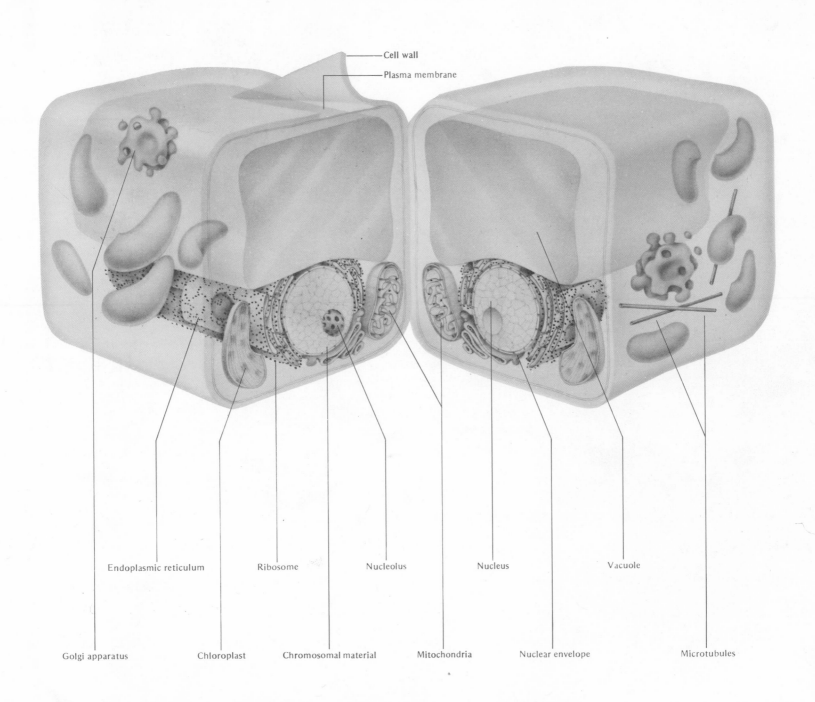

Cell wall

Plasma membrane

Endoplasmic reticulum   Ribosome   Nucleolus   Nucleus   Vacuole

Golgi apparatus   Chloroplast   Chromosomal material   Mitochondria   Nuclear envelope   Microtubules

BIOLOGY
A HUMAN APPROACH

Out of the dreaming past, with its legends of steaming sea and gleaming glaciers, mountains that moved and suns that glared, emerges this creature, man—the latest phase in a continuing process that stretches back to the beginnings of life. His is the heritage of all that has lived; he still carries the vestiges of snout and fangs and claws of species long since vanished; he is the ancestor of all that is yet to come.

   Do not regard him lightly—he is you.

DON FABUN, The Dynamics of Change, Prentice Hall, Englewood Cliffs, New Jersey, 1967.

IRWIN W. SHERMAN AND VILIA G. SHERMAN

University of California at Riverside

BIOLOGY
A HUMAN APPROACH

SECOND EDITION

OXFORD
UNIVERSITY
PRESS

NEW YORK 1979

Copyright © 1975, 1979 by Oxford University Press, Inc.
Third printing, 1980
Library of Congress Cataloging in Publication Data

Sherman, Irwin W.
  Biology : a human approach.

  Includes index.
  1. Biology. 2. Human biology. I. Sherman,
Vilia G., joint author. II. Title.
QH308.2.S5 1979   574   78-6078
ISBN 0-19-502439-7

Printed in the United States of America

Designed by Gayle Jaeger

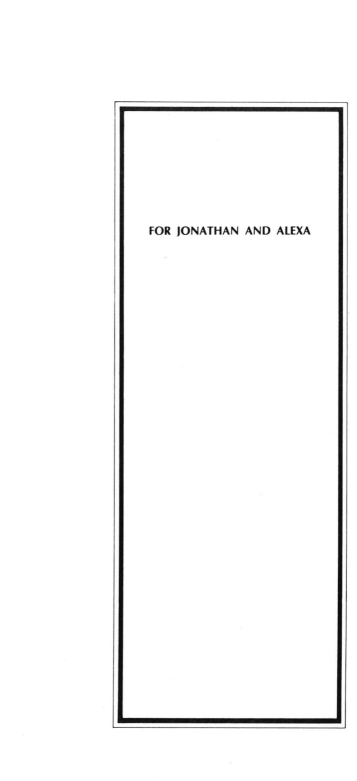

FOR JONATHAN AND ALEXA

PREFACE TO THE SECOND EDITION

The wide acceptance of the first edition of *Biology: A Human Approach* has provided us with a great sense of satisfaction. We are pleased to know that the book is being used by so many students and that instructors have found our view of teaching biology in harmony with their own. Yet over the last few years, as readers have told us of their views, we knew portions of the text could be revised and improved upon. This second edition is our attempt to respond to these challenges.

The major changes in this edition have involved a condensation and simplification of the chemistry found in Part I (The Cell), the addition of a totally new chapter, Animal and Human Behavior, to Part II (The Human Organism), and the rewriting and amplification of topics that have changed markedly in the past few years (for example, genetic engineering, environmentally induced cancers, and immunology). At the end of each chapter there is a summary, a list of key words that are boldfaced and ordered as they are defined in the text, and topics for review and discussion. No chapter, figure, or figure legend in need of improvement was left untouched. At the same time, we felt that change for its own sake was not constructive; thus material that we felt was satisfactory remains unchanged.

As in the first edition, the index may also be used as a glossary. Key terms are boldfaced where they are defined in the text; the page number where their definition may be found is likewise boldfaced in the index, providing a quick and easy reference. We hope that students will find this method of indexing and the lists of key words added to the end of each chapter to be useful study aids. For example, after they read a given chapter, students may wish to run down the list of key words. If they can successfully define all key words for themselves, they are on the way to achieving a good grasp of the chapter's content. Definitions of unknown terms can be checked quickly by looking back through the chapter, since the key words are listed in order as they appear in the chapter; or definitions may be easily found by turning to the index.

We would like to thank Clay Sassaman and Charles Taylor for their assistance in updating the genetics and ecology chapters.

As before, any errors remain our own.

I.W.S.
V.G.S.

February, 1978
Riverside, California

This book is intended for use by undergraduate students who are likely to pursue careers in fields other than the biological sciences. Because this will perhaps be both their introductory and final course in the subject, we want to present to those students not only biological concepts but related information that will better enable them to understand themselves and the world in which they live. We have tried to write a book that is clear, enjoyable, and relevant to the personal life of students majoring in the arts, humanities, and social sciences, but not a watered-down major's text.

As implied by the book's title, we have selected the human organism as its central theme. Here we treat biology—the study of life—in its modern and often controversial aspects. Hopefully, biology framed in this way has both meaning and excitement rather than being an abstract stumbling block serving merely to satisfy the breadth requirements of the liberal arts major. The human approach to biology has immediate personal relevance for most students because humans have an inherent interest in themselves. By design, we have omitted detailed considerations of the life histories of plants or animals, and plant and animal anatomy. On the other hand, the text goes more deeply into areas relevant to the student's present and future. Basic biological concepts are illustrated by reference to areas of immediate human concern, such as cancer, abortion, human genetics, human ecology, transplantation immunity, and genetic engineering. When a topic is mentioned, it is given sufficient depth to make exposure meaningful. It is our hope that a human approach to biology will generate sufficient interest in biological phenomena and how these are investigated that, although formal coursework may end here, in depth studies will continue thereafter in an informal way.

The chapters are arranged in a sequential order, the study of advanced areas depending on an understanding of more elementary concepts. Thus, the book progresses from The Cell (Part One) to The Human Organism (Part Two) and ends with a view of man from an ecological perspective, The Human Population (Part Three). Within each section there is also a progression from simple to complex levels so that what goes before is used as a basis for later discussions. However, the text treatment is flexible enough to permit diverse sequences of chapters, so an array of readings can be accommodated to courses of varying length.

To demonstrate the human implications of biological processes, most chapters begin with a photograph, a news article, or a short excerpt from the popular literature. These are designed to pique reader interest and to point up the timeliness of the subject matter. The chapter proper, the academic core of the text, attempts to stress the importance of biological facts to the human condition—where we are and how we might possibly change. At intervals in each chapter, relevant asides are presented in boxes. For example: why cyanide kills; curling of hair and the shape of protein molecules; transplanting a gene; human mammary tumors; smog, smoking, and emphysema. Boxed material serves as motivational cement, holding firm the reader's attention to the text's core material.

Since the book is written for undergraduate college students regardless of their previous scientific background, it seemed important that all students have some of the basic tools of biology at hand. The role of chemistry in biology cannot be minimized. The simple chemistry essential to an understanding of today's biology is presented in an appendix to the text proper so that chapter continuity and student interest are not sidetracked by a "chemistry hurdle" at the book's beginning. Reading the chemistry appendix at appropriate points in the text rather than all at once will serve to reduce the activation-energy barrier of this subject.

This book spans its subject from the level of molecules to populations, and no single person can claim excellence in all of these diverse specialties. In the preparation of the text we have been most fortunate in having the able assistance of colleagues who read and gave criticism of these pages so that the text was as error-free as possible. No doubt errors still remain; the fault for these is ours. We particularly want to thank: William L. Belser, Roger D. Farley, Robert W. Gill, Robert L. Heath, Richard L. Moretti, Donald I. Patt, E. Crellin Pauling, Timothy Prout, Rodolfo Ruibal, Vaughan H. Shoemaker, William W. Thomson, Irwin P. Ting, and Linda Tanigoshi.

We hope the fruits of our labors are enjoyed with the same delight that we derived by writing about and teaching biology in this way.

CONTENTS

PROLOGUE

PART I. THE CELL

1 GENESIS: THE ORIGIN OF LIFE 5

1-1 What is life? 5
1-2 Spontaneous generation 6
Box 1A Vitalism versus mechanism 7
Box 1B The methods of science 8
1-3 History of the early earth 10
1-4 The primitive soup 12
1-5 The primitive organism 14
Box 1C Creation: producing life in the laboratory 1b
1-6 Heterotrophs and autotrophs 17
Box 1D The origin of life: an alternate view 19
Box 1E Does life originate abiotically today? 19
Summary 20

2 CELL STRUCTURE: THE ORGANIZATION OF LIFE 23

2-1 What is a cell? 23
Box 2A Thereby hangs a tail: a tale of resolution 28
2-2 Cell specialization: tissues and organs 48
2-3 Cells and organisms 48
Box 2B Seeing the unseen: viruses 52
Summary 54

3 OF MOLECULES AND CELLS 57

3-1 The nature of the living system: the chemicals of life 57
3-2 Carbohydrates: cellular fuel 58
Box 3A Through the looking glass: symmetry of molecules 60
3-3 Lipids: fuel, structure, and coordination 68
3-4 Proteins: fabric of the cell 66
3-5 Enzymes: proteins that make haste 70
Box 3B Curling hair and the shape of protein molecules 71
3-6 Vitamins and cofactors: enzyme helpers 74
Box 3C Beneficial poison: sulfanilamide 76
3-7 Nucleic acids: informational molecules 77
Summary 77

4 HARVEST OF CHEMICAL ENERGY: CELLULAR RESPIRATION 81

4-1 What is energy? 81
4-2 Energy currency: ATP 84
4-3 Chemical energy harvest 85
4-4 The energy cascade 86
4-5 Getting along without oxygen 86
Box 4A Beer, bread, and muscle: moledular changes in pyruvic acid 88
4-6 Anaerobic respiration: cost accounting 89
4-7 Getting along with oxygen 89
4-8 Aerobic respiration: cost accounting 92
Box 4B Why cyanide kills 92
4-9 Where is ATP used? 93
4-10 Chemical energy harvest: evolution and cellular location 94
Summary 94

5 HARVEST OF LIGHT ENERGY: PHOTOSYNTHESIS 97

5-1 All flesh is grass 97
5-2 The energy cascade revisited 98
5-3 Light and chlorophyll: light phase 99
5-4 Carbon dioxide and sugar formation: dark phase 102

5-5 Cellular location and evolution of
photosynthesis 102
Box 5A Weed killers 103
5-6 Efficiency and implication of photosynthesis 103
Summary 104

6 COMMUNICATION BETWEEN CELL GENERATIONS: NUCLEIC ACIDS 107

6-1 Inheritance: general features 107
6-2 Inheritance: chemical structure 108
6-3 Genetic code: four-letter alphabet, three-letter
words 112
Box 6A An inborn error: PKU and idiocy 113
6-4 Copying and translating the genetic code: RNA
and protein 115
Box 6B The library of heredity 116
6-5 Implications of the genetic code 121
6-6 Genetic engineering: recombinant DNA 122
6-7 Prospects for the refabrication of man 123
Box 6C Redesigning bacteria 124
Summary 124

7 CELLS IN ACTION: DUPLICATION AND CONTROL MECHANISMS 129

7-1 Cell duplication 129
7-2 Chromosome structure 136
7-3 Control of cellular activity 138
Summary 148

8 CANCER: CELLS OUT OF CONTROL 151

8-1 What is cancer? 151
8-2 Carcinogens: cancer-causing agents 153

Box 8A The cancer screen 157
Box 8B Human mammary tumors 158
8-3 The causes of cancer: some hypotheses 159
Summary 162

PART II. THE HUMAN ORGANISM

9 PERPETUATION OF LIFE I: SEX CELL FORMATION AND FERTILIZATION 167

9-1 Sexuality and meiosis 167
9-2 The male reproductive system 170
9-3 The female reproductive system 174
Box 9A Sperm banks 176
9-4 Coitus (copulation) 182
9-5 Fertilization 183
9-6 Contraception 185
Box 9B Detection of ovulation 186
Box 9C Pregnancy: signs and symptoms 187
9-7 Control of the female reproductive cycle:
the pill 188
Summary 189
Box 9D Sterilization for both sexes 190

10 PERPETUATION OF LIFE II: HUMAN DEVELOPMENT 195

10-1 From egg to embryo 195
10-2 The fetus 202
10-3 The placenta: lungs, kidneys, and digestive
tract for the human fetus 206
10-4 Parturition (birth) 208
Box 10A From water to land: a baby is born 210
10-5 The mammary glands and lactation 211

10-6 Birth defects 213
Box 10B Abortion 216
Summary 217

11 PROCESSING NUTRIENTS: DIGESTION 221

11-1 Why do we eat? 221
Box 11A A peephole into the stomach 222
11-2 Intake apparatus: food processing 223
11-3 Digestion: food degradation 223
11-4 Absorption: food uptake 224
11-5 Defecation: waste expulsion 229
11-6 Nutrition: the use of processed nutrients 229
11-7 Trouble along the alimentary canal 231
Box 11B Ulcers: the price of success 232
Summary 234

**12 INTERNAL TRANSPORT:
CIRCULATORY SYSTEM 237**

12-1 Why do we need a circulatory system? 237
12-2 What does the circulatory system do? 238
12-3 The heart: a remarkable pump 240
Box 12A Artificial pacemakers 246
Box 12B Heart attack 248
12-4 Cardiovascular plumbing at work 249
Box 12C The Death of Lenin 256
Summary 255

13 BLOOD AND IMMUNITY 259

13-1 Blood: a tissue that flows 259
13-2 Against the foreign peril: immunity 264
**Box 13A Counting the reds and whites: barometer of
disease 266**
Box 13B Putting antibodies to work 272
Box 13C New organs for old: transplantation 275

Box 13D Wayward antibodies: allergy 276
Summary 277

14 GASEOUS EXCHANGE: RESPIRATION 281

14-1 Why do we need a respiratory system? 281
14-2 The respiratory tree 283
14-3 Alveolus: rendezvous of blood and air 285
Box 14A Smog, smoking, and emphysema 288
14-4 How do we breathe? 290
14-5 Control of breathing 291
14.6 Low pressure: high pressure 296
Box 14B Why can't we breathe underwater? 296
Summary 298

15 CHEMICAL BALANCE: THE KIDNEY 301

15-1 What does the kidney do? 301
Box 15A What is urea? 302
15-2 Structure and function of the kidney 303
15-3 Regulation of fluid and salt balance 309
15-4 Discharge of urine 311
Box 15B The artificial kidney machine 312
Summary 314

16 CHEMICAL COORDINATION: HORMONES 317

16-1 The nature of hormones 317
16-2 The human hormonal orchestra:
endocrine glands 318
Box 16A Giants and dwarfs 322
Box 16B Rickets 328
Box 16C The critical cortex 334
16-3 Management of the endocrine orchestra 336
16-4 The molecular activity of hormones 337
Box 16D The hypothalamus and history 338
Summary 341

17 ELECTROCHEMICAL COORDINATION: NERVES 346

17-1 What does the nervous system do? 346
Box 17A 1984 Revisited? 348
17-2 The neuron: signal cell of the nervous system 349
17-3 The nerve impulse: an electrochemical message 352
17-4 The nerve impulse: production and conduction speed 354
17-5 Signals from neuron to neuron: the synapse 355
Box 17A Bugging the brain 356
17-6 Integrated circuits of neurons: the nervous system 360
17-7 Neuronal circuits in action: reflexes 365
17-8 Drugs and the nervous system 366
Summary 370

18 BUILT-IN SIGNAL CORPS: SENSE ORGANS 373

18-1 Sensory listening posts 373
18-2 Smell: detecting chemical signals in the air 374
18-3 Taste: detecting chemical signals in solution 375
18-4 Sight: receiving light signals 376
18-5 Hearing: receiving sound signals 381
Box 18A Unwanted sound: noise pollution 383
18-6 Balance and equilibrium 384
18-7 Touch, temperature, and pressure: skin senses 385
Summary 385

19 ANIMAL AND HUMAN BEHAVIOR 389

19-1 Why behave? 389
Box 19A Purpose is in the mind of the beholder 391
19-2 Toward understanding behavior 392
19-3 Simpler types of behavior 393
Box 19B Skinner's utopia: panacea, or path to hell? 394

Box 19C Genes über alles? 396
19-4 More complex types of behavior 398
Box 19D Is man an instinctive killer? 399
19-5 Where and how does learning occur? 405
19-6 Memory 406
19-7 Social behavior 408
Box 19E The life of the honeybee 410
19-8 Prospects for the future 413
Summary 414

20 MOVEMENT: SKELETON AND MUSCLE 417

20-1 The skeleton: tower of support and protection 417
20-2 Motion: muscle 422
Box 20A Your muscles: use 'em or lose 'em 433
Box 20B Paralytic poisons
Summary 436

21 STAYING IN PLACE: HOMEOSTATIC CONTROLS 439

21-1 Keeping things in order or maintaining the steady state 439
21-2 Body-temperature control 440
Box 21A Fever: resetting the body's thermostat 443
Summary 444

PART III. THE HUMAN POPULATION

22 GENES IN INDIVIDUALS: HUMAN INHERITANCE 449

22-1 Communication between parent and offspring 449
22-2 Genes, chromosomes, and sex-cell formation 450

Box 22A Preparing chromosomal pictures 451
22-3 Phenotype and genotype 454
Box 22B Mendel and his peas 456
22-4 Transmission of genes 458
22-5 Multiple alleles: blood will tell 463
Box 22C Hemophilia and history 466
22-6 Polygenes and skin color 468
Box 22D The Charlie Chaplin paternity suit 469
22-7 Chromosomal abnormalities 471
Summary 478

**23 GENES IN POPULATIONS:
AN INTRODUCTION TO EVOLUTION 483**

23-1 Sickle cell: a molecular disease 483
23-2 Our changing genes: mutation 484
23-3 Genes in populations 486
**Box 23A Ill winds blow over the *Lucky Dragon*:
radiation as a mutagen 489**
23-4 Natural selection 491
23-5 Eugenics: improving our genes by selection 494
Summary 496

24 THE EVOLUTION OF MANKIND 499

24-1 Man's place in nature 499
24-2 What is a man? 501
24-3 Darwinism: from finch to man 508
24-4 The road to man 512
24-5 The races of man 520
24-6 What is man's evolutionary future? 523
Summary 524

25 MAN AND THE ENVIRONMENT 527

25-1 What's ecology? 527
25-2 Energy flow 528
25-3 Food chains and ecological pyramids 530
25-4 The danger of accumulated chemicals 533
25-5 Recycling matter 537
25-6 Homeostasis and succession in the ecosystem 544
25-7 Agriculture and unbalanced ecosystems 547
Box 25A Nuclear garbage 549
Summary 550

26 ABOARD THE SPACESHIP EARTH 553

26-1 The inhabitants of the spaceship earth and how
they grew 553
26-2 Implications of population growth on
spaceship life 561
Box 26B Last chance to keep off the grass? 564
Box 26B Thermal pollution: we can't break even 571
26-3 Limits on population growth:
strategy for survival 573
Summary 576

EPILOGUE 579

APPENDIX

A SIMPLE CHEMISTRY FOR BIOLOGY 583

**B MEASURING UP OR OUR SYSTEM IS DYING
INCH BY CENTIMETER 589**

INDEX 591

PROLOGUE

Man is a threatened species. The twin specters facing him are over-population and unbridled technology—both self-induced.

The double threat is aimed most directly at man's environment. As the United States strives to accommodate more human beings than it has ever had to serve before, increased demands are placed on our natural resource bank. Our surroundings become increasingly crowded, noisy, and soiled.

. . . Buffeted by the elements and beset by other life forms, man has always stubbornly insisted on exercising every option open to him. Does he still run his own show today: Or has he finally stumbled upon two forces—population and technology—that he considers too sacred to tamper with: Is he still convinced that the roaring crescendo from babies and bulldozers is the sweet music of progress?

Does he confuse technology with science? Will he continue to accord to the jackhammer the same revered status as the test tube? Or will he recognize in time that the tools he uses to rip up mountains and destroy estuaries must be extensions of his mind as well as his muscle? Will he see that science must remain free, since it is the *search* for truth, but that technology is only a means of *applying* truth—and that these applications need the control and balance of wisdom and a concern for posterity?

Man stands at a fork in his environmental road to the future.

The two arms of the signpost do not state categorically, "Man—Master of Himself" and "Man—An Extinct Species," but it is increasingly apparent that the direction he takes now will move him rapidly along the path toward one or the other destination.

Let us look closely for a moment at this creature who pauses at the crossroads and clamors for attention with our own voice. Who is he? Where has he come from and how has he made the journey this far?

From "Man . . . An Endangered Species?" U.S. Department of the Interior Conservation Yearbook No. 4 (Washington, D.C.: U.S. Government Printing Office, 1968).

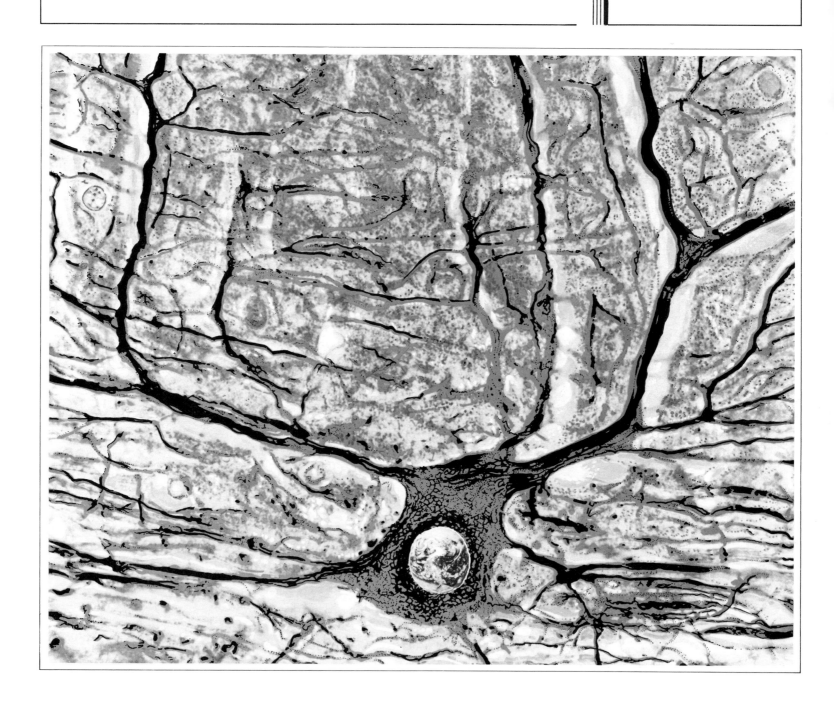

THE CELL

Some say the spaceship earth is doomed. There is only a short time before there will be little oxygen to breathe or food to eat; the effluents of the inhabitants will soon overcome them. Time has run out, and no matter what we do, nothing can save us. This pessimistic view of our future may become a reality in a very short time, but let us imagine for the moment that there is still time to recover the ship and its inhabitants. How can the earth be saved from self-destruction?

Recovery tactics must depend upon a clear understanding of the nature of the earth and its occupants—the substances which compose this self-contained system and how they interact with each other. It would be helpful, indeed desirable, to understand the most fundamental elements that make up this system. In the physical and chemical world the fundamental unit is the atom, and it is impossible to understand physical and chemical events without recognizing what atoms are like and what they do. In the biological world, our world, the cell is the fundamental unit, and similarly we cannot comprehend biological phenomena without an appreciation of the cell, its structure and operation. It is elemental that many of the problems confronting earthly inhabitants depend upon the nature of the cells which make up earthlings. An understanding of the cell is basic to understanding ourselves and all life.

Is there life on Mars?

On July 20, 1976, as the sun broke over the Martian horizon, a spindly three-legged spacecraft sat silently on a dry, barren landscape. Suddenly, a silvery arm stretched out from the Viking Lander and scooped up a heap of reddish soil. The Viking, after an 11-month journey in space, successfully responded to commands from controllers on the Earth 200 million miles away. The robot arm retracted, twisted its wrist, and dropped the soil into an opening on the top of the spacecraft. The search for extra-terrestrial life had begun.

How does one look for life that may be very different from our own? How can one be sure that observations reveal living processes rather than nonbiological activity? More than 10 years ago, a Space Study Board agreed that anything organized to draw nourishment from the environment and reproduce itself should be considered "life."

The most obvious way to seek life on Mars is to look for it, and the Lander's cameras are being used to see if anything moves or looks suspiciously biological. Another way is to test for the presence of organic molecules. On board Viking, a tiny laboratory is housed in a one-foot cube weighing only 30 lbs. It is crammed with 140,000 electronic components—including 122,000 transistors, 40 thermostats, 3 tiny ovens, bottled radioactive gases, a small xenon lamp to simulate sunshine, and a pocket-sized chromatograph to analyze the chemical components of the soil. Three separate experiments have been performed.

One of the experiments sought to learn if anything in the soil assimilates Martian air (mainly carbon dioxide) to form carbon-containing material. The soil sample was exposed to carbon dioxide whose carbon was radioactive, and incubated for up to 5 days under simulated Martian sunlight. The soil was then tested to see if it became radioactive by incorporating that carbon. This would provide evidence that organisms are growing on Mars. The tests did not find organic molecules; however, small amounts of such carbon-based compounds could go unnoticed.

Living things on Earth use nutrients and release waste products and gases, a process called metabolism. To test for this, a Martian soil sample was moistened with a substance scientists have named "chicken soup," a nutrient broth rich in vitamins and amino acids and containing radioactive carbon. The sample was incubated to see if anything consumed the "soup" and released radioactively labelled products. A rapid release of carbon dioxide occurred. This could have been the result of microbes metabolizing the "soup," or simply the chemical activity of soil compounds called peroxides.

The third test submerged a soil sample in a liquid nutrient for 12 days in an atmosphere of helium, krypton, and carbon dioxide. At intervals, the miniaturized lab sampled the atmosphere in the chamber for hydrogen, nitrogen, oxygen, methane, and carbon dioxide—gases generally produced by living organisms. The sample released oxygen far more rapidly than plants usually do; such a reaction could be biological or the result of decomposition of soil peroxides.

The results of these three experiments were equivocal. The official NASA position stated that it was impossible to say that there was or was not life on Mars; however, physicist Robert Jastrow said unofficially, "Although the Viking experiments have contradictory elements, they seem to indicate that life, or some process closely imitating life, exists on Mars today." Further experiments with Martian soil are planned.

Should the billion-dollar Viking project find even the most primitive organisms, it will help confirm what many scientists suspect: Life is not unique to Earth, and it is probably commonplace throughout the universe.

GENESIS: THE ORIGIN OF LIFE

1–1 WHAT IS LIFE?

The business of biology is life. But what is life? What do we really mean when we say that an object is living or nonliving? How do we distinguish between the inanimate and the dead? How shall we recognize living matter on other planets (see facing page), and when can we say that a man is dead and can serve as a donor for organ transplants? It is necessary for us to develop some notion of what life is. We shall find in the process that this is no easy task and no definition of life is completely satisfactory; we can nevertheless make an attempt.

For most of us it is easier to recognize life than to define it. We are aware that we are alive and that someday we shall be dead. Animals and plants are alive; earth, fire, water, and air are not alive. Living things have certain characteristics, none of which by itself is sufficient to define them as being alive, but which, when taken together, enable us to distinguish them from the nonliving. The capacities for growth, maintenance and repair, reproduction, movement, responsiveness, change—these are the properties of the living. But how are these characteristics of life different from the growth of crystals, the division of raindrops, the swift movement of a mountain stream, the response and change in a piece of wood as it is consumed by fire? Let us examine the characteristics of living things one by one in greater detail.

Growth, maintenance, and repair

It is easy for us to recognize living things; however, it is important to recognize that life is not a thing or a substance. **Life** is a property possessed by individuals characterized by the capacity to perform a series of highly organized interacting processes that occur within a definite structural framework. In order to continue these processes, living systems must obtain materials from the environment, utilizing and altering these substances for the synthesis, maintenance, and repair of their own structures and eliminating those materials that are no longer useful. The flow of materials through the living system is called **metabolism.** The tests for life performed on Martian soil by the Viking Lander attempted to detect metabolism in various ways (see facing page).

The metabolic processes of life require an energy supply for their continuance which involves the performance of work, and living systems are capable of converting, transporting, and storing energy. They perform these energy transformations with the aid of organic catalysts called **enzymes.**

One of the end results of all these transformations is an increase in size, or **growth**—from the inside out. Thus, the growth of living systems differs from that of nonliving ones such as crystals, which grow by the addition of material from the outside and do not transform the added material during the process. Moreover, a crystal is unable to repair or maintain itself except under special conditions.

Reproduction

Living systems do not only grow, convert, transport, and store energy—they are also capable of self-duplication, or **reproduction.** Many nonliving systems are capable of reproduction, however. Crystals divide, as do streaming raindrops, and fire can be reproduced by incandescent sparks. How does reproduction in living systems differ from this kind of reproduction? The method of reproduction in living systems is far more complex, exact, and specific than in nonliving• ones. Living systems can make identical copies of themselves. Moreover, the process of self-duplication is especially important in living sys-

tems because it includes a capacity for **mutation,** or change, and these changes are perpetuated exactly in subsequent generations of the organism involved. The changes in form which may occur during the division of a raindrop or a crystal or the reproduction of fire are not subsequently perpetuated.

A system that cannot change to meet the needs imposed by changes in its environment is at a severe disadvantage, which may eventually prove fatal; in other words, a system incapable of mutation is unable to **evolve** (change with time). Living things exist in endless variety and complexity, and they are able to cope with the demands of the environment because of their ability to mutate, adapt, and evolve.

Responsiveness and movement

Living systems are not static; they are capable of responding to changes in both the external and internal environment, not only from generation to generation but within generations. If somebody pricks your finger with a pin, your initial response is probably to withdraw your finger. The capacity to respond to stimuli is related to the fact that living systems are self-regulating. A particular stimulus induces an appropriate response, but the response may not always be the same. After withdrawing your finger, you might take the pin from the offending party, you might hit him with your fist, or you might simply walk away. Responses are mediated by the living system. A crystal, when it is placed in the right conditions, will always grow; likewise fire and rain are at the mercy of outside controls. Living systems to a greater or lesser extent can control the end results of stimuli themselves.

The dynamic changes characteristic of the living world are often evidenced as motion. Sometimes the motion is on a slow time scale

so that the change is not immediately apparent (as in plants), and at other times it takes place quite rapidly and is easily recognizable (as in animals). In many living systems the motion takes place over such short distances that it is not perceived by the human eye, but there are conditions where motion covers great distances and is easily observable. Like the capacity for responsiveness, the capacity for movement that living systems have is self-regulating; in nonliving systems it generally is not. A raindrop must fall, and a fire must rage or die.

Thus, **life** as we have defined it (and the definition is to a large extent arbitrary) involves a series of highly organized interacting processes that form a system that is potentially capable of perpetual change.

What is the advantage of defining and describing living systems in these mechanistic terms (Box 1A)? First, it avoids mystery, and second, it rules out the concept of the existence of a vitalistic force such as entelechy or the "soul of life." The reason for avoiding such ideas is that they preclude a scientific examination of the living system; simply stated, such a characterization of life defies testing by scientific methods (Box 1B) and is of little value in biological inquiry.

1–2 SPONTANEOUS GENERATION

If we agree that life can be defined in mechanistic terms, then we can further ask the question: How did life originate? Somewhere in our educational experience, most of us have come across the doctrine of **spontaneous generation,** that is, the sudden appearance of living things from something nonliving. This view held dominion over much of man's thinking up until the seventeenth century, and it was supported by some of the greatest and clearest scientific minds of the time. The supporters of this view cited examples and gave recipes for

BOX 1A Vitalism versus mechanism

Throughout the history of biological thought, there has raged the controversy whether or not living phenomena can be described in terms of chemistry and physics. The **vitalists** (adherents to the philosophy of vitalism) take the view that living phenomena cannot be included under the heading of chemistry and/or physics but rather that there is a kind of directing force or spirit resident within the living organism. Vitalists believe that this life force is beyond human comprehension and that a distinct and inviolable barrier exists between the living and the nonliving world. According to the vitalists, the origin of life could be explained only as a result of divine creation.

In contrast, the **mechanistic** view of life states that living phenomena can be described by chemistry and physics; the line of demarcation between the living and the nonliving is not sharply defined, and no vitalistic spirit or soul directs the living organism. The mechanists suggest that living phenomena can be examined and tested by the methods of science (see Box 1B).

Vitalism is an ancient view of living phenomena, but it reappears even in modern times. A distinguished physicist, W. Elasser, in 1958 wrote of "biotonic phenomena," that is, "phenomena in the organism that cannot be explained in terms of mechanistic function." And in 1963 P. Mora suggested: "Living entities, at all levels in almost all their manifestations, have something of a directed, relentless, acquiring and selfish nature, a perseverance to maintain their own being and a continuous *urge* to dominate their surroundings, to take advantage of all possible circumstances, and to adjust to new conditions." Perhaps the clearest refutation of vitalism would be the creation of a living organism by completely synthetic means out of simple chemical elements. Disturbing and convincing as such an experiment might be, no doubt a resourceful vitalist would claim that even this system was taken over by a life force. Vitalistic views, old as well as new, replete with misconceptions and defying examination by objective criteria, are of little value in helping us to understand biology.

As you read this book, you will see that most of the processes having to do with life that we do understand can be explained in terms of chemistry and physics. Those we do not yet understand are potentially explicable in these terms. Superstition and vitalism were early attempts to explain away things that could not be understood at the time because physics and chemistry had not yet advanced far enough.

the production of living material from the nonliving: sweaty shirts stored with wheat in a dark place were supposed to give rise to mice, the hairs of a horse's tail when placed in water produced worms, and decaying meat gave rise to maggots (fly larvae). The strangest part about these recipes was that they really appeared to work!

In 1668 an Italian physician and poet named Francesco Redi (Figure 1–1) performed a simple but classic test that shook the foundations of the doctrine of spontaneous generation. Where others were content to observe nature and suggest imaginative explanations for various phenomena, Redi was not content merely to observe natural phenomena as they occurred, but he set out to test ideas and to arrange some of the components of nature so that analysis of phenomena could be made; in short, he did an **experiment.** Redi arranged three jars of decaying meat; one of the jars he covered with gauze, another was covered with parchment, and the third was left uncovered. Flies were attracted to the meat samples in the gauze- and parchment-covered jars, but could land only on the meat in the open jar; in this jar maggots developed, but not in either of the others. Decaying meat in itself did not give rise to maggots, he concluded. It was necessary for flies to land on the meat and deposit their eggs, which subsequently hatched and gave rise to maggots.

This simple refutation of the generation of life from substances such as rotting meat held sway for only a short time. In 1675 a Dutch linen merchant with a penchant for grinding magnifying lenses, Anton van Leeuwenhoek (Figure 1–2), found that his lenses showed living microscopic creatures in rainwater. Broth too would give rise to all sorts of living creatures if one waited for a while. The resourceful defenders of spontaneous generation argued that although one could not get worms, flies,

BOX 1B The methods of science

Science is the product of man's attempt to understand himself and the world in which he lives; it embodies knowledge about the natural world and ourselves, and it is organized in a systematic fashion derived from experimentation and observation. Scientific knowledge often allows the prediction of specific events and in some cases even permits control of them; scientifically determined facts are repeatable and capable of verification. Science admits "we do not know the answer" and tries to discover it by a variety of repeatable and testable methods rather than by fabricating myths which cannot be tested.

There is no one scientific method. Discoveries of a scientific nature are made in a variety of ways, and the methods employed are also quite diverse. The methods of science are not magical or practiced exclusively by scientists working in white lab coats and surrounded by exotic chemicals and apparatus, but represent a rational approach to understanding the universe. Science assumes that the world about us is knowable.

Scientific method is often described in an idealized fashion; such a formal description bears only a slight resemblance to reality, but it serves to emphasize some of the important elements of scientific inquiry. First, it is important to recognize that a phenomenon exists which demands inquiry. Second, an attempt is made to state the problem in very specific terms. Third, a working **hypothesis** is formulated that attempts to explain the phenomenon. Fourth, the phenomenon is investigated by observation or by experimentation or by both, and evidence is gathered and recorded—the testimony of raw data. Fifth, the data are analyzed, and the working hypothesis is reexamined in the light of the evidence provided by the data. Sixth, a conclusion is drawn from the experimental observations about whether the working hypothesis is correct or must be reformulated and reexamined. If it is concluded that the hypothesis is correct, predictions can be made from it. If the hypothesis stands the test of verification and predictability, it becomes a part of the established body of knowledge — a **theory.**

Perhaps no real scientist ever goes through the steps cited above, and obviously such a brief discourse does not take into account the trials, errors, intuitions, and leaps of imagination that are a part of scientific inquiry. Nevertheless, the scientific method represents a powerful tool because of its rigor. Many of the procedures of science are identical to our own daily reasoning, but most of us fail to apply the method stringently in formulating and testing the hypothesis, and when the observations do not satisfy it, we fail to begin anew by reformulating the hypothesis.

Scientific method is a powerful tool for inquiry, but it is not without its limitations. We should not be surprised that sometimes scientific "facts" change or that the practitioners of science harbor decidedly unscientific notions. Since science deals with observable phenomena, its range is limited by our senses or by the machines which sense for us. Refinements in techniques and methods of observation often alter our view of nature and necessitate appropriate modifications in hypotheses and conclusions. Some have therefore characterized science as "the study of errors slowly corrected." The strength of scientific inquiry is that it is self-correcting. Although the scientific method attempts an accurate description of natural phenomena, there is considerable documentation of its mistakes. Scientists are human, with human feelings, failings, and prejudices; on occasion these interfere with their reasoning. This may be particularly true when a scientist engages in discussions outside his specialty and neglects the rigorous proof used in his own area of competence. Science can and should be judged apart from the people who practice it; by its very nature science is a productive and important contributor to our knowledge and well-being.

and rats from the inanimate world, one could get living microorganisms from nonliving broth, rainwater, hay infusions, and other such materials. The first attack on this microlevel doctrine of spontaneous generation came from the Italian scientist Lazzaro Spallanzani (Figure 1–3). In 1767 he discovered that if he boiled meat broth in a flask and then sealed the neck of the flask, no microorganisms developed. If, however, the flask neck was broken, in a short time the broth swarmed with microscopic forms of life. Boiling the broth (sterilization) prevented the growth of microorganisms; contamination of it with unheated broth or another substance apparently provided the source of the living creatures. Thus, life did not arise spontaneously in the meat broth, but came from an outside source. However, the adherents of the doctrine of spontaneous generation were also imaginative and resourceful. They argued that not only did heat destroy the life in the broth, but

that sealing the flask prevented the vital force of life from entering the flask and bringing forth life. The issue remained in doubt for another century, for it was well known that many organisms required oxygen, a so-called vital principle, and the absence of oxygen in a sealed flask could be expected to inhibit the spontaneous development of living organisms.

In 1862 the French chemist Louis Pasteur (Figure 1–4a) performed an experiment that was both simple and elegant. Pasteur contended that microorganisms (bacteria, yeast, and protozoa) not only cause disease in man and animals but give rise to decay and change in organic materials such as broth. In the absence of these microorganisms, he said, no change in the broth should occur. Pasteur placed some meat broth in a flask and boiled it until it was sterile; then he drew out the neck of the flask so that it was formed into the shape of an S (Figure 1–4b). He did not seal the neck,

and air could freely communicate with the broth after passing through the twisting neck of the flask. The long curving neck of the flask trapped the airborne microorganisms and prevented contamination of the broth. In spite of the ease of access of the vital principle to the broth, the flasks remained sterile. However, if Pasteur tipped the flask so that some of the broth ran into the bend of the S and then back into the flask, the broth became contaminated with microorganisms. Obviously, something more than broth plus air was necessary to produce life. Kept at the Pasteur Institute in Paris, some of the uncontaminated flasks were sealed to prevent evaporation of the contents, and they remain sterile to this day. The doctrine of spontaneous generation never recovered from the mortal blow provided by this simple experiment of Louis Pasteur.

Such experiments suggest the improbability of life, a complex of organization, arising in a

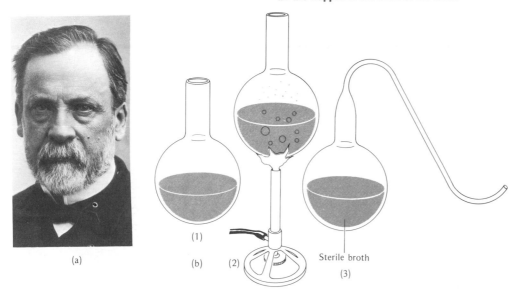

FIGURE 1–4 (a) Louis Pasteur (1822–1895). (Courtesy of The Bettmann Archive, Inc.) (b) Pasteur's experiments disproved the doctrine of spontaneous generation. Broth placed in a flask (1), is heated, sterilized (2), and the neck of the flask drawn out into the shape of an S (3). Such flasks remain sterile though air communicates with the broth through the opening in the S-shaped neck. Microorganisms in the air are trapped in the bend of the neck.

(a)   (b)   (1)   (2)   Sterile broth   (3)

TABLE 1–1 Relative abundance of some of the major elements found in the universe and the human body

Element	Relative abundance in atoms %	
	Universe	Human body
Hydrogen	90.79 ⎫	60.3 ⎫
Carbon	9.08	10.5
Nitrogen	0.0415 ⎬ 99.8	2.42 ⎬ 98.7
Oxygen	0.0571 ⎭	25.5 ⎭
Sodium	0.00012	0.73
Magnesium	0.0023	0.01
Aluminum	0.00023	
Silicon	0.026	0.00091
Phosphorus	0.00034	0.134
Sulfur	0.00091	0.132
Chlorine	0.00044	0.032
Potassium	0.000018	0.036
Calcium	0.00017	0.226
Iron	0.0047	0.00059

short time before our eyes, from unorganized organic solutions. We know that today living things perpetuate themselves—roses produce roses, dogs produce dogs, people produce people—but where did it all begin? How did life originate in the first place? Perhaps "spores" came from outer space, and upon arrival on this planet flourished and evolved into the creatures that we now recognize as plants and animals. Space exploration convinces us that this is a highly improbable possibility; the kind of "spores" that could have given rise to life as we know it would be unable to survive the lethal radiations and the low temperatures of outer space. The trip from space to the earth would be disastrous for living substances such as we have on this planet.

Since living things today appear to be produced only by other similar living things, and are not generated in a short space of time from nonliving materials, we are drawn almost inescapably to the conclusion that life arose on this planet at some distant period in time. Furthermore, it probably arose in a spontaneous man-

ner from nonliving materials. Thus, the proponents of the doctrine of spontaneous generation may not have been completely wrong after all. They were right about the place, but wrong about the time and the mechanism. Let us go back in time to the period when the earth was being formed and gain some understanding of the conditions that could have fostered the emergence of life as we know it.

1–3 HISTORY OF THE EARLY EARTH

If your body were reduced to its constituent elements, we would find that 99% of it consisted of carbon, hydrogen, nitrogen, and oxygen (Table 1–1); the remainder would be composed of trace quantities of the elements sodium, magnesium, phosphorus, chlorine, potassium, iron, and some other metals. The chemical composition of the universe, derived from analyses of the sun, stars, meteorites, and the earth's crust (Table 1–1), demonstrates that the most abundant elements are the very same ones that make up the bulk of living matter. Is

this sheer coincidence? Most scientists think not. The relative abundance of elements in the universe and within our own bodies suggests a common origin. To understand the relationship, we must briefly examine the nature of the universe.

If we had a time machine, we could travel backward in time to the remote past and view the beginning of our solar system. Such a machine unfortunately does not exist, so the solution to the mystery of the origin of our sun and its nine planets must be based on our knowledge of the phenomena that go on in the universe today, most of which are clearly observable through our telescopes.

Where did our solar system come from? How did it arise? The theory that is most popular today and is supported by most **cosmologists** (scientists interested in the origin of stars and planets) is the **primordial cloud** theory. Like other stars, our sun and all the planets were probably formed from a cloud of primordial gas and dust. There are numerous such clouds in space, some of which are today in the process of forming stars and solar systems. The thinly distributed primordial cloud that ultimately gave rise to the planets and the sun eventually became thicker and more condensed by the gravitational attraction of the molecules composing the cloud; the cloud was in constant motion owing to rotational and gravitational forces, and eventually assumed the shape of a flat revolving disc; matter condensed as a series of concentric rings within the cloud. The hub of this spiral cloud had the greatest concentration of matter, and surrounding the center were smaller masses of matter; the center of the condensing cloud formed the sun, and the peripheral condensations that encircled it formed the planets (Figure 1–5). Each planet became distinctive in character depending on its distance from the central mass and the nature of its raw materials. Over the millions of years the heavier elements composing each planet, mostly nickel and iron, settled to the center and formed the core. Lighter substances such as granite and basalt migrated near the surface. Surrounding this was a surface mantle or envelope of lighter gases, mostly hydrogen and helium. The gaseous matter of the sun began to condense further; hydrogen atoms began to undergo fusion into helium, a process that produces a reaction not unlike an H-bomb exploding. With the beginning glow of the sun's thermonuclear furnace, the darkness gave way to light. The radiant heat of the newborn sun drove off the residual gases that surrounded the newly formed planets; after several hundred million years the planets had most of their mass boiled away, and what remained were virtually naked sun-warmed cores—the planets of today. According to recent estimates, astronomers believe that the formation of the Milky Way took place approximately 10 to 15 billion years ago, and our solar system was formed about 5 billion years ago.

Time of the order of billions of years is

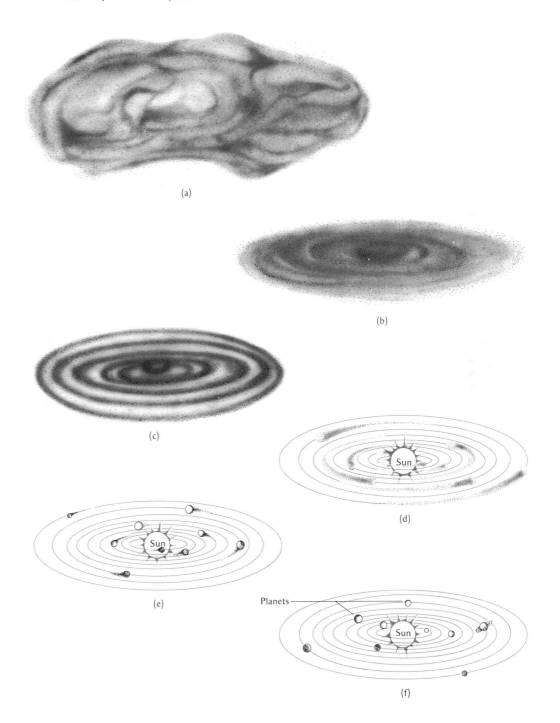

(a)

(b)

(c)

(d)

(e)

Planets

Sun

(f)

sometimes a difficult concept to grasp so let us use an analogy.[1] Let us assume that the history of the earth can be written in a series of 10 volumes: each volume contains 500 pages, and therefore each page represents a million years. The age of the earth and of our solar system is 10 times 500 million, or 5 billion years. If we could read the history of the earth in these volumes, we would find the first 6 volumes (3 billion years) to describe the lifeless period (Figure 1–6). What was it like?

According to astronomers, the primeval solar cloud contained principally hydrogen and helium, and present in smaller quantities were carbon, nitrogen, oxygen, and iron. In this interstellar cloud the carbon and nitrogen and some oxygen could combine with the hydrogen to form methane (CH_4), ammonia (NH_3), and water (H_2O). Note that methane is a simple hydrocarbon—an **organic compound.**[2] Thus the atmosphere of the early earth contained organic materials long before there were any organisms. Evidence that tends to confirm these theories comes from spectroscopic analysis of the atmosphere of the planets of our solar system that are farthest from the sun today, planets whose residual gases were not completely driven off by the heat of the protosun. Their atmospheres are similar to that suggested for the protoearth: the atmospheres of Jupiter and Saturn contain mostly methane, hydrogen, and ammonia, and the atmospheres of Uranus and Neptune are largely methane. The atmosphere of the early earth was a **reducing atmosphere;** that is, it contained much more hydrogen than today in combination with other elements. In contrast, the atmosphere of the earth today is **oxidizing,** with carbon present

---

[1] Modified from A. I. Oparin.
[2] An organic compound is a substance that is a constituent of organisms. In a broader sense it can be any compound in which the element carbon is combined with hydrogen (hydrogenated). Chemically speaking, the carbon is relatively reduced.

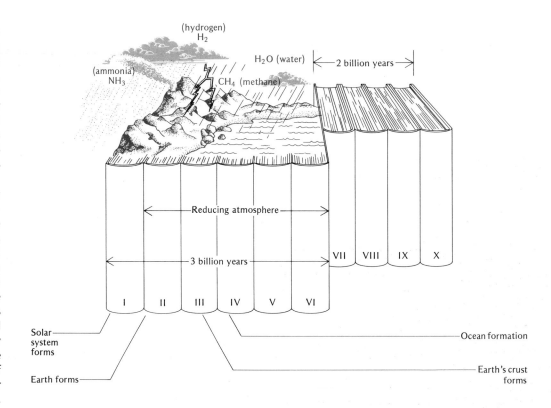

FIGURE 1–6 The history of the earth: formation of the reducing atmosphere.

as carbon dioxide (CO_2—0.03%), nitrogen as N_2 (78%), oxygen as O_2 (21%), and hydrogen as H_2O. The story of how the reducing atmosphere of the earth became an oxidizing one is bound up with the story of life itself, as we shall see. Radioisotope studies suggest that a reducing atmosphere surrounded the earth for 2 to 3 billion years; in the history of the earth this phase could be read in Volumes II through VI (Figure 1–6).

The atmosphere of the early earth contained considerable quantities of water vapor, and as the planet cooled, this vapor condensed and torrents of rain fell. The low places became the ocean basins, the rain that washed down the mountain slopes carved out valleys and river beds. Slowly, salts and minerals, dissolved from

the rocks by the rain, accumulated in the newly formed oceans of the early earth, which had a consistency described by some as like that of pea soup. In the history of the earth, this stage could be read in Volumes III and IV (Figure 1–7).

1–4 THE PRIMITIVE SOUP

An event of momentous importance took place on the primeval, foul-smelling planet earth some 2.5 billion years ago—the first living organisms appeared. How did this happen? Speculation and science have merged into an imaginative attack on this question and have produced some startling results. In 1924 the Russian biochemist A. I. Oparin published a

FIGURE 1–7 The history of the earth: formation of
the primitive soup.

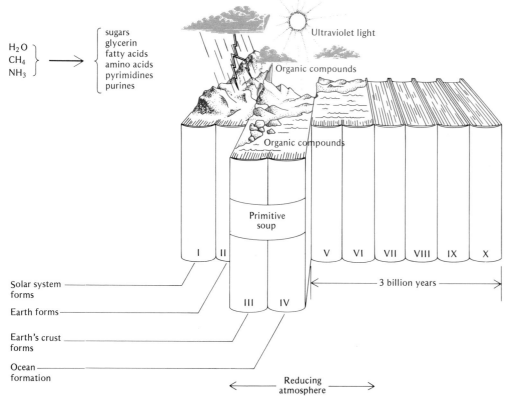

$\left.\begin{array}{l} H_2O \\ CH_4 \\ NH_3 \end{array}\right\} \longrightarrow$ sugars
glycerin
fatty acids
amino acids
pyrimidines
purines

Ultraviolet light

Organic compounds

Organic compounds

Primitive
soup

I II   V VI VII VIII IX X

III IV

Solar system
forms

Earth forms

Earth's crust
forms

Ocean
formation

3 billion years

Reducing
atmosphere

little book entitled *Origin of Life* (English edition, 1936). Disappointed with the lack of progress in understanding how life originated on the planet earth, he pointed out that scientific inquiry had reached a no-exit position by limiting studies of possible processes to those that exist on the earth today. Oparin emphasized the differences between the atmosphere of the early earth and that of the earth today, and suggested that spontaneous organic synthesis could only occur under the former conditions.

What kinds of compounds can be produced from the gases that made up the atmosphere of the early earth, and what sources of energy are needed for their synthesis? Since we cannot go back in time, some uncertainty exists about

the nature of the sources of energy present on the primitive earth; the sources we know of that are available today and that might also have been available then are ultraviolet radiation (from the sun), electrical discharges (from lightning), cosmic rays (raining down from outer space), radioactivity (from radioactive elements), and heat of volcanic origin. The relative importance of each of these in the prebiological synthesis of organic materials is unknown. Given the nature of the primitive earth's atmosphere and the available sources of energy, can we reproduce the same situation now? By mimicking the conditions present on the early earth in the laboratory today, can we in fact produce organic compounds of the complexity found in living organisms, or even

life itself? Just such a challenging problem was given to Stanley L. Miller, then a graduate student at the University of Chicago, by his major professor, Nobel laureate Harold C. Urey. In 1953 Miller described an ingenious experiment which sought to simulate the conditions of the early earth. In his experiments, Miller used a sterile, airtight glass apparatus in which he circulated methane, ammonia, hydrogen, and water (Figure 1–8). As the gases passed a set of tungsten electrodes, they were subjected to an electrical discharge; the circulation of gases was maintained by boiling the mixture in one limb of the apparatus and condensing it in the other. After 1 week the condensed liquid was removed and analyzed. What had originally been pure water now contained a number of complex organic compounds including urea, vinegar (acetic acid), and some very simple amino acids. **Amino acids** are the building blocks of proteins and basic constituents of living organisms. The yield of amino acids was surprisingly high. This demonstration created considerable excitement and produced a flurry of experiments along similar lines with similar results.

The amino acids synthesized in the laboratory under primitive-earth conditions are the same amino acids that are most abundant in the proteins of the organisms we see aound us. Nevertheless, the early reports of the synthesis of amino acids from gases such as ammonia, methane, and hydrogen were received with skepticism. After all, it was an established fact that "living" molecules such as amino acids are formed only by living organisms, and the doubters suggested that the apparatus used was contaminated with microorganisms. Miller, however, showed not only that his apparatus was sterile but that no living microorganisms could survive the temperature of the boiling water. In addition, amino acids were never formed when oxygen or carbon dioxide was present in the apparatus or in the absence of

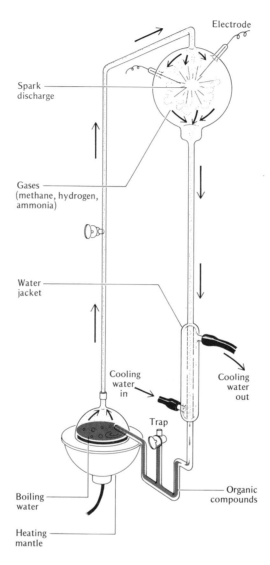

FIGURE 1–8 Stanley Miller's spark-discharge apparatus. The boiling water provides water vapor that carries the gas mixture upward, and the gases are then subjected to a spark (electrical discharge). Afterward the vapors are condensed by a cool-water jacket and returned to the boiling flask.

an electrical discharge; amino-acid synthesis required a reducing atmosphere, not an oxidizing one, and a source of energy.

Living organisms contain many compounds besides amino acids and protein, however. Therefore, it is necessary for us to account for the formation of other organic compounds. The hereditary material of all organisms consists of nucleic acids. Could these compounds too have been formed on the early earth under the conditions that existed at the time? The answer is yes. If a solution of ammonia is heated in the presence of cyanide gas, the purine base adenine (a component of nucleic acids) is produced; variations on this experimental scheme produce other nucleic-acid precursors. Sugars are important components of living systems, and these can by synthesized in a similar way from a dilute solution of formaldehyde in the presence of hot clay.

We can also envisage and demonstrate in the laboratory the **polymerization** (joining) of these small organic molecules into larger ones—for example, amino acids into proteins and purines into nucleic acids. The experiments of Fox and his collaborators are of particular interest. These workers took a mixture of dry amino acids, and by heating them to 150°–180° C were able to link them to form a small protein chain; when water was added to this mixture, small spherules called **proteinoids** developed (Box 1C).

It is clearly possible that in the ancient seas of the early earth, synthesis and accumulation of organic compounds took place. However, organic molecules are labile entities and are quickly destroyed by oxidation and decay. How did organic materials accumulate, and why weren't they destroyed? Since neither oxygen nor microorganisms that effect decay were present on the primitive earth, the organic molecules were able to persist for very long periods of time. The seas of the earth some 3.0 to 3.5 billion years ago contained a dilute

mixture of organic molecules; some scientists estimate that the waters could have been as much as 1% solids. Thus, we can visualize conditions under which a **primitive organic soup** could exist for a long time on the surface of this planet.

1–5 THE PRIMITIVE ORGANISM

The warm seas of the primitive earth doubtless contained a rich array of organic molecules, but how did their concentration increase to such a level that chemical reactions could proceed at a reasonable rate? Several mechanisms may have operated. The simplest way may have been by evaporation. In the lakes, lagoons, and pools of the slowly cooling earth we would expect evaporation to occur and the concentrations of solids to increase. Another mechanism could have been by the adsorption of organic substances on the surfaces of claylike materials.

Perhaps the most ingenious mechanism is the one originally suggested by Oparin—the formation of colloidal droplets (**coacervates**). Colloids are produced by the interaction of organic polymers in solution; under the appropriate conditions, drops of the organic compound are formed that become sharply set off from the watery solution that surrounds them. The boundary between the droplet and its surroundings is formed by a shell of rigidly oriented water molecules (Figure 1–9). Such droplets are capable of reacting chemically with their surrounding medium. Coacervate formation is a method of concentrating highly polymerized substances from very dilute solutions. (Gelatin in water at a concentration of 0.001% can form coacervate droplets in which the concentration of gelatin in the droplet is 10% or a 10,000-fold increase over the concentration in the surrounding medium.) In the laboratory it is possible to produce coacervate droplets containing protein and nucleic acids, and it is possible to produce these at ordinary temperatures. The fate of these droplets depends on the nature of their internal composition as well as on the array of substances in the surrounding medium. Some droplets are able to concentrate materials better than others and thus interact differently with the environ-

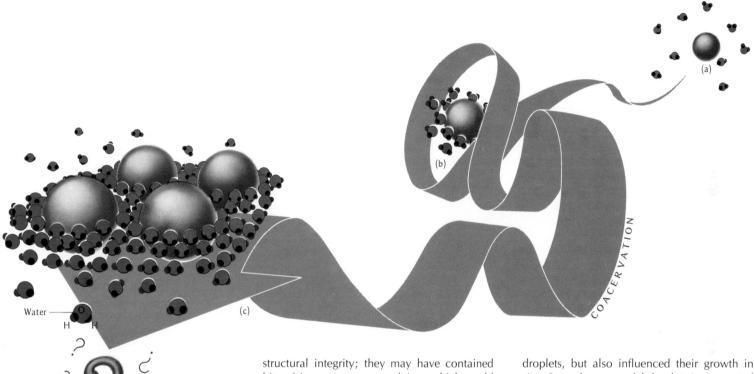

FIGURE 1–9 How colloidal droplets become organized. (a) Colloidal droplet with a diffuse orientation of water molecules. (b) Concentration of water molecules around colloidal droplet. (c) Coacervate, a combination of colloidal droplets with a tight boundary of water molecules.

Water — O
H   H

ment that produced them. Once favorable combinations of materials are established, coacervates take on a dynamic condition. They persist for a short while in a dynamic state, but when equilibrium is established with the surrounding medium, they become static. The coacervate drops, like the proteinoid spherules, are not models of a living system (Box 1C).

We can imagine that in the primitive oceans the surface film of some droplets might have broken down faster than it could be replenished; these droplets must have disintegrated and returned to the primeval soup. Perhaps other droplets were able to maintain their

structural integrity; they may have contained bits of iron, copper, or calcium which could have acted as catalysts so that changes in the droplets occurred at a more rapid rate and the droplets were able to persist in a dynamic state. In time some of these metals may have become permanently attached to the protein and functioned as enzymes do today. (See Chapter 3.) Rates of processes within the droplet were thus further increased, and some droplets could derive materials from the environment at a greater rate than the other droplets. Droplets now were characterized by individuality. Those that could obtain materials from the environment more efficiently were favored by the selective process. If a particular droplet took up materials faster than its neighbor, the neighbor would not fare as well and perhaps would not survive. These faint beginnings of **natural selection** (survival of the fittest) not only contributed to the preservation of the most efficient

droplets, but also influenced their growth in size. Soon the successful droplets incorporated materials that could be laid down as a membrane just below the layer of oriented water molecules; this conferred even greater capacity for differential selection of materials entering the droplet. Soon the growing droplet was so big that the physical forces in the membrane could no longer hold it together. It fragmented, and there resulted two or more drops with stable and environmentally interacting components. The division of these droplets was governed by mechanical forces (for example, waves or increased size), and the droplets fragmented in much the same fashion as droplets of oil in water are broken up by shaking. Those droplets that were able to maintain themselves—to grow and to reproduce—were the ones that survived.

There is not a hard and fast line of demarcation between the structural organization of

BOX 1C

CREATION: Producing Life in the Laboratory

Spheres of living protein made under conditions similar to early earth

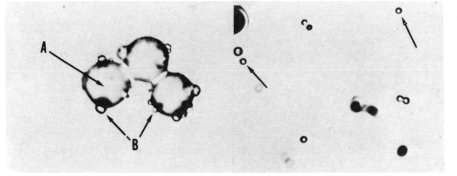

THE "PROTEINOID" (A) SHOWS THE SPONtaneous production of buds (B) as the proteinoid microspheres stood in the mother liquor. When the mother liquor was warmed, "buds" were released from the parental proteinoid microspheres. (Courtesy of S. Fox.)

In the beginning, there existed on the earth the basic chemicals essential for life—but no life, no living thing capable of reproducing itself. How then did life arise on the early earth?

For a long time, the idea that lifeless molecules could combine into a living substance capable of reproducing itself (spontaneous generation) intrigued both scientists and philosophers, but evidence obtained from carefully controlled experiments always seemed to discredit the idea. Recently, however, there has been renewed interest in a modern theory of spontaneous generation.

Dr. Sidney W. Fox and his colleagues at the Institute of Molecular Evolution in Miami have reported a series of exciting experiments: they have been able to synthesize the stuff of life in a laboratory flask. Previously, other workers had shown that if a mixture of gases (methane, ammonia, and water) simulating the atmosphere of the primitive earth is subjected to electric discharge, duplicating on a laboratory scale the lightning bolts that crashed through the atmosphere of the lifeless planet, these simple substances combined into more complex molecules such as amino acids. While not living things themselves, amino acids are the building blocks of proteins, which form the basic structural material of living things. Dr. Fox and his colleagues attempted to form assemblages of these amino acids into proteins and thus continue the biological chain reaction toward life itself. By simply heating dry amino acids in the laboratory, chemical blending occurred and proteins were formed. Upon the addition of water these thermally produced proteins gave rise to little spheres which Dr. Fox calls "proteinoids." The proteinoids resemble certain bacteria in size and shape, and when viewed under the electron microscope they are so similar to bacteria that their true origins are not recognized. "Expert audiences tend to identify sections of bacteria as sections of proteinoid microspheres and vice versa," Dr. Fox said.

To Dr. Fox, these proteinoids suggested other similarities to the earliest forms of life. Proteinoids can take up materials from surrounding solutions (their "mother liquor"), they can multiply by a kind of "budding," grow by accretion, and have enzymatic activity. In many of their properties they are very similar to contemporary living cells. The original chain reaction that produced living forms may have been triggered by volcanic eruptions, or perhaps the heat of hot springs. "A well-known region of appropriate geologic character for the formation of proteinoids," Dr. Fox said, "is Yellowstone National Park." The laboratory-produced proteinoids demonstrate how primordial life could have arisen in the absence of parental or ancestral systems on a lifeless earth.

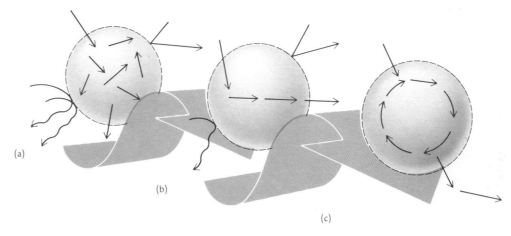

FIGURE 1-10 The development of metabolism in the protoorganism. (a) At first, reactions of organic materials are sporadic. (b) Later they become linked into a series. (c) Finally, the chain reactions branch and form cycles.

coacervate droplets, proteinoid spherules, and the earliest organisms; we can recognize in each of them some of the properties that fit our definition of life. Our own origins are in the substance of the successful line of coacervate droplets. The primitive organisms reproduced by mechanical fragmentation, and the daughter droplets initially obtained their constituents by chance. The constitution of the daughter droplets was dependent on the materials they received. Today all organisms have a more precise hereditary mechanism, mediated through the nucleic acids. It would appear that the protoorganisms did not have their composition and inheritance controlled by nucleic acids, but it is conceivable that in the colloidal droplet some nucleic acid was present, and when this material was able to reproduce itself and control the chemical reactions in the droplet, a protoorganism, resembling in organization the living creatures we see about us today, came into being. How this complex interaction and control came about remains unknown. Since time is the hero of this story, there was no doubt sufficient opportunity for these interactions to take place.

Living systems are characterized by a high degree of organization. The protoorganism may have been capable of obtaining materials from the environment and by virtue of substances within its body may have been able to alter these materials. The chemical alterations (reactions) occurring in present-day organisms, known as metabolism, are mediated by special proteins with catalytic properties (enzymes). Enzymes do not make impossible reactions occur, but they make possible reactions proceed at a faster rate. The primitive organisms no doubt had inefficient catalysts, but with time these became more efficient and more specific. Later, the hereditary materials (nucleic acids) specified (encoded) the formation and structure of the catalysts. We know virtually nothing about the details of the metabolism of

these early organisms, but we assume that the reactions which were originally sporadic became linked into a series, and as these chains of reactions became longer, some became branched and some became cycles (Figure 1–10). Ability to select materials entering the organism, to trap molecules for synthesis, to control rates of reactions, and to obtain energy were all acquired, but we do not know how. It is obvious that improvement in the metabolic machinery of the living system must have involved the development of coordination within the organizational framework; it appears remote that a perfect system developed in a single instant.

We can now consider the self-reproducing and adapting droplet to be a protoorganism, and natural selection acquires its biological meaning—differential reproduction. In our history of the earth, it is in Volumes IV and V that coacervates and protoorganisms make their appearance (Figure 1–11).

1–6 HETEROTROPHS AND AUTOTROPHS

It is now commonly believed that the earliest organisms on the planet earth were **heterotrophs** (*hetero*—other, *troph*—feeder; Gk.).

Heterotrophic organisms require complex organic nutrients such as proteins, amino acids, and sugars for their survival—you are a heterotroph. The primitive organisms living in their organic soup presumably had adequate supplies of these substances, which were continually being formed by lightning, ultraviolet radiation, and so on, acting on the gases of the reducing atmosphere. As these organisms increased in numbers, they must have depleted the soup at a rate faster than the organic molecules could be replenished. Eventually the primitive organisms had to compete for materials, and some may even have fed on other droplets. As droplets could not maintain themselves, they returned to the soup and thus provided materials for other growing and reproducing droplets. As the nutrients disappeared from the primitive soup, those organisms that could synthesize needed materials from some other, perhaps simpler and more plentiful, substances present in the soup had an enormous advantage. In such a way we can imagine the stepwise development of metabolic pathways.

Heterotrophy was the primary nutritional mode of life, but life would have ceased long ago if all nutrition had remained heterotrophic.

The spontaneous production of organic compounds would have been insufficient to maintain a growing and reproducing population indefinitely, and eventually much of the available organic material would have been tied up in the bodies of the heterotrophic organisms. In order to obtain energy to drive the chemical reactions going on inside them, the heterotrophic organisms degraded (broke down) energy-rich molecules from the soup such as sugars (carbohydrates). When carbohydrates are degraded, carbon dioxide and water are released, along with energy. In time the reducing atmosphere changed and contained considerable quantities of carbon dioxide gas just as it does today. Such carbon dioxide in the atmosphere must have screened out some of the high-energy ultraviolet light and reduced the **abiotic synthesis** (a synthesis without organisms) of organic molecules. The decisive step that enabled life to continue on the planet earth was the development of organisms capable of utilizing simple compounds such as carbon dioxide and water for the synthesis of complex organic molecules. This was made possible by the presence in these organisms of pigments that could absorb and make available to them the energy of the sunlight, together with enzymes necessary to catalyze the chemical reactions. Organisms capable of gathering light energy for the production of organic molecules from simple compounds such as carbon dioxide and water are called **autotrophs** (*auto*—self, *troph*—feeder; Gk.), and the process is called **photosynthesis** (literally "building with light"). All green plants today are photosynthetic autotrophs. An important byproduct of their autotrophic activity is molecular oxygen. The appearance of photosynthesis about 2 billion years ago led eventually to the accumulation of oxygen, and this helped transform the reducing atmosphere into an oxidizing one. It is estimated that at present the entire

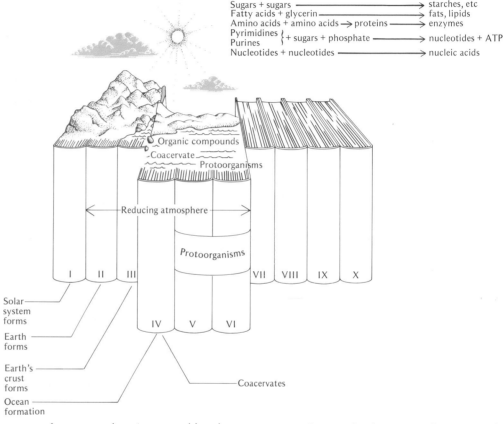

FIGURE 1–11 The history of the earth: coacervates and protoorganisms.

oxygen of our atmosphere is renewed by photosynthetic activity once every 2,000 years. The layer of oxygen that accumulated and surrounded the planet earth was acted upon by solar radiations and gave rise to a layer of ozone that efficiently screened out most of the lethal effects of the sun's ultraviolet radiation. By this means, the energy source that perhaps played so important a role in the spontaneous generation of life on the primitive earth was effectively reduced so it could not destroy that life, and ultimately permitted living organisms to leave the protective filtering waters of the oceans and colonize the land.

Up to the time that free oxygen first appeared in the atmosphere of the primitive earth, the energy needed to perform all the chemical reactions (metabolism) that went on inside the droplets and first organisms was obtained from the organic molecules of the primitive soup. During the breakdown of these molecules within the organisms, energy was released. If oxygen is utilized during these breakdown reactions, more energy per molecule of fuel material can be released than without oxygen. Thus the development of photosynthesis had important implications in regard to energy yield in metabolism for both heterotrophic and au-

BOX 1D The origin of life: an alternate view

An alternate view to the colloidal droplet theory of the origin of life suggests that the primitive soup contained proteins or nucleic acids that were capable of replicating themselves. Later, these molecules became surrounded by a shell of other materials and eventually were enclosed by a semipermeable membrane (Figure 1–12a–c). This view is attractive to some degree because it allows for the development of a precise mechanism for inheritance and change. At present, there is no satisfactory explanation for the establishment of a system that can make identical copies of itself (as does the hereditary material of every living thing) and that can control the manufacture of various kinds of proteins. As the simplest reproducible combination of protein and nucleic acid existing today, **viruses** provide a possible link to these ancient self-reproducing particles (Figure 1-12d); they can thus be considered to be modern representatives of protoorganisms. If the ancestral viruslike particles were free molecules of nucleic acid and could use the organic molecules of the soup to replicate these nucleic acids, then they could have existed as free-living forms so long as the primordial soup and source of material persisted. Modern viruses, however, are obligate parasites and require a living cell to reproduce themselves; incapable of self-reproduction in the absence of another living organism, they are in fact considered by many to be nonliving. Viruses are at the threshold of life. As we learn more about them (Chapter 2), we shall see what a tenuous distinction there is between the living and the nonliving.

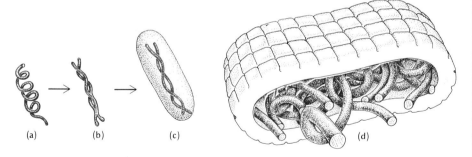

(a)   (b)   (c)   (d)

FIGURE 1–12 Hypothetical scheme of the steps involved in the formation of the first viruslike organism. (a) Single strand of nucleic acid. (b) Synthesis of a sister strand of nucleic acid from materials of the organic soup; this permits self-duplication. (c) The double strand of nucleic acids becomes enclosed in a semipermeable membrane. (d) A cowpox virus sliced to reveal an inner core of nucleic acid surrounded by a protein jacket. Note its similarity to the protoorganism (c).

BOX 1E Does life originate abiotically today?

Does spontaneous generation of organic molecules and perhaps protoorganisms occur today? The answer is probably not. The oxidizing atmosphere of the earth today quickly destroys labile organic molecules, so any that might form could not persist for very long. In addition, living organisms by their production of oxygen and carbon dioxide have effectively removed one of the energy sources (ultraviolet light) for the synthesis of organic molecules by abiotic means. If by some rare chance such primitive droplets were to form, they would soon be fed upon by organisms that populate the earth today.

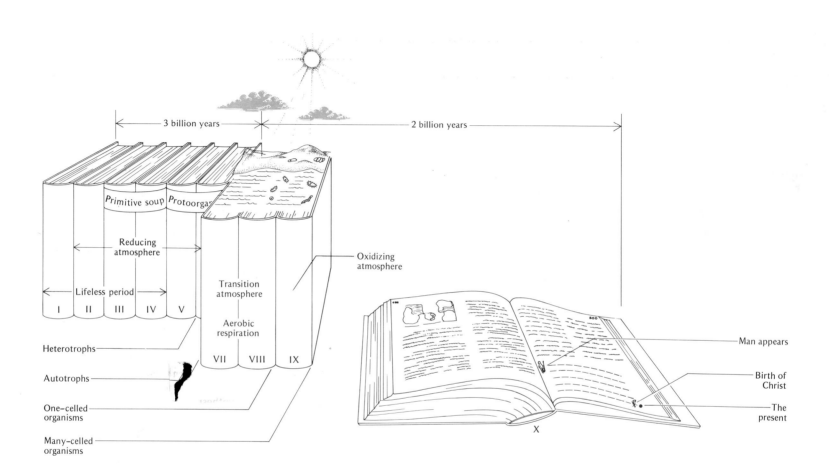

totrophic organisms. The development of autotrophy was further important because it permitted heterotrophs to maintain themselves by utilizing the autotrophic organisms or their products as nutrients, instead of depending on the ever-diminishing supply of organic nutrients in the soup. To this very day we, as heterotrophic organisms, ultimately depend on green-plant photosynthesis for all our food (Chapter 5).

In the history of the earth (Figure 1–13), heterotrophs would appear late in Volume V and autotrophs in Volume VI (over 2 billion years ago). What took five volumes (2.5 billion years) to develop flourishes in abundance and variety on the planet earth in Volumes VII through X (2 billion years). To carry our historical analogy a step further, let us assume that there are 500 words per page in each of the volumes; each word represents 2,000 years. Man appears 50 words from the end of the last page in Volume X; the birth of Christ is the last word on the last page, and we are but a speck on the final period!

SUMMARY

1. Characteristics of living things include the capacities for: growth, maintenance and repair, reproduction, responsiveness, and change.

2. Life can be defined as a property possessed by individuals characterized by the capacity to perform a series of highly organized interacting processes that take place within a definite structural framework.

3. Biologists are usually mechanists, not vitalists. They believe that biology can be understood in terms of physics and chemistry rather than by mystic or spiritual explanations.

4. To understand and investigate biology, rigorous scientific methods are used. A working hypothesis is repeatedly tested, and if the results of repeated experiments can be predicted accurately, the hypothesis is proved and becomes a theory and part of our knowledge.

5. The spontaneous generation of life was a vitalist idea that was refuted by the experiments of Redi, Spallanzani, and Pasteur. Today, living things only come from other living things.

6. The first living things probably arose in the primitive seas of the early earth, which contained a rich array of organic molecules. A variety of theories exists to account for this.

7. Our solar system formed 5 billion years ago from a condensing primordial cloud. The early reducing atmosphere of the earth contained no oxygen; it consisted mostly of methane, ammonia, and water. High-energy ultraviolet light from the sun and other energy sources such as lightning and cosmic rays permitted the abiotic synthesis of complex organic molecules from these simpler substances.

8. Laboratory experiments by Miller and Urey mimicking the conditions of the early atmosphere produced complex organic compounds from an array of simple ones by subjecting them to alternate heat and condensation and an electrical discharge.

9. Fox and Oparin have shown how the formation of proteins and coacervates could have occurred in the primitive oceans and perhaps led to the emergence of competing protoorganisms.

10. The earliest organisms on the earth were probably heterotrophs. Subsequently, photosynthetic autotrophs appeared.

11. The metabolic wastes of autotrophs, specifically oxygen, produced an oxidizing atmosphere. Oxygen destroyed labile organic molecules so quickly that no further abiotic generation of life could occur.

12. An alternative view on the origin of life suggests that the first living organisms were self-replicating molecules resembling modern viruses.

KEY WORDS

life
metabolism
enzymes
growth
reproduction
mutation
evolve
responsiveness
vitalists
mechanists
hypothesis
theory
spontaneous generation
experiment
cosmologists
primordial cloud
organic compound
polymerization
proteinoids
reducing atmosphere
oxidizing atmosphere
amino acids
primitive organic soup
coacervates
natural selection
heterotrophs
abiotic synthesis
autotrophs
photosynthesis
viruses

TOPICS FOR REVIEW AND DISCUSSION

1. What is metabolism? What aspects of metabolism were the Viking Lander experiments designed to detect?

2. Explain why the results of the Viking Lander experiments were equivocal.

3. What is scientific method? Why is it necessary?

4. Today, all heterotrophs depend ultimately on autotrophs for their food. Explain why heterotrophs were probably the first living things, and how they survived initially without autotrophs.

5. Why is autotrophy essential for continued life on earth?

6. Why does the spontaneous generation of life not occur today?

7. How was the "primitive soup" formed? Why are modern oceans not a "soup" of organic materials?

8. What are the characteristics of living things? Is there one characteristic above all others that distinguishes a living thing from a nonliving thing?

9. How could we recognize life on other planets, or even beyond our solar system?

10. What energy sources were available for the synthesis of organic compounds on the early earth, and of what importance was the earth's reducing atmosphere?

11. Describe some laboratory experiments that illustrate the conditions on the early earth that might have contributed to the origin of life.

12. Contrast the views of vitalists and mechanists.

13. Discuss the pros and cons of the following statement: Viruslike particles were the earliest organisms.

14. What was the meaning of "survival of the fittest" to the protoorganism?

University Days

I passed all the other courses that I took at my University, but I could never pass botany. This was because all botany students had to spend several hours a week in a laboratory looking through a microscope at plant cells, and I could never see through a microscope. I never once saw a cell through a microscope. This used to enrage my instructor. He would wander around the laboratory pleased with the progress all the students were making in drawing the involved and, so I am told, interesting structure of flower cells, until he came to me. I would just be standing there. "I can't see anything," I would say. He would begin patiently enough, explaining how anybody can see through a microscope, but he would always end up in a fury, claiming that I could *too* see through a microscope but just pretended that I couldn't. "It takes away from the beauty of flowers anyway," I used to tell him. "We are not concerned with beauty in this course," he would say. "We are concerned solely with what I may call the *mechanics* of flowers." "Well," I'd say, "I can't see anything." "Try it just once again," he'd say, and I would put my eye to the microscope and see nothing at all, except now and again a nebulous milky substance—a phenomenon of maladjustment. You were supposed to see a vivid, restless clockwork of sharply defined plant cells. "I see what looks like a lot of milk," I would tell him. This, he claimed, was the result of my not having adjusted the microscope properly, so he would readjust it for me, or rather, for himself. And I would look again and see milk.

I finally took a deferred pass, as they called it, and waited a year and tried again. (You had to pass one of the biological sciences or you couldn't graduate.) The professor had come back from vacation brown as a berry, bright-eyed, and eager to explain cell-structure again to his classes. "Well," he said to me, cheerily, when we met in the first laboratory hour of the semester, "we're going to see cells this time, aren't we?" "Yes, sir," I said. Students to right of me and to left of me and in front of me were seeing cells; what's more, they were quietly drawing pictures of them in their notebooks. Of course, I didn't see anything.

"We'll try it," the professor said to me, grimly, "with every adjustment of the microscope known to man. As God is my witness, I'll arrange this glass so that you can see cells through it or I'll give up teaching. In twenty-two years of botany, I—" He cut off abruptly for he was beginning to quiver all over, like Lionel Barrymore, and he genuinely wished to hold onto his temper; his scenes with me had taken a great deal out of him.

So we tried it with every adjustment of the microscope known to man. With only one of them did I see anything but blackness or the familiar lacteal opacity, and that time I saw, to my pleasure and amazement, a variegated constellation of flecks, specks, and dots. These I hastily drew. The instructor, noting my activity, came back from an adjoining desk, a smile on his lips and his eyebrows high in hope. He looked at my cell drawing. "What's that?" he demanded, with a hint of a squeal in his voice. "That's what I saw," I said. "You didn't, you didn't you *didn't!*" he screamed, losing control of his temper instantly, and he bent over and squinted into the microscope. His head snapped up. "That's your eye!" he shouted. "You've fixed the lens so that it reflects! You've drawn your eye!"

CELL STRUCTURE: THE ORGANIZATION OF LIFE

2–1 WHAT IS A CELL?

In the living world, the fundamental unit is the cell, and all the processes of life occur within cells. Cells may be broken up by grinding, or they may be extracted to yield particles that perform some of the life processes, but it is only within a unit such as the cell that all the processes of the living system can occur. A **cell** may be defined as the standard unit of biological activity, bounded by a membrane, and able to reproduce itself independently of any other living system. All living organisms, large and small, plant and animal, fish and fowl, man and microbe, are made up of cells. All cells are basically similar to each other, having many structural features in common. It is one of the marvels of nature that the endless variety of living organisms should all be constructed of or depend upon such basically similar units.

How can we recognize a cell when we see one? What are universal features of cells? Exactly what do we mean when we say that the cell is the standard unit upon which all living systems are based? Let us take the last question first, and in attempting to answer it we shall find answers to the first two questions. We can more easily understand the concept of the cell by making an analogy and comparing living organisms to buildings; the rooms of buildings are analogous to the cells of organisms.

Both rooms and cells have boundaries with exits and entrances—rooms have walls, floors, ceilings, doors, and windows, for example, while cells have walls or membranes with pores of various sizes. Both rooms and cells come in a variety of sizes and shapes, with various contents; each kind of room or cell has its own particular use, function, or specialty. Buildings may be composed of only one room or of many rooms. In the same way, organisms may be composed of only one cell, when we

(a) ANIMAL CELL

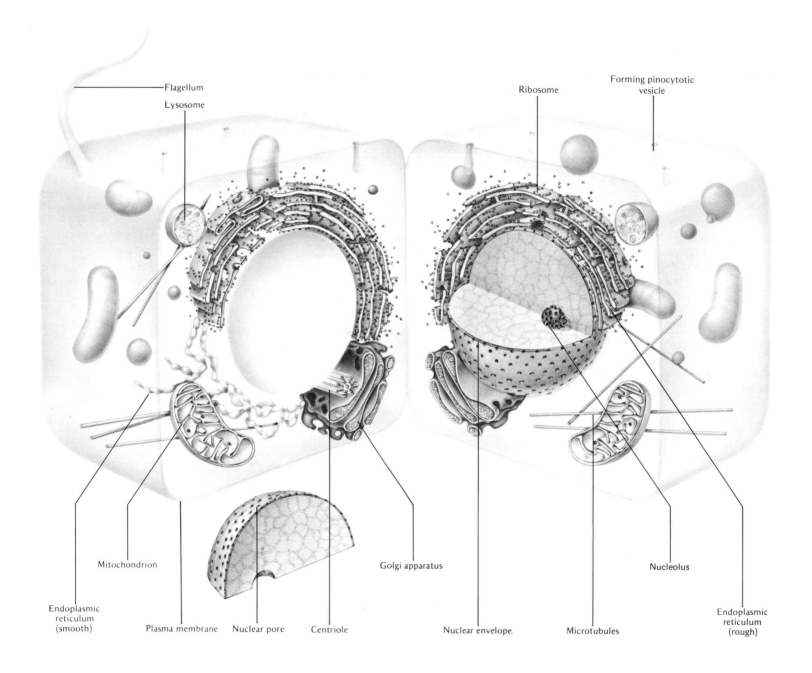

Flagellum

Lysosome

Ribosome

Forming pinocytotic vesicle

Mitochondrion

Endoplasmic reticulum (smooth)

Plasma membrane

Nuclear pore

Centriole

Golgi apparatus

Nucleolus

Nuclear envelope

Microtubules

Endoplasmic reticulum (rough)

FIGURE 2–1 Architecture of a generalized or typical (a) animal and (b) plant cell, as revealed by the electron microscope.

(b) PLANT CELL

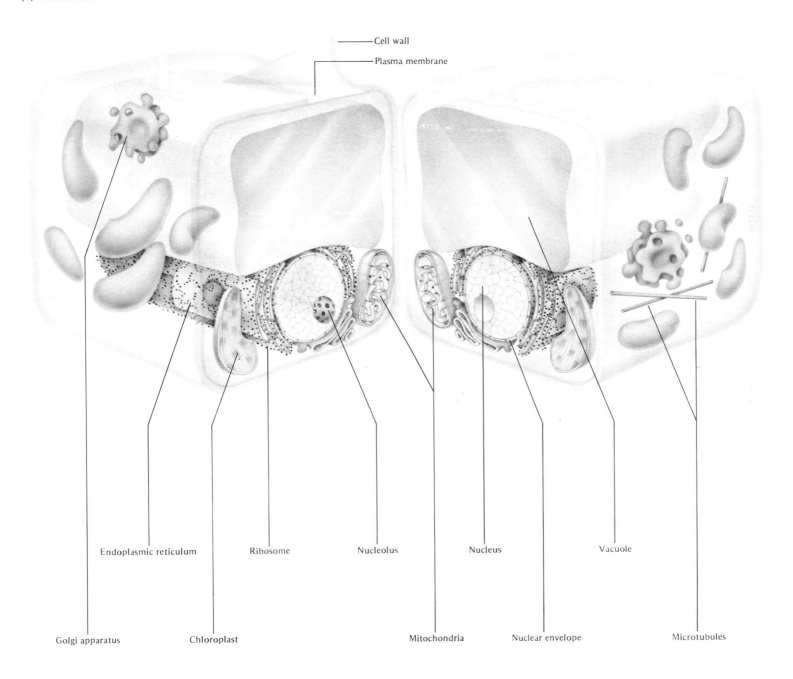

Cell wall

Plasma membrane

Endoplasmic reticulum

Ribosome

Nucleolus

Nucleus

Vacuole

Golgi apparatus

Chloroplast

Mitochondria

Nuclear envelope

Microtubules

describe them as being **unicellular,** or of many cells when we say they are **multicellular.** By combining rooms of various kinds, we can construct a variety of buildings—homes, apartments, schools, offices, and so on. Similarly, different organisms are constructed of a variety of cell types. Just as there is no building without rooms, so there is no life without cells.

Could you describe a typical room to someone who for some reason had never seen one before? What would you include and what would you leave out? Your final description would probably end up by being somewhat all-inclusive, but not applicable to any room that actually exists. It would serve as a kind of map or guide, enabling its user to recognize different features of different rooms as he encountered them, but he would be unlikely to find any existing room exactly as you had described it. In the same way, descriptions of the typical cell end up by being rather generalized; some features will be universal, some will not. The "typical" cell is a convenient aid, but no "typical" cell actually exists.

Like rooms, cells have specialized roles, and correlated with this special function they are specialized in their form. Before examining specialized cells, however, let us construct a typical cell so that we have a guide on our journey through the cells of an organism. Figure 2–1 is a diagrammatic rendition of a typical plant and animal cell.

Cells are separated from their external environment by an interface or plasma membrane. Everything inside the plasma membrane is sometimes referred to as protoplasm, consisting of the jellylike cytoplasm (*cyto*—cell, *plasma*—thing; Gk.) and various structures collectively known as organelles, including the membrane-bound nucleus with its contained nucleoplasm (*nuci*—nut, *plasma*—thing; Gk.). Each organelle represents a highly specialized compartment or submodule in which particular functions of the cell are localized. Continuing our analogy between cells and rooms, we can think of the organelles as items of furniture or equipment (such as stoves, refrigerators, sinks) that are localized in the room and specialized for specific functions. Just as articles of furniture have a form that is easily recognizable and closely related to their function, cellular organelles also show a form related to their function.

The plasma membrane: cell boundary

Although the early microscopists conceded that cells were probably bounded by a membrane, they were unable to see such a structure. The reason was that the thickness of the plasma membrane (0.075 to 0.1 micron (μm)[1] is outside the limits of resolution of the light microscope (Box 2A, page 28). In spite of the inability of the light microscope to resolve the plasma membrane, considerable information had accumulated about its nature and composition before the development of improved microscopes. Information about the chemical composition of the membrane was obtained from the ghosts of red blood cells; **ghosts** are the membrane remnants of the red cell after it has been placed in water, then washed and extracted with a variety of solutions (Figure 2–2). Chemical analyses of these ghosts indicated that the membrane consists of lipid (fatty material) and protein. How are the lipid and protein molecules arranged in the membrane?

Recent studies suggest that in the plasma membrane, and other membranes as well, the bulk of the protein lies external to a double layer of lipid (fatty) molecules; the water soluble

---

[1] Appendix B contains the conversion of metric system units of measure to the English system and vice versa.

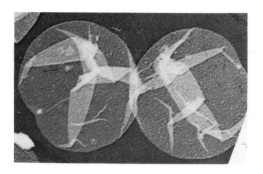

or polar ends of the lipid molecules face outward while the water insoluble ends face each other. Some proteins are embedded in the lipid layer and others extend all the way through it. Further, both the lipids and the proteins have considerable freedom of movement, and this is responsible, in part at least, for some of the membrane's dynamic properties. According to this view, the plasma membrane is like a thick sea of lipid in which protein icebergs float (Figure 2–3a).

Viewed with the electron microscope, the plasma membrane appears as two dark bands separated by a light band (Figure 2–3b); the thickness of the membrane is 75 Angstrom units (Å). Membranes that have this appearance and dimension in electron micrographs have been called **unit membranes,** a term first used by J. D. Robertson.

Gelatin, protoplasm, and cytoplasm

The internal contents of the cell are sometimes referred to as **protoplasm.** Protoplasm has been considered by some authors to be the substance of life, but as we have already observed, the cell is the unit of life and a constituent part of the cell is not by itself alive. Another serious reservation to the use of the term protoplasm is that it suggests a uniformity of composition, and nothing could be further from the truth.

FIGURE 2–3 The plasma membrane. (a) The fluid mosaic model of the membrane; (b) as revealed by the electron microscope. (Courtesy of J. David Robertson.)

FIGURE 2–4 Molecules in action. Owing to inherent thermal agitation molecules are always moving. This is what we really mean by heat. (a) Colloidal droplets have bound water and are jostled to and fro by bombarding water molecules and repulsive electrical forces on the surface of the colloidal droplets. (b) The path of a single droplet.

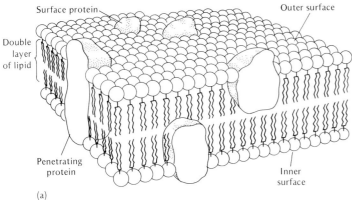

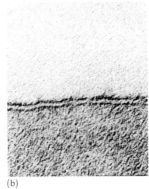

Surface protein

Outer surface

Double layer of lipid

Penetrating protein

Inner surface

(a)

(b)

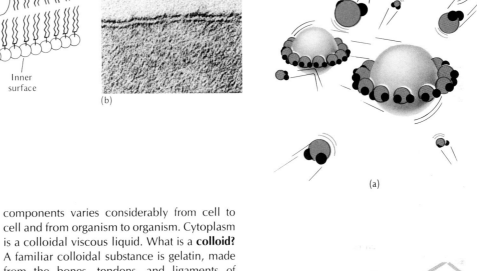

(a)

Start

Finish

(b)

Our own protoplasm differs from the protoplasm of the cells of the dog, the horse, the cat, and the ameba; it is different from the plasms of all other organisms. Your protoplasm is different from that of any other living creature; it is this which confers a uniqueness upon every organism, large or small. Protoplasm varies in its composition and properties not only between organisms but within organisms; the protoplasm of a red blood cell differs from that of a bone cell, a brain cell, and a liver cell. We shall use the term protoplasm because it is a convenient descriptive word to indicate the material inside a cell; however, do not think of it as a defined chemical entity or as characterized by a life of its own, because this is nonsense.

The protoplasm of the cell bounded by the plasma membrane but lying outside the nucleus of the cell is called the **cytoplasm.** Chemically, cytoplasm consists of proteins and lipids, some carbohydrates, minerals, salts, and a great deal of water (70%–90%). The proportion of these components varies considerably from cell to cell and from organism to organism. Cytoplasm is a colloidal viscous liquid. What is a **colloid?** A familiar colloidal substance is gelatin, made from the bones, tendons, and ligaments of cattle by treatment with boiling water to solubilize some of the constituents. It is a protein, and when placed in warm water, the protein molecules become finely and uniformly dispersed in the water and remain so because of their electrostatic charge and because they are being bombarded by the constant activity of the water molecules (Figure 2–4). Because of their charge, the protein (gelatin) molecules bind water molecules; as the particles start to settle out in response to gravitational forces, the gelatin molecules are brought closer together, the like electrical charges on their surfaces cause repulsion, and they move apart once again. Thus, the activity of the water molecules and the electric charge on the molecule of protein maintain the molecules of gelatin in a dispersed phase, and settling does

FIGURE 2–5 The conversion of a colloid from a solution of sol state (a) to a solid or gel state (b). In the sol the particles are randomly dispersed, whereas in the gel there is an orderly network.

not occur. The size of colloidal particles is on the order of 0.001 μm. Gelatin has another property that is familiar to the household chemist: it can exist as a solid or a liquid. Place some warm dissolved gelatin in the cold and it sets—it gels. The change from the liquid state to the solid state is called a **sol-gel** (solution-solid) **transformation.** In the sol state the gelatin molecules are dispersed randomly in the fluid medium; in the gel state the molecules become more rigidly oriented so that a spongy network results, and the water is trapped in the interstices of this network (Figure 2–5). The sol-gel transformation shows the change from a liquid, free-flowing situation to one in which there is considerable resistance to flow, rigidity, and elasticity. Anyone who has witnessed the movement of an ameba has seen how the animal travels forward by producing a fluidlike extension, gelating the sides of the extension so that it forms a tubular, fingerlike projection (the pseudopod), and then pulling along its rear end by a gellike contraction. The colloidal properties of gelatin are generally similar to the colloidal properties of the cytoplasm. The capacity of the cytoplasm to form a semisolid material as well as to exist in a liquid state allows the cell to maintain its shape and also provides a degree of plasticity. Changes in cell form are in part related to these sol-gel transformations of the colloids of the cytoplasm.

Colloids need not consist of a solid such as gelatin in water, but can be formed of two liquids. If we place some light oil in water and shake vigorously, it is possible to disperse the oil droplets and obtain a stable suspension of these. In such a case we have produced an **emulsion**—colloidal droplets of a liquid in another liquid. Homogenized milk is a commonplace emulsion in which the fat globules are so small and evenly dispersed that the cream does not rise to the top and separate out. The colloidal material of the cytoplasm of

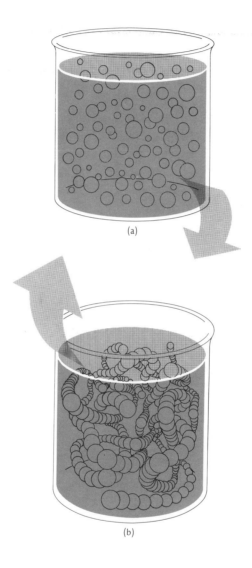

(a)

(b)

The power to distinguish as separate two objects that lie close together, such as lines, dots, or the letters of words, is known as the power of **resolution.** The quality of any optical system, our own included, depends on its resolving power. For example, as you look at the words in the mouse's tale, you are able to see those at the top more clearly than those at the bottom. If the resolving power of your eye is low, you are unable to read the words at the bottom of the tale unaided—the letters merge and become indistinguishable from one another. If your vision is 20/20, you can see two letters, lines, or dots as being separate if the distance between them is 0.1 mm (4/1000 in.) or greater. If the letters, lines, or dots are closer than this, they are indistinguishable from each other even if you have perfect vision.

Resolution and magnification are two different things. By the combination of their lenses microscopes enlarge **(magnify)** objects for us; however, if the optical system employed does not permit objects to be resolved into separate entities, no matter how many times they are enlarged they will still be blurred. Indeed the measure of the performance of an optical system such as a microscope is its enhanced power to record fine detail (increased resolution); without this, higher magnification is worthless, since it merely makes a small blur into a big blur. Resolving power is limited by the kind of illumination used. Objects that are closer to each other by less than one half the wavelength of their illuminating light cannot be clearly distinguished as separate entities. Objects or details of objects that are smaller than 2750 Å (or 0.275 μm) cannot be resolved by white (visible) light as the source of illumination (Appendix B). To increase resolution beyond this level an illuminating source with a wavelength shorter than that of white light must be used. When electrons (generated by a cathode-ray tube) are energized or propelled by a charge of 50,000 V of electricity, they have a wavelength of 0.05 Å. The electron microscope, using electrons as the "light", never achieves its maximum resolution owing to limitations imposed by materials used in construction, but it does have a resolving power of 2 A. To gain some idea of the new dimensions revealed by microscopes, let us become quantitative: the human eye can see objects as small as 100 μm, the light microscope permits us to view objects as small as 0.2 μm, and the electron

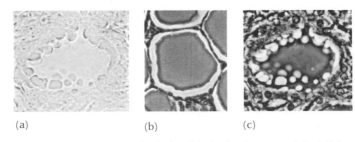

(a)　　　　　　　　(b)　　　　　　　　(c)

FIGURE 2–6 Cells from a dog's thyroid gland. (Courtesy of Carl Zeiss, Inc.) (a) Unstained. (b) Stained. (c) Unstained but viewed with the phase-contrast microscope.

microscope reveals objects as small as 0.001 μm. Looked at in another way, the light microscope has improved our vision 500 times, and the electron microscope has improved it 10,000 times.

Magnification and resolution are not the only factors involved in the study of cells. Under a conventional light microscope, most living cells appear transparent and almost structureless. It is imperative to be able to distinguish the different parts of the cell from one another and from their surrounding materials. Contrast between the various parts of the cell can be achieved by killing the cell (also called **fixing**) in such a way that minimum damage to its parts occurs and then selectively **staining** (dyeing) certain parts of it (Figure 2–6b). Such techniques are employed in both light and electron microscopy. However, when cells are fixed and stained, there is always the possibility that they may be altered in such a way that the structures seen in a prepared cell are not in fact present in the intact, living cell, but are produced by the techniques employed in the preparation of the cell for examination. A special type of light microscope called the **phase-contrast microscope** does permit living cells to be viewed with sharp contrast between their parts without staining and fixing (compare Figures 2–6a and c). With a combination of fixation, staining, phase contrast, and electron microscopy techniques, a rather accurate and more complete view of the cell can be achieved.

Quotation from Lewis Carroll, *Alice's Adventures in Wonderland,* 1865.

cells is mostly protein molecules and fat globules.

The cytoplasm also contains materials larger than colloidal particles, and such particles do not remain stabilized. If left to gravitational forces, these particles would eventually settle out. Systems that include such materials are referred to as **suspensions.** Sand in water is a common example of a suspension; if kept agitated, the sand particles remain suspended, but if left undisturbed, they soon settle to the bottom of the container. Suspensions of crystals and other particulate materials are commonly found dispersed in the colloidal and watery solutions of the cytoplasm.

Although many of the properties of the cell are closely linked to the colloidal state of matter, there exists within the cell material in **solution,** that is, material dissolved more or less evenly in the water of the cell. The size of such particles may be the dimensions of individual atoms or molecules or small clusters of these and is usually less than 0.001 μm. Molecules of this dimension cannot be seen with the microscope, do not settle out due to gravitational forces, cannot be removed by filtration, and are kept dispersed by the constant agitation of the water molecules. When you add sugar to your coffee or prepare a glass of salt water, you produce a solution of these substances; similar solutions are found in our own cells. In addition, gaseous materials such as carbon dioxide and oxygen are dissolved in the cytoplasm of the cell; these are often the least conspicuous components of the cytoplasm.

Action at the interface: plasma-membrane function

The plasma membrane is a dynamic part of the cell. In addition to performing the obvious function of containing the cell components, it regulates the flow of materials into and out of the cell; all substances entering or leaving the cell must pass across the plasma membrane.

The chemical and physical properties of a particular substance determine its ability to penetrate the plasma membrane. Amino acids (the subunits that make up proteins) penetrate moderately well, but proteins penetrate poorly. Fatty acids (components of fats) penetrate well, but fats penetrate poorly. Table sugar (sucrose) does not penetrate cell membranes, but a simple sugar such as dextrose (glucose) penetrates membranes readily. Salts, which ionize, vary in their ability to penetrate membranes, but usually the negatively charged ions pass across the membrane better than the positively charged ones. In general, molecules that have no electrical charge, that are soluble in lipid, and that are small in size, penetrate the membrane most readily.

If we were to place a small drop of blood on a slide with a drop of water and observe the blood cells under the microscope, we should see that the red blood cells would swell and perhaps even burst; conversely, if we placed a drop of blood in a drop of seawater, we should find that the red blood cells would shrink in size and wrinkle. How can we explain such phenomena? In order to understand the properties and the role of the plasma membrane in the swelling and shrinking of cells, it is necessary to examine some of the physical and chemical phenomena that affect the movement of materials from one place to another.

Diffusion. In any situation where molecules (or particles) are free to move, they move from a region of greater concentration to a region of lesser concentration. The movement is net movement. That is, any given particle could be moving in any direction at any particular time; an individual particle could be moving toward the center of concentration, but on the average the movement will be away from the center. Given time, the particles eventually distribute themselves evenly within the space available; at this point the system is in equilibrium. The particles are still moving, but there is no net change in the system; that is, the particles remain evenly distributed. The net movement of molecules from a region of high concentration to a region of low concentration is called **diffusion.**

Gas, liquid, and solid are the three common states of matter. Diffusion is fastest in gases and slowest in solids. Diffusion is due to the inherent heat energy of molecules and atoms, as a result of which they are always in constant motion except at absolute zero (−273°C). In all cases, the rate of diffusion increases with an increase in temperature. Some familiar examples of the rates of diffusion may be helpful. A woman enters a room and the fragrance of her perfume penetrates the air very quickly. The air currents carry the volatile perfumed oils quickly from the source throughout the room, but even in the absence of these currents distribution of the fragrance is rapid. A crystal of writing ink placed on the bottom of a container of water slowly diffuses through the water if there is little disturbance of the water by agitation or heat; left undisturbed the coloration of the water by the ink to produce uniform dispersion may take years, but it eventually will occur. A crystal of writing ink placed on an ice cube moves even more slowly, but its movement can be accelerated by melting the ice. It is obvious that the rate of diffusion in the various media depends on the arrangement of the atoms and the molecules in the medium; in a gas the molecules are capable of free movement, in a liquid the molecules are more tightly held together, and in a solid the movement is very restricted. The process of diffusion plays

FIGURE 2–7 Osmosis: the diffusion of water through a semipermeable membrane. (a) Initially, the sac contains a watery solution of gelatin, sucrose, and salt. The sac is permeable to salt and water only. Because the concentation of water molecules is greater outside the membrane, more water mole-cules collide with the outer face of the sac and enter it; the converse is true for the salt. (b) Hence, there is a net movement of water into the sac, the sac swells owing to the influx of water, and the volume of fluid in the container remains constant.

an important role in the entry and distribution of molecules in the living cell.

Osmosis. Some molecules can pass through the plasma membrane by diffusion and others cannot. In this sense the membrane is considered to be **semipermeable** or differentially permeable. Sausage casing is an example of a semipermeable membrane and can be used as a convenient experimental substitute for the plasma membrane. If we placed a watery solution of table salt, table sugar, and gelatin in a sac made of sausage casing and then suspended this sac in a jar containing pure water, we should find that the membrane is permeable to water and salt, but not to sugar and gelatin. This means that water molecules and salt molecules can freely pass into and out of the sac, whereas the sugar and gelatin are retained within the sac. More salt molecules pass out of the sac than into it because there are more salt molecules inside than outside; conversely, more water molecules pass into the sac than in the other direction because there are more water molecules per unit volume outside the sac. In both instances, there is a net movement of particles from a region of higher concentration to one of a lower concentration owing to the purely physical phenomenon of diffusion (Figure 2–7). In time, the fluid content inside the sac will increase owing to the movement of water into it; the amount of water in the jar will decrease by an equal amount. Since the volume of the sac is limited and some of it is obviously occupied by molecules other than water (salt, sugar, and gelatin) there is a limit to the amount of water that can enter the sac. The increasing amount of fluid within the sac exerts pressure against the sac membrane until, even though the concentration of salt and water is not the same on each side of the membrane, a point of equilibrium is reached, and no further change of volume

Start

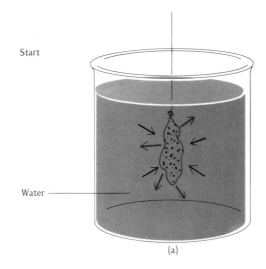

Water

(a)

12 hours later

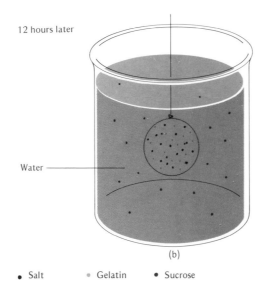

Water

(b)

● Salt ● Gelatin ● Sucrose

occurs. The diffusion of water through the semipermeable membrane is called **osmosis;** the pressure exerted by the water in the sac at equilibrium is called the **osmotic pressure.** At equilibrium the concentrations of sugar, salt, and gelatin in the sac will have become somewhat lower than originally; that is, the solution has become more dilute, and salt can be detected outside the sac by simply dipping a finger into the jar and tasting the fluid that once was pure water. Sugar and gelatin cannot be detected outside the sac since the membrane is impermeable to these substances.

The osmotic pressure of a solution is independent of the kinds of particles but proportional to the number of particles per unit volume. Thus, the higher the osmotic pressure of a particular solution, the greater the number of particles per unit volume in it and the greater the tendency for water to flow into it.

In our sample of blood placed in water, the blood cells became swollen (Figure 2–8) for the same reason as in the case of the sac described above. Our blood cells normally exist in a moderately dilute salt solution (as you know if you have ever placed a cut finger in your mouth), and this solution has the same osmotic pressure as the inside of red blood cells. The cells are in an **isosmotic medium** (osmotic pressure of the external environment is equal to that inside the cell). When a red cell is placed in water, a medium **hypoosmotic** to blood cells (osmotic pressure less on the outside than inside), there are fewer salt particles outside the cell, and water molecules move into the cell, dilute its contents, and stretch the plasma membrane (Figure 2–8). A red blood cell placed in seawater shrinks because the medium is **hyperosmotic** to the blood cell (osmotic pressure greater on the outside than inside), water passes out of the cell, and the membrane collapses (Figure 2–8).

Against the tide: active transport

Charged molecules do not penetrate the plasma membrane readily. Nevertheless, in spite of the fact that potassium ions (K^+) are higher in concentration inside the cell and sodium ions (Na^+) are higher in concentration outside the cell, sodium ions can pass out of the cell and potassium ions can pass into the cell; that is, they can move against their concentration gradients. These passages apparently defy the principles of diffusion we talked about in our previous discussion of concentration gradients; therefore, some mechanism other than diffusion must be operational. The cell has to expend energy or do work in order to transport materials against a concentration gradient. This process is known as **active transport** and involves the movement of materials across the plasma membrane by the expenditure of cellular energy.

Since the plasma membrane is slightly permeable to sodium, sodium moves into the cell by diffusion, but instead of continuing until equilibrium is established, the sodium which enters is rapidly expelled. In so doing, the cell is working against a concentration gradient and in effect is pumping the sodium out; the system is therefore called the **sodium pump** and may perhaps be crudely likened to a bilge pump bailing water from a leaky ship. The situation for potassium is just the reverse; it is retained at a higher concentration within the cell. The fact that energy is required for the system to work is easily demonstrated, since if we poison the cell with cyanide (a respiratory inhibitor) or remove dextrose (a source of cellular energy) then the cell leaks potassium and sodium enters freely until equilibrium is established. All cells engage in active transport, and they accumulate many kinds of materials. For example, sugars and amino acids are moved into the cell by active transport.

What is the mechanism of active transport?

FIGURE 2–8 (top) The swelling or shrinkage of a red blood cell depends on the osmotic pressure of the solution in which it is suspended.

FIGURE 2–9 (bottom) Active transport as described by a hypothetical model. Carrier substances in the membrane combine with molecules to be transported across the membrane. Material is moved uphill, against the concentration gradient, by the expenditure of energy.

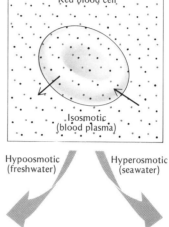
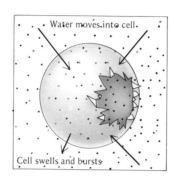
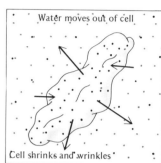
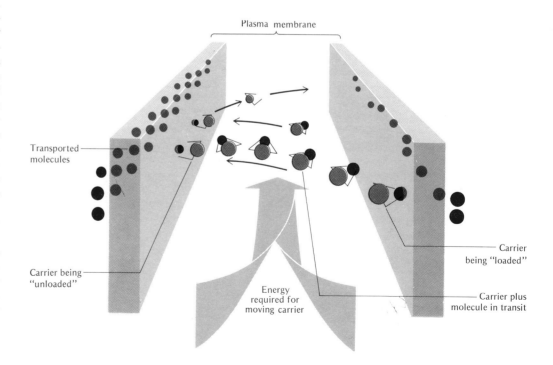

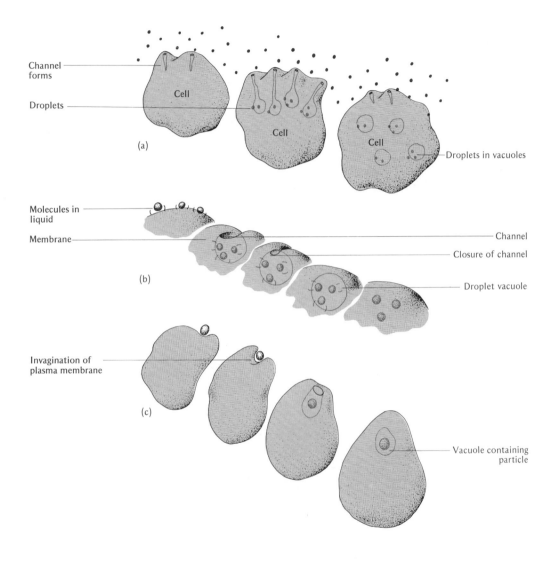

FIGURE 2–10 Pinocytosis and phagocytosis. (a) Cell drinking (pinocytosis) in a living cell. (b) Pinocytosis in schematic detail. (c) Cell eating (phagocytosis).

Channel forms

Droplets

Cell

Cell

Cell

Droplets in vacuoles

(a)

Molecules in liquid

Membrane

Channel

Closure of channel

Droplet vacuole

(b)

Invagination of plasma membrane

(c)

Vacuole containing particle

One theory suggests that the molecule to be transported across the membrane is bound to a material in the membrane called a carrier (Figure 2–9). The carrier and the molecule move across the membrane, and at the inner surface of the membrane the transported molecule is released and enters the cytoplasm. Now the carrier acquires energy at the inner face of the membrane and returns to the outer surface to pick up another molecule for transport. Conversely, the carrier could also move molecules from the inside to the outside.

Cell drinking and cell eating: pinocytosis and phagocytosis

The processes of diffusion, osmosis, and active transport are not the only mechanisms whereby the cell can acquire materials from the external environment. In 1931 W. H. Lewis made a remarkable discovery while observing some human cells he had grown in a sterile medium outside the body. He observed that the cells produced fingerlike projections at the surface, and within these projections fine channels developed. Small amounts of fluid from the medium flowed into these channels, and at the base these were pinched off to form tiny droplets, each enclosed by a portion of the plasma membrane (Figure 2–10a, b). In time the fluid was released from the membrane-enclosed vesicle and became mixed with the cytoplasmic materials. Lewis called the phenomenon **pinocytosis,** which in Greek means "cell drinking." In this manner the cell is capable of taking in soluble molecules that ordinarily would have difficulty in penetrating the membrane. The cell can, as one author has said, "take them (the molecules) by the scruff of the neck and pull them pinocytotically into its interior." The phenomenon of pinocytosis is an energy-dependent process, highly selective for certain substances such as salts, amino acids, and certain proteins, all of which are water-soluble. This process, which occurs widely in a variety of cells, could account for some of the phenomena described by some workers as active transport, since both situations show selectivity and energy dependence. The phenomenon of pinocytosis may be of considerable medical importance. For example, suppose it were necessary to treat a group of cells with a drug that ordinarily does not penetrate the plasma membrane owing to its size or its electrical charge. If these cells were treated with the drug made up in a solution containing a pinocytosis-inducing material such as albumin, then the drug would enter in piggyback fashion along with the albumin and could do its job inside the cell.

Phagocytosis, or "cell eating," is in many respects similar to pinocytosis, but whereas in pinocytosis the material engulfed by the plasma membrane is in solution, in phagocytosis the material is in particulate form. Our own white blood cells react to foreign materials such as bacteria, cell debris, or even a damaged cell by engulfing the material and surrounding it in a vesicle lined by an infolding of the plasma membrane (Figure 2–10c). Once inside the phagocytic cell, cytoplasmic enzymes are secreted into the vesicle by the cell, and these effectively degrade the material into a harmless form; the smaller molecules then pass across the membrane of the vesicle and enter the cytoplasm.

Microspecialists: cytoplasmic organelles

The word **organelle** means "little organ" and is obviously not to be taken literally. The early microscopists, who used relatively poor microscopes to study the cell, had vivid imaginations, and some of the structures they observed they described as "hearts," "stomachs," and "kidneys." Today it is recognized that the objects these early cell explorers discovered are not true organs, but subcellular structures within the cell. However, since organs (composed of many cells) are the functional units within our bodies and these subcellular structures are functional units within cells, the name organelle has been retained.

Endoplasmic reticulum and ribosomes: transport, storage, and synthesis. In 1945 Keith Porter found that cells have a system of internal membranes which form an extensive network in the cytoplasm, and he called this system the **endoplasmic reticulum.** The endoplasmic reticulum is visible in great detail with the electron microscope (Figure 2–11) and consists of an extensive network of membrane-enclosed spaces.

The membranes of the endoplasmic reticulum may appear smooth along their outer surface. However, sometimes the outer surface is studded with small particles called **ribosomes,** and in this case the endoplasmic reticulum has a coarse appearance and is spoken of as rough. The rough endoplasmic reticulum is found with greater frequency and abundance in cells which are actively synthesizing protein. The manufacture of proteins in the cell is associated with the ribosomes, which are dense particles containing protein and ribonucleic acid (RNA). Ribosomes need not always be associated with the endoplasmic reticulum, but may exist free or in clusters in the cytoplasm; these clusters are called **polyribosomes.**

The endoplasmic reticulum, by virtue of its extensive branching, functions in transport; in some cases, the endoplasmic reticulum accumulates large masses of protein and acts in a storage capacity. Of considerable importance is the fact that the endoplasmic reticulum provides a kind of structural framework or a compartmentalization of the cytoplasm. The presence of these sheets and tubules, lined by membrane, provides a tremendous surface area within the cell, and if certain areas of the membrane are specialized for a particular function, the cell shows local patterns of synthesis or breakdown of substances.

The Golgi apparatus: secretion. In 1898 the Italian scientist Camillo Golgi observed a series of vesicles in the nerve cells of a barn owl. These structures, which could be stained with certain stains containing silver and which accumulated other dyes, were subsequently found to exist not only in nerve cells but in a number of other cells. The vesicles came to be called the **Golgi apparatus,** after their original discoverer.

Since the structures could be seen at the level of the light microscope only with the help of stains and dyes, some workers contended that they were artifacts produced by the microscopists and their dyes and did not really exist in the living cell. The electron microscope has revealed that Golgi was right and his detractors and critics were wrong. The Golgi apparatus as disclosed by the electron microscope (Figure 2–12) consists of a group of parallel membranes in flattened saclike configurations (cisternae). The individual sac (cisterna) in the stack of Golgi membranes is usually thin in its center and dilated or distended at the edge. At the edge of the Golgi membranes, vesicles of various sizes are seen. Golgi membranes are found in both plant and animal cells.

Controversy regarding the Golgi apparatus has moved from existence to function. Some electron microscopists believe the Golgi apparatus is a part of the endoplasmic reticulum and that it may be a site at which new membranes for the endoplasmic reticulum are manufactured. Other workers suggest that its primary function is synthesis and/or storage of materials. In certain cells, such as those of the pancreas, the Golgi membranes participate in the elaboration and packaging of digestive enzymes. Vesicles containing the digestive en-

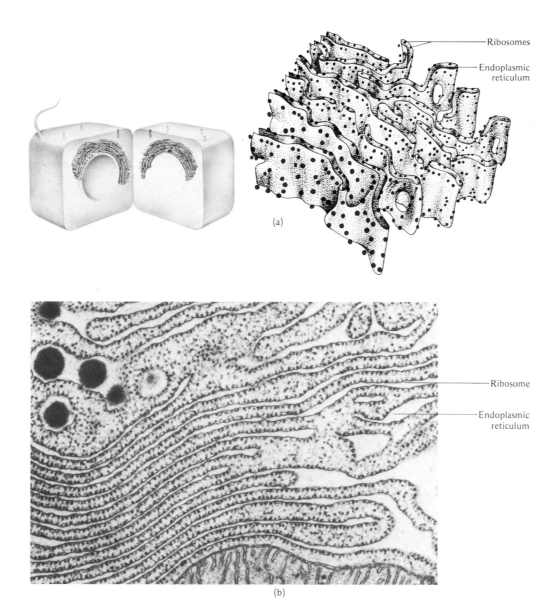

FIGURE 2–11 The endoplasmic reticulum. (a) Reconstruction. (b) As revealed by the electron microscope. (Courtesy of K. R. Porter.)

Ribosomes

Endoplasmic reticulum

(a)

Ribosome

Endoplasmic reticulum

(b)

zymes leave the surface of the Golgi stack, migrate through the cell, and are secreted into the intestine. Recent evidence indicates that lysosomal enzymes are packaged and secreted by Golgi membranes. Much remains to be learned about the Golgi apparatus and its special cellular roles.

Mitochondria: the power house. Mitochondria are microscopic bodies that can be stained with certain dyes. They were described in 1890 and called bioplasts ("life germs") because they were thought to be the smallest living things. In 1897 these structures were renamed mitochondria (singular, mitochondrion; *mito*— thread, *chondrion* granule; Gk.) and recognized as true organelles of the cell. Mitochondria vary in number and shape: some are rods, others are granules, and still others are filaments (Figure 2–13). There may be a single mitochondrion in a cell, or a thousand or more. They are widely distributed in almost all cells and function in cellular metabolism. Carbohydrates, fats, and proteins, which provide the metabolic fuel for the machinery of the cell, are broken down by enzymes of the cytoplasm to smaller molecules that can enter the mitochondrion. Through a series of complex chemical reactions, together called respiration (discussed later), these organic molecules are oxidized to carbon dioxide and water, and the energy released during the reactions is trapped for useful work by the cell. Approximately 90% of the cell's energy is obtained through the chemical reactions that take place in the mitochondrion. It is for this reason that the mitochondrion is sometimes called the powerhouse of the cell.

In living cells viewed under the light microscope, mitochondria appear as lively little organelles that tend to aggregate where energy is required; they move back and forth, change their shape, and fragment to give rise to new mitochondria. The activity is probably due not to inherent movement but to the jostling by other particles and the streaming activity of the cytoplasm. Although it is great fun to watch the active particles under the light microscope, it is with the electron microscope that greater detail of their structure can be observed. A mitochondrion is bounded by two membranes; the outer membrane is smooth and encloses the organelle, while the inner membrane is thrown into a series of folds or pleats (Figure 2–13). These pleats, called mitochondrial crests or **cristae,** extend well into the mitochondrion itself and provide either a series of shelflike partitions or a mass of fingerlike projections. Cristae are most abundant in cells with high metabolic activity, and it is here that most of the respiratory activities of the cell take place.

Chloroplast: energy fixation. The food we eat and the oxygen we breathe are produced by organelles called **chloroplasts.** Chloroplasts ("green bodies") are found in green plants; they are disc-shaped bodies about 5–8μm in diameter and up to 3μm in width, and they contain a green pigment called **chlorophyll.** From 1 to 300 chloroplasts are found in each green plant cell, and new chloroplasts are produced by division of other chloroplasts. The complex chemical processes of **photosynthesis** take place in the chloroplast, where the energy of sunlight is trapped and utilized for the synthesis of complex organic materials (carbohydrates) from simple inorganic molecules (carbon dioxide and water).

There is considerable structural organization to a chloroplast, and this is best revealed by the electron microscope (Figure 2–14). Each chloroplast is surrounded by two membranes that enclose its contents and separate it from the cytoplasm. The internal portion of the chloroplast consists mainly of two parts: a fluid matrix (**stroma**) surrounding a complex membrane system. The membrane system of the chloroplast generally consists of a series of

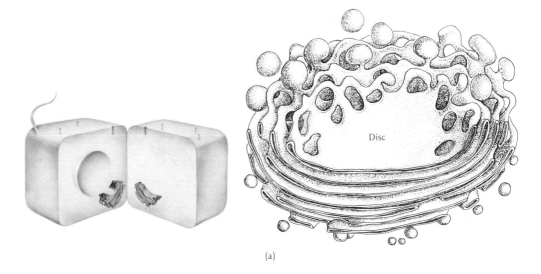

Disc

(a)

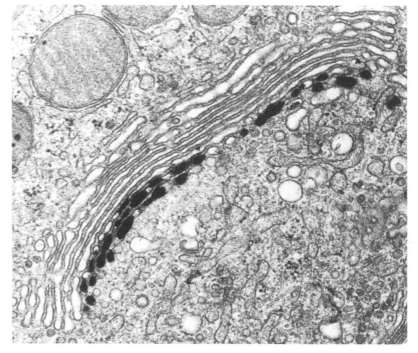

(b)

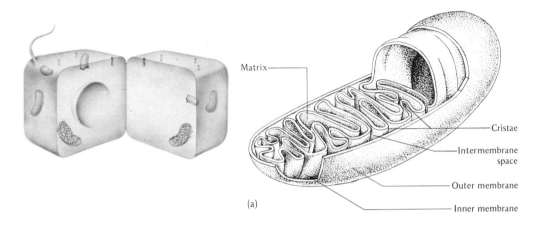

Matrix

Cristae

Intermembrane space

Outer membrane

(a)

Inner membrane

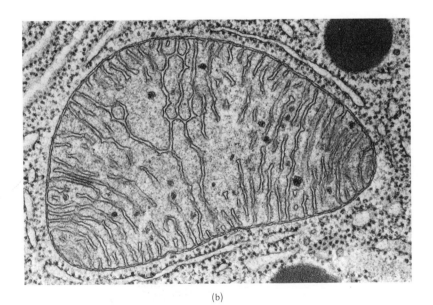

(b)

multilayered fluid-filled discs (**grana**) resembling a stack of coins or pancakes and a system of closed flat sacs (**lamellae**) extending throughout the chloroplast and connecting the grana. The grana are so-called because they appear as small green grains under the light microscope, and it is here that chlorophyll, carotenoids, lipids, and proteins are associated into units that can convert the radiant energy of the sun into biologically usable forms of chemical energy; the grana are also the site of oxygen production. The chemical energy made in the grana is used in the stroma for the fixation of carbon dioxide into carbohydrate. It is clear that even within the chloroplast itself there is a division of labor.

Chloroplasts do not develop in the dark, and so plants that have been grown in the dark appear colorless or pale green. However, if one puts such a plant in the light, there is a rapid elaboration of the grana and chlorophyll, and greening occurs. Chloroplasts do not develop in all plant cells, and root tissues even when exposed to the light ordinarily do not green up. (Carrot, pea, and morning glory roots are exceptional cases.)

Higher plants contain a variety of intracellular bodies called **plastids.** Usually we distinguish two types of plastids: **chromoplasts** ("colored bodies") and **leucoplasts** ("white or colorless bodies"). Chloroplasts belong to the chromoplast group. Other kinds of chromoplasts give many flowers and leaves their colors of yellow, orange, or red; these do not show the extensive membrane development seen in the chloroplast, but are membrane-bound vesicles containing pigment. Leucoplasts serve as food-storage depots for the cell and contain oil, starch grains, and protein. The edible part of the potato plant consists of starch-laden leucoplasts.

The color changes in ripening fruit and in leaves during the autumn are due to the decline of chloroplasts and the unmasking of other chromoplasts. In some cases the chloroplasts themselves turn yellow or red because there is a decline in the chlorophyll content and an unmasking and synthesis of the yellow-orange carotenoids present in the grana. When bananas, tomatoes, or sweet peppers ripen, they change from green to yellow or from green to red as the chlorophyll is broken down and the carotenoids originally present become dominant. The reasons that underlie such changes remain obscure.

Lysosomes: suicide bags. Lysosomes were discovered very recently (1952) and have been found only in animal cells. We are still somewhat uncertain about their exact function, but we know that they are rich in hydrolytic enzymes, which are capable of breaking down and destroying a number of important cellular constituents. Lysosomes are separated from the cytoplasm by a single unit membrane; under normal conditions the enzymes within remain inactive and cannot contact the cellular materials, which they would soon destroy. However, if the lysosomal membrane becomes damaged, the enzymes are released and result in the destruction of the cell. It seems puzzling that the cell manufactures the seeds of its own destruction and then isolates them; however, there may be a sensible explanation for this. In some cases cell death is a natural event in development. For example, when a tadpole changes into a frog, a number of bodily alterations occur. Perhaps the most dramatic of these is the loss of the tadpole's tail. As is evident to every small child who has ever collected pollywogs, the tail does not simply drop off, but is reabsorbed by the developing adult frog. During the resorptive process, the tail cells disintegrate, and part of their demise is due to the activity of the lysosomes. The end result of the digestion and reabsorption of the tadpole's tail is a tailless adult frog.

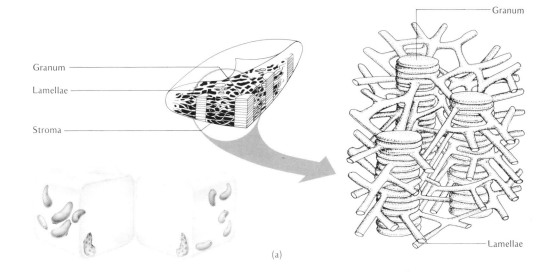

(a)

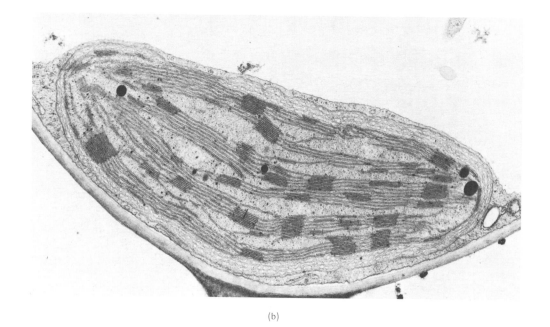

(b)

Lysosomes also play a role in cellular digestion. Bacteria and other foreign bodies are phagocytized by the ameboid white blood cells of our body and contained in membrane-enclosed spaces called vacuoles; lysosomes migrate to the surface of these vacuoles, the membranes of the vacuole and the lysosome fuse, and the lysosomal enzymes enter the vacuole (Figure 2–15). In such a manner, the foreign substances are broken down in the vacuole by the enzymes contributed by lysosomes and, once degraded, are utilized by the cell.

Lysosomes may also play a role in drug action. Vitamin A destroys lysosomal membranes, and when animals receive excess vitamin A, a condition known as vitamin-A intoxication results, which is associated with spontaneous fractures and other lesions in bones and cartilage. Cortisone has the opposite effect; it stabilizes the lysosome and this property may account for the anti-inflammatory effect of this drug. The other roles of these newly discovered organelles await further research.

Vacuoles: inner space. The cell may contain fluid-filled spaces surrounded by a membrane, called **vacuoles.** We have already discussed the formation of vacuoles in animal cells during the processes of pinocytosis and phagocytosis.

Plant cells have more prominent vacuoles; in young plant cells the vacuoles are many and they are rather small, but as the plant gets older, these vacuoles fuse to form a large, conspicuous central vacuole (Figure 2–1). The hydrostatic (fluid) pressure of the vacuole forces the cytoplasm to the periphery of the cell, and there it remains as a thin layer closely pressed against the plasma membrane. The vacuole of plant cells contains primarily water and a variety of other substances together called **cell sap;** because cell sap has a higher osmotic pressure than the external medium, water

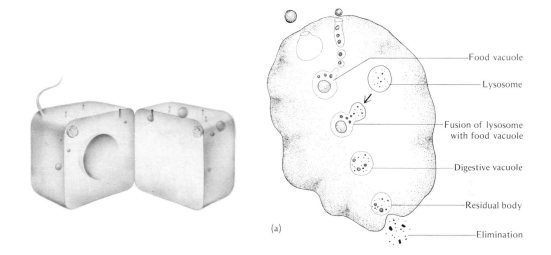

Food vacuole

Lysosome

Fusion of lysosome with food vacuole

Digestive vacuole

Residual body

Elimination

(a)

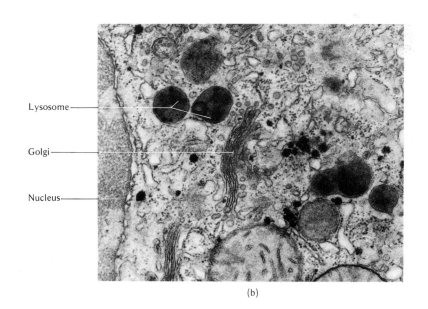

Lysosome

Golgi

Nucleus

(b)

moves into the cell and the cell becomes turgid. It does not burst because it is surrounded by a rigid cell wall (Figure 2–23). The turgid nature of the plant cell contributes to the strength of certain plant stems and the crispness of vegetables such as celery and lettuce. Limp celery and wilted lettuce are produced when large amounts of water are lost from the cells of the plant. The cells lose their rigid form and then become limp in much the same way that a tire of your automobile goes flat when it loses air. The plant cell stores a number of important substances in the watery fluids of the vacuole, and these include amino acids, proteins, salts, sugars, and the red pigment anthocyanin. The red color of roses, red onions, and beets is due to the presence of anthocyanins in the vacuolar fluid.

Microtubules and microfilaments: Minimuscles: Cellular movements involve two kinds of rodlike structures: microtubules and microfilaments (Figure 2–16). Microtubules are capable of rapid assembly and disassembly and are primarily composed of the protein tubulin. **Microtubules** are the structural framework of cilia and flagella; in the mitotic spindle, microtubules act to move the chromosomes during cell division.

The threadlike **microfilaments** are smaller in diameter than microtubules and do not contain tubulin. They are often associated with the inner surface of the plasma membrane where they occur in bundles and sheets. The musclelike contractions of microfilaments are involved in cell movement, in changes of cell shape and in cytoplasmic streaming.

Centrioles, cilia, and flagella: motion. As early as 1887, microscopists found that their stains showed a dark central body just outside the nucleus; it was given the name **centriole** ("central body"). This densely staining granule plays an important role in the division of animal cells, a function that will be discussed in

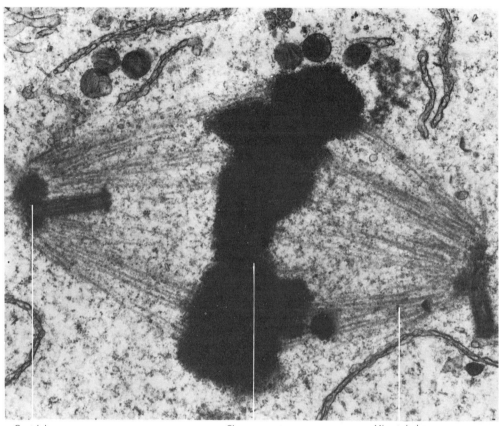

Centriole   Chromosomes   Microtubules

FIGURE 2–17 The centriole. (a) Reconstruction. (b) Electron microscope view of centriole cut in cross and longitudinal section. (Courtesy of K. R. Porter.)

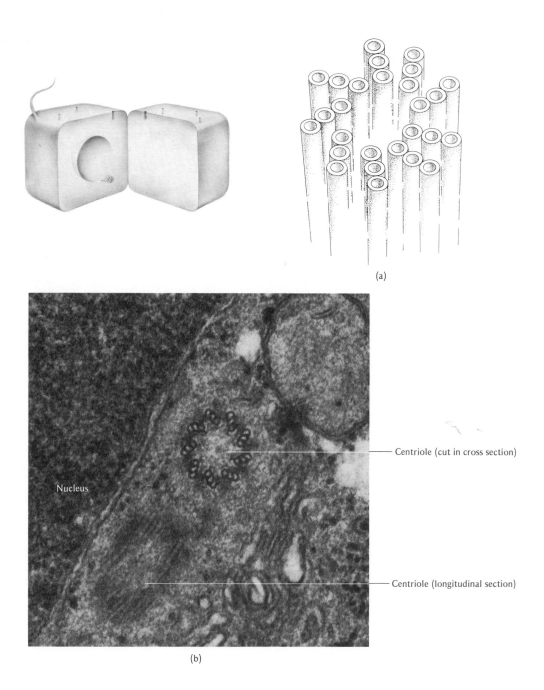

(a)

Nucleus

—— Centriole (cut in cross section)

—— Centriole (longitudinal section)

(b)

Chapter 7. Centrioles are not indispensable during cell division, however; the cells of higher plants contain no centrioles, and yet are still capable of dividing properly. When the centriole is studied with the electron microscope we can see that it consists of a circlet of nine microtubules, each of which is further subdivided into three smaller tubules (Figure 2–17). If we reconstruct a three-dimensional picture from the two-dimensional thin sections viewed with the electron microscope, we find that the nine triplets of the centriole form a short, hollow cylinder. Centrioles are self-replicating, but their chemical nature and exact function remain obscure.

The surfaces of many cells have short hairlike or long whiplike appendages that move fluid across the surface of the cell. If the cell is free to move on its own, the appendages can propel the cell in a watery medium (Figure 2–18). The hairlike appendages are called **cilia** (*cilium*—eyelash; L.), and the whiplike appendages are called **flagella** (*flagellum*—whip; L.). The cells of the trachea (windpipe, see Chapter 14) are covered with cilia that move mucus across the delicate surfaces so that foreign materials such as soot and tobacco tars do not lodge directly on the plasma membrane (Figure 2–18a). In the human female, the cells of the Fallopian tubes of the reproductive system (Chapter 10) are lined with cilia that move the egg from the ovary along the tube and toward the uterus (Figure 2–18b). Spermatozoa produced in the testes of the human male are motile because of the activity of their lashing tails, which are really flagella (Figure 2–18c).

What do cilia and flagella have to do with centrioles? In cells that bear cilia and flagella, the centriole replicates itself, and the copies migrate to the cell surface, where they become basal bodies that in turn give rise to the cilia and flagella (Figure 2–18). Every cilium and every flagellum has the same structure when

FIGURE 2–18 Cilia and flagella. (a) Cilia line the surface of tracheal cells. (b) Cilia line the cells of the
Fallopian tube. (c) Sperm tail (flagellum). (d) Cross
section of the flagellum as seen with electron microscope. (Courtesy of D. Friend.) (e) Diagrammatic
reconstruction of a cilium (or flagellum).

viewed with the electron microscope. Each
cilium and flagellum is covered by the plasma
membrane, and internal to this is a ring of nine
pairs of microtubules surrounding two central
tubules (Figure 2–18). The basic structure of
these organelles is often referred to as a nine-
plus-two arrangement of tubules. Only the
cylinder of nine tubules continues below the
cell surface (Figure 2–18), and there it forms
the basal body, which appears structurally
identical to the centriole. The ability to move
rhythmically or to beat is an inherent property
of cilia and flagella, and even when detached
from the cell, they can be made to move.
However, the mechanism by which the microtubules produce such contractions is unknown
at the present time.

Nucleus: control center. The most conspicuous feature of a cell viewed with a microscope
is the nucleus. This organelle was first described
from the cells of a flowering plant in 1833 by
Robert Brown, but it was soon recognized that
both plant and animal cells contain a nucleus.
The nucleus, as we shall see, is the control
center of the cell, and it contains the genetic
material that determines the characteristics of
the cell and its offspring. The nucleus is quite
obvious in living cells, but it becomes even
more recognizable if we add a stain such as
iodine, which binds strongly to substances in
the nucleus called **d**eoxyribo**n**ucleic **a**cid
(DNA) and protein. The DNA and protein are
organized into structures called **chromosomes**
(*chroma*—color, *some*—body; Gk.). In a non-
dividing cell the chromosomes are in an extended and diffuse state (Figure 2–19), their
most active metabolic condition.

The nucleus is a relatively large structure,
spherical in shape and separated from the
cytoplasm by a nuclear envelope. The electron
microscope reveals that the nuclear envelope
really consists of two membranes and that the
membranes have pores that appear to connect

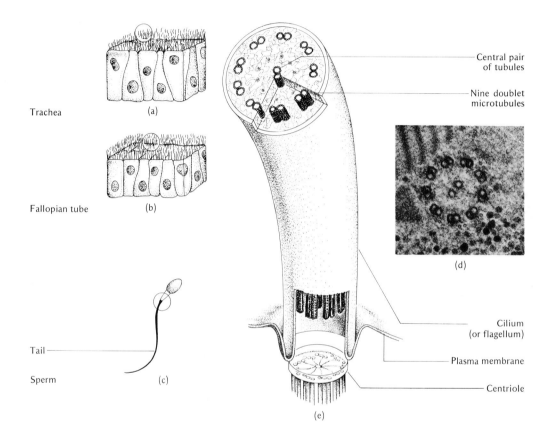

Trachea

(a)

Fallopian tube

(b)

Tail

Sperm

(c)

Central pair
of tubules

Nine doublet
microtubules

(d)

Cilium
(or flagellum)

Plasma membrane

Centriole

(e)

FIGURE 2–19 The nucleus. (a) Reconstruction. (b) As revealed by the electron microscope. (c) Detail showing nuclear pores. (Electron micrographs (b) and (c) courtesy of K. R. Porter.)

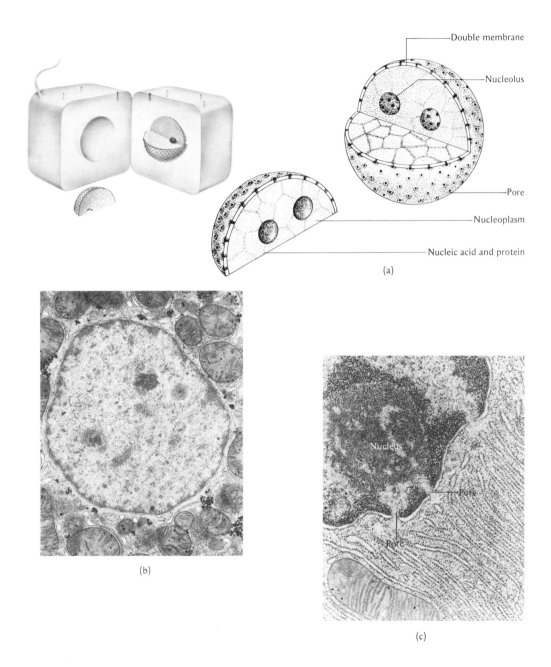

Double membrane

Nucleolus

Pore

Nucleoplasm

Nucleic acid and protein

(a)

(b)

Nucleus

Pore

Pore

(c)

the inside of the nucleus (**nucleoplasm**) with the cytoplasm (Figure 2–19b). In addition to lumps and strands of chromosomes, the nucleoplasm contains a small, dense, roughly spherical body known as the **nucleolus** ("little nucleus"). Under the light microscope, the nucleolus appears to contain a relatively fluid phase as well as a more solid or dense phase; the dense phase appears dark and consists of clusters of ribosomes, which are rich in RNA; the lighter-appearing, more fluid areas of the nucleolus are made up of tiny granules that are presumably protein. The nucleolus is formed by particular chromosomes that have an active region called the **nucleolar organizer,** and it is here that ribosome precursors are manufactured and accumulated. The dense accumulation of these ribosomal components surrounding the nucleolar organizer forms the major part of the nucleolus as seen under the microscope. The ribosomal components pass out of the nucleolus region, through the nuclear envelope, and into the cytoplasm, where they are assembled and function in the manufacture of proteins. The nucleolus is absolutely fundamental for the growth of the cell. If there is no nucleolus, ribosomes are not synthesized; then there is no protein synthesis, and without protein synthesis the cell cannot grow or maintain itself. The nucleolus may be thought of as the cell's pacemaker, since any change in the activity of the nucleolus will result in a change in the growth rate of the cell.

The nucleus is the control center of the cell, and its role as such can easily be demonstrated. If you cut an ameba in half with a fine glass needle so that one half contained the nucleus and the other half contained only cytoplasm, you would find that the half containing the nucleus could grow and reproduce itself, whereas the portion consisting exclusively of cytoplasm could not maintain itself and would eventually disintegrate (Figure 2–20a and b).

FIGURE 2–20 Surgery on the ameba. (a) A fine glass needle is used to cut the ameba in halves, one piece with a nucleus and one without. (Courtesy of the Carolina Biological Supply Company.) (b) If left in this condition, only the piece containing the nucleus survives. (c) to (g) If the nucleus from one ameba is transplanted into the cytoplasm of another enucleate ameba, it survives. The donor, without a nucleus, dies.

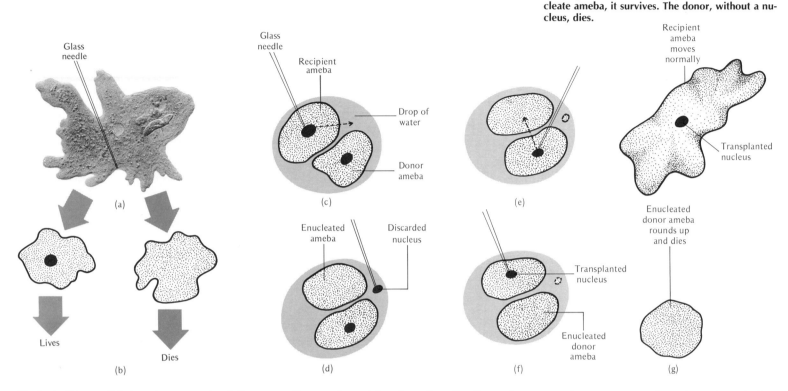

Similarly, it is possible to restore normal activity to an ameba that lacks a nucleus by transplanting a nucleus from another ameba (Figure 2–20c–g). Lacking a nucleus, the donor ameba rounds up and dies.

Our own mature red blood cells lack a nucleus and cannot reproduce themselves; their life span is limited to about 120 days, after which they disintegrate and are replaced by new ones developed from nucleated juvenile cells found in the bone marrow and spleen. Although the nucleus directs the activities of the cell and the cell cannot survive for long without it, it is not independent of the cytoplasm, for an isolated nucleus devoid of cytoplasm cannot survive either. It is obvious that there must be dynamic interplay between nucleus and cytoplasm, and for coordinated function a cell must have both.

The nucleus determines the direction of development a cell will take. This can be illustrated by an elegant experiment with an unusual one-celled plant called *Acetabularia*. *Acetabularia* is an umbrella-shaped marine alga that grows to a length of about 1 in., but consists of a single cell with a single nucleus (Figure 2–21). Because of its large size, surgical manipulations are relatively easy. Two kinds (species) of *Acetabularia* differ in their external appearance; in one the umbrella portion (cap) has a smooth edge (*A. mediterranea*), whereas in the other species it is fringed (*A. crenulata*). In both *A. mediterranea* and *A. crenulata,* the cap is supported by a stem, and the nucleus is contained in its rootlike base, called the rhizoid. If we removed the caps from some specimens of both kinds, we should find that the caps would regenerate; they would look just like the

caps that were present in each specimen before surgery. However, if we removed the caps from fresh specimens of both species, and then grafted the rhizoid of one species onto the capless stem of the other, we should find that the cap that grew back on any specimen would look like the cap of the cell that donated its nucleus (Figure 2–21c). If these caps were then removed, the caps that regenerated would still be identical to those of the species from which the transplanted nucleus came. It seems that the nucleus exercises its control by secreting messages into the cytoplasm; upon arrival in the cytoplasm, the message is converted into information that details the type of cap to be formed.

How does the nucleus send out these directive messages? In our description of the nuclear envelope we noted that it is periodically inter-

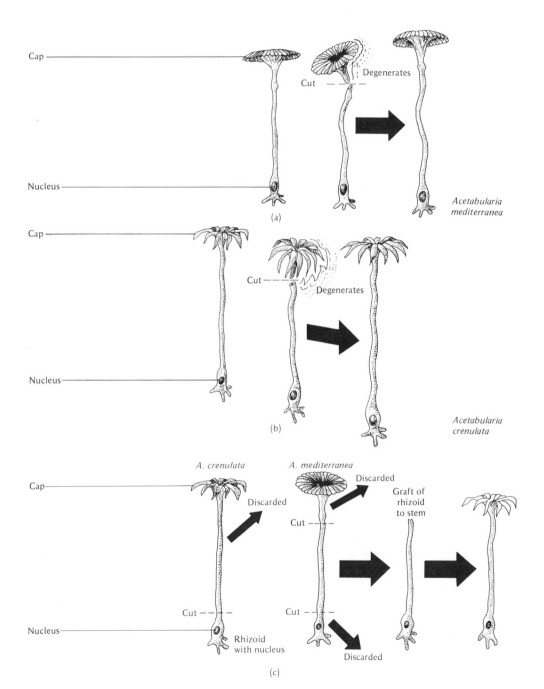

FIGURE 2–21 **The nucleus controls the direction of the development. (a) and (b) If the cap of** *Acetabularia* **is amputated, the nucleus directs the regeneration of a species-specific cap. (c) Grafting of the nucleate rhizoid of** *A. crenulata* **onto the stem (with cytoplasm only) of** *A. mediterranea* **produces a cap with the characteristics of** *A. crenulata.*

Cap

Cut --- Degenerates

Nucleus

Acetabularia mediterranea

(a)

Cap

Cut --- Degenerates

Nucleus

Acetabularia crenulata

(b)

A. crenulata   *A. mediterranea*

Cap   Discarded

Discarded

Graft of rhizoid to stem

Cut ---

Cut ---

Nucleus

Rhizoid with nucleus

Discarded

(c)

rupted along its surface by pores. These pores are not simple holes, but are often filled with a plug that probably confers specific properties of selectivity upon the nuclear envelope, enabling it to determine the type of materials that cross into the cytoplasm; chemical messages of a specific nature do in fact pass through these pores. There is another way in which the nucleus can communicate with the cytoplasm. The electron microscope has shown that the outer membrane of the nuclear envelope is continuous in some areas with the membranes of the endoplasmic reticulum. In this way, the materials of the nucleus may be provided with a direct connecting passage to a portion of the cytoplasm (Figure 2–19). The nature of the nuclear message and the manner of its expression in the cytoplasm will be considered in another chapter.

Extracellular substances: glue, support, and protection. The cell is a delicate entity. Although the plasma membrane confers a certain amount of protection and support upon the contents of the cell, it is in itself a rather fragile component. Many cells are covered with one or more external substances that give the cell support, protection, and elasticity, and provide for adhesiveness and for the retention of water. Most of these coating substances consist of large molecules such as protein and carbohydrate fabricated from smaller molecules within the cell.

If the cells of our bodies had no adhesiveness, then no doubt we would exist as loose lumps, constantly coming apart. Obviously, cells must have a type of intercellular glue to hold them together; there are two kinds: **hyaluronic acid** and **chondroitin sulfate.**

Hyaluronic acid is a carbohydrate; when combined with protein, it is a viscous jellylike material with no definite form. Its viscosity and adhesiveness depend on the amount of calcium present. This material acts as glue and binds cells together. For example, it is possible to convert a live sponge into a mass of separate cells which will not reaggregate by placing it in seawater free of calcium. The human egg is surrounded by numerous small cells glued to each other by hyaluronic acid, making the egg virtually impregnable. However, the hyaluronic acid can be dissolved by an enzyme, hyaluronidase, which is present in the heads of human spermatozoa. This permits penetration of the egg by the sperm. Hyaluronic acid may function in other capacities: in the fluid of our bone joints, it acts as both a lubricant and a shock absorber; in the fluid of our eyes (vitreous and aqueous humors), it helps give the eye a fixed shape.

Chondroitin sulfate is also a carbohydrate, and when combined with protein it forms a rather rigid gel. It is produced and secreted by cartilage cells and forms the intercellular matrix (substance) of cartilage (gristle); it provides for architectural support in the human body. When cartilage contains fibrous materials such as collagen, increased flexibility and rigidity result, a situation not unlike that of steel-reinforced concrete. The cartilage of our ears, nose, and windpipe, and much of the skeleton of an infant are good examples of the occurrence of chondroitin sulfate in this capacity.

Fibrous protein elements such as **collagen** (commercial gelatin), **reticulin,** and **elastin** are also found in the interstices of cells. Collagen is a long fibrous protein with a distinct pattern of dark and light bands (Figure 2–22); it confers properties of rigidity and firmness on muscles and tendons. Elastin and reticulin do not show banding in their protein fibers. Elastin, as its name suggests, is elastic in nature and is present in the skin and in the major blood vessels. Reticulin, closely related to collagen, also helps to bind cells together and give organs their shape; it is well developed in the skin.

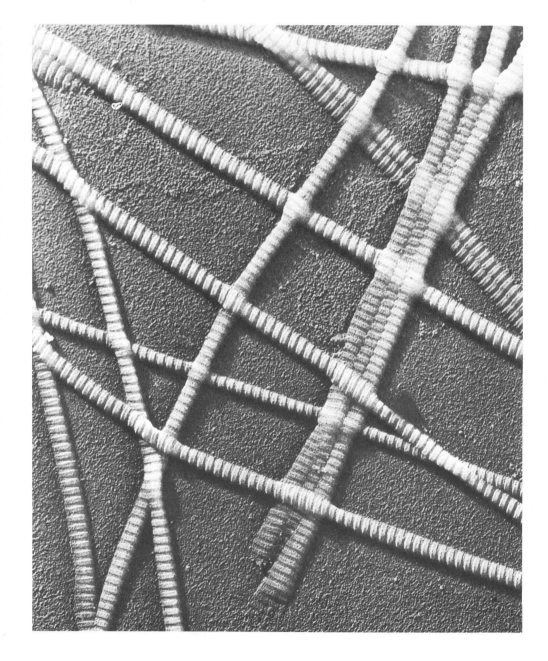

FIGURE 2–22 Collagen fibers enlarged 50,000 times by the electron microscope. (Courtesy of Jerome Gross.)

FIGURE 2–23 Plant cells and wall formation.

Finally, it should be mentioned that the mucus secretions of cells provide a means of protection against the loss of excessive amounts of water and in some cases help to trap foreign materials. For example, fine particles of dust and the like are trapped in the mucus secretions of the cells of our nasal passageway; by means of cilia which protrude into the mucus layer and beat toward the throat, the mucus and the trapped particles are removed and eventually swallowed. Such a mechanism is extremely efficient in cleansing and filtering the air we breathe and ensures that relatively few particulate materials reach the delicate surfaces of our lungs. Sometimes the copious amounts of mucus produced during a bacterial or viral infection such as the common cold become troublesome and continue as the familiar postnasal drip; here too, the function of the mucus is to remove the source of irritation.

Plants also have supportive and protective materials outside the plasma membranes of their cells. It is these substances that provide most of our fuel, clothing, building materials, food, and innumerable other conveniences which make our lives more comfortable. The most striking difference between plant and animal cells is that plant cells are surrounded by a nonliving cell wall. The cell wall may consist of three parts: an intercellular cement called the middle lamella, a primary cell wall, and a secondary cell wall (Figure 2–23). The middle lamella is of jellylike consistency and is formed between adjacent cells during division; it is rich in **pectin,** the cement that binds the plant cells together. During ripening, the fruit becomes soft (as any shopper knows when she pokes a ripe tomato or melon) owing to the dissolution of the pectin and the loosening of the cells. Pectin is commonly added to jams and jellies to ensure that these materials set or jell. The primary wall is found in all plant cells,

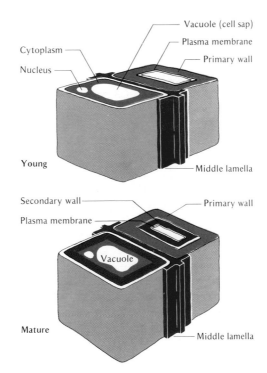

lies between the plasma membrane and the middle lamella, and is generally formed during early development of the cell. It is thin, quite flexible, and forms the outer covering of the cells of fruits, roots, and fleshy stems. When growth of the cell ceases, certain mature cells form a secondary cell wall between the primary wall and the plasma membrane. It may be thick or thin, hard or soft, colorless or colored. The secondary cell wall may be as much as 10 μm thick, and it gives the plant cell strength and mechanical support. For us, the secondary cell walls of plants provide lumber, cotton, flax, and hemp. The major component of the cell wall is **cellulose,** a large molecule made up of long chains of sugar (glucose) molecules; derivatives of cellulose are used in the manufacture of rayon, cellophane, and plastics. Present

in the walls of some plant cells is **lignin,** a complex, nonfibrous molecule unrelated to glucose that forms between the fibrils of cellulose. The arrangement of the cellulose fibers provides for elastic strength, with the harder resistant material, lignin, acting as a rigid support; in many respects it is like reinforced concrete used in the construction industry—the cellulose acts like the steel rods in providing flexibility, while the cementlike lignin is effective in resisting pressure. Balsa wood is an example of cell walls that lack lignin; it is a soft, brittle wood.

Membranes: compartmentalization and specialization. One of the most striking features of cellular organization is the way in which organelles are constructed of membranes. Plasma membrane, nuclear membrane, Golgi apparatus, endoplasmic reticulum, mitochondria, chloroplasts, and lysosomes are all membranous structures. The membranes of these organelles at first glance appear similar when viewed with the electron microscope, and this structural unit has been emphasized by the application of the descriptive term unit membrane. However, with increasing resolution and newer techniques of electron microscopy, differences in membrane structure are beginning to be shown. These differences are not entirely unexpected, since the various organelles composed of these membranes do differ greatly in function.

Membranes increase the internal surface area and compartmentalize the cell so that considerable specialization can be effected within specific regions of the cell. It is apparent that the cell is not an amorphous blob characterized by randomness and chaos, but a highly structured and well-organized living system. Its structural organization depends intimately upon the use of membranes in the formation of organelles.

2–2 CELL SPECIALIZATION: TISSUES AND ORGANS

We cannot see the cells of which we are composed in any detail with the naked eye, yet the most casual glance reveals that these cells are not all the same. For example, each of us recognizes that the cells on the surface of our skin are quite different from those that appear when we cut through the skin and bleed. Just as there are no typical buildings composed of typical rooms, so organisms are not composed of uniformly structured typical cells. According to their function, cells vary in their shape, size, and the degree of development of their constituents. Thus, the cells that make up our muscles are elongate and capable of contraction whereas the cells that cover the various surfaces of the body, both external and internal, have a characteristic shape that may be cubical, columnar, or platelike so that they fit together and function in lining and protection. Cells that occur together and have a similar structure and function are called **tissues.** Table 2–1 details the structure and function of the major tissues that are found in the human body.

Groups of tissues may be organized to form structural and functional units known as **organs.** The kidney is an organ composed of highly specialized epithelial and connective tissues; our skin is not a simple tissue but a complex organ consisting of epithelial, muscular, nervous, and connective tissues. Organs do not function apart from one another, but form an integrated whole, a smoothly functioning organism, but some groups of organs have particularly integrated functions and are referred to as **organ systems.** Thus, the digestive system includes such organs as the mouth, esophagus, stomach, intestines, liver, and pancreas, and the respiratory system is made up of organs that

include the ribs, lungs, diaphragm, and so forth. It should be emphasized that the functional attributes of organ systems reflect the organs and tissues of which they are composed and these in turn are only products of their specialized cells. What makes a cell specialize and become, for example, a red blood cell (with no nucleus and lots of pigment) rather than a skin cell (with nucleus and little or no pigment) during the course of its development from a nonspecialized embryonic cell is another question entirely, and one that lies at the frontiers of modern biology.

2–3 CELLS AND ORGANISMS

From the time of Aristotle, man has attempted to categorize the kinds of living creatures around him. The simplest classification of organisms was the system of two kingdoms—plants and animals. Classified as **plants** were non-motile, photosynthetic autotrophs with cellulose cell walls. Classified as **animals** were generally motile non-photosynthetic heterotrophs without cellulose cell walls.

However, during the early 19th century it was recognized that the two kingdom system was in difficulty, especially when it came to the classification of one-celled organisms. In the unicellular creature the body is not divided up into tissues and organs, but its organelles are the functional equivalent of organs found in multicellular organisms. Various authors suggested that the single-celled organisms be placed in a kingdom all of their own—neither plant nor animal—and that it be given the name **Protista** (proto—first; Gk.). For a time the three kingdom system of plants, animals and protists, was in popular usage; however as knowledge about the Protista accumulated it became apparent that these one-celled organisms had two very different types of structural

organization. One group, the **prokaryotes,** typified by the bacteria, rickettsiae, and blue-green algae, lack a well-defined, membrane-enclosed nucleus and other organelles. The fine filaments of nucleic acid lie free in the cytoplasm and no nuclear membrane can be seen, nor can one identify an organization of fibrils corresponding to chromosomes. (The term prokaryote comes from the Greek words: pro—before and karyon—nucleus). The cytoplasm contains no mitochondrion, endoplasmic reticulum or Golgi apparatus, but is richly endowed with ribosomes. Photosynthesis and respiration are localized in the plasma membrane and its local invaginations. Flagella, when present, do not show microtubules arranged in the 9 + 2 pattern. These prokaryotes, it has been proposed, should be placed in a kingdom called **Monera.**

The remaining protists, the **eukaryotes** (eu—true; Gk.), included organisms such as the green, yellow-green, and golden algae, protozoans and diatoms. These have a membrane-bound nucleus and other organelles. Nucleic acid is combined with protein to form chromosomes which are enclosed by a nuclear membrane. Photosynthesis and respiration are localized in the chloroplast and mitochondrion. Flagella and cilia, when present, have a 9 + 2 arrangement. Currently, such eukaryotic unicellular organisms are classified in the kingdom Protista.

Most plants and animals, including ourselves, are multicellular. They are, like the Protista, eukaryotic in their cell structure, having a membrane-enclosed nucleus and organelles. As multinucleate heterotrophs with cellulose cell walls, the fungi form a kingdom of their own. The major kingdoms for these many-celled eukaryotes are: Animalia, Plantae, and Fungi. The diversity of all these living creatures is a reflection of the plasticity of the cell.

BOX 2B Seeing the unseen: viruses

As human beings we sometimes doubt that what we cannot perceive with our senses, and in particular see with our eyes, really does exist. Viruses are a case in point. Although the existence of viruses has been known for a great many years because of the diseases they produce in animals and plants, they are invisible with the light microscope. Indeed, because of their small size they pass through the finest porcelain filters, making it impossible to collect them by filtration. Before the advent of the electron microscope their structure was very poorly characterized.

Viruses come in various shapes and sizes and cause a variety of diseases (more than 500 animal viruses have been identified and described so far), but they all have the same basic composition: an outer protein coat and a central core of nucleic acid. Viruses do not have a plasma membrane and, therefore, according to our definition of a cell, viruses are not cells. Some biologists would go a step further and suggest that they are not alive: they have no metabolic machinery of their own, they contain no raw materials for making a new virus, and they can be crystallized just as a salt can be crystallized, and still remain infective when the crystals are dissolved and come in contact with a suitable host cell.

How do these nonliving noncells exist? Viruses cannot be grown on nonliving substances, but must have a living cell to supply the raw materials and the necessary enzymatic machinery for their reproduction. The virus provides the blueprint for new virus particles (in the form of nucleic acid, the genetic material) and directs the living host cell to perform the actual synthetic operation of viral reproduction. Although it is not exactly clear how all viruses enter cells and take over the host cell's machinery to manufacture new viruses, it is quite clear how the viruses of

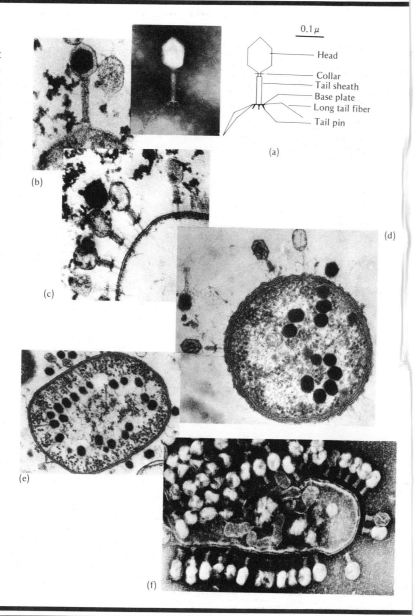

0.1 μ

Head
Collar
Tail sheath
Base plate
Long tail fiber
Tail pin

(a)
(b)
(c)
(d)
(e)
(f)

FIGURE 2-24 The life cycle of a bacteriophage. (Courtesy of Lee D. Simon.) (a) Bacteriophage. (b) Bacteriophage attached to bacterial cell. (c) Bacteriophage injects DNA into bacterial cell. (d) Reproduction of bacteriophage heads. (e) Production of bacteriophage heads and tails. (f) Lysis of bacterial cell and release of bacteriophage.

TABLE 2-1 A morphological classification of human tissues

Type	Function	Form and location	Appearance
I. Epithelial	Cells that form a continuous layer and cover external and internal surfaces; protect, repair, and regulate the passage of substances across themselves; may be absorptive or secretory.		
	A. Covering and lining May be stratified (layered), ciliated or keratinized.	i. Squamous Flattened, thin irregularly shaped cells. Line the body cavities, blood and lymphatic vessels. Stratified and keratinized squamous epithelium forms the outer layer of the skin.	
		ii. Cuboidal Cube-shaped cells. Line kidney tubules. Cover ovary.	
		iii. Columnar Elongate pillar-shaped cells with nucleus usually located near the base. Often ciliated at outer surface. Widely distributed as linings of ducts, digestive tract, etc.	Cilia
	B. Glandular Cells specialized to secrete substances such as milk, wax, perspiration, sex cells or hormones. Cuboidal or columnar in shape.	i. Exocrine Glands with a duct to the outside surface, e.g. sweat and mammary glands.	Duct Secretory cells

TABLE 2-1 (Continued)

Type	Function	Form and location	Appearance
		ii. Endocrine Ductless glands that secrete their products directly into the blood stream, e.g. thyroid, pituitary.	Secretory cells Blood vessel
II. Connective	Cells whose intercellular secretions (matrix) support and hold together the other cells of the body.		
		A. Fibrous Widely distributed cells with supportive carbohydrate and flexible protein secretions including collagen and elastic fibers. Form tendons and ligaments and wrap many organs, muscles, and nerves.	Fibers Connective tissue cell Matrix
		B. Supporting	
		i. Cartilage Cells secrete a firm, rubbery matrix. Support embryonic skeleton and parts of adult.	Cartilage cell Cartilage (matrix)
		ii. Bone Calcium salts secreted into cartilage imparting great strength. Supports skeleton and protects nervous system.	Blood vessel Bone (matrix) Bone cells

TABLE 2-1 (Continued)

Type	Function	Form and location	
	C. Blood forming tissue	Blood, including plasma, red cells, white cells, and platelets, derived from cells resembling connective tissue.	Cells Plasma (matrix)
III. Nervous	Elongate cells or neurons specialized for the expression of irritability and conductivity.	i. Central nervous system: brain and spinal cord ii. Peripheral nervous system: peripheral nerves, ganglia, autonomic nervous system (sympathetic and parasympathetic nerves), and sense-organ nerve endings	Nerve endings Nucleus Cell body
IV. Muscle	Elongate, cylindrical or spindle-shaped cells containing contractile fibers. Perform mechanical work by contraction.	i. Cardiac Branching fibers with many nuclei and striations. Involuntary. Located only in the walls of the heart.	Nuclei
		ii. Skeletal Elongate cylindrical fibers with many nuclei and striations. Under voluntary control. In general attached in bundles to bones of skeleton.	Cross striations Nuclei
		iii. Smooth Spindle-shaped uninucleate cells with no visible striations. Involuntary. Generally occur in sheets in walls of viscera, digestive tract, etc.	Nuclei

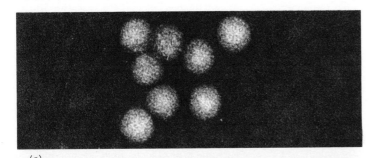

(a)

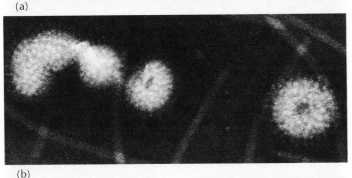

(b)

(c)

bacteria take over. The viruses of bacteria were discovered in 1917 by the French scientist F. d'Herelle, who noticed them destroying his cultures of dysentery bacteria. He named them **bacteriophages** because they appeared to eat the bacteria (*phago* —eat; Gk.). In actuality the viruses had caused the bacteria to burst or **lyse.**

The life cycle of a bacteriophage, or phage virus, is illustrated in Figure 2-24. The bacteriophage is tadpole-shaped, and it attaches to the cell wall of the bacterium by its tail. The phage nucleic acid (DNA) is then injected into the host cell, and its protein coat remains outside. Once inside, the phage DNA takes over the metabolic machinery of the cell; for the first 10–15 minutes (called the **eclipse period**) nothing is readily apparent inside the phage-infected bacterium, but then virus particles appear in the cytoplasm of the bacterial cell. In time (20–30 minutes) the entire host cell is filled with phage particles containing a protein coat and a nucleic-acid core. Eventually the bacterial cell bursts, and the phage viruses (100 or more) are liberated and capable of infecting new bacteria. The entire cycle may take less than half an hour.

Viruses are obviously parasites, but in some cases there is no evidence of injury to the host organism. Most of the viruses we are aware of cause effects that we call disease and therefore are regarded as pathogenic. Viruses are responsible for human maladies such as rabies, smallpox, influenza, polio, the common cold, fever blisters (cold sores), shingles, encephalitis, infectious hepatitis, mumps, measles, yellow fever, and viral pneumonia. Some cancers are caused by viruses, but there is no unanimity among scientists in thinking that all cancers are caused by virus particles. (A discussion of cancer is found in Chapter 8.) Some of the viruses that afflict us are shown in Figure 2-25 as they appear with the electron microscope. You are seeing what was first seen only 30 years ago.

FIGURE 2-25 Viruses that cause human disease as revealed by the electron microscope. (Courtesy of Robley C. Williams and the Virus Laboratory, University of California, Berkeley.) (a) Poliovirus x 300,000, (b) influenza (spheres) and potato virus (elongate rods) x 150,000, (c) smallpox virus x 100,000.

SUMMARY

1. The basic unit of life is the cell. Cells are bounded by membranes and capable of self-reproduction.

2. The cell or plasma membrane is composed of a double layer of lipid molecules with proteins embedded in or extending through it.

3. The cytoplasm of the cell consists of a viscous, colloidal complex of proteins, lipids, carbohydrates, minerals, and salts. The sol-gel properties of the cytoplasm are in part responsible for maintenance of and changes in cell shape. Other materials in the cell exist in solution as emulsions and suspensions.

4. The plasma membrane regulates the flow of materials into and out of the cell. Regulation is a product of the physical and chemical properties both of the membrane and of the materials that cross it.

5. Entrance to and distribution of molecules in the cell are largely governed by the laws of diffusion and osmosis.

6. Osmotic pressure is a measure of the concentration of particles in a solvent enclosed by a semipermeable membrane.

7. Active transport of materials across the cell membrane against their concentration gradient requires the expenditure of energy. By this means, cells may accumulate or get rid of particular substances.

8. Cells may also acquire liquids and solids by pinocytosis ("cell drinking") and phagocytosis ("cell eating").

9. The cell is compartmentalized by the endoplasmic reticulum, an intracellular network of membranes which functions in transport and storage. It may be studded with RNA-containing ribosomes, important in protein synthesis.

10. The Golgi apparatus consists of a stack of saclike membranes or cisternae. It plays a role in secretion.

11. Mitochrondria are membrane-bound structures that are the site of cellular respiration and energy release.

12. Chlorophyll-containing chloroplasts, found in green plants, are the site of photosynthesis—the fixation of energy from the sun into complex organic molecules (carbohydrates). Chloroplasts develop in the light; they consist of layers of fluid-filled grana and flat lamellae surrounded by a fluid stroma or matrix. Other plant plastids include chromoplasts, which give plants their color, and leucoplasts, which store oil, starch, and protein.

13. Lysosomes contain hydrolytic enzymes and are important in cellular digestion.

14. Vacuoles are fluid-filled areas, surrounded by a membrane. The prominent central plant cell vacuole keeps the cytoplasm peripheral. The plant contains a watery cell sap with a high osmotic pressure that keeps the cell turgid.

15. Microtubules and microfilaments are important in cellular movement: microtubules in hairlike cilia and whiplike flagella and chromosome movements in cell division (the mitotic spindle): microfilaments affect changes in cell shape and play a role in cytoplasmic streaming.

16. The centriole, just outside the nucleus, is important in cell division. Its microtubular structure is similar to the arrangement in cilia and flagella.

17. The nucleus is the control center of the cell and contains the genetic material. The nucleus contains the chromosomes, composed of deoxyribonucleic acid (DNA). An active region of particular chromosomes organizes the nucleolus, which is important in ribosome synthesis and therefore in cellular growth. The nucleus and cytoplasm communicate across the nuclear membrane and via nuclear pores. The nuclear membrane is also continuous with the endoplasmic reticulum in some areas.

18. Cellular adhesiveness and support are provided by hyaluronic acid and chondroitin sulfate, collagen, reticulin, and elastin. Mucus secretions lubricate and protect against dehydration and trap foreign materials. In plant cells, support is provided by the primary and secondary cellulose cell wall, and sometimes by woody lignin and other materials. Plant cells are bound together by a jellylike middle lamella containing pectin.

19. Resolution is the power to distinguish between objects, whereas magnification is the enlargement of objects. Both are important in microscopy. Resolution depends on illumination. The electron microscope improves the resolution of objects closer together than one half of the wavelength of white light, using an electron beam as a source of illumination.

20. Tissues are composed of groups of cells with similar structure and function. Epithelial tissues may be covering and lining or glandular. Connective tissue may be fibrous, supporting, or blood forming. Nervous tissue forms the nervous system, and muscle tissue provides for movement.

21. Organs are structural and functional units composed of groups of tissues. Groups of organs form organ systems.

22. Viruses are noncellular, submicroscopic parasites consisting of a nucleic acid core and a protein coat. They invade cells and take over their genetic machinery, redirecting it to make new viruses. They cause many human diseases. Bacteriophages, much used in research, are bacterial viruses.

KEY WORDS

cell
unicellular
multicellular
ghosts
unit membranes
protoplasm

cytoplasm
colloid
sol-gel transformation
emulsion
suspensions
solution
resolution
magnify
fixing
staining
phase-contrast microscope
diffusion
osmosis
semipermeable membrane
osmotic pressure
isosmotic medium
hypoosmotic
hyperosmotic
active transport
sodium pump
pinocytosis
phagocytosis
organelle
endoplasmic reticulum
ribosomes
polyribosomes
Golgi apparatus
mitochondria
cristae
chloroplasts
chlorophyll
photosynthesis
stroma
grana
lamellae
plastids
chromoplasts
leucoplasts
lysosomes
vacuoles
cell sap
microtubules
microfilaments

centriole
cilia
flagella
nucleus
chromosomes
nucleoplasm
nucleolus
nucleolar organizer
hyaluronic acid
chondroitin sulfate
collagen
reticulin
elastin
pectin
lignin
cellulose
tissues
organs
organ systems
viruses
bacteriophages
lyse
eclipse period
plants
animals
Protista
prokaryotes
Monera
eukaryotes

TOPICS FOR REVIEW AND DISCUSSION

1. Why is the cell considered to be the basic unit of life?
2. Is the statement that viruses are nonliving noncells nonsense? Defend the statement.
3. How are cilia, flagella, and centrioles related to one another?
4. How have developments in optical instrumentation increased our understanding of the cell?
5. How is the structure of the cell (plasma) membrane related to its function?
6. What are the cellular roles of the Golgi apparatus, the endoplasmic reticulum, and lysosomes?
7. Give experimental evidence for the statement: the nucleus is the control center of the cell.
8. How is the form of cells related to their specific roles in the body?
9. Describe the three-dimensional appearance of the organelles in a typical cell.
10. Contrast the structure of the plant cell with an animal cell. Are there similarities? Differences?

Kellogg's Special K is labeled in accordance with federal standards for nutrition labeling as established by the U. S. Food and Drug Administration.

These toasted flakes made from rice and wheat are fortified with eight important vitamins and iron.

NUTRITION INFORMATION PER SERVING

SERVING SIZE: One ounce (1¼ cups) Special K in combination with ½ cup vitamin D fortified whole milk.

SERVINGS PER CONTAINER: 7

	SPECIAL K	
	1 oz.	with ½ cup whole milk
CALORIES	100	180
PROTEIN	6 gm	10 gm
CARBOHYDRATES	20 gm	26 gm
FAT	0 gm	5 gm

PERCENTAGE OF U.S. RECOMMENDED DAILY ALLOWANCE (U.S. RDA)

	SPECIAL K	
	1 oz.	with ½ cup whole milk
PROTEIN	10	15
VITAMIN A	25	25
VITAMIN C	25	25
THIAMINE	25	25
RIBOFLAVIN	25	35
NIACIN	25	25
CALCIUM	*	15
IRON	25	25
VITAMIN D	10	25
VITAMIN B₆	25	25
FOLIC ACID	25	25
PHOSPHORUS	4	15
MAGNESIUM	4	8
ZINC	4	4
COPPER	2	2

*Contains less than 2 percent of the U.S. RDA for this nutrient.

INGREDIENTS: Rice, wheat gluten, sugar, defatted wheat germ, salt, nonfat dry milk and malt flavoring with vitamin A, ascorbic acid, thiamine (B₁), riboflavin (B₂), niacinamide, vitamin D, pyridoxine (B₆), folic acid and iron added. BHA and BHT added to preserve product freshness.

MADE BY KELLOGG COMPANY
BATTLE CREEK, MICHIGAN 49016, U. S. A.
©1965 BY KELLOGG COMPANY
® KELLOGG COMPANY

Mfd. under U.S. Patent No. 2,836,495

THIS PACKAGE IS SOLD BY WEIGHT, NOT VOLUME. SOME SETTLING OF CONTENTS MAY HAVE OCCURRED DURING SHIPMENT AND HANDLING.

3

OF MOLECULES AND CELLS

A living system such as our own body consists of about 6 dollars' worth of shelf chemicals. This low price does not reveal the true value of such a marvelous machine, for we and all living things are much more than a heap of granular materials that can be purchased in the local pharmacy or supermarket. Nevertheless, we *are* composed of chemicals—substances that obey the same physical laws whether they are a part of a living or a nonliving system— and it is virtually impossible to comprehend how life operates without some understanding of these chemicals.

Before becoming unduly discouraged by a chapter devoted to chemistry, remember that chemicals are a part of daily life. We consist of chemicals, and the energy we need to run our bodies is obtained from chemicals. We replenish our supply each day from the food we eat. Most mornings we sit down for breakfast, open a package of breakfast cereal, and eat some of the contents. What we fail to observe is that the "breakfast of champions" or the "snap, crackle, pop" is loaded with chemicals. The side panel of the cereal box tells some of the story. This cereal, it says, contains proteins, fats, carbohydrates, vitamins, minerals, and so on. In some cases, it designates how much of each chemical we require each day and what percentage of our minimum daily requirement for certain chemicals the cereal provides. Does everybody really need all these chemicals every day? What are these chemicals, and why do we need them? Where do these materials come from? How are we able to utilize them?

Just as we need not be mechanics to understand the basic operation of automobiles, so we need not be chemists to understand the basic workings of the living cell. The rudiments presented here are designed as a guide to the fundamental components of each and every living entity. This text cannot teach you all the chemistry required for a complete understanding of the workings of the living system, but it can give you some insight into the roles played by chemicals in the living organism—in you and me. We shall confine our discussion to the five major organic molecules of life: (1) carbohydrates, (2) fats or lipids, (3) proteins, (4) vitamins, and (5) nucleic acids.

Those who have no background in chemistry or who require a brief refresher course should consult Appendix A which provides a summary of the simple chemistry necessary for understanding this and subsequent chapters.

3–2 CARBOHYDRATES: CELLULAR FUEL

Carbohydrates, commonly known as sugars and starches, are the primary source of chemical energy needed to run the cells of our bodies; in addition they, or their carbon skeletons, form part of the building materials of a variety of cellular constituents. Carbohydrates are the cheapest foods available, and for this reason they probably represent a large fraction of the average diet. The poorer the economic situation, the higher the proportion of carbohydrate in the diet.

In the economy of the cell, carbohydrates serve as fuel. The carbohydrates used directly by the cell for the production of energy are generally sugars; because of their small molecular size and their solubility they can readily pass across the plasma membrane, enter the cellular "engine," and upon combustion yield useful energy. However, since sugars can pass into the cell across the plasma membrane with such ease, they can also pass out of the cell again just as easily. If they were not used at once by the cell, they might pass out and be lost. If our cells had to depend for energy on

soluble and easily lost sugars alone, a continuous supply of these molecules would be necessary. How does the cell prevent loss of sugars, and how does it avoid being without a ready supply of these fuel molecules?

In order to store fuel, the cell converts the fuel into a form that cannot pass across the membrane and therefore cannot exit. The cell stores carbohydrates in the form of large and insoluble molecules such as glycogen or starch. However, the cellular engine cannot use these large molecules as fuel so, when needed, they must be converted back into the smaller soluble sugars. In this manner, the cell has conveniently solved the problem of having both the storage and usable forms of carbohydrate within the confines of its plasma membrane.

How is starch converted back into usable sugar? The conversion of available fuel into storage form and vice versa by living cells is based on the fact that all carbohydrates are composed of simple units—building-block compounds of small molecular size. These can be linked together (**polymerized**) to yield large, complex molecules; in turn, the large molecules can be broken apart again to yield their constituent building blocks when required. Very simply, the situation is as follows. We eat carbohydrates such as starch. In the digestive tract (mouth, intestine) these foods are broken down into soluble units of small molecular size which can pass across the plasma membrane and be taken up by the cells of the body. Once inside the cell, the simple molecular units can be polymerized into storage products that are of large molecular size. When the cell needs energy, the storage products can be broken down into metabolically manageable units once again, and these are then metabolized to yield energy or they may be fabricated into other products required by the cell (Figure 3–1).

The mechanisms involved in the synthetic (building up) and degradative (breaking down)

processes of carbohydrate metabolism can be understood with the aid of some simple chemistry; this is what follows.

Single sugars: monosaccharides

Sugars are the simplest of carbohydrates, but in chemical terms what exactly are carbohydrates? As their name implies, they are "hydrates of carbon"; that is, they contain both carbon and water, and for every carbon atom in a carbohydrate there are, with few exceptions, 2 hydrogen atoms and 1 oxygen atom. We can therefore write the chemical formula of a carbohydrate as CH_2O, but since there are always more than 2 carbon atoms present, we must modify this formula as $(CH_2O)_n$, where n is a number greater than 2.

Monosaccharides (*mono*—one, *saccha*—sugar; Gk.), or single sugars, are the simplest kind. The simplest monosaccharides contain at least 3 carbon atoms and may contain as many as 10. Thus, we may write their formulas as follows:

$(CH_2O)_3$ or $C_3H_6O_3$

$(CH_2O)_4$ or $C_4H_8O_4$

$(CH_2O)_5$ or $C_5H_{10}O_5$

$(CH_2O)_6$ or $C_6H_{12}O_6$

and so on, up to

$(CH_2O)_{10}$ or $C_{10}H_{20}O_{10}$

These sugars are relatively small molecules and are called **trioses** (*tri*—three), **tetroses** (*tetra*—four), **pentoses** (*penta*—five), and **hexoses** (*hexa*—six).[1] Both trioses and pentoses

---

[1] The names of sugars usually end in -*ose*; this may help you recognize them as you read on.

FIGURE 3–1 The cellular "engine." In the cell, molecules of fuel enter through pores and can be used directly, but for storage they are combined to form complex chains (they are polymerized). As such they cannot be burned in the "engine" because they are too large to enter. However, storage molecules can be broken down, and release of one unit at a time permits entry to the "engine."

FIGURE 3–2 Structural formulas of the common monosaccharides—glucose, fructose, and galactose—shown in the ring form.

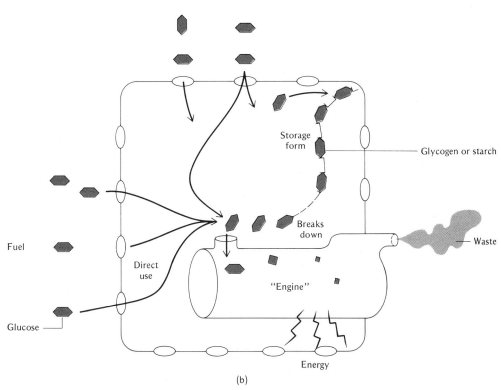

(b)

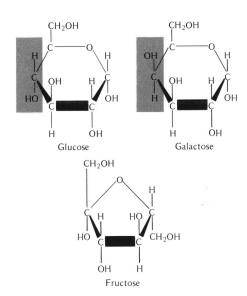

results in convulsions and unconsciousness, and perhaps even in death. Foods rich in glucose (Figure 3–3) provide quick energy because the glucose dissolves easily in water and is of a molecular size that can easily pass across the plasma membrane and into the cell, where it is metabolized.

Double sugars: disaccharides

In nature the single sugars are rare. More commonly the sugars of beets, honey, milk, and other foods consist of 2 monosaccharide molecules bonded together chemically as a **disaccharide** (di—two, saccha—sugar; Gk.), or a double sugar molecule (Figure 3–5). Common table sugar (cane sugar or maple sugar or beet sugar) is called **sucrose,** and it is a double sugar consisting of the combination of glucose and fructose; sucrose has the formula $C_{12}H_{22}O_{11}$. Although sucrose is soluble in water, its molecular size prevents it from entering the living cell; to be used by the cell, it must first be broken apart in the presence of water **(hydrolysis),** as shown in the following equation:

play important roles in the living cell, and these we shall treat in a later chapter; here we shall confine our discussion to the hexoses, or 6-carbon sugars, that act as fuel and building blocks for more complex carbohydrates.

Glucose (also known as **dextrose;** see Box 3A), fructose, and galactose are the most common hexose sugars; they are represented by the chemical formula $(CH_2O)_6$ or $C_6H_{12}O_6$. Although the molecular formulas for these sugars are exactly the same, their three-dimensional configurations are not. Molecules of the same chemical composition but different in their structural organization, and consequently in their chemical properties, are called **structural isomers.** The structural formulas (Figure 3–2) show these different geometries; at first

glance the differences appear slight, but they produce significant differences in both chemical and physical properties. For example, fructose is twice as sweet as glucose, and glucose is much sweeter than galactose.

Glucose (dextrose) is the most common single sugar (monosaccharide) found in the body; it is the most important fuel sugar for our cells. Normally glucose is present in the blood at a concentration of about 0.15%, and this level is rather rigidly regulated by a complex mechanism involving the nervous system, liver, pancreas, adrenal glands, and pituitary gland. Brain cells are particularly sensitive to glucose concentrations, and below a certain level their function is impaired; it is for this reason that a severe drop in blood glucose level often

$$C_{12}H_{22}O_{11} + H_2O \longrightarrow C_6H_{12}O_6 + C_6H_{12}O_6$$
sucrose   water   glucose   fructose

Hydrolysis of sucrose takes place in the small intestine during digestion and yields monosaccharide units—glucose and fructose; these pass across the intestinal wall, enter the blood stream, and then enter the body cells. Another double sugar is **lactose** (milk sugar); lactose is found only in milk, and although it has the same chemical formula as sucrose, it has quite different properties, as it is composed of the monosaccharides glucose and galactose. Maltose, sucrose, and lactose are structural isomers that are made up of different monosaccharide isomers. Milk provides energy because of its lactose content (about 2%–6%), but lactose, like sucrose, cannot be directly utilized by the body cells because of its molecular size. It too must first be digested, and the resultant products of hydrolysis serve as chemical fuel. In the production of beer, the disaccharide malt sugar **(maltose)** is a common starting material; upon hydrolysis of maltose, 2 molecules of glucose are liberated and then fermented to yield alcohol and carbon dioxide, the active ingredients of beer.

Having considered the breakdown of disaccharides into monosaccharides, let us consider the reverse—the synthesis of disaccharides from monosaccharides. The cell chemically fuses the monosaccharide units to form disaccharides, a process known as **condensation.** A molecule of water (a hydroxyl, OH, from one monosaccharide and a hydrogen, H, from the other) is formed in the process, and the monosaccharides are connected by an oxygen atom shared between them (Figure 3–5).

Many sugars: polysaccharides

In the same way that cells make disaccharides from monosaccharides by condensation reac-

BOX 3A Through the looking glass: symmetry of molecules

Substances in the universe exist in two optical forms called **stereoisomers*** which are mirror images of each other; the two forms are called D and L, because of their spatial configurations: right-handed or left-handed (D for *dextro*—right; L., and L for *levo*—left; L.). Sugar molecules come in D- and L-forms; when sugars are synthesized in the laboratory, the two kinds are produced in equal amounts, but it is one of nature's curiosities that only the D-form appears in living organisms. The glucose we utilize is the dextro- or right-handed form, and is therefore sometimes called **dextrose** (Figure 3-3). If you glance at the illustration of the structure of glucose and dextrose (Figure 3-4), you will see that the D-form

---

*Do not confuse stereoisomers with structural isomers, which are molecules with the same chemical composition but a different structure, such as glucose and fructose (Figure 3-2).

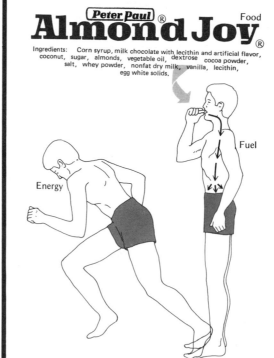

FIGURE 3-3 Food into fuel. Dextrose or D-glucose is directly absorbed in the intestine and can be directly utilized by the cell. (Courtesy of Peter Paul, Inc.)

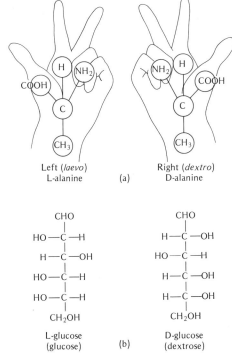

L-glucose
(glucose)   (b)   D-glucose
                   (dextrose)

Left (*laevo*)   Right (*dextro*)
L-alanine   (a)   D-alanine

FIGURE 3–4 Mirror molecules. Note that although these molecules consist of the same atoms, the molecules cannot be superimposed one on top of the other. (a) The amino acid alanine. (b) Glucose and dextrose.

(dextrose) is the mirror image of the L-form (glucose), and there is no way of rearranging the atoms so that one form can be converted into the other. Similarly, all amino acids exist in both the right- and left-handed forms (except for the simplest, glycine). Remarkably, all the proteins found in living material contain only the left-handed form of the amino acids. Nobody knows why almost all naturally occurring sugars are right-handed and why we and other organisms contain only left-handed amino acids. Is there something special about the L-form of amino acids or D-sugars? So far as we can tell, none. Molecules that are left-handed or right-handed have exactly the same chemical and physical properties except when they interact with living organisms.

If, like Alice in "Through the Looking Glass" you were suddenly to become a mirror image of yourself there would not be much noticeable change in you; you might part your hair on the opposite side of your head, your heart would be on the right side instead of the left, you would write with your opposite hand and so on. Except for some of these slight anatomical differences you would look much the same as usual. You would breathe oxygen in and exhale carbon dioxide, and all of your body functions could be carried out except one—you would be unable to eat ordinary food. As soon as you tried to eat natural food, you would find it impossible to digest it; the geometry of the food molecules would be all wrong for you. You could be maintained only on a diet of laboratory manufactured D-amino acids and L-sugars. You could not even reproduce yourself unless there were other "mirror humans" like yourself.

The living system can be compared with a machine composed of interchangeable parts held together by nuts and bolts. If the machine parts were held together with two kinds of bolts having a left- or a right-handed thread, then two separate kinds of bolts and nuts would be required to put the machine together and the nuts and bolts and machine parts would not be interchangeable. Clearly, the construction and operation of a machine requiring two separate kinds of noninterchangeable fasteners would be very inefficient. In the same way, if both left- and right-handed molecules were present in the living cell, two metabolic systems would be required to handle them. Proteins are very specific in their configuration for attachment to right- or left-handed molecules, so that, just as it is impossible to fit a glove for the left hand correctly on the right hand, an improper fit of enzyme and substrate molecules in the cell would result in a malfunctioning metabolism. The design of a machine that is simple, efficient, and effective must involve the selection of right-handed or left-handed threads; which system is adopted does not matter so long as one system is used throughout. In the living machine, the cell, we can assume that once the pattern of left-handed amino acids and right-handed sugars was established in the earliest organisms (probably by chance) the subsequent introduction of the opposite stereoisomer, the mirror image, would have fouled things up structurally. Since the emergence of life, living things have used only one molecular form: right or left, not both.

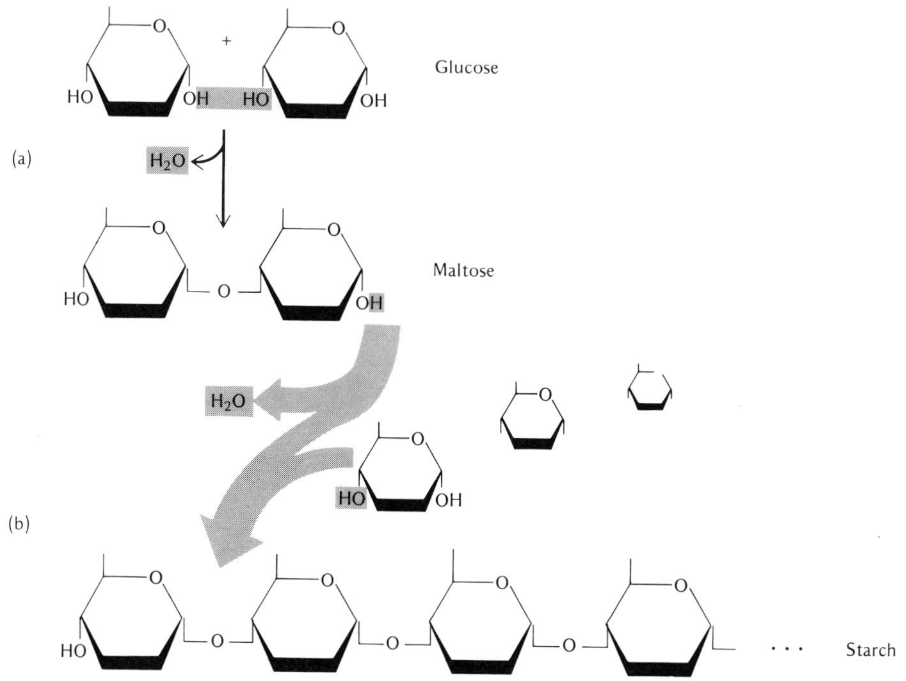

FIGURE 3–5 The formation of disaccharides (double sugars) and polysaccharides (many sugars). Only the reactive OH groups are indicated here. (a) By linking two glucose (monosaccharide) units, the disaccharide maltose is formed. (b) The addition of other glucose units produces an elongated chain, the polysaccharide starch.

tions, they can make long chains of sugar molecules called **polysaccharides,** or multiple sugars (*poly*—many; *saccha*—sugar; Gk.). **Starch** is formed of long chains of glucose molecules and is an example of a polysaccharide (Figure 3–5b). Starch is poorly soluble, but when it is hydrolyzed (Figure 3–6c), soluble and directly utilizable glucose is released. Starch is the main storage product of many plants (the potato tuber, the carrot, the seeds of wheat and corn) and provides a major source of glucose when we eat these plants. The large molecular size and insolubility of starch admirably suit it to be a storage substance. We are able to store carbohydrate within our bodies, principally in liver and muscle; however, we store it not in the form of starch, but as the polysaccharide glycogen, sometimes called animal starch. **Glycogen** is also a polymer of glucose; it differs from starch in the way in which the glucose molecules are joined together (compare Figures 3–6b and 3–6c). Unlike most starches, glycogen is soluble, but it is of such large molecular size that once fabricated inside the cell it cannot leave; however, when glucose is required by other cells of the body, the stored glycogen in the liver is broken down into glucose and passed into the blood, from where it is carried to the cells requiring it.

Some polysaccharides, such as cellulose, function in a structural capacity rather than as a fuel reserve. **Cellulose** is abundant in the cell walls of plants, where it functions for support and protection. Like starch, it is composed of long chains of glucose molecules. It differs from other polysaccharides, such as starch and glycogen, in the manner by which the glucose units are joined together (Figure 3–6a). If we could digest cellulose, it would provide a rich source of glucose. Perhaps, unfortunately, we do not have the proper digestive enzymes to degrade it; cattle, on the other hand, harbor

microorganisms in their stomachs capable of digesting cellulose to useful glucose molecules, and this enables them to derive glucose from hay and grass. For us, grass and hay do not provide food; if we were to eat them, they would provide only roughage. Paper is made of plant materials and is largely cellulose; thus, these pages could provide a useful meal for a cow, but for you they provide food only for the mind.

3–3 LIPIDS: FUEL, STRUCTURE, AND COORDINATION

Lipids include fats and fatlike molecules such as steroids. They are components of membranes, serve as fuel, act as chemical messengers (hormones), and as fat deposits in the skin provide good insulation against loss of body heat. Lipids are generally or partly insoluble in water, but they are soluble in organic solvents such as alcohol, ether, acetone, and chloroform.

Fats

Fats, like carbohydrates, consist of the elements carbon, hydrogen, and oxygen, but the ratio of hydrogen to oxygen is greater than 2:1. The larger hydrogen content of fats allows a greater degree of **oxidation** (removal of hydrogens) and correspondingly more energy per molecule than carbohydrates. Most fats provide twice as many calories per pound as do carbohydrates, and thus are a concentrated source of metabolic fuel. Fats may be taken in directly in the diet,

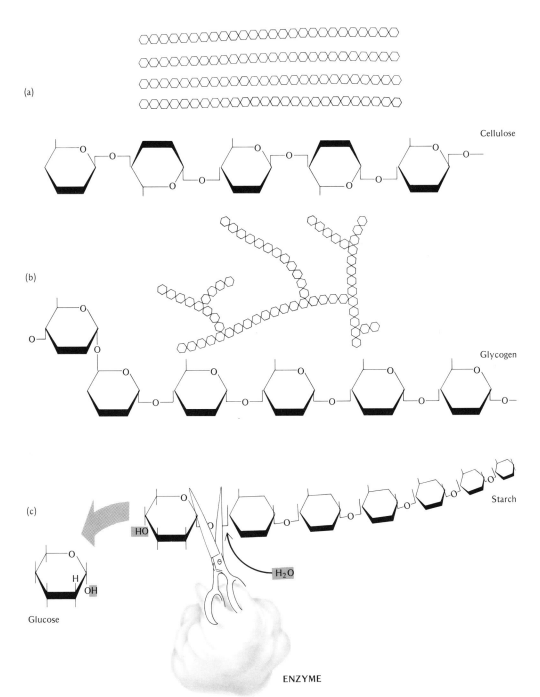

(a)

Cellulose

(b)

Glycogen

(c)

Starch

HO

H₂O

Glucose

H
OH

ENZYME

or carbohydrate and other nutrients may be converted into fat and stored as fat by the body cells. As we age, we ordinarily become fatter; young men average 13% of their body weight as fat; the amount of fat in the bodies of older men around age 50 increases to about 20%.

True fats, including oils and waxes, are composed of two kinds of molecules: (1) an alcohol (usually glycerol, sometimes called glycerine) and (2) fatty acids. The **glycerol** molecule has a backbone of 3 carbon atoms, and attached to each is a hydroxyl group (OH) (Figure 3–7). A **fatty acid** consists of a hydrocarbon chain and like all organic acids contains an acidic group, the carboxyl group (COOH) (Figure 3–7). When glycerol reacts with a fatty acid, an H from the glycerol and an OH from the carboxyl end of the fatty acid combine to form water. This chemical combination of glycerol and fatty acid is another example of a condensation reaction; we have already seen that this kind of reaction is important in the formation of complex carbohydrates. The linkage of the fatty acid and the glycerol (technically called an **ester linkage**) can occur at three places, since there are three OH groups in the glycerol molecule (Figure 3–7).

Fats differ in their properties because different kinds of fatty acids may be linked to glycerol to form a fat. The fats we eat may contain fatty acids with backbones of from 4 to 24 carbon atoms, but the most commonly occurring fatty acids are those with 16 and 18 carbon atoms. The hydrolysis of fats, which occurs in the intestine, is accomplished by the enzyme lipase (derived from the term lipid), which breaks the ester linkage, liberating glycerol and fatty acids (Figure 3–8).

The fatty acids that are chemically bound to glycerol vary not only in the length of their carbon chains, but also in the amount of their saturation. Current advertising tells us that polyunsaturated fats are better for our health

FIGURE 3–7 (top) The formation of a neutral fat involves the bonding of three fatty acids to a molecule of glycerol.

FIGURE 3–8 (bottom) Hydrolysis of fats and their entry into the circulatory system. Hydrolysis.

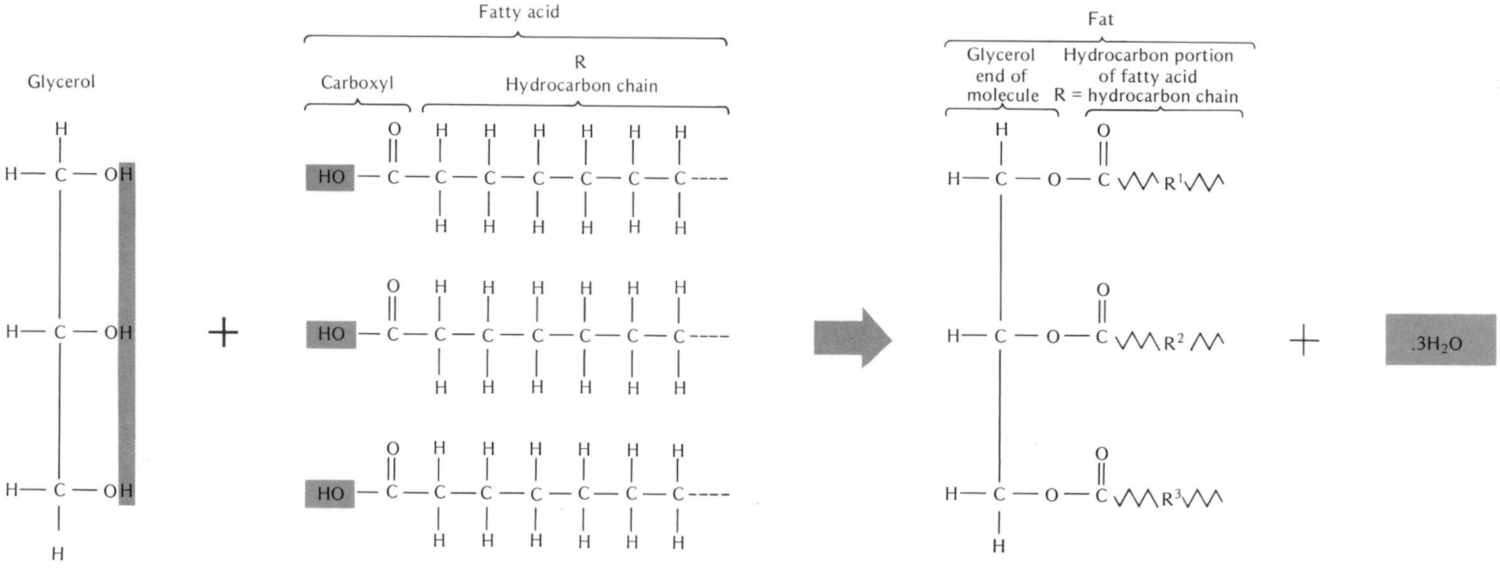

than saturated ones, but what do these terms actually mean? If the carbon atoms in the hydrocarbon portion of the fatty acids have most or all of their bonds filled with hydrogen atoms, these molecules are described as **saturated fats.** This type of fat is solid at room temperature. If a large proportion of the carbon atoms in the hydrocarbon chain of the fatty acid is not bonded to hydrogens, the **unsaturated** condition is obtained; these fats are liquid at room temperature and are commonly called oils. Most vegetable oils are unsaturated; they can be converted by **hydrogenation** (chemically adding hydrogens) to saturated fats such as the familiar vegetable shortening (for example, Crisco) or margarine (from cottonseed oil).

Atherosclerosis, a disease of the arteries, is caused by accumulation and deposition of lipids, chiefly cholesterol (a steroid) in combination with fatty acids and proteins, as

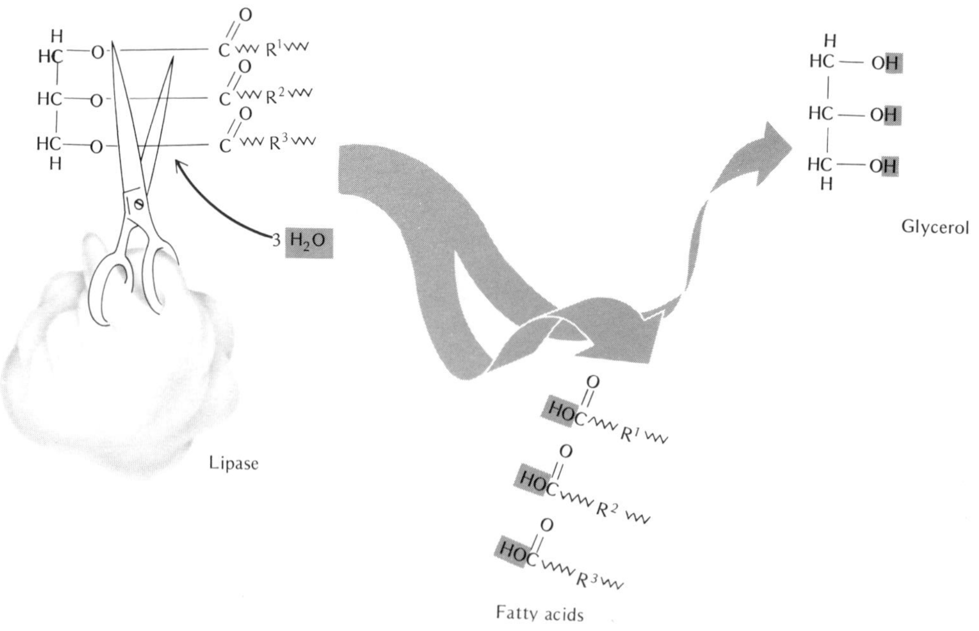

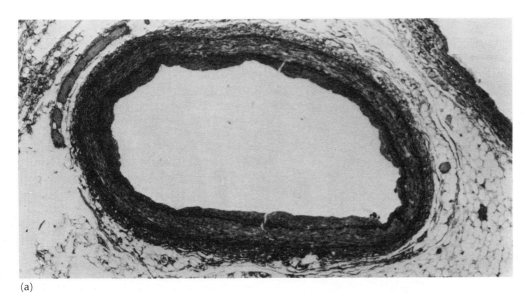

(a)

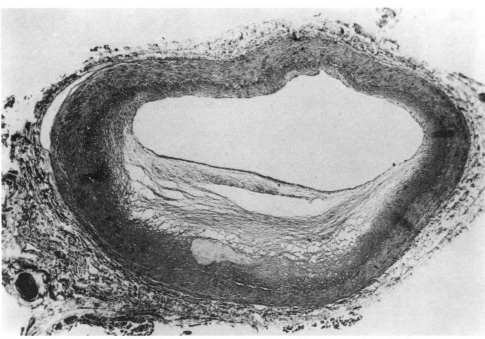

(b)

plaques in the arterial walls (Figure 3–9). Such deposits are common during and after middle age, when the percentage of body weight that is fat is higher, and the disease is promoted by diets rich in lipids. Polyunsaturates may in fact be better for one's health than saturated fats because they depress the synthesis of cholesterol and are not deposited in the walls of arteries to the same extent as are saturated fats. However, factors other than diet play a role in atherosclerosis. For example, blood cholesterol levels are increased as a result of smoking, stress, and lack of exercise.

Phospholipids

Phospholipids are important structural components of the plasma membrane and other membrane systems of the cell. As their name indicates, they are lipids that contain the element phosphorus; one of the three fatty acids linked to the glycerol is replaced by a group containing phosphorus and nitrogen (Figure 3–10a). This makes the molecule highly charged, in contrast to the fats previously discussed, which are therefore sometimes called neutral fats. This charged portion of the molecule is soluble in water, but the uncharged portion is not, resulting in a polarity. In the plasma membrane (Figure 3–10b), the water-soluble portion of the phospholipid is external, and the water-insoluble ends are internal and face each other.

The best known of the phospholipids are the lecithins and cephalins. Lecithin is found in egg yolk, brain, yeast, liver, and wheat germ; brain tissue is rich in cephalin.

Steroids

Steroids are classed as lipids, but they are not fats; they are considered to be lipids because of their solubility in organic solvents. Steroids have a complex structure of four interlocking rings of carbon atoms (Figure 3–11a). Probably the most familiar steroid is **cholesterol** (Figure 3–11b), which is important in membrane structure, in atherosclerosis, and as the building block for many hormones (Figure 3–11c). Simple addition or deletion of hydrogens, oxygens, and carbons on the basic ring structure creates profound functional change and is indicative of the parsimony of cellular chemistry. For example, the beard and deep voice of a man depend on minor changes in the steroid molecule; a slight molecular alteration is associated with a high-pitched voice, absence of a beard, and well-devoloped breasts (Figure 3–11c).

3–4 PROTEINS: FABRIC OF THE CELL

All living cells and all living organisms contain proteins. Although 60%–70% of the body consists of water, most of our tissues contain between 10% and 20% protein, the exact amount varying with the individual tissue. Proteins are essential to living systems; the word itself comes from Greek roots meaning "preeminence." If carbohydrates and lipids are principally the fuels of the living machine, then protein can be considered the structural material of which the machine is made. Furthermore, since all enzymes, without which life could not exist, are proteins, the preeminent position of proteins is well justified. Not all proteins are enzymes; some function in myriad other roles: gas transport, immune reactions, chemical regulation, and structural support. Common and important proteins found within the body include hemoglobin (the red pigment in red blood cells which carries oxygen), keratin (a fibrous protein that forms nails and hair), fibrin (a fibrous protein involved in clotting), collagen (perhaps the most abundant protein of all, the principal fibrous protein of skin, tendons, ligaments, bone, cartilage, etc.), in-

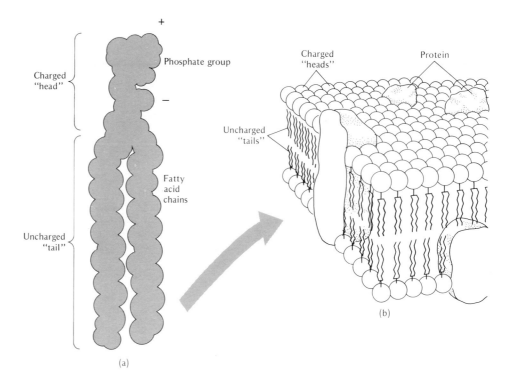

FIGURE 3–10 Phospholipids. (a) The form of lecithin, a phospholipid. (b) Phospholipid and plasma-membrane structure.

sulin (the hormone regulating blood sugar level), gamma globulin (containing antibodies), oxytocin and vasopressin (hormones that, respectively, regulate uterine contractions during childbirth and the diameter of blood vessels).

Protein structure

Proteins are organic molecules that contain nitrogen in addition to carbon, oxygen, and hydrogen; the fundamental building blocks of proteins are the **amino acids.** Amino acids are nitrogenous compounds having an amino (NH_2) group, a carboxyl (COOH), group and another characteristic group of atoms often referred to as an **R group,** since it is variable.

We can, therefore, write the generalized formula:

$$NH_2{-}\overset{\displaystyle R}{\underset{\displaystyle H}{\vphantom{|}C}}{-}COOH$$

Twenty amino acids occur in nature, and they differ from each other in the structure of the R group. They combine in various ways to form different proteins in much the same fashion as the letters of the alphabet can be arranged to form a variety of words; amino acids are the alphabet of protein structure, and all the proteins that exist in nature, whether they be of cabbage or of king, are constructed from this

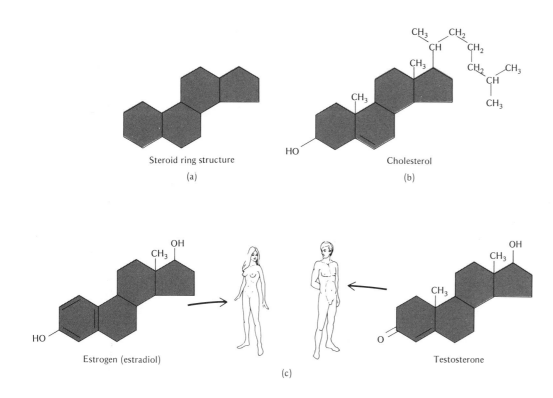

FIGURE 3–11 The structure of some common steroids: (a) Basic steroid ring structure. (b) Cholesterol. (c) Testosterone (male hormone) and estrogen (female hormone).

Steroid ring structure

(a)

Cholesterol

(b)

Estrogen (estradiol)

Testosterone

(c)

is called a **peptide bond** (−CONH; Figure 3–12). The linking together of amino acids is much like attaching "pop-it" beads together to form a long chain. The joining of up to 100 amino acids results in a **polypeptide chain,** and generally speaking, 100 or more linked amino acids are called a **protein.** Proteins are generally large molecules having molecular weights of thousands to millions. Hemoglobin, for example, has a molecular weight of 64,000; insulin has a molecular weight of 5,000; and the protein of a virus weighs about 40,000,000. By comparison, water has a molecular weight of 18, glucose a molecular weight of 180, and table sugar (sucrose) a molecular weight of 342. Size and function in proteins are not related: vasopressin (the hormone controlling the diameter of blood vessels) contains only nine amino acids, whereas gamma globulin (the antibody molecule) contains hundreds of amino acids; both are indispensable for normal body function.

The variety of proteins existing in nature is due both to the kinds of amino acids composing the peptide chain and to the particular sequence in which they are arranged. On occasion, a single variation in an amino acid out of a chain of hundreds of amino acids makes the molecule behave entirely differently. To illustrate this point, let us once again use our alphabet analogy. If a sequence of letters spells out the word *good,* then a single alteration in one letter could change the entire meaning— for example, *gold* or *food* or *goof.* Sometimes an alteration can produce a word that is meaningless, such as *lood.* For example, the hemoglobin in red blood cells is a protein composed of a chain of 287 amino acids. Some persons have a variant kind in their blood cells called sickle-cell hemoglobin. When the red blood cells of persons so affected are deprived of oxygen, the cells assume a sickled shape and can no longer function normally. This striking

universal 20-letter alphabet. Insofar as proteins are concerned, all living things are not equal. The proteins of a person differ from those of a dog, cat, elephant, and baboon. They differ from those of other living creatures because of a different arrangement of amino acids. This is what we mean by **species specificity.**

Since proteins may contain several hundreds or thousands of amino acids, it is obvious that the number of possible combinations of the 20 kinds of amino acids is almost infinite; thus, there are endless possible kinds of proteins. The possible number of proteins of average size (containing about 500 amino acids) is so large that it would be expressed as the number 1 followed by 600 zeros! To understand the

magnitude of the situation, imagine taking the 26 letters of the alphabet and counting the number of words that can be constructed. Additionally, to make the comparison equivalent to the structure of a protein, you would have to make words that contained 500 letters.

Proteins are long, complex chains of amino acids linked end to end. The connecting link between two amino acids is formed by a chemical bonding of the carboxyl group (COOH) of one amino acid with the amino group (NH₂) of another; its formation (a condensation reaction) involves the loss of a water molecule between the two amino acids (an OH from the carboxyl group and an H from the amino group), and the bond thus formed change in the properties of a red blood cell containing sickle-cell hemoglobin is produced by a change in only 1 amino acid out of a total of 287!

The protein chain of amino acids formed by peptide bonding makes up the so-called **primary structure** of the protein. Few proteins, however, exist as a straight chain, and most polypeptide chains are coiled or twisted to yield a springlike or helical form (Figure 3–13). The most common kind of coiling arrangement is called an **alpha helix;** in its precise geometry there is a turn of the helix every 3.6 amino acids (5 turns involve 18 amino acids). The helix is stabilized by the formation of hydrogen bonds between successive turns of the spiral; the hydrogen bonds are formed between the C = O group of one amino acid and the NH group of another (Figure 3–13). The spiral-staircase arrangement of the polypeptide chain is often called the **secondary structure** of the protein.

Proteins undergo other contortions, for the helical polypeptide chain can be further twisted to yield a globular or an elongated shape (Figure 3–13). This folding of the alpha helix into a variety of shapes produces what is called the **tertiary structure** of the protein. Just as the secondary structure of the protein is maintained by hydrogen bonds, so too is its tertiary structure.

The secondary and tertiary structures of a protein are easily destroyed by harsh environmental conditions such as heat, excessive acidity, chemical reagents, and radiation. If a protein solution (for example, the white of an egg) is subjected to high temperature (boiling), the helical arrangement is destroyed and the protein becomes **coagulated** or **denatured** (Figure 3–14). In this condition, although the amino-acid sequence in the protein is unchanged, the spatial orientation of the molecule

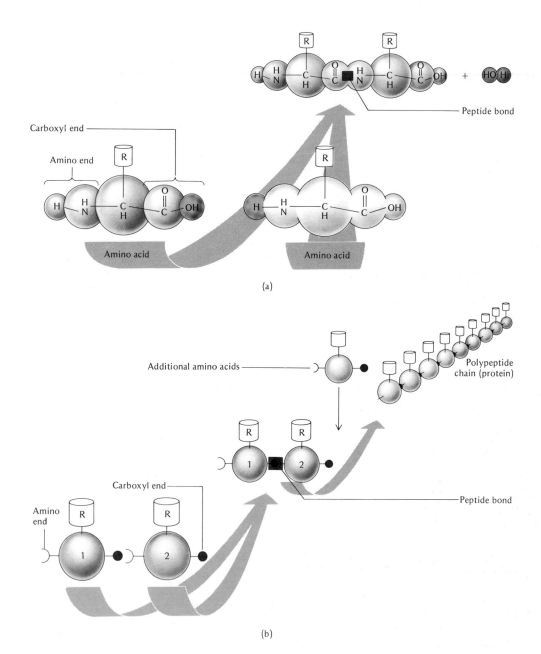

Carboxyl end

Amino end

Amino acid

Amino acid

Peptide bond

(a)

Additional amino acids

Polypeptide chain (protein)

Amino end

Carboxyl end

Peptide bond

(b)

FIGURE 3–13 The formation of the secondary and tertiary structure of a protein involves coiling of the polypeptide chain.

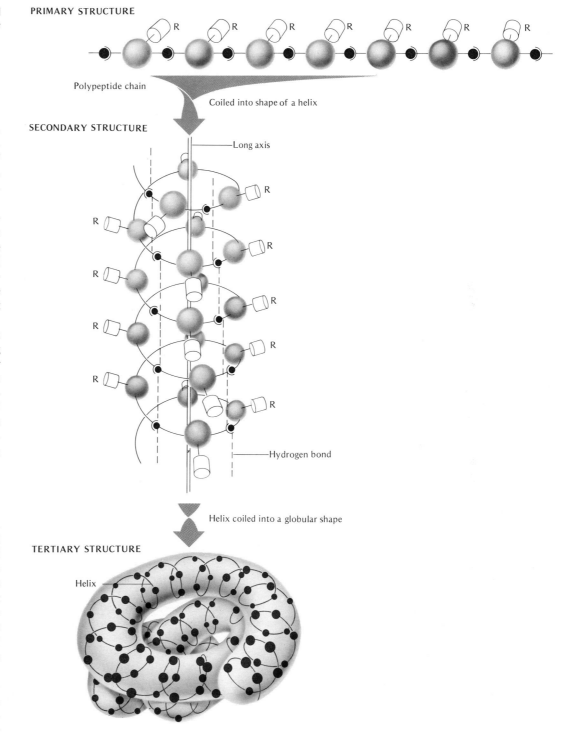

PRIMARY STRUCTURE

Polypeptide chain

Coiled into shape of a helix

SECONDARY STRUCTURE

Long axis

Hydrogen bond

Helix coiled into a globular shape

TERTIARY STRUCTURE

Helix

is severely altered; the helix does not re-form spontaneously, and the molecule loses its original properties. This process of denaturation of proteins may also occur by drying or by exposing the solution to acid or alkaline conditions.

Denaturation may be reversible or irreversible. If the diffuse or uncoiled condition can be returned to the coiled configuration, the solubility and other properties of the normal (or native) state are obtained, and this is reversible denaturation, whereas if the native state is not reestablished (as in coagulation) it is irreversible denaturation.

The structure of proteins is important in a practical sense because it tells us something about their properties (Box 3B). The reason for refrigerating blood plasma and other protein solutions becomes apparent; the three-dimensional configuration of proteins is maintained by relatively weak hydrogen bonds that are easily disrupted by elevated temperatures. Consequently the geometry as well as the function of the proteins are destroyed by heat.

Conjugated proteins

Many proteins consist entirely of amino acids, but some include other components (called **prosthetic groups**) and are called **conjugated proteins.** Conjugated means bound, and in conjugated proteins, the protein portion is bound to a prosthetic group. One of the most familiar of all conjugated proteins is hemoglobin. Hemoglobin consists of four polypeptide chains wound about four ring structures called heme, each of which bears an atom of the metal iron (Figure 3–15). The heme is the prosthetic group. The globular, protein portion of the molecule is called globin, and by its conjugation (binding) with heme it becomes hemoglobin. Other conjugated proteins con-

FIGURE 3–14 (top) Denaturation of a protein. First, the weak interactions maintaining the tertiary structure are destroyed; then there is disruption of the hydrogen bonds to produce a random coil.

FIGURE 3–15 (bottom) Protein and prosthetic-group organization in hemoglobin. The molecule is folded in a complicated way, and the iron-containing heme group (called the prosthetic group) is embedded in one region. Four such protein chains, each with a heme, form a single hemoglobin molecule.

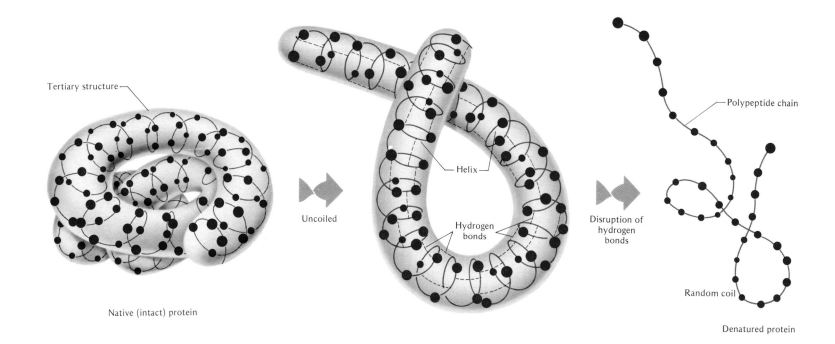

Tertiary structure

Native (intact) protein

Uncoiled

Helix

Hydrogen bonds

Disruption of hydrogen bonds

Polypeptide chain

Random coil

Denatured protein

taining heme are the cytochromes, important in the energy transformations of the cell.

Other examples of conjugated proteins are nucleoproteins (nucleic acid plus protein), found in the nucleus and ribosomes; glycoproteins (carbohydrate plus proteins), present on the surface of erythrocytes and responsible for blood types; and lipoproteins (lipid plus protein), which are structural elements in all cellular membranes.

3–5 ENZYMES: PROTEINS THAT MAKE HASTE

An old adage states: "Haste makes waste." However, in all cells hasty chemical reactions make life possible. Before we see how cells are able to speed up chemical changes, let us briefly consider the conditions under which chemical reactions themselves take place.

Before molecules can react to form new chemical configurations, their atoms must be so oriented that electrons may be lost, gained, or shared. The sharing and rearrangement of electrons form the basis of all chemical change. However, spontaneous electron alterations do not commonly occur in nature: if they did, matter in the universe would be very unstable and constantly undergoing change.

The bringing together of molecules so that atomic collisions can occur and electron sharing can take place requires energy. That is, to effect the molecular collisions which allow chemical reactions to occur, there must be an increase in the amount of energy in the reacting substances. This is called **activation energy.**

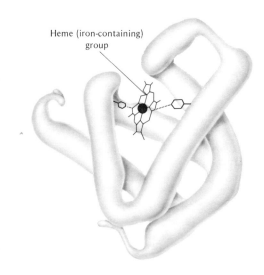

Heme (iron-containing) group

BOX 3B Curling hair and the shape of protein molecules

The major protein of hair and wool is keratin. Long protein molecules are lined up alongside each other to form a fibrous arrangement and bound together by numerous cross linkages that give the hair its elasticity and strength (Figure 3-16a). About 10% of all the amino acids forming keratin are cysteine which contains sulfur. These amino acids form sulfur-sulfur linkages called **disulfide bridges,** which link the protein chains. The position of a disulfide bridge contributes to the straightness or curl of hair (Figure 3-16). To change the arrangement of the bonds, one must break the bridges and re-form them in another configuration. This is what is done in a permanent-wave treatment. If the keratin fibers in hair are forcibly coiled and subjected to heat of over 200°F, the strained sulfur-sulfur linkages snap, and new ones re-form on cooling; the hair remains permanently curled. Such treatment often damages the hair itself, and so the cold-wave or home-permanent-wave treatment, which does not involve high temperatures and is less harmful, has largely replaced the hot-wave permanent. Permanent-wave kits for home use contain a strong, odoriferous chemical, which is applied to the hair; the chemical contains a so-called reducing substance that breaks the disulfide bridge. The hair is then curled or straightened with rollers, and a neutralizer is applied. The neutralizer reestablishes the disulfide bridges in the configuration of the rollers, and the protein chains are reoriented. The normal configuration of hair proteins and the position of the disulfide bridges are determined by hereditary constitution, so when new hair grows it must, in turn, be treated.

Disulfide bridges also help to stabilize other proteins or polypeptides. For example, insulin is composed of two chains held together by disulfide bridges (Figure 3-16b); even small molecules such as vasopressin and oxytocin, with only nine amino acids, are formed into a ring by the disulfide bridge (Figure 3-16c). A contorted protein chain achieves additional stability by the strong binding produced by the sulfur-sulfur linkage.

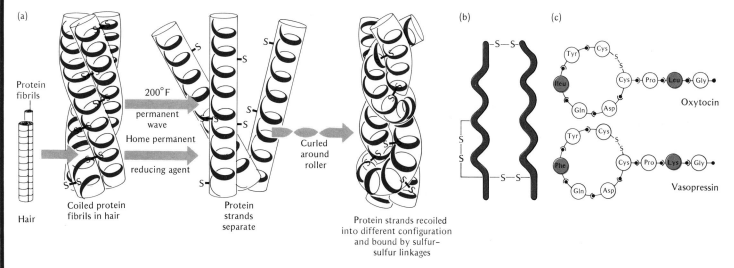

FIGURE 3–16 Disulfide bridges, (a) Hair curling. (b) Insulin, a hormone, contains two polypeptide chains held together by disulfide bridges. (c) Oxytocin and vasopressin, hormones, each consists of nine amino acids; six of these are linked via disulfide bridges. Note that the two hormones differ in only two amino acids (shaded).

FIGURE 3–17 Skiing and activation energy. (a) Skiing downslope requires energy input to reach the hilltop (energy barrier). Once this is reached, the skier is activated and can easily ski downhill without additional energy. (b) Ordinary chemical reaction. The energy of the reactants is insufficient to carry them over the energy barrier, and so they do not react until additional energy is supplied (activation energy) to allow movement across the energy barrier. (c) Enzyme-catalyzed reaction. Required energy of activation is lowered, permitting more molecules to cross the energy barrier and the reaction to proceed at a faster rate.

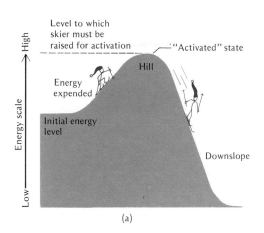

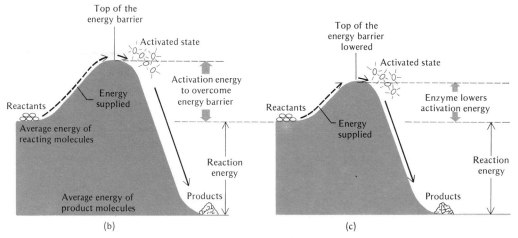

Once molecules have acquired the necessary activation energy, the chemical reaction can begin. In other words, each chemical reaction has an energy barrier; the higher the barrier, the more stable are the substances involved, and the less likely they are to react. We can compare the situation to a skier who must get to the top of a hill in order to ski down the other side. He starts at the bottom of the hill, and to get to the top he must expend energy; a strong push of his arms on his ski poles, and the skier moves up the slope. The energy expended to get the skier to the top of the hill is equivalent to the activation energy necessary to overcome the energy barrier in a chemical reaction (Figure 3–17a and b). The skier at the top of the hill can be considered to be in an activated state and can easily go down the opposite side of the hill; the barrier has been hurdled. In just the same way, chemical reactions can proceed only when the reacting molecules have acquired the energy of activation. Thus, a molecule may never react or change unless activation energy is supplied to it. If this happens, the molecule is lifted over

the top of the energy hill and rolls down the other side. In rolling down, it is chemically altered.

How can we influence the activation energy of molecules so that rates of chemical reactions can be increased? We can visualize a number of ways of making molecules more reactive. For example, we can increase the concentrations of the reactants so that the frequency of molecular collision is increased, or we can increase the energy of the reactants by heating the materials, or we can add an accelerating substance called a **catalyst.** Catalysts are substances that affect the rate of a reaction without themselves being changed in the process. Catalysts are not magic substances. They do not make impossible reactions possible, but they do accelerate the rates of reactions that would occur spontaneously (but very slowly) in their absence. Thus, catalysts have similar effects upon reactants as do temperature and increased concentration. Catalysts are widely used in industry to reduce the energy of activation of reactants. They lower the energy barrier, and chemical reactions proceed rapidly at ordinary

temperatures with reduced concentrations of reactants. Iron filings and finely divided platinum are common industrial catalysts. We can demonstrate the catalytic effects of iron filings quite easily. For example, hydrogen peroxide (used in bleaching hair or as an antiseptic) is a naturally unstable material which breaks down slowly at room temperature to yield oxygen and water. Its decomposition can be accelerated by warming the solution or by adding a pinch of a catalyst such as iron filings. As the iron filings are added the hydrogen peroxide froths; the liquid turns almost white as oxygen is rapidly liberated. The iron filings, however, remain unchanged.

How do catalysts produce their effects? One mechanism is by providing a site of adsorption for the reactants—a sort of meeting place or rendezvous so that the reactants can find each other more easily and are more likely to meet. Molecular contacts are thus more frequent and assured, facilitating molecular rearrangements (Figure 3–18). Other factors, still obscure, probably also play a role in catalytic activities. Catalysts are economical materials for produc-

FIGURE 3–18 How a catalyst works. (a) Molecules react chemically if they collide; however, such collisions are rare. (b) If an adsorption site for the molecules is present, collisions occur more easily. Catalysts provide adsorption sites, thus facilitating chemical reactions and increasing their rates.

FIGURE 3–19 The interaction of enzyme and substrate. (a) Enzyme and substrate have geometries that allow them to fit together. (b) Enzyme and substrate form a complex. (c) Product is liberated, and enzyme is free to participate with a new molecule of substrate.

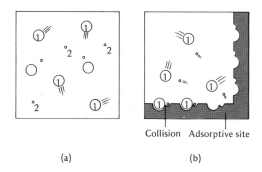

Collision   Adsorptive site

(a)   (b)

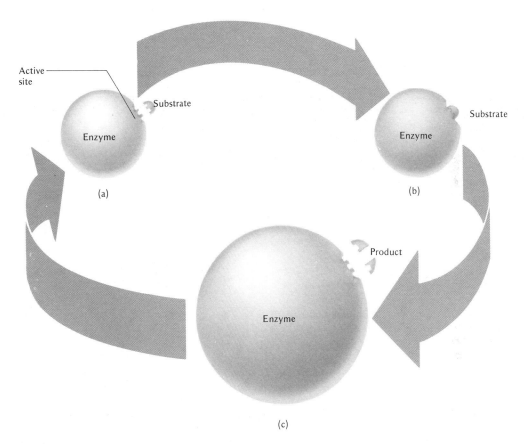

ing chemical change; they are usually effective in very small quantities, and since they are unchanged by the chemical reactions they effect, they can be used over and over again.

Now let us consider how the cell makes chemical reactions occur at a quickened pace. Clearly heat cannot be used, since it would coagulate (irreversibly denature) the cellular proteins; the concentrations of reactants cannot be increased beyond a certain level because many of these key materials are unavailable in large quantities; and the addition of substances such as platinum or iron filings is quite out of order. Living cells hasten chemical reactions by employing certain kinds of proteins as catalysts—**enzymes.** All enzymes are proteins, and they are involved as catalysts in almost all the chemical reactions that take place in the living organism. Without enzymes, life would not be possible.

How do enzymes work? Enzymes operate in much the same fashion as other catalysts; that is, they lower the amount of energy necessary to start the reaction and thus assist molecules in jumping over the activation-energy barrier (Figure 3–17c). To do this an enzyme must initially combine with its **substrate** (the substance[s] acted upon by the enzyme) and form a chemical complex (Figure 3–19a and b); in forming the complex, the enzyme steers the

electrons of the substrate into deformed orbits, orienting them so they can be shared between reactants, and then the chemical reaction occurs. Once the reaction has been completed, the enzyme is released and is free to participate once again with new substrate (Figure 3–19c).

Unlike inorganic catalysts, enzymes are specific in their action and often can catalyze only a single reaction. The degree of their specificity is probably dependent upon the structure of the protein. Current thinking holds that the substrate, which is usually a smaller molecule,

fits into a specific area on the surface of the large-protein enzyme. The region of attachment is called the **active site** of the enzyme. Since an enzyme must have the proper geometry for its active site to fit the substrate properly, it becomes clear why enzymes are easily inactivated by heat, acid, and alkali. Such treatments denature the protein and destroy the three-dimensional structure of the enzyme; the active site and the substrate cannot fit properly, and because of the misalignment no chemical reaction takes place.

The specificity of enzyme and substrate suggests mechanisms whereby enzymes may be **poisoned.** Poisoning occurs when the enzyme is prevented from affecting the substrate even though they are in proximity. If molecules resembling the substrate are introduced into a living cell (or to an isolated enzyme), but these molecules cannot be acted upon by the enzyme, binding of the mimic compound (called an **analog** or **antimetabolite**) to the enzyme may occur, and then the enzyme is prevented from carrying out its function with its normal substrate (Figure 3–20). If the binding process is irreversible, the enzyme is permanently poisoned. Fluoride acts as a poison for the enzymes involved in glucose breakdown. The nerve gases, agents of chemical warfare, operate by blocking the enzymes involved in the transmission of the nerve impulse and thus produce their paralyzing and lethal effects.

Over a thousand different enzymes are known, and undoubtedly many more exist that have not yet been identified. Each enzyme is involved in a rather specific chemical reaction. The names given to enzymes generally end in -ase and are derived either from the substrate on which they work or from the kind of reaction in which they are involved. For example, lipase acts on lipids, amylase attacks starch (amylum—starch; L.), and sucrase attacks sucrose (table sugar). This uniformity of enzyme nomenclature is convenient, logical, and of recent vintage. Some of the first enzymes discovered escaped this systematic approach and were given names that end in -in; these have been so commonly used that they are retained to this day. Trypsin and pepsin, which digest proteins, and ptyalin (present in saliva), which digests starch, are examples of these oldtime enzymes.

Enzymes are organic catalysts; like industrial catalysts, they are effective in very small amounts, they are relatively unchanged by the

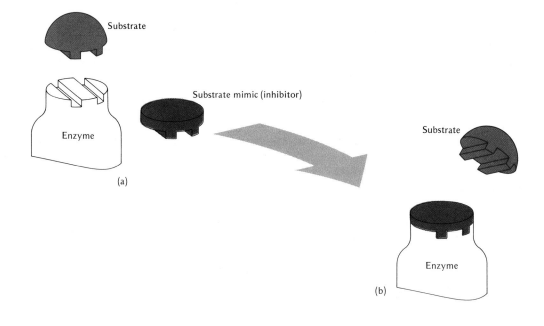

FIGURE 3–20 Enzyme inhibition. (a) Inhibitor molecule is mixed with the enzyme. (b) Enzyme-inhibitor complex forms and prevents the natural substrate from attaching to the enzyme; no reaction occurs since the enzyme is blocked, or poisoned.

chemical reactions they effect, and they can be used over and over again; this is highly advantageous to the cellular economy, since only a minute amount of each required enzyme need be synthesized by the cell, and the organism need not expend energy and materials for the synthesis of enzymes at a rate proportional to the reactions they effect. Life depends on enzyme function, and when enzymes are inactivated, death may eventually be the outcome.

3–6 VITAMINS AND COFACTORS: ENZYME HELPERS

Vitamins

Sailors on extended voyages used to be prone to the disease scurvy, characterized by skin hemorrhage, bleeding gums, extremely tender joints, and great pain upon movement. An English surgeon named Lind demonstrated that food alone would cure scurvy and that the disease could easily be prevented by drinking lemon and lime juices. Since British sailors were required to drink lime or lemon juice daily as a preventive measure, they were nicknamed "limeys." In 1870 sailors of the newly created Japanese navy fell ill with a painful and crippling disease that paralyzed the legs, affected the heart, and produced loss of appetite and severe constipation. The disease, called beriberi, was cured by having the sailors eat less polished rice and more wheat, meat, and vegetables. In 1914 a debilitating disease, pellagra, was widespread in the southern United States. Persons affected showed a roughening

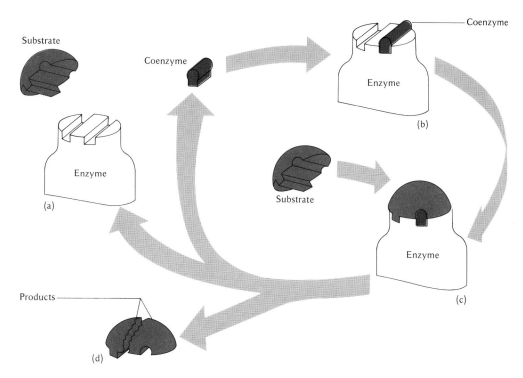

of the skin, diarrhea, and dementia. Dr. Joseph Goldberger claimed that the disease was not spread by infection, and to prove the point he and his wife and 22 volunteers swallowed capsules containing the blood, feces, and urine of pellagra patients. None of them developed pellagra. A diet of wholegrain cereals quickly cured the afflicted.

Miracle cures for scurvy, beriberi, and pellagra? No, the diseases were obviously related to diet; today it is well-known that these diseases are the result of deficiencies in specific dietary materials called **vitamins.**

Vitamins, the miracle workers in deficiency diseases, are molecules that must be acquired from food because they cannot be manufactured by the body. A glance at the side panel of a cereal package (page 56) indicates that we require these materials in very small quantities, what we might call catalytic amounts. Do vitamins play a role as catalysts? Yes, but in a special way. Vitamins are misnamed because they are not "vital amines" as was once thought, but represent a diverse group of chemical compounds. However, all vitamins have a similar function: within the body they are converted into substances known as **coenzymes** which are essential participants in a number of enzyme-catalyzed reactions. Not all enzyme reactions involve coenzymes, but in those that do, the characteristics of vitamin deficiency are produced by the absence of required coenzymes.

How do enzymes and coenzymes differ? Coenzymes are not proteins, they are of much smaller molecular size, and they do not ordinarily lose their normal properties and activity when subjected to treatments which inactivate (denature) most proteins. How do coenzymes work? They may act as adapter molecules, facilitating binding of enzyme and substrate, or removing one of the products of the chemical reaction, thus making the reaction tend to proceed in one direction only.[2] Like enzymes, coenzymes are not consumed in the reaction and are effective in small quantities. However, unlike enzymes, they are generally not as specific with respect to substrate. The relationship of coenzyme and enzyme to catalytic function is illustrated in Figure 3–21.

In the treatment of disease considerable gains have been made from the simple recognition that coenzymes function in a wide variety of enzyme-catalyzed reactions (Box 3C). As we have observed, enzyme function can be inhibited or enzymes can be poisoned by substances that resemble the natural substrate (Figure 3–20); this type of inhibition is quite specific, since enzymes themselves ordinarily act in restricted catalytic capacities. On the other hand, a coenzyme inhibitor causes a more generalized inhibition or poisoning because it affects all enzymes requiring that coenzyme. For example, folic acid is a vitamin and coenzyme required by all our cells for the synthesis of nuclear materials. A deficiency of folic acid prevents the operation of enzymes involved in nuclear synthetic reactions, and eventually cell growth and division are inhibited. Certain drugs (such as aminopterin) structurally resemble folic acid, but cannot serve as the coenzyme; they are called folic acid analogs. When cancer patients, particularly those with leukemia, are treated with such drugs, there is remission of the disease. The remission is due to the nature of the greedy cancer cell, which avidly takes

---

[2] Refer to Appendix A for reasons for this.

BOX 3C Beneficial poison: sulfanilamide

The vitamin p-aminobenzoic acid (PABA) is required by many bacteria for the synthesis of the coenzyme folic acid. Folic acid is a coenzyme that is required by all organisms, including ourselves. Unlike bacteria we cannot take PABA and make folic acid; we require a supply of preformed folic acid and obtain it from our diet. It may be a matter of semantics, but for bacteria PABA is a vitamin, and for us folic acid is a vitamin.

Since bacteria require PABA, anything that will interfere with its utilization will ultimately prevent the manufacture of the critical coenzyme folic acid. Sulfanilamide, a sulfa drug, discovered in the 1930s was found to be effective in controlling bacterial growth; it was especially valuable in the treatment of pneumonia. Why does sulfanilamide kill bacteria but not humans? The answer lies in the similarity of structure of PABA and sulfanilamide (Figure 3-22). When bacteria are presented with the sulfa drug, they are "fooled" into incorporating this material into their folic acid, but this sulfa-bearing folic acid molecule is completely functionless. Enzyme reactions in the bacterial cell requiring folic acid cannot be carried out, and so the bacteria die. Since we do not manufacture folic acid from PABA there is little chance that the cells of our body will utilize the sulfa drug. (However, note that an overdose could ultimately kill the cells of the patient as well as those of the bacteria.)

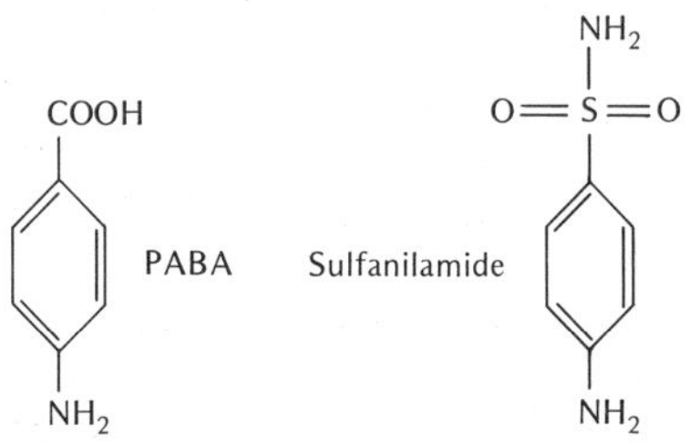

FIGURE 3-22 The structure of PABA and sulfanilamide.

up the drug; the drug inhibits synthesis of nuclear materials and causes destruction of the cancer cells. Aminopterin also affects the requirements of normal cells for folic acid, but to a much lower degree, since normal cells divide at a much slower rate. Hence, adequate precautions must be taken not to overdose the patient; the effectiveness of drug treatment depends on the delicate balance between administered dose and the requirement of the cancer cells for folic acid.

Vitamins must be supplied regularly in the diet because the body cells cannot manufacture these key materials, and they are broken down and excreted during cell metabolism. This means that although vitamins are required in very small amounts, any deficiency in the diet can cause a disruption in cellular function and the symptoms of disease. If the diet is well balanced, there is no need for vitamin supplements, but if the diet is poorly balanced, then vitamin supplements may be helpful. Vitamin pills are relatively expensive, and the paradox is that "those who need vitamin pills can't afford them, and those who can afford them don't need them."[3] The best advice on vitamin pills is: Buy them if you can afford them, but if your diet is balanced and varied you can save your money.

Minerals

We are conscious that minerals in the soil are necessary for good plant growth, and we assure healthy plants by supplementing garden soil with iron, phosphorus, and magnesium. We may not yellow like the leaves of a plant when suffering from a lack of iron or magnesium, but

[3] Quoted in Garrett Hardin, *Biology: Its Principles and Implications,* 2nd ed., San Francisco, Freeman, 1966, p. 494.

we certainly do show some ill effects from deficiencies of mineral elements. The cereal packet (page 56) indicates that the cereal supplies certain amounts of trace elements—magnesium, cobalt, iron, manganese, zinc, and copper.

The minerals we obtain in our food are not degraded by digestive juices, but are simply freed from their complex associations with food components and absorbed across the intestinal wall to enter the bloodstream. Unlike carbohydrates, fats, and proteins, minerals supply no energy, but they play important roles in the structure of the body, and in trace quantities they perform essential metabolic functions. For example, calcium and phosphorus are major structural components of bones and teeth, and iodine is necessary for the synthesis of thyroid hormone. Some minerals function as part of the prosthetic group of conjugated proteins. For example, iron is a part of the prosthetic group (heme) of both blood hemoglobin and the cytochrome enzymes. Minerals may also function as cofactors of enzymes. Cofactors may function much like coenzymes in assisting the enzyme to bind its substrate.

3–7 NUCLEIC ACIDS: INFORMATIONAL MOLECULES

Like proteins, **nucleic acids** are generally large molecules. They have two principal functions in the cell. First, they are the materials that record the hereditary information that every living thing carries with it; second, they translate this information so that they can direct the cell's activities. The latter fucntion is mediated through their direction of the synthesis of proteins, and in particular proteins with an enzymatic function. We shall discuss the structure and function of these very important molecules at greater length in Chapters 6 and 7.

SUMMARY

1. The organic molecules that make up living things fall into five major categories: carbohydrates, fats or lipids, proteins, vitamins, and nucleic acids.

2. *Carbohydrates* (sugars and starches) have the basic composition $(CH_2O)n$. They are the primary source of cellular energy.

3. Polymerization of carbohydrate units into larger storage molecules prevents them from leaving the cell, where they can later be degraded and used as needed. Monosaccharides contain from 3 to 10 carbon atoms. The commonest are 6-carbon, hexose sugars, with the formula $(CH_2O)_6$ or $C_6H_{12}O_6$. Structural hexose isomers include glucose (dextrose), fructose, and galactose. Of these, glucose is the most common in cells, and its level in the blood is rigidly controlled.

4. Glucose occurs only in the right-handed or D-form (stereoisomer) in living things, and is known as dextrose. (Proteins in living things contain only the left-handed or L-isomer of amino acids).

5. Two monosaccharide units bonded together form a double sugar or disaccharide. Disaccharides are formed by condensation reactions with the loss of a water molecule. Conversely, they can be broken apart during digestion with the addition of a water molecule, a process called hydrolysis. Common disaccharides are table sugar or sucrose (glucose + fructose), milk sugar or lactose (glucose + galactose), and malt sugar or maltose (glucose + glucose).

6. Long chains of sugar molecules linked by condensation reactions are called polysaccharides or multiple sugars. Starch, glycogen, and cellulose are formed from long chains of glucose molecules. Polysaccharides are usually insoluble in water and are important storage and structural components of cells.

7. *Lipids* include fats and steroids; insoluble in water but soluble in organic solvents, they are important fuels, storage materials, and components of membranes and of hormones.

8. Like carbohydrates, fats contain CHO, but the ratio of hydrogen to oxygen is greater than 2:1. Thus fats contain more hydrogen than carbohydrates, which permits a greater degree of oxidation by removal of hydrogen; therefore fats provide about twice as much energy as the same amount of carbohydrate.

9. True fats are composed of an alcohol (usually glycerol, with 3C) and fatty acids.

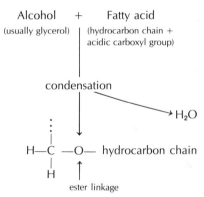

The link between the alcohol and fatty acid is called an ester linkage.

10. The ester linkage of glycerol to fatty acid, each with its long hydrocarbon chain and carboxyl group, can occur at three places because there are 3 hydroxyl groups in glycerol.

11. Digestion of fats is by hydrolysis with the help of the enzyme lipase.

12. Saturated fats are solid at room temperature; all the carbons in their hydrocarbon chains are filled with hydrogens. Unsaturated fats contain less hydrogen and are not solid at room temperature.

13. Phospholipids contain a phosphorus and nitrogen group in place of one hydrocarbon chain, which polarizes the molecule. Phospholipids are important constituents of membranes.

14. Steroids consist of 4 interlocking carbon rings and are important constituents of hormones, membranes, and cholesterol. Cholesterol causes atherosclerosis when deposited on arterial walls in combination with fatty acids from fats. A diet high in saturated fats appears to speed this process.

15. *Proteins* contain C, H, and O plus nitrogen. They are built of amino acids whose formula is:

$$NH_2-\underset{\underset{H}{|}}{\overset{\overset{R}{|}}{C}}-COOH$$

The R group is variable.
There are 20 kinds of amino acids in nature, which form polypeptide chains and proteins by condensation reactions among them:

$$H_2N-\underset{\underset{H}{|}}{\overset{\overset{R}{|}}{C}}-COOH \quad + \quad H_2N-\underset{\underset{H}{|}}{\overset{\overset{R}{|}}{C}}-COOH$$

$$\searrow H_2O$$

$$N_2H-\underset{\underset{H}{|}}{\overset{\overset{R}{|}}{C}}-\underset{\text{peptide bond}}{CONH}-\underset{\underset{H}{|}}{\overset{\overset{R}{|}}{C}}-COOH$$

16. Up to 100 amino acids joined together = a polypeptide chain. Longer chains = protein.

17. Primary structure of a protein = straight polypeptide chain.
Secondary structure = helical coiling (alpha-helix) with subsidiary H-bond formation.
Tertiary structure = twisting of helically coiled chain into globular form with subsidiary H-bond formation.

18. Denaturation (coagulation) = uncoiling of tertiary or secondary structure and breaking of the H-bonds. May be caused by heat, excess acidity, chemical reagents, radiation etc. It may be reversible.

19. Subsidiary bonds such as sulfur-sulfur links between molecules of the R group stabilize proteins and can produce changes in protein geometry.

20. Some molecules exist in two geometric forms, called stereoisomers, which are mirror images of each other. In living things, carbohydrates are all right-handed (D-form), and amino acids are all left-handed (L-form).

21. Conjugated proteins have a nonprotein prosthetic group attached, and include nucleoproteins, glycoproteins, lipoproteins, and metal-containing proteins.

22. All enzymes are proteins and are cellular organic catalysts. Enzymes combine with their substrates, making it possible for substrate and reactant to share electrons. Following the reaction, the enzyme is released. Enzymes are reaction-specific, unchanged by the reactions they catalyze, and effective in minute quantities. Enzymes can be poisoned by mimics called analogs or antimetabolites.

23. Vitamins, components of coenzymes, facilitate enzyme-catalyzed reactions. The body cells cannot manufacture vitamins so they must be supplied in the diet. Deficiency of particular vitamins causes various diseases related to the lack of the vitamin coenzyme.

24. Minerals obtained in the diet are important in structural components of the body; others function as part of the prosthetic group of conjugated proteins or as cofactors of enzymes.

25. Nucleic acids record hereditary information and translate it to direct the cell's functions.

KEY WORDS

carbohydrates
polymerized
monosaccharides
trioses
tetroses
pentoses
hexoses
glucose
dextrose
fructose
galactose
structural isomers
disaccharide
hydrolysis
sucrose
lactose
maltose
condensation
polysaccharides
starch
glycogen
cellulose
lipids
fats
oxidation
glycerol
fatty acid
ester linkage
saturated fats
unsaturated fats
oils
hydrogenation
stereoisomers
phospholipids
steroids
cholesterol
amino acids
R group
species specificity
peptide bond
polypeptide chain

protein
primary structure
alpha helix
secondary structure
coagulated
denatured
tertiary structure
disulfide bridges
prosthetic groups
conjugated proteins
activation energy
catalyst
enzymes
substrate
active site
poisoned
analog
antimetabolite
vitamins
coenzymes
cofactors
minerals
nucleic acids

TOPICS FOR REVIEW AND DISCUSSION

1. How is the structure of a protein related to its function?
2. What is denaturation?
3. Compare and contrast the structure of carbohydrates with that of fats.
4. Explain the role of condensation reactions in the formation of a polysaccharide.
5. What is the functional significance of hydrolysis reactions to living organisms?
6. What is activation energy and how does it relate to enzyme function?
7. Which amino acid stereoisomer is commonly found in living systems? Why don't both stereoisomers occur in living organisms?
8. Why is it that carbohydrates tend not to be species specific whereas proteins generally are?
9. What is the relationship of vitamins to cofactors?
10. By what mechanism(s) do enzymes work?
11. What is the difference between saturated and unsaturated fats? Of what importance is this to human health?
12. Discuss the use of molecular building blocks in living systems.

YOUR USE OF ENERGY FOR ONE DAY*

Activity	Hours spent in activity	Energy used (Calories per pound per hour)	Total Calories used per pound (Calories × hours)
Asleep	8	0.4	3.2
Lying still, awake	1	0.5	0.5
Dressing and undressing	1	0.9	0.9
Sitting in class, eating, studying, talking	8	0.7	5.6
Walking	1	1.5	1.5
Standing	1	0.8	0.8
Driving a car	1	1.0	1.0
Running	½	4.0	2.0
Playing ping-pong	½	2.7	1.3
Writing	2	0.7	1.4
TOTAL	24	—	18.2

Total Calories used per pound   18.2
Weight in pounds   × 115.0
Total Calories expended for the day   2,093.0

* Adapted from Helen S. Mitchell *et al.*, *Cooper's Nutrition in Health and Disease*, 15th Ed., Philadelphia, J. P. Lippincott, 1968. Used by permission of the publisher.

HARVEST OF CHEMICAL ENERGY: CELLULAR RESPIRATION

4-1 WHAT IS ENERGY?

The human heart is about the size of a clenched fist. In a life span of 70 years it beats 2.5 billion times and pumps 40 million gallons of blood. The heart is indispensable to one's very being, for in the pumping of blood, oxygen and food are carried to every cell of the body, and by the return circuit of the bloodstream, cellular wastes and carbon dioxide are removed. The heart, we humans, and indeed all our cells are composed of matter and, as such, occupy space and have both weight and density. **Matter** is something we can perceive through our senses—we can feel it and see it. However, it takes more than matter to constitute a living cell, an organ such as the beating heart, or an entire organism like one's self. What makes matter in the living system go, or maintain itself, is **energy.** The human heart has an energy demand, and this varies with its work load. For example, no mathematics, chemistry, or physics is required to recognize that the heart of an obese individual requires more energy to operate effectively than that of a person of more normal girth. The same size heart has to pump blood a greater distance in the overweight individual; it has been estimated that for every pound of fat there are 3 miles of blood vessels, and to pump blood over these distances, the heart needs more energy in a person with more pounds of fat. It is not profitable to be overweight. Similarly, the energy requirements of all our cells depend directly on their work load. Energy is quite different from matter: energy has no mass, occupies no space, has no density; yet we know it exists. We refer to energy frequently when we say, ''I wish I had his energy,'' or, ''Atomic energy has changed the course of modern history.'' What is this thing called energy?

   Energy can be measured by the effects it produces on matter and is defined as the capacity to do work. **Work** can be thought of as any process requiring energy. You can see that this kind of circular reasoning does not get us very far, but it does indicate that there is a close relationship between work and energy. The greater the amount of energy available, the greater the amount of work that can be performed; the more matter to be moved, for example, the greater the energy required to move it.

Energy is considered to exist in two states: potential and kinetic. **Potential energy** is stored or inactive energy; the stored energy confers the capacity for work, but no work is actually being done. **Kinetic energy,** on the other hand, is energy in action—the energy used during the performance of work. Let us consider an example. If your automobile can go 15 miles per gallon of gasoline, we can say that the potential energy of 1 gallon of gasoline for your car is 15 miles of travel, and when the car is actually moving, the energy contained in the fuel is being released and is in a kinetic state.

Energy can take a variety of forms: potential energy may be chemical, electrical, or atomic; kinetic energy may be in the form of motion, light, or heat. The forms of energy are interchangeable; for example, electrical energy can be converted into light and heat (as when electricity flows through the tungsten filament in a light bulb), and light can be changed into electrical energy (as occurs in an exposure meter). Just as there are different forms of energy, so too there are different kinds of work: mechanical, chemical, osmotic, and electrical. The forms of energy and their uses are everyday occurrences. Let us briefly examine one of these: the starting of the automobile engine. When you activate the starter, the stored chemical energy in the battery is converted into electrical energy, which produces a spark at the spark plugs; this ignites the gasoline vapors,

and by the process of combustion (burning) the gasoline's potential chemical energy is released as kinetic energy in the form of heat, light, and an expanding gas. (Heat, you may recall, is simply the random motion of molecules.) The kinetic energy of the expanding gas moves the pistons, producing mechanical work, and the car moves.

What has energy to do with life processes? As you sit and read these words, your cells are in ceaseless activity. All living cells depend on the release and use of energy for the maintenance of the living state. Living systems are highly organized, and to maintain this order, energy must be fed into the system. When a living cell can no longer obtain, release, and use energy it is dead. However, no machine or living system can produce energy out of nothing; it can only change energy from one form to another. Automobile engines obtain energy and produce work from fuels such as gasoline or diesel oil. Living machines—cells—obtain, release, and use energy for growth, maintenance, and repair from the energy-rich molecules that constitute their food. Like the automobile engine, cells are transducers of chemical energy and perform work, but as we shall see there are significant differences between cells and gasoline engines as energy-exchange systems.

In both living and nonliving things, transformations of energy are governed by a number of constant factors which we formalize by stating them as laws. The principal laws dealing with energy conversions are called the **Laws of Thermodynamics** ("heat-energetic" laws). The first law, often called the Law of Conservation of Energy, states that when energy goes from one form to another no energy is gained or lost. In other words, the total energy of the universe remains constant: energy does not perish, it is transformed. The Second Law of

Thermodynamics states that during an energy transformation the amount of usable energy is reduced and some of the energy is not available to do work. Succintly stated, **entropy**[1] (the tendency toward randomness and disorder in the universe) increases.

Let us consider some examples. If we put a fixed amount of fuel into a gasoline engine, we obtain a certain amount of mechanical work, but a fraction of the fuel energy is lost to the surroundings in the form of heat. Proof of this is easily obtained by holding your hand over a running engine. By saying energy is lost, we do not mean the energy disappears, but only that it is unable to produce work. The total available energy of the gasoline liberated by combustion is the sum of the work of the pistons of the engine and the heat that is unavailable for work. (This satisfies the First Law of Thermodynamics.) The conversion of one form of energy into another is never 100% efficient, and some of the energy is liberated in a useless form. (This satisfies the Second Law of Thermodynamics.)

Now let us consider the living machine. A living cell is an open thermodynamic system; that is, it must continually be supplied with energy in order to maintain itself. When the energy supply ceases, the cell ceases to work; it becomes disorganized, and death ensues. A convenient way of looking at the energy relationships in the living system is illustrated in Figure 4–1. The organized state of the living cell is represented by level A; to maintain this level at a steady state, energy must constantly be fed in, because the system is open and there is a certain loss that must be compensated. The level will decline and the system will approach level B, a state of disorganization (entropy will

---

[1] Entropy is a randomized state of energy that is unavailable to do work.

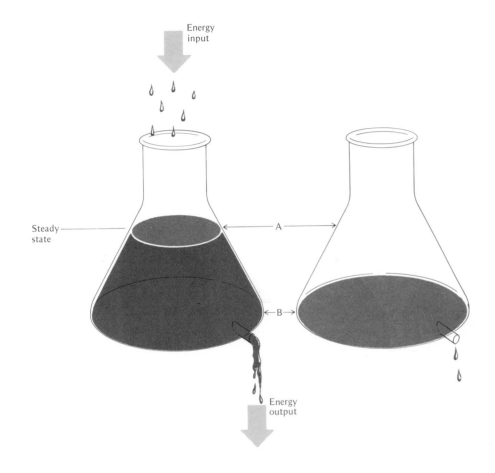

Energy
input

Steady
state

A

B

Energy
output

water 1° C — from 14.5° to 15.5° C.)[2] In a living system that burned glucose and liberated 700 Calories in a single burst, there would result a sharp rise in temperature owing to the sudden release of heat. If a man ate 0.5 lb of glucose and all the energy was liberated in a single burst, his body temperature would rise about 10° C, from 37° to 47° C. From what you already know about proteins and cells, it is obvious that such a temperature rise would coagulate most proteins. In effect, the single-step liberation of heat energy from food in a human would mean self-incineration with every meal.

Living cells are essentially **isothermal;** that is, there are no significant changes in the temperature of cells throughout their life. Unlike heat engines cells must obtain their energy from the chemicals that are their fuel so that large bursts of heat do not occur. The mechanism of trapping energy and storing it involves breaking the energy-rich chemical bonds of molecules; a portion of the energy obtained from these molecules is trapped in the chemical bonds of other molecules. Energy is thus redistributed among reacting molecules rather than liberated all at once in a free form. The energy obtained in this way is called **bond energy,** for obvious reasons, and it is obtained in stepwise fashion, that is, during a sequence of chemical reactions. The remainder, unavailable for cellular work, is dissipated into the

increase), if energy is not brought into the system. The continual energy loss from the living, open thermodynamic system is represented predominantly by heat, which is dissipated and unavailable for work to maintain the organized state. Energy received in the form of food cannot be utilized with 100% efficiency for work, and at every energy transformation there is a reduction in the usable energy of the system—usually in the form of heat.

We have said that carbohydrate in the form of glucose is one of the fuels of the living cell. How does the cell convert glucose into energy? When the fuel of an automobile engine is burned, energy is liberated in a single quick burst. If we took 0.5 lb of glucose, burned it, and measured the heat released, we would obtain about 700 Calories. (A Calorie is a unit measure of heat. It is the amount of heat necessary to raise the temperature of 1 kg of

[2] A calorie (c) is the amount of heat required to raise the temperature of 1 gm of water 1°; this unit is so small that most nutritionists prefer to use the larger calorie, written Calorie (capital C), which is a kilocalorie or 1,000 times greater than the small calorie. We shall use Calorie, the common way of writing kilocalorie, and drop the *kilo.* A Calorie is roughly equivalent to the work available in 0.10 horsepower; put another way, 100 Calories are needed to keep a 100-W light bulb burning for 1 hour.

FIGURE 4–2 (top) The dry-cell battery. Chemical energy is transformed into heat and light.

FIGURE 4–3 (bottom) ATP: its role in the cell can be likened to that of a storage battery.

environment as almost imperceptible amounts of heat. Heat is liberated in step with the chemical reactions, and in this way the release of large amounts of heat and energy at one time is avoided and waste is minimized.

4–2 ENERGY CURRENCY: ATP

The fires of life burn with a cool flame. The fuels of the living cell are the chemicals that make up its food, and energy is obtained by breaking the bonds of these various chemicals. The released energy is trapped and stored in other chemicals in the cell and can be released again when needed. What are these chemicals? How does the cell make all these energy conversions?

The oxidation of carbohydrates and other energy-rich compounds (such as lipids and proteins) provides the main source of energy for many cells and organisms. **Oxidation** is a process by which oxygen is added to a molecule or hydrogens and electrons are removed. In spite of the name, therefore, oxygen is not always necessary for an oxidative reaction to take place. Since hydrogens may be removed from molecules during an oxidative reaction, these reactions are sometimes called **dehydrogenations.** When oxygen is added to a substance, it must be obtained from somewhere, and when hydrogens or electrons are removed, they have to go some place. Thus, for every oxidation reaction there must be a corresponding reduction reaction. **Reduction** in the chemical sense does not mean a diminution in size, but indicates the removal of oxygen or the addition of hydrogens or electrons to a substance. (Remember the reducing atmosphere of the primitive earth and the important role it played in the formation of organic compounds?) The following equation illustrates the general form of an oxidation-reduction reaction: the

substance X is being oxidized and Y is being reduced; X is the reductant (it supplies the hydrogens), and Y is the oxidant (it receives the electrons or hydrogens).

electron donor	electron acceptor	oxidized substance	reduced substance	

$$XH_2 + Y \rightleftharpoons X + YH_2 + energy$$

energy rich	energy poor	energy lost	energy gained	

In the human body carbohydrate serves as fuel and is oxidized to carbon dioxide and water; during the process energy is liberated. The simplified formula for this reaction is:

electron donor	electron acceptor	oxidized substance	reduced substance	

$$CH_2O + O_2 \rightarrow CO_2 + H_2O + energy$$

carbo-hydrate	oxygen	carbon dioxide	water	

The carbohydrate (CH_2O) corresponds to the XH_2 molecule (plus an O), and the molecular oxygen corresponds to the Y; the carbohydrate becomes oxidized, not by a loss or gain of oxygen but by a loss of hydrogens. The hydrogens derived from the carbohydrate are passed to the molecule of oxygen and chemically reduce the latter, and a molecule of water is formed. The carbohydrate without its hydrogens yields the molecule carbon dioxide.

The significant feature of this oxidation-reduction shuttle system is that when an electron (or a hydrogen) flows from a reducing system to an oxidizing system, energy is liberated. As electrons are transferred, they fall from a higher energy state to a lower one. In a storage battery or a dry cell, the electrons flow from one electrode to the other when the elctrodes are connected by a wire, and energy is produced in the form of electricity; this energy can be made to produce heat or light or to perform work (Figure 4–2). In much the same manner, the oxidation of foodstuffs liberates energy;

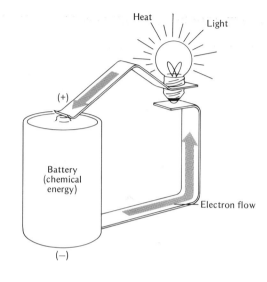

Heat
Light
(+)
Battery (chemical energy)
Electron flow
(−)

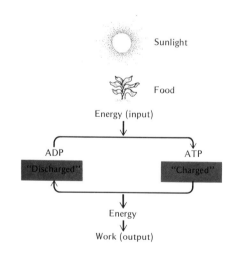

Sunlight
Food
Energy (input)
ADP "Discharged"
ATP "Charged"
Energy
Work (output)

FIGURE 4-4 The ATP molecule. The terminal phosphate bonds are energy-rich. When these bonds are broken, more calories are released than if other bonds in the molecule were broken.

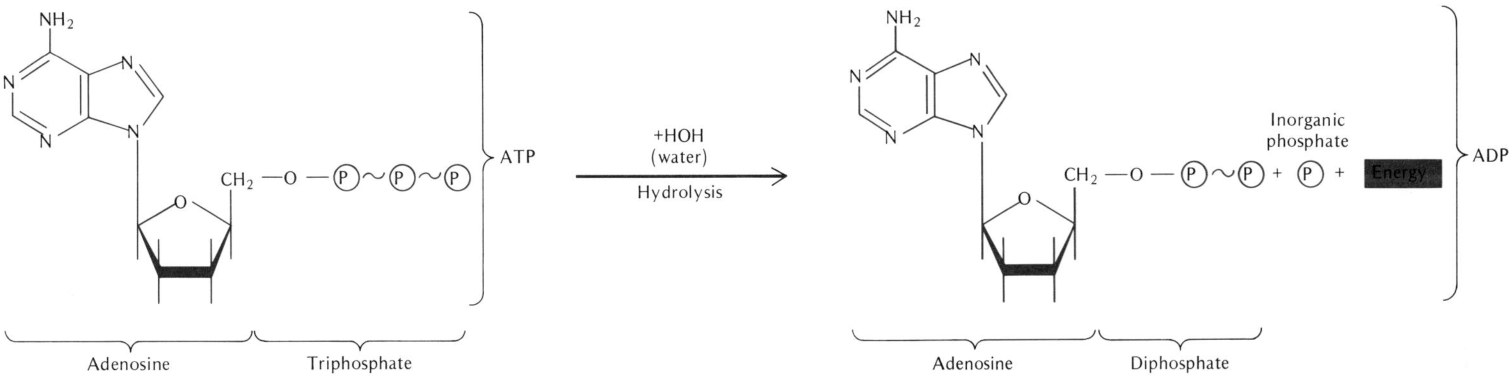

such energy is trapped and conserved in the form of a high-energy phosphate compound called adenosine triphosphate (ATP)[3] and can later be released for work. ATP can be likened to the "charged" form of the storage battery with energy available for cellular work (Figure 4-3). Just as a battery runs down and becomes discharged when the chemicals in the electrodes no longer donate electrons, so too there is a "discharged" form of ATP called adenosine diphosphate (ADP). The molecule of ATP has three phosphate groups; ADP has only two. The link between the second and third phosphate of ATP releases a great deal of energy when it is broken and is sometimes spoken of as a high-energy bond (Figure 4-4). It is of primary importance in conserving the energy that is released by oxidation of foodstuffs. When phosphate and energy are added to ADP, it becomes ATP and is thus "energized."

_____

[3] Other energy-rich compounds in the cell also trap energy (for example, $NADPH_2$, $NADH_2$, GTP, and CoA); however, we shall omit detailed consideration of these and focus attention primarily on ATP.

The ATP↔ADP reaction shuttles energy back and forth between energy-requiring and energy-yielding reactions of the cell; ADP is charged to ATP during the oxidation of foodstuffs, and ATP is discharged to ADP during the performance of cellular work (Figure 4-4).

Most energy-dependent reactions within the cell are tied to ATP. What is the advantage of this? A similar question is: What is the advantage of money or currency? Our society could, as do many primitive societies, exist perfectly well on a barter system. The payment for goods and services could easily be made by other goods and services. If, for example, you wanted to purchase a loaf of bread and a pound of meat, you could do a chore as payment. If you didn't have the time to do the work, you could pay off the debt by giving the shopkeeper a gallon of oil or some other item. However, if the shopkeeper had no desire for the oil, you could not pay for the bread and meat. Clearly, if there were some universally exchangeable and therefore universally desirable object, transactions could be carried out more conveniently and efficiently. In commerce, the universally exchangeable material is money;

in the cell, the currency is ATP. Energy is liberated from ATP as needed, and in this way the cell can conveniently and efficiently pay for most of its work. By having such a common currency for energy transactions, considerable flexibility of energy supply is attained. For example, the body can satisfy its energy requirements by the oxidation of carbohydrates, proteins, and fats (and on occasion even by peculiar nutrients such as alcohol); all these can provide energy for cellular work. The living machine is thus capable of using a variety of fuels, roughly equivalent to an automobile engine that could run on oil, wood, gasoline, coal, or gas. The automobile engine lacks such versatility, but living organisms have it; this is largely achieved because ATP is the single mediator between the energy suppliers and the energy users. The food or fuel does not directly drive the energy-requiring processes of the cell; this is handled by ATP.

4-3 CHEMICAL ENERGY HARVEST

In what manner do the cells of our bodies obtain their energy? How is the harvest of energy accomplished? We have indicated that we are dependent upon a supply of food (carbohydrates, lipids, proteins) as a source of energy. These are our fuels; however, we cannot manufacture these for ourselves and ordinarily require other organisms to make them for us. We consume and then digest (break down) these organisms (plants and animals) or their products; some of the digested materials are then stored within our bodies as food reserves. (The process of digestion is the subject of Chapter 11.) Foodstuffs are rich in potential energy, and our cells release this energy and use it for growth, maintenance, reproduction, and other life functions.

How do living things release the potential energy of digested materials? In human beings, as in most organisms, energy is released from the various digested foodstuffs by oxidation. Although many substances may be oxidized to yield energy, we shall confine our discussion to glucose, a material that is our prime source of energy.

The breakdown of glucose in the living cells of the body is carried out by a complex series of chemical reactions, each catalyzed by a specific enzyme. The early sequence of steps in the oxidation of glucose is completely independent of oxygen and is called **anaerobic respiration.** Some organisms can carry out the entire sequence without oxygen; such organisms can, therefore, live in the absence of oxygen and are called **anaerobes** (an—without; aero—air; Gk.). The tetanus bacterium and yeast are examples of anaerobes. However, we and many other organisms cannot complete the extraction of energy from foodstuffs without oxygen; in such organisms the sequence of reactions that constitutes anaerobic respiration is followed by a sequence of reactions which depends on the availability of molecular oxygen. The sequence is called **aerobic respira-**

tion, and organisms like ourselves in whom aerobic respiration is the rule are called **aerobes.** Oxygen is involved only in the final steps of aerobic respiration, but all the preceding steps indirectly depend on it because they could continue for only a short time if oxygen were not available to participate in the terminal reactions of the process. We obtain the oxygen used by our cells in aerobic respiration by breathing, and we can live for only a short time when breathing stops; without oxygen, too little ATP is generated, the processes in our cells become disorganized, and the cells are damaged beyond repair. In many parts of the body the cells can be restored to normal function if oxygen should become available before total death has occurred, but after only 2 minutes without oxygen, irreparable brain damage occurs. This is because brain cells require a rich supply of ATP, which comes principally from glucose via aerobic respiration.

4–4 THE ENERGY CASCADE

Before we examine the sequence of events that occurs within our cells during anaerobic and aerobic respiration, let us first understand what can be called the **energy cascade.** A flowing river has a great deal of kinetic energy that can be useful if some way can be found to harness it. If the river is dammed to prevent it from flowing, a reservoir is created and the river's energy is converted into potential energy. This energy can be harvested by allowing the water to spill over the dam and flow downhill, and by making it do useful work on the way, such as driving waterwheels or electrical generators. To get the energy out of the reservoir, the sluice gate of the dam must be opened; this requires energy (a kind of pump priming, to get things going), and then during the water's fall, energy and work are obtained. The amount of energy

obtained from this system depends on the initial volume of the reservoir, the quantity of water permitted to flow out of the dam, and the vertical distance traveled by the water. Energy for work can no longer be obtained when the water has completed its downhill run and reached the lowest level. Allowing the water to flow downhill in a single sharply declining slope supplies a large, but unmanageable, spurt of energy. Greater control is achieved if a series of sluice gates or dams is included, creating a number of cataracts. At each cataract, a waterwheel or electrical generator is positioned. By opening and closing the sluice gates, energy is released in manageable steps rather than in a single big rush (Figure 4–5).

This cascade of water, producing energy in stepwise fashion, provides us with a simple model somewhat analogous to the release of energy in the cell. The cellular regulators (gatekeepers) of energy are specific proteins called enzymes. In the living cell molecules that are rich in potential energy are activated by means of enzymes to get them over the barrier of chemical reactivity, and then degraded in stepwise fashion to release small, measured amounts of energy. Just as it is impossible to convert all the kinetic energy of the flowing water into mechanical work, since some is lost through friction, so there is some wasted energy in the conversions that take place in the cell. The process of energy conversion is always less than 100% efficient (the Second Law of Thermodynamics).

4–5 GETTING ALONG WITHOUT OXYGEN

When your cells are working, glucose, serving as the fuel for this work, is rapidly consumed. The degradation of energy-rich glucose yields energy, and the process may or may not involve oxygen. The breakdown of glucose and the

FIGURE 4-5 The energy cascade. Water in the reservoir is allowed to leave and flow downhill. At certain places there are sluice gates and waterwheels. The flowing water can move the wheels to produce mechanical work, and the rate of flow is controlled by opening or closing the sluice gates.

FIGURE 4-6 A schematic representation of the anaerobic respiration of glucose.

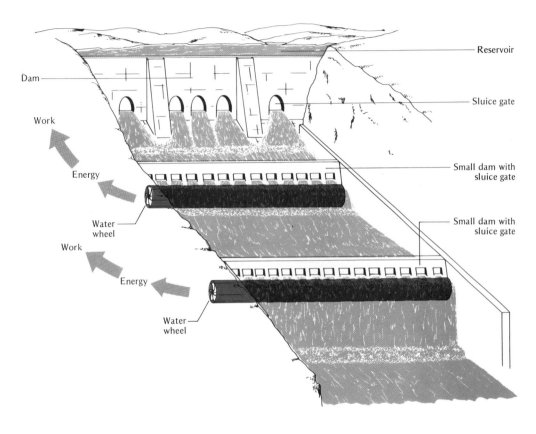

Reservoir

Dam

Work

Sluice gate

Energy

Small dam with sluice gate

Work

Water wheel

Small dam with sluice gate

Energy

Water wheel

Glucose

Stage 1: Glycolysis

removal of hydrogens occurs in stepwise fashion, and the energy released is trapped in the form of ATP. If oxygen is not present, the oxidation of glucose is by anaerobic respiration, as follows.

Glucose, a 6-carbon compound, is broken down to 2 molecules of pyruvic acid, a 3-carbon compound, by 11 enzymatic steps (Figure 4–6). Glucose is prepared initially for its molecular dismemberment by the attachment of phosphate groups, donated by ATP. This is significant, because the adding of phosphate (**phosphorylation**) to glucose converts it from a substance that can leave the cell to one that cannot pass across the plasma membrane. Now the glucose molecule is trapped inside the cell, where cytoplasmic enzymes can alter its molecular architecture. The phosphorylation of glucose also boosts it to a higher energy level (activates it) and this renders it more reactive chemically: the sluice gates have been opened. Two molecules of ATP are used to activate the glucose initially, and during the following steps, there is the synthesis of 4 molecules of ATP; the net yield to the cell is 2 ATP molecules (Figure 4–6).[4]

During the degradation of glucose, hydrogens are removed and transferred to an acceptor molecule, the coenzyme **NAD** (**n**icotinamide **a**denine **d**inucleotide).[5] Two molecules of $NADH_2$ are produced.

The end product of anaerobic respiration, the 3-carbon fragment known as pyruvic acid, is subsequently converted into lactic acid or ethyl alcohol and carbon dioxide, which can leave the cell (Box 4A).

---

[4] This value pertains only if pyruvic acid is subsequently converted to lactic acid or alcohol.

[5] The coenzyme NAD is formed from the vitamin niacin; a deficiency of niacin results in a decrease in the amount of NAD, a slowing of anaerobic degradation of glucose, and production of the disease known as pellagra. The situation is analogous to the gatekeeper of the energy cascade not allowing water to exit and flow downhill; energy release is thus prevented.

BOX 4A Beer, bread, and muscle: molecular changes in pyruvic acid

Pyruvic acid plays a pivotal role in reactions that take place in such diverse occurrences as muscular work, the manufacture of alcohol, and the baking of bread. The fact that the fermentation of sugars produces alcohol and carbon dioxide has been known for thousands of years, but the cause of the phenomenon was discovered only about a century ago. Under anaerobic conditions, the yeast cell utilizes glucose and performs enzymatic conversions similar to those we have discussed. The pyruvic acid is then further degraded to a 2-carbon compound, acetaldehyde, and eventually this forms ethyl alcohol plus carbon dioxide (Figure 4–7). The enzyme that carries out this split of pyruvic acid is called **carboxylase,** and it requires the coenzyme **cocarboxylase** (the new name is **t**hiamine **p**yro**p**hosphate, or **TPP**), a derivative of vitamin B_1 (thiamine).

When our muscle cells are heavily exercised, they run out of oxygen, and glycogen is utilized anaerobically (more accurately, glycogen is converted into phosphorylated glucose molecules); up to a point, the molecular dismemberment of glucose takes place in the same way as in yeast cells. The principal difference between the process in fatigued muscle and in yeast is the product formed: yeast produces alcohol and carbon dioxide, and muscle produces lactic acid. In both cases there is a reduction (acetaldehyde ⟶ ethyl alcohol; pyruvic acid ⟶ lactic acid) and the donor of hydrogens is reduced NAD ($NADH_2$) formed earlier during anaerobic respiration (Figure 4–8). This step makes the product of glucose degradation its own hydrogen acceptor, and at the same time NAD is regenerated for continued participation earlier in the sequence. In yeast, the process is called **fermentation,** and it is responsible for the production of alcoholic beverages such as beer; the carbon dioxide is used in the baking of bread (it causes the rising of the dough). In muscle the process is called **glycolysis,** because glycogen is the original fuel molecule. The products of glycolysis (lactic acid) and of fermentation (carbon dioxide and ethyl alcohol) leave the cell easily and become a part of the external cellular environment. A twist of enzymatic fate has determined that lactic acid rather than alcohol

be poured into the blood when our muscles are exercised. Were our muscle enzymes slightly different, heavy physical effort might be both popular and intoxicating!

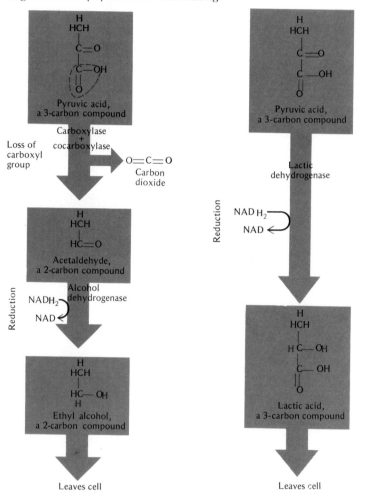

FIGURE 4–7 (left) The fermentation of sugar by yeast.

FIGURE 4–8 (right) Anaerobic respiration of pyruvic acid by muscle.

4–6 ANAEROBIC RESPIRATION: COST ACCOUNTING

The efficiency of any machine is measured not by the rate at which the fuel is degraded, but by the amount of energy that is made available for work. A cell is not simply a machine for burning sugar, any more than an automobile is just a machine for burning gasoline. If breaking down fuel molecules into small size for disposal were the only object of combustion, it would be simpler and cheaper for cells to burn fuel by some other means. The object of breaking down a fuel is to get energy for work—in an automobile, to get the car moving; in the cell, to produce ATP for storage of energy and ultimately cellular work. The more efficient the automobile engine, the less fuel it consumes in moving the car from place to place; the more miles per gallon, the happier the automobile owner. Similarly, the more efficient the cell, the less fuel consumption required for its work load.

How efficient is the anaerobic degradation of glucose in providing a cell with energy in the form of ATP? To evaluate efficiency, it is necessary to do some simple arithmetic. The amount of energy potentially available to the cell from glucose can be estimated by completely burning the glucose in a device called a **bomb calorimeter** and measuring the caloric output. The calorie tables that you use to determine how much you should or should not eat have been worked out by using a bomb calorimeter or a similar device. If glucose is burned in this device, it is found that 1 mole[6] of glucose (180 gm) yields approximately 700 Calories of energy. The burned glucose in the calorimeter ends up as carbon dioxide and water. When 1 mole of ATP is hydrolyzed to

ADP, about 10 Calories are released. By anaerobic respiration of 1 mole of glucose, the cell obtains a net yield of 2 ATP or approximately 20 Calories; the total available was 700 Calories, and so the cell obtains about 3% of the total energy available ($20/700 \times 100 = 3\%$). Looked at another way, the breakdown of 1 mole of glucose in anaerobic respiration yields 2 moles of lactic acid. A mole of lactic acid burned in a calorimeter yields approximately 325 Calories and produces carbon dioxide and water. The cell has potentially available to it 700 Calories per mole of glucose, and it produces lactic acid which has a caloric value of 650 Calories. Thus, when glucose is broken down to lactic acid, the amount of energy available (as determined by calorimetry) is 50 Calories ($700 - 650$). The cell has trapped 20 Calories in ATP and lost 30 Calories; it is about 40% efficient as an energy-trapping machine ($20/50 \times 100$). By comparison, the modern steam-generating plant and the automobile engine are only 25%–30% efficient.

Let us now do a cost accounting of the degradation of glucose under anaerobic conditions. The energy-rich glucose may be thought of as a large denomination of currency, let us say a 5-dollar bill. In order for the cell to pay for goods and services, that large bill must be exchanged for smaller change, of the order of 10 cents. The 10-cent pieces are used by the cell to pay all outstanding debts. The 10-cent piece is the only currency used; dollar bills or other amounts must be made up out of multiples of 10 cents. The commission for converting the large bill into small change is about 60 cents on the dollar. By financial standards it appears exorbitant, but for the cell it is the only means of operation. If 1 mole of glucose represents a 5-dollar bill, the cell only gets 20 cents' worth of usable currency from each 5 dollars' worth of glucose; it loses about 30 cents due to the exchange rate (as heat,

which cannot be used for paying debts), and about 4 dollars and 50 cents is not converted into small change. It would appear that the cell using anaerobic respiration has let a great deal of money slip through its fingers by producing pyruvic acid, lactic acid, or ethyl alcohol.

The efficiency expert might say that this system of generating small change from larger currency is a highly inefficient way to run an economy. Why does the cell not dispense with all the enzyme reactions that make ATP from glucose and form ATP directly from inorganic materials? As simple as this scheme sounds, it would mean that the cell would have to handle a lot more material in the form of ATP in order to gain an amount of energy equal to that contained in a molecule of glucose. It would be equivalent to the cell's obtaining all its fuel energy in the form of dimes instead of dollars. (Just imagine receiving your paycheck in dimes!) Glucose does not provide energy in an immediately available form, but it is compact, easy to transport, and quite convenient to handle. The ATP-generating system using energy-rich molecules is, after all, quite economical and efficient for the cell.

4–7 GETTING ALONG WITH OXYGEN

During anaerobic respiration, pyruvic acid forms either alcohol (as in yeast cells) or lactic acid (as in fatigued muscle) (Box 4A), but when oxygen is plentiful, these products are not formed and pyruvic acid has another fate. As we have seen, this 3-carbon compound is still quite energy-rich, and it is apparent that a cell which loses such a molecule also loses a potential source of ATP. If a cell were to exist only on anaerobic respiration, then another molecule of glucose must be consumed for the cell to derive another 2 molecules of ATP. In discarding pyruvic acid or other related molecules (lactic acid, alcohol), the anaerobic cell

---

[6] A mole is the molecular weight in grams.

is throwing away potential chemical energy; if the pyruvic acid could be further degraded, the cell could obtain additional amounts of energy without taking in any more glucose. This is what happens during aerobic respiration, and the pyruvic acid is broken down to carbon dioxide and water. The principal yield of energy to the cell occurs during this phase. Clearly, as the name aerobic implies, the degradation of pyruvic acid involves oxygen.

During aerobic respiration, pyruvic acid is split to produce a 2-carbon fragment (called an **acetyl group**) and carbon dioxide (Figure 4–9). Simultaneously, the carbon dioxide passes out of the cell and the 2-carbon fragment is linked to a coenzyme called **coenzyme A,** producing a molecule called **acetyl coenzyme A** (Figure 4–9).

It is apparent that one third of the carbon atoms of the glucose have become carbon dioxide; the remaining two thirds are now in the form of the acetyl group of acetyl coenzyme A. The acetyl group is oxidized to carbon dioxide, by a cyclic series of enzyme-catalyzed reactions called the **Krebs,** or **citric acid cycle** (Figure 4–9).

When acetyl coenzyme A enters the Krebs cycle, the acetyl group is joined by a 4-carbon compound (oxaloacetic acid) to form a 6-carbon compound (citric acid), and the coenzyme A molecule is released in the process (Figure 4–9). The coenzyme A liberated by the acetyl group entering the cycle is available for ferrying more acetyl groups to the enzymes of the Krebs cycle (Figure 4–9). By a series of enzyme-catalyzed reactions, the citric acid is broken down to oxaloacetic acid, the starting material, which can then pick up another acetyl group; in the process, the original acetyl group has been broken down to yield 2 molecules of carbon dioxide. The Krebs cycle provides a continuous mechanism for breaking down acetyl groups. During the operation of the

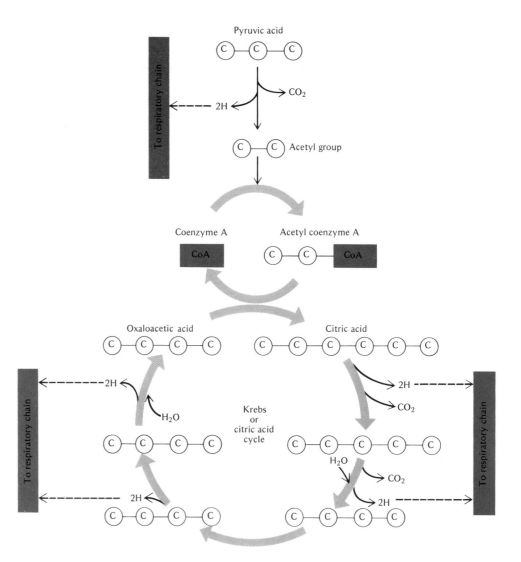

FIGURE 4–10 The respiratory chain.

cycle, the acetyl group also loses its hydrogens; they are linked to molecular oxygen and turned into water through the operation of a series of reactions called the **respiratory chain** (Figure 4–10).

The respiratory chain consists of an energy cascade involving a series of enzyme-coenzyme reactions that pass hydrogens down the line eventually to reduce oxygen and form water (Figure 4–10). For each turn of the Krebs cycle, there are four dehydrogenation steps, each involving a pair of hydrogens. Thus, at each turn of the Krebs cycle, four pairs of hydrogens are produced. Each pair of hydrogens moving down the respiratory chain generates 3 molecules of ATP; thus for every turn of the cycle 12 ATP molecules are generated. In addition to the release of hydrogens for the production of energy via the respiratory chain, the cell derives another benefit from the Krebs cycle, and that is the production of building blocks. Some of the compounds formed during the cycle can be used to manufacture amino acids and other essential cellular components; these intermediate products can be siphoned off at different points in the cycle. In the same way, molecules not needed for biosynthesis can be fed into the cycle at different places, allowing the cell a greater diversity of usable foodstuffs. The Krebs cycle can be thought of as the metabolic hub of the cell.

Now let us examine some of the details of the flow of energy through the respiratory chain that enable the cell to obtain energy currency, ATP. When pyruvic acid is split to yield acetyl coenzyme A and carbon dioxide, it also loses a pair of hydrogens; these hydrogens, as well as those from the Krebs cycle acids, are passed to the coenzyme NAD, which becomes reduced ($NADH_2$) (Figure 4–10). The $NADH_2$ passes its hydrogens to another acceptor, the coenzyme **FAD** (**f**lavin **a**denine **d**inucleotide, a derivative of the vitamin riboflavin); the

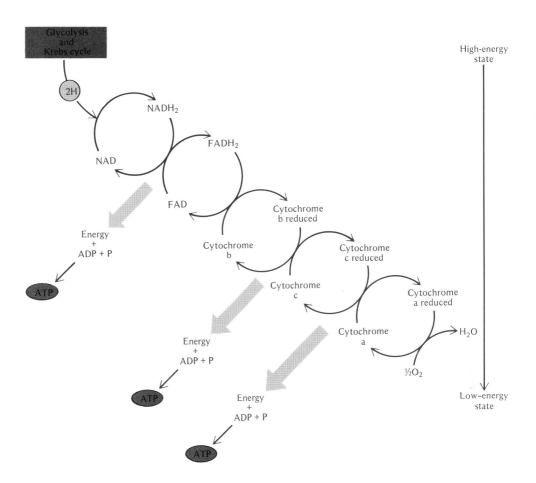

transfer of hydrogens again is an oxidation-reduction reaction. The FAD bearing hydrogens ($FADH_2$) can pass the hydrogens to a series of enzymes called the **cytochromes.** The cytochromes are conjugated proteins containing iron in a ring structure (heme) and resemble hemoglobin. There are three types of cytochromes, and the last one donates its hydrogens to molecular oxygen (Figure 4–10). The respiratory chain consists of NAD, FAD, and the cytochromes.

At this point you may wonder why the cell goes through such a series of complicated reactions. The answer is that with each respiratory-chain transfer (an oxidation-reduction reaction) the hydrogen is lowered from a higher energy state to a lower energy state; at some of these transfers, ATP is formed. We might analogize the transfer of hydrogens down the respiratory chain to the movement of water down the energy cascade, releasing measured amounts of energy along the way.

In summary, let us write our general equation for oxidation-reduction:

$$XH_2 + Y \longrightarrow X + YH_2$$

In the respiratory chain, pyruvic acid or the acids of the Krebs cycle are XH_2, and Y is NAD; the next reaction would be:

$$\overset{\text{NADH}_2\quad\text{FAD}}{YH_2 + Z} \longrightarrow \overset{\text{NAD}\quad\text{FADH}_2}{Y + ZH_2}$$

$$\downarrow$$

$$\text{ATP}$$

in the next series of steps:

$$\overset{\text{FADH}_2\quad\text{cytochrome}}{ZH_2 + Cy} \longrightarrow \overset{\text{FAD}\quad\overset{\text{reduced}}{\text{cytochrome}}}{Z + CyH_2}$$

$$\downarrow$$

$$\text{2ATP}$$

and finally:

$$\overset{\overset{\text{reduced}}{\text{cytochrome}}\quad\text{oxygen}}{CyH_2 + O_2} \longrightarrow \overset{\text{water}\quad\text{cytochrome}}{H_2O + Cy}$$

The overall equation for the aerobic degradation of glucose is:

$$\underset{XH_2}{C_6H_{12}O_6} + \underset{Y}{6O_2} \longrightarrow \underset{X}{6CO_2} + \underset{H_2Y}{6H_2O}$$

4–8 AEROBIC RESPIRATION: COST ACCOUNTING

The degradation of pyruvic acid by the enzymes of the Krebs cycle and the transfer of hydrogens along the respiratory chain yield energy that is available for cellular work; in the process the pyruvic acid becomes carbon dioxide, and the hydrogens removed from it become water. Just how efficient is aerobic respiration? Each molecule of pyruvic acid going to acetyl coenzyme A produces 1 molecule of $NADH_2$, and by passing the hydrogens through the cytochromes 3 ATP molecules are formed. Since 2 pyruvic acid molecules are formed per molecule of glucose, the net yield to the cell is 6 ATP molecules. In each revolution of the Krebs cycle there are generated 12 ATP molecules (four dehydrogenations and 3 ATP molecules produced per dehydrogenation); since there are two turns of the cycle for the 2 molecules of acetyl coenzyme A formed from the original molecule of glucose, the net yield is 24 molecules of ATP. Now recall that during anaerobic respiration 2 molecules of $NADH_2$ were produced. If these hydrogens are passed down the respiratory chain, 3 ATP molecules are generated per pair of hydrogens for a total of 6 ATP molecules. Now for the grand total: 6 ATP from pyruvic acid going to acetyl coenzyme A, 24 ATP from the Krebs cycle–respiratory chain sequence, and 6 ATP from the glycolysis–respiratory chain sequence. The total yield from aerobic respiration is 36 ATP molecules; to this must be added the 2 ATP from anaerobic respiration, and thus the cell gains 38 ATP molecules from the complete oxidation of 1 molecule of glucose.

Now for some simple arithmetic to calculate efficiency. We noticed previously that the total energy available in 1 mole of glucose was about 700 Calories. By both anaerobic and aerobic respiration, the cell traps 38 ATP molecules, which is equivalent to 380 Calories. The efficiency of the combined anaerobic and aerobic respiratory scheme is, therefore, 54% ($380/700 \times 100 = 54\%$). This efficiency is considerably better than any heat engine yet invented (the efficiency of the average heat engine is about 30%). If we could manufacture an engine employing principles similar to those incorporated in the design of the cell, shortages of electricity and heat and problems of transportation would be alleviated; there would be plenty of energy for all.

BOX 4B Why cyanide kills

Cytochromes are capable of reacting with cyanide. The lethal effect of the gas chamber (cyanide gas is generated by the addition of sodium cyanide pellets to an acid such as hydrochloric acid) results from the fact that the iron of the cytochrome binds these materials irreversibly. In this way the cytochrome enzymes are poisoned and incapable of acting in the capacity of transferring electrons. Aerobic cells unable to obtain energy from glucose become disorganized and die. It is for this reason that cyanide kills.

4–9 WHERE IS ATP USED?

ATP provides the driving force for most cellular
activities, including muscular and nervous activity, the movement of cilia and flagella, active
transport, and a wide variety of energy-dependent metabolic processes. We shall describe some of these events more fully later
(Chapters 17 and 20).

When materials have to be moved into or
out of cells against a concentration gradient,
energy is required. In some cases, this is called
uphill transport. The pumping of sodium out
of the interior of a cell is an ATP-requiring
process; if ATP is withdrawn from participation
(by a poison, for example) the process stops.
The work of bailing out the water that enters
the cell by osmosis requires ATP. In this manner, ATP is a participant in the work of the
cells of our kidneys, which maintain our water
balance.

Nerve cells perform electrical work, and here
again ATP functions in the movement of
charged molecules (called **ions**) across the
plasma membrane, thus generating an electrical potential; the movement of this electrical
charge is the basis of transmission of nerve
messages. If the ATP system is poisoned by a
substance such as nerve gas, paralysis results
because electrical messages cannot be generated.

The fabrication of molecules, that is, the
formation of new chemical bonds, is dependent
upon ATP. In green plants, ATP is formed
during the so-called light phase of photosynthesis and subsequently this is used for the
synthesis of carbohydrate (sugar) from CO_2. It
is this carbohydrate upon which we all depend
for energy.

Thus, in the economy of life, the coin of the
realm is ATP (Figure 4–11). ATP is the common
intermediate of exchange in reactions that
require energy and those that yield energy.

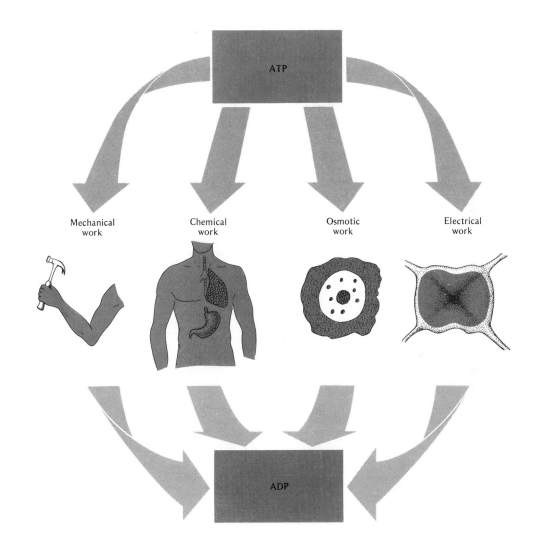

ATP

Mechanical
work

Chemical
work

Osmotic
work

Electrical
work

ADP

FIGURE 4-12 The fractionation of liver.

4-10 CHEMICAL ENERGY HARVEST: EVOLUTION AND CELLULAR LOCATION

All cells, whether they be anaerobic or aerobic, have the anaerobic scheme for extracting energy from glucose. It would appear, therefore, that during the evolution (development) of cellular life, anaerobic organisms arose first and the aerobic respiratory system was later added on to the anaerobic system. This line of reasoning fits well with our thoughts about the origin of life, since the atmosphere of the early earth was a reducing one and aerobic organisms could not function in it; later, once oxygen became available, anaerobic organisms must have acquired the ability to use oxygen in cellular respiration. Because of this, their metabolism became more efficient, and they were able to conserve greater quantities of energy (in the form of ATP) for every molecule of energy-rich material they fed upon.

Cells are specialized transducers of chemical energy, but where in the cell is the energy factory located? How does the machinery for energy production fit inside the cell? We have seen that there is considerable division of labor within cells, that some cellular functions are restricted to specific structures or compartments of the cell known as organelles. The energy-yielding processes that occur inside every cell are also compartmentalized. If you put some liver in a blender and whirled it at high speed so that it became homogenized, and then spun it in a centrifuge, you would obtain two fractions—a watery solution and a solid pellet (Figure 4-12). The enzyme-catalyzed sequences of anaerobic respiration occur in the soluble portion of the cytoplasm, in the liver fraction that you separated into solution. Although the enzymes are probably organized, they do not occur in a fixed, observable geometric configuration. The enzymatic events that take place during aerobic respiration do

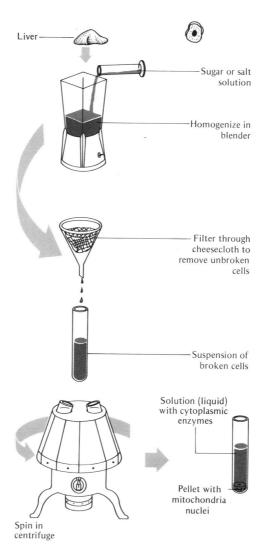

Liver

Sugar or salt solution

Homogenize in blender

Filter through cheesecloth to remove unbroken cells

Suspension of broken cells

Solution (liquid) with cytoplasmic enzymes

Pellet with mitochondria nuclei

Spin in centrifuge

not occur in the soluble cytoplasmic portion of the cell, but are arranged in a fixed spatial orientation in a special organelle, the mitochondrion. These mitochondria would be found in the pellet portion of your fractionated liver sample. The mitochondrion has been called the powerhouse of the cell because it is here that the major harvest of energy from glucose is obtained. You have seen pictures of the mitochondrion as it is revealed by the electron microscope (page 37). The matrix tains the Krebs-cycle enzymes, whereas the enzymes of the respiratory chain are located on the inner mitochondrial membranes, the mitochondrial crests or cristae. Imagine the systematic degradation of a molecule of pyruvic acid; the 15 enzymes of the sequence are located in sequence in the mitochondrion, and the electrons and hydrogens are removed in a stepwise fashion. The enzymatic sequence fits the geometric structure of the mitochondrion beautifully. We can compare the situation to the assembly line of a factory; the enzymes and coenzymes are lined up and in bucket-brigade fashion, they remove electrons and hydrogens and pass them along to where they eventually reach molecular oxygen; in the process ATP is produced and made available to the entire cell.

SUMMARY

1. Energy is the capacity to do work. Potential energy is stored energy; kinetic energy is energy in action. All living cells depend on the release and use of energy from food for maintenance of the living state.

2. The Laws of Thermodynamics state: I. When energy goes from one form to another, no energy is gained or lost. II. During energy transformations, the amount of *usable* energy is reduced.

3. Living cells are open thermodynamic sys-

tems: to maintain an organized, steady state, energy must be supplied. Living cells obtain energy for growth, maintenance, and repair by the stepwise release of energy from organic molecules. During this process, heat is dissipated into the environment.

4. Oxidation (= dehydrogenation) of energy-rich compounds such as carbohydrates liberates energy. For every oxidation reaction (addition of oxygen) there is a corresponding reduction reaction (removal of hydrogen). When an electron (hydrogen) flows from a reducing system, energy is released.

5. In living systems, energy is stored primarily in the high-energy bonds of ATP, which can be discharged to ADP when energy is needed.

6. The oxidation of glucose to provide energy in living cells is called respiration. It occurs in two stages: anaerobic and aerobic.

7. During the anaerobic respiration of glucose, it is broken down from a 6-carbon compound to two molecules of pyruvic acid, a 3-carbon compound, by 11 enzymatic steps. During this process, hydrogens are removed and there is a net yield of 2 molecules of energy-rich ATP to the cell.

8. Pyruvic acid may form alcohol (fermentation of yeast), lactic acid (glycolysis of muscle), or it may be further metabolized in the presence of oxygen (aerobic respiration) to produce more energy-rich ATP, carbon dioxide, and water. Carbon dioxide and water are energy-poor molecules and freely leave the cell.

9. During aerobic respiration, pyruvic acid is first broken down to carbon dioxide and an acetyl group. The acetyl group is itself metabolized via the Krebs cycle to carbon dioxide. During both sets of reactions, hydrogens are released; the hydrogens are transported by a series of reactions called the respiratory chain, and ultimately combine with molecular oxygen to form water. During the transport of hydrogens down the respiratory chain, energy is released and stored as ATP.

10. For every mole of glucose metabolized in respiration, the cell obtains 38 energy-rich molecules of ATP from ADP, a 54% efficiency rate. (The average heat engine is only 30% efficient.)

11. The reactions of anaerobic respiration occur in the soluble cytoplasm. Aerobic respiration occurs in the mitochondrion.

KEY WORDS

matter
energy
work
potential energy
kinetic energy
Laws of Thermodynamics
entropy
calorie
Calorie
isothermal
bond energy
oxidation
dehydrogenation
reduction
adenosine triphosphate (ATP)
adenosine diphosphate (ADP)
high-energy bond
anaerobic respiration
anaerobes
aerobic respiration
aerobes
energy cascade
pyruvic acid
phosphorylation
nicotinamide adenine dinucleotide (NAD)
bomb calorimeter
mole
carboxylase
thiamine pyrophosphate (TPP)
fermentation
glycolysis
acetyl group
coenzyme A
acetyl coenzyme A
Krebs or citric acid cycle
respiratory chain
flavin adenine dinucleotide (FAD)
cytochromes

TOPICS FOR REVIEW AND DISCUSSION

1. Explain how potential and kinetic energy differ from each other.
2. How do the Laws of Thermodynamics govern the operations of living cells?
3. Why is ATP called the "energy currency" of the cell?
4. Compare anaerobic and aerobic respiration.
5. Why can yeast cells survive anaerobically?
6. What is the significance of enzymes in the release of energy by cells?
7. Explain the similarities and differences between glycolysis and fermentation.
8. What is energy and how is it measured?
9. Describe the cellular compartmentalization of aerobic and anaerobic respiration.
10. How are oxidation-reduction reactions involved in the release of energy from cells?

Reprinted by permission Sawyer Press, Los Angeles, California.

HARVEST OF LIGHT ENERGY : PHOTOSYNTHESIS

5–1 ALL FLESH IS GRASS

Man, the dominant species of the planet earth, can change the course of rivers, build mountains out of rubble, create machines that think, split the atom, and travel through outer space. Surely nothing on this planet is more impressive than man's capacity to think creatively and alter the environment to suit his needs. However, in spite of his dominion over much of the planet, man is totally dependent upon other organisms for all his food.

For example, in an average day's meals you might consume the following:

Breakfast: Eggs, orange juice, toast, coffee, butter, milk
Lunch: Hamburger, salad, french fries, milk
Dinner: Roast beef, potatoes, carrots, peas, ice cream, coffee

We can divide these foods into two categories: those that came from plants and those that were derived from animal materials, the latter being eggs, milk, and meat products. The chickens, cows, and beef-on-the-hoof from which the latter products came obtained their nourishment, in turn, from plant materials. When we sit down to a juicy steak or drink a glass of creamy milk, we are consuming nothing more than converted plants. All animals that feed on other animals are in the same situation. A chain or pyramid of food organisms exists, linked by producers and consumers, and headed by the flesh eaters; all food pyramids ultimately are based on plant material of one kind or another (Figure 5–1). Many food pyramids, such as the one we have described for ourselves, exist in nature, and some of them are so complex that they form webs rather than pyramids. This is the subject of Chapter 24. The point to be made here is that ultimately we and all other animal life depend on plants

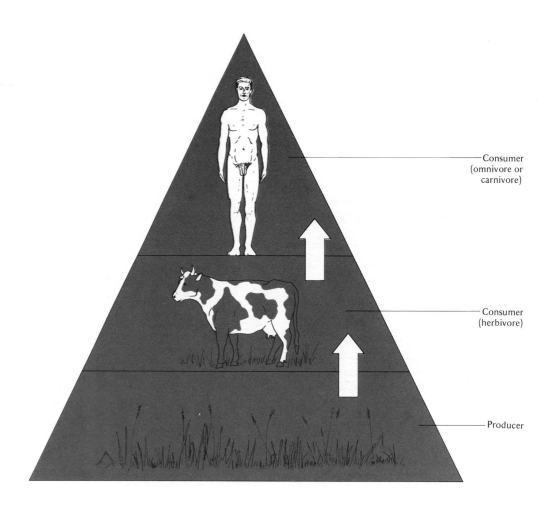

FIGURE 5–1 A simple food pyramid. This food pyramid involves several transformations of chemical energy. The cow eats the grass, and the chemical fuel of the plant is used to form the chemicals of the cow's flesh and milk. Man drinks the milk and eats the flesh (meat), thus obtaining chemical fuel. Since at each transformation there is a loss of energy, it takes a greater amount of the supplier to support an equal amount of consumers (For example, 10,000 lb of grass to support 1,000 lb of cow to produce 100 lb of human).

Consumer (omnivore or carnivore)

Consumer (herbivore)

Producer

to produce our food or the food of organisms upon which we feed. Everything we eat is thus either plant material or derived from plant material. The author of Isaiah recognized this when he wrote: "All flesh is grass" (Isa. 40:6). But what do plants depend on for their food?

Plant cells, green-plant cells to be more exact, combine the simple inorganic substances carbon dioxide (CO_2) and water (H_2O) to form complex organic carbohydrates such as sugar or starch ("CH_2O"). They obtain the energy needed for this process from sunlight. This process, upon which all life on this planet depends, is called **photosynthesis** (literally, "putting together with light"). As a byproduct of such activity, the photosynthetic plant liberates molecular oxygen.

If all flesh is grass, then the amount of flesh that the earth can support is dependent on the amount of energy trapped by the grass (green plants) through photosynthesis. The photosynthetic process is enormous in magnitude. Each year 690 billion tons of carbon dioxide and 280 billion tons of water are combined by photosynthesis to produce 500 billion tons of carbohydrate. About 80% of this production occurs in the sea. Some 90% of the total carbohydrate goes into materials that cannot be used as food, such as wood, and some is used by the plants themselves as their energy source. Even so, the total sugar yield available for use by other organisms is about 4 billion tons each year; this amounts to about 1.3 tons of sugar per person per year or 7 lb of sugar per person per day. As an additional bonus there is the liberation of about 500 billion tons of life-giving oxygen into the atmosphere. Every 2,000 years the atmospheric oxygen of our planet is completely renewed by the activities of green plants.

The population of the earth continues to grow at an enormous rate. Calculation of the number of people the planet can support must involve some recognition of the photosynthetic process. Can the energy-trapping efficiency of photosynthesis be increased? Can we increase photosynthetic food production? Will the capacity of green plants to carry out photosynthesis finally determine the number of persons who can live on the planet earth? Before we can attempt to answer these questions, we must have some appreciation of the mechanisms of photosynthesis itself.

5–2 THE ENERGY CASCADE REVISITED

By the oxidation of complex foodstuffs such as sugar, living things obtain chemical energy (in the form of ATP), lose energy (in the form of heat) and in the process form carbon dioxide and water, energy-poor molecules. We have seen that the energy is released in gradual steps rather like water flowing from a reservoir down through a series of controllable sluice gates.

(See Chapter 4, Figure 4–5.) Consequently, the reservoir is slowly and continually being depleted, for its resources are not unlimited. In order to maintain a store of potential energy, it must have its water supply replenished. There must be uphill transport of water, and this requires a massive input of energy from the outside. In nature, water is cycled continually from reservoir to ocean or sea and back to the reservoir; the heat energy of the sun evaporates water from the oceans and seas, and via precipitation (rain, snow, etc.) the reservoir can be refilled (Figure 5–2a). The reservoir can also be refilled by mechanical means, such as a pump (Figure 5–2b). As long as it is replenished, the reservoir can continue to flow downhill and create a cascade of energy. In the same way, cells and organisms require an energy reservoir for their work. The energy reservoir of life would run dry if only the oxidation of food took place, and life could no longer continue. How is life's energy reservoir replenished?

The energy of sunlight in photosynthesis acts in an analogous way to the sunlight that evaporates water from oceans and seas or the electric pump that lifts water to a reservoir. The tremendous input of solar energy via photosynthesis enables CO_2 and H_2O to be chemically fused into an energy-rich molecule ("CH_2O") which acts as a reservoir or energy source for life (Figure 5–3). In this way, photosynthesis makes the energy cascade operate on a continuing basis and permits the continuity of life.

5–3 LIGHT AND CHLOROPHYLL: LIGHT PHASE

The sun, that thermonuclear furnace which warms us and lights our way, is the ultimate energy source for man and every living creature on the earth. Our sun is like a gigantic hydrogen

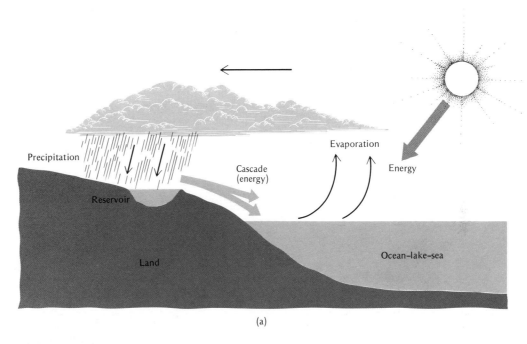

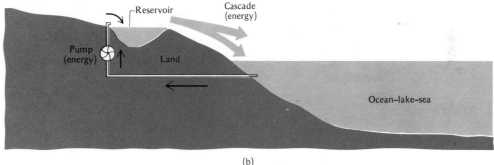
(b)

bomb in which hydrogen is converted into helium; in the process, radiant energy is released as light and heat.

Light behaves as if it were made up of particles or radiant energy packets, called **photons** or **quanta.** Light energy is converted into chemical energy by photosynthesis, but first it must be absorbed by plant pigments, principally the green pigment **chlorophyll** located in the chloroplast. (This phase of photosynthesis

is thus called the light phase.) There are two kinds of chlorophyll: chlorophyll *a* and chlorophyll *b*, which differ slightly in their molecular structure (Figure 5–4) and in the way in which they absorb visible light. It is believed that only chlorophyll *a* is active in photosynthesis. Other pigments present in a leaf (carotenoids, phycobilins, and chlorophyll *b*) help to gather light energy and pass it on to chlorophyll *a*.

FIGURE 5–3 A highly schematic view of the energy cascade in the biological world.

FIGURE 5–4 The molecular structure of chlorophyll. In chlorophyll a, $X = -CH_3$ whereas in chlorophyll b, $X = -CHO$.

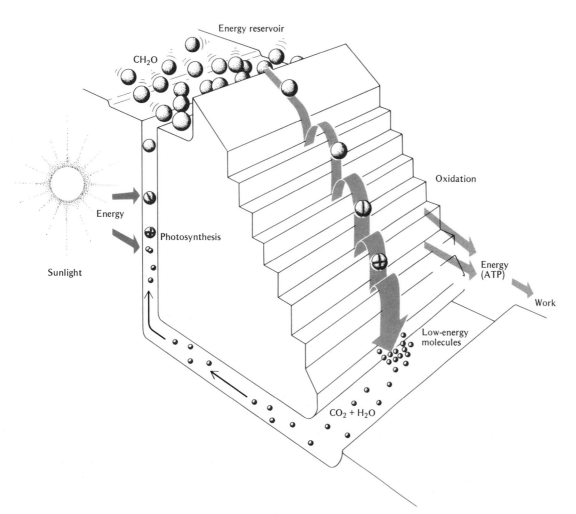

The ability of a pigment to absorb light depends on its molecular architecture and in particular on the arrangement of the electrons within its atoms. Certain pigments, by virtue of their structure, possess particularly mobile electrons that are associated not with individual atoms but with the molecular system as a whole, and chlorophyll is one of these. When light strikes chlorophyll a in the chloroplast, a particle of light energy (a photon) is absorbed, and an electron is raised to a higher energy level; this electron is then passed on to neighboring molecules, and thus energy is transferred from chlorophyll molecule to chlorophyll molecule. The freely migrating electrons react with molecules called **electron acceptors.** Upon receiving an electron, an electron acceptor becomes reduced and in turn donates its electrons to a series of other electron acceptors. In bucket-brigade fashion, the electrons flow from one compound to another, as electricity flows in a wire. Eventually, light energy is traded for

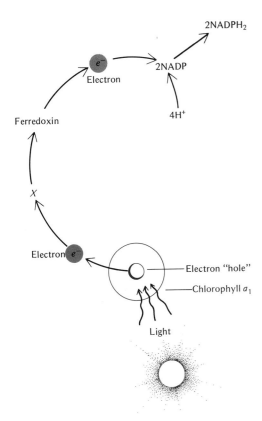

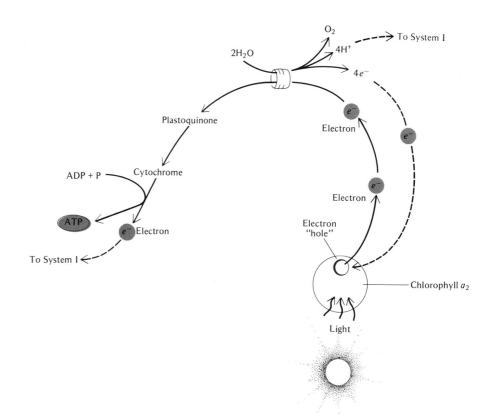

chemical energy, by a process called **noncyclic photophosphorylation.** Noncyclic photophosphorylation occurs in two phases or systems, and both involve chlorophyll a. In **system I,** chlorophyll a_1 is "excited" by photons of light, and electrons are boosted to a high energy level; the electrons are passed to some unknown intermediate material X, then to the electron acceptor ferredoxin. The ferredoxin in turn passes the electrons on to the coenzyme NADP, which combines with hydrogen ions (H^+) in the medium to become reduced ($NADPH_2$). (**NADP** stands for **n**icotinamide **ad**enine **d**inucleotide **p**hosphate, and is a derivative of niacin; it is not identical to, or exchangeable in function with, its coenzyme cousin, NAD, but both play almost identical roles in cellular metabolism.) The situation is shown schematically in Figure 5–5.

The chlorophyll a_1 of system I, excited by light energy, has generated $NADPH_2$, but in the process it has lost its electrons; now electron-deficient, this "hole" must be filled with another ground-state electron. It is believed that replacement comes from another chlorophyll molecule a_2; this is called **system II.** In this system the absorption of light energy by the chlorophyll produces an "excited" state, and in some manner (as yet unknown) water is split into hydrogen (H^+) ions, electrons, and molecular oxygen:

$$2H_2O \longrightarrow 4H^+ + 4e^- + O_2$$

water   hydrogen   electrons   oxygen
                    ions

The chlorophyll a_2 electrons are carried downhill through a series of electron acceptors (plastoquinone and cytochrome), eventually reaching the ground state and filling the "hole" in the chlorophyll a_1 of system I. This downhill transit of electrons is not only a "digging of an electron hole" to "fill an electron hole," for during the process ATP is generated from ADP and inorganic phosphate. The electrons to replace those lost from the chlorophyll a_2 of system II are derived from the splitting of water. Now the chlorophyll of system II has recovered ground-state electrons and is no longer "excited." The hydrogen ions, also produced by splitting water, are used for the formation of $NADPH_2$ in system II. The reactions of system II are shown schematically in Figure 5–6.

So far, we have seen how the energy of light is trapped and conserved in the high-energy bonds of ATP. However, we said that one of the main results of photosynthesis was the production of carbohydrate. How does this occur? We shall examine this process.

5-4 CARBON DIOXIDE AND SUGAR FORMATION: DARK PHASE

The chemicals ATP and $NADPH_2$, generated rapidly during the light-driven phase of photosynthesis, are used to produce carbohydrate in a more leisurely fashion, by a process that does not require light and is called the **dark phase** of photosynthesis. (It can go on in the light or the dark.)

In the dark phase a 5-carbon phosphorylated sugar (called **r**ibu**l**ose **p**hosphate or **RuP**), already present in the chloroplast, receives another phosphate group from ATP to form a molecule with two phosphates, one on each end of the molecule (called **RuDP** for **r**ibulose **d**i**p**hosphate). Now carbon dioxide (from the air and in solution in the cell sap) is joined to this diphosphorylated 5-carbon sugar, and the resulting 6-carbon molecule is split immediately to produce 2 molecules of a 3-carbon compound (**PGA** or **p**hospho**g**lyceric **a**cid) (Figure 5–7). By the energy supplied from ATP and the reducing power of $NADPH_2$ (derived from the light reactions), the PGA becomes modified into a 3-carbon sugar (**PGAL** or **p**hospho**g**lyceraldehyde). These two 3-carbon molecules are then combined to give a 6-carbon sugar with two phosphates (hexose diphosphate). The 6-carbon sugar (hexose) with its two phosphates is then altered and added on to a starch molecule,[1] and in this manner can serve as a reserve energy fuel (Figure 5–7).

---

[1] Glucose in a free state is not directly produced in photosynthesis.

FIGURE 5-7 Simplified reactions of the dark phase of photosynthesis.

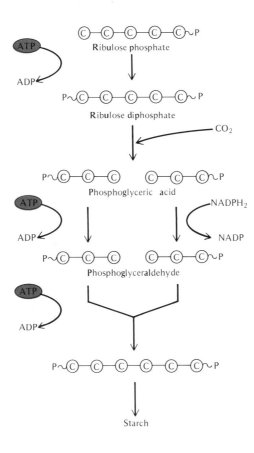

It should be noted that for every 3 molecules of carbon dioxide used, 6 molecules of PGAL are produced, but only 1 of these will be converted into starch. The remaining 5 molecules of PGAL are converted into 3 molecules of the starting material, the 5-carbon phosphorylated sugar (RuP). The 5-carbon sugars can then react with more carbon dioxide to initiate another cycle of dark reactions; if the regeneration of the 5-carbon sugar does not occur, the sequence of reaction stops.

Thus, in the dark phase of photosynthesis, carbon dioxide, which is a relatively oxidized

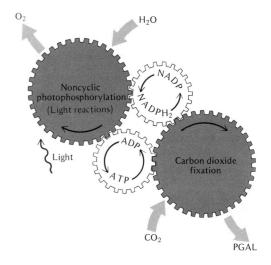

molecule, is chemically reduced to form "CH_2O," carbohydrate, at the expense of the energy of ATP and the reducing power of $NADPH_2$. Note that in the production of carbohydrate within the chloroplast the intermediates are all phosphorylated and, therefore, do not leak out of their place of manufacture; by being directly converted to starch rather than a freely permeable product such as glucose, the carbohydrate is conveniently stored within the chloroplast. When energy is needed for cellular work, the starch is enzymatically converted to glucose or fructose, and via oxidation energy is made available to the plant cell. (Thus, plants respire too!)

The process of photosynthesis can be seen as a mechanical model (Figure 5–8).

5-5 CELLULAR LOCATION OF PHOTOSYNTHESIS

The processes of photosynthesis take place in the chloroplast. The chloroplast is compartmentalized, and the light and dark reactions take place in different parts of it (Figure 2–14). Chlorophyll is incorporated in the grana in a series of membranes stacked one on top of the other, resembling a pile of coins. The chlorophyll molecules are arranged in layers in these stacks of membranes, presumably sandwiched between layers of protein and lipid. The chlorophyll is thus rigidly oriented, and it behaves much like a crystal. When light (photons) strikes the chlorophyll in the grana, the electrons are free to move in the crystal and in their cyclic path generate ATP, $NADPH_2$, and oxygen. Thus the grana are solar power plants. The dark reactions take place in the stroma of the chloroplast, the region devoid of membrane stacks. In the stroma are the enzymes necessary for the fixation of carbon dioxide into carbohydrate. The reducing agent, $NADPH_2$, and the energy source, ATP, which are generated in the grana, move into the stroma and there participate in the synthesis of carbohydrate.

5–6 EFFICIENCY AND IMPLICATIONS OF PHOTOSYNTHESIS

Let us examine the efficiency of the photosynthetic process as it relates to us. The surface of the earth receives annually about 500×10^{18} Calories of solar energy. (The bomb dropped on Hiroshima, which was equivalent to 20,000 tons of TNT, yielded only 0.00002×10^{18} Calories, equivalent to the amount of solar radiation received on 1.5 square miles of the earth's surface in 1 day.) About one third of the total annual solar radiation reaching the earth is expended in the evaporation of water (from the surfaces of oceans, lakes, ponds, etc.), which leaves about 150×10^{18} Calories available for photosynthesis. Plants convert less than 1% (actually closer to 0.34%) of this available energy into the chemical energy of carbohydrate. Even so, this amounts to 1.5×10^{18} Calories stored by land plants alone, and if we

calculate only the fertile areas (about 20% of the land area of the earth), the stored energy equals 0.3×10^{18} Calories. Of this, man uses only about 1% or 0.003×10^{18} Calories annually—about 2×10^3 (2,000) Calories per day for each of the approximately 3 billion persons presently on the earth. Based on these values, it should in theory be possible to support a population 100 times greater than we now have—that is, 300 billion persons.

This argument has obvious fallacies. First, many of the photosynthetically produced Calories in the previous statistics are useless to us as food. For example, an acre of pine trees is useful to man in many ways, but not directly as food. Second, to be of any use, the enormous annual production of photosynthetic energy must be harvested. Our means of harvesting are not good. Optimistic assumptions indicate that, under present farming conditions, about 1 acre of land is required to support one person. The total potentially arable (meaning harvestable) land of the earth is about 7 billion acres. The term "potentially arable" is misleading, as much of this land cannot really be farmed because of problems of irrigation, clearing, low yields, and the exorbitant costs of making it produce. Hence, even supposing all this acreage could be farmed efficiently in the next few years, the absolute maximum number of persons the earth can support is 7 billion.

The population of the earth is increasing at a rate of 1.8%–2.0% yearly, and we can expect it to at least double within 40 years, and to have reached this 7 billion limit by the year 2010. However, as public-health measures are improved, the death rate will go down, and therefore the rate of population increase will rise. Furthermore, since almost 40% of the population of the world is now under 20 years of age, we can expect that within the next decade as these persons reach childbearing age the rate of population increase will again rise. Thus we see that the number of years before the population doubles will be reduced, and we shall reach 7 billion long before 2010— probably by 1985 or 1990. (See Chapter 26.) Of course, the more people there are, the less food there will be to go around, and the greater the probability of malnutrition or starvation for everyone. Since there is an unequal distribution of populations and food resources, starvation is already occurring in many countries of the world, especially in underdeveloped countries, and the United States is not immune to the problem. (See Chapter 26.)

What can be done about it? One way is to increase food production so that less than 1 acre of arable land is needed to feed one person adequately. Much has been done already, with the use of fertilizers, insecticides, etc. However, the effects of these chemicals have not been totally beneficial, because they are not easily degraded and they accumulate in the environment to dangerous levels (Chapter 25).

Another way to increase food production is to produce more efficient photosynthetic plants; again, something along these lines has already been done. The Nobel Peace Prize for 1970 was presented to Dr. Norman E. Borlaug, whose efforts in producing higher-yield wheat and rice plants have greatly contributed to feeding the world. When we understand the genetic mechanism of photosynthesis in detail, we may be able to breed superefficient plants and perhaps even create a manmade chloroplast to substitute for the living green-plant cell.

Even if we could feed more than 7 billion persons, it is questionable how pleasant life would be under such crowded conditions (Chapters 25 and 26). There is one other way out of our dilemma: to decide what number of persons can comfortably live on the earth and then find a way to restrict the population to that number (Chapter 26). For now, suffice it to say that it is clearly not the energy-conversion system of the photosynthetic plant, the amount of carbon dioxide, or the availability of sunlight that will limit the earth's population. Rather the number of persons the earth can support will be restricted by the amount of space man needs for living. No conceivable increase in food supply can long keep pace with the current growth rate of world populations; what happens after 1985 depends on the political, religious, social, and economic considerations that affect measures which must be taken now.

SUMMARY

1. All life depends on the sun.

2. Plants trap solar energy by using it to combine CO_2 and H_2O into carbohydrate; oxygen is liberated into the atmosphere as a waste product. This process is called photosynthesis.

3. Animals obtain their energy in turn from plants or animals that feed on plants.

4. Photosynthesis is essentially the opposite of respiration:

Photosynthesis:

$6CO_2$ + $6H_2O$ + solar energy
carbon   water
dioxide
$$\longrightarrow C_6H_{12}O_6 + 6O_2$$
carbohydrate   oxygen

Respiration:

$C_6H_{12}O_6$ + $6O_2$
carbohydrate   oxygen
$$\longrightarrow 6CO_2 + 6H_2O + \text{energy}$$
carbon   water
dioxide

5. Solar energy is trapped as ATP in the light phase of photosynthesis (Figure 5–9). The green pigment chlorophyll, present in the chloroplast, is excited by a photon; in the process an electron is raised to a higher energy level and is passed along a chain of electron acceptors, including ferredoxin and NADP. This series of reactions is called system I.

6. At the same time, chlorophyll a_2 in a second system (system II) is excited by light energy (photons); this energy is used to split water into H^+ ions, electrons, and oxygen, which is released. The hydrogen ions go to system I to reduce NADP to $NADPH_2$. The electrons are passed along a series of acceptors (plastoquinone and cytochrome) and replace the original electrons in systems I and II.

7. During the passage of electrons from system II to system I, ATP is formed from ADP and inorganic phosphate.

8. The reduced $NADPH_2$ and ATP formed in the light phase of photosynthesis are subsequently used during the dark phase to form carbohydrate (sugar) from CO_2.

9. In the dark phase, the energy of ATP is used to phosphorylate ribulose phosphate (RuP) and convert it to a 6-carbon sugar (ribulose diphosphate, RuDP) using CO_2 as the carbon source ($5C \rightarrow 6C$). The 6-carbon sugar is then split to 2 molecules of PGA (phosphoglyceric acid). More ATP energy and the reducing power of $NADPH_2$ (from the light reactions) convert the PGA into PGAL (phosphoglyceraldehyde). Two such 3-carbon molecules are then combined into a diphosphorylated 6-carbon sugar (hexose disphosphate) which is altered and added onto a starch molecule.

10. The light reactions occur in the grana of the chloroplast. The dark reactions occur in the stroma.

11. The earth receives 500×10^{18} Calories of solar energy annually. Land plants convert 1.5×10^{18} Calories of this. Man uses only 0.03×10^{18} Calories of this annual solar input. However, the earth cannot support a vast increase in population because our means of harvesting are not good and our methods of farming the 7 million potentially arable acres cannot be improved much more. The maximum limit to population is probably about 7 billion, predicted to occur about 2010.

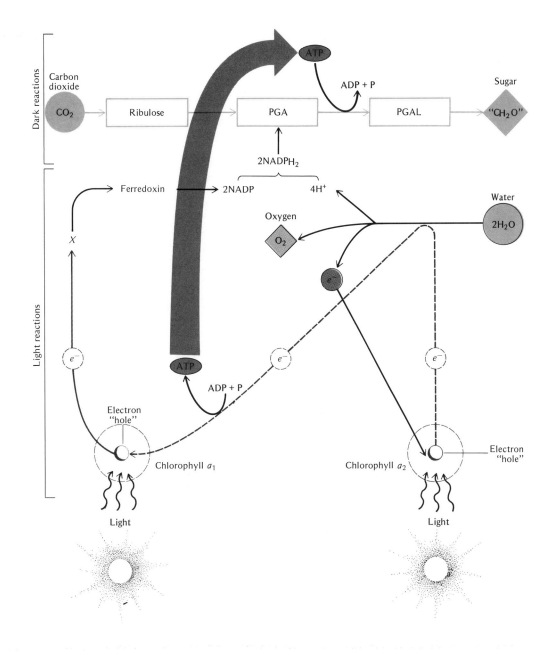

FIGURE 5–9 A summary of the light and dark reactions of photosynthesis (reactants in circles and products in diamonds).

KEY WORDS

photosynthesis
photons
quanta
light phase
chlorophyll
electron acceptors
noncyclic photophosphorylation
system I
nicotinamide adenine dinucleotide
   phosphate (NADP)
system II
dark phase
ribulose phosphate (RuP)
ribulose diphosphate (RuDP)
phosphoglyceric acid (PGA)
phosphoglyceraldehyde (PGAL)

TOPICS FOR REVIEW AND DISCUSSION

1. What is the role of light in the process of photosynthesis?
2. How can a knowledge of photosynthesis be used for the improvement of crop yields?
3. What is the biological meaning of the statement: "All flesh is grass"?
4. What functional roles are served by plant pigments?
5. Describe in some detail the light and dark reactions of photosynthesis.
6. Does the oxygen produced during photosynthesis come from water or from carbon dioxide?
7. Contrast cellular respiration with photosynthesis.
8. How does the Second Law of Thermodynamics relate to photosynthesis?
9. How does photosynthesis keep the energy cascade operating?
10. Why are green plants green? Of what significance is this?

TINKERING WITH LIFE
Genetic Manipulation and Morality

By AMITAI ETZIONI

The acceleration of biological engineering has been urged before Congress by Nobel Laureate Dr. Joshua Lederberg. He has called for the establishment of a National Genetics Task Force to increase the momentum of efforts aimed at unlocking the genetic code of man. Such a breakthrough in biology could lead to the prevention of many illnesses whose origin is wholly or partially in the genetic code.

There is much to be said in favor of such a task force. But it ought to be accompanied by a task force on the social and moral consequences of genetic manipulation. The imminent breakthroughs in biology may affect man as much or more as he was affected by previous revolutions in engineering and physics: the imposition of a new set of capacities, of freedoms, of choices society must make, of evil it can inflict.

Gene manipulation may also allow man to tamper with biological elements which heretofore had to be accepted, including the sex of children to be conceived, their features and color, and ultimately their race, energy levels, and perhaps even their IQ's. Thus, what may start as the biological control of illnesses could become an attempt to breed supermen. While this may appeal to some, think about the agonizing problems if man has to act as the creator and fashion the image of man.

What supermen will the national task force order? Blond or brown, white or black? Highly charged or low-keyed? More males? And, who will make all these decisions—the parents shopping for genes in the supermarket, again expecting society to pick up the bill for the aggregate effects of individual decisions? Or, a Government agency, a task force?

Before such guiding of scientific efforts can be effectively applied to the new genetics, we must have a clearer notion of the moral and social choices involved in the biological revolution and the mechanisms by which science can be guided without being stifled.

To this end, I suggest that at least 1 per cent of the $10 million a year requested for National Genetics Task Force be set aside to explore the options genetic engineering is about to impose on us.

Is the Genie Out of the Bottle?

The article above was extracted from one written in 1970. Since that time, a whole new field of genetic research has opened up, causing enormous and stormy public debate. The technique that has many people so worried is known as recombinant DNA research. It involves the splicing of hereditary material from one organism into that of another, and is said to be so simple, a beginning graduate student can do it. Essentially, new organisms are created, breaching natural barriers among species, shortcutting evolution and bringing biologists into new and untested territory. One public official has compared the breakthrough to that obtained by physicists when they split the atom.

The possibilities are almost infinite. One, suggested by the humorist Russell Baker, involves combining the genes of man and ape to produce more hairy customers for depilatory makers! To date, however, research has been performed only with bacteria—the lowly E. coli, a normal resident of the human intestine. Genes from any other organism can be added to the genetic material of bacterial cells, and are reproduced along with the bacterial genes.

The possibilities for both good and evil are immediately obvious. Since bacteria reproduce rapidly, they might be persuaded to produce large quantities of insulin directly for diabetics, or they might be enlisted to turn out vitamins or antibiotics for the drug companies. They might also be persuaded to produce substances causing new and interesting diseases, however. What if a scientist, studying a gene thought to produce human cancer, added it to a bacterial cell that later escaped from its laboratory flask and invaded the intestines of the general population? What would be the effects of a bacterial insulin factory on normal human metabolism? Some scientists, in efforts to unlock genetic secrets, are performing "shotgun" experiments, where strips of genetic material from a chosen organism are spliced randomly and inserted into E. coli, just to see what they do. While helping to elucidate the fine genetic structure of an organism, such experiments are blind to the extent that the scientist has no idea what will be produced until after the experiment has been done. Might he make a new "Andromeda strain"? Can the human mind master what the human hand has made?

To eliminate or reduce the chances of an accidental epidemic produced by an escaped strain of laboratory manufactured E. coli, the National Institutes of Health and the Government have worked to produce a set of guidelines for such research. It seems that the sentiments expressed in Dr. Etzioni's 1970 article are more pertinent today than ever before.

6

COMMUNICATION BETWEEN CELL GENERATIONS: NUCLEIC ACIDS

6-1 INHERITANCE: GENERAL FEATURES

"Chip off the old block" is an expression that indicates the relationship between parent and offspring. It simply suggests that offspring generally resemble their parents, that like begets like. Humans give birth to humans, cats give birth to cats, and dogs give birth to dogs; this is the way of nature, and the phenomenon occurs (at least to date) without exception. The phenomenon is more complicated than this: not only do cats always produce other cats and dogs other dogs, but black cats tend to produce black kittens and brown dogs brown puppies. Tall humans tend to have tall children. Not only this, but when liver and skin cells reach a certain size and stage of development, they divide and produce more liver and skin cells, respectively. Nowhere does it occur within our bodies that on occasion one of our cells makes a mistake and slips in a dog cell or a cat cell, or even a human skin cell among the cells of the liver. Cells pass on their specific and characteristic messages from one generation to the next with great fidelity; otherwise chaos would reign within us.

Although these phenomena may not surprise you because they are everyday occurrences, they are remarkable because they imply the existence of a system capable of accurately retaining information and passing it on with equal accuracy from generation to generation. In our culture, information is recorded in books, on records, or on tapes, and is stored and available to anyone who has access to it. During the building of a house, a builder works from a plan—a blueprint prepared by an architect for his client. In much the same way, the cells of the body record and store information about the kinds of cells they are, and when new cells arise, this information can be translated and is available to those cells in a form very like a blueprint. A new cell can be constructed using the directions of the blueprint.

Genetics, the study of inheritance, deals with the transfer of information (or passing on of traits) from parent to offspring, whether this be at the level of the cell or of the organism. In the present context we shall explore the chemical nature of the genetic blueprint (so-called genes) and show how that material performs its function in making us what we are. In the next chapter we shall deal with the mechanisms whereby cells are able to duplicate themselves, and much later (Chapters 22 and 23) we shall see how human traits are inherited.

Considerable fear and excitement surround discussions of the nature of the inherited message, for once we understand how the genetic blueprint is constituted and how it can be read and translated, we should be able to manipulate it (see facing page). The implications of this are far-reaching. Will this knowledge be used to change defective traits into beneficial ones? It is not difficult to see that a child born with an inherited defect, such as cystic fibrosis, muscular dystrophy, or diabetes, would be better off if this defect could have been prevented by the manipulation of his parents' genetic material prior to his conception. If taken further, the subject of genetic manipulation raises interesting questions. Could we and shall we create an entire race of superhumans with predictable and "desirable" characteristics? Who is qualified to decide what is desirable and what undesirable in a human being? Understanding the chemical nature of inheritance may provide a "second genesis," but it is fraught with the same kinds of social and political questions that surrounded the suggested programs of eugenics that were popular not too long ago (breeding for beneficial traits in human populations, just as cows are bred to produce more milk, poodles for more interesting hair patterns, and so on). This chapter will detail the mechanism of inheritance; the use to which this information is put rests with all of us as individuals and as citizens.

6-2 INHERITANCE: CHEMICAL STRUCTURE

The disease pneumonia is caused by a bacterium called *Diplococcus pneumoniae*. When grown in the laboratory on a plate of nutrient jelly or agar, the bacteria multiply rapidly and form colonies that appear smooth and shiny to the naked eye. The smooth kind of *D. pneumoniae* is the normal, virulent disease producer. The smooth appearance is the result of the presence of mucilaginous capsules, which probably prevent the bacteria from being engulfed (phagocytized) by the white blood cells of the body and enable them to multiply and cause the symptoms of the disease. Some strains of *D. pneumoniae* have lost the ability to produce the mucilaginous capsule; when grown on agar, the colonies appear rough. The cells that form rough colonies are not virulent and do not cause pneumonia because they are easily ingested and destroyed by the white blood cells. Each strain of bacterium breeds true: rough strains give rise to rough strains and smooth strains to smooth strains. In 1928 F. Griffith studied these strains of *D. pneumoniae* and obtained some puzzling results. If he injected living cells of the rough strain into a mouse, the mouse lived; if he injected living cells of the smooth strain into another mouse, the mouse died (Figure 6-1). When he killed the bacteria (he heated the broth in which they were growing to 60° C) and then injected each strain into mice, the mice did not become infected (Figure 6-1). So far, there are no puzzles. However, for the sake of completeness Griffith injected into a mouse a mixture of living rough cells and some killed smooth cells; unexpectedly, the mouse died (Figure 6-1).

The cause of death was pneumonia, and bacteria recovered from the body multiplied on agar and formed smooth colonies. Obviously, there had been a transformation of rough cells into smooth cells, and Griffith postulated that an agent had been given off by the dead smooth cells which was taken up by the living rough cells; this agent was transmissible and conferred heritable properties upon the recipient. The agent was called **transforming principle.** What was this mysterious material?

In 1944 O. T. Avery, C. M. Macleod, and M. McCarty carefully collected and purified transforming principle; they found that it was a chemical compound called **deoxyribonucleic acid (DNA).** It has since been found that this chemical is the material universally responsible for inheritance—recording and transmitting the characteristics of a cell or an organism from generation to generation.[1]

What is the structure of DNA, and how does it work? What properties must DNA have in order to qualify as the chemical substance of hereditary material? It must have two fundamental properties: first, it must have the ability to duplicate itself, otherwise it could not be passed undiluted from generation to generation; and second, it must be able to contain all the information necessary for directing cellular activities that must be carried from one generation to the next.

Chemical analysis of DNA shows it to belong to a class of molecules known as the nucleic acids. Each DNA molecule is made up of a series of building blocks called **nucleotides** (more specifically, **deoxyribonucleotides**), each of which contains three major constituents: (1) phosphate, (2) a 5-carbon sugar called **deoxyribose,** and (3) a nitrogenous base (Figure

---

[1] There exist viruses which have ribonucleic acid (RNA) as their genetic material; these are exceptional cases.

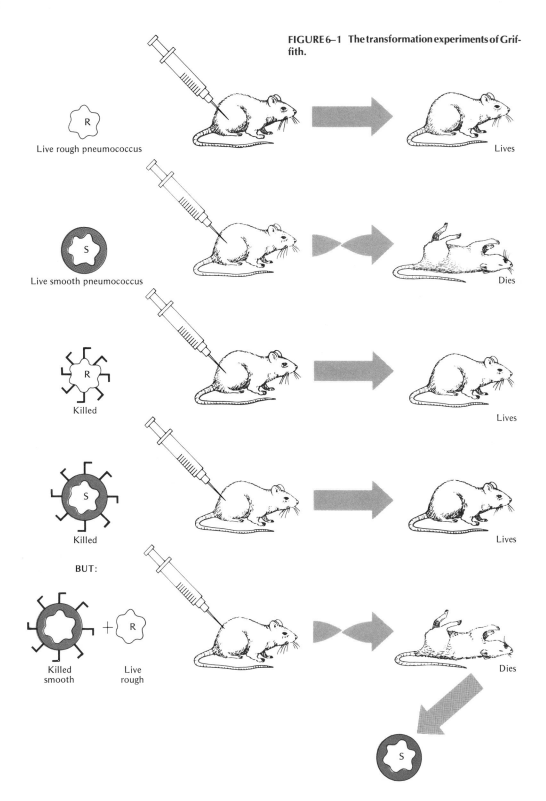

FIGURE 6–1   The transformation experiments of Griffith.

Live rough pneumococcus

Lives

Live smooth pneumococcus

Dies

Killed

Lives

Killed

Lives

BUT:

Killed   Live
smooth   rough

Dies

FIGURE 6–2 Structure of a nucleotide (deoxyribonucleotide), with detailed structure of the sugar deoxyribose.

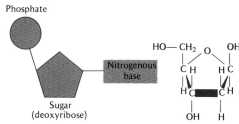

TABLE 6–1   The percentage of nitrogenous bases present in DNA

Source of DNA	Adenine	Guanine	Cytosine	Thymine
Human				
Sperm	29%	18%	18%	30%
Thymus	28	19	16	28
Liver	27	19	*	*
Ox				
Thymus	26	21	18	25
Liver	26	21	17	24
Spleen	26	20	*	*
Yeast	31	17	18	33

* These determinations were not made by Chargaff. What do you think the values should be?

6–2). The phosphate and sugar portions of the nucleotide are a constant feature, but the nitrogenous bases come in four varieties: **adenine, guanine, thymine,** and **cytosine** (Figure 6–3). Adenine and guanine are structurally similar, consisting of two fused rings (Figure 6–3), and are called **purines.** Thymine and cytosine, having a single ring, are structurally alike but differ from the purines, and are called **pyrimidines** (Figure 6–3).

In 1950 Erwin Chargaff purified DNA from a variety of sources and measured the amounts and kinds of nitrogenous bases present in each type of DNA. The data from some of his analyses are shown in Table 6–1. Three facts become apparent from examination of Chargaff's data. First, within the limits of experi-

FIGURE 6–3 The nitrogenous bases found in DNA.

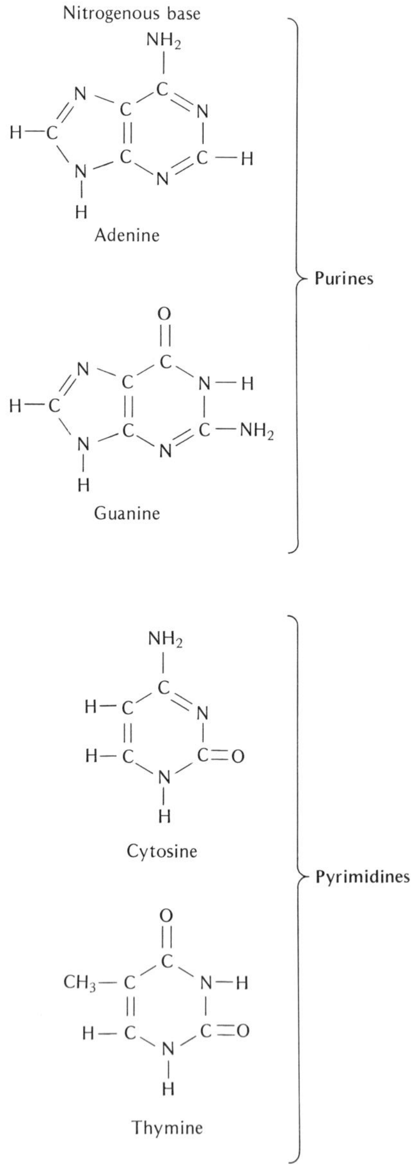

mental error, the proportion of the bases from the tissues of any particular species is the same; for example, human tissues all have about 28% adenine, 19% guanine, 28% thymine, and 19% cytosine. Second, the composition of DNA varies from species to species; the DNA of the ox is different from the DNA of yeast. Third, the percentage of purines (adenine plus guanine) equals, within the limits of experimental error, the percentage of pyrimidines (thymine plus cytosine); moreover, the amount of adenine equals the amount of thymine and the amount of guanine equals the amount of cytosine. Let us now substitute the letters A, T, G, and C for the nitrogenous bases that begin with those letters and abbreviate the discovery of Chargaff as A = T and G = C, and A + G = T + C.

How are the nucleotides arranged to form a molecule of DNA? As we have already mentioned in our discussion of proteins, the spatial or three-dimensional appearance of a molecule is often related to the functions of the molecule. Thus, in order to comprehend the functional properties of DNA, it is necessary to understand the spatial arrangement of nucleotides in the molecule. As we shall see, it is the three-dimensional nature of the DNA that is the key to how it really works. As with proteins, which you recall are composed of long chains of amino acids, DNA consists of nucleotides linked in a long strand. Are the nucleotides of the DNA molecule arranged to form a single strand, two strands, three strands, or four strands? Are the strands twisted, and how are the nitrogenous bases oriented? The problem of the structure of DNA was solved in 1953 by the efforts of M. H. F. Wilkins, James D. Watson, and Francis Crick. Wilkins was an x-ray crystallographer; that is, he studied the arrangement of atoms in crystals by bombarding the crystal with x rays. The way in

Nitrogenous base

Adenine

Guanine

} Purines

Cytosine

Thymine

} Pyrimidines

which the x rays are scattered by the atoms in the crystal (**diffraction pattern**) gives information on the spatial arrangement of atoms forming the crystal. (Similarly, the manner in which x rays are scattered by your bones, muscles, and skin gives information on the density of these tissues and on the arrangement of atoms within them.) Wilkins examined purified crystalline DNA by this method and found that there were three spacing arrangements (**periodicities**) in the crystalline DNA; these were 3.4Å, 20Å, and 34Å. Why did the diffraction pattern always indicate this repetitive spacing? It became a jigsaw puzzle to arrange these pieces of information into a meaningful picture. The puzzle was solved elegantly by Watson and Crick. Aware of the molecular structure of the nitrogenous bases and of their organization in the nucleotide, Watson and Crick built scale models of DNA, using the structure and size of the purines and pyrimidines and the dimensions of the DNA crystal. They arranged the nucleotides in a linear sequence, the backbone of the chain being formed of repetitive sequences of sugar-phosphate-sugar-phosphate; furthermore, they proposed that the DNA molecule consisted of two long chains, each with a sugar-phosphate backbone. The position of the bases proved somewhat troublesome, but soon it became apparent that adenine could only bond with thymine (by two hydrogen bonds) and guanine could only bind with cytosine (by three hydrogen bonds). Such a specific pairing of a purine with a pyrimidine would account for the identical amounts of adenine-thymine and cytosine-guanine. Furthermore, when the two nitrogenous bases were joined as A-T or G-C they were 20Å in width. Watson and Crick proposed that the nucleotides were arranged lengthwise forming two long chains; the backbone of each chain consisted of sugar and phosphate groups,

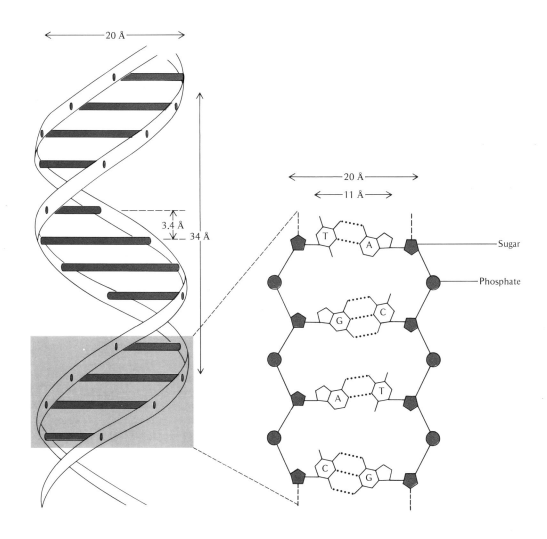

FIGURE 6–4 Organization of the DNA double helix as proposed by Watson and Crick. The uprights of the helix consist of alternating sugar-phosphate groups, and the rungs are formed by pairs of nitrogenous bases held together by hydrogen bonding.

20 Å

20 Å

11 Å

3.4 Å

34 Å

Sugar

Phosphate

T — A
G — C
A — T
C — G

and the two chains were linked by the inward-pointing bases (Figure 6–4). It then became obvious that the bases formed the rungs of a ladder, and the uprights consisted of sugar and phosphate arranged in an alternating sequence; the distance between successive bases (rungs) was 3.4Å. The 34Å periodicity that Wilkins had found in crystalline DNA is 10 times 3.4Å, and Watson and Crick proposed that the ladder was twisted in the form of a helix (similar to a spiral staircase), with a full twist (360°) at every tenth base. The Watson-Crick model for DNA was complete. They called it the double helix, since it was made of two strands of nucleotides wound around one another (Figure 6–4).

One of the properties that DNA must have in order to function in inheritance is the ability to duplicate itself. The Watson-Crick model provides for this. One strand of the DNA is complementary to the other, since when adenine occurs in one strand, thymine occurs in the opposite strand, and guanine in one strand is matched by cytosine in the other strand. The helical strands separate by the paired bases moving apart; this "unzipping" of the strands, resulting from reduced hydrogen bonding between the bases, allows a new complementary strand to be formed from nucleotides and other materials present in the cell, under the direction of a specific enzyme called **DNA polymerase** (Figure 6–5). All the information necessary for arranging these bases in a linear sequence to complement the original strands of DNA is provided for by this mechanism of **complementary base pairing.** For example, if the strand of DNA contains adenine, a nucleotide containing thymine will pair with it; and if the next base in the DNA strand is guanine, the nucleotide which will pair must contain cytosine; and so on. In this manner, the old strands of DNA direct the sequence or order of nucleo-

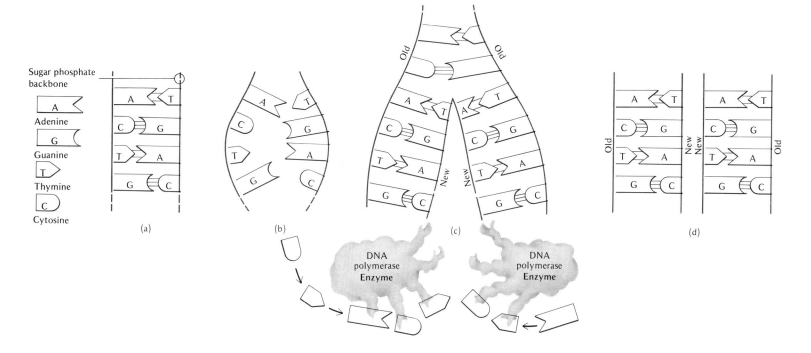

tides that form the new sister strands; the new strand is a complementary copy of the original strand. The formation of two new chains identical to the original two chains is thus effected, but in addition and most importantly, the exact sequence of base pairs in the original strands will have been faithfully reproduced (Figure 6–5d). This process of DNA duplication is called **replication.**

6–3 GENETIC CODE: FOUR-LETTER ALPHABET, THREE-LETTER WORDS

As we have seen, the double helical strands of DNA have built in to them all the information necessary for their own duplication; by means of base pairing, the old strand builds onto itself a new and complementary DNA strand. Thus,

DNA satisfies one of our criteria for being the genetic material. How does the DNA carry information from one generation to the next? How can the bases of the DNA provide all the necessary information to build an entirely new cell? It becomes obvious that somehow the DNA must contain a **code,** a symbolic message that directs the cell to manufacture what the code specifies.

Before we explore the nature of the code itself, let us briefly look at inheritance. Studies with a variety of organisms amply demonstrate that enzyme synthesis depends upon specific entities within the nucleus known as **genes.** In 1940 G. W. Beadle and E. L. Tatum proposed that for every gene there is a specific enzyme; the concept became known as the **one gene–one enzyme hypothesis.** This hypothesis states that

genes exert their control over cellular metabolism by controlling the synthesis of enzymes; since enzymes control all the reactions that take place in the cell, the form and function of the cell depend directly upon enzyme function, and therefore upon gene action (Box 6A). To illustrate this very simply, let us consider an example. In order to have brown pigment in your hair there must occur a series of chemical reactions in the cells that produce a pigmented hair. The reaction sequence can be represented by: $A \rightarrow B \rightarrow C \rightarrow D \rightarrow$ brown pigment. The enzymes that catalyze this sequence of reactions can be represented as a, b, c, d. Thus, the reaction can be represented:

$$A \rightarrow B \rightarrow C \rightarrow D \rightarrow \text{brown pigment}$$
$$\uparrow \quad \uparrow \quad \uparrow \quad \uparrow$$
$$a \quad b \quad c \quad d$$

BOX 6A An inborn error: PKU and idiocy

Phenylketonuria (PKU) is a rare hereditary disease of man. About 1 in 15,000 infants have this disease, which is characterized by severe mental retardation, the daily excretion in the urine of large amounts of phenylpyruvic acid, and high concentrations of the amino acid phenylalanine in the blood plasma. Normal individuals metabolize phenylalanine to tyrosine, but individuals with PKU lack the enzyme that catalyzes this reaction, and instead the blood plasma contains high levels of phenylalanine (Figure 6-6). In other words, in a child with the symptoms of PKU, the gene that codes for the enzyme that catalyzes the conversion of phenylalanine to tyrosine is faulty or missing altogether. Because of this, a derivative of phenylalanine, phenylpyruvic acid, accumulates; this substance is deleterious to brain cells and results in mental retardation or idiocy.*

Knowledge of PKU, its genetic basis and the nature of the enzymatic deficiency, provides a means for the prevention of mental retardation due to this disease. At birth each child is tested for the PKU trait; urine is examined for phenylpyruvic acid, or a drop of blood is placed on a card saturated with chemicals to ascertain the level of phenylalanine. If urine and blood are high in phenylpyruvic acid and phenylalanine, the child bears the defective genes for PKU. Idiocy in the bearer of PKU trait is easily prevented by controlling the diet.

Human beings are unable to synthesize eight amino acids (so-called **essential amino acids**)† which are necessary for protein synthesis, and these must be obtained from our dietary proteins.

Phenylalanine is one of these. In a child with the PKU trait it is necessary to prevent the production of phenylpyruvic acid, derived from phenylalanine in the diet and responsible for damage to the brain. The treatment of PKU is to maintain the dietary level of phenylalanine just sufficient for protein synthesis, but low enough to prevent a buildup of phenylpyruvic acid. Once brain development has proceeded beyond the sensitive stages and there is no longer any chance that phenylpyruvic acid will influence mental capacity, a normal diet can be instituted.

The simple recognition of the relationship of genes to enzymes in PKU has virtually eliminated idiocy produced by this disease in the United States.

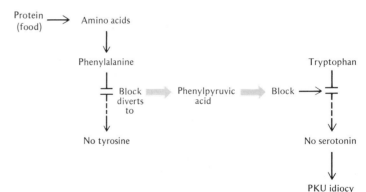

(a) Normal pattern

(b) PKU

FIGURE 6-6 The metabolic basis of phenylketonuria (PKU). (a) Normal metabolism of phenylalanine. (b) Abnormal metabolism of phenylalanine in a person with PKU.

---

* Phenylpyruvic acid blocks the conversion of the amino acid tryptophan to serotonin; serotonin, a transmitter substance in the brain, is essential for normal mental processes.

† Because valine, leucine, isoleucine, lysine, methionine, threonine, phenylalanine, and tryptophan cannot be synthesized from other diet-derived molecules or the various intermediates of metabolism, they are called essential amino acids. The nonessential amino acids (glycine, alanine, serine, glutamic acid, glutamine, tyrosine, aspartic acid, asparagine, cysteine, histidine, proline, and arginine) can be made in the body using metabolic intermediates and the essential amino acids as amino donors. "Essential" is an unfortunate term, since all amino acids must be present in the cell for protein synthesis to take place. (Only plants are capable of synthesizing all the amino acids from substances simpler than the amino acids themselves.)

When all the enzymes are present, brown pigment is produced, as long as substance A is present in the body. If substance A is lacking, then we get no brown pigment and the hair is colorless; and if enzyme a, b, c, or d is lacking, we get no brown pigment.

Genes direct the production of enzymes. What do genes have to do with DNA? Let us say for the time being that specific regions of the DNA helix present in the cell are the stuff of which genes are made; in a simplistic way, a portion of the DNA molecule equals a gene. (How much DNA makes up one gene is a complex question that we shall allow to rest for the time being.) The question that remains is: How does the DNA direct the synthesis of enzymes? Enzymes, as we have seen, are proteins, and so the question can be rephrased: How do we get from DNA to a protein?

This brings us once again to our problem of coding. The DNA of most organisms exists as a long, unbranched, double helical strand localized within the nucleus. In each strand of the double helix there are four different kinds of molecules—the nitrogenous bases. It must be that the sequence of bases determines the nature of the information available to the cell. But what do we need to code for, and what is the product of the sequence of nitrogenous bases? If the DNA that composes a gene or genes makes a protein, the sequence of bases must code for the subunits of a protein—the amino acids. There are 20 different kinds of amino acids present in most proteins; therefore, the base sequence must be able to code for at least 20 different amino acids. How many bases, of which only four kinds exist, are needed to code for 20 amino acids? To put it another way, let us think of the nitrogenous bases as the letters A, G, T, and C. How many words can we construct using these four letters? If we use one letter for each word, we can

Singlet code (4 words)					Doublet code (16 words)				Triplet code (64 words)			
									AAA	AAG	AAC	AAT
									AGA	AGG	AGC	AGT
									ACA	ACG	ACC	ACT
									ATA	ATG	ATC	ATT
									GAA	GAG	GAC	GAT
									GGA	GGG	GGC	GGT
A					AA	AG	AC	AT	GCA	GCG	GCC	GCT
G					GA	GG	GC	GT	GTA	GTG	GTC	GTT
C					CA	CG	CC	CT	CAA	CAG	CAC	CAT
T					TA	TG	TC	TT	CGA	CGG	CGC	CGT
									CCA	CCG	CCC	CCT
									CTA	CTG	CTC	CTT
									TAA	TAG	TAC	TAT
									TGA	TGG	TGC	TGT
									TCA	TCG	TCC	TCT
									TTA	TTG	TTC	TTT

make only four words (A, G, T, and C); if each word is composed of two letters, and four letters are available, we can make 16 words (AA, AG, AT, AC, GT, GG, GC, and so on); and if we make three-letter words and have four letters available, we can compose 64 words (Figure 6–7). Clearly, if the genetic code were to use only one- or two-letter "words" it could not code for 20 amino acids, but three-letter "words" (or "words" composed of three bases) can code for more than is required. The genetic code, it has been established, is a **triplet code;** that is, it consists of sequences of three nitrogenous bases. The nature of the triplet code words has been completely deciphered. This is shown in Figure 6–8. As you can see, there is some redundancy in the code, since some combinations signify the same amino acid; for example, the DNA code words which specify alanine are CGA, CGG, CGT, and CGC. We have a similar situation in the English language, where several different words can be used to specify the same object: house, home, dwelling, residence. Codes in which several different code words describe the same thing are called **degenerate codes;** the genetic code is a degenerate code.

It has been found that the genetic code is universal; that is, the same triplet stands for the same amino acid in all organisms from viruses to man. This suggests that the genetic code probably evolved only once. The differences among viruses, bacteria, men, and all living organisms are due not to differences in the kinds of nitrogenous bases found in their DNA, but to variations in the sequential order of these bases. It is much like using the letters of the alphabet to construct different words. The letters used are always the same; however, the words formed and their meaning (message) may differ depending on the manner in which the letters have been arranged. Similarly, different arrangements of triplets (letters) along the length of the DNA strand will result in a varying amino-acid sequence (words) in a protein (sentence); it is possible to produce an almost infinite number of triplet arrangements in DNA, and thus it is possible to have an

FIGURE 6–8 The genetic code deciphered. (a) DNA code words. (b) RNA code words (codons).

Second base in DNA triplet

First base	A	G	T	C	Third base
A	AAA / AAG phenylalanine; AAT / AAC leucine	AGA / AGG / AGT / AGC serine	ATA / ATG tyrosine; ATT / ATC "terminator"	ACA / ACG cysteine; ACT "terminator"; ACC tryptophan	A G T C
G	GAA / GAG / GAT / GAC leucine	GGA / GGG / GGT / GGC proline	GTA / GTG histidine; GTT / GTC glutamine	GCA / GCG / GCT / GCC arginine	A G T C
T	TAA / TAG / TAT isoleucine; TAC methionine	TGA / TGG / TGT / TGC threonine	TTA / TTG asparagine; TTT / TTC lysine	TCA / TCG serine; TCT / TCC arginine	A G T C
C	CAA / CAG / CAT / CAC valine	CGA / CGG / CGT / CGC alanine	CTA / CTG aspartic acid; CTT / CTC glutamic acid	CCA / CCG / CCT / CCC glycine	A G T C

(a)

Second letter in mRNA triplet

First letter	U	C	A	G	Third letter
U	UUU / UUC phenylalanine; UUA / UUG leucine	UCU / UCC / UCA / UCG serine	UAU / UAC tyrosine; UAA / UAG "terminator"	UGU / UGC cysteine; UGA "terminator"; UGG tryptophan	U C A G
C	CUU / CUC / CUA / CUG leucine	CCU / CCC / CCA / CCG proline	CAU / CAC histidine; CAA / CAG glutamine	CGU / CGC / CGA / CGG arginine	U C A G
A	AUU / AUC / AUA isoleucine; AUG methionine	ACU / ACC / ACA / ACG threonine	AAU / AAC asparagine; AAA / AAG lysine	AGU / AGC serine; AGA / AGG arginine	U C A G
G	GUU / GUC / GUA / GUG valine	GCU / GCC / GCA / GCG alanine	GAU / GAC aspartic acid; GAA / GAG glutamic acid	GGU / GGC / GGA / GGG glycine	U C A G

(b)

endless variety of amino-acid sequences in proteins. Species-specificity has as its basis the different amino-acid sequences in a protein (that is, dog hemoglobin differs from human hemoglobin in a number of amino acids), but it must be emphasized that this is ultimately a reflection of the specific arrangement of the three-letter words in the hereditary blueprint, DNA (Box 6B).

6–4 COPYING AND TRANSLATING THE GENETIC CODE: RNA AND PROTEIN

The genetic code of the DNA is localized within the nucleus of the cell; the nucleus is in a sense a library of blueprints, but the blueprints cannot be taken out of the library. How then does the information encoded in the DNA reach the rest of the cell? The manufactured product specified by the code is a protein, and the apparatus for protein synthesis is in the cytoplasm. Thus, a number of problems must be solved by the cell before it can utilize the information contained in the DNA.

First, a means must be available for copying or transcribing the genetic code and transporting this copy of the code to the cytoplasm, where the information can be put to use. Second, the coded message must be deciphered; if the code is in a cryptic or symbolic language, a key to the code will be needed in order to translate its contained information. Let us now introduce the transcriber and the translator of the genetic code: ribonucleic acid or **RNA.**

RNA, like DNA, consists of three molecular constituents: (1) phosphate, (2) a 5-carbon sugar called **ribose,** and (3) a nitrogenous base. As with DNA, these constituent groups together form a nucleotide, or more specifically a **ribonucleotide** (Figure 6–9). Although RNA, like DNA, contains four different nitrogenous bases,

FIGURE 6–9 The general structure of a ribonucleotide. For comparative purposes, the structural differences between the sugars of a ribonucleotide (ribose) and a deoxyribonucleotide (deoxyribose) as well as between the unique nitrogenous bases (uracil and thymine) are shown.

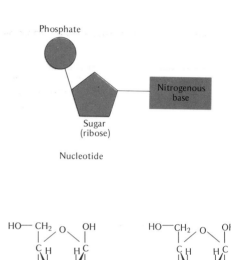

Nucleotide

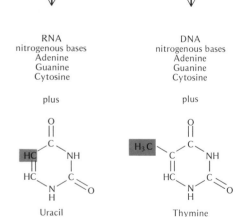

BOX 6B The library of heredity

The microscopic bearers of hereditary characteristics, which we call genes, are long chain molecules. They may be thought of as scrolls upon which the genetic language is written, the four letters of whose alphabet correspond to atomic groups of four different kinds.

A particular sequence of letters in a portion of the molecular chain constitutes, as it were, a file in the library of heredity, and controls one or more characteristics, whilst each complete chain can be faithfully reproduced and passed on to a succeeding generation.

This hereditary filing system becomes progressively more voluminous as one passes from the lower to the higher forms of life.

In a small virus the scroll is of the order of 0.003 mm in length and contains as many letters as half a page of newsprint (8,000).

In the T₂ bacteriophages (another kind of virus extensively used in research), the information capacity of a 300 page paperback novel (say 500,000 letters) is packed into 0.1 mm of molecular chain.

In the mammal the uncoiled chain would span a full meter, and is able to accommodate some 3,000 million letters, that is roughly 500 times the letter count of the Bible.

An interesting feature is that in the case of the virus it has been shown that changing one of the letters of its genetic language—corresponding to making one printing error in its hereditary file—would make the file unreadable.

R. HOUWINK, *The Odd Book of Data,* New York, Elsevier, 1965, pp. 28–29.

---

only three of these are the same as those of DNA. RNA contains the bases adenine, guanine, cytosine, and uracil. **Uracil** is a pyrimidine almost identical to thymine, and in RNA the uracil replaces the thymine of DNA.

There are three important distinctions between DNA and RNA: (1) RNA contains ribose, a slightly different sugar from that found in DNA (Figure 6–9); (2) RNA is usually single-stranded in contrast to double-stranded DNA; and (3) RNA is distributed throughout the cell, whereas DNA is found (almost) exclusively within the nucleus.[2]

A look at the nucleotides of RNA suggests that the base sequences are similar to those of DNA and could in themselves represent a language or code written in four letters. This RNA alphabet of nitrogenous bases is exactly like that of DNA, except for the substitution of uracil for thymine. Even with this one-letter substitution, the bases of DNA can be copied or transcribed by RNA bases, and in so doing the genetic message is unchanged. In the cell the transcription or copying of code words of the DNA is made by a specific kind of RNA called **m**essenger **RNA (mRNA).**

In order for the sequence of nucleotides in the DNA to be read and copied, the DNA strands must separate; a region of the DNA double helix "unzips" and a sequence of bases

---

[2] Mitochondria and chloroplasts contain small amounts of DNA, which may code for specific mitochondrial and chloroplast proteins

FIGURE 6–10 A schematic representation of transcription. (a) DNA strands separate, exposing a sequence of bases. This serves as a guide for the synthesis of a complementary strand of mRNA. The nucleotides of the incipient RNA are bonded together by means of RNA polymerase. (b) DNA strands zip closed, and the strand of mRNA leaves the nucleus.

is exposed (Figure 6–10); only one of the exposed strands of the DNA serves as a guide for the synthesis of a complementary strand of mRNA. Using the mechanism of specific complementary base pairing, the nucleotides of the incipient mRNA, which are free in the nucleus, line up and match the DNA nucleotides: mRNA-uracil with DNA-adenine, mRNA-guanine with DNA-cytosine, mRNA-cytosine with DNA-guanine, and mRNA-adenine with DNA-thymine. The base-pairing mechanism assures that the sequence of nucleotides in the DNA strand is followed exactly by the sequence in which the nucleotides of the incipient mRNA are lined up. Once aligned, the nucleotides are bonded together with the aid of the enzyme **RNA polymerase,** to form a chain of mRNA. The mRNA molecule detaches from the strand of DNA and leaves the place of synthesis in the nucleus: the DNA strands now "zip" closed (Figure 6–10). In this manner the nucleotide sequences of DNA, corresponding to **genes,** are copied in the nucleotide sequences of mRNA; the genetic code has been rewritten or copied by mRNA without any change in the wording: the process is called **transcription.**

As we have already seen, the language of the genetic code, contained in the DNA molecule, is written in three-letter words using an alphabet of four letters. The mRNA, using an alphabet of four letters, only one of which is different from the alphabet of DNA, faithfully copies these three-letter code words. The four letters of the DNA alphabet are capable of making 64 different three-letter code words, or triplets, and it is easy to see that in the mRNA the four-letter code can also make 64 three-letter words; the 64 code words of the mRNA are called **codons** (Figure 6–8).

The problem of getting the genetic information out of the nucleus without the release of DNA has been solved by the manufacture of a strand of mRNA. Messenger RNA is ap-

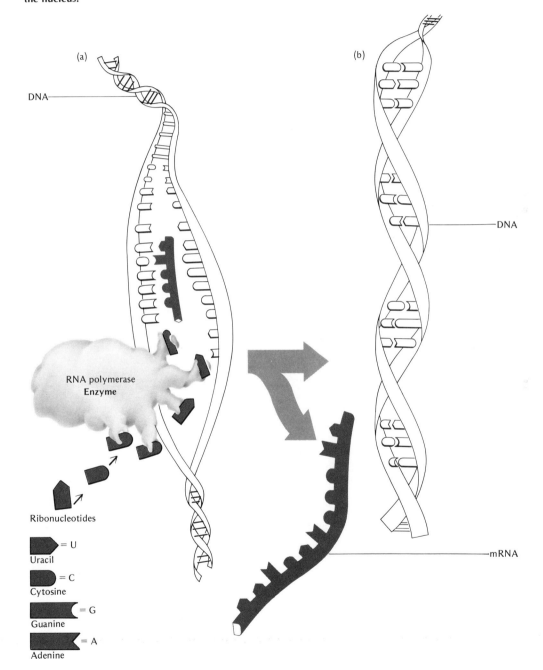

propriately named, for it not only copies the genetic code, but also moves out of the nucleus, where it was synthesized, and enters the cytoplasm. In the cytoplasm the mRNA attaches to a particle, the **ribosome.** The ribosome contains protein and a special kind of RNA called **r**ibosomal **RNA (rRNA).** The ribosome is the site within the cell where proteins are synthesized.

How is the coded message of the mRNA translated into a protein? It is clear that before a protein can be synthesized it is necessary to move the constituent amino acids into position, arrange them in the proper sequence, and forge the peptide bonds between adjacent amino acids so that a protein is produced. This cannot be accomplished by the amino acids themselves, since they will not bond directly to the mRNA; the process depends upon an adapter-type molecule that will pick up an individual amino acid and transfer it to the mRNA; this adapter is called **t**ransfer **RNA (tRNA).** Transfer RNA is a molecule of RNA which is twisted in such a manner that it forms a cloverleaf pattern (Figure 6–11). There are at least 20 kinds of tRNAs present in the cytoplasm, and each one binds to a specific amino acid at its free end (Figure 6–11). At the bend in the cloverleaf-shaped molecule of tRNA there is a specific triplet region of nucleotides which is complementary to a particular triplet (codon) of the mRNA; the specific triplet of the tRNA is called an **anticodon.**

We have introduced all the participants in the assembly of proteins; now let us see how they function. Each amino acid, free in the cytoplasm, is activated by ATP and an enzyme, and then this activated amino acid is transferred to one of the prongs of the cloverleaf-shaped tRNA, which is specific for that particular amino acid. The tRNA carries the activated amino acid to the ribosome, which is composed

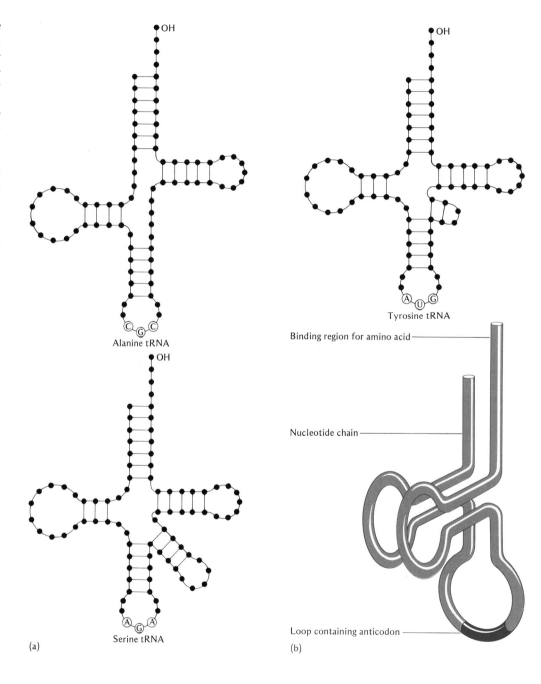

FIGURE 6–11 Transfer RNA, a single-stranded molecule that exists in the cytoplasm. (a) Cloverleaf pattern of three tRNAs from yeast showing the free end for binding to an amino acid and the bend with a region for binding to mRNA (anti-codon). (b) Three-dimensional representation of tRNA.

Alanine tRNA

Serine tRNA

Tyrosine tRNA

Binding region for amino acid

Nucleotide chain

Loop containing anticodon

(a)

(b)

FIGURE 6–12 The ribonucleic acids (RNAs) of the cell.

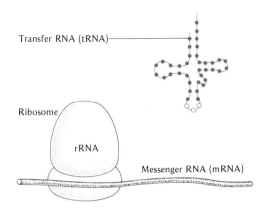

Transfer RNA (tRNA)

Ribosome

rRNA

Messenger RNA (mRNA)

of two globluar regions, one larger than the other (Figure 6–12). Protein synthesis actually begins when a single strand of mRNA attaches to the smaller region of the ribosome. By means of specific base pairing, the triplet region of the tRNA (anticodon) becomes attached to the complementary triplet region of the mRNA (codon) (Figure 6–13a). The tRNA, carrying its amino acid, is held in position on the larger portion of the ribosome; in this manner the ribosome acts like a jig or a vise, holding the mRNA in position and guiding the tRNA anticodon into proper alignment for coupling to the codon of the mRNA. The next step involves the movement of a second tRNA with its amino acid to the ribosome and the specific coupling of its anticodon region with the next mRNA codon. Thus, when the two adjacent codons of the mRNA are read, two tRNAs are present on the ribosome, properly aligned and attached to the mRNA and each bearing an activated amino acid. With the aid of two enzymes and an energy-rich molecule of GTP (similar to ATP but containing guanosine instead of adenosine), a peptide bond is formed between the adjacent amino acids, and the bonding of the first tRNA

to the mRNA and the amino acid is weakened (Figure 6–13b). The ribosome then moves along the strand of mRNA as the third codon is read, and the first tRNA, now minus its amino acid, becomes discharged from the mRNA (Figure 6–13c). A new tRNA carrying its activated amino acid can now attach to the third codon of the mRNA. The movement of the ribosome along the strand of mRNA produces a sequential reading of codon after codon. With each reading an amino acid is brought into position, a peptide bond is formed between adjacent amino acids, the tRNA in the first position is released, and as the next codon is read there is room for another tRNA with its amino acid to be bound to another mRNA codon further down the line. The free tRNAs move away from the ribosome and can be used over and over again to pick up other activated amino acids. By this one-by-one addition of amino acids, a long free polypeptide chain is formed (Figure 6–13d); its sequence of amino acids is determined by the codon sequence of the mRNA and the proper alignment of the complementary anticodons of the tRNA. In this manner, a nucleotide sequence of mRNA becomes the amino-acid sequence of a protein. This process is called **translation.**

The final one or two codons of the mRNA are **chain terminator** codons and signal the end of the message (Figure 6–7); the final amino acid in the polypeptide chain is released from the last tRNA, the tRNA is released from the mRNA, the ribosome detaches from the mRNA, and a complete free polypeptide chain results. Since molecules of mRNA are often very long (thousands of nucleotides in length), only a small portion of the molecule is in contact with the ribosome at any one time. Frequently other ribosomes attach at the beginning of the mRNA, so that the message is being read by several ribosomes in succession;

this creates a chain of ribosomes held together by a strand of mRNA, and such a group is called a **polyribosome** or **polysome** (Figure 6–14).

In summary, let us once again look at the genetic code and the synthesis of proteins. The hereditary information of all living cells is contained in molecules of DNA; a sequence of nucleotides in the DNA (a gene) carries all the necessary information for the fabrication of the gene product, a protein. A sequence of three nucleotides on the DNA strand codes for an amino acid—this triplet genetic code is specific and universal. In addition, the DNA molecule is self-duplicating and contains all the information necessary for making additional copies of itself. DNA is ordinarily restricted to the nucleus of the cell, and we can consider that the coded hereditary blueprint resides in a library of heredity, but the blueprint does not circulate. It is as if the master blueprint were so precious that it is kept confined so that it cannot be lost. In order for the coded message of the DNA to be put into operation, it is necessary to get the message to the cytoplasm of the cell, where the site of protein synthesis is located. Since the DNA is locked in the nucleus, the coded blueprint is copied, and only the copy of the blueprint leaves the nucleus to enter the cytoplasm. Messenger RNA copies a single strand of the DNA and faithfully transcribes the triplets that make up the genetic message. In the cytoplasm the mRNA attaches to a ribosome; another kind of RNA (tRNA) binds to a specific amino acid and carries it to the ribosome. At the ribosome the copied blueprint of the mRNA is read, and the message is translated into an amino-acid sequence. In this manner the DNA of the genetic code becomes transcribed into a molecule of mRNA and this in turn is translated into a polypetide or protein.

FIGURE 6–13 Diagrammatic view of peptide-bond formation and the role of the ribosome in translation. (a) A strand of messenger RNA (mRNA) attaches to the small subunit of the ribosome, and by means of specific base pairing the triplet region (anticodon) of the transfer RNA (tRNA) becomes attached to the complementary region of the mRNA (codon). The tRNA carrying an amino acid is held in position on the large subunit of the ribosome. A second tRNA bearing an amino acid then moves to the ribosome, and its anticodon region binds to the second codon of the mRNA. (b) A peptide bond is formed between the adjacent amino acids, and the bonding of the first tRNA to the mRNA is weakened. (c) The first tRNA minus its amino acid leaves the ribosome as the ribosome moves along the strand of mRNA. (d) Another tRNA carrying an activated amino acid can now attach to the third codon of the mRNA. Thus, a sequential reading of codon after codon is produced by the movement of the ribosome along the mRNA; with each reading an amino acid is brought into position by the tRNA, a peptide bond is formed between adjacent amino acids, the tRNA in the first position is released, and as the next codon is read, there is room for another tRNA bearing an amino acid.

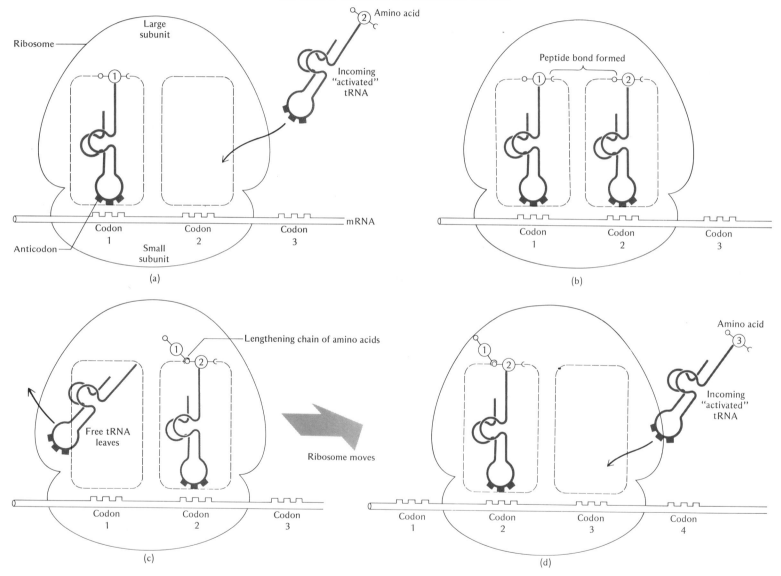

FIGURE 6–14 The ribosome: assembly site of polypeptide chains. (a) Electron microscope view of isolated ribosomes, with inset showing ribosome at higher magnification. (Courtesy of Masamichi Aikawa.) (b) Ribosomes move (in direction of the small arrow) along a strand of mRNA, reading and translating the message into a polypeptide chain. A chain of ribosomes held by a strand of RNA is called a polysome (polyribosome).

(a)

(b)

Complete polypeptide chain

Growing polypeptide chains

tRNA

Free ribosome about to attach to mRNA

Messenger RNA

Polysome

Free ribosome

6–5 IMPLICATIONS OF THE GENETIC CODE

So far we have suggested that the nucleotide sequences of DNA code for amino acids; via mRNA and tRNA amino acids are assembled in specific ordered arrangements dictated by the DNA and enzymatically linked to form proteins. Originally we restricted our discussion of genes and the genetic code to a specific group of proteins, enzymes, because it is in enzyme activity that gene function is most clearly observed. The one gene–one enzyme hypothesis of Beadle and Tatum is true, but as you will already have surmised, it should be modified and broadened in its scope to include more than proteins with enzymatic activity. Recent evidence indicates that a better statement of the genetic code would be one gene–one polypeptide chain. (Remember that a polypeptide chain is a subunit of a protein; Chapter 3.) Let us return to an example of a protein that is not an enzyme but that illustrates the nature of the genetic code and its importance in the function of the body.

There exists in human populations a variant of normal hemoglobin called sickle-cell hemoglobin. **Sickle-cell hemoglobin** differs from normal hemoglobin in only 1 amino acid among the more than 286 that compose the protein chain of hemoglobin. If we look at 5 amino acids in the normal hemoglobin sequence, we find the following:

-threonine-proline-glutamic acid-glutamic acid-lysine-

Looking at the same region in a person with sickle-cell hemoglobin, we find:

-threonine-proline-valine-glutamic acid-lysine-

The difference between the two hemoglobins is the substitution of valine for glutamic acid in sickle-cell hemoglobin. To see how such a small change produces such a profound effect, let us look at the nucleotide sequence in the DNA that codes for both these amino-acid sequences. The DNA code words for valine are CAT and CAC; those for glutamic acid are CTT and CTC. Essentially, the difference between valine and glutamic acid (as far as the nucleotide sequence in DNA is concerned) is a change in the middle nucleotide; when A is substituted for T in the triplet, glutamic acid is transcribed and translated as valine. It is apparent that one way of producing a genetic change (**mutation**) in a cell or an organism is by the simple exchange of one nucleotide for another. In Chapter 23 we shall discuss in some detail changes in our genetic constitution (mutations) and their effect on our body functions.

How do nucleotide substitutions occur? One type of error might be a typographical error. It is conceivable that during the replication of the nucleotide strands of DNA, when the strands separate, a rearrangement of the hydrogen atoms of the nucleotide might occur, resulting in faulty hydrogen bonds; this would permit binding of the wrong nucleotide in the new chain being synthesized. For example, instead of A binding T, perhaps because of defective hydrogen bonding it would bind C. When the next round of replication occurred, the chain with the C would pair with a G; the original parent strand would pair with a T. Thus, there would arise a DNA molecule in the next generation which had AT substituted by GC. Other errors might occur due to unnatural effects. For example, there is evidence that mutations can arise in double-stranded DNA as a result of exposure to x rays, ultraviolet rays, cosmic rays, and certain chemical agents such as nitrous acid; such agents alter either the base sequence (by cutting out a base) or the base itself.[3] Thus, mutational changes can be considered to result, at least in some cases, from changes in the nucleotide sequence of DNA. A major hazard of radiation (to be discussed in Chapter 23) is the hazard to the genetic code itself. As typographical errors rarely improve a sensible printed message, so too random changes in the base sequence are rarely favorable. Many mutations are lethal, and so we never see them and they are not perpetuated. If the mutations are not lethal, their favorable or unfavorable nature often depends on the environment and the genetic constitution in which the gene functions, so that a hard-and-fast rule is impossible.

An important implication of the Watson-Crick model for DNA and the mechanism of protein synthesis is that it tells us why men differ from mice, cats, dogs, and elephants. We manufacture human proteins and not any other kind because of the specific order of the nucleotides in our DNA. We may eat a variety of proteins in our diet, but after digestion and absorption, the building blocks of amino acids are fabricated within our cells into our very own proteins because each cell of our body contains very specific coded directions in the DNA. It can be said that the more similar the order of the bases, the more closely related are the organisms; in identical twins the DNA base sequences are all identical.

---

[3] Nitrous acid works as a mutagen by converting the pyrimidine cytosine to uracil. This means that in the DNA strands GC, for example, is converted to GU; when replication occurs one of the daughter strands will contain GC, but the other will contain UA. In a second round of replication this mutant strand will give rise to a daughter molecule with AT and UA. The error or alteration in bases is perpetuated.

6–6 GENETIC ENGINEERING: RECOMBINANT DNA

We said earlier that a cat is a cat and a dog is a dog because they respectively have cat DNA and dog DNA. What if one could combine dog and cat DNA? How would it be expressed? Would we create a new organism?

Recently, a new technique called **recombinant DNA** research has made it possible to splice the DNA sequences of organisms such as dogs and cats and to insert them into the DNA sequences of the bacterium *E. coli*. The foreign DNA is then replicated along with the bacterial DNA (Box 6C). The scientific community has been deeply divided about the risks and morality of such research (page 106).

The technique produces new organisms, essentially shortcutting nature. One scientist has claimed that recombinant DNA research irreversibly counteracts the evolutionary wisdom of millions of years. Certainly, if gene combinations for such diseases as bubonic plague, smallpox, typhoid, polio, and cancer were inserted into bacteria that were then let loose on an unsuspecting population, the results would be disastrous. It is worth remembering, however, that this same "evolutionary wisdom" gave us the gene combinations for these diseases in the first place! At the same time, most organisms are uniquely fitted to their environments, and to tamper with their genetic blueprints may invite disaster.

What then is the promise of recombinant DNA research?

First, by carving up the genetic material of particular species and inserting the genes, one at a time, into *E. coli*, it is possible to study the organization and expression of genes as well as the function of different genes, something that up to now has required long and costly breeding programs, most often with limited results. With understanding comes control, possibly of genetic diseases such as hemophilia and diabetes. We could construct, say, bacterial strains that produce massive amounts of antibodies, antibiotics or hormones, or we could develop bacteria that enable all plants to draw their nitrogen from the air rather than from costly fertilizers. Already, bacteria have been manufactured that can produce the human brain hormone, somatostatin. Somatostatin inhibits the secretion of pituitary growth hormone. The researchers who first isolated somatostatin needed almost half a million sheep brains to produce 5 mg of the substance, an amount that can be produced rapidly by only 2 gallons of altered bacterial culture.

With all this promise, what are the dangers?

Reengineered *E. coli* could escape from the lab and multiply in their normal environment—the human intestine. No one knows the effects such bacteria would have because, as far as we know, it has never happened. To minimize the risks, a crippled strain of *E. coli* has been developed for use in such research. This strain cannot live outside the lab or even in the presence of sunlight. It cannot survive in the human intestine or human serum, and is easily destroyed by household detergents. A new set of governmental guidelines has been developed which will, it is hoped, make it illegal to perform this research under any but the most stringent, safe conditions. To ban such research totally may be shortsighted: where would we be today, for example, if we had decided much earlier that there ought to be no research on communicable diseases because an epidemic might be let loose? Despite predictions of imminent disaster as a result of research on recombinant DNA, the fact remains that in laboratories over the past few years, billions of bacteria have received "foreign DNA" from viruses, fruit flies, frogs, and mammals, and no

"Andromeda strain" has yet been produced. Further, bacteria in nature have long been exposed to DNA from human cells, such as the lysed cells of our gut and those in decomposing corpses, yet there is no evidence that such natural recombinant DNA experiments have threatened our lives. Fortunately, nature tends to bury her mistakes; abnormal combinations usually fail to survive.

Legal, social, and moral implications aside, someone, somewhere, is bound to perform recombinant DNA research, just because it is possible. The prospect of a Nobel prize to a scientist, or the economic incentive for a drug company, may prove irresistible, making any moratorium ineffective. Further, we have no control over research in other countries, and to date there are no international regulations under consideration by any government. Provision of stringent regulations in the U.S. means that such research is encouraged to proceed with all due caution. Should someone, somewhere produce a new "Andromeda strain," we shall then perhaps be in a better position to cope with it.

6–7 PROSPECTS FOR THE REFABRICATION OF MAN

We have deciphered messages written in the DNA strands and can isolate single genes; now we are on the road toward learning to spell new words, words of our own creation—in effect, writing new genetic messages. Can we change the pattern of the machinery that predestines our being?

The idea of correcting a human genetic defect by introducing a new piece of DNA into a cell has been obvious on theoretical grounds and has been much discussed in recent years. The problem, which thus far has been almost insoluble, is to get that DNA in a position to

express itself—to get it unharmed to the site of action in the human cell, the nucleus. Direct injection of DNA would not work because enzymes in the human body would promptly destroy the injected materials. However, observations in the 1930s suggested that viruses might be used as packages and delivery vehicles. DNA of infecting viruses could become a temporary or permanent part of the cell's nuclear DNA, and thus a cell of a particular genetic constitution could receive additional genetic information. Can we induce cells to incorporate new genetic messages in this way? Could the DNA of a virus be so reconstructed that it could carry vital information, such as the code for the enzyme phenylalanine hydroxylase, the missing enzyme in PKU (Box 6A), and thus correct this hereditary disease? Some scientists speculate that this may occur in the near future. Viruses hold out tremendous potential for repairing or replacing defective genes and effecting one-shot cures of hereditary diseases.

At the same time, things are not as simple as they seem. Responsible scientists warn that the possibilities of genetic intervention (genetic engineering) in man are not without limitations. The number of diseases (for example, hemophilia, PKU, and diabetes) and normal traits (for example, eye color) that are controlled by a single gene (relatively short lengths of DNA) is small and forms a special class of the genetic spectrum. Most human traits are **polygenic;** that is, they depend on the interaction of many genes (long stretches of DNA) and they vary continuously rather than producing an all-or-none effect. For example, instead of a trait such as skin color being represented as white or brown, there exists a continuum of shades of tan between these extremes. Especially important is the fact that behavioral traits, such as intelligence and temperament, as well as physique are highly polygenic. Man probably has hundreds of thousands of genes for polygenic behavioral traits compared with the few hundred recognizable monogenically (single gene) controlled ones. Polygenic inheritance is difficult to study and control because so many combinations are possible and gene interactions are extremely frequent. Also it should be emphasized that, singly or in clusters, genes can determine the range of potential for a given trait in an individual, but past and present environments may determine the nature of its expression within that range. In other words, having a genetic trait does not assure expression of that trait; the conditions for its manifestation must be appropriate. (A simple example: although the cells of your skin have the gene for making hemoglobin, they do not manufacture it.)

There is another fly in the ointment when it comes to genetic manipulation of the human being; namely, that redirection often cannot be accomplished. For example, although genes direct the development of the brain and its billions of constituent nerve cells, the insertion of a new piece of DNA following the establishment of the brain could not alter that organization; neither could learning processes be altered by means of a shot of DNA because they depend so much on previous occurrences, such as experience and conditioning (Chapter 19). The transfer of genetic defects to produce debilitation for political or military purposes also seems more in the realm of science fiction than science. Most monogenic diseases produce behavioral aberrations whose usefulness to a tyrant is hard to conceive; even if such a malevolent end result is desired, it would seem that a genetic bomb could not easily and secretly be introduced to mass populations and affect already developed individuals.[4]

---

[4] It should be noted that any society wishing to direct

Sensational pronouncements about the dangers of genetic manipulation and the Promethean predictions of unlimited control are without scientific evidence; blueprinting a human personality is not on the horizon. Genetic intervention could be used in a limited way for improvement of the human condition, and while dangers may exist, it is necessary to evaluate and regulate them in a rational way.

SUMMARY

1. Genetics, the study of inheritance, deals with the transfer of information from parent to offspring.

2. Griffith found in 1928 that rough *D. pneumoniae* can be altered into a lethal strain by transforming principle from a killed smooth strain. Avery, McCleod, and McCarty purified transforming principle in 1944 and found it to be DNA (deoxyribonucleic acid).

3. DNA is universally responsible for recording and transmitting the genetic blueprint.

4. DNA is composed of deoxyribonucleotides, composed of a sugar (deoxyribose), linked to a phosphate and to one of four nitrogenous bases: adenine, guanine (purines), thymine, or cytosine (pyrimidines).

5. In 1950, Chargaff found that in any DNA: (a) the proportion of bases from any given species was always the same; (b) the composition varied between species; and (c) the amounts of A = T and G = C, and A + G = T + C.

---

the evolution of its genetic composition has an alternate solution to genetic engineering, and that is selective breeding. This method has been used for thousands of years to produce better stocks of animals and plants. It was the basis of the eugenics movement, which faltered because humans could not be induced to mate and breed by directive. Inbreeding, selection, and eugenics are discussed in Chapter 23.

BOX 6C Redesigning bacteria

The development of the recombinant DNA technique ushered in a new era of genetic engineering—with all of its promise and possible peril. The lowly organism that currently plays the largest role in the process is the *E. coli* bacterium. This microbe—a laboratory derivative of a common inhabitant of the human intestine—lends itself to being engineered because its genetic structure has been so well studied. In the first step of the process, scientists place the bacterium in a test tube with a detergent-like liquid. This dissolves the microbe's outer membrane, causing its DNA strands to spill out in a disorderly tangle. Most of the DNA is included in the bacterium's chromosome, in the form of a long strand containing thousands of genes. The remainder is found in several tiny, closed loops called plasmids, which have only a few genes each and are the most popular vehicles for the recombinant technique.

After the plasmids are separated from the chromosomal DNA in a centrifuge, they are placed in a solution with a chemical catalyst called a restriction enzyme. This enzyme cuts through the plasmids' DNA strips at specific points. It leaves overlapping, mortise-type breaks with "sticky" ends. The opened plasmid loops are then mixed in a solution with genes—also removed by the use of restriction enzymes—from the DNA of a plant, animal, bacterium or virus. In the solution is another enzyme called a DNA ligase, which cements the foreign gene into place in the opening of the plasmids. The result of these unions are new loops of DNA called **plasmid chimeras** because, like the Chimera—the mythical lion-goat-serpent after

which they are named—they contain the components of more than one organism.

Finally, the chimeras are placed in a solution of cold calcium chloride containing normal *E. coli* bacteria. When the solution is suddenly heated, the membranes of the *E. coli* become permeable, allowing the plasmid chimeras to pass through and become part of the microbes' new genetic structure. When the *E. coli* reproduce, they create carbon copies of themselves, new plasmids—and DNA sequences—and all. Thus they become forms of life potentially different from what they had been be-

fore—imbued with characteristics dictated not only by their own *E. coli* genes but also by genes from an entirely different species.

FIGURE 6–15 Gene transplantation in *E. coli* bacteria. (a) Plasmid is isolated from *E. coli* bacterium. (b) Plasmid undergoes enzymatic cleavage. (c) Foreign DNA segment is prepared. (d) Foreign DNA is inserted into plasmid. (e) The plasmid chimera is introduced into a fresh *E. coli* bacterium. (f) The "engineered" *E. coli* bacterium reproduces itself.

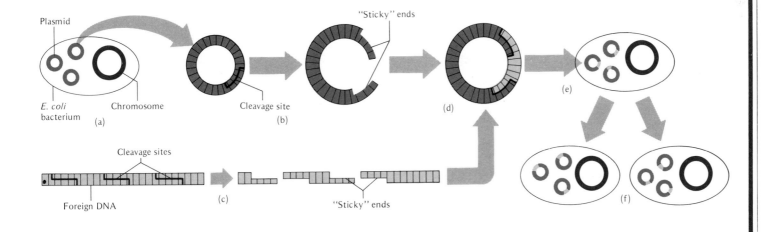

6. In 1953, Watson and Crick proposed that the nucleotide bases of DNA are arranged in a linear sequence. Each DNA molecule consists of two such sequences, and the bases of each strand bond with each other in a complementary way so that A always bonds with T and G with C. The bonded bases form the rungs of a ladder, with sugar-phosphate uprights. The entire DNA molecule is twisted helically at every 10th base.

7. DNA replicates itself by unzipping and matching complementary base pairs, which are bonded together by DNA polymerase.

8. Specific regions of the DNA helix code for, and direct, the synthesis of proteins. These regions are the genes.

9. A specific sequence of three bases codes for a specific amino acid. Each triplet is called a codon. The triplet code is degenerate because more than one triplet codes for the same amino acid. (There are 64 possible triplet combinations and 20 amino acids in most proteins.)

10. The code is universal—the same triplet codes for the same amino acid in all organisms.

11. Many diseases are genetic. Phenylketonuria (PKU) for example is caused by a fault in the genetic code for an essential enzyme that metabolizes phenylalanine in the diet. Build up of phenylpyruvic acid, derived from phenylalanine, causes mental retardation in the young. It can be prevented by controlling the diet.

12. The genetic message is contained in the DNA of the nucleus; proteins are synthesized in the cytoplasm. The genetic instructions in DNA are copied, or rather transcribed, into messenger ribonucleic acid (mRNA), which resembles DNA except that the sugar is ribose and uracil replace thymine. mRNA then travels to the cytoplasm.

13. Messenger RNA (mRNA) transcribes one strand of DNA by specific complementary base pairing, with the aid of RNA polymerase. The mRNA moves from nucleus to cytoplasm and attaches to a ribosome.

14. Transfer RNA is a cloverleaf-shaped molecule in the cytoplasm. Each of the 20 amino acids is carried by a specific kind of tRNA molecule. At the bend of the cloverleaf of tRNA is a triplet region of nucleotides (anticodon) which is specifically complementary to a triplet region (codon) of the mRNA).

15. Activated by ATP, cytoplasmic amino acids are picked up by the prongs of their specific cloverleaf tRNA, carried to the mRNA on the ribosome where the anticodons match and bond with their complementary codons. Peptide bonds are forged enzymatically between adjacent amino acids on the ribosome, which moves along the strand of mRNA "reading" the codons. This process is called translation. Following translation of the genetic message into polypeptide chain, the tRNAs are discharged and can be reused.

16. The final codons of the mRNA are chain terminators and signal the end of the message and of the polypeptide chain.

17. Long strands of mRNA may be read by more than one ribosome at once. Such associations are polyribosomes, or polysomes.

18. One gene codes for one polypeptide chain.

19. The sequence of triplets (codons) in the DNA determines the sequence of codons in the RNA and hence the kind of protein made. Species specificity and individuality are determined by codon sequencing.

20. Changes in the genetic message, or mutations, occur when nucleotide base substitutions or deletions occur.

21. Recombinant DNA research, or gene splicing, has given us the capacity to add a small number of genes from any organism to bacterial cells (*E. coli*), and hence to reproduce them in large quantities. The potential for solving biological and medical problems is offset by the novel risks presented by this research and has led to the development of government guidelines.

22. Prospects for changing the genetic code of organisms other than bacteria are remote since most traits are polygenic, and it is difficult to change an organism once it is fully developed. Correction of some genetic diseases may be accomplished in the future by using viruses as delivery vehicles for the insertion of genetic messages into the nuclei of human cells, however.

KEY WORDS

genetics
transforming principle
deoxyribonucleic acid (DNA)
diffraction pattern
periodicities
nucleotides
deoxyribonucleotides
deoxyribose
adenine
guanine
thymine
cytosine
purines
pyrimidines
complementary base pairing
DNA polymerase
sequence
replication
code
genes
one-gene–one-eyzyme hypothesis
triplet code
degenerate code
essential amino acids
ribonucleic acid (RNA)

ribose
ribonucleotide
uracil
messenger RNA (mRNA)
RNA polymerase
genes
transcription
codons
ribosome
ribosomal RNA (rRNA)
transfer RNA (tRNA)
anticodon
translation
chain terminator
polyribosome or polysome
sickle-cell hemoglobin
mutation
recombinant DNA
plasmid chimeras
polygenic

TOPICS FOR REVIEW AND DISCUSSION

1. How is DNA able both to duplicate itself and to carry information?
2. How did transformation in bacteria lead to the discovery that DNA is the hereditary material?
3. What is the hereditary basis of sickle-cell hemoglobin?
4. Discuss the validity of the statement: One gene–one enzyme.
5. What is the relationship of codon to anticodon?
6. Explain the significance of the universality of the genetic code.
7. Describe the cause of phenylketonuria (PKU) and how it may be "cured."
8. During gene expression, the genetic code is transcribed and translated. Describe these molecular events.
9. What is genetic engineering?
10. What is the Watson-Crick model for the structure of DNA? How can it be confirmed experimentally?
11. Name two different kinds of RNA and list their functions.
12. Describe the role of polyribosomes in the manufacture of protein.

Cloning: babies to order?

During the last few years, science fiction has explored a new theme —human cloning. The infiltration of modern Europe by clones of Hitler and the cloning of a powerful and wealthy industrialist are just two examples of the scenarios produced by this theme. What is human cloning? Is it based on fact, and if so, what are the implications for the future?

Reproduction begins with the fertilization of the female egg by the male sperm. The sperm and the egg each contain only one set of chromosomes; following the union of sperm and egg, the offspring is endowed equally by both parents and equipped with a complete double set of chromosomes. But there is another possible way for life to begin. Because the nucleus of every body (somatic) cell derived from the fertilized egg has a complete double set of chromosomes, the nucleus of every body cell contains all the genetic information required for the formation of an entire individual. If body cells could be isolated and made to divide and develop, asexually produced offspring with only one parent would result. This procedure is already being used successfully with other species (plants, fruit flies—and more significantly, frogs); it is called *cloning*.

More than a decade ago at the University of Oxford, J. B. Gurdon produced a clonal frog. Taking an unfertilized frog egg, he destroyed its nucleus with ultraviolet radiation and replaced it with the nucleus of an intestinal cell from another frog. Suddenly finding itself with a double set of chromosomes, the egg began to divide and develop, tricked into starting the reproductive process. The egg developed into a genetic twin of the frog from which the intestinal cell nucleus was taken. The baby frog inherited nothing from the "mother" frog, though, since her ability to pass on her traits was eliminated along with the nucleus of her egg.

Would such a technique work with humans? Some believe so. An egg could be taken from a female and its nucleus could be destroyed and replaced with a nucleus from a donor whom we'll call "J." The egg would then be reimplanted in the uterus of another woman. Even though its nucleus had been destroyed, it could nevertheless divide because it had received the proper start —a full set of chromosomes. The child resulting from that process would have only one parent—"J"—and would be, in fact, a carbon copy of "J" but a generation removed.

What might the consequences of human cloning be? Widespread human cloning could revolutionize human society. Families would no longer exist and sexuality would be divorced from reproduction. The male sex might even become redundant. Parenthood would be completely altered as a concept and actuality. Human diversity, presently insured by sexual reproduction, could be the first casualty. Entire communities of people might look exactly alike and have the same potential. "Clones and clonishness" could replace nationality and race. Would everyone be allowed to replicate, or would society choose to permit only its most "valuable" members to clone? Who would set the parameters for this activity?

When a single cell—a fertilized egg or any other body cell— divides, the genetic makeup of cells derived from it is identical to that of the original parent cell. The manner by which genetic information is transferred undisturbed is the subject of this chapter, and if human cloning is in the foreseeable future, it is imperative that we understand the nature of the process.

7

CELLS IN ACTION: DUPLICATION AND CONTROL MECHANISMS

7–1 CELL DUPLICATION

In the last chapter, we saw how a molecule of DNA is able to reproduce itself, and how it contains within it information for directing the synthesis of proteins. Now we shall see how the DNA is packaged within the cell and how synthesis of DNA and of protein is involved in cellular reproduction—one of the fundamental properties of life.

Mechanics of cell division

It is a paradox of nature that living things change, yet stay the same. Viewed over a life span, we can recognize that we grow old and die, but on a shorter time scale, say minutes or hours, we notice hardly any difference in our appearance—we are the same. Or are we? Critical examination shows that we are in fact not the same as we were an instant ago. Our bodies are in constant activity, and change is unending. The cells of which we are composed are restless, they reproduce, they grow, they specialize, and they die. Our cells must work hard to maintain the status quo, for they must replace those that are lost from wear and tear. No cell lives forever, and every second that passes sees the death of some 50 million body cells; during that same period 50 million new cells are produced to take their place. Cells, like people, vary in the duration of their life span; an intestinal cell lives for 36 hours, then it dies and passes out of the body; a white blood cell lives for about 2 weeks, a red blood cell for about 4 months, and a nerve cell for 60 years or more.

What kind of delicate balance operates within our bodies to preserve our integrity? How is it that the body's active cells, which may number in the millions of millions, are capable of regulating themselves? And how is it that every so often some cells lose their

FIGURE 7–1 Mitosis in the living cell. Shown here are
stages in the reproduction of cells from the African
blood lily. (Courtesy of Andrew S. Bajer.)

regulatory controls and, no longer subject to
ordinary restrictions, become lawless cancer
cells?

Let us consider our cellular origins and see
how we became what we are today—an organized population of diverse cells. We began
life as a single cell, and through a series of
dramatic events, that cell became divided into
many smaller cells. Later, by utilizing externally
supplied raw materials, the cells grew in size,
again they multiplied, and soon there were
thousands of cells; as the process continued
cells propagated their kind, and eventually
there were millions of cells: the cells became
organized into distinctive nerve cells, skin cells,
liver cells, and so on. How did the original
cells recognize that multiplication was their
role? When did the cells receive the signal to
stop dividing continuously, and instead to just
keep pace with cellular death? Why and how
did our cells specialize? Before we can answer
questions such as these, we must understand
a basic activity of the living cell—its reproduction. Cells reproduce themselves by dividing.
Cell division is a commonplace occurrence, in
that it happens with great frequency; however,
the process is intricate and beautiful to observe.
As we shall see, the division of a cell is a
remarkable event.

As the living cell prepares to divide, the
nucleus, which is ordinarily translucent, begins
to cloud up, and within it distinct threadlike
bodies called **chromosomes** appear (Figure
7–1). The nuclear membrane disintegrates, the
chromosomes, which are now free, split lengthwise, and one set migrates to one pole of the
cell while the other moves to the opposite pole.
Each set then is reestablished within a nucleus,
and the cytoplasm divides producing two new
daughter cells, each with a full set of chromosomes.

Thus, the division of the cell involves a
complex pattern of events and is not merely a

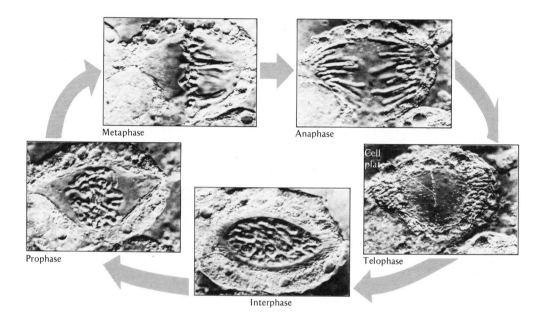

Metaphase

Anaphase

Prophase

Interphase

Cell
plate

Telophase

mechanical fragmentation of the parent cell to
form daughter cells. The process can be looked
upon as a cycle of events or steps, each of
which depends or is conditional upon what
precedes it. The entire process is, for the sake
of convenience, divided into stages, but it must
be remembered that these stage designations
are somewhat arbitrary, and during division
one stage merges into another to form a continuum. We can compare the cycle of cell
division to a dramatic play that is divided into
a series of scenes or acts to focus attention on
a particular situation. A detailed view of cell
division, like a view of a continuously running
play, can begin at any time, but for simplicity
let us begin with an undividing **(resting)** cell.
This cell, like most cells, is in fact doing all
sorts of things, for in the living state cells are
never at rest; the only thing the cell has not

begun to do is divide. Such a cell is described
as containing an **interphase nucleus;** that is,
the nucleus is between the phases of activity
that make up cell division. At this stage, the
nucleus is large and nucleoli are evident. The
internal contents of the nucleus show a fine
network of stainable material, called **chromatin**
(*chrom*—color; Gk.), but there is little discern-
ible structure to it (Figure 7–2). Near the nucleus
is a small, deep-staining body called the **centriole.** The centriole is often not visible in the
living cell. The interphase nucleus of the resting
cell is not inactive, but quite the opposite; it
is during this stage that protein and nucleic-
acid synthesis occur. Interphase may be looked
upon as a preparation stage for the process of
cell division, and it occupies the major portion
of the cycle of cellular reproduction.

When the cell enters the initial stage of cell

FIGURE 7–4 Coiling and uncoiling of chromosomes. The appearance of the chromosome depends on the degree of coiling; in the wire model shown, the strand is more apparent with the larger coil diameter and fewer turns (right).

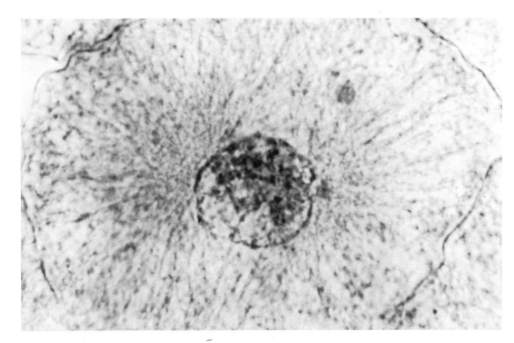

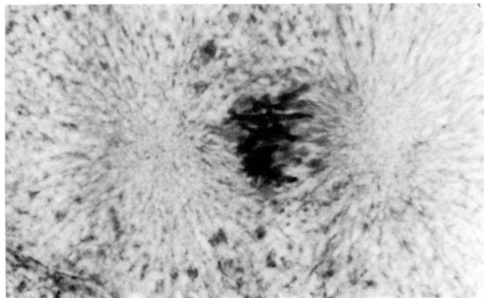

division, called **prophase** (pro—first; L.), the dynamic processes within the nucleus become visible (Figure 7–3). The chromatin material undergoes a striking change in organization. The diffuse network of chromatin forms dense threads, and the nucleus begins to resemble a carelessly rolled ball of yarn; the chromatin is not a single continuous thread, but consists of discrete and separate threads that are easily stained with a variety of dyes. These are the chromosomes ("colored bodies"; chroma—color, soma—body; Gk.). The increasing visibility of the chromatin and its distinct longitudinally double appearance in the chromosome are due to the coiling and consequent shortening of the chromatin threads. You can visualize the process that is occurring in the nucleus by thinking of the chromatin as a wire coiled into the shape of a bedspring (Figure 7–4); as it goes from an elongated, barely visible entity to one that is short and thick, it becomes clearly visible. The larger the diameter of the coil, the more infrequent the number of turns, and the more apparent the chromosome becomes. Why does the chromosome coil and shorten during cell division? One suggested reason is that in the shorter, more compact form the chromosome can be moved about more easily without becoming hopelessly tangled and snarled. Now we might turn the question around and ask: Why does the chromosome not remain coiled even during interphase? Again the answer is speculative: it could be that the synthetic activity of the chromosome is directly related to the amount of exposed surface, and since during interphase synthetic activity is maximal the uncoiled form, presenting the greatest exposed surface, is its

FIGURE 7–5 (top left) Chromosome structure. When magnified with the light microscope, a chromosome is seen to consist of two threads called chromatids, held together by the centromere. It is believed that each chromatid is made up of a helically wound DNA molecule and protein. (From *DNA and Chromosomes* by E. J. Du Praw. Copyright © 1970 by Holt Rinehart and Winston, Inc. Reprinted by permission of Holt Rinehart and Winston, Inc.)

FIGURE 7–6 (bottom) The cell in metaphase. (Whitefish blastula photograph courtesy of Carolina Biological Supply Company.)

functional condition. Careful inspection of the chromosome shows that each actually consists of two spiral filaments, called **chromatids,** closely associated along their length but not actually fused (Figure 7–5); the chromatids are held together at a specific region, the **centromere.** During prophase the originally distinct nucleoli of interphase disappear, and the nuclear membrane also breaks down and disappears.

Now there is the appearance of activity within the cytoplasm. The centriole divides into two daughter centrioles, and these separate. Radiating outward from each centriole are protein fibrils, called **asters** or **astral rays,** and between the centrioles there is another group of fibrils, the **spindle,** arranged in the shape of two cones lying base to base. The centrioles, along with their asters, migrate in opposite directions describing a semicircle, until they reach the poles of the cell. With the dissolution of the nuclear membrane, the chromosomes lie free in the center **(equator)** of the cell and orient themselves in a characteristic manner in this location. The appearance of the spindle and the positioning of the chromosomes on the equator of the cell[1] begin the stage called **metaphase** (*meta*—middle; Gk.) (Figure 7–6). The metaphase chromosomes are clear and distinct, and at this point it is quite easy to count the number of individual chromosomes. Ordinarily, to make certain that the chromosomes remain arrested at this stage so that they can be counted easily, the dividing cell is treated with the drug colchicine. This prevents the aggregation of the spindle proteins, and without a spindle the chromosomes are incap-

---

[1] Actually it is the centromeres that are precisely positioned on the equator, while the chromosomal arms may extend in any direction (Figures 7–1 and 7–6).

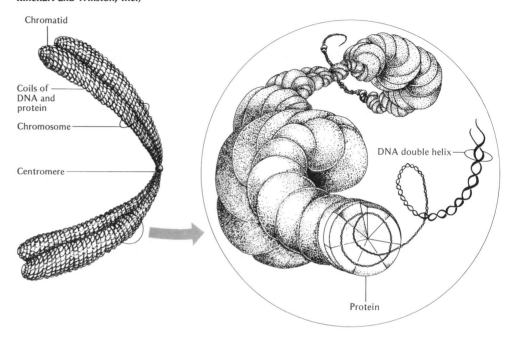

Chromatid

Coils of DNA and protein

Chromosome

Centromere

DNA double helix

Protein

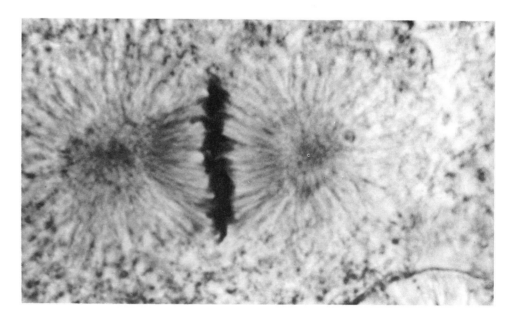

FIGURE 7–7 (top) Chromosomes of the human female. (Courtesy of W. R. Centerwall.)

FIGURE 7–8 (bottom) The cell in anaphase. (Whitefish blastula photograph courtesy of Carolina Biological Supply Company.)

able of moving away from their position on the cell equator. Such a colchicine-arrested condition of the metaphase chromosomes from the cells of a human female is shown in Figure 7–7. Notice how each centromere differs in its location along the chromosome, conferring a distinctive appearance upon each chromosome; thus, each chromosome can be recognized by the length of its chromatid arms on either side of its centromere.

The metaphase chromosomes, lying on the equatorial plate of the cell, become connected to the spindle fibers at the centromere. Each centromere divides, and as it does so the fibers of the spindle contract and the chromatids separate and move toward opposite poles of the cell. The separate chromatids, each with its own centromere, are now called **daughter chromosomes,**[2] and their separation signals the start of the stage called **anaphase** (*ana*—further or later; Gk.).

Anaphase chromosomes, dragged through the cytoplasm by the contracting spindle fibers, often appear as rods or are V or J-shaped. The shape each chromosome takes is related to the position of the centromere along its length and is produced as a passive response to the contractile force of the spindle. The situation is much like the arm movement of a string-manipulated puppet. As the puppeteer pulls on the string attached to the puppet's elbow, the arm moves and is bent into the shape of a V, whereas a tug on the string attached to the wrist of the puppet moves the arm as a stiff rod; the shape assumed by the arm is a passive response to the pull of the string. At the end of this chromosomal dance, two dense groups of chromosomes are located at opposite ends of the cell (Figure 7–8); this concludes anaphase.

---

[2] In determining the number of chromosomes, we count the number of centromeres present in the nucleus, not the number of strands.

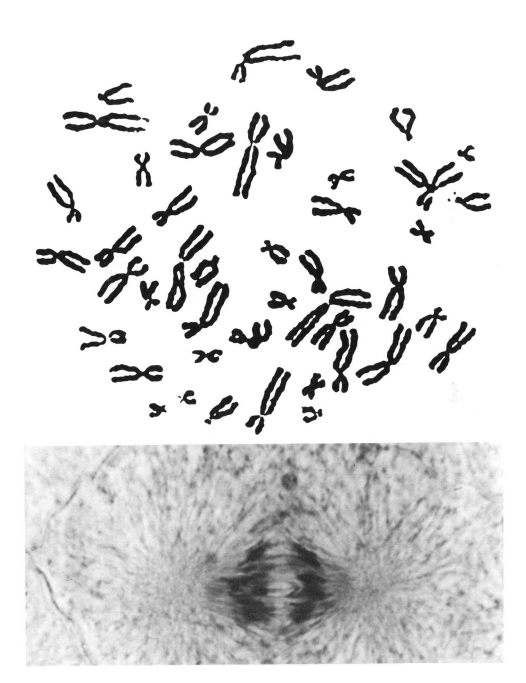

The next phase, **telophase** (*telo*—final; Gk.), begins when the movement of the chromosomes ends. During this phase (Figure 7–9), the nuclear membrane is reconstructed from endoplasmic reticulum remnants, the chromosomes uncoil to become slender chromatin threads, the spindle disintegrates, and the nucleoli appear at specific regions along the length of certain chromosomes (**nucleolar organizer** regions). The events of telophase are essentially the reverse of those that took place during prophase, but there is an added dimension; segmentation and separation of the cytoplasm (**cytokinesis**) occur. The cytoplasm constricts along a plane perpendicular to the spindle, and a process known as **furrowing** (or **cleavage**) divides the cytoplasm into two as if the cell were being constricted by a tightly drawn string around its equator.

This process of cell division, or strictly speaking, nuclear division, is called **mitosis;** it comes from the Greek word for "thread" (the prefix *mito* is already familiar to you from the name of the threadlike body of the cytoplasm, the mitochondrion or powerhouse of the cell). The precise mitotic ballet of the chromosomal threads assures equal distribution of nuclear materials between the daughter cells, and the chromosome number is reproduced with great fidelity. The cycle of division begins in a parent cell with a certain number of chromosomes, and the cycle ends with two daughter cells having exactly the same number of chromosomes as the parent cell did (Figure 7–10).

We have been describing mitosis as it occurs in animal cells. The process in plant cells is basically the same; the differences are related to the fact that plant cells differ from animal cells in certain aspects of their anatomy. For example, plant cells lack a centriole but do form a spindle. Obviously a centriole as a visible entity is not an absolute requirement for the organization of the spindle fibers. Plant

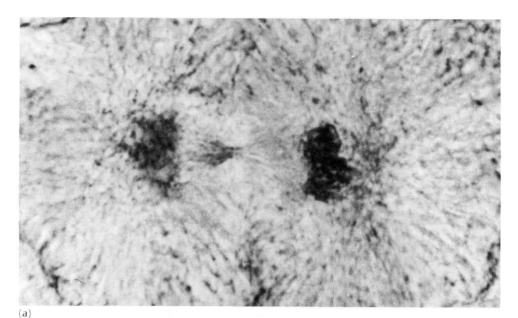

(a)

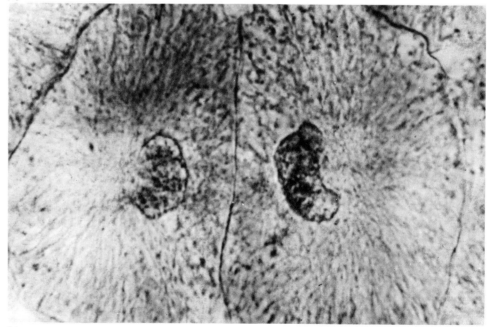

(b)

cells are surrounded by a rigid, nonliving cellulose cell wall, and therefore furrowing is impossible. Instead, in telophase of plant cells, the division of cytoplasm occurs by the laying down of a cell plate between the daughter nuclei (Figure 7–1).

Timing of cell division

The division of a cell during mitosis is like a continuously running play. How long does it take for the cell to perform a complete cycle of division? The answer to this question is: Quite variable, depending on the particular cell type. Some cells, such as nerve cells and muscle cells, never divide again once they are formed; some bacteria divide every half hour. Furthermore, the cells of the human body may divide at different rates at different times. For example, if a portion of your skin is removed by accident or surgery, the remaining cells begin to divide rapidly to replace the lost tissue; at other, more normal times, the division rate of skin cells is much slower. A human connective-tissue cell takes about 20 hours to undergo mitosis. It is evident from Figure 7–11 that most of a cell's time is spent in interphase (90%) and it may take as little as 45–90 minutes for the cell to go from prophase to telophase of one division. Interphase, the longest phase of a cell's life, can be subdivided into three periods: a gap (G_1) after telophase, a time of DNA synthesis (S), and then another gap (G_2) just before prophase commences. Exactly what goes on in the nucleus during the gap phases is obscure; however it is presumed that those metabolic events which must precede prophase occur during these times.

Implications of cell division

Cells, like humans, progress through a series of transitions from birth until death. Cells di-

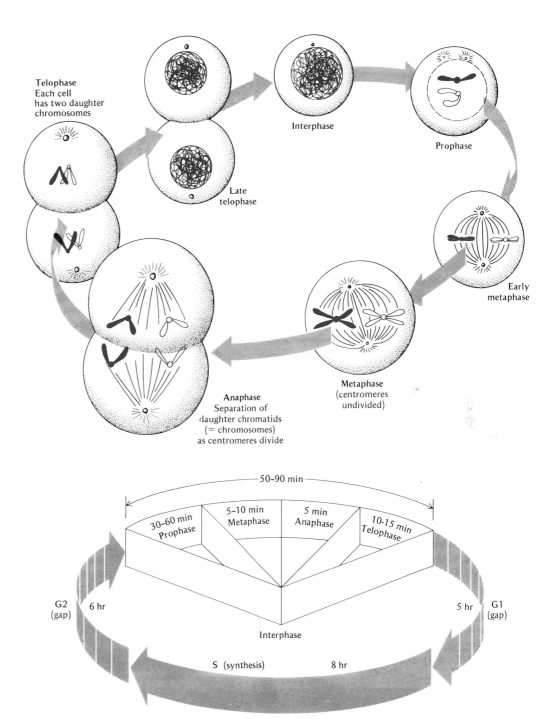

FIGURE 7–10 (top) Diagrammatic representation of mitosis in an animal cell.

FIGURE 7–11 (bottom) The cell cycle in a human connective tissue cell.

vide, grow, fabricate new materials from the raw stuff of their environment, and divide again; in this way they obtain a freshness and vitality—their substance is constantly renewed. As soon as a cell reaches a certain age, it divides and begins again; the aging process is thus avoided or diminished. A cell that retains the ability to divide is always young; through this mechanism, life to a degree thwarts time. In contrast, cells that do not divide eventually become senescent and die.

The division of cells and their increase in numbers and size by growth provide the raw material for differentiation and division of labor within the organism of which they are a part. The cells of our bodies are grouped into units called tissues and organs, which perform varied but specialized functions. The differentiation of a cell toward a specific role is the beginning of its eventual death, since the most highly differentiated cells do not divide; a return to youth and cell division may come about only by a loss of the differentiated state (called dedifferentiation). The paradox of cellular life is that youth belongs to the generalized cell, but the specialized cell, which is probably much more efficient for its job, is destined to die.

Mitosis is the mechanism whereby continued genetic similarity of cells is assured. Although mitotic mistakes do occur, they are rare. For example, if one or more chromosomes did not attach to the spindle for some reason, such as loss of the centromere, it might lag behind during division and end up in the wrong daughter cell. The centrioles might divide abnormally, producing abnormal division centers. As a result of these mitotic mistakes, the daughter cells would contain unequal numbers of chromosomes, which could lead to death or to abnormal patterns of cellular behavior. Some human disorders, such as Down's syndrome

(mongolism) are produced by the unequal distribution of particular chromosomes during the cell divisions that are preliminary to formation of the egg cell. If mistakes were a part of the natural process of cell division and the products of division all varied in their constitution, our bodies would consist of a random assortment of cells. By assuring the similarity of inherited characteristics in each and every cell derived from the original fertilized egg, it is possible to construct organisms like ourselves, which have both uniformity and specialty in function. The mechanism by which cells of uniform genetic character become differentiated is poorly understood and remains one of the most intriguing of all biological problems. We shall speculate on some aspects of the situation later in this chapter.

7–2 CHROMOSOME STRUCTURE

Packaging the DNA

Chromosomes contain DNA, but they also contain proteins called **histones** and **protamines**.[3] The proteins play a role in the organization of chromosome structure and function, but do not carry hereditary information; that function is restricted to the DNA. As we saw in Chapter 6, the Watson-Crick model for DNA shows that it is a double helix, and the genetic code is embodied in the nucleotide sequence of the DNA. The problem that now confronts us is: How are the DNA and protein organized in the chromosome?

It has been shown that the chromatid consists of two complementary strands, suggesting that a chromatid is in fact a chain of DNA. However,

---

[3] In prokaryotic cells such as bacteria, the chromosome contains only DNA.

the matter is not as simple as this. Some indication of the packaging problem is evident from Figure 7–12, and the statistics in Table 7–1. The nucleus of a human cell contains about 3 billion nucleotide pairs distributed among 46 chromosomes. Thus each chromosome, composed of two chromatids, contains an average of 60 million nucleotide pairs, and the double helix twists once for every 10 nucleotide pairs—6 million times! If a chromatid consisted of a single linear chain of DNA, then the DNA chain of a virus would be 60 μm long, that of a bacterium would be 1 mm in length, and that of a man would be almost 1 in. long. The total length of the DNA chains in each cell of a human being would be about 1 yd! When we consider that all this DNA must be packaged into cells with diameters many times smaller than the length of the DNA chains, we can begin to appreciate the nature of the problem (Figure 7–12). It seems unthinkable that DNA strands inches long could ever uncoil, as they must each time transcription or replication occurs. Furthermore, when fully extended, each chromatid is 100 times thicker and only one-ten thousandth as long as it ought to be if it consisted of a linear DNA chain. How then are DNA chains packaged into chromosomes?

Staining methods and light microscopy have revealed that each chromosome thread consists of transverse alternating dark and light bands, which presumably represent the many genes arranged along its length (Figure 7–13). It can also be seen that the strand of material forming a chromatid is very tightly wound in a helical coil, which may wind up on itself like a coiled telephone cord twisted into a series of secondary coils (Figure 7–5). This would in some measure account for the amount of DNA that can be accommodated in such a small space as that afforded by the nucleus of a cell. It is

FIGURE 7–12 Extracted DNA from a variety of organisms compared to scale with the structure into which the DNA is organized. (a) A virus (60-μm DNA). (b) A bacterium (1-mm DNA). (c) A fruit fly (16-mm DNA).

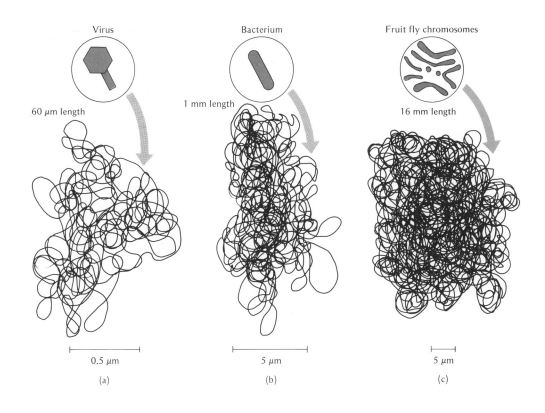

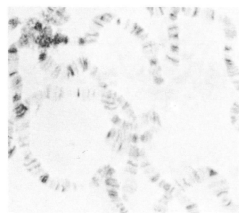

Virus

60 μm length

0.5 μm

(a)

Bacterium

1 mm length

5 μm

(b)

Fruit fly chromosomes

16 mm length

5 μm

(c)

impossible by light or electron microscopy to see how the DNA and protein molecules are arranged in a chromatid; therefore, we are uncertain of their exact organization, but biologists with considerable ingenuity have proposed models that seem to solve the packaging problem. Any proposed model of the chromatid must be reconciled not only with the previously described statistics and observations, but with the molecular organization of the DNA as suggested by the work of Watson and Crick.

One rather speculative model of the structure of the chromosome and chromatid duplication is shown in Figure 7–14. According to this model, a chromosome consists of a long, tightly wound coil of DNA packed in protein (Figure 7–14a). At duplication the DNA helix unwinds, permitting synthesis of a complementary DNA strand and formation of sister chromatids (Figure 7–14b). The sister chromatids coil to form a visible chromosome at metaphase (Figure 7–14c). At anaphase the sister chromatids separate, becoming the chromosomes distributed to the daughter cells during telophase. The

TABLE 7–1 The genes in different organisms

Organism	DNA nucleotide pairs per cell	Number of genes (estimated)
φ 174 (virus)	5,500	12
T4 (virus)	2×10^5	450
Escherichia coli (bacterium)	6×10^6	13,300
Drosophila (fruit fly)	6×10^6	13,300
Frog	7×10^{10}	120,000,000 (?)
Man	3×10^9	6,000,000*

* Although 6,000,000 are possible, only 1,487 genetic traits have been catalogued. From E. Carlson, *The Gene: A Critical History*, Philadelphia, Saunders, 1966, table 1, p. 268.

FIGURE 7–14 (facing page) A model of chromosome duplication. (a) Before replication the chromosome consists of a long, coiled double helix of DNA surrounded by protein. (b) During DNA synthesis (S phase) the DNA helix unwinds, permitting replication and formation of sister chromatids. (c) At metaphase the sister chromatids coil to form a visible chromosome. (Modified from *Cell and Molecular Biology* by E. J. Du Praw, New York, Academic Press, 1968.)

sequence of events would begin once again during interphase.

Implications of chromosome structure

As we have seen, the hereditary materials or **genes** of all organisms are organized into structures known as chromosomes. What exactly is a gene and what is the significance of the organization of genes into chromosomal structures? Let us say that a gene is a nucleotide sequence of the DNA which codes for a specific protein product. The number of genes an organism has is to some extent related to the complexity of the organism. The number of genes of a variety of organisms is represented in Table 7–1. It is obvious that a small virus has fewer proteins, less DNA, fewer nucleotides, and fewer genes than a larger and more complex virus. Similarly, the intestinal bacterium *Escherichia coli* has about 20 times as many genes as a complex virus, and when we come to our own DNA we see that we have enough DNA in each of our cells to provide for 6 million genes. Are we really a thousand times more complex than a bacterium? Probably we are, but notice also that DNA content and degree of complexity do not always go hand in hand; for example, the cells of the fruit fly *Drosophila* have the same number of nucleotides as those of *E. coli,* and the cells of some frogs have 20 times the DNA content of human cells.

How many genes do each of our cells contain? To approach the solution to this question, we could catalog all the human traits that have been shown to be controlled by a single gene; a recent catalog of such traits indicates the number to be 1,487. Of course, this value is the lower limit for the number of human genes, since there are a great many genetic traits that are undescribed; estimates for the total number range from 5,000 to 15,000 (with little good evidence for either). Thus, there is a sharp discrepancy between the potential number of genes a human cell could have (6 million) and the number it may actually have (thousands); such a discrepancy indicates that we have a great deal of extra DNA in every cell, and to some biologists this suggests that a considerable fraction of our DNA is redundant, consisting of duplicated genes. The functions of all of these duplicated messages have as yet been worked out in only a few instances.[4] The whole mechanism is highly complex, and discussions of genetic engineering and suggestions for manipulation of hereditary material involving the insertion of linear stretches of DNA into cells may be further away than the headlines in newspapers would have us believe. We must not only work out the mechanism by which genetic insertions can be made, but determine how to eliminate specific portions of the message already in the cell. These and many other problems will have to be solved before a special-to-order human can be manufactured.

The calculation that there are 6 million genes in every human cell, or even the conservative estimate of thousands of genes, makes it plain that if each gene were localized in a discrete particle, and each had to segregate at cell division, a haphazard and chaotic state would be produced every time a cell divided. The organization of genes into a linear array on the chromosome (Figure 7–13) minimizes the number of segregating units, increases the efficiency of segregation by diminishing the possible loss or gain of individual genes, and permits gene interaction. This is not to say that the process would be impossible with thousands of individually separate gene bodies, but it would probably be much less efficient. A cell obtains many advantages from its structural organization, including the organization of genetic material into chromosomes; a more independent and less organized condition of the genetic material would be less advantageous for the cell and the organism.

7–3 CONTROL OF CELLULAR ACTIVITY

Seeing gene activity

It has been said that we are what our enzymes make us. Each of us is a product of the subtle activities of the many enzymes that control our cellular processes. Ultimately these cellular activities depend on the nature of the genetic material and its manner of expression. As we have seen, the genetic material is composed of nucleotide units of DNA—genes—which are organized into discrete nuclear structures called chromosomes. Therefore, it seems logical to look for evidence of the activities of genes in or on the chromosomes. Can we with our own eyes directly observe the activities of submicroscopic entities such as genes? The answer, quite surprisingly, is yes.

We are fortunate that there exist in nature certain organisms with cells that have giant chromosomes, for without such structures it would be impossible to see gene activity directly. Our own chromosomes are much too small for studies of chromosome function, and even the electron microscope with its increased powers of resolution is of no help. However, in the cells of the salivary glands of certain flies, **giant chromosomes** are formed by the repeated division of chromatids without nuclear division. After 9 or 10 multiplication cycles, there may be up to 2,000 chromatids.

---

[4] For example, in the fruit fly there are 130 copies of the genes involved in making the ribosomes; in the chick there are 100 copies; in a salamander 1,000 copies, and in *E. coli* 5 copies. Similar redundancy has been found for transfer RNA, where there are 13 copies in the chromosomes of the fruit fly.

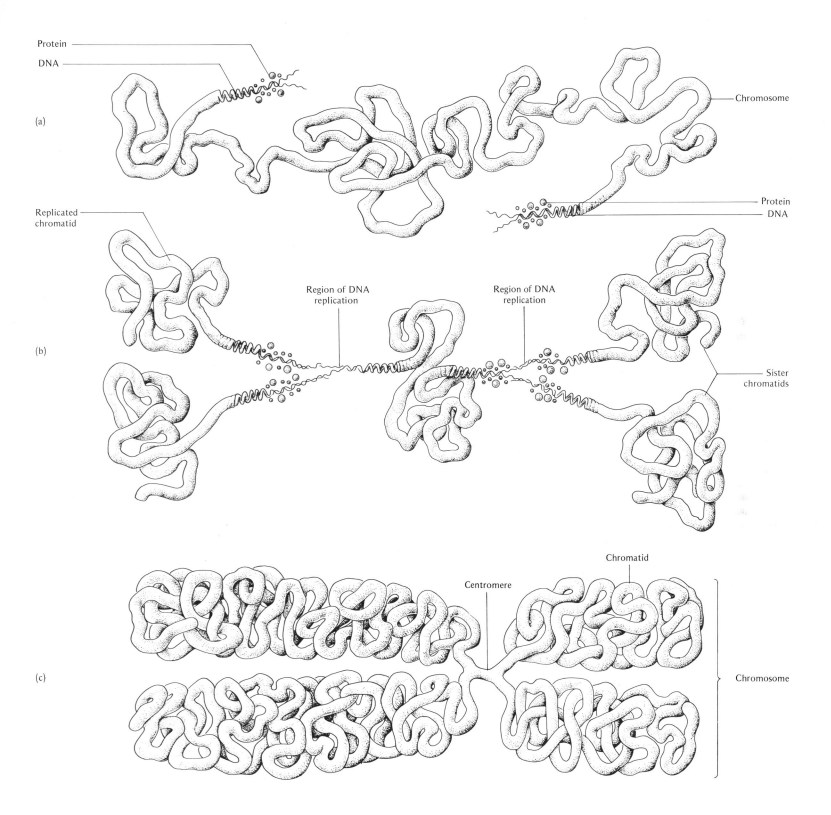

Protein

DNA

(a)

Chromosome

Replicated
chromatid

Protein

DNA

Region of DNA
replication

Region of DNA
replication

(b)

Sister
chromatids

Centromere

Chromatid

(c)

Chromosome

These do not separate, but remain in parallel association so that there results a giant chromosome resembling a multistranded cable (Figure 7–15). At full size they are 100 times thicker and 10 times longer than ordinary chromosomes like our own. These large chromosomes provide us with a better view of chromosome organization and behavior than do the normal-sized chromosomes of ordinary cells, because not only their size but their activities are magnified. Studied under the light microscope, the giant salivary-gland chromosome appears to have transverse bands; the dark bands can be shown to contain DNA and histone (a protein) in high concentration, whereas in the light bands these substances are present in lower concentrations. Are these bands the genes themselves? Investigators studying these giant chromosomes found that particular regions of certain chromosomes at various times in the life of a fly (not the same fly, of course) showed a reversible change in their circumference. These inflated regions were called **chromosomal puffs.** The puff pattern of chromosomes changed during the life cycle of the fly, and this was interpreted to mean that different regions of the chromosome were active. In addition, puffing could be induced by applying certain hormones to the fly. What sort of activity was taking place at the chromosomal puff? Studies using special strains and radioactive materials have shown that the puffs are regions of uncoiling, and in the vicinity of the puff significant amounts of RNA can be detected (Figure 7–15). By the use of radioactive uridine, which is a specific precursor molecule for RNA but not for DNA, it is possible by special techniques to find radioactivity in the region of the puff, but not in the giant chromosome itself. It appears that RNA is manufactured on the DNA template, and this is interpreted to be messenger RNA (mRNA). The RNA leaves the region of the puff, migrates out

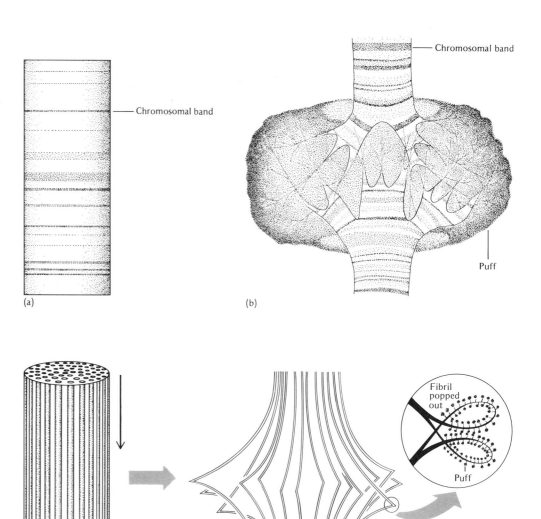

Chromosomal band

Chromosomal band

Puff

(a)

(b)

(c)

Fibril popped out

Puff

Fibrils "pop out"

of the nucleus, enters the cytoplasm, and interacts with the ribosomes to make new protein. Puffing can be inhibited by preventing nucleic-acid synthesis, but not by blocking protein synthesis. There seems to be little question that puffing represents the activity of certain genes at the level of the chromosome; gene activity thus represents the DNA-directed synthesis of RNA.

The activity of genes, visible under the light microscope, is not just an odd case found in certain cells of certain flies. In the primary occytes (early stages in the formation of the egg) of amphibians (frogs, salamanders, etc.) reptiles, and birds, the nuclei contain very large chromosomes called **lampbrush chromosomes.** These consist of a main longitudinal axis from which arises a series of loops that makes the chromosomes appear fuzzy and gives them the appearance of a bottle brush or lampbrush (Figure 7–16). Unlike salivary-gland chromosomes, there is no increase in the number of strands. The loops represent lateral extensions of the chromosome and are rich in protein and RNA, whereas the axis is rich in DNA. Loop formation is a reversible process, which is variable. The similarity to puffing is obvious.

Control mechanisms

Now that we have morphological evidence for gene activity, we can ask the question: Are gene activities one-way avenues of communication out of the nucleus with the rest of the cell playing a passive role, or is it possible to influence gene activity? If the latter is the case, what turns the genes on and how can synthesis of protein, the gene product, be controlled? There are thousands of genes in every cell of the body, and theoretically each is capable of providing the information necessary for the synthesis of a protein. Yet it is quite apparent

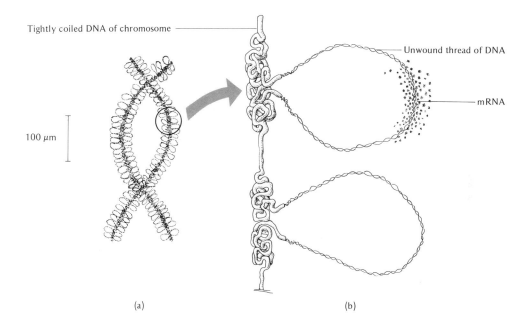

Tightly coiled DNA of chromosome

100 μm

Unwound thread of DNA

mRNA

(a)　　　　　　　(b)

that not all genes are active at the same time, and in the case of certain cells some genes are never active. For example, every cell in the body has the gene for making hemoglobin, and yet only certain cells (juvenile red blood cells) make this particular protein; similarly all our cells have the gene for producing the protein of hair, keratin, but only specific cells produce this protein. Furthermore, a gene may be active for a period of time, then inactive, and perhaps reactivated again after this. Such on-off genetic switching may continue for the entire lifetime of the cell. Thus, the genes of every cell are regulated in their pattern of activity; behind their regulation lies a complex system of controls.

By the regulation of gene or enzyme activity, it should be possible to control the destiny of a cell and ultimately the entire organism. By turning protein and especially enzyme production on or off, or by controlling the action of

genes responsible for enzyme synthesis, it should be possible to regulate cell growth and specialization. Some of the most intriguing problems in biology center on understanding the mechanisms that control cellular activities. At present, the state of our knowledge concerning this is quite incomplete. Nevertheless, let us now examine what is known about some of these control mechanisms.

Mechanical systems. Frequently in this text we have compared cells to machines, and although such an analogy is often useful in helping us to understand the activities of cells, it is well to recognize its limitations. Living things, cells included, are machines with some remarkable properties, such as the ability to duplicate themselves, to change, and to perpetuate such changes. However, when we come to the regulatory mechanisms that operate within a living organism, the similarity between cells and machines is striking. Cells,

FIGURE 7–17 A mechanical feedback circuit: the thermostat. (a) When the bimetallic strip is cooled, it moves the plate downward; this completes the electrical circuit, and the heating system goes on. When contact is broken by an elevation in temperature, the circuit is broken and the heating system is shut off. (b) The room temperature oscillation in a thermostatically regulated heating system. (c) The negative feedback loops in temperature control of room temperature. (From *Biology: Its Principles and Implications* by Garrett Hardin, 2nd ed., San Francisco, W. H. Freeman and Company. Copyright © 1966.)

like many man made machines, consist of an array of structures that work in a coordinated fashion; processes are integrated and the metabolic maelstrom is organized so as to produce a smoothly running living system. For a cell to regulate its pattern of activity, there must be a systematic organization of the control systems. What is that organization? Two types of control systems are found in cells, and in principle these greatly resemble those of machines: (1) a rigid or **fixed control system,** that is, a system with a preset, unchanging pattern; and (2) a **feedback control system,** that is, a self-correcting and oscillating system.

The first kind of control is seen in a machine such as the automobile engine. The combustion of fuel in the cylinder moves the pistons, the piston rod moves the crankshaft, and the latter moves the drive shaft. The organization is rigid in such a machine, in that movement of the piston always produces a similar response in the drive shaft. When we consider the duplication of the genetic material, we can see how it resembles such a rigid control system. DNA is duplicated in a fixed manner according to the rules of complementary base pairing. When errors (mutations) occur, they are usually not corrected, and these lead to a new message; from then on, the new system is itself duplicated. The duplication of DNA is therefore an example of a rigid or fixed control system.

In contrast a feedback control system is self-correcting. The response to a given command or stimulus is quantitative, specific, and automatic. Such systems operate within the living cell, but before we study these, let us briefly examine a mechanical one, the thermostat.

When we set the thermostat in a room to a specific temperature (set point), we are initiating a regulatory mechanism (Figure 7–17a). When the temperature drops below the set point, a metallic strip in the thermostat moves, makes a complete electrical circuit, and the

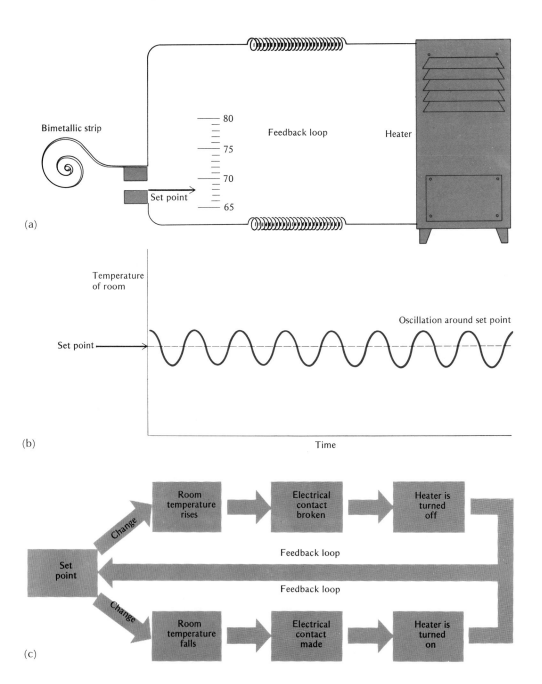

heating system is turned on. When the room warms to the temperature set on the thermostat, the circuit is broken, and the heating system is turned off. The essential feature of this system is that it is represented by a closed loop (or circle); such a closed loop is called a feedback system (Figure 7–17b). Information about this system (in this case, temperature) is fed back to change the response. This is called a **negative feedback** because it compensates for, or negates, any change in the system (the temperature remains constant); negative feedback systems are error-actuated (that is an error in the temperature of the room sets in operation the corrective process) and tend to stabilize the system. (Note that the set point can be moved to produce another system for stability.) A system that is controlled by such feedback controls, which tends automatically to correct for errors, is far more flexible and less vulnerable to damage than a rigid one. Moreover, it is economical to operate because it is self-correcting and does not require the intervention of a separate control.

There is, as you might imagine, an opposite kind of feedback which would result in a lack of correction by changes in the system. This is a **positive feedback.** It produces an unstable situation and is sometimes called a **runaway feedback.** As an example, let us suppose that in the thermostat when the circuit was broken the heating system was turned on; the temperature in the room rises, but this causes no change in the thermostat, and the temperature rise continues indefinitely. Another mechanical representation of negative and positive feedback controls is illustrated in Figure 7–18.

The mechanical control systems just discussed are quite simple compared with the controls found in the living cell. The brief survey of metabolic pathways (Chapters 4 and 5) and of the synthesis of proteins and nucleic acids (Chapter 6) convinces us that in order for

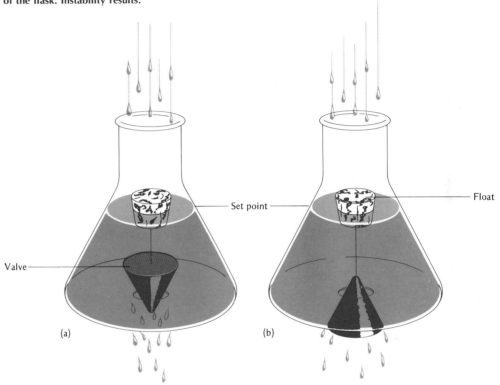

FIGURE 7–18 A mechanical model of feedback control systems. (a) Negative feedback. As water enters the flask, the valve automatically compensates for the inflow and maintains a constant level. (b) Positive feedback. As water enters the flask, the float is raised; this closes the valve and the water rises above the set point. Similarly, if water evaporates, this lowers the float; the valve opens, and water runs out of the flask. Instability results.

cells to maintain their status quo, the traffic in chemicals must move in an orderly fashion, and there must be certain priorities established that channel energy into certain activities and not others. A metabolically regulated cell can function without interruption during periods of nutritional feast or famine and can adjust to adverse or injurious situations—the cell maintains what may be called its set point or **steady state.** It is important to recognize that both change in the performance of a pattern of metabolic activity (alteration of the set point) and maintenance of the steady state (fixed set point) operate in the same way: effects act back upon the cause. Signals are received by cellular receptors, and these signals are trans-

lated into biological activity; typically it is found that small mobile molecules are the signals and large molecules are the receptors. The on-off feedback system of the cell (or the thermostat) usually shows a delay in response and then a slight overshoot, and there results an oscillation around the set point. Let us now look at these feedback controls of the living cell.

Enzyme induction. Much of the detailed information we have about enzyme regulation in cells comes from studies of microorganisms, but many biologists feel that what goes on in such creatures may be similar to what occurs in the cells of higher organisms. However, at present we have little direct evidence that this

FIGURE 7–19 The operon and enzyme induction.
(a) Operator off. Repressor protein attached to operator gene blocks the beginning of transcription by
RNA polymerase; no synthesis of mRNA at the operon. (b) Operator on. Inducer attaches to repressor,

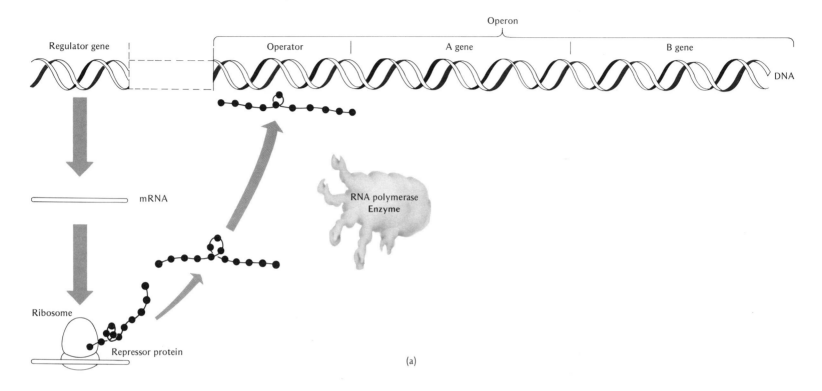

(a)

optimistic notion is true. In the bacterium *E. coli*, it is possible to stimulate the synthesis of certain enzymes by environmental factors. If *E. coli* is grown in a medium containing glucose, it has only traces of the enzyme β-galactosidase (an enzyme that catalyzes the breakdown of lactose into glucose and galactose); however, if the bacteria are grown with lactose in the medium, then the enzyme activity may be increased over a thousand times. It is a "clever" bacterium that makes the right kind of enzyme when the right nutrient substance (food) is available. How does the bacterium "know" what kind and how much enzyme to produce? Is it due to an activation of the enzyme; that is, does the lactose make the enzyme itself more active, or does it affect the amount of the enzyme produced? It has been shown that lactose acts as an inducer for the

synthesis of the enzyme; the lactose affects the rate of manufacture of the enzymatic protein, and new protein can be detected within the cell only 3 minutes after the inducer has been added to the medium. When the inducer is removed, the synthesis of the enzyme ceases. Here we have an example of an environmental substance that turns specific genes on and when the substance is lacking these genes are switched off; it is a negative feedback system. Not all enzymes exhibit these properties; many are produced and are active regardless of the presence or absence of substrates (inducers). Such enzymes are called **constitutive** because they are always present in the cell at a fixed level—the genes that code for them are always turned on and, therefore, the enzymes are a permanent part of the cell's normal constitution. They are part of a rigid control system.

How is the induced enzyme genetically controlled? The production of β-galactosidase is actually controlled by two genes: one (*A*) codes for the amino-acid sequence of the enzyme itself, and another (*B*) codes for a permease that moves lactose into the cell. An alteration in the nucleotides (mutation) of any of these genes (so-called **structural genes**) changes the structure of the enzyme but does not affect the rate of enzyme synthesis. The rate of synthesis is controlled by a third gene called the **regulator** (*R*); if this gene mutates, then enzyme synthesis is continual, and the inducer, lactose, need not be present; in other words the loss of gene activity in the regulator gene (*R*) causes enzyme synthesis to be switched to the on position (alternatively, it can be thought of as not switched off; the reason for this will become apparent momentarily). How does the

changes its shape, and releases it from the operator gene. RNA polymerase attaches to DNA, and mRNA synthesis begins. The translation of mRNA at the ribosome gives rise to enzymes.

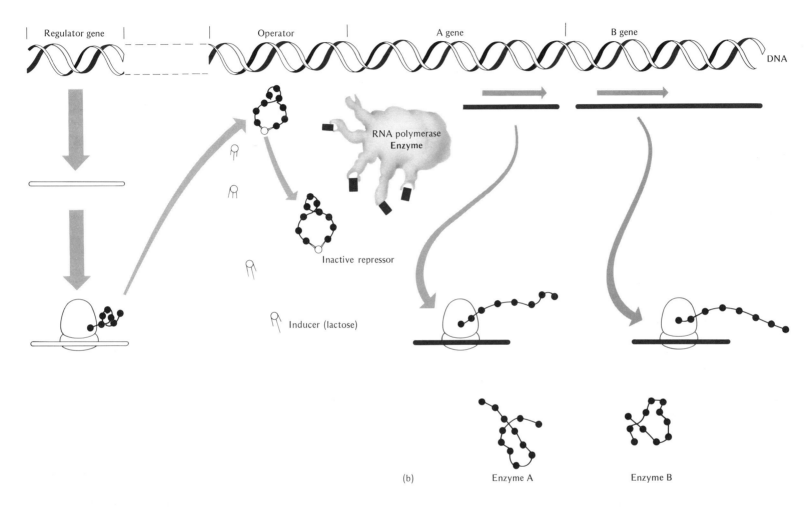

| Regulator gene | Operator | A gene | B gene |

RNA polymerase Enzyme

Inactive repressor

Inducer (lactose)

(b)

Enzyme A   Enzyme B

regulator work? In 1961 F. Jacob and J. Monod proposed that the regulator gene (R) produces a substance called **repressor,** and that this material diffuses through the cell and shuts off the genes (A and B) responsible for the synthesis of β-galactosidase and the permease. They hypothesized that the repressor substance does not act on the structural genes (A and B) directly, but binds to a special region of the chromosome located next to the A and B genes. This special region they called the **operator**

(designated as O), and it can be thought of as a switch that turns on genes A and B. The genes O, A, and B are called the **operon.** According to the Jacob-Monod model, this region of the bacterial chromosome would look as follows:

$$\cdots - \boxed{R} - \boxed{O} - \boxed{A} - \boxed{B} - \cdots \text{chromosome}$$

$$\underbrace{}_{\text{operon}}$$

Let us see what the operon model predicts. When repressor is produced by the R gene, it

binds to the DNA at the operator gene (O), and this switches off the activity of the adjacent genes, A and B—the operon is repressed. (The mechanism of switching off the operator is that when repressor binds to the operator gene, it prevents RNA polymerase from initiating transcription.) However, when lactose is present, repressor binds to it, and by virtue of a change in the shape of the repressor molecule, it falls off the operator gene (Figure 7–19). Now the operator is switched on, the gene (A + B) is transcribed (via RNA polymerase), mRNA is made, β-galactosidase and permease are formed, and lactose is utilized by the cell.

Recently, the Jacob-Monod model of the operon has been experimentally substantiated; normally, only 10 molecules of repressor substance exist in a cell—hardly enough to detect experimentally. Using special techniques that involve tricking the cell into producing much larger quantities of repressor substance than normal, it has been possible to isolate the substance; it appears to be a protein of high molecular weight (about 150,000). Repressor substance is very specific in its binding to a specific region of the bacterial chromosome; it turns off only the lactose operon. However, the mechanism by which the repressor recognizes the appropriate nucleotide sequence of the DNA is not yet understood.

It is quite possible that not all repressors control the activity of genes in the same manner. For example, it is conceivable that there are molecules that stimulate the activity of genes (**promoters**) and thus act in a positive sense to turn genes on. The operon model provides us with a situation whereby switching on of genes can alter the functional properties of cells; if the product of a gene were a division protein, it could even initiate cell division. The possibilities for future understanding of differentiation and division of cells through repressors and the like are quite encouraging.

Through repressor substances and enzyme induction, as well as another mechanism to be discussed shortly, the chemical machinery of the cell is self-regulating, and the amount of energy used in the manufacture of enzymes (a major cellular expenditure because a different enzyme is required for every reaction) is effectively and economically controlled; in a frugal and systematic way the cell produces only the kinds and amounts of enzyme mole-cules necessary to metabolize the amount of available nutrients (in this case, lactose).

Feedback inhibition. In the previous discussion we have seen how chemical reactions within the cell can be genetically regulated by controlling the kinds and amounts of enzymes produced. There is another mechanism of cellular control that involves regulation of enzymatic activity in direct response to environmental situations; this is not directly controlled by the genes.

Several years ago it was found that when *E. coli* was grown in the presence of an excess of the amino acid isoleucine the cells stopped producing this amino acid. It was known that the synthesis of this amino acid required several intermediate materials, and the conversion of these into the final product (isoleucine) was controlled by specific enzymes. This situation is illustrated in Figure 7–20.

It was also shown that when excess isoleucine is present it directly affects only the first enzyme in the series; all the other enzymes are unaffected. Moreover, no other amino acid can turn off the system, and the inhibition does not even require an intact cell, for inhibition can be produced by an extract of the cells containing only the enzymes and totally lacking any nucleic acids.

Here we have another example of a negative feedback control system. When the supply of isoleucine is adequate, the cell no longer continues its manufacturing processes; the control mechanism is simply to turn off the first enzyme in the sequence, and since this prevents the formation of subsequent materials in the pathway, the whole production line is brought to a standstill. The mechanism is as follows: isoleucine combines with the first enzyme to produce an enzyme-isoleucine complex, and this prevents the enzyme from binding to its natural substrate (a); as a result the enzyme is inhibited in its activity. The binding of isoleucine is not at the active site of the enzyme (that is, the same site as the enzyme would bind to its natural substrate [a]), but nevertheless it does interfere with the usual binding of this enzyme with substance (a). The reason for this can be visualized as shown in Figure 7–20b. If the level of isoleucine diminishes, isoleucine is no longer bound to the enzyme, and then (a) can once again bind to the active site of the enzyme and set in motion the reaction sequence to produce more isoleucine.

Such a control mechanism adjusts itself quickly and automatically to the changing needs of the cell and such a control mechanism requires no additional expenditure of cellular energy once the first enzyme is made. This control system is perhaps the ultimate in efficiency, since it requires no energy for its operation. (Consider what the effects of this would be in an industrial process!)

Regulation of gene action in higher organisms. How are our genes regulated? If we knew the answer to this question, all problems of differentiation, cancer, and the like could be understood and perhaps controlled. At this writing there is no conclusive answer as to how the genes in the cells of higher organisms operate; the previously discussed models of gene and enzyme regulation are largely based on experiments with bacteria, and it seems reasonable to suppose that differences in control mechanisms exist between bacteria and man. The bacterial cell interacts with its environment in a precise way. Fluctuations in the environment in which bacteria live are likely to occur quite suddenly, and the bacterium must be able to cope with them; conditions are feast or famine, and the control switches must be turned on or off rather rapidly. Although in the multicellular organism genes are also switched on and off, it is rare that they are

FIGURE 7–20 Feedback inhibition. (a) The biosynthesis of isoleucine requires five enzymes, and the end products of the pathway act as a regulatory signal to inhibit the first enzyme in the pathway. (b) The mechanism by which isoleucine inhibits or restores activity to the first enzyme. The first enzyme in the pathway has two sites for binding to its natural substrate (a) and the end product, isoleucine (b). When the end product (b) is bound, it changes the configuration of the enzyme so that it cannot bind to the substrate, and this effectively stops enzyme activity in the pathway. Activity can be restored by release of the end product.

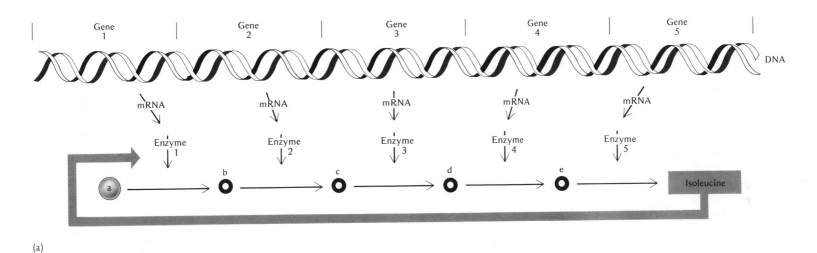

(a)

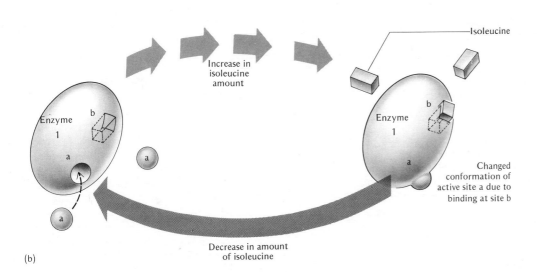

(b)

switched on and off with such rapidity. The multicellular organism seems to be buffered against fluctuations in the environment, and examples of enzyme induction are unusual. In more complex organisms it may be that environmental stimuli trigger internal changes, which in turn affect the gene itself. For example, production of hormones may be affected by environmental factors, and it is suspected that certain hormones act to turn genes on and off. During development, many genes are turned on and others turned off simultaneously. In many of these cases, once the direction of differentiation (specialization) has been taken, the cell is committed to a certain path and reversibility is rare. Based on very crude and incomplete knowledge, it has been postulated that when genes are switched on in the cells of higher organisms, there may be a simultaneous switching off of other genes; that is, they become irreversibly repressed and never again can be turned on. Another possibility is that many genes are operational, but for some enigmatic reason the messages of the mRNA are not translated. Which of these possibilities is true remains a problem for the future, as does the exact mechanism whereby a specific gene is selectively activated by molecules inside or outside the cell. Some biologists suggest that the histone proteins control gene activity, but how such proteins recognize specific genes and influence them remains unknown. Control of genes in higher organisms requires a system of molecules that can recognize a specific group of nucleotides in the DNA and influence them; such molecules have not yet been discovered.

SUMMARY

1. Mitotic cell division involves the replication and separation of the chromosomes and the production of two cells with identical sets of genetic information to the parent cell. It occurs in several definable stages.

2. The undividing or resting cell contains an interphase nucleus with a diffuse network of stainable chromatin and an obvious nucleolus. Protein and nucleic acid synthesis occur during this stage.

3. During prophase, the chromatin threads coil and shorten and are increasingly stainable and visible as chromosomes. Each chromosome consists of two closely associated chromatids, held together at the centromere. The nuclear membrane and nucleoli break down.

4. During metaphase, the centriole divides and separates, forming the spindle and astral rays. The chromosomes orient themselves at the equator of the cell, and individual chromosomes can be recognized according to the location of their centromeres, by which they attach to the spindle.

5. At anaphase, the centromeres divide and the chromatids separate. Pulled by contracting spindle fibers, the chromatids move to opposite poles of the cell.

6. During telophase, the nuclear membrane is reconstituted, the chromatids, now called chromosomes, uncoil, and nucleoli appear at the nucleolar organizer regions of certain chromosomes. In animal cells, cytokinesis divides the cytoplasm by cleavage.

7. Plant cells lack a centriole, and have a rigid cellulose cell wall. Cytokinesis occurs by the laying down of a cell plate rather than by cleavage.

8. Cells spend most of their time in interphase, which can be divided into three phases: G_1, S, and G_2. DNA synthesis occurs during the S phase, and the chromatids are doubled by the time of the next cell division.

9. Chromosomes consist of DNA and structural proteins (histones and protamines). A chromatid consists of two complementary strands of tightly coiled DNA, which coils on itself and is packed in protein.

10. Genes are nucleotide sequences of DNA that code for specific proteins. The number of genes an organism has is related to its complexity. Most cells seem to have many duplicated redundant genes. The linear organization of genes on chromosomes increases the efficiency of replication and segregation and permits gene-gene interaction.

11. Fly salivary glands have cells with giant chromosomes formed by the repeated division of chromatids without nuclear division. Their banded structure reflects functional differences, possibly at the level of the gene. Chromosomal puffing reflects uncoiling and RNA synthesis. The primary oocytes of amphibians, reptiles, and birds contain large lampbrush chromosomes with lateral loops where RNA synthesis occurs, similarly reflecting gene activity.

12. DNA replication is a fixed control system with no feedback.

13. Maintenance of the steady state in a cell involves negative feedback systems comparable to the thermostat.

14. The regulation of gene activity and the switching on and off of genes often involves negative feedback systems. In *E. coli*, lactose induces synthesis of the enzyme β-galactosidase which digests it. The enzyme is not produced unless lactose is present. (Enzymes that are produced all the time, regardless of the presence or absence of the substrate, are constitutive enzymes.)

15. Induced enzyme synthesis is controlled by more than one gene. In β-galactosidase synthesis in *E. coli*, gene A codes for the amino-acid sequence of the enzyme, gene B for a permease that moves lactose into the cell, and gene R (regulator) for the rate of synthesis. R produces minute amounts of a repressor substance that binds to an operator gene (O) next to A and B. This binding switches off the activity of genes A and B. Genes O, A, and B are called the operon.

16. When repressor binds to O it prevents RNA polymerase from initiating transcription. When lactose is present, repressor binds to it rather than to O so the operator is switched on, and A and B are transcribed. Hence, β-galactosidase and permease are made, and lactose is used by the cell.

17. The production of a substance by a cell may be inhibited by the presence of adequate amounts of that substance in the cellular environment with no genetic control. Feedback inhibition of the production of the amino acid isoleucine by *E. coli* occurs when isoleucine combines and thus interferes with the first enzyme of a sequence that normally produces isoleucine. This prevents it from binding to its substrate and stops the production sequence. Production does not resume until isoleucine levels drop.

18. Control of genetic switching on and off in higher organisms is not yet well understood.

KEY WORDS

cell division
chromosomes
resting cell
interphase nucleus
chromatin
centriole
prophase
chromatids
centromere
asters or astral rays
spindle
equator
metaphase
daughter chromosomes

anaphase
telophase
nucleolar organizer
cytokinesis
furrowing or cleavage
mitosis
histones
protamines
genes
giant chromosomes
chromosomal puffs
lampbrush chromosomes
fixed control system
feedback control system
negative feedback
positive (or runaway) feedback
steady state
enzyme induction
constitutive enzymes
structural genes
regulator gene
repressor substance
operator
operon
promoters
feedback inhibition

TOPICS FOR REVIEW AND DISCUSSION

1. Describe the stages of mitosis. What is the significance of mitotic cell division?
2. Can gene activity be seen? How?
3. What is the meaning of negative feedback? How does it operate in cellular metabolism?
4. Describe the operon theory.
5. Why may mitosis be thought of as cellular immortality?
6. Explain the relationship of genes to chromosomes.
7. Illustrate the chromosomal movements during mitotic division of a cell containing four chromosomes.
8. What is the relationship of chromatid to chromosome?
9. How can semiconservative chromatid duplication be demonstrated?
10. Describe the packaging of DNA into chromosomes.
11. What would happen to a cell's chromosomes during mitosis if a centromere were lost? If a centriole were lost?
12. What is the difference between plant and animal cell mitosis?

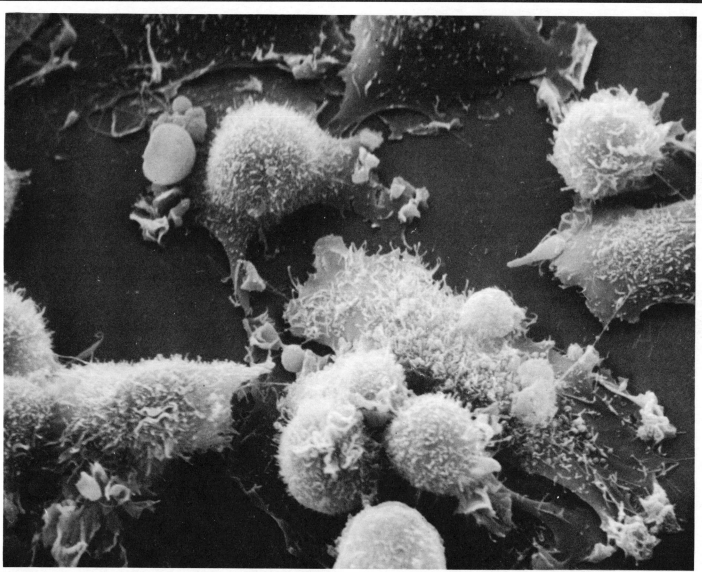

These are "lawless" cancer cells. As their cytoplasmic fingers stretch out and claw their way into the surrounding normal tissues, death and destruction are all that remain.
(Courtesy of Keith R. Porter)

8

CANCER: CELLS OUT OF CONTROL

8–1 WHAT IS CANCER?

Cancer is a ubiquitous killer. Worldwide, 2 million people each year die of cancer, and in the United States more than 0.25 million deaths are annually attributed to this disease. In morbidity statistics for the United States, cancer ranks second as the principal disease causing death, surpassed in importance only by the cardiovascular diseases. It has been estimated that about 50,000 man-years are lost annually because of disabilities caused by cancer, and $12 billion are lost yearly in goods and services. Cancer is indiscriminate; it affects young and old, male and female, rich and poor. What is cancer and how can it be recognized?

One of the inherited tendencies of living matter is to multiply. Cellular growth and multiplication are dynamic processes, but in our bodies such events are under some sort of restraining influence; unchecked or unlimited cell division ordinarily does not occur. For example, if a wound is made in the skin, let us say a cut, some striking changes in the pattern of cell growth take place (Figure 8–1a). The cells that surround the gaping wound detach themselves, become ameboid, and move into the wound area; on the way, these cells begin to divide. The cells continue the process of division until the entire wound is filled. Now cell touches neighboring cell, the proliferation of cells ceases, and the wound is healed. Similarly, when a tissue or an organ is in the process of growing, the cells multiply, but once a certain size is reached the multiplication process ceases.

Suppose a few of our skin cells, especially those that divide to replace the cells continually being lost at the surface, escape from controlling factors and the daughter cells continue to multiply; eventually a cluster of cells is produced (Figure 8–1b). Similar uncontrolled cell division in any part of the body could result in

FIGURE 8–1 Patterns of skin growth. (a) Wound healing. (b) Cancer.

an abnormal functionless lump or mass of cells—a **cancer** or **tumor** (the terms are synonymous). The cancer cell is often described as being **neoplastic** (*neos*—new; *plasma*—a thing formed; Gk.), since it has been transformed from a cellular condition of rest with the synthetic machinery turned off to a condition of continuous growth and multiplication.[1] Cancer cells contribute nothing to the tissue from which they arose; they often interfere with and damage neighboring normal cells and monopolize the food supply.

To most people, cancers are "rampant, predatory, savage, ungovernable, greedy cells—flesh destroying itself." To a biologist, the cancer cell is one with an inherited capacity for autonomous growth; that is, it determines its own activities irrespective of the laws that precisely regulate the growth of all normal cells. To a clinical physician, a cancer is a malignancy that is ultimately lethal. The clinical physician distinguishes tumors, or cancers, by their appearance and their manner of distribution, designating them as **benign** (harmless) or **malignant** (deadly). The property of autonomy is expressed in both benign and malignant tumors; how then do they differ from one another?

Normal cells of the body tend to respond to the presence of abnormal objects such as tumors by attempting to surround them in an envelope of tissue, thus isolating them and cutting them off from the rest of the body. If a tumor is surrounded by a connective-tissue capsule and is thus restricted in its growth, it is called a benign, or harmless tumor (Figure 8–2a). Of themselves, benign tumors do not kill, but they can interfere with normal function;

[1] The time spent in mitosis by a cancer cell is the same as that of normal cells, but the frequency of cell division is greater for the cancer cell.

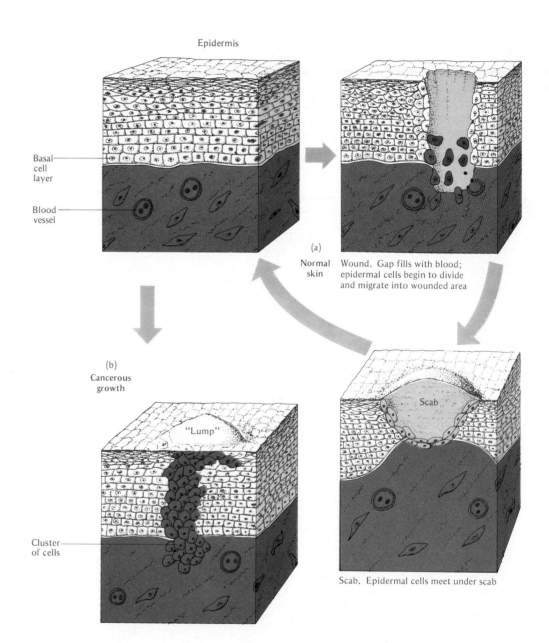

Epidermis

Basal cell layer

Blood vessel

(a)

Normal skin

Wound. Gap fills with blood; epidermal cells begin to divide and migrate into wounded area

(b)
Cancerous growth

"Lump"

Cluster of cells

Scab

Scab. Epidermal cells meet under scab

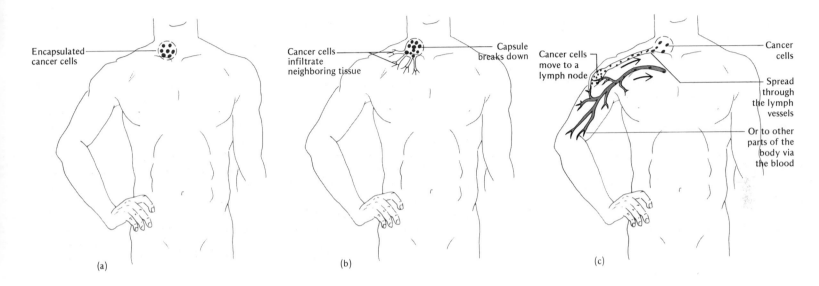

FIGURE 8–2 Clinical diagnosis of cancer. (a) Benign. (b) Invasive. (c) Metastatic.

Labels for (a): Encapsulated cancer cells

Labels for (b): Cancer cells infiltrate neighboring tissue; Capsule breaks down

Labels for (c): Cancer cells move to a lymph node; Cancer cells; Spread through the lymph vessels; Or to other parts of the body via the blood

for example, they can obstruct the bowel. Benign tumors can be cured because they can be removed by surgical means; warts are examples of benign tumors, caused by a virus.

Sometimes benign tumors give rise to another kind, called an **invasive tumor.** In such a condition the normal tissue cannot keep pace with the rapidly dividing cancer cells, and the capsule breaks down. The cancer cells break out of the connective-tissue capsule, grow outside the capsule, and infiltrate the surrounding tissues (Figure 8–2b). Such a condition is more difficult to handle and may cause a diseased condition; this is a malignant tumor. The malignant invasive tumor may eventually become surrounded by blood vessels and lymphatics; the tumor or cancer cells may then enter these vessels and be transported to other parts of the body (Figure 8–2c). This is a malignant **metastatic tumor** and is extremely difficult to treat by surgical procedures. The cancer cells may eventually lodge in the lungs, liver, or kidney,

or be held in the lymph nodes; there they may proliferate and cause further damage and destruction to the vital organs of the body. Whole-body treatment with drugs or irradiation must be instituted to control such a malignant metastatic cancer.

It is apparent from the previous discussion that the cancer cell differs from ordinary normal cells in its relationship with neighboring body cells. Ordinarily, normal cells are confined within specific tissues and maintain themselves in a precise and orderly architectural pattern; the cancer cell does not remain confined to the tissue from which it originated, but moves into the surrounding tissues and there continues its proliferative ways. Many cancers show this invasive tendency, and in this way they "claw" their way through the body, injuring and killing other cells. To the early physicians who first observed these cellular tendencies, the hardened body of tumor tissue resembled the body of a crab, and the rapidly dividing cancer cells

projecting out of this mass and invading surrounding tissues appeared like the crab's claws; it was imaginative, if perhaps not quite logical, to call the disease "cancer" after the Greek word *karkinos,* meaning "crab."

8–2 CARCINOGENS: CANCER-CAUSING AGENTS

Agents capable of causing cancer are called **carcinogens;** they come from a variety of sources and can be classified into three basic groups: chemical, physical, and viral.

Chemical agents

Occupational. The first occupational cancer caused by a chemical was described in 1775 by the English physician Percival Pott as "a scrotal cancer of chimney sweeps." The frequency of skin cancer, particularly in the area of the scrotum, was greater in sweeps who had started as climbing boys early in childhood, yet the symptoms only appeared in sweeps who had been in the profession for 20 years or more. Pott surmised that the tumors had originated from soot which accumulated and persisted longer in the scrotal region due to poor hygiene. The clinical observation of the relationship between soot and scrotal cancer resulted in a set of safety regulations, in the form of bathing requirements. As a result, the incidence of scrotal cancer among sweeps showed a sharp decrease. It was not until 150 years later that Pott's clinical observations were confirmed experimentally. In 1932, Kennaway and his colleagues isolated a specific chemical from coal tar, 3:4-benzpyrene, and this alone induced cancerous lesions.

Workers in the dye industry also show a high incidence of bladder cancer, and the carcinogen involved is β-naphthylamine, which is used as the starting material for the synthesis of many dyestuffs. Petroleum workers too show a higher incidence of cancers than the general populace, but the agents involved have not yet been identified. Asbestos, nickel, uranium, chromates, and arsenic have all been shown to be involved in cancerous lesions.

Dietary. Recent evidence indicates that we are surrounded by numerous dietary agents that may be carcinogenic. A potent carcinogen popularly known as "butter yellow" (N-dimethylaminobenzene) used to be routinely added to butter to make it more yellow and presumably more attractive to the consumer; fortunately, the quantity used for coloring was quite low and the practice has now been prohibited. Certain other food additives, such as cyclamates and saccharin, have been implicated in bladder cancer in mice, and older case studies showed that the hormone diethylstilbestrol, used for caponizing fowl, could also produce animal cancers. Dietary factors in

some countries are associated with or implicated in cancer of the esophagus and the stomach. In Ceylon and India, cancers of the mouth and throat are common and are related to the habit of chewing mixtures of tobacco, betel nut, and lime. The evidence at present, as with most cancer research on humans, is strictly statistical, but diet itself does represent a potential source of carcinogens.

Hormonal. Within our own bodies there may be agents potentially capable of inducing cancer. Particularly suspect is the female hormone estrogen. Long-term application of estrogen to the skin has caused cancers in experimental animals, and the continued application of estrogen in guinea pigs causes uterine tumors; mice can be induced to produce pituitary tumors by estrogen administration. For humans, the effects of hormones on cancer development are not as clear-cut, but recent evidence indicates that women taking the contraceptive pill show a higher incidence of breast and uterine cancer than women who do not take it; the pill contains synthetic steroids that have effects similar to those of estrogen and progesterone. In men under treatment with estrogen it is not unusual to find an enlargement of the breasts, and some of these show precancerous lesions. It is suspected, but by no means proved, that estrogen may play a significant role in the development of certain forms of cancer.

Atmospheric pollution. Whenever hydrocarbons such as gasoline, coal, oil, gas, or other fuels are incompletely burned, a variety of compounds are produced that have been identified as carcinogenic. In the atmosphere over most large cities, a dense blanket of air pollutants causes eye irritation, difficulty in breathing, and a range of other discomforts. In every month in every square mile of Manhattan more than 1,520 lb of tar settle down, a small fraction of the 176 tons of solids that rain down

on the inhabitants each month. The soot of the air contains the powerful carcinogen 3:4-benzpyrene, and the amount of that chemical increases during the winter with increased burning of fossil fuels. The frequency of lung cancer is directly related to the density of the population, indicating that atmospheric pollution may play a significant role. With continued increase in population and an increased utilization of fossil fuels, there seems little doubt that the larger amounts of carcinogens produced will enhance the incidence of cancer.

Smoking and lung cancer. At the present time, the side panel of every package of cigarettes manufactured in the United States carries the statement: "Warning: The Surgeon General Has Determined That Cigarette Smoking Is Dangerous to Your Health." Much experimental evidence for a variety of laboratory animals indicates that inhaled cigarette smoke can be carcinogenic. Cigarette smoke contains the compound **dimethylbenzanthracene** (DMBA), shown in animal tests to be a powerful carcinogen.[2] For example, in experiments where dogs smoked cigarettes and extracts from cigarette smoke were painted on the backs of mice, tumors developed. Critics of these tests maintain that the results cannot be extrapolated to humans. They may be correct, but continually accumulating evidence indicates that the critics are probably wrong. No direct experiments using cigarette-smoke carcinogens can be performed on humans, so that the evidence is primarily statistical. Be that as it may, the statistics show that smoking is hazardous to one's health. If every smoker in America

---

[2] Cigarette smokers can take little consolation from the fact that the roasting of meat leads to the formation of carcinogens such as 3:4-benzpyrene, since the glandular portion of the stomach (in test animals) is insensitive to this kind of carcinogen.

FIGURE 8–3   Smoking and cancer. (a) Mean death rate from lung cancer for cigarette smokers and their cigarette consumption. (b) Mean death rate from lung cancer for cigarette smokers who have stopped smoking for varying periods of time. (Courtesy of *Journal of the American Medical Association* **and American Cancer Society, Inc.)**

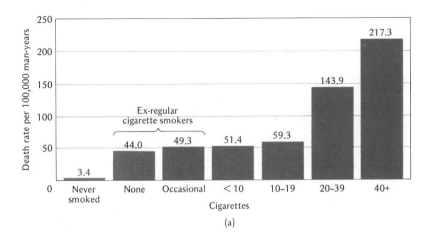

(a)

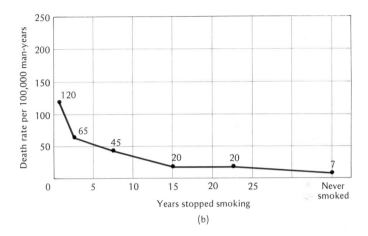

(b)

stopped smoking today, over 60,000 premature deaths from a variety of cancers could be prevented. Since such warnings are likely to go unheeded, we can expect 63,000 persons to die annually from lung cancer alone; more persons will die from lung cancer than from automobile accidents!

The data in Figure 8–3 show that the longer an individual smokes, the greater the possibility of developing lung cancer. The most important factor in the development of lung cancer is that the risk increases with the amount of smoke inhaled and the duration of the smoking habit. Cigarette smoking also bears a causal relationship to other disorders such as Buerger's disease (an inflammation of the arteries), emphysema, chronic bronchitis, and heart disease. These correlations do not prove a cause-and-effect relationship, but they do suggest a strong link between the smoking of cigarettes, cancer, and related diseases.

Over the last 40 years, overwhelming evidence has accumulated that Americans, and others elsewhere around the world, are increasingly filling their environment with carcino-

genic chemicals. The World Health Organization currently estimates that environmental agents are responsible for more than 75% of human cancers. A National Cancer Institute map shows the areas of the United States with the highest death rates from lung, liver, and bladder cancer (Figure 8–4); these are the same regions that have the greatest amount of chemical pollution. Yet there is hope: if the main causes of cancer are indeed environmental, then potentially this disease could be prevented (Box 8A).

Physical carcinogens

Ultraviolet light and skin cancer. One of the diversions considered fashionable in modern society is to lie in the radiant sunshine for hours at a time, covered with some kind of greasy substance; at the end of this time the sunbather hopes to have acquired a deep brown suntan. The effect of the sun's ultraviolet light upon the skin of most humans is to produce first a redness that may be accompanied by a soreness; then if the dose is not

severe enough to cause blistering and peeling, the red fades and is replaced by a brown coloration—the much desired tan. The redness is produced by the dilation of blood vessels creating increased blood flow in the area, and the skin becomes red and warm. In the tanning process, ultraviolet light causes the deeper layers to produce the pigment melanin, which migrates to the more superficial layers (Figure 8–5). If the skin is not reexposed, the tan fades. In many respects the effect of ultraviolet light represents a type of inflammation, and commonly there is some damage to the skin cells. Such injury results not only in the elaboration of melanin but in the thickening of the skin itself by a proliferation of cells and a greater deposition of keratin.[3] Does this ultraviolet-

---

[3] Microscopic examination of skin from the face and arms of fair-skinned but overtanned individuals shows that the cells resemble those taken from the skin of very old men and women (cells from unexposed areas such as the buttocks retained a youthful appearance). Sunburning and suntanning tend to age the skin prematurely.

light-induced aging and cellular proliferation eventually lead to cancer? There is some uncertainty and doubt regarding this question, but certain factors suggest that this may indeed be the case. It is estimated that 10% of human cancers are related to exposure to ultraviolet rays; most of these are cancers of the skin, and 95% of such cancers are curable if treated in time—while they are still benign and before they metastasize. Skin cancers occur more frequently on the parts of the body that are habitually exposed to sunlight, skin cancer is more prevalent among populations dwelling in the sunnier regions of the globe, and skin cancer appears more frequently in light-skinned than in dark-skinned persons and more frequently among outdoor than indoor workers. Mice can be induced to produce skin cancers by treatment with low doses of ultraviolet light for 5 days a week over 161 days; man has 70 years or so for exposure and therefore a greater likelihood of developing skin cancer (the life span of a mouse is about 2 years). If the effects of sunlight are cumulative, as experiments with mice suggest, the risk of skin cancer in man due to the effects of sunlight increases with age and is very great indeed.

X radiation. X rays are much more energetic than ultraviolet rays, and as one would expect, are much more effective in producing skin cancer. The early radiologists, unaware of the carcinogenic effects of x rays, developed severe carcinomas of the hands after repeated brief exposures to low doses of x rays. The frequency of leukemia among radiologists is 10 times greater than among other physicians. The leukemia incidence among the survivors of Nagasaki and Hiroshima is 5 times that of the rest of the Japanese population; the incidence of breast cancer among the female survivors is also increased (approximately 3 times).

X rays, by virtue of their energy, break molecules down into charged particles called

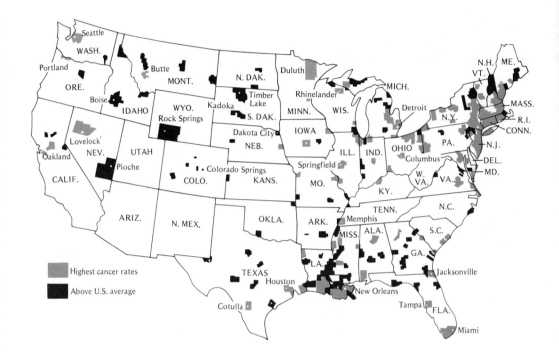

FIGURE 8–4 (top) Mortality among white males suffering from lung, liver and bladder cancer (1950–1969). The national average is 174 per 100,000 yearly. (Reprinted by permission from TIME, The Weekly Newsmagazine; Copyright Time Inc. 1975.)

Highest cancer rates

Above U.S. average

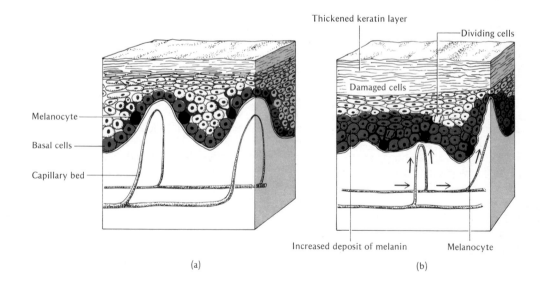

FIGURE 8–5 (bottom) Changes in the skin induced by sunburn. (a) Before sunburn. (b) After sunburn, there is increased blood flow in the capillary bed (arrows), greater deposition of melanin, death of epidermal cells, increased cell division, and a thickening of the keratin layer.

BOX 8A The cancer screen

Medical scientists have been trying for decades to identify those chemicals that are carcinogenic. That task is not easily accomplished. During a lifetime we are exposed to literally millions of chemical compounds, both natural and synthetic. To screen each of these compounds for their carcinogenic properties would require extensive animal testing, involving many years and costing millions. However, there may be a way to speed up the process. A simple and inexpensive cancer screening test was developed by Bruce Ames, a biochemist at the University of California. The Ames Test is based on the theory that carcinogens damage nucleic acids, causing mutations, and thereby inducing cancer. Bacteria do not get cancer but are vulnerable to mutations. Ames used a mixture of four mutants of the bacterium *Salmonella typhimurium* that were unable to make the amino acid histidine; such mutants do not grow unless histidine is added to the agar on which they are placed. When exposed to agents causing mutations (mutagens), these strains may mutate further; some of these mutations, by chance, will repair the original genetic defect, making their possessor normal. If this occurs, the bacteria will grow on agar containing no histidine and colonies will form on the agar surface. The colonies show up as white dots and can easily be counted. The more powerful the mutagen, the more colonies there are that will grow in the absence of histidine. Thus, colony numbers give some measure of the potency of the mutagenic agent. Of 174 known carcinogens tested, 157 (90%) were mutagenic in the Ames test. Out of a group of 108 so-called non-carcinogens, 14 were mutagenic. Some of these mutagens may well cause cancer. For example, a food additive called furylfuramide that was used in Japan for 8 years had been shown to be noncarcinogenic in animal tests; however, recent studies show it to be both mutagenic and carcinogenic. Furylfuramide is positive in the Ames test.

All evidence indicates that certain chemicals and radiation in the environment damage DNA; over a lifetime this cumulative damage is probably the cause of most human cancers. The Ames mutagenicity test is a cheap way of determining whether a chemical compound is likely to be dangerous to humans. It may also enable us to tell whether an individual who has been exposed to a suspected carcinogen is carrying a lethal time bomb within him. A physician suspicious of such a condition could take a sample of the patient's urine, add it to a culture of *Salmonella,* and see if it is mutagenic. If detected early, treatment could be initiated. Mutagenic screening could make it possible to protect the public against hazardous substances before and not after they have been in widespread use.

ions. These highly reactive ions stimulate mitosis or destroy cells, depending on the dose and the tissue involved. How these effects are produced, and what factors trigger normal cells to divide and retain these proliferative properties, remain speculative, but mutations may be involved. We shall return to this later in the chapter.

Viruses and cancer

As early as 1910 Peyton Rous found evidence that a fibrous connective-tissue tumor (sarcoma) in the breast muscle of chickens was due to the presence of a virus (Rous sarcoma virus, or RSV), but this was lightly dismissed as a peculiarity of chickens. Later, the evidence made such criticisms less valid. In 1932 Rous, Beard, and Shope found that a nonmalignant skin cancer in wild rabbits was caused by a virus (Shope papilloma virus);[4] when the purified virus was transferred to other rabbits, it metastasized and eventually killed the animal. Today it is well recognized that viruses are responsible for a great many cancers. (In fact Rous was awarded the 1966 Nobel Prize in medicine for his 1910 work on RSV.)

At the present time uncertainty exists about whether viral invaders or viral residents already present within human cells are the causative agents of human cancers. The principal difficulty is this: the mere presence of viral particles in cancer cells does not necessarily *prove* viruses to be the cause of the condition. For example, it is conceivable that the viruses are harmless cellular occupants, and it is quite by

---

[4] Perhaps the best known of all human cancers that is viral in origin is the nonmalignant skin tumor known as the **common wart**. Electron microscopy reveals the presence of viruses in wart tissues; isolation of these viral particles and introduction into a susceptible host produce warts. (Handling toads has nothing to do with the transmission of warts.)

chance that they occur within these cells; or it could be that viruses invade cells after the cells have been transformed into cancer cells, and so have not contributed to the transformation process itself. A convincing demonstration that a virus does induce cancer is required, and certain criteria (called **Koch's postulates**) must be satisfied before we can say that proof positive has been obtained:

1. The virus must be isolated from the cancer.
2. The virus must be inoculated into a healthy animal and produce the disease.
3. The same virus must be reisolated from this experimentally induced cancer.

It is obvious that to apply such methods to the direct study of human cancers is almost impossible, and therefore other techniques have been employed. The transmission of human cancers to animals (other than man) and to human cells grown outside the body in tissue-culture systems has provided a great deal of information on the nature of the relationship between viruses and human cancers, but the results are still far from conclusive. Presently, it is suspected that viruses may be involved in human cancers such as Burkitt's lymphoma (a tumor that destroys the jaw), breast cancer (Box 8B), leukemia, and Hodgkin's disease (a cancer of the spleen and lymph nodes). However, the assumption held by some scientists that all cancers are caused by viruses, and that carcinogens such as chemicals and irradiations are effective merely because they activate the virus, is presently without foundation.

One of the problems involved in investigating virally induced cancers is that the behavior of viruses may depend on the host or the particular tissue that is invaded; in one tissue a virus may have no ill effects (it is simply a passenger virus); in another it may cause lethal effects, and in still others it may induce malignancy. To illustrate this let us look at a particular case.

BOX 8B Human mammary tumors

Breast cancer is the most common type of cancer in women in the United States: there are 65,000 new cases each year, and 25,000 deaths annually. In human breast cancer the earliest symptom is a lump in the breast; most commonly these are non-malignant, or benign, tumors. Breast cancer risk increases with age, and the tumor seldom occurs in young women. As the cancer develops, it tends to spread to the lymphatics and then to other parts of the body; it is at this stage that the cancer is dangerous. Detection of cancer in the early stages is simple and effective. By self-examination of the breast more than 90% of breast cancers are discovered.

There is a tendency for breast cancer to occur more frequently in certain females, and this makes it likely that a genetic predisposition for it may exist (the genetic factor, if it exists, does not appear to be strong). Human breast cancers show a hormone dependency. As early as 1896 G. Beatson, a British physician, found that surgical removal of the ovaries in patients with breast cancer caused a remission of the tumor. However, since breast cancer occurs more frequently among women who have never married and those who have not had children, the mechanism of hormonal action in human breast cancer remains an enigma. In mice, tumors of the mammary gland are also influenced by hormonal factors, but it has been conclusively shown that mammary tumors in mice are caused by a virus that is passed on in their mothers' milk. Although it seems to be different from the situation in the mouse, it may be that some human breast cancers are also viral in origin.

Treatment for breast cancer is ordinarily by surgery, but sometimes radiation and hormone therapy are employed. Cancer of the breast also occurs in men (the frequency is 1 case of male breast cancer for 100 cases of female breast cancer). The most common treatment for male breast cancer is, likewise, surgical removal.

In summary, we can say that we know little about the events that initiate human breast cancer, but we do know that hormones play an important role in the development and growth of mammary tumors. It is conceivable that the effect of hormones on the breast tissue produces an environment conducive to tumor formation; it is also possible that a combination of genetic and hormonal factors alone can induce tumor formation. Which, if any, of these is the causal mechanism remains uncertain at present.

In the early 1950s, Sarah E. Stewart and Bernice E. Eddy isolated a virus that caused salivary-gland tumors in mice; when reisolated and grown in tissue culture, the viruses multiplied; and upon reinjection into newborn mice, the viruses produced 20 different kinds of cancers in addition to the one found originally only in the salivary gland. This broad-spectrum virus was named **polyoma virus.** Other experiments using tissue cultures showed that on occasion polyoma virus not only induced cells to become cancerous but resulted in cell death. What is responsible for these many faces of the polyoma virus?

The explanation may in part relate to the protein coat that surrounds the virus. The viral protein coat readily provokes an **immune response** by the host (that is, specific proteins called **antibodies** develop in the blood of the host upon contact with foreign proteins and the antibodies neutralize or inactivate the virus); when viruses are contracted in small quantities, the animal builds an immunity to the virus. In nature, polyoma virus is spread among mice in the saliva and in the feces, and the young gradually build an immunity to it.[5] Antibodies are also passed to the offspring via the mother's milk. This acquired immunity probably accounts for the low incidence of spontaneous tumors in mice. Experimentally, massive quantities of virus can be produced in tissue cultures, and then these can be injected into highly susceptible animals. In young animals death often results because the defense system is unable to cope with the virus, but in older, more resistant animals the same virus may allow the host cells to live and cause them to proliferate; in this latter case, malignant cancers are produced.

---

[5] Laboratory experimenters working with polyoma virus also build antibodies to it.

How does the polyoma virus cause cancer? Electron microscopy shows the polyoma virus to be a round particle with a diameter of 2.7 Å, made up of a protein coat surrounding a core of DNA. When polyoma viruses enter a susceptible cell, the cell has two possible fates: (1) death, as a consequence of virus production, or (2) transformation into a cancer cell without the formation of new viruses. In both cases, viral particles are taken up by the host cell into small sacs, these sacs accumulate around the host-cell nucleus, some of the viruses lose their protein coats, and the naked viral DNA enters the nucleus. In the lethal infection, within 48 hours the host-cell nucleus becomes filled with viral particles; the progeny of the infecting viruses eventually fill the host cell and cause it to burst, and the released viral particles can begin a new round of infection. However, on rare occasions the transformation of the host cell into a cancerous condition by polyoma virus occurs. Viral particles are not seen in the transformed cell; rather the naked viral DNA is inserted into the DNA of the host-cell chromosomes. These viral genes usurp and corrupt the host cell, making it act in a malignant way. How the viral DNA usurps the prerogatives of the host cell is still incompletely understood, but some information is at hand.

Polyoma viral DNA carries about six to eight genes (it has 5,000 nucleotide base pairs); these genes are actively responsible for the production of a cancer cell. One of the viral genes probably codes for a protein that is a regulator (a promoter or repressor substance) and this turns on or off specific genes of the host cell or the virus. (Genetic regulators are discussed in Chapter 7.) For example, the viral genes that code for viral coat protein are silent, or off, since once a cell is transformed, viruses with protein coats are never formed. On the other hand, the virus does turn on the cellular enzymes involved in DNA synthesis, presumably

at the transcription level, thus causing both host-cell and viral genes to be duplicated, and the cell to divide. In addition, viral genes direct the production of a specific coat on the host cell; this change in the host-call surface may be responsible for the altered relations of the cancer cell with its neighbors.

Why do viruses induce cancer? The induction of cancer by some viruses may be a byproduct of viral functions that have developed for their own multiplication. Because the virus is small, it cannot contain much genetic information and must therefore exploit the synthetic machinery of the cell to achieve its own replication. The viral activities induce cancer because they are similar to the activities and mechanisms by which cells regulate their own replication. In other words, viruses enter cells in order to reproduce their kind. To do this, they must turn on the DNA-replicating machinery of the host cell and redirect it to make viral materials. A side effect of continued DNA replication is likely to be continued cell division and proliferation—cancer.

8-3 THE CAUSES OF CANCER: SOME HYPOTHESES

What is the difference in
these lethal shapes
that makes them deathful deviates
from quiet harmony
with those around?
CHAUNCEY D. LEAKE

Cancer, as we have seen, is a descriptive term for a galaxy of diseases having a common biological pattern: the breakdown of the regulatory controls that precisely govern the growth of cells within the body. Cancers also show another property: an escape from normal cell-cell contact relations. Let us examine this loss of "stick-togetherness" of cancer cells and see how it contributes to their ability to infiltrate

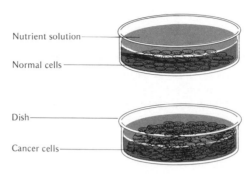

Nutrient solution

Normal cells

Dish

Cancer cells

normal tissues, detach, migrate, and metastasize to distant sites.

If a small piece of your skin were removed and placed in a glass dish containing a nutrient solution, the cells would settle down and attach to the glass surface. For a time, if there were very few cells, there would be rapid cell division, and the skin cells would spread out to form a uniform sheet of cells one cell thick (Figure 8-6). Having reached the edge of the dish, and now with each cell touching another, cell division would slow down. The sheet of cells would continue to live if supplied with nutrients, and when cells died they would be replaced by limited divisions in certain cells of the sheet. In this manner the continuity of the single layer of cells would be maintained. It appears that cell contact inhibits cell division; we have already seen this type of growth pattern in wound healing. A similar type of experiment can be performed with a cancer of the skin, but here the results would be quite different. Once the cancer cells were placed in the glass dish containing nutrient solution, they would divide and spread out in a single layer in the same way as normal cells, but division would not cease at this point. Rather, growth and division would continue without limit, and the cells would pile one upon the other producing a lump of cells several layers deep. There appears to be no boundary limiting the growth of the cancer cell. Division of cancer cells can occur in the living organism (in vivo) or in a glass dish (in vitro) over and over again; it occurs not only with skin cancer but with cancers derived from liver, nerve, kidney, bone, and so on. All cancers show the same sort of phenomenon: uncontrolled growth and loss of normal contact relations.

This phenomenon, where cells cease to divide when they touch one another, is called **contact inhibition.** It is not clear what factors are responsible for contact inhibition, but it

may be related to a greater stickiness of normal cells so that their contacts are firm; the cancer cells, in contrast, are less adhesive and can separate and pile upon one another. This local regulation of growth makes good sense, for inside our own bodies or the body of any other multicellular organism there is a limited amount of space; thus it would be a great disadvantage if cells continued to divide once the space was filled—there would be no room for them. At the same time, if for some reason a space appears, as for example in a wound, then division can begin and the space is filled up. Stickiness of cells is also rather specific: liver cells adhere to liver cells, and kidney cells to kidney cells, but cancer cells show no preferences. Thus, we can begin to see why cancer cells are able to invade and maintain themselves in a variety of tissues irrespective of their tissue of origin.

One of the puzzling things about cancer concerns the diversity of physical, chemical, and biological agents that are capable of producing what appears to be the same end result: an inherited pattern of cell growth that is abnormal, unregulated, permanent, and autonomous, accompanied by a loss of cellular contact inhibition. When we begin to ask, How is cancer caused? rather than, What is cancer?

we move from an area of descriptive science to the realm of speculative science. A plethora of hypotheses concerning the mechanisms of cancer formation is available: each hypothesis has its merits and each has its limitations. We will discuss three of them here.

Hypothesis: mutations cause cancer

Since cancer is usually a permanent (irreversible) change, it has been suggested that cancers develop as a result of mutations. It is well known that physical agents such as ultraviolet light and x rays as well as certain chemical agents are both carcinogenic and mutagenic (capable of producing mutations). Hermann Muller, a Nobel laureate in genetics, suggested that the carcinogenic properties of these agents are due to the accumulation of changes in individual genes (**mutations**) in certain cells of the body, and that these reduce the cell's susceptibility to inhibition of cell division. This hypothesis also states that several mutations are necessary for inducing the cancerous state, and as such fits well with what we know about the slow rate of development of most cancers and their increased incidence with advancing age. The older a person gets, the mutation hypothesis holds, the greater the chance that some of the body cells could accumulate these mutations, thereby becoming transformed into cancer cells. The essential mutations are those directly involved in cell division and could have as their basis the removal of an essential protein, such as the repressor substance that regulates the expression of genes involved in mitosis. Thus, the mutational hypothesis attempts to correlate chemical and physical agents with the capacity for tumor induction. Some investigators suggest that these agents could also activate a latent virus and thus could play a role in virally induced cancers. Once the carcinogen induces such a change, it no longer has to be present to perpetuate the tumorous condition; since the change affects the genetic material, it is permanent, and autonomous growth results.

What evidence is there to support the view that mutations are involved in cancer? The death rate from cancer of the large intestine increases about 1,000 times between ages 20 and 80, but significantly most of this increase occurs after age 40 (Figure 8–7). The relationship between age and death rate shown in the graph (Figure 8–7) is logarithmic, and can be explained by the hypothesis that it takes both time and a series of mutations to produce a cancer. The slope of the line suggests about five mutations are required to produce a cancer, and the incubation period for a cancer may encompass a lifetime. Thus, the incidence of lung cancer may be related not to the number of cigarettes an individual smokes today, but to the number he or she smoked 20 years ago. Similarly the scrotal cancer of chimney sweeps described by Pott was initiated during childhood and took many years for its full development.

If mutations cause cancer, is it due to normal errors in DNA replication or are environmental mutagens responsible? Current evidence implicates the environment. Cancer of the stomach is more common in Japan than in the United States whereas for cancers of the breast, prostate, and bladder the reverse is true. When Japanese leave Japan and come to live in the U.S. these differences are lost within two generations despite the fact that such immigrants tend to marry within the group. The time factor, the changed environment, and the rather constant genetic background tend to implicate environmental mutagens.

What is the mechanism involved in mutagenicity? The basic mechanism of mutation is either by base-pair substitutions or frameshift mutations. In the base-pair substitution the

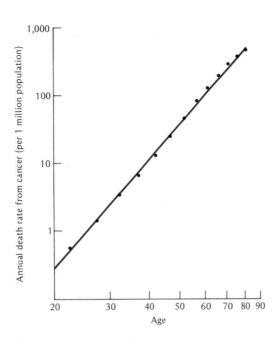

mutagen resembles the normal nucleotide base so that it is mistakenly inserted in the DNA during replication. This substitution alters base pairing; the coding for an amino acid would then be incorrect, producing an abnormal protein (or a shortened polypeptide if the new triplet coded for termination). Frameshift mutations occur when the mutagen slips between the base-pair "rungs" of the DNA, and by distortion of the molecule—a stretching or shortening—the precise reading of the triplets is upset. As a consequence, bases are read in an entirely new sequence of triplets, and the protein produced may be either novel or functionless. No matter what the mechanism, if mutations are responsible for many human cancers, then prevention and eradication should be possible.

Hypothesis: selective gene activation causes cancer

Recently, investigators have suggested that gene alterations (addition or deletion of genes) need not occur in order to produce the cancerous state; all that is necessary is a more or less permanent shift in gene activity from a condition of cellular rest to a condition characteristic of the cell during the time of division. This hypothesis suggests that the neoplastic state is fundamentally a more or less permanent switch in the pattern of cellular syntheses. What is the basis for such thoughts? Although most tumors show properties of irreversible and autonomous growth, not all tumors do; in fact, in some rare cases tumors show a reversibility and become normal. Such a condition, proponents of the gene-activation hypothesis argue, could not occur if the genetic constitution of the cell were altered during transformation from the normal to the neoplastic state—that is, if the cell acquired a gene for cancer.

The control mechanisms and materials involved in normal cell division are incompletely understood. We have little idea of the nature of the key compound that normally triggers cell division and changes the metabolic and synthetic pattern of the cell from the resting state to the dividing state. We do not understand how selective gene action is accomplished so that cells synthesize new species of mRNA and turn on growth and division. The search for and elucidation of the functional role of division-regulating proteins remains one of the most exciting tasks ahead for biologists. However, it has been suggested that carcinogens act in a similar way to substances that trigger division in the normal cell, but they act at the wrong time, and the new patterns of synthetic activity (characteristic of dividing cells) they produce are self-perpetuating (difficult to reverse). According to A. Braun, the reorientation of synthetic pathways from a resting to a dividing cancer cell can be visualized as follows:

The carcinogenic agent temporarily inactivates a repressor molecule. Inactivation of this repressor allows the operon which it normally inhibits to become active, and this newly activated operon can, in turn, indirectly inhibit production of the original repressor. Once this stable shift in gene activity occurs, the presence of the carcinogenic agent is no longer necessary to maintain the new steady state. Although the switch in gene activity is stable and can be perpetuated by cell division, it is theoretically reversible. Any agent capable of temporarily repressing the newly activated operon would allow the original steady state to be resumed again.

A. BRAUN, *The Cancer Problem: A Critical Analysis and Modern Synthesis,* New York, Columbia University Press, 1969.

Additionally, the selective gene activation hypothesis suggests that once the detailed regulatory mechanisms involved in normal growth and division are understood, then the growth pattern of the cancer cell will also be solved. In other words, when we understand what normally turns a cell from a resting state into a dividing state and how it accomplishes this, we shall be closer to understanding how other substances can do the same thing. At that point, it is conceivable that by artificial manipulation of nuclear function, cancer could be cured.

Hypothesis: viruses cause cancer

The hypothesis that cancer is caused by viruses is not really a hypothesis, since there already is considerable evidence for the viral origin of certain kinds of cancer. However, questions persist: Are all cancers induced by viruses? How do viruses contribute to the transformation of a normal cell into a neoplastic one?

We have considered two cancer hypotheses: one in which there was change in, or deletion of genetic information (mutations), and another in which there was no change in the genetic constitution (selective gene activation). Now let us consider a third hypothesis, involving the addition of new genetic information resulting from viral infection.

When viruses infect a host cell, their nucleic acids (genes) enter the nucleus to become integrated into the chromosomes of the host cell; in this manner new genetic information is introduced into the host cell. The introduced viral genes could exert their cancer-inducing effects in three possible ways: (1) viral genes could act as a direct source of genetic information, leading to the production of new and specific products directly responsible for the establishment and maintenance of the tumorous state; (2) products of viral genes could exert an indirect effect by causing the permanent activation of certain genes in the host cell, particularly those involved in cell growth and division; and (3) viral genes could derepress (activate) host-cell functions by means other than those involving transcription and translation of the viral genetic message. Which of these is operational in virally induced cancers is not clear at present.

The attractive aspect of the viral hypothesis for cancer induction is its suggestion that viral genes are incorporated into host-cell chromosomes and are replicated along with the host-cell genes at each cell division. Since the viral DNA is associated with, or incorporated into, the DNA of the transformed host cell, it is often impossible to demonstrate viral particles in a cancer cell. This conveniently explains why cancer cells show no viral particles even when examined with the electron microscope.

The hypotheses presented here by no means exhaust all the possibilities, but they do offer attractive suggestions for future work. Perhaps by the time you read this we may have a better view of how viruses, mutations, gene activation, and other carcinogens act to product the cancerous state.

SUMMARY

1. Cancer is one of the principal diseases causing death in the United States.

2. A cancer or tumor is composed of neoplastic cells in a state of autonomous growth and multiplication.

3. Clinically, cancers may be benign or malignant. Malignant tumors arise from benign tumors which become invasive. If invasive cancer cells are transported to other parts of the body by blood or lymph, they are metastatic.

4. Carcinogens are agents capable of causing cancer; they may be chemical, physical or viral.

5. Chemical carcinogens include many environmental substances, thought to be responsible for more than 75% of human cancers; examples are soot (3:4-benzpyrene), dyes (β-napthylamine), asbestos, nickel, chromates, and arsenic; dietary additives such as cyclamates and saccharin; hormones, including some synthetic estrogens and progesterones; air pollutants, including fossil fuel and coal tar derivatives such as 3:4-benzyprene; and cigarette smoke (DMBA).

6. The Ames test is a screen for environmental carcinogens. It depends on the mutation rate of a strain of *Salmonella* bacteria.

7. Physical carcinogens include the ultraviolet rays of the sun and x rays.

8. Some viruses are carcinogenic. The effect of a cancer-causing virus may vary with the tissue invaded. For example, polyoma virus causes a broad spectrum of cancers in newborn laboratory mice, possibly owing to lack of antibodies and absence of an immune response in these animals.

9. Polyoma virus invades the host-cell nucleus. The virus may cause cell death as a consequence of virus production, or it may transform the cell into a cancer cell; then viral DNA is replicated along with host-cell DNA, and the host-cell surface is changed, altering cell-cell interactions.

10. Cancer cells show a loss of contact inhibition.

11. There may be a genetic factor in human breast cancer, although it seems to be hormone dependent. Mammary tumors in mice are caused by a virus that is passed on in their mothers' milk.

12. Cancer may be caused by cumulative mutations. Mutagens may have their effect by altering the DNA itself (base-pair substitutions or frameshift mutations) or by activating a latent virus.

13. Carcinogens may selectively and permanently activate the genes that switch cell division on, or repress those that keep it switched off.

14. Viruses may cause cancer by adding genetic information that causes cell division; by activating host-cell genes responsible for cell division; or by activating cell division in some other way.

KEY WORDS

cancer
tumor
neoplastic
benign
malignant
invasive tumor
metastatic tumor
carcinogens
Ames test
mutagens
Koch's postulates
polyoma virus
immune response
antibodies
contact inhibition
mutations

TOPICS FOR REVIEW AND DISCUSSION

1. **How may viruses cause cancer?**
2. **Why is it difficult to determine the role of viruses in the establishment of human cancers?**
3. **Describe the varied behavior of the polyoma virus and its significance to cancer.**
4. **Describe how environmental carcinogens are related to the incidence of cancer.**
5. **What are some of the ways that scientists use to screen potential carcinogens?**
6. **What is loss of contact inhibition?**
7. **How may somatic mutations be related to the development of cancer?**
8. **How is cancer defined biologically? Clinically?**
9. **How can an understanding of normal cell division aid in an understanding of cancer?**
10. **The cancerous condition is usually a permanent, heritable change in a cell. What does this tell us about the mechanism of induction?**

WHAT IS MAN?

A self-balancing, 28-jointed adapter-base biped; an electrochemical reduction plant, integral with segregated stowages of special energy extracts in storage batteries for subsequent actuation of thousands of hydraulic and pneumatic pumps with motors attached; 62,000 miles of capillaries . . .
"The whole, extraordinary complex mechanism guided with exquisite precision from a turret in which are located telescopic and microscopic self-registering and recording range finders, a spectroscope, etc.; the turret control being closely allied with an air-conditioning intake-and-exhaust, and a main fuel intake . . .

R. BUCKMINSTER FULLER,
Nine Chains to the Moon, J. B. Lippincott Co., 1938

THE HUMAN ORGANISM

What is a human being? Anthropologists regard humans as the tool-making descendants of an apelike stock and the makers of culture. Psychologists view man as a behavioral system capable of learning, memory, and reason. To sociologists we are of interest because of the institutions and societies we form. Poets see the human organism as the abode of the soul, and for artists humans are forms of transcendent beauty. To a biologist the human organism is an anatomically complex, highly adaptable self-reproducing animal who can accumulate experience and use it to alter his environment and his fate.

The machinery of the human body can be most conveniently studied by taking it apart piece by piece, examining each part in detail, and then putting the parts together again and determining the way in which they function to make a smoothly operating whole. The parts of the body are known as organs, and organs that function together constitute systems. The chapters that follow describe the various systems of the human body, which number about eight, depending upon how they are classified. The classification is merely a didactic convenience, and there is nothing fixed in any arrangement; one may split or lump organs into systems in a variety of ways. It is important to remember that all the body's systems interact and the simplest of bodily functions requires more than one system. A malfunction in one system can lead to breakdown or dysfunction of others.

Study of the human organism has intrinsic interest for us, because we are the subject of our own study. Moreover, it is of considerable importance for all of us to know what we are and how we came to be, for in understanding these we hold the key to our future. As far as we know, man is the only creature to have developed on earth who is able to stand apart, as it were, and to examine himself in his own setting. No other animal has altered its environment to the degree that civilized man has. To a large extent, all other creatures are at the mercy of their particular environment and adapt themselves to it. Man adapts the environment to himself. The implications of this are, of course, enormous, and the more each of us understands about himself and his relationship to the world in which he finds himself, the better equipped he will be to adjust himself to that world and that world to him. The admonition, "know thyself," is the prime reason for studying the human organism. As a consequence we may be more reasonable in our relations with other human beings, other forms of life, and our total environment.

Test-Tube Babies Seem Nearer

Science is making the relationship between sex and procreation ever more tenuous. The contraceptive pill prevents conception and legalized abortions make it simple to terminate unwanted pregnancies. Now word comes to us that the day is not far off when the union of sperm and egg to produce a fertilized egg and its development into a human embryo will all take place within the confines of a test tube. The methods are simple and straightforward. About 30–32 hours beforehand the woman is given an injection of a hormone that stimulates the maturation of eggs in the ovary. Removal of the eggs is achieved by making a small incision into the abdomen, locating the ovary, and sucking out the eggs with a fine tube. The eggs are then placed in a nutrient broth for 1 to 4 hours before being exposed to sperm cells obtained from the husband. Embryos that have developed to the 16- or 32-cell stage may be reintroduced into the womb. Although implantation of a human embryo by such methods has not as yet been achieved, it seems that the day is not too far off.* Ultimately fertilization and development may be completed outside a woman's body—test-tube babies in the literal sense.

*Indeed, in 1978 prophesy may have become reality: Drs. Steptoe and Edwards of England reported that the world's first test-tube baby—a 5 pound 12 ounce girl—was born to Mrs. Lesley Brown. The infant was, it was claimed, conceived by having Mr. Brown's sperm fertilize an egg removed from Mrs. Brown's ovary: two days later this was implanted into the womb, where development took place until birth.

PERPETUATION OF LIFE I: SEX-CELL FORMATION AND FERTILIZATION

9–1 SEXUALITY AND MEIOSIS

"And this," said the Director opening the door, "is the Fertilizing Room." . . . "These," he waved his hand, "are the incubators." And opening an insulated door he showed them racks upon racks of numbered test-tubes. "The week's supply of ova. Kept," he explained, "at blood heat; whereas the male gametes," and here he opened another door, "they have to be kept at thirty-five instead of thirty-seven. Full blood heat sterilizes."

Still leaning against the incubators he gave . . . a brief description of the modern fertilizing process; spoke first, of course, of its surgical introduction— "the operation undergone voluntarily for the good of Society, not to mention the fact that it carries a bonus amounting to six months' salary"; continued with some account of the technique for preserving the excised ovary alive and actively developing; . . . referred to the liquor in which the detached and ripened eggs were kept; and . . . showed them how this liquor was drawn off from the test-tubes; . . . how the eggs which it contained were inspected for abnormalities, counted and transferred to a porous receptacle; how (and he now took them to watch the operation) this receptacle was immersed in a warm bouillon containing free-swimming spermatozoa—at a minimum concentration of one hundred thousand per cubic centimetre, he insisted; and how, after ten minutes, the container was lifted out of the liquor and its contents re-examined; how, if any of the eggs remained unfertilized, it was again immersed . . . ; how the fertilized ova went back to the incubators; where the Alphas and Betas remained until definitely bottled; while the Gammas, Deltas and Epsilons were brought out again after only thirty-six hours, to undergo Bokanovsky's Process.

"Bokanovsky's Process," repeated the Director. . . . "One egg, one embryo, one adult—normality. But a bokanovskified egg will bud, will proliferate, will divide . . . and every bud will grow into a perfectly formed embryo, and every embryo into a full-sized adult. Making ninety-six human beings grow where only one grew before."

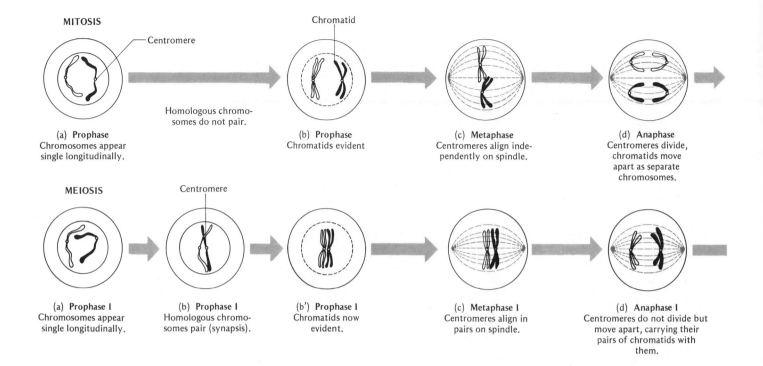

MITOSIS

(a) **Prophase**
Chromosomes appear single longitudinally.

Homologous chromosomes do not pair.

(b) **Prophase**
Chromatids evident

(c) **Metaphase**
Centromeres align independently on spindle.

(d) **Anaphase**
Centromeres divide, chromatids move apart as separate chromosomes.

MEIOSIS

(a) **Prophase I**
Chromosomes appear single longitudinally.

(b) **Prophase I**
Homologous chromosomes pair (synapsis).

(b′) **Prophase I**
Chromatids now evident.

(c) **Metaphase I**
Centromeres align in pairs on spindle.

(d) **Anaphase I**
Centromeres do not divide but move apart, carrying their pairs of chromatids with them.

So it went in the Central London Hatchery and Conditioning Centre as described in Aldous Huxley's 1932 novel, *Brave New World.* Even in such a soulless, streamlined Eden, human reproduction could be mechanized but the human donors could not be completely abandoned. Bokanovsky's Process produced 96 carbon-copy adults from one fertilized egg, but to get the fertilized egg required sperm and eggs from adult individuals.

Bokanovsky's Process—that is, the mitotic division of a fertilized egg—provides for the continuity of life. All our cells, numbering in the million of millions, are derived from such mitotic divisions. Thus, each of us is essentially a complex organization of carbon-copied cells, for mitosis ensures that genetic information

remains more or less constant from cell generation to cell generation (Chapter 7). Identical human twins or triplets are the result of the separation of cells that are mitotic products of a single fertilized egg—to use Huxley's term, identical twins are derived by Bokanovsky's Process.

We replace lost cells in our body by mitosis, and we grow from a single cell to 100 million million (10^{14}) cells by the same process, but we do not reproduce ourselves by this method. Even in *Brave New World,* Bokanovsky's Process could not make an Alpha from an Epsilon or vice versa; that required the combination of a sperm with an egg. Why? The answer lies in our sexuality. Bacteria and one-celled plants and animals are asexual; that is, they usually

do not produce sex cells, and all the offspring come from a single parent. Most other living creatures on this planet, ourselves included, reproduce by sexual means. Sexually reproducing organisms produce sex cells that fuse in pairs to form a new cell, the **zygote,** and from this the many-celled organism eventually develops. The fusion of sex cells **(gametes)** is called **syngamy,** or **fertilization;** thus, sexually produced progeny have not one but two parent cells. It is of critical importance to recognize that when fusion of gametes takes place the nuclei of the gametes also fuse, and as a result there is an intermingling of two separate sets of genetic messages. (More about this in Chapter 22.)

Recall that the messengers of inheritance in

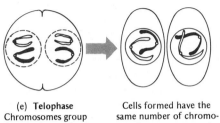

Chromatid replication would occur during interphase before another division took place.

(e) **Telophase**
Chromosomes group into nuclear region.

Cells formed have the same number of chromosomes as were present in the parental cell.

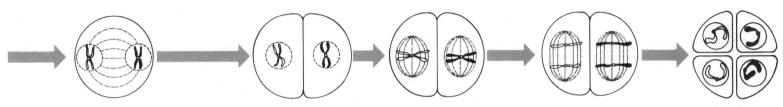

(e) **Telophase I**
Chromosomes group into nuclear region. (Each nuclear region contains a chromosome with two chromatids.) The cells formed contain half the number of chromosomes present in the parental cell.

(f) **Prophase II**
No replication of genetic material prior to this, hence no new chromatids are formed. Therefore, each cell is haploid, containing half the number of parental chromosomes.

(g) **Metaphase II**
Centromeres arranged on a new spindle.

(h) **Anaphase II**
Centromeres divide for the first time; sister chromatids (now separate chromosomes) separate and move to opposite poles.

(i)
Four cells, each with a single chromosome or half the number present in the parental cell.

our cells are called **genes** and that these are localized in discrete nuclear bodies called **chromosomes** (Chapter 7). Chromosomes move from one cell generation to the next during cell division and are the vehicles of inheritance. Let us consider our own chromosomal condition for a moment. Each cell of the body contains 46 chromosomes, and if we assume that our gametes (sperm and egg) arise by mitosis, then the chromosome number of our gametes would also be 46, since mitosis maintains a constant chromosome number. When fertilization occurred, the zygote would contain double the number of chromosomes— 92. Such individuals would also produce eggs and sperm containing 92 chromosomes, and the resulting zygote would contain 184 chro-

mosomes. In each succeeding generation the chromosome number would double if mitosis were the only type of cell division. In 10 generations, about the time since the United States declared its independence, the number of chromosomes would increase from 46 in the cells of first-generation individuals to 23,332 in the cells of members of the tenth generation. Clearly, this is not what occurs.

It is obvious that in a sexually reproducing population of organisms, the gametes cannot be formed by simple mitotic division. Sexually reproducing organisms like ourselves produce gametes by a process that includes a special type of cell division through which the number of chromosomes in the daughter cells is reduced. This process is called **meiosis.** Meiosis

encompasses two divisions but involves only one replication of the chromosomes; in this way the number of chromosomes in each daughter cell is halved. Let us examine the process of meiosis and compare it with the process of mitosis by looking at the dance of a single pair of chromosomes in each type of nuclear division (Figure 9–1).

A number of important differences between mitosis and meiosis are apparent. At the outset of both types of cell division the number of chromosomes is the same, and the chromosomes occur in pairs. Members of each pair of chromosomes are identical in size and shape, and are called **homologous chromosomes.** Cells having two of each type of chromosome are said to be **diploid** and have the 2*n* number of chromosomes (n stands for the number of different types of chromosomes). At the start of both meiosis and mitosis, the cells are diploid, and in the example illustrated the 2n number is 2 (n = 1). First, in meiosis homologous chromosomes pair (this is called **synapsis**); in mitosis this does not occur. Second, in meiosis there are two cell divisions without replication of the genetic material between divisions; in mitosis there is one division. Third, mitosis reproduces the chromosome number exactly in each daughter cell; in meiosis the number in each is halved. At the end of mitosis the chromosome number is 2n (diploid); in meiosis the products have only a single member of each chromosome pair present, and the chromosome number is n (**haploid**). In the illustrated example, the cells derived from the two divisions of meiosis have a single chromosome, while each of the cells derived from the mitotic division have two chromosomes.

A human body cell, such as a skin or liver cell, contains 46 chromosomes, and whenever it divides mitotically the daughter cells also have 46 chromosomes—the diploid (2n) number. If a body cell containing 46 chromosomes divides meiotically, the products of such a nuclear division will have 23 chromosomes, the haploid (n) number.

Where in the human body does meiosis take place? In sexually reproducing organisms such as ourselves, the body consists of two basic kinds of cells which are called **somatic cells** and **germ cells.** The germ cells divide meiotically to give rise to sex cells or gametes, the sperm and eggs. The germ cells are usually located in organs called **gonads.** Indeed, the only place in the body where meiosis occurs is in the gonads. Somatic or body cells do not contribute to the germ line, and although they may assist in the act of reproduction in higher organisms, they never contribute directly to the

cells of a new individual. Somatic cells reproduce themselves solely by mitosis and in this manner they provide for growth in bulk or increase in body-cell number.

Meiosis enables us and all other sexually reproducing creatures to reproduce our kind without multiplying the number of chromosomes in the cells of succeeding generations. More important, it provides a mechanism whereby the genetic heritage of two individuals may be reassorted and recombined in new ways in a new individual. The result is that every generation produced by sexual means has new genetic possibilities. Meiosis provides a bridge between succeeding generations— parents passing on to offspring a minute fraction of their own matter. Somatic cells are mortal; they die with the individual, but through the germ cells, eggs and sperm, we gain a kind of physical immortality.

9–2 THE MALE REPRODUCTIVE SYSTEM

Producing sperm and semen

The human male reproductive system is designed for the production of gametes (**spermatogenesis**) and their transfer to the female, where fertilization will take place. It is also concerned with the production of the chemical messages (hormones) that control development of the male secondary sexual characteristics— for example, hair on the face and body, deepening voice, body size and shape. Figure 9–2 illustrates the male reproductive system, which consists of the following: (1) the testes, paired male gonads (gonos—seed; Gk.) contained within the scrotum, (2) a duct system leading from the testes to the penis, including the seminiferous tubules and vas deferens, (3) accessory glands, and (4) a copulatory organ, the penis.

Each **testis**[1] consists of two functional parts: one produces endocrine secretions (endo— within, inside; Gk.), the male sex hormones (**androgens**) which circulate in the blood, and the other produces an exocrine secretion (exo— out; krinein—to separate; Gk.) which leaves the body, the **spermatozoa** or **sperm.** At maturity the testis is a solid, egg-shaped structure (Figure 9–2b) that contains highly coiled tubules, the **seminiferous tubules,** enmeshed in connective tissue and blood vessels. In each testis there are approximately 800 tubules, whose combined length is about 0.5 mile. In the fetus and up to the time of puberty, the seminiferous tubules are represented by solid cords of cells. These cords contain primordial germ cells (**spermatogonia**) and nurse cells (**Sertoli cells**) that will eventually nourish the developing spermatozoa. Some cord cells also secrete male hormones that markedly affect the maleness of the developing child.

At puberty the young male child becomes an adolescent, and these are times of turbulent change until the boy becomes a man. The changes depend upon a small, pea-sized gland, the pituitary gland, suspended by a short stalk from the base of the brain. The pituitary gland receives signals from the brain and secretes hormones that travel throughout the body via the blood. (The pituitary gland and hormones are discussed in Chapter 16.) Two hormones secreted by the anterior portion of the pituitary trigger the testis to activity: **follicle-stimulating hormone** (FSH) and **interstitial-cell-stimulating hormone** (ICSH). FSH causes proliferation of the germ cells in the seminiferous cords, and sperm

---

[1] The singular of testes is testis. The diminutive form of testis is testicle; the two words are synonymous. For some reason the popular literature commonly uses testicle, whereas the scientific literature uses testis.

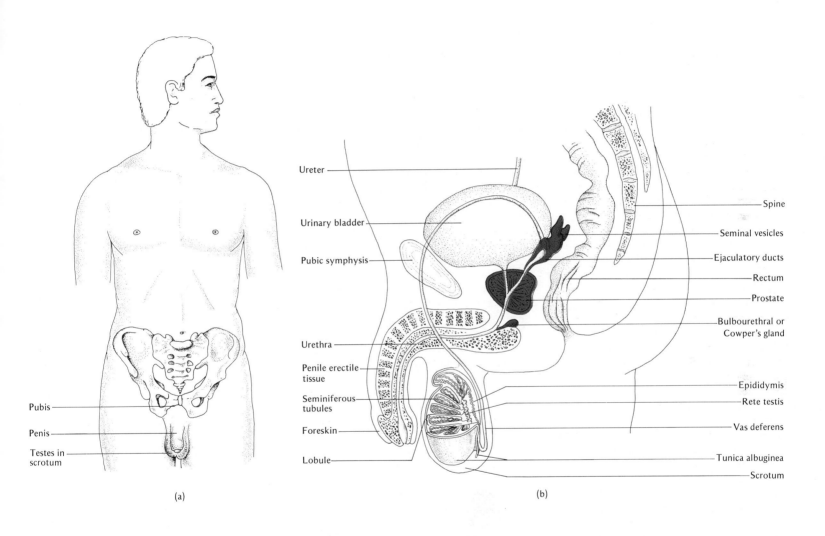

FIGURE 9–2 The male reproductive system. (a) External anatomy. (b) Detailed view of male pelvic organs shown in cross section.

Ureter

Urinary bladder

Pubic symphysis

Urethra

Penile erectile tissue

Seminiferous tubules

Foreskin

Lobule

Pubis

Penis

Testes in scrotum

Spine

Seminal vesicles

Ejaculatory ducts

Rectum

Prostate

Bulbourethral or Cowper's gland

Epididymis

Rete testis

Vas deferens

Tunica albuginea

Scrotum

(a)

(b)

formation begins. ICSH stimulates the secretion of male hormones (androgens, notably **testosterone**) by certain cells called interstitial cells that lie between the seminiferous cords. The androgens initiate development of those characteristics we recognize as adolescent, such as deepening voice, development of a beard, and muscular body shape.

The most important effect of ICSH, FSH, and testosterone is the initiation of spermatogenesis. The seminiferous cords enlarge in length and diameter, they hollow out, and sperm formation from spermatogonia begins. The processes involved are meiosis, maturation, and differentiation. In the human male, sperm formation requires about 14 days—2 days for spermatogenesis plus 12 days for passage through the male reproductive tract. Details of the process are shown in Figures 9–3, 9–4, and 9–5.

Mature sperm are liberated into the seminiferous tubule and make ready for their journey through a half mile of highly convoluted tubing (Figure 9–2b). In some as yet unknown way the sperm move from the tubule to straight collecting ducts, then they enter a network of smaller tubules **(rete testis);** eventually these unite to form about 20 collecting tubules, and from here the sperm move to the coiled ducts of the **epididymis.** The epididymis functions to remove fluids and cell debris accompanying the sperm, and in addition serves to increase the fertilizing capacity of sperm. (Sperm removed directly from the seminiferous tubules have a reduced capacity for fertilizing eggs compared with those taken from the epididymis.) Sperm are routinely stored (up to a month) in the epididymis. The tubules of the epididymis are continuous with the long, thick-walled, relatively straight **vas deferens.** The muscular vas deferens passes upward from the scrotum and enters the abdominal cavity through the **inguinal canal.** (The inguinal canal,

located in the groin, is the opening through which the testis descended from the abdomen at birth, and this is routinely checked by a physician for an inguinal hernia.) The vas deferens continues upward and across the surface of the bladder, then turns downward and crosses under the bladder. Eventually the two vasa deferentia join the tube leaving the bladder and enter the penis as a common duct called the **urethra.** Shortly before ejaculation the part of the urethra that lies between the points of entry of the vasa deferentia and Cowper's glands receives, in addition to sperm cells, fluid secretions from three sets of accessory glands—the two **seminal vesicles,** the **prostate gland,** and the two Cowper's **(bulbourethral) glands.** These secretions, together with the sperm cells, constitute the **semen** (seed; L.). The climax of sexual excitation in the male occurs when the accumulated semen is forcibly ejaculated by spasms or contractions of the muscle layers that surround and are adjacent to the urethra. By a remarkable system of engineering, urine and semen never pass through the urethra at the same time.

In older men it is not uncommon for the prostate to enlarge sufficiently to obstruct the urethra and to make passage of urine difficult; this calls for surgical removal of the gland. In elderly males the prostate is also a common site of cancer.

Seminal fluid, a complex liquid material, contains fructose, citric acid, amino acids, choline, enzymes, prostaglandins, and mucus, and is slightly alkaline. The tiny sperm cells can store very little food themselves and are entirely dependent upon extrinsic nutrients for use in respiration to provide the ATP necessary to keep the flagella lashing to and fro. Most of the energy for sperm motility is derived from the metabolism of fructose, a unique property of sperm, since most cells use glucose as their

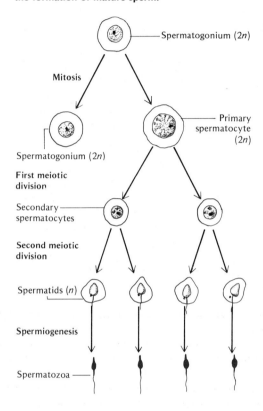

metabolic fuel. Seminal fluid also plays other roles; it lubricates the passageways through which the sperm pass during ejaculation, it helps protect the sperm from the acidity of the female vagina, and it serves as a medium for carrying sperm to the vagina. Immediately after ejaculation, the sperm are motionless, as they were before ejaculation, because the seminal fluid is thick and mucuslike in consistency, but within 30 minutes the fluid thins (owing to proteolytic enzymes contained in the fluid), and the sperm become highly motile—swimming as fast as 3 in. per hour. Sperm deposited

FIGURE 9–4 Spermatogenesis in the human male. (a) The seminiferous tubules are the factory for sperm production. Below the connective-tissue sheath are the spermatogonia. In layers approaching the lumen of the tubule, consecutive stages of developing sperm can be seen. Mature sperm are nearest the lumen. (b) Spermiogenesis (spermatid maturation). The spermatid loses most of its cytoplasm.

The nucleus shrinks and becomes the sperm head. The Golgi aggregates to form a cap, the acrosome, which helps the sperm to penetrate the egg. The centrioles migrate behind the nucleus, where one gives rise to the axial filaments of the sperm tail (flagellum). The mitochondria move to the middle piece providing power for the lashing tail.

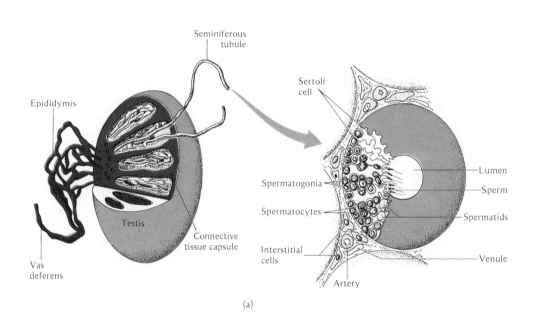

(a)

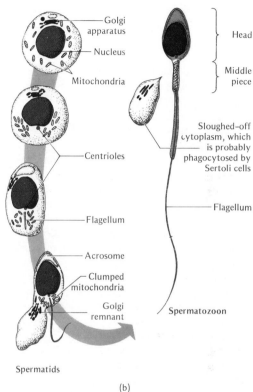

(b)

in the vagina can remain motile for as long as 2 hours; the entire trip along the female canal from the cervix to the Fallopian tube (a distance of 5–7 in.), where fertilization takes place, may take less than 70 minutes. Figure 9–6 illustrates the route taken by the sperm.

Human ejaculate (semen) may be about a teaspoonful in amount (3–4 ml), although the quantity varies from 1 to 11 ml. Each milliliter contains about 100 million sperm, so that there are about 300–400 million per ejaculate. There is, however, considerable variation in the actual number. Some 3 or 4 males out of every 100

are sterile or infertile. Infertile males have fewer than 150 million sperm per ejaculate, and more than 25% of these sperm are abnormal in appearance (all ejaculates contain some abnormal-looking sperm). The causes of male sterility are variable and include infections (such as mumps, typhus, or gonorrhea), ionizing radiation, and congenital defects. It may be difficult to realize why sterility should result when so many sperm are present, since it takes only a single sperm to fertilize the egg, but it is believed that the others are necessary to assist that sperm in reaching the egg.

If the male is continuously producing sperm, what happens if there is no ejaculation? That is, what happens to sperm that are not utilized? There are indications that unused sperm are resorbed by the tubules so that ejaculation after a period of rest usually produces a more or less constant volume of semen.

External genitalia: scrotum and penis

The most obvious parts of the male reproductive system are the penis and scrotum. Shortly before birth the testes descend from the abdomen and

FIGURE 9–5 Diagrammatic reconstruction of the human sperm as revealed by the electron microscope (side view).

enter the scrotal sac by way of the inguinal canals. The **scrotum** is a wrinkled sac of skin with a midline seam that indicates its double origin in the embryo. In the small child, the inguinal canal may remain open and the testes may be withdrawn into the abdomen on occasion (for example, when it gets cold.) Normally, the canal is closed by the growth of connective tissue, and testicular retraction becomes impossible; however, if it fails to close properly, a loop of the intestine may drop into the scrotum via the enlarged ring of the inguinal canal and cause obstruction and pain in the groin. Even when the inguinal canal is properly closed, it is a point of weakness and easily ruptures. An open inguinal canal is called an **inguinal hernia** and is quite common in human males. The weight of the abdominal organs rests on the lower abdomen, and thus when this area is subjected to excessive strain—as by lifting a heavy object, for example—the probability for development of an inguinal hernia is increased. Such hernias are uncommon in four-legged mammals, in which the weight of the organs is more evenly distributed.

A major function of the scrotal sac is to keep the testes cool, or cooler by 0.5–4.5°C than the rest of the body. Spermatogenesis will not proceed at normal body temperatures, so the scrotum acts as a thermoregulator: when the temperature surrounding the testes is too cold the scrotal muscles contract bringing the testes closer to the body; another set of scrotal muscles causes the scrotal skin to shrink and pucker. Increased environmental temperatures cause these scrotal muscles to relax. It is crucial for the testes to descend prior to adolescence; if they do not, they atrophy, and sterility results. The condition of undescended testes is called **cryptorchidism** (*cryptos*—hidden, *orchis*—testis; Gk.). (The name of the familiar orchid plant is derived from the same Greek word, because

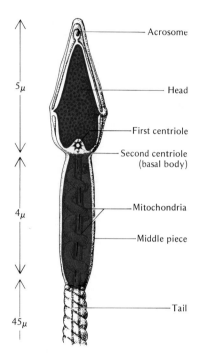

in some species the bulbous roots resemble human testes). Cryptorchidism often can be treated by surgery or administration of hormones.

The **penis** hangs in front of the scrotum and is the copulatory organ used for transferring sperm to the vagina of the female reproductive tract. At its tip is the slitlike opening of the urethra. The head of the penis, the **glans** (acorn; L.), is covered with a fold of thin skin attached at the neck region; this is the **foreskin** or **prepuce,** and it is this portion that is surgically removed at circumcision. Under normal circumstances, the penis is limp or flaccid, but in order to function as a copulatory organ it must become rigid and erect. Penile erection de-

pends upon a spinal reflex (Figure 9–7a) and can be induced by physical and/or psychic stimuli. Most carnivorous mammals such as cats, dogs, and seals have a stiffening bone, the os penis, but primates like ourselves have no such supporting element. The tissues of the human penis are so organized that there are two large cavernous beds of blood vessels above the urethra, the **corpora cavernosa,** and a smaller bed surrounding the urethra, the **corpus spongiosum** (Figure 9–7b). Under appropriate reflex stimulation, the cavernous blood spaces fill rapidly with blood. This increases the fluid pressure, compresses the veins, and prevents blood from leaving the penis. As a result of this engorgement the penis enlarges, stiffens, and becomes erect; it can then be introduced into the vagina. (There is no relation between penis size and virility or sterility, although a considerable mythology has developed about this.)

Ejaculation occurs when the penis is erect. Wavelike (peristaltic) contractions of the smooth muscles of the testes spread to the epididymis, vasa deferentia, seminal vesicles, and prostate gland, and at the same time the entrance from the bladder to the urethra is closed by a sphincter. Muscles in the penis itself contract, and sperm and seminal fluid are discharged. The entire process can take as little as a few seconds.

9–3 THE FEMALE REPRODUCTIVE SYSTEM

Producing eggs

The human female reproductive system is organized (1) to produce eggs **(oogenesis),** (2) to receive sperm and provide a milieu where fertilization of the egg by the sperm can take place, (3) to provide for the development of

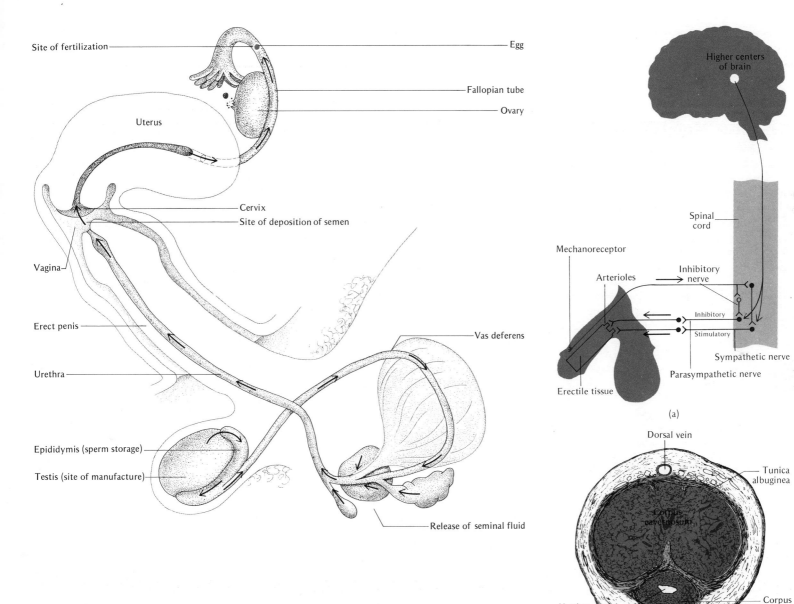

FIGURE 9–6 Coitus: the route taken by a sperm from its origin (testis) to the place where it fertilizes the egg (Fallopian tube).

FIGURE 9–7 (a) Reflex pathway for ejaculation. Reflex can be initiated from higher centers or these centers can inhibit arousal; mechanoreceptors in the penis can also initiate the reflex. (b) Cross section through the penis.

Site of fertilization

Egg

Fallopian tube

Ovary

Uterus

Cervix

Site of deposition of semen

Vagina

Erect penis

Urethra

Epididymis (sperm storage)

Testis (site of manufacture)

Vas deferens

Release of seminal fluid

Higher centers of brain

Spinal cord

Mechanoreceptor

Arterioles

Inhibitory nerve

Inhibitory

Stimulatory

Sympathetic nerve

Parasympathetic nerve

Erectile tissue

(a)

Dorsal vein

Tunica albuginea

Corpus cavernosum

Urethra

Corpus spongiosum

(b)

BOX 9A Sperm banks

From the Day of Deposit—A Lien on The Future

Ever since 1776, when Lazzaro Spallanzani found that human sperm could be frozen and then thawed out and revived, biologists have predicted a day when a man might have his semen put into cold storage and, perhaps after his death, used to beget a child.

That day is here now, not just as a laboratory curiosity but as a commercially available sperm banking service that brings much closer the prospect of controlled breeding programs to produce superior members of the human species.

Some experts predict that enlightened couples may come to prefer having a child by the frozen semen of a modern-day Einstein or Lincoln rather than by the husband. Others, equally expert, say that, on a large scale, such a program could actually redound to the disadvantage of the species by diminishing genetic variability and, therefore, adaptability to changing environments.

Whatever the consequences, it appears that we are soon to learn them, for two private corporations announced last week that they plan to open frozen sperm banks in New York City within the next few weeks. One, Genetic Laboratories, Inc., already has facilities in Minneapolis and St. Paul and plans a branch in Chicago. The other New York concern, the Iatric Corporation, said it also plans to open a sperm bank in Baltimore.

Both companies will offer their services to anyone who wishes to have quantities of his semen specially treated, frozen and kept at 321 degrees below zero until wanted for use. At that time, which may be many years later, the sperm bank will thaw out the semen and hand it over to a physician, who will place it in the uterus of the woman intended to be the mother of the child.

At this point, the procedures of artificial insemination are similar to those currently used with freshly collected semen. Last year in the United States, some 25,000 children were born following insemination of the mother with fresh semen from an anonymous donor.

Frozen semen, on the other hand, has been used by a handful of researchers over the last 17 years to initiate nearly 400 successful pregnancies. The rate of miscarriages and birth defects has been somewhat lower than normal.

With sperm banks, the use of frozen semen can be expanded to offer the woman not just the semen of a donor the doctor chooses, usually on the basis of superficial physical characteristics, but the semen of a broad range of outstanding men, living and dead, who have contributed, or sold, their semen to the bank.

For their part, the men who are setting up the commercial sperm banks wish to avoid the question of whose semen shall be used for what purposes. They intend to exercise no control over the use of the semen, except to say that it shall remain the property of the client and his to use as he sees fit.

Some 85 per cent of the clients in the two existing sperm banks are men who stored quantities of semen before undergoing vasectomies. Proprietors of the new sperm banks believe that the current boom in such male sterilizations (750,000 last year, as compared with 100,000 in 1969) will increase as men learn there is a method by which they may still have children.

Beyond the medical and scientific questions are many legal and moral issues to be resolved. If a man undergoes a vasectomy in the belief that he may have children by his frozen semen, and then the sperm bank folds, what recourse has he? If artificial insemination programs are to use sperm banks for eugenics purposes, who shall determine which men exemplify worthwhile characteristics—and, indeed, just what these characteristics are?

Many such questions have already come up with respect to artificial insemination with fresh semen. There the questions have been dealt with by avoiding them. That procedure is usually carried out in the utmost privacy, with no records or legal sanction. Now that sperm banks are going commercial, the questions again ask for answers.

BOYCE RENSBERGER, *The New York Times,* Aug. 22, 1971. © 1971 by The New York Times Company. Reprinted by permission.

FIGURE 9–8 The female reproductive system. (a) External anatomy. (b) Internal anatomy.

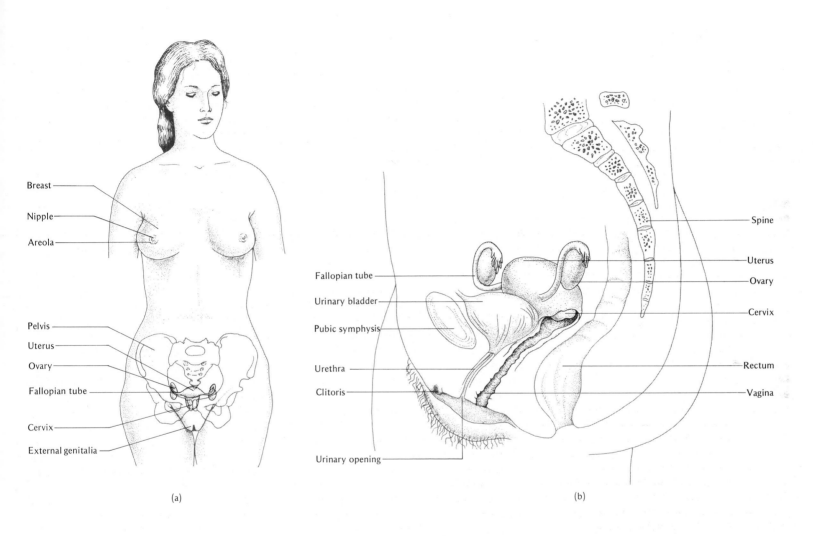

Breast

Nipple

Areola

Pelvis

Uterus

Ovary

Fallopian tube

Cervix

External genitalia

(a)

Fallopian tube

Urinary bladder

Pubic symphysis

Urethra

Clitoris

Urinary opening

Spine

Uterus

Ovary

Cervix

Rectum

Vagina

(b)

the embryo, and (4) to produce hormones to control the development of the secondary sex characteristics such as pubic hair, the breasts, and the feminine form. These are much more complex tasks than mere sperm production and ejaculation, and this is reflected by the greater complexity of the reproductive organs of the human female. The essential organs of the female system are (1) a pair of gonads (the ovaries); (2) a duct system including uterine tubes called the Fallopian tubes, the uterus (womb), and a copulatory organ (the vagina); and (3) the external genitalia (vulva) (Figure 9–8). The mammary glands (breasts) are also part of the female reproductive system.

The **ovaries** are almond-shaped glands about 1.5 in. long, lying completely within the body and located on each side of the pelvic cavity (Figures 9–8 and 9–9). Just as the testes are designed for sperm and hormone production, so the ovaries are designed for the production of eggs (ova) and hormones. Like the testes, the ovaries contain germs cells, but these are not arranged in tubules as in the testes; instead they are distributed in a cortical layer of varying thickness (Figure 9–10a). The ovaries are attached to the back wall of the pelvis by a fold of connective tissue called the **broad ligament** (Figure 9–9). Along the top edge of the broad ligament and close to the ovary lie the **Fallopian (uterine) tubes,** which connect to the uterus. They are called Fallopian tubes after the sixteenth-century anatomist Gabriel Fallopius (1523–1563), who described them. Curiously he did not recognize their true function, but thought they were chimneys for the escape of "sooty humors" from the uterus. The end of the Fallopian tubes closest to the ovary is fringed with projections that appear to caress the surface of the ovary. The Fallopian tubes provide a passageway to the uterus for the egg after it has been liberated from the ovary. Fertilization takes place in the Fallopian tubes; the fertilized egg travels down to the uterus itself and becomes implanted in the uterine wall, where development of the embryo takes place.

During infancy and childhood the immature egg cells exist as primary **oocytes** numbering 2 million at birth, but reduced by age 21 to 300,000. The ripening of the primary oocytes begins at puberty, usually between ages 9 and 17 (the average is 12.5 years). Like the male, at puberty the female child becomes an adolescent, and many physical and psychological changes take place. The most obvious signal of the onset of puberty is the first menstrual discharge of blood **(menarche)** from the vagina. The first menstrual flow, or period, marks the beginning of a female's reproductive life.

Menarche signals the beginning of cyclic changes in the ovary. Oocytes that were quiescent are now roused to activity. Each month (on the average, every 28 days) an **ovarian cycle** takes place: an egg matures in the ovary and is released and carried down the Fallopian

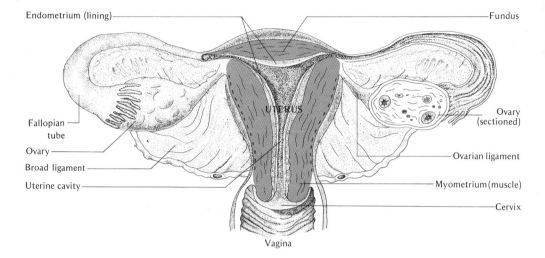

tubes. The uterus undergoes changes in its lining to prepare for pregnancy; if the egg is not fertilized, the uterine lining breaks down and is shed as a bloody discharge **(menstruation).**

During the reproductive years the monthly liberation of eggs from the ovary results in a continued decline in oocyte numbers, and at **menopause** (about age 45), when menstruation ceases, oocyte development ceases, the remaining oocytes degenerate, and a woman's reproductive life ends. Although ovulation ceases, sexual life continues and frequently improves. The psychological and physiological changes that characterize menopause do not end abruptly, but may last years. They appear to be unique to human females, for there is nothing to compare with them in wild animals. Primitive human females probably never lived long enough to experience menopause. Improved health care and civilization have made menopause the obvious sign of the termination of a woman's reproductive activity.

The ovarian cycle

The events of an ovarian cycle are shown in Figures 9–10 and 9–11. The signal for oocyte development and for the attendant change from child to woman is a rise in FSH, secreted by the anterior lobe of the pituitary gland (Figure 9–11). (Probably the brain itself activates the pituitary, but why this occurs at this particular time is still unknown.) Several oocytes begin to ripen in one or both ovaries under the influence of FSH. Each oocyte is surrounded by a nest of epithelial cells constituting a follicle. Different follicles mature at different rates; some are slower and others faster. A few of the follicles show a greatly expanded fluid-filled cavity and appear like inflated balloons. The enlargement is rapid and the follicle becomes many times its original size, almost 0.5 in. in diameter. At this inflated stage the follicle protrudes slightly and appears as a swollen blister on the surface of the ovary; it is called a **Graafian follicle** (Figure 9–10b) after the seventeenth-century

FIGURE 9-10 (a) Cross section of the mature ovary showing the events of the ovarian cycle. (b) Schematic representation of the ovarian cycle (oogenesis). (Modified from M. B. V. Roberts, *Biology: A Functional Approach.* Copyright © 1971 by M. B. V. Roberts. Published by The Ronald Press Company, New York, 1972. Used with permission of The Ronald Press Company and Thomas Nelson & Sons Limited.)

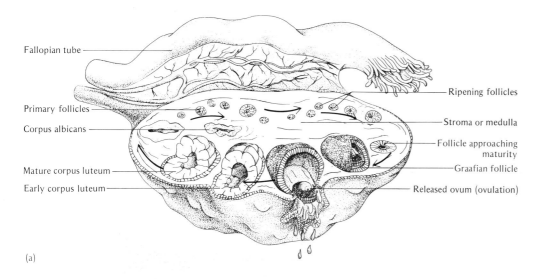

Fallopian tube

Primary follicles

Corpus albicans

Mature corpus luteum

Early corpus luteum

Ripening follicles

Stroma or medulla

Follicle approaching maturity

Graafian follicle

Released ovum (ovulation)

(a)

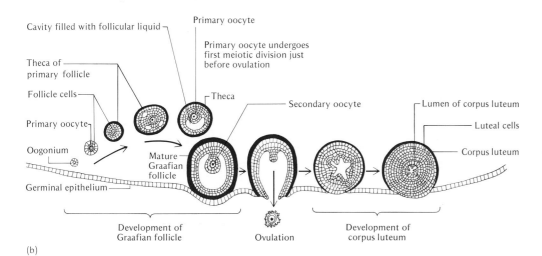

Cavity filled with follicular liquid

Primary oocyte

Primary oocyte undergoes first meiotic division just before ovulation

Theca of primary follicle

Theca

Follicle cells

Primary oocyte

Oogonium

Germinal epithelium

Mature Graafian follicle

Secondary oocyte

Lumen of corpus luteum

Luteal cells

Corpus luteum

Development of Graafian follicle

Ovulation

Development of corpus luteum

(b)

Dutch embryologist, Regnier de Graaf (1641–1673), who first described it.

By the time the follicle develops its cavity, the primary oocyte has undergone its first meiotic division, producing a large secondary oocyte and a tiny polar body. This secondary oocyte contains a haploid number of chromosomes. The other set of chromosomes is extruded into the polar body. The unequal division of cytoplasm in the first meiotic division assures that the secondary oocyte will have a large amount of cytoplasm, needed for subsequent development (Figure 9–13).

For further maturation the secondary oocyte requires another pituitary hormone—**luteinizing hormone** (LH) (Figure 9–11). The LH of the female is the same as ICSH of the male. Under the stimulating influence of FSH and LH, the follicle cells begin to secrete ovarian sex hormones, **estrogens** (Figure 9–11). Thus, FSH and LH both initiate maturation of the ovarian follicles and elicit ovarian-hormone production. The ovary, like the testis, is both an endocrine and an exocrine gland.

The mature (Graafian) follicle contains the oocyte embedded in a small mound of cells (Figure 9–10), the **cumulus oophorus** (literally "heap holding the egg"). The mature follicle is now ready for ovulation, that is, the expulsion of the secondary oocyte from the follicle. The cause of the rupture of the follicle wall is not known; it may be increased fluid pressure. A change in the balance of LH and FSH (LH suddenly rises) brings about a stretching of the follicle and a loss of elasticity, the follicle ruptures, the oocyte is ripped away from the follicle cells, and fluid oozes out. Some believe such changes are produced by an ovulatory enzyme in the extracellular environment. The shed secondary oocyte is surrounded by a thick jellylike layer (the **zona pellucida**) and a loose covering of cells (the **corona radiata**).

As a rule only one follicle ripens sufficiently and releases its oocyte each month. On occasion, however, two mature, and if both liberate their oocytes the possibility of multiple fertilizations and multiple births exists. The reason why one follicle liberates its oocyte and another does not is unknown, but those follicles that do not shed their oocytes regress, die, and are replaced by scar tissue; no other trace of them remains.

The secondary oocyte is liberated into the abdominal cavity, but its time there is quite brief. Only a few millimeters away lie the finger-like processes of the Fallopian tube, covered with cilia. The cilia beat in waves toward the interior of the tube, and this current sweeps the egg into it. In less than a few hours the oocyte enters the Fallopian tube and starts to move toward the uterus. The viability of the oocyte is probably about 24 hours, and if not fertilized during this time, it dies.

After ovulation the follicular cavity collapses and becomes filled with cells. The remains of the follicle, now called the **corpus luteum** ("yellow body" because de Graaf noted that color in the cow ovary; in the human ovary it is cream-colored), develops rapidly during the next 4 days (Figure 9–10). By 9 days it is solid and almost 1 in. across. Under the influence of LH, the corpus luteum begins to secrete the hormones estrogen and progesterone. The progesterone and estrogen circulate in the blood, arrive at the pituitary, and act on it to suppress production of FSH (Figure 9–11). Thus, no new follicles are stimulated at this point, and egg production is arrested. Progesterone also acts on the uterine lining, causing its glands to secrete glycogen and inhibiting its contraction. In the event that fertilization does not occur, the corpus luteum becomes inactive in 13 or 14 days and is transformed into a fibrous mass of scar tissue, the **corpus albicans** (Figure 9–10).

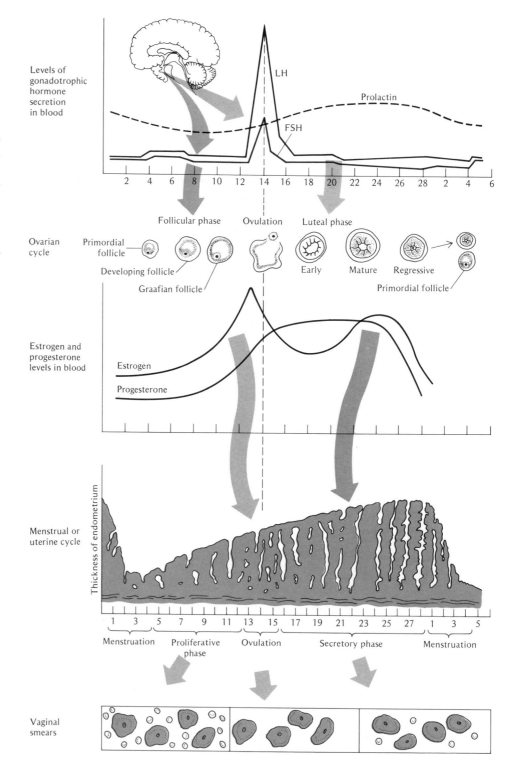

In about 2 months there is hardly a trace of the scar tissue left.

About 7 females out of 100 are infertile. The causes of infertility may be infection (most commonly gonorrhea), which blocks the Fallopian tubes, or failure to ovulate because of either insufficient hormone or an excessively thick capsule on the ovary. Sometimes these conditions can be remedied by hormone treatment or surgery. Hormone treatment involves the use of FSH or some synthetic steroid to stimulate follicular growth; occasionally, more than one follicle is stimulated, several oocytes are released at the same time, and multiple fertilizations and births result.

The uterus and the uterine cycle

The ovarian cycle of egg maturation and release is intimately related to the events taking place in the uterus, collectively described as the **uterine** or **menstrual cycle.** Menstruation and ovulation are different events occurring in different organs at different times, but are linked to each other by hormonal action.

The **uterus** is a hollow muscular organ about the size and shape of a pear (Figures 9–8 and 9–9). It is loosely anchored to the bony pelvis by ligaments, and is suspended in the pelvic cavity between the bladder in front and the rectum in the rear. The upper part of the uterus, the body, is mostly muscle, and it is this portion that enlarges greatly during pregnancy. The smaller, lower end of the uterus, the **cervix,** points downward toward the vagina. It can readily be felt by inserting a finger into the vagina; it feels like the tip of the nose. The uterus is where the developing child, first as an embryo and then as a fetus, is nourished and sheltered. The lining of the uterus (**endometrium**) is elaborately prepared for implantation of the fertilized egg.

Menstruation is a time of uterine bleeding, accompanied by a regression and shedding of the endometrium. It usually indicates that the egg has not been fertilized, the preparations of the uterus for implantation are in vain, and the entire lining is torn down and discarded. (For obvious reasons, menstruation has been described as the "weeping of a disappointed uterus.") The building up and tearing down of the endometrium constitute the menstrual cycle. It is a continuum that runs for about a lunar month or 28 days (*menses*—lunar, Gk.) before repeating itself (Figure 9–11). Cycles have no beginning or end, but for convenience we can designate four stages in the menstrual cycle. We begin with a landmark in the cycle: the start of bleeding.

Stage 1. Menstruation. The beginning of menstruation or bleeding is designated as day 1 of the cycle. The bleeding (often referred to as a period) lasts on the average for 4 days (range 3–7 days); at the end of this time the endometrium is only 1–2 mm thick and is rather smooth (Figure 9–11).

Stage 2. Proliferation. Under the influence of FSH, ovarian follicles begin to develop and to produce estrogen. As the estrogen level increases, endometrial repair begins and the thickened uterine lining acquires blood vessels (Figure 9–11). The proliferative stage lasts 8–10 days.

Stage 3. Ovulation. On or about day 14 of a standard cycle a rise in LH causes the ovarian follicle to rupture, releasing the egg. This is called ovulation.

Stage 4. Secretion. Estrogen and progesterone (produced by the corpus luteum) act upon the endometrium to prepare it for the fertilized egg. Blood vessels in the uterine wall become more abundant, and glycogen accumulates in the glands of the endometrium. The secretory phase lasts about 13–14 days.

On about the twentieth or twenty-first day of the uterine cycle, the corpus luteum receives a message, in the form of a hormone, indicating whether the egg has been fertilized and implanted in the endometrium. If implantation has not occurred, the corpus luteum starts to degenerate on day 22 or 23, and the progesterone and estrogen levels drop. The endometrium begins to break down, fragments are shed along with mucus from the uterine glands, blood oozes from the denuded endometrium, and on the average 1–8 oz of blood are lost. (Menstrual blood is distinctive in that it does not clot, probably because it lacks fibrinogen and thrombin, the clotting factors.) The decreased production of progesterone and estrogen causes removal of the block on FSH production by the pituitary. New FSH production initiates new ovarian and menstrual cycles.

The vagina

The **vagina** is a tubular canal about 4–6 in. in length, directed upward and backward to connect the uterus with the vestibule of the vulva (Figures 9–8b and 9–9). Like the uterus, it is situated between the rectum and the bladder. The vagina has muscular walls and is ordinarily folded to form a collapsed tube. It serves as a copulatory organ, receiving the shaft of the penis, and during childbirth it functions as a part of the birth canal. The vaginal walls contain no glands and normally are kept moist by a slightly acidic secretion of the cervix; a bacterium aids in maintaining this state. If acidity is not maintained, the vagina becomes prone to infection. Cells from the vaginal fluid are useful in determining the time of ovulation (Figure 9–11) and as a diagnosis of cervical (uterine) cancer.

The opening of the vagina into the vestibule may be partially covered by a mucous mem-

FIGURE 9–12 The external genitalia of the female.

brane, the **hymen.** The absence of a hymen is no more proof of sexual promiscuity than its presence is proof of virginity. Frequently it is absent in women who have never had sexual intercourse, and it may persist in women who have.

In the vaginal vestibule mucus glands, called **Bartholin's glands** (homologous to the bulbourethral glands of the male), secrete a viscous fluid and are stimulated to activity during sexual play.

External genitalia

The external genitalia of the female (Figure 9–12), collectively called the **vulva,** consist of the **mons veneris** (pubis), the **labia majora** and **minora,** the **clitoris,** and the **vestibule.** These organs play a minor role in the reproductive process per se, but do function in foreplay and the sexual act (coitus). The labia majora are fatty skin folds which pass backward from the mons veneris and in postpubertal females are covered with hair. The labia minora are two fleshy folds of skin which enclose the vestibule. The vestibule of the vagina is a cleft between the labia minora, and situated within are the hymen, and the openings of the vagina and the urethra (leading from the bladder).

The clitoris is a pea-shaped projection of erectile tissue well supplied with nerves and blood vessels; located at the upper end of the labia minora it is partially covered by them. The clitoris is the homolog of the male penis, is highly sensitive, and is important in sexual stimulation and satisfaction. (It does not contain the urethra.)

9–4 COITUS (COPULATION)

Courtship is the prelude to **coitus** (also called **copulation** or **sexual intercourse**). In humans

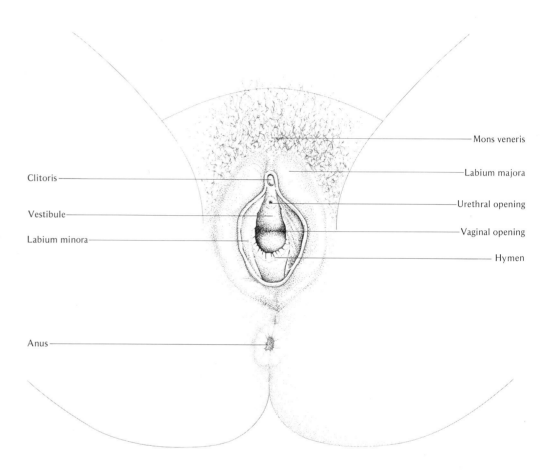

Clitoris

Vestibule

Labium minora

Anus

Mons veneris

Labium majora

Urethral opening

Vaginal opening

Hymen

the courtship process varies and may run the gamut of sensory experiences including touch, taste, smell, and sight—all usually pleasurable. The sexual foreplay of courtship serves one primary biological function: arousal and preparation for coition.

The result of sensory stimulation is tumescence—swelling of the breasts and genitals as they become engorged with blood. The nipples become erect, as do the clitoris and penis. The

erection of these organs occurs because blood is forced through the arteries faster than it can be drained by capillaries and veins. In the female, the cervix begins to secrete a fluid that lubricates the vagina. At first, individual droplets line the vaginal wall, and these then coalesce to form a uniform surface coating. The production of lubricating fluid, the first physiological evidence of the female's response to sexual stimulation, is involuntary and results from the marked dilation of blood vessels and massive localized vasocongestion in the region of the vagina. Occasionally, vaginal muscles contract and cause this vaginal fluid to squirt out (this is not ejaculation, for which it is sometimes mistaken). The inner two thirds of the vagina dilates and the uterus contracts rapidly and irregularly. The labia minora swell and Bartholin's glands in the vaginal vestibule secrete a small amount of fluid. In both male and female, the heart beats faster and the blood pressure rises.

Erection of the penis facilitates insertion into the vagina.

The clitoris serves both as a receptor and a transformer of sensual stimuli, and stimulation of the clitoris is the major cause of female **orgasm.** During sexual stimulation, the clitoris becomes enlarged and erect and then the entire clitoral body withdraws beneath a sheath of skin, the clitoral sheath. This retraction is reversible, and if there is a reduction in sexual activity it will return to its unstimulated overhanging position. Because of clitoral retraction the penis rarely comes in direct contact with it during sexual intercourse. However, secondary stimulation may occur via the sensitive labia minora and the vagina. The onset of female orgasm is indicated by contractions in certain organs: at first the labia minora and outer third of the vagina show long spasmodic contractions lasting 2–4 seconds, and then contractions about 1 second apart occur; this phase is followed by less regular contractions of the uterus and the muscles of the pelvic floor. The strength and number of orgasmic contractions vary with the individual. No particular orgasmic reaction in the breasts or clitoris has been observed. During orgasm, the muscular contractions are accompanied by a diffuse feeling of pleasure. Hearing, vision, taste—all the senses are diminished or lost

during orgasm, and it seems to pervade the entire body.

It has long been believed by many women and their physicians that clitoral and vaginal orgasms were distinctly different from each other. Indeed, in the Freudian view the human female progressed during sexual maturation from a maximal response to clitoral stimulation to a maximal response to vaginal stimulation. Recent anatomical evidence shows that vaginal and clitoral orgasms are not physiologically distinguishable; the organs involved in orgasm respond exactly the same way to effective sexual stimulation of any kind.

The orgasmic phase is explosive and brief, usually lasting 3–10 seconds. Female orgasm corresponds to ejaculation in the male, but there is no fluid expelled in the female. Whether orgasm in the female facilitates conception is unknown, but it is not essential for conception to occur. In many women orgasm must be learned, and some women never experience it in spite of frequent intercourse. Emotional and psychological factors have profound effects. Aside from ejaculation, the female differs from the male in two respects with regard to orgasm: (1) the female is capable of several orgasms, one immediately after the other, and (2) the orgasmic experience can be maintained for a relatively long period of time.

In the male, orgasm and ejaculation occur at the same time. Ejaculation consists of two distinct activities: (1) semen is delivered from the sperm ducts to the urethra by contraction of smooth muscles in the vasa deferentia, seminal vesicles, and prostate gland (this is often called **emission**); and (2) seminal fluid is forcibly expelled by contraction of muscles surrounding the urethra. The process is controlled by a spinal reflex located in the lower back (Figure 9–7), although higher centers can alter the response, and ejaculation can occur

without tactile stimulation of the penis (influenced by erotic thoughts, petting, and so on). Higher centers may also inhibit erection or ejaculation under certain conditions.

Following orgasm there is a return to the unstimulated condition. In the male, detumescence of the penis occurs rapidly, and it becomes flaccid. After orgasm the male experiences a refractory period, that is, a resistance to sexual stimulation. Those events that elicited excitement no longer do so; in fact, they may even be irritating and distasteful. The relaxation of the penis after ejaculation-orgasm precludes a second intromission for a varying period of time.

9–5 FERTILIZATION

Ovulation takes place between the tenth and the sixteenth days of the average menstrual cycle. The egg is quickly caught on the fringed edges of the Fallopian tube, and through muscular and ciliary activity the egg is whisked into the tube's upper portion (ampulla). In order for fertilization to occur, sperm must be introduced at about this time. Ejaculation during copulation sends almost 0.5 billion sperm into the vagina. The alkaline seminal fluid is mixed with acidic cervical secretions that lubricate the vagina; sperm must first survive this acid bath and then must swim upstream to the farthest reaches of the Fallopian tube, almost 6 in. away, to contact the egg (Figure 9–6). By means of the lashing motion of the whiplike tail, the sperm move against the ciliary currents in the Fallopian tubes which are directed downward toward the uterus. Within seconds after ejaculation, sperm are in the cervical canal, and they continue to swim upstream against the prevailing current. The Fallopian tubes have many folds and blind passages, and the number of swimming sperm is reduced all along the

Would you be more careful if it was you that got pregnant?

way. Probably only 1 million sperm ever reach the uterine lumen, and by the time they enter the Fallopian tubes only thousands remain. Perhaps less than a few hundred actually arrive at the vicinity of the egg in the upper part of the Fallopian tube.

Sperm are fragile and can survive for about 48 to 72 hours inside the female reproductive tract. Indeed, if they have not found and fertilized the egg by then, they perish. The egg is even more sensitive and cannot survive for more than about 24 hours. For fertilization to take place, coitus must occur less than 72 hours before the egg reaches the Fallopian tube or else the sperm will not be viable; if coitus occurs more than 24 hours after the egg reaches the uterine tubes, the egg will no longer be viable. Thus, there is a critical maximum period of 72–96 hours during a monthly ovarian cycle during which coitus must occur if an egg is to be fertilized.

The sperm, numbering about 100, arrive in the vicinity of the egg about 1–2 hours after coitus. The sperm, which are relatively small (they are 54 μm long, but only 5 μm of this length encompasses the head), quickly surround the enormous (150 μm) egg. The egg itself is enclosed by a layer of cells, the corona radiata, and prior to entry the sperm must first disperse these (Figure 9–10). Each sperm contains in its caplike acrosome the enzyme hyaluronidase; upon release of this enzyme, the hyaluronic acid cement which holds the cells of the corona radiata together is broken down and the cells of the corona become dispersed. Now the remaining sperm confront another barrier, the jellylike zona pellucida. A single sperm moves through the zona pellucida and reaches the plasma membrane of the egg. The plasma membranes of the sperm and the egg fuse, and the contents of the sperm enter the egg. As soon as this occurs, the egg becomes impermeable to all other sperm; they are barred from entry.

As the sperm nucleus enters, the cytoplasmic activity of the secondary oocyte becomes violent. The oocyte completes its second meiotic division and expels the second polar body (Figure 9–13). The remaining egg nucleus and the sperm nucleus, called **pronuclei,** approach each other, lose their enclosing nuclear membranes, and merge. The merging of pronuclei is called syngamy, or fertilization, and this occurs within 12 hours after sperm penetration. The fertilized egg is now called a zygote, and it contains the diploid number of 46 (23 pairs) chromosomes. Now the cytoplasm of the zygote becomes less agitated, and the period of embryo development **(embryogenesis)** begins.

9–6 CONTRACEPTION

Anything that interferes with fertilization of the egg can be termed **contraception,** and anyone who deliberately avoids conception during sexual intercourse is employing a contraceptive measure. There are about half a dozen ways that conception can be avoided:

1. Limiting time of intercourse to a period during the female reproductive cycle when the egg is not available for fertilization
2. Preventing sperm from being deposited in the vagina
3. Preventing sperm from entering the cervix
4. Preventing the release of sperm in the seminal fluid
5. Preventing ovulation or entry of the egg into the Fallopian tube
6. Interfering with egg transport in the Fallopian tube or preventing implantation of the fertilized egg.

Let us consider the effectiveness of each of the six methods.

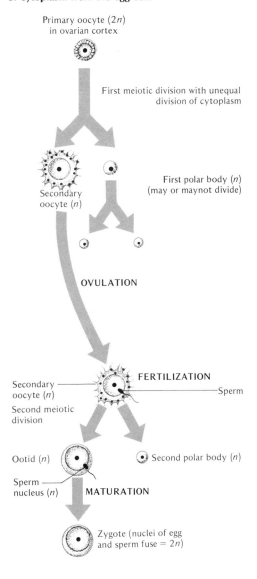

FIGURE 9–13 Meiotic events in oogenesis. A primary oocyte that undergoes oogenesis produces one functional ovum and three (or two) minute, functionless polar bodies. The unequal cytoplasmic division during the two meiotic divisions ensures that the ovum is well endowed with the cytoplasmic components essential for development of the fertilized egg. In this way, chromosome reduction occurs without loss of cytoplasm from the egg cell.

Primary oocyte ($2n$)
in ovarian cortex

First meiotic division with unequal
division of cytoplasm

First polar body (n)
(may or maynot divide)

Secondary
oocyte (n)

OVULATION

FERTILIZATION

Secondary
oocyte (n)
Second meiotic
division

Sperm

Ootid (n)

Second polar body (n)

Sperm
nucleus (n)   MATURATION

Zygote (nuclei of egg
and sperm fuse = $2n$)

1. The **rhythm method** is quite simple: couples do not engage in sexual intercourse when a fertilizable egg is present in the Fallopian tubes. In theory, the egg is fertilizable for only a brief time (72–96 hours) and, therefore, by avoiding intercourse during an unsafe period—2.5 days before and after ovulation—conception can be avoided. This is the ideal situation, but for many women there is no regularity of ovulation. Detection of the timing of ovulation presents the most difficult aspect of using the rhythm method. Analysis of the cervical mucus and measurement of body temperature are often used to determine the danger or unsafe period (Figure 9–14 and Box 9B). How safe is the rhythm method? It is estimated to be 74% effective; that is, of 100 women using this method for 1 year, 26 will become pregnant.[2]

2. Sperm may be prevented from deposition in the vagina by a variety of natural and artificial methods. The oldest is **coitus interruptus,** or withdrawal of the penis prior to ejaculation; this is mentioned in the Old Testament (*Gen.* 38:9), was practiced by the ancient Hebrews, and is still widely practiced. The method requires self-control by the male and is 78% –85% effective. It is generally unacceptable to Americans, who believe it limits full sexual expression by both partners. Even the smallest amount of semen deposited in the vagina can cause pregnancy, and many couples employing coitus interruptus do not realize that it is so failure-prone.

The **condom** is an effective artifical male contraceptive (83%–88% effective). Invented in the 1700s in the court of Charles II, it is a rubber sheath 7–8 in. in length worn over the erect penis during intercourse. The sheath cap-

[2] In the absence of any birth-control procedures, the pregnancy rate is about 40%.

BOX 9B Detection of ovulation

Aside from the liberation of oocyte and follicular fluid during ovulation, there may be some hemorrhage. The hemorrhage, if excessive, may cause a local inflammation, which may in turn cause low abdominal pain, called **mittelschmerz** (middle pain; Ger.) because it occurs in the middle of the month between menstrual bleedings.* If it does occur it is an indication that ovulation has taken place.

Changes in the cervical mucus occur during the ovarian cycle, and two tests indicate the time of ovulation. In the **fern test** a drop of cervical mucus is placed on a microscope slide and allowed to dry. If the donor is in the middle of her cycle, the mucus dries in a fernlike pattern. In the **spinnbarkheit** test mucus from the cervix is placed between the thumb and the forefinger and is drawn out to form a thread. Mid-cycle mucus usually stretches to 4 in.; mucus obtained at other times stretches less.

Some women find that ovulation can be detected by changes in body temperature. The body temperature is recorded daily and ovulation is indicated by a drop in temperature (Figure 9–14). After ovulation there is a rise in temperature, and this is maintained during the rest of the menstrual cycle. It is important to note that this method tells the safe period *after* ovulation, not before.

None of these tests is foolproof and there is considerable variation from woman to woman.

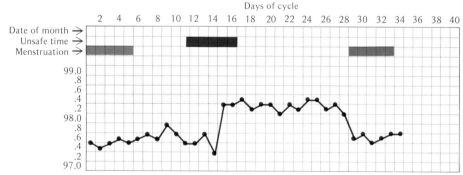

FIGURE 9–14 Body temperature and the ovarian cycle.

*Some physicians believe mittelschmerz is not caused by an irritation from blood, but is a result of spasmodic contractions of the Fallopian tubes.

tures the sperm in the sealed end and prevents their deposition in the vagina. The condom is effective if it does not rip or slip off during intercourse. Condoms are made of latex rubber, can be purchased without prescription, and are frequently used to prevent the contraction of venereal diseases such as gonorrhea and syphilis. (Premature ejaculation in the male, which is not uncommon during the early months of marriage, may be alleviated by wearing two condoms. This substantially reduces the sensitivity of the glans penis and makes ejaculation occur later in sexual intercourse.)

3. Sperm may be blocked from entering the cervical canal by a variety of artificial methods: douche, spermicidal solutions or foams, and the diaphragm.

The **vaginal douche** is perhaps the oldest of contraceptives; however, it is also the least effective (40%–66%). Shortly after ejaculation, the vagina is flushed with a spermicidal solution in order to kill spermatozoa. Since sperm enter the cervix 90 seconds after deposition in the vagina, a quick dash to the bathroom is required to prevent sperm from moving out of the vaginal region. The method is judged unacceptable from the point of view of sexual expression.

Spermicidal solutions, foams, suppositories, jellies, or creams contain chemicals that kill sperm and make them incapable of penetrating the cervical mucus. The jellies and creams are inserted in the vagina with a special applicator, the foams and suppositories are inserted without an applicator. The chemical methods are about 80% effective. Their prime disadvantage is that they must remain in place for 1–6 hours after intercourse; oozing of the spermicide from the vagina may be objectionable.

The **diaphragm** is a mechanical device that fits over the cervix and prevents sperm entry (86%–88% effective). It is usually used in conjunction with a spermicide; it must be kept

BOX 9C Pregnancy: signs and symptoms

Pregnancy lasts 9.5 lunar months (38 weeks or 266 days) after conception. Physicians usually calculate the date of birth by adding 14 and 266, and tell the mother that the baby will be born 280 days after the first day of the last menstrual period (the figure 14 signifies that fertilization occurred on the fourteenth day of the menstrual cycle.). Another frequently used method for calculating the date of confinement (birth) is to add 7 days to the first day of the menstrual cycle, substract 3 months, and add 1 year. If the last menstrual period began on June 17, the delivery date would be March 24 of the following year (June 17 + 7 = June 24 − 3 months = March 24 + 1 year). It should be noted that the calculation of date of confinement is a rough approximation; less than 10% of normal pregnancies run 280 days exactly. There is less than a 50-50 chance that the child will be born within a week of the due date, and the odds are 1 out of 10 that labor will occur 2 weeks later.

Although conception and implantation take place about 7 days after ovulation, the prospective mother is not aware of her condition until 2 weeks later. During the third week after fertilization, the telltale sign of pregnancy becomes apparent: **amenorrhea,** a cessation of menstruation. Her awareness is heightened by a tingling sensation in her breasts, more frequent urination, and constipation. Nausea and vomiting—**morning sickness**—appear in about two thirds of pregnant women. The cause of morning sickness is uncertain, but may be due to stretching of uterine tissues. A hormone appears in her urine, called **chorionic gonadotrophin,** which is produced by the cells of the chorion.* Secretion of this hormone causes the corpus luteum to continue to secrete progesterone; thus the endometrium is maintained in its thickened, highly vascularized state. Detection of chorionic gonadotrophin in a urine sample is the basis of most pregnancy tests.

The pregnancy test most commonly employed today is an immunological one (sold commercially as Pregnosticon). The urine from the possibly pregnant female is placed in contact with a reagent, consisting of an antiserum to human chorionic gonadotrophin. If the urine contains gonadotrophin, it reacts with the antiserum and the combination precipitates as a white, flocculent material. The test results may be read within 2 hours and because of the test's sensitivity and high degree of specificity, pregnancy can be detected as early as 6 days after the missed period.

Later in pregnancy the effects of hormonal changes and the developing fetus become more obvious; the breasts enlarge, veins are more prominent, the nipples and areolae become more deeply pigmented. The oil glands around the areolae enlarge and look like tiny bumps. The pressure of the enlarging uterus may be distressing; the shifting of the center of gravity by an extra 20 lb forward can cause unpleasant sensations of falling forward. Cravings for foods may develop, and there may be a change in attitude. The so-called mask of pregnancy **(chloasma),** a deep coloration of the face may develop, as well as dark lines **(striae)** on the abdomen. Most of this will disappear after childbirth. The pituitary gland doubles in size, the thyroid, parathyroids, and adrenals grow larger. Engorged and dilated blood vessels are visible in the skin. Varicose veins and hemorrhoids often accompany or follow pregnancy. Edema and swelling are common, and there is a sluggish return of blood from the extremities to the heart.

*The outermost covering of the developing embryo (Chapter 10).

FIGURE 9–15 Insertion and position of the IUD. (a) Insertion of IUD into syringelike device. (b) Insertion of IUD into uterus. (c) IUD in place. (Modified from Robert Demarest and John Sciarra, *Conception, Birth and Contraception: A Visual Presentation,* **New York, McGraw-Hill, 1969.**)

in place for 6 hours after intercourse and can be worn for 24 hours. The diaphragm must be fitted by a physician and refitted after each pregnancy. If inserted improperly, it will not perform satisfactorily.

4. Prevention of sperm release into the semen involves **vasectomy** (sterilization; Box 9D).

5. Prevention of egg release or transport to the uterus involves the oral **contraceptive pill** (Section 9–7) or **tubal ligation** (sterilization; Box 9D).

6. Prevention of implantation is apparently effected by **intrauterine devices** (IUD). These are 98% effective. The IUD is a small object (loop, ring, spiral, or bow) made of steel or plastic, which is inserted by a physician into the uterus (Figure 9–15). The physician inserts the IUD by straightening it in a tubelike instrument, pushing this through the cervix, and then pushing the IUD out the end of the tube. Free in the uterus, the IUD assumes its original shape. Usually the IUD is provided with nylon threads that protrude through the cervix so that the woman can check whether the device is still in place. Since the IUD does not alter the hormonal balance and produces fewer undesirable side effects, it offers advantages over the pill. However, there may be discomfort, bleeding, cramps, and spontaneous expulsion. Usually the IUD cannot be used until after the first pregnancy. The exact mechanism by which the IUD prevents conception is unknown. Some physicians believe it interferes with implantation of the embryo into the uterine wall.

9-7 CONTROL OF THE FEMALE REPRODUCTIVE CYCLE: THE PILL

The complete understanding of how hormones influence the ovarian cycle provided the necessary background for developing a method of suppressing ovulation, the oral contraceptive

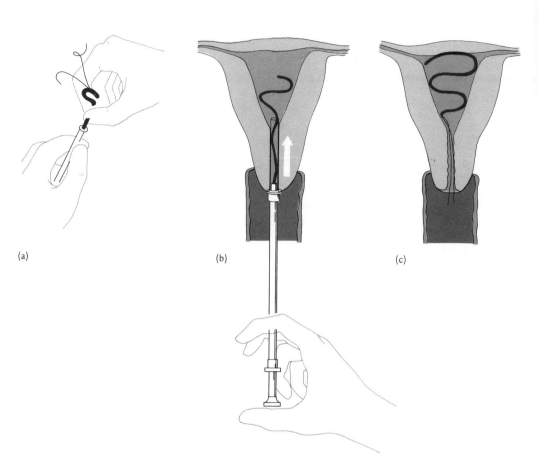

(a) (b) (c)

commonly called the pill. The contraceptive pill is 97%–99% effective; that is, out of 100 women taking the pill for 1 year, less than 3 will become pregnant. According to the best estimates there are between 12 and 15 million women in the world today using oral contraceptives by their own choice. Some 8 million women in the United States not only choose the oral contraceptive, but find its use pleasant and easy.

How does the oral contraceptive pill work? Recall from our discussion of the normal ovarian cycle that ovulation is triggered by a change in the relative levels of FSH and LH and that the secretion of these hormones is inhibited by estrogen and progesterone (Figure 9–11). If the levels of estrogen and progesterone could be adjusted sufficiently to interfere with the normal change in relative levels of the two pituitary hormones, ovulation could be inhibited artificially. Estrogens such as estradiol and progesterone are ineffective when taken by mouth (they are degraded during digestion), so they must be injected to produce an effect. Synthetic estrogens and progesterones (progestins) have been developed that are not so degraded and act like the natural hormones; that is, they inhibit pituitary production of FSH and LH. Theoretically, estrogen alone could suppress FSH secretion, and indeed it does, but estrogen alone will not suppress midcycle production of LH; in fact it stimulates LH secretion. Therefore, if there were developing follicles, an oral contraceptive that possessed only estrogenic activity would allow the possibility of an escape ovulation. Thus, the contraceptive effect of a pill employing the estrogenic method alone would be uncertain. The oral steroid contraceptive procedures most commonly used are the combination method and the sequential method. In the combination method a pill containing estrogen and progestin is taken for 20–21 days after menstruation (day 5). The period usually starts 1–3 days after the last pill is taken. (The progestin maintains the endometrial lining, and when its production is diminished menstrual bleeding results.) This establishes an artificial 28-day menstrual cycle. In the sequential technique 14–16 days of estrogen are followed by 4–6 days of estrogen and progestin.

Side effects from taking the pill are not uncommon. Indeed, it would be unfair to say that all the questions about its long-term effects are settled. For example, estrogens have widespread effects on body tissues aside from those involved in reproduction. They affect protein and carbohydrate metabolism, they play a role in calcium balance, they affect water and salt retention, and they act on the adrenal and thyroid glands as well as the pituitary. Some women taking the pill experience water retention and weight gain. In large doses, progestins have effects similar to those produced by male hormones. The contraceptive pills usually balance these steroids and reduce the amount of synthetic hormone to a minimum. Still there are women taking the pill who experience acne, headaches, nausea, and tenderness of the breasts. Some critics of the pill contend that long-term estrogen treatment may induce breast cancer (Chapter 8), and there are indications of enhanced clot formation in women taking the pill. In some women the pill causes the veins to dilate abnormally causing slow blood flow, and this may in turn produce a sludging of blood and clot formation. Moreover, the pill may change the chemistry of the clotting reaction itself. A clot that develops in the veins **(thrombophlebitis)** may break loose from its point of origin, travel through the body (as an **embolus**) and lodge in the blood vessels of the lungs or the brain. Many such pulmonary (lung) and cerebral (brain) embolisms are fatal. The use of oral contraceptives increases by six to eight times the risk of being admitted to a hospital for a thromboembolism. Between the ages of 20 and 34 years, the annual death rate per 100,000 population (1968 figures) was 0.8 for non-pill-users and 1.3 for pill-users. In the 35–44 age group the pill-user mortality was 3.4. This statistic must be balanced against the dangers of pregnancy itself. Today, in a developed country, a woman's chances of dying during pregnancy are 1 in 3,000. Although death is a tragic event and the risk of it is not to be dismissed lightly, the risk of complications from the pill is in fact low or lower than that from other drugs and is equivalent to the risk of death that a woman runs in her own kitchen in the same period of time. Clearly, the decision to use oral contraceptives should be based on individual choice after consultation with a physician.

SUMMARY

1. Meiosis is the method of cell division by which germ cells (gametes) are produced in the gonads. Following meiosis, the gametes are haploid. The diploid number of chromosomes is restored at fertilization.

2. Meiosis involves two cell divisions. During the first reduction division, synapsis of homologous chromosomes is followed by a halving of the chromosome number. This first division is followed by a mitotic division. Four haploid cells are thus produced from one diploid cell.

3. Male gametes, or spermatozoa, are produced from spermatogonia in the seminiferous tubules of the paired testes by spermatogenesis: this involves meiosis, maturation, and differentiation. At puberty, the pituitary gland produces FSH and ICSH, which stimulate both spermatogenesis and testosterone production in `the testes. Testosterone and other male hormones produced in the testes control the development of secondary sex characteristics.

4. Mature sperm pass via a series of collecting ducts and tubules and the coiled epididymis to the vas deferens. The two vasa deferentia join with the tube leaving the bladder and enter the penis as the urethra.

5. The seminal vesicles, prostate and Cowper's (bulbourethral) glands add secretions to the sperm cells, constituting semen or seminal fluid.

6. The testes normally hang in the scrotal sac. In prepubertal males, they may be retracted via the inguinal canal, which closes after puberty. An inguinal hernia is caused by the dropping of an intestinal loop into the scrotum. Cryptorchidism is due to undescended testes.

7. The muscular penis is the male copulatory organ. The glans (head) is covered by a prepuce (foreskin). Erection occurs in response to stimulation and depends on the filling of the corpora cavernosa and corpus spongiosum with blood. During ejaculation, the entrance from the bladder to the urethra is closed by a sphincter.

8. Female gametes or ova are produced from oocytes in the paired ovaries by oogenesis, involving meiosis and maturation.

9. FSH produced by the pituitary signals the onset of menarche and the development of secondary sex characteristics. It also signals the onset of oogenesis.

10. The oocytes develop in Graafian follicles. Unequal division of the cytoplasm at each meiotic division produces one mature ovum and two (or three) polar bodies from each primary oocyte.

11. The secondary oocyte requires LH from the pituitary for maturation. FSH and LH induce the follicle cells to produce estrogens.

12. The changing balance of FSH and LH produces ovulation of the oocyte, which is surrounded by the cumulus oophorus. When

a secondary oocyte is shed, it is picked up by the Fallopian tubes and moved toward the uterus.

13. Influenced by LH, the follicle remnants, as the corpus luteum, secrete estrogen and progesterone. In turn, the latter suppress FSH production and hence further egg production. Progesterone also acts on the uterine lining to prepare it for implantation. In the absence of fertilization, the corpus luteum is transformed to a corpus albicans and eventually is resorbed.

14. The uterus consists of the uterine body and cervix. A uterine cycle begins with menstruation, that is, with breakdown of the endometrium; this is followed by endometrial proliferation under the influence of estrogen and progesterone produced by the ovarian follicles and the corpus luteum. If fertilization does not occur following ovulation on about day 14, the corpus luteum degenerates on about day 22 or 23. Progesterone and estrogen levels drop and menstruation recommences about day 28.

15. The muscular vagina is the female copulatory organ. Its entrance may be partly covered by the hymen. Vestibular Bartholin's glands secrete a lubricant during sexual intercourse. Externally, the female vulva consists of the mons veneris (pubis), labia majora and minora, clitoris, and vestibule.

16. During coitus, sexual arousal causes vasocongestion that results in the erection of nipples, clitoris, and penis and produces secretions from the various glands. Sexual orgasm and emission coincide in the male. Vaginal and clitoral orgasm are physically indistinguishable in females. Males experience a postorgasmic refractory period.

17. Fertilization occurs high in the Fallopian tube. The fertilizing sperm penetrates the corona radiata and zona pellucida, and the egg becomes impenetrable to other sperm. The egg and sperm plasma membranes fuse, the oocyte

---

BOX 9D Sterilization for both sexes

One drastic way to practice birth control is by means of sterilization. The surgeon's knife is now being sought by increasing numbers of Americans, both men and women, who want to be sure that they will have no more babies. Sterilization operations for men and women are based on the same strategy: cutting the tubes that carry the sex cells on their paths toward junction and conception.

Because of anatomical differences, the male operation is the simpler. After injecting a local anesthetic, the surgeon makes an incision about half an inch long on one side of the scrotum, draws out one vas, and cuts out a section up to an inch long. He usually cauterizes the remaining stumps of the vas and ties them shut with nonabsorbable thread. The surgeon then sutures the small wound and repeats the procedure on the other side.

Because patients have some discomfort for two or three days after a vasectomy, the Margaret Sanger Research Bureau in Manhattan schedules all such operations on Fridays: thus the patient will be able to return to work on Monday. The vasectomy patient undergoes no hormonal changes, and if he has fully understood the operation beforehand he should have no emotional problems. His capacity for sexual relations may even be increased, because he no longer fears conception. His sperm, trapped in the testicles, are reabsorbed, and eventually his body manufactures fewer of them. However, some sperm are left "in the pipeline" at the time of operation, so for the next six to twelve acts of coitus a contraceptive must be used. Most surgeons require that their patients return after four to twelve months and leave a semen sample for analysis to make sure that neither vas deferens has joined itself up again.

For women there are a variety of surgical procedures. The most obvious is used on the woman who is having a baby by caesarean section, and has decided that this will be her last. Since her abdomen is already open, the obstetrician simply reaches in for the Fallopian tubes, ties them off and severs them—much as the urologist does in a vasectomy. Most surgeons also remove part of the tube. This procedure is called tubal ligation.

Equally common is the operation on a woman who has just given birth to a baby normally. Within 36 hours after the delivery, the surgeon makes a three- or four-inch incision in her lower abdomen to reach the tubes. The surgical wound is almost healed by the time the woman goes home with her baby.

In recent years, especially in Britain and Europe, gynecological surgeons have been seeking means of reaching and severing the Fallopian tubes without making a long pelvic incision. They have succeeded with the aid of the laparoscope, a tube containing a "light pipe," less than half an inch in diameter. The techniques vary in detail. At Johns Hopkins Hospital in Baltimore, Dr. Clifford R. Wheeless makes two incisions less than half an inch long just below the navel. Through one, after blowing in carbon dioxide to separate the organs, he inserts the laparoscope to locate a tube. Through the second he inserts the electric cautery and a tiny surgical knife. The operation, under general anesthesia, takes about 30 minutes and allows the patient to leave the hospital the same day.

In a modified version of the operation, Dr. Alvin Siegler of New York's Downstate Medical Center makes his first incision for the laparoscope so close to the navel that no separate scar will be visible, then inserts the cautery and knife through two punctures not much bigger than those made by a heavy-gauge hypodermic needle.

The simplest development in sterilization of women, requiring only local anesthesia, is reported by Dr. Martin Clyman at Manhattan's Mount Sinai School of Medicine. He has designed special instruments that enable him to operate through an incision little more than an inch long in the vaginal wall, reaching and tying off both tubes in about ten minutes. This vaginal approach leaves the patient with no visible scars, and she can go home in 24 to 48 hours after the operation.

Why do couples prefer sterilization to the long-term use of contraceptives? "Because they know that failure rates from most forms of contraception are too high." one doctor says. "Or they are afraid of side effects from the Pill, or they have aesthetic objections to having to remember to insert something at what is emotionally the wrong time."

Rates of Reversal. One nagging question that still deters many men from seeking sterilization: Is the operation reversible? The answer is that in some cases, after either male or female sterilization, fertility can be restored by a reverse operation to rejoin the severed tubes. The success rate of these procedures is disputed. Some physicians put it as high as 80%; most think 30% is more realistic. But the question seldom arises. Most urologists' records show that not more than 1% or 2% of their male patients have ever asked for a reverse operation.

completes its second meiotic division, the egg and sperm pronuclei fuse, and fertilization is complete. The zygote is now diploid.

18. Conception can be avoided by preventing the sperm from reaching the egg. Methods include use of: rhythm, coitus interruptus, condoms, vaginal douche, spermicides, diaphragm, vasectomy, tubal ligation, intrauterine devices, and oral contraceptive pills.

19. Ovulation is normally triggered by changes in FSH and LH levels. Secretion of FSH and LH is inhibited by estrogen and progesterone. Contraceptive pills depend on oral administration of synthetic estrogen and progesterones; they are not without side effects.

20. Ovulation can be detected by a variety of methods, including the fern and spinbarkheit tests, or by monitoring changes in body temperature. Mittelschmerz indicates ovulation in some women.

KEY WORDS

zygote
gametes
syngamy
fertilization
genes
chromosomes
meiosis
homologous chromosomes
diploid
synapsis
haploid
somatic cells
germ cells
gonads
spermatogenesis
testis
androgens
spermatozoa (sperm)
seminiferous tubules

spermatogonia
Sertoli cells
follicle-stimulating hormone (FSH)
interstitial-cell-stimulating
   hormone (ICSH)
testosterone
rete testis
epididymis
vas deferens
inguinal canal
urethra
seminal vesicles
prostate gland
bulbourethral glands
semen
seminal fluid
scrotum
inguinal hernia
cryptorchidism
penis
glans
foreskin (prepuce)
corpora cavernosa
corpus spongiosum
oogenesis
ovaries
broad ligament
Fallopian (uterine) tubes
oocytes
menarche
ovarian cycle
menstruation
menopause
Graafian follicle
luteinizing hormone
estrogen
cumulus oophorus
zona pellucida
corona radiata
corpus luteum
corpus albicans
uterine or menstrual cycle

uterus
cervix
endometrium
vagina
hymen
Bartholin's glands
vulva
mons veneris
labia majora
labia minora
clitoris
vestibule
coitus
copulation (sexual intercourse)
orgasm
emission
pronuclei
embryogenesis
contraception
rhythm method
coitus interruptus
condom
vaginal douche
spermicidal solutions
diaphragm
vasectomy
contraceptive pill
tubal ligation
intrauterine device (IUD)
thrombophlebitis
embolus
mittelschmerz
fern test
spinnbarkheit
pregnancy
amenorrhea
morning sickness
chorionic gonadotrophin
chloasma
striae

TOPICS FOR REVIEW AND DISCUSSION

1. What is the relationship of meiosis to sexual reproduction?
2. What are the biological advantages of sexual reproduction (meiosis) as contrasted with asexual reproduction?
3. Describe spermatogenesis. Compare and contrast the process with oogenesis.
4. How are the structural attributes of the sperm and egg related to their functional roles in reproduction?
5. Discuss the hormonal basis of the contraceptive pill.
6. What are some of the methods used for detecting ovulation and pregnancy?
7. Describe the functional roles of the external genitalia.
8. Contrast vasectomy with castration.
9. Describe the ovarian and uterine cycles.
10. Describe the nature and possible causes of infertility in humans.
11. Discuss the exocrine and endocrine functions of the testes or the ovaries.

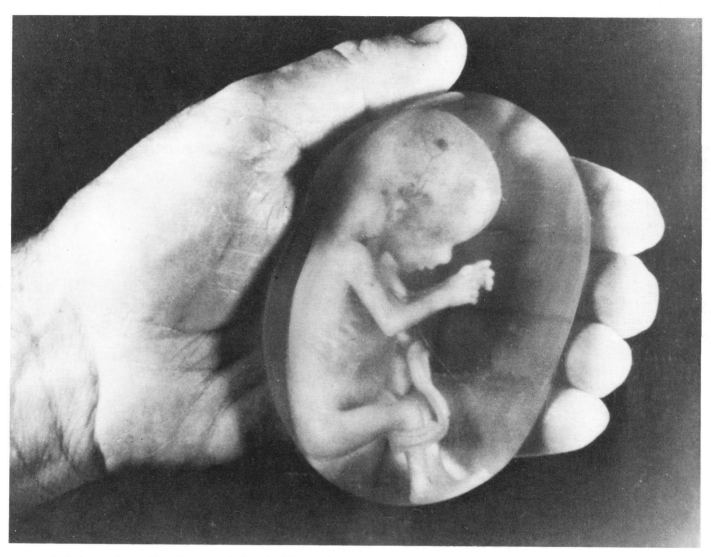

A 12-week-old human fetus still within its amniotic sac held in an obstetrician's hand.
(Courtesy of Roberts Rugh and Landrum B. Shettles.)

PERPETUATION OF LIFE II: HUMAN DEVELOPMENT

10–1 FROM EGG TO EMBRYO

Each of us represents a kind of miracle. We began in an "unfelt, unknown and unhonored instant when a minute, wriggling sperm plunged headlong into a mature ovum or egg."[1]

The fertilized egg cell, or zygote, is one of the largest cells in the body with a high ratio of cytoplasm to nuclear material. Following fertilization, the zygote divides mitotically into a number of smaller cells, a process known as **cleavage** (Figures 10–1 and 10–2). The main functions of cleavage are to compartmentalize the components of the cytoplasm and to reduce the amount of cytoplasm relative to nuclear material in the enormous zygote. This is accomplished by a reduction in cell size through subdivision and the formation of new nuclear material. Although the total mass of the zygote is not substantially reduced during cleavage, the arrangement of its components is radically altered. Cleavage lasts until other developmental processes beside mere subdivision begin to assume importance (Figures 10–3 and 10–4). These processes involve the movement of the cells of the subdivided zygote relative to one another; the cells in the different regions interact and affect each other's fate. From this time onward, the developing organism is referred to as an **embryo.** After implantation in the wall of the mother's uterus the embryo begins to obtain nourishment (Figure 10–5). Only then does an increase in total size of the developing organism begin to occur.

Let us now trace these events in more detail by looking at a series of remarkable photographs that show human development from the moment of fertilization until the seventh week, when the embryo begins to assume a clearly recognizable human form (Figures 10–6—10–11).

---

[1] M. S. GILBERT, *Biography of the Unborn*, New York, Hafner, 1963.

FIGURE 10-1 The moment of fertilization. The mature egg, barely visible to the naked eye, is about the size of the period at the end of this sentence and is 2,000 times larger than the sperm that fertilizes it. At the moment of fertilization shown here, sperm and egg fuse, producing a zygote weighing one twenty-millionth of an ounce. In 30 days this zygote will increase its size 40 times and its weight 3,000 times. (Courtesy of Landrum B. Shettles.)

FIGURE 10-2 Cleavage: one cell into two. Cleavage of the zygote into two cells takes place in about 36 hours. Surrounding the cells is the transparent zona pellucida, which helps to hold the cells together. (Courtesy of Carnegie Institution of Washington.)

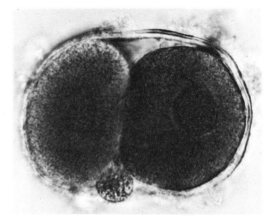

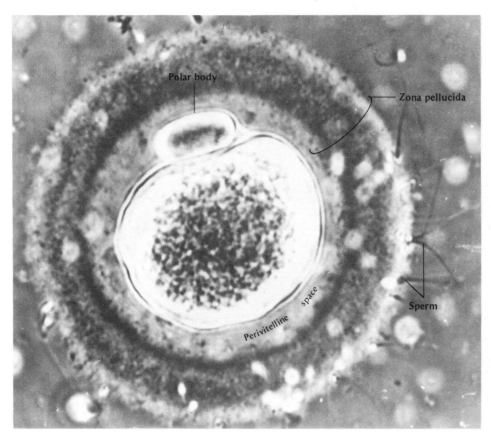

FIGURE 10–3. (top) Cleavage and morula formation. By the third day there is a clump of 16–32 cells—the morula (literally "mulberry")—enclosed by the zona pellucida. The total mass of the morula is not greater than that of the egg, since cleavage has simply divided the egg into smaller and smaller cells. (Courtesy of Carnegie Institution of Washington.)

FIGURE 10–4 (below left) The blastocyst. At 4 days fluid enters through the zona pellucida, and the solid morula becomes the hollow blastocyst (blastula). The cells at one pole form a thickened mass, the internal inner cell mass, and this will ultimately contribute cells to the embryo as well as the extraembryonic membranes. The exterior blastocyst cells (trophoblast) function in implantation and contribute to the chorion. (Courtesy of Carnegie Institution of Washington.)

FIGURE 10–5 Implantation of the blastocyst. After 7–9 days the blastocyst, free of its zona pellucida, becomes implanted in the endometrial lining. (Courtesy of Carnegie Institution of Washington.)

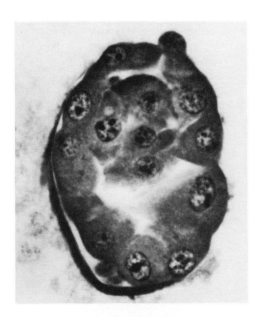

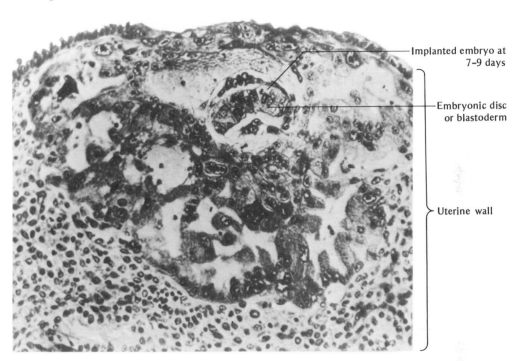

Implanted embryo at 7–9 days

Embryonic disc or blastoderm

Uterine wall

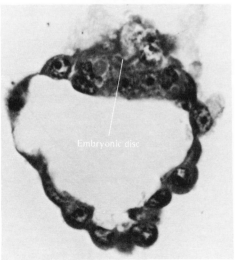

Embryonic disc

FIGURE 10–6 (left) A two-layered embryo: the second week. The 2.5-week embryo with its two layers of cells, the ectoderm and the endoderm. The ectoderm will form the skin and nervous system and the endoderm the digestive tract and its derivatives. The bulge in the ectoderm, called the primitive streak, marks the long axis of the embryo, and it is along this line that the spinal column will form.

FIGURE 10–7 (right) A three-layered embryo: the third week. The 3-week embryo contains three layers: ectoderm, endoderm, and a middle layer—mesoderm. The ectoderm thickens and contains a groove and begins to fold its edges in such a way that it forms the neural tube, which will ultimately give rise to the brain, spinal cord, and the entire nervous system. (Courtesy of Carnegie Institution of Washington.)

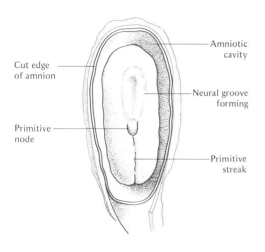

Cut edge of amnion

Primitive node

Amniotic cavity

Neural groove forming

Primitive streak

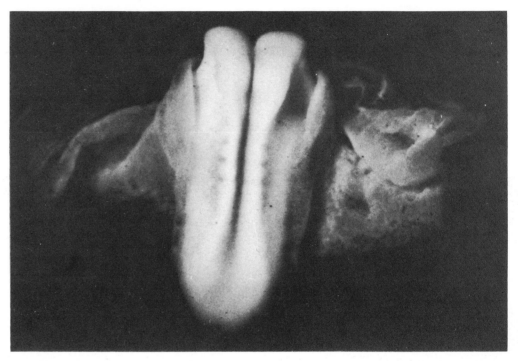

Actual size

FIGURE 10–8 (left) Organ systems develop: the fourth week. A 4-week embryo is about the size of a pinkie nail. The neural tube has closed, the brain begins to develop, and the heart appears. Segmentation shows up in the 30 pairs of mesodermal blocks (somites) that will form muscle and the vertebrae of the spinal column. (Courtesy of Carnegie Institution of Washington.)

FIGURE 10–9 (right) At 5 weeks of development the embryo is almost complete, although its size and weight are about the same as those of an aspirin tablet. About one-third of the embryo is head, and this is bent forward (cranial flexure) in such a way that it almost touches the rudimentary tail. The heart bulges outward from the body and pulsates, forcing blood through the circulatory channels. At this time the heart is a simple tube, and only later will chambers develop. The arm and leg buds are formed. It is at this stage that the tranquilizer thalidomide had its tragic effects, producing flipper limbs. Gill clefts and the outline of the eyes appear. (Courtesy of Roberts Rugh.)

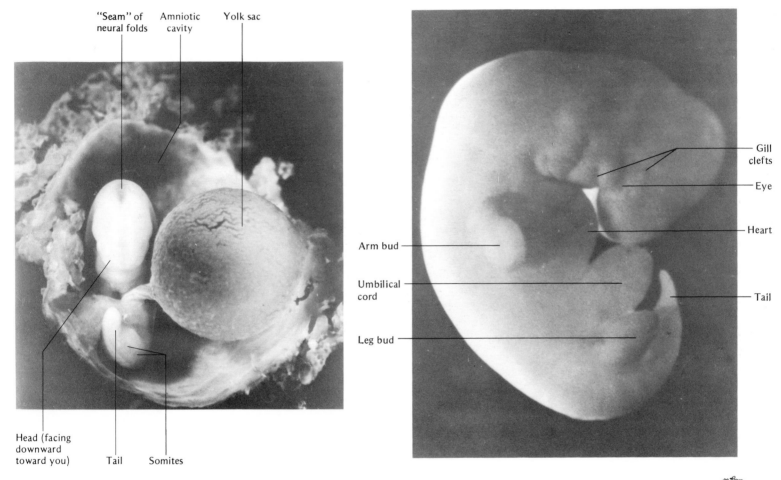

"Seam" of neural folds Amniotic cavity Yolk sac

Head (facing downward toward you)   Tail   Somites

Actual size

Arm bud

Umbilical cord

Leg bud

Gill clefts

Eye

Heart

Tail

Actual size

FIGURE 10-10 At 6 weeks the eyes become pigmented, the jaw is well formed, and the gill clefts have almost disappeared. The head still curls downward, almost touching the tail. Muscle blocks (somites) are clearly seen. The embryo is now about 0.5 in. long, and the earliest reflexes appear. The heart is partitioned into chambers, the liver begins to manufacture red blood cells, the diaphragm is formed, and the intestines and liver are so enlarged that they bulge outward. The germ cells have completed their migration from the yolk sac and are settled into the gonad region. (Courtesy of H. Nishimura and Roberts Rugh.)

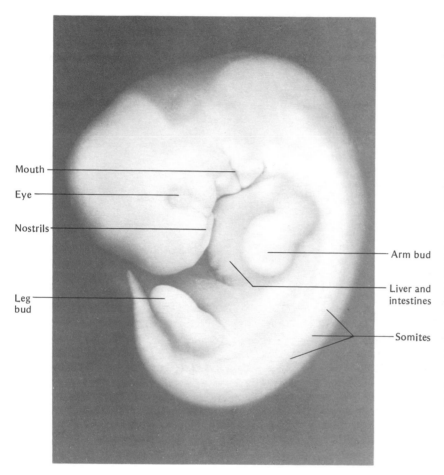

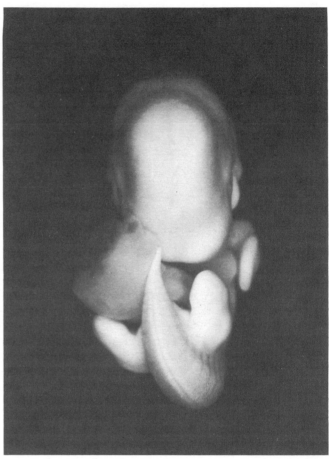

Mouth

Eye

Nostrils

Leg
bud

Arm bud

Liver and
intestines

Somites

Actual size

FIGURE 10–11 By the seventh week the embryo begins to assume a more human form. The rudimentary tail has diminished in size; the ears and eyes are developing rapidly. The fingers and toes are present, and the development of the face is evident. One of the temporary gill pouches becomes part of the ear (Eustachian tube). The embryo is about 1 in. long, and because of the large head appears top-heavy. (Courtesy of E. Ludwig and Roberts Rugh.)

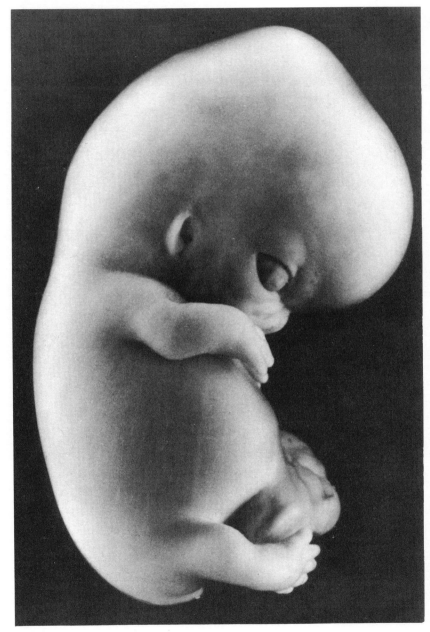

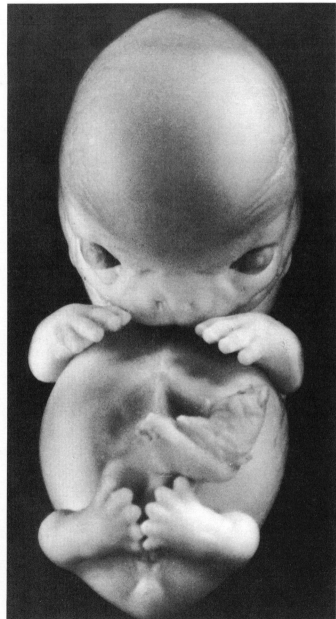

10-2 THE FETUS

By the eighth week of development, the ground plan of differentiation has been established and all the human features are evident (Figure 10–12). No longer called an embryo, the developing child has become a **fetus.** One and a quarter inches in length from the top of its enlarged head to its buttocks, it weighs 1/30 oz (0.059 gm) and appears top-heavy because of the large head; indeed, the head composes almost one half of the fetus. The eyes, slitlike in appearance, have moved from the sides of the head and are directed forward, the jaws are almost fully developed, as is the nose; there are small earlobes and the face is now unmistakably human. The mouth region forms and the alimentary canal runs downward from this. The arms and legs are gracefully proportioned, the hands well formed, the fingers and toes distinctly human, and simple reflexes occur. The skeleton, which was once cartilage, now begins to harden with the deposition of bone. The abdomen bulges outward like a potbellied stove, and the external sexual organs can clearly be seen.

Now the fetus grows (Figures 10–13—10–15). Growth depends on nutrition from the mother via the placenta, the umbilical cord, and the fetal blood vessels (Section 10–3).

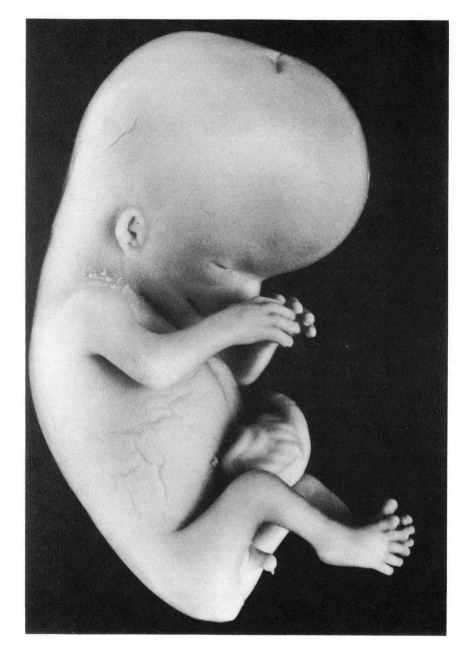

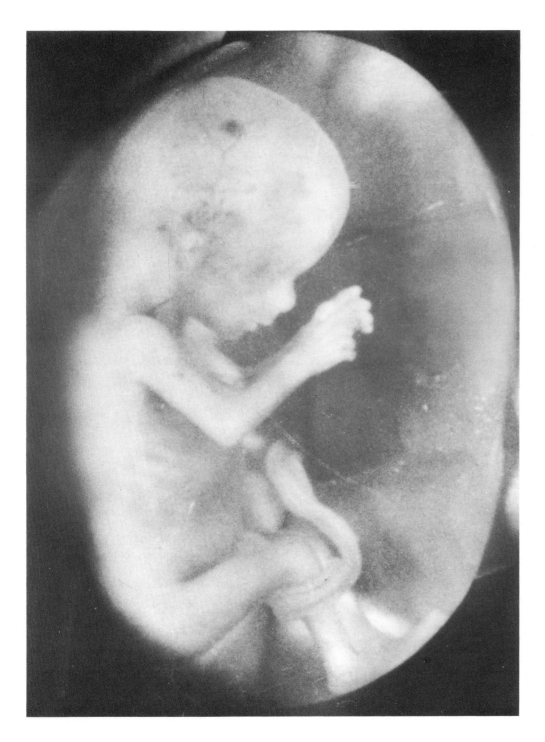

FIGURE 10-13 The third month. Between weeks 9 and 12 the fetus grows from 1.5 to 3 in. in length. The head no longer bends forward, and the fetal posture becomes upright. Weight goes from 1/7 oz. to 1/2 oz. Fingernails, toenails, and hair appear. The face becomes more human in appearance because the eyes (which are closed) are moved forward; tooth buds and bones form. Blood is now manufactured in the bone marrow instead of in the spleen and liver. A fetal pulse can be heard; the skeleton and muscles take on human contours. The male fetus is recognizable by its external penis, and the female forms a uterus and vagina. The kidneys begin to function. The fetus is sensitive to touch, and the motions of breathing and eating become coordinated and purposeful. Individuality in fetal facial expressions appears. Now the mother can feel her enlarging uterus just below her navel. (Courtesy of Roberts Rugh and Landrum B. Shettles.)

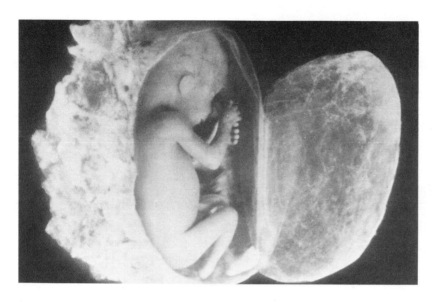

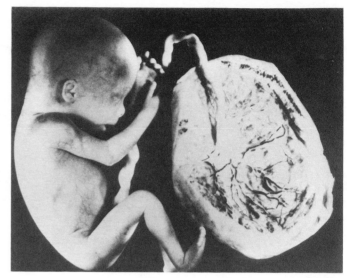

FIGURE 10-14 The fourth month. During the fourth month there is a rapid burst in fetal growth: now the fetus grows to 6 in. and weighs 4 oz. Posture becomes more erect, fingerprints appear, and the fetus shows a startle reflex. It moves, stretching its arms and legs, but ordinarily these movements are not felt by the mother. If prematurely delivered, it would curl its fingers and toes when tickled on the palms or soles. The heart beats 120–160 times a minute, twice the rate of the mother's heart. The lungs are fully formed, but are collapsed and functionless. The fetal membranes surrounding the fetus are shown. (Courtesy of Carnegie Institution of Washington.)

FIGURE 10-15 The fifth month. The fetus, 8 in. long and weighing 0.5 lb, is shown attached to the placenta by the umbilical cord. Gripping reflexes develop now. Fetal movements, such as kicking and turning, can easily be felt by the mother. The fetus may hiccup and sleep. Suspended in a quart of amniotic fluid, the fetus is in a state of weightlessness. The body is covered by a fine down called lanugo. Although the baby appears capable of a great deal if delivered prematurely at this time, it cannot survive more than a few minutes, since its lungs, skin, and digestive tract are unprepared for independent existence. The heart beats rapidly, and the sounds are easily recognized by a physician listening in with a stethoscope on the mother's abdomen. (Courtesy of Roberts Rugh.)

The sixth month

The fetus averages 12 in. in length and weighs 1.5 lb. It appears to be a miniature human. The internal organs occupy their normal positions, and the abdomen no longer bulges outward so much. Sucking of the thumb and swallowing of amniotic fluid occur. The eyelids separate, and eyebrows and lashes form. The oil glands in the skin produce a protective cheeselike covering called the **vernix caseosa.** The intestine becomes filled with a green pastelike material **(meconium),** which is derived from the breakdown of red blood cells and debris from the liver, gallbladder, and pancreas; it remains there until birth. The hands develop a strong grip. Scalp hair becomes abundant. The first bones develop (breastbone), and cartilage in the limbs is slowly replaced. Survival is usually not possible for the baby delivered after only 6 months of fetal life, since body temperature cannot be regulated and the lungs and digestive system cannot function on their own.

The seventh month

By the seventh month the child has at least a chance of independent survival. (Premature births are the cause of 50% of infant mortality in the United States.) If born at this time, there would be a 10% chance that the premature infant (preemie), given proper care in an incubator, could live. Essentially an incubator is an artificial womb: warm moist sterile air aids in protection and survival. The baby's nervous system has developed sufficiently so that breathing is under control, swallowing is reg-

ular, and body temperature is regulated. The fetus is about 1 ft in length and weights 2–3 lb. The downy fur coat, **lanugo,** which was more plentiful in the sixth month, is now found only on the back and the shoulders. Movements continue to be random, and there is thumb-sucking. The brain enlarges greatly, its surface becomes furrowed, and functional localization in the brain becomes evident. Sexual differences also manifest themselves: in the male, the testes descend into the scrotum.

The eighth month

During the eighth month the rate of weight gain slows down, the fetus weighs 4–5 lb and is 13 in. long. Fat tends to accumulate under the skin, and the fetus appears chubby—a change from his wrinkled condition in previous months. The fetus can perceive light and taste sweet substances, but remains deaf. If born at this time, the probability of his survival rises to 70%. Fat deposits help to regulate the body temperature of the eight-month preemie, but he still has difficulty in breathing because the air sacs of the lungs are not prepared for gas exchange. In addition, the digestive tract is immature. For this reason premature babies are given supplemental oxygen, but they can live on stored nutrients for a while.

The ninth month

Fetal activity diminishes during this month because all the available space within the uterus is occupied, and the fetus simply has no room to move. The fetus is 19 or 20 in. long and weighs between 6 and 8 lb. Boys may

weigh more than girls. Shortly before delivery the fetus turns so that its head rests in the pelvic basin—a comfortable position for the fetus and a safe position for delivery. The movements of the fetus in this position are even more restricted than before. The cheesy coating, the vernix caseosa, becomes dislodged and muddies the amniotic waters, the intestines are filled with green meconium. Fingernails and toenails grow so long that they require trimming shortly after birth, the gums are ridged, the eyes are almost always blue because pigmentation has not formed. This requires a few weeks of exposure to light. The baby acquires disease-resisting antibodies from the maternal blood supply, which will give it temporary immunity against certain diseases that could adversely affect the newborn child. Immunities acquired in this way usually persist for the first 6 months or so of life, and then gradually wear off as the baby begins to develop its own immune reactions. The hormones of the mother that cause her breasts to enlarge in preparation for lactation also cause the unborn child's breast to become firm and protrude. The newborn child may even secrete a few drops of milk (witches' milk). A few days after birth, when the effects of maternal hormones wear off, breast enlargement in the infant subsides and further secretion of milk ceases.

A week or two before delivery, the growth of the fetus slows down as the placenta regresses and becomes fibrous. Blood clots appear indicating that placenta blood vessels have degenerated; this prevents the mother's blood from reaching the fetal capillaries. The placenta begins to fail and the fetus must prepare to take care of itself.

FIGURE 10-16 The uterus and embryo at 7 weeks.

10-3 THE PLACENTA: LUNGS, KIDNEYS, AND DIGESTIVE TRACT FOR THE HUMAN FETUS

A chick, a frog, or a fish embryo survives apart from the mother because its egg contains a rich food supply—yolk. A human embryo has no such abundant resource and, therefore, requires considerable maternal care.

The first source of maternal nourishment is the glycogen-rich secretion of the endometrium (uterine milk). The egg implants by burrowing into the endometrium, where it derives its nourishment from then until birth. Since this elaborately prepared layer is shed at birth, it is given the name **decidua** (*deciduus*—falling off; L.). As early as the first week of development, the embryo begins to form an enclosing membrane, the **amnion,** within which it secretes amniotic fluid. This bag of water cushions and protects the early embryo. The trophoblast cells also burrow deeper into the endometrial lining. About 5–6 days after attachment to the uterus the trophoblast has eroded spaces (**lacunae**) in the endometrium, giving it a spongy appearance (Figure 10–5). Blood from the mother fills these spaces and bathes the growing trophoblast. The outermost membrane of the embryo (the **chorion**) begins to form fingerlike projections called **chorionic villi,** which project into the lacunae. By the third week the chorionic villi contain blood vessels that absorb nourishment from the mother's blood and carry it into the blood vessels of the developing embryo. These blood vessels are called the **allantoic arteries** and **veins,** and they run from the body of the developing child via a stalk, the **umbilical cord,** and connect up with the chorionic blood vessels (Figures 10–16 and 10–17).

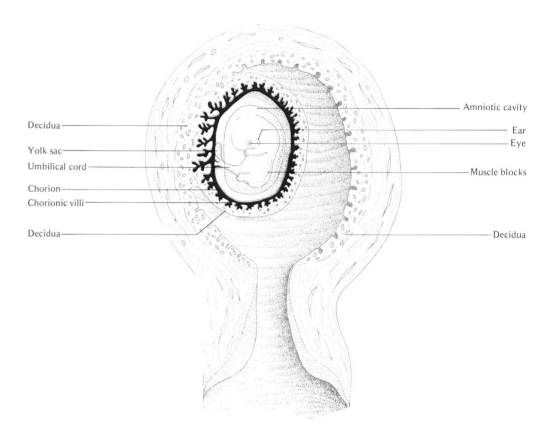

Decidua

Yolk sac

Umbilical cord

Chorion

Chorionic villi

Decidua

Amniotic cavity

Ear

Eye

Muscle blocks

Decidua

The **placenta** then develops. It is a disc-shaped mass of tissue that has a double origin. The fetal portion consists of the branched chorionic villi (called the **chorion frondosum**), and the maternal portion consists of a part of the decidua. It is principally via the chorionic villi that most of the exchanges between mother and fetus take place. The placenta provides for gaseous transfer, for elimination of wastes, and for the fetus's supply of predigested nutrients.

In a sense, it is lungs, kidneys, and digestive tract for the fetus during life in the womb. How does the placenta perform these functions?

The fetus floats in an almost weightless state in a fluid-filled sac (the amnion), but it is connected to the placenta from the very beginning by the umbilical cord (Figure 10–17), a conduit carrying two fetal arteries and one fetal vein. The umbilical cord provides a connection to life-support systems provided by the

FIGURE 10-17 (a) Section through the uterus showing fetal membranes, umbilical cord, and placenta,
(b) Detail of circulation of maternal and fetal blood
through the placenta.

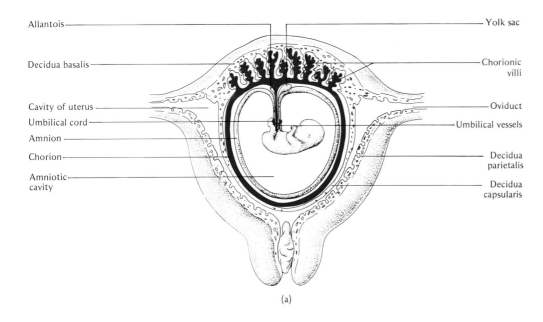

Allantois

Decidua basalis

Cavity of uterus

Umbilical cord

Amnion

Chorion

Amniotic
cavity

Yolk sac

Chorionic
villi

Oviduct

Umbilical vessels

Decidua
parietalis

Decidua
capsularis

(a)

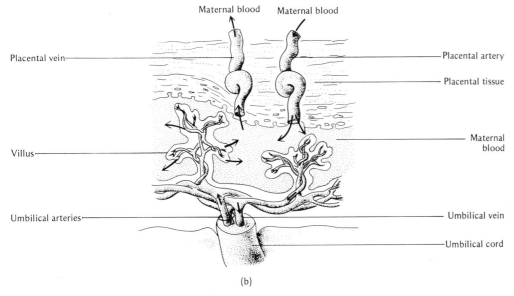

Maternal blood   Maternal blood

Placental vein

Villus

Umbilical arteries

Placental artery

Placental tissue

Maternal
blood

Umbilical vein

Umbilical cord

(b)

mother. In the fetus, the umbilical cord grows
from a few inches to a length of up to 4 feet.
Its blood vessels are cushioned in a substance
known as **Wharton's jelly,** which prevents them
from kinking. Just prior to birth about 300 qt
of blood are moved across the placenta each
day, each round trip between fetus and mother
taking only 30 seconds. The waste products of
the fetus are carried away by the umbilical
arteries and delivered to the chorionic villi,
from which they diffuse into the mother's
blood. The cleansed blood returns to the fetus
via the single umbilical vein, laden with nutrients and oxygen picked up from the mother's
blood. Maternal blood enters the placenta via
the uterine arteries, and these open into an
intervillous space. The spaces in turn are
drained by uterine veins.

   Thus, in the placenta, blood of the fetus and
the mother circulate in distinct and separate
channels; normally the two are never in direct
contact. Materials are exchanged across the
placenta by diffusion across capillary membranes, and the membranes keep out relatively
large molecules (such as proteins) but permit
smaller ones to pass. To facilitate exchange,
the surface area of the placenta is quite large,
amounting to many square yards. The placenta
functions for nutrition, respiration, and excretion; acts as a barrier against disease organisms;
and secretes hormones to maintain the endometrium. All these functions (except hormone
secretion) are accomplished by purely physical
means and require no special mechanisms
other than semipermeable capillary walls. The
growth of the placenta is phenomenal: by 1
month it covers 20% of the uterus, and by 5
months 50%. When the baby is delivered, the
placenta is shed as the afterbirth and weighs
1.5 lb.

10-4 PARTURITION (BIRTH)

After 40 weeks in the womb, the baby is ready for birth **(parturition).** The fetus in the amniotic sac has been surrounded by amniotic fluid for 9 months. The fluid capsule acts as a shock absorber and enables the fetus to grow without restriction. The uterus in turn has increased to 16 times its normal size, and its elastic walls have been stretched to capacity. Now, approximately 280 days after the last menstrual period, labor begins. **Labor** brings about the expulsion of the fetus from the uterus; it will mean departure from the watery world and the beginning of a new life on land (Box 10A).

The factors that initiate labor are obscure. It begins as a series of rhythmic movements in the uterine muscles and those in the abdomen. These muscular contractions, called **labor pains,** produce discomfort beginning in the lower abdomen and spreading to the back and thighs. As a consequence of the tremendous forces created by the muscular contractions of the uterine wall and the fluid pressure of the amniotic fluid, the cervix begins to be forced open. The dilation or opening of the cervix is the first stage in labor and permits the baby to enter the birth canal. A pink vaginal discharge (show) is also indicative of cervical dilation. The amnion, placed under 25–30 lb of pressure by the uterine muscles, may rupture, and the breaking of the waters occurs. Once the uterine contractions are occurring at 10-minute intervals, it is time to set off for the hospital (Figures 10-18—10-20).

FIGURE 10-18 First stage of labor: dilation of the cervix. This stage takes the longest time: 8–20 hours for the first baby and 3–8 hours for the second or later babies. During pregnancy the cervix has remained closed; now it must be opened to a diameter of 4 in. to permit the baby to pass through. The cervix also becomes softer during pregnancy, and the uterine muscles rearrange themselves to produce maximum expulsive force. As the body of the uterus contracts, considerable pressure is placed on the cervix, and it dilates and effaces itself. The uterus, cervix, and vagina become a single canal, the birth canal. Although the birth canal, like the uterus, is elastic, the baby's head itself acts as a wedge, widening the birth canal and slowly moving downward. If there is difficulty at this stage of labor, the child may be delivered by Cesarean section, an incision into the abdomen and uterus.

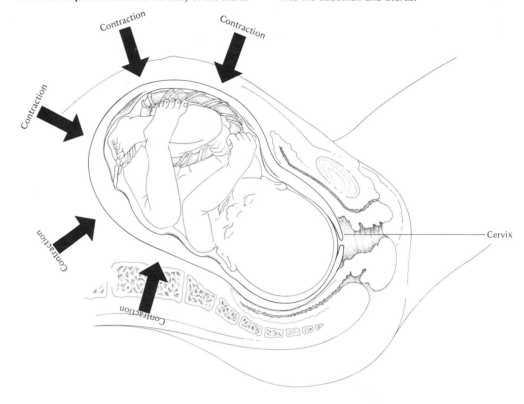

Contraction
Contraction
Contraction
Contraction
Contraction

Cervix

FIGURE 10-19 (facing page) Second stage of labor: delivery of the baby. The uterus, cervix, and vagina become a single canal, the birth canal, and the child moves outward through it. Now the mother bears down with each contraction. (This is involuntary and is stimulated by distension of the vagina and vulva by the fetus's head.) Usually the duration of this stage is shorter in women who have given birth previously and in those with strong uterine musculature. It usually lasts between 20 minutes and 2 hours. Moving downward through the birth canal, the newborn is rotated by the bones of the mother's pelvis. The obstetrician often guides the child at the delivery stage. Abnormal positions of the baby's limbs and the umbilical cord can complicate labor, but most physicians are well versed in handling these nonroutine births. As the child passes down the birth canal, his head may be pressed out of shape, but this produces no damage to the underlying brain, since the bones of the skull are pliable and separated by sutures which allow some degree of overlap. A few days after birth, the head resumes its normal shape. Uterine contractions usually cease with expulsion of the baby. The uterus now shrinks in size, and by shearing forces at its base the placenta is separated. (Courtesy of Roberts Rugh.)

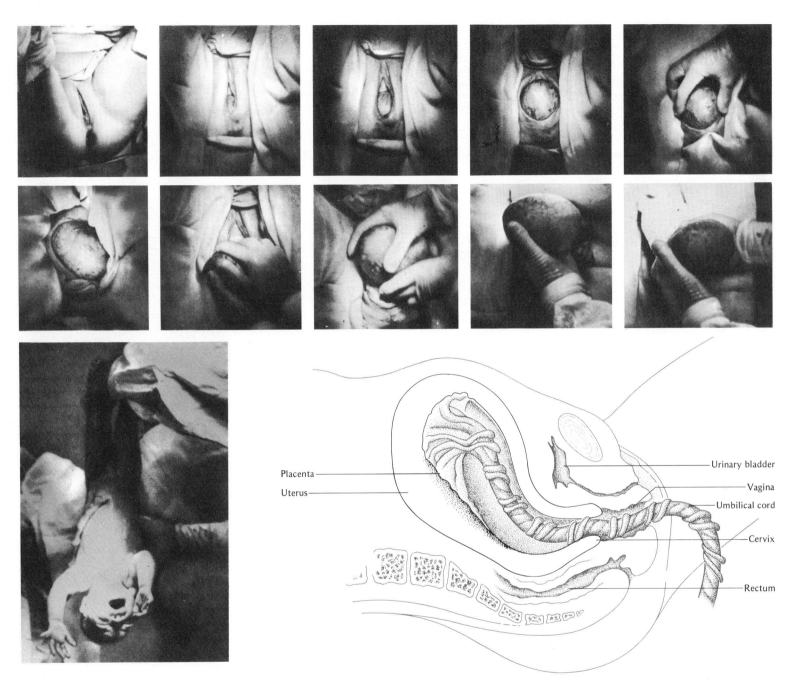

| Placenta |
| Uterus |
| Urinary bladder |
| Vagina |
| Umbilical cord |
| Cervix |
| Rectum |

FIGURE 10–20 (above) Third stage of labor: delivery of the placenta. About 15 minutes after the baby is delivered, the placenta and its associated membranes, together called the afterbirth, are expelled. Usually this is effected by the mother. The blood loss during this stage is about 200 ml, and is minimal because the firm contractions of uterine muscles serve to constrict blood vessels. Blood loss can be further lessened by intromuscular administration of Pitocin (an oxytocic drug), which helps the uterine muscles regain their tone. The umbilical vein collapses and the cord is tied. Life on land begins. (See Box 10A.)

BOX 10A From water to land: a baby is born

We emerge from water to the land at birth and announce the event by screaming. The change from fetus to newborn is dramatically seen in the circulatory system. If changes in the circulatory system did not occur, our chances for survival would be poor indeed. As we live warm and protected within the womb, our lungs are collapsed and idle. The fetal heart contains a by-pass for the lungs: blood entering the right side of the heart from the body can pass directly to the left side through an opening called the **foramen ovale,** the oval window (Figure 10–21a). To-gether with another short circuit in the fetal circulation, the heart-lung bypass ensures that only 10% of the fetus's blood is sent to the lungs, 55% goes to the placenta for oxygenation, and 35% is circulated through the blood vessels. The placenta acts as the fetal liver as well as lungs, and blood that enters the fetus via the umbilical vein is sent on a bypass around the liver.

At birth the circulatory routing must be changed and the short circuits removed. The newborn infant can no longer de-pend on the placenta for food and oxygen; now lungs and liver assume important functional roles, and blood must be routed through these organs. Ordinarily, the shock of birth triggers the first breath of life. If not, the traditional slap on the baby's bottom provides a sensory stimulus that triggers the respiratory center of the brain to initiate breathing (Chapter 14). Failing this, the umbilical blood vessels close following delivery of the placenta, and there is a sharp rise of carbon dioxide in the blood; this triggers the respiratory center into activity. As a result blood flow between the heart and the lungs is increased, this equalizes the pressure on the heart, and the foramen ovale in the heart closes. Closure of the umbilical blood vessels results in increased flow from the body organs through the liver, the other heart-lung short-circuit vessel closes, and the land-type pattern of cir-culation takes over (Figure 10–21b).

If by chance there is some slipup and the heart-lung bypasses do not close, not enough blood is routed to the lungs, and a "blue baby" results. Surgery can be used to seal the foramen ovale and tie off the short-circuit vessel, restoring normal land-type circulation.

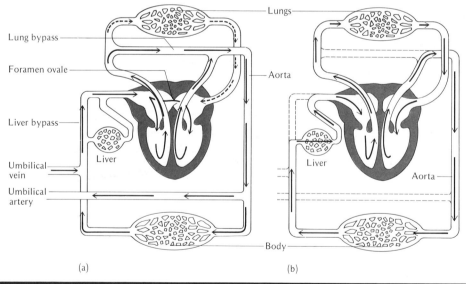

FIGURE 10–21 (a) Human fetal circulation (diagrammatic). (b) Circulatory system following birth (diagrammatic).

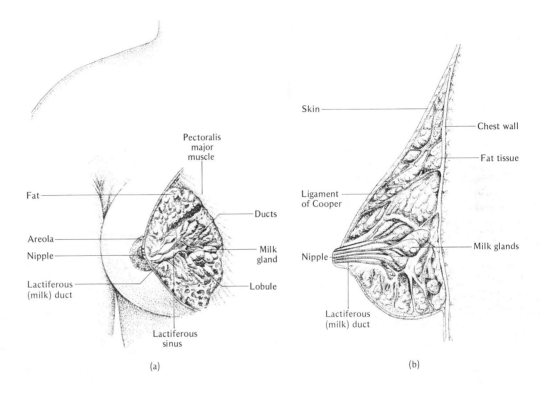

FIGURE 10-22 (a) External and internal anatomy of the mature female breast as seen in front view. (b) Side view of the internal anatomy of the breast.

10-5 THE MAMMARY GLANDS AND LACTATION

The **mammary glands** (breasts) are accessory reproductive organs in females (Figure 10-22). Prior to puberty the breasts in the male and female are similar. At about 10 years of age a girl's breasts begin to enlarge owing to the influence of estrogen and progesterone.

Estrogen causes duct development and increase in the supporting connective tissue, whereas progesterone is responsible for growth of 15 to 20 glandular lobes of tissue embedded in fat and connective tissue (Figure 10-22). The absence of significant quantities of these hormones in the male makes his breasts remain rudimentary. Growth continues until late adolescence, when the breasts assume a firm and rotund condition. After puberty the breasts enlarge slightly and then return to the normal size in cyclic fashion during each menstrual cycle because of the slight rise in progesterone and estrogen (Figure 10-23). At puberty the nipples become more sensitive, as does the pigmented area around each nipple (the **areola**). Each breast lobule is drained by a **lactiferous duct** that runs to the nipple.

Just prior to their termination beneath the areola, the ducts dilate into **lactiferous sinuses** that serve as reservoirs for milk. The actual size of the breasts bears little relation to the amount of mammary tissue present. In a mature non-lactating woman there may be no more than a spoonful of mammary tissue; the rest of the breast is occupied mostly by fat and connective tissue. These contribute to the ornamental features of the female body, not the primary function of the breast. At menopause, a woman's breasts are deprived of their hormonal stimulation, there is contraction and shrinkage of the connective tissue and the lobes, and as a result her breasts sag and lie closer to her chest.

Early in pregnancy there is an alteration in the breasts that is more marked than during menstruation—the lobes become more complex and larger and the areolae darken. Sebaceous (oil) glands in the areolae enlarge, and their oily secretion keeps the nipples supple and prevents the skin from cracking. Growth is brought about by ovarian and placental hormones (estrogen and progesterone). However, **lactation** (milk secretion and release from the mammary gland) is suppressed by these same hormones. At parturition, the levels of estrogen and progesterone decline with the expulsion of the placenta. This decline removes the inhibitory influence and permits the mammary gland to be stimulated. Stimulation for lactation comes from the action of the pituitary hormones prolactin, vasopressin, and oxytocin (Figure 10-24). Milk is not produced until a day or two after parturition, and the first secretion is a watery, yellowish-white fluid called **colostrum.** Colostrum has the same composition as milk but contains little or no fat; it serves as the baby's first food.

Although milk secretion begins about 2 days after the child is born, the milk remains stored within the gland until the breast is suckled. Sucking the nipples triggers a **letdown reflex** and the flow of milk. The letdown reflex is a complex affair (Figure 10-24). As the baby suckles the nipple, the nerves in the nipple are excited and impulses are transmitted to the

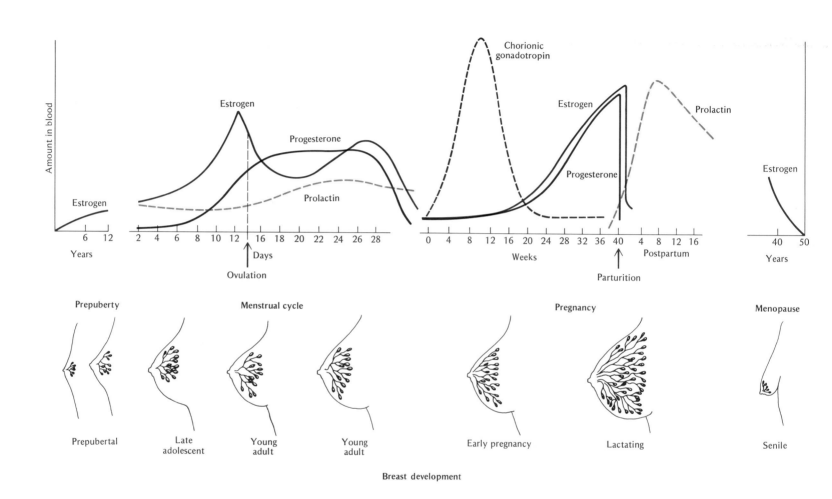

FIGURE 10-23 Hormonal changes during a human female's lifetime and their relation to breast development.

Breast development

hypothalamus of the brain, which releases the hormones oxytocin and vasopressin. The actual release of these hormones occurs via the posterior lobe of the pituitary gland, and when the two hormones arrive at the breasts, they trigger the contraction of the small muscles that surround the milk-engorged lobules of the mammary glands. As a result, the glands are squeezed, and milk is moved into the main ducts; the negative pressure caused by the sucking brings milk out of the holes in the surface tip of the nipple. The more milk is removed, the more hormones are released, and the more milk is produced; thus the suckling infant regulates his own supply. Once established, milk production can continue for many months, and once suckling is instituted, the letdown process may take less than a minute.

If the mother does not suckle her young, the mammary glands do not receive local stimulation, milk formation stops within 1 or 2 weeks, and the milk stored in the gland, if not expressed manually, is resorbed. Lactation is a delicately balanced process, and psychogenic

FIGURE 10-24 Sucking reflex control of the milk letdown in the mammary gland.

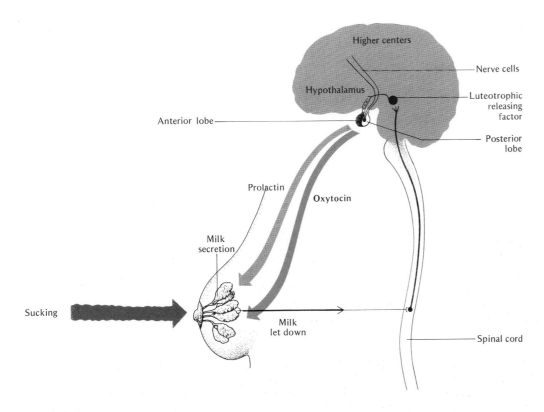

TABLE 10–1 The composition of milk

Component	Approximate concentration (gm/100 ml)	
	Human milk	Cow's milk
Water	88.5	87.0
Lactose (milk sugar)	7.0	4.8
Lipid	3.3	3.5
Protein		
Casein	0.9	2.7
Lactalbumin and other proteins	0.4	0.7
Minerals		
Potassium	0.041	0.150
Calcium	0.030	0.120
Phosphorus	0.013	0.095
Sodium	0.011	0.050

10-6 BIRTH DEFECTS

The boy shown in Figure 10–25 is a victim of thalidomide, a "harmless" sedative taken by his mother during her second month of pregnancy. His plight tragically illustrates the intricate complexities of human development and shows how interference at some crucial stage can produce defective offspring.

Development and growth can go awry to produce Siamese twins, hydrocephalics with balloon-shaped heads, children with six fingers or harelip or clubfoot, fetuses without brains or with split spines. How does this happen?

The processes of development are not always perfect, but usually the final product of embryonic development fits a statistical description that we call normal. (**Normality** simply means that the variable in question most frequently fits within a certain range—for example, normal height for a human adult male is between 5 ft 2 in. and 6 ft 4 in.; outside this range, the individual is considered to be abnormal.) When tissue interactions go awry, as indeed they often do, a functional or a structural abnor-

stimuli can automatically inhibit oxytocin release and prevent letdown. Because of this possibility, nursing mothers are advised to remain calm during lactation.

Milk production by the mother may amount to 1.5–2.0 qt each day. Lactation in humans may continue for as long as the infant is suckled. The production of prolactin during this time usually inhibits the menstrual cycle. It is important to know that a nursing mother can become pregnant, and nursing does not constitute a reliable contraceptive method.

The composition of milk in cows and humans is shown in Table 10–1. Cow's milk is richer in protein than human milk and curdles in a human infant's stomach.

mality may result. These developmental aberrations are probably more common than we believe. In 1964 Columbia-Presbyterian Medical Center studied 6,000 infants from the fourth month of fetal life through the first year after birth and tabulated all the malformations. Of the babies born alive (by far the majority of abnormal embryos never reach full term), 4% had severe disorders such as heart defects, mental retardation, and imperfect closure of the neural folds resulting in a neural canal open to the exterior of the body **(spina bifida).** Extrapolated to the country at large, this means that over 250,000 seriously abnormal children are born in the United States each year, 700 each day, or 1 every 2 minutes.

Most of the abnormalities occur prior to birth, for the embryo is the most vulnerable stage of development. The embryo passes through a series of developmental stages at any of which it can be changed from normal to abnormal. Birth defects are generally caused by three conditions: (1) defective genes, (2) abnormal arrangements of chromosomes, and (3) teratogens. **Teratogens** are environmental agents that act on the embryo, directly or indirectly, during its intrauterine existence to produce malformations. We shall discuss the first two causative elements in Chapter 22. Here we shall concern ourselves with the teratogens that are responsible for 80% of the distortions (so-called **congenital aberrations**) in the architectural plan of development.

Thalidomide and other drugs

The thalidomide tragedy of the 1960s points up how birth defects can be induced by subjecting the pregnant mother to an environmental hazard. During 1960–1962 in West Germany numerous reports of deformed infants appeared in newspapers around the country.

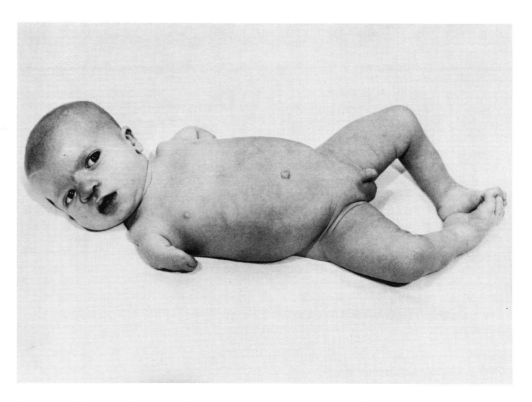

FIGURE 10-25 A thalidomide victim. (Courtesy of W. Lenz.)

The arms of these babies had failed to grow and their hands extended almost directly from their shoulders, producing a seallike flipper (Figure 10–25). Careful sleuthing revealed that the condition was not genetic, but that the mothers of the flipper-limbed infants (and there were hundreds of cases in 1961, and as many as 6,000 by 1962) had taken a sedative preparation containing thalidomide. The drug was used in cough syrups and was recommended for patients with morning sickness. Little is known about the metabolism of thalidomide and how it produces its effects, but it seems to act on the embryo between the twenty-eighth and forty-second days of development, a time when the future limbs of the embryo are forming. The limbs first appear as microscopic buds at 10 days, and by 42 days they can easily be identified by eye in the 1-in. embryo (Figure 10–9). The arms develop first and then the legs; this may be the reason that the arms are malformed more frequently than are the legs. Thalidomide does not produce limb abnormalities in mice, rats, or rabbits. In November 1961 the German manufacturer stopped production of the drug, but the thalidomide victims continue to suffer to this day. Americans were spared a thalidomide disaster because the drug was not approved by the U.S. Food and Drug Administration. (The few American babies deformed by thalidomide were damaged by pills brought in from Europe.)

TABLE 10-2   Drugs and pregnancy

Drugs taken by mother	Possible effects on fetus or newborn
Androgens (male hormones)	Masculinization of fetus
Anesthetics	Depression of fetus, asphyxia
Antimetabolites	Hydrocephalus, cleft palate, eventual infertility
Antihistamines	Abortion, malformations
Aspirin	Persistent truncus arteriosus, abnormal heart
Barbiturates	Depressed breathing, drowsiness for 6 days after birth
Chloramphenicol (an antibiotic)	Gray spinal syndrome, death, leukemia, damage to chromosomes
Cortisone acetate	Anomalies, cleft palate (?) withdrawal syndrome
Demerol	Brain damage(?)
Diuretics and potassium-depleting agents	Polycystic kidney disease
Erythromycin	Liver damage(?)
Estrogens (female hormones)	Malformations, hyperactivity of fetal adrenals
Gantrisin	Hyperbilirubinemia (a blood disorder)
Heroin and morphine	Convulsions, tremor, neonatal death
Insulin shock	Fetal death
Lead	Anemia, hemorrhage, abortion, lead line on gums
LSD (d-lysergic acid diethylamide)	Chromosomal anomalies, deformity
Nicotine (from smoking)	Stunting, accelerated heartbeat, premature birth, organ congestion, fits and convulsions
Norethisterone or Enovid	Androgenesis, masculinization
Oral progestogens	Secondary sex anomalies(?)
Androgens	Masculinization
Estrogens	Feminization
Phenacetin	Anemia, death
Phenobarbital in excess	Asphyxiation, brain damage(?), neonatal bleeding, death
Sodium seconal	Electrical depression of brain waves
Stilbesterol	Virilization of females
Streptomycin	Damage to auditory nerve
Sulfonamides	Jaundice, kernicterus
Thalidomide	Hearing loss, abnormal appendages, death
Tranquilizers	Retarded development(?)
Thiazine (a diuretic)	Thrombocytopenia, polycystic kidney disease
Vaccination against influenza	Anemia
Vaccination against smallpox	Fetal vaccinia
Vitamin K analogs in excess	Hyperbilirubinemia, jaundice in premature infants

From R. Rugh and L.B. Shettles, *From Conception to Birth*, New York, Harper & Row, 1971.

Other drugs that can cause malformations when taken during pregnancy are listed in Table 10-2. The best advice for a pregnant woman who is contemplating taking drugs is, don't.

Disease and birth defects

The uterus, which harbors the developing embryo, is a protected place indeed, but it is not entirely safe. As we have seen, drugs such as thalidomide do penetrate the placental barrier and can produce malformations. The pregnant mother can also transmit diseases and toxins to her unborn offspring, and such agents can be teratogenic. The agents that cause smallpox, chickenpox, measles, mumps, scarlet fever, tuberculosis, syphilis, influenza, and rubella (German measles) are all capable of crossing the placenta. During the first month of development rubella often produces eye cataracts; in the second month heart malformations are likely; and in the third month fever may result in deafness of the fetus. The nervous system begins its formation about the third week, but continues its development long after birth. Thus, for a considerable period of time the structure and function of the embryonic and fetal nervous system can easily be affected by disease agents, toxins, and viruses. Table 10-3 lists some dangerous maternal diseases and possible effects on the embryo and fetus.

X-rays and birth defects

Hereditary defects from ionizing radiations such as x-rays occur because sperm and egg genes are altered (more about this in Chapter 23). Congenital defects can develop in the fetus as a result of direct exposure of the pregnant mother to x-rays. Dividing cells are more sensitive to ionizing radiation than are nondi-

BOX 10B Abortion

Abortion refers to the interruption of pregnancy before the fetus can live on its own outside the uterus. (To have a minimum chance of survival, the fetus must be 7 months old.) Premature expulsion of an embryo or nonviable fetus constitutes abortion, and 96% of all such are spontaneous. Spontaneous abortions **(miscarriages)** are probably due to abnormalities of fetus or placenta, and about 10% of pregnancies end in spontaneous abortion, the majority occurring in the first 3 months.

Prior to 1967, millions of American women underwent illegal abortions, and hundreds of these died as a result of incompetent bunglers performing the operations under unsanitary conditions. Recently however, the status of abortion in the U.S. has changed dramatically. At first only a dozen states provided for legal abortion, but by 1973 the Supreme Court ruled that abortion could not be denied to any women in her first 3 months of pregnancy; after 3 months, regulation by states was permissible if it related to maternal health. Although the availability of safe, legal abortions has significantly reduced the number of deaths due to illegal operations, the practice is not without controversy. Some opponents say abortion destroys a potential and sacred life, penalizes the poor and the unmarried, induces psychiatric problems, and promotes irresponsibility and sexual promiscuity. The proponents say it is necessary to curb over-population, protect against the delivery of an abnormal child, and precludes adverse effects for the mother who desires no more children. At present a vast gulf exists between science and religion, and between law and morality; and the abortion laws yet to be drafted must satisfactorily bridge that gap.

How is a therapeutic abortion performed, and how safe is it? This depends on the length of the pregnancy. During the first 3 months, **suction** or **dilation and curettage** (D and C) are most frequently used. Suction or vacuum aspiration is performed under local anesthesia, the cervix is distended or dilated sufficiently to permit introduction of a suction tube, and the uterus is emptied by gentle suction. The dilation process may produce some cramps, but the suction is quick and usually quite painless. The entire procedure, including preparation, takes less than 20 minutes, and the patient can return home after several hours' rest. In over 54,000 patients aborted by suction, there was not a single fatality due to the method.

Performance of a D and C involves dilation of the cervix and the insertion of a curette (a wide loop of steel attached to a special handle) to scrape the inside of the uterus to remove the implanted embryo and the surrounding endometrial tissue. General anesthesia is often used because somewhat greater discomfort is produced by the dilation. The procedure takes 20 minutes and the patient may be admitted to the hospital for an overnight stay. When properly done, there is less than a pint of blood lost.

Beyond 12 weeks of pregnancy the risk and difficulty of abortion increase. Most physicians find that the risk of hemorrhage during the thirteenth to sixteenth weeks precludes abortions by suction or D and C, but by the seventeenth week a **saline abortion** can be employed. The procedure takes from 1 to 3 days. Under local anesthetic, a long hollow needle is inserted through the abdominal wall into the amniotic sac, a small amount of amniotic fluid is withdrawn and replaced with a concentrated solution of salt water. The salt injection ends fetal life within hours, labor contractions are induced, and the fetus plus placenta are expelled within 24 to 72 hours after the injection. There is little discomfort during the saline injection, and the labor is quite similar to a normal delivery, though not as intense. The risk of complications by saline abortion is three times greater than with the procedures used for early abortions. Beyond 20 weeks most physicians will not attempt a saline abortion.

Two other methods for later abortion are a small **Cesarean section** (cutting through the walls of the abdomen and uterus to remove the fetus) and **hysterectomy** (total removal of the uterus). Both are considered major surgery, may cause complications, and are ordinarily performed only when the mother's health demands (for example, diseased uterus).

The death rate for legal abortions averages less than 8.2 fatalities per 100,000 cases. This means that early abortion is far safer than childbirth and saline abortions are somewhat safer than full-term delivery. Cesarean section and hysterectomy incur greater risks than normal childbirth. Notwithstanding the safety of legal abortion, it remains the least desirable method of birth control.

TABLE 10–3 Maternal diseases and the fetus

Maternal infection	Possible effects on fetus or newborn
Chickenpox or shingles	Chickenpox or shingles, abortion, stillbirth
Congenital syphilis (*Treponema pallidum*)	Miscarriage
Coxsackie B virus	Inflammation of heart muscles
Cytomegalovirus (salivary-gland virus)	Small head, inflammation and hardening of brain and retina, deafness, mental retardation, enlargement of spleen and liver anemia, giant cells in urine from kidneys
Hepatitis	Hepatitis
Herpes simplex	Generalized herpes, inflammation of brain, cyanosis, jaundice, fever respiratory and circulatory collapse, death
Influenza A	Malformations
Mumps	Fetal death, endocardial fibroelastosis, anomalies
Pneumonia	Abortion in early pregnancy
Poliomyelitis	Spinal or bulbar poliomyelitis, acute poliomyelitis of newborn
Rubella (German measles)	Anomalies; hemorrhage; enlargement of spleen and liver, inflammation, of brain, liver, and lungs; cataract; small brain; deafness; various mental defects; death
Rubeola	Stillbirth, abortion
Scarlet fever	Abortion in early pregnancy
Smallpox	Abortion, stillbirth, smallpox
Syphilis	Stillbirth, premature birth, syphilis
Toxoplasmosis (protozoan parasite infection)	Small eyes and head, mental retardation, water on the brain (encephalitis), heart damage, fetal death
Tuberculosis	Fetal death, lowered resistance to tuberculosis
Typhoid fever	Abortion in early pregnancy
Western equine, encephalitis	Encephalitis, idiocy

From R. Rugh and L. B. Shettles, *From Conception to Birth*, New York, Harper & Row, 1971.

viding cells. Since the major part of development involves cell division and specialization, it is no wonder that it takes less x-ray exposure to damage a fetus than an adult. The embryo is most sensitive during the second to the sixth week, because this is the period of fastest embryonic growth when cell division of the germ layers is most active. Based on the experiences of the victims of Hiroshima and Nagasaki, we know that ionizing radiation causes leukemia, cataracts, microcephaly, and other malformations. Seven out of eleven Hiroshima children were born retarded if their mothers were less than 20 weeks pregnant and were less than 0.5 mile from ground zero. Those whose mothers were further away showed deformed hips, eyes, and hearts.

Laboratory experiments have shown that a minimum exposure of 10 roentgens will affect the fetus. Usually a single x-ray exposure involves 0.1 roentgen, but fluoroscopy may involve greater doses. When exposure to 10 roentgens or more during the first 6 weeks of pregnancy is necessary (for example, when radiation therapy must be used to treat a pelvic tumor), a therapeutic abortion may be recommended by the physician. If nonessential x-ray examinations were forbidden for women during the 9 or 10 days following menstruation (the fertile period in the cycle), radiation-induced congenital abnormalities would be virtually eliminated.

SUMMARY

1. After fertilization the zygote divides mitotically into a number of smaller cells; this process constitutes cleavage. Cleavage lasts until other developmental processes besides mere subdivision begin to occur.

2. By the 16–32 cell morula stage, the fate of each cell is committed.

3. The morula hollows out to a blastocyst at 4 days. The polar blastoderm cells will form the embryo; the remaining trophoblast cells will form embryonic membranes.

4. Following implantation of the blastocyst, the embryo increases in size and cell-cell relationships change. Embryonic development begins.

5. Embryonic development

Weeks	Stage of embryonic development
2–5	Ectoderm, endoderm, and primitive streak appear.
3	Mesoderm and neural tube appear.
4	Neural tube closes; organ systems begin to develop; brain, heart, and somites appear.
5	Tubular heart pumps blood; cranial flexure becomes acute; limb buds form; gill clefts and outline of eyes appear.
6	Jaws form; gill clefts disappear; eye pigment, diaphragm, gonads, and reflexes appear. Heart is partitioned; liver and intestine enlarge; cranial flexure is apparent.
7	Tail diminishes; eyes and ears, fingers and toes, and inner ear are well formed.

6. Fetal development At 2 months, the embryo is well developed and is now called a fetus.

Months	Size	Stage of fetal development
2	1.5 in., 1/7 oz	Eyes directed forward; jaws, nose, earlobes, limbs, external genitalia, mouth, and alimentary canal are well formed. Skeleton hardens; simple reflexes occur. Blood made in fetal liver and spleen. Growth ensues.
3	3 in., 1/2 oz	Cranial flexure disappears. Fingernails, toenails, hair, tooth buds, and bones appear. Blood manufactured in bone marrow. Fetal pulse apparent; kidneys function; coordinated motion improves.
4	6 in., 4 oz	Posture erect; startle and other reflexes appear; fingerprints appear, refinements in development continue.
5	8 in., 0.5 lb	Gripping reflexes appear; fetal movements and rapid heartbeat apparent. Lanugo covers fetus. Lungs, skin, and digestive tract well developed but fetus incapable of independent existence.
6	12 in., 1.5 lb	Thumb sucking, swallowing, and vernix caseosa appear; meconium fills intestine; hair, bones well developed.
7	12 in., 2–3 lb	Chance of survival following birth is 10%. Breathing, temperature regulation, and swallowing possible, though systems immature. Lanugo reduced. Brain enlarges and differentiates; testes descend.
8	13 in., 4–5 lb	Growth slows. Fat accumulates; temperature regulation capacity improves; light and taste perception develop. Lungs and digestive tract immature; no sound perception.
9	19–20 in., 6–8 lb	Activity diminishes; fetus turns; acquires immunities; development to prenatal state completed. Placenta and umbilical cord regress.

7. The embryo implants into the endometrium (decidua), and the protective amnion forms. The trophoblast erodes lacunae; the chorion (outermost membrane) forms villi and blood vessels. Chorion and amnion form the allantoic sac; the embryo is suspended in amniotic fluid. Two allantoic arteries and one vein run along the umbilical cord to and from the embryo, protected by Wharton's jelly.

8. The chorionic villi (chorionic frondosum) + decidua = placenta. The placenta has a large surface area and functions in fetal respiration, excretion, and nutrition. It also secretes hormones and bars many maternal disease organisms. Placental exchange occurs by diffusion, with no mixing of maternal and fetal blood.

9. At 40 weeks, cervical softening, dilation and effacement, and muscular uterine contractions (labor pains) signal imminent parturition. The amnion ruptures; the baby moves into the birth canal and is born. Placenta and membranes (afterbirth) follow.

10. At birth, the fetal umbilical vessels, foramen ovale, and heart-lung bypass close. Rising CO_2 levels trigger the respiratory center and breathing ensues. Postnatal circulatory patterns take over.

11. Female mammary glands develop in response to progesterone and estrogen, which also suppress lactation. Levels of these hormones drop with placental expulsion. Prolactin, vasopressin, and oxytocin stimulate lactation. Milk flow is preceded by colostrum. A letdown reflex is triggered by sucking on the nipples. Prolactin may also inhibit menstruation during lactation.

12. Miscarriages and/or birth defects may be caused by defective genes; by chromosome abnormalities; or by teratogens such as drugs, viruses, and x rays.

13. Pregnancy termination (abortion) may be performed by suction, D and C, saline abortion, Cesarian section, or hysterectomy.

KEY WORDS

cleavage
embryo
zona pellucida
morula
blastocyst
embryonic disc or blastoderm
trophoblast
ectoderm
primitive streak
mesoderm
neural tube
somites
cranial flexure
fetus
vernix caseosa
meconium
lanugo
decidua
amnion
lacunae
chorion
chorionic villi
allantoic arteries and veins
umbilical cord
placenta
chorion frondosum

Wharton's jelly
intervillous space
parturition
labor
labor pains
birth canal
Cesarean section
foramen ovale
mammary glands
areola
lactiferous duct
lactiferous sinuses
lactation
colostrum
letdown reflex
spina bifida
teratogens
congenital aberrations
miscarriages
suction
dilatation and curettage (D and C)
saline abortion
hysterectomy

TOPICS FOR DISCUSSION AND REVIEW

1. Discuss the role of cleavage in development.
2. What are the differences between the embryo and the fetus?
3. List the functions of the placenta.
4. Discuss methods of therapeutic abortion.
5. How may birth defects arise?
6. What structural changes take place in the circulatory system at birth?
7. Describe the events taking place during the first eight weeks of embryonic life.
8. What are the feedback controls involved in lactation?
9. Describe the various stages of labor.
10. To what organs do the embryonic ectoderm, endoderm, and mesoderm give rise?

"Why should I eat it? I got my daily essential Vitamin A and all the thiamine and riboflavin I need in the fortified Dynaflakes and irradiated milk I had for breakfast."

PROCESSING NUTRIENTS: DIGESTION

11–1 WHY DO WE EAT?

We all recognize what food is. Just the thought of food may make the mouth water. But why is it necessary for us to eat? Most of us would respond by saying, "We eat to live." Intuitively, each of us recognizes that his life depends upon a supply of food, and we all know that we have a continuing and constant demand for matter and energy (Chapters 3 and 4). From our food we obtain the materials we need for growth, repair, and energy. As Walter de la Mare put it:

It's a very odd thing—
As odd as can be—
That whatever Miss T. eats
Turns into Miss T.[1]

Our own cells as well as those of Miss T. require that the nutrients or regulatory substances contained in food be available in the tissue fluids in a soluble form. However, the food we eat is a complex material that is often insoluble. Therefore, the problem presented to the body with breakfast, lunch, or dinner is to break down these complex foodstuffs into small soluble molecules that can pass across the plasma membranes of the cells of the digestive system. The process by which our food is broken down from complex insoluble substances into simple soluble ones is called **digestion.** This is a special function of the organ system known as the **digestive tract** (also called the **gut, alimentary canal,** or **gastrointestinal [GI] tract**). Once the food is digested, the digestive tract functions in another capacity: it permits absorption of the soluble substances and makes them available to the cells of the

---

[1] From "Miss T" by Walter de la Mare. Reprinted by permission of The Literary Trustees of Walter de la Mare, and The Society of Authors as their representatives.

BOX 11A A peephole into the stomach

On June 6, 1812, in a small community at the Canada-Michigan border, Alexis St. Martin accidentally discharged his musket and tore a large gaping hole in the left side of his abdomen. Dr. William Beaumont, an army surgeon stationed at nearby Fort Mackinac, examined the young man within 30 minutes of the accident and treated St. Martin for his wound, but held out little hope that the 18-year-old boy would recover. St. Martin was young and vigorous, and he not only survived but recuperated remarkably well. However, his wound healed peculiarly: a tunnel 2.5 in. wide remained open in his side, and this led through the skin and muscle of the abdominal wall directly into the stomach. Food and drink constantly exuded from the perforation unless prevented by a compress and a bandage.

St. Martin's **fistula** (abnormal opening) provided Dr. Beaumont with a remarkable opportunity to peep into the digestive system, and he retained the young man as a servant in his home. St. Martin performed a variety of household duties; he fathered children and enjoyed good health and vigor as do men in general.

Periodically, Dr. Beaumont used the boy for experiments on the digestive processes by introducing materials into the stomach via the hole, sealing the opening with a bandage to prevent leakage, and later removing the materials from the stomach to evaluate the changes. Here is his first recorded experiment:

August 1, 1825. At 12 o'clock, M., I introduced through the perforation, into the stomach, the following articles of diet, suspended by a silk string, and fastened at proper distances, so as to pass in without pain—viz.:—a piece of high-seasoned à la mode beef; a piece of raw, salted, fat pork; a piece of raw, salted, lean beef; a piece of boiled, salted beef; a piece of stale bread; and a bunch of raw, sliced cabbage; each piece weighing about two drachms; the lad continuing his usual employment about the house.

At 1 o'clock, P.M., withdrew and examined them—found the cabbage and bread about half digested: the piece of meat unchanged. Returned them into the stomach.

At 2 o'clock, P.M., withdrew them again—found the cabbage, bread, pork, and boiled beef all cleanly digested, and gone from the string; the other pieces of meat but very little affected. Returned them into the stomach again.

At 3 o'clock, P.M., examined again—found the à la mode beef partly digested: the raw beef was slightly macerated on the surface, but its general texture was firm and entire. The smell and taste of the fluids of the stomach were slightly rancid; and the boy complained of some pain and uneasiness at the breast. Returned them again.

W. Beaumont, *Experiments and Observations on the Gastric Juice and the Physiology of Digestion,* Pittsburgh, Allen, 1833.

St. Martin and Beaumont collaborated off and on for 8 years; St. Martin remained healthy and outlived Beaumont by 27 years, dying at the age of 76. Today we know much about the function of the digestive tract, but it was a remarkable set of circumstances that permitted Beaumont a glimpse into the stomach for the first time 150 years ago.

body via the body fluids. Finally, the digestive tract eliminates the food residues that cannot be digested or absorbed.

11–2 INTAKE APPARATUS FOOD PROCESSING

The digestive tract is a continuous tube running from the mouth to the anus (Figure 11–1a); if stretched out, it would measure nearly 30 ft in length. In life, however, **muscle tonus** (a sustained state of partial contraction) shortens the length of the digestive tube to between 12 and 15 ft. The walls of the tube have four layers: the innermost lining (the **mucosa**) consists of epithelial cells; the layer beneath this (the **submucosa**) is connective tissue with fibers, blood vessels, and nerve endings; and the outermost layers are muscle with the inside layer having a circular orientation of the fibers and the outer one a lengthwise arrangement (Figure 11–1b). The mucosal lining of the digestive tract secretes a slimy mucous material, and in some regions it also produces digestive juices. This mucus lubricates the tube, facilitates the movement of food along its length, protects the delicate epithelial cells against the abrasive substances in the food, and provides a coating that resists the action of enzymes in the digestive juices.

Although the digestive tract is often called one's "innards," in a very real sense, any substance present within it is not in the body at all. The digestive tube goes through the body and is surrounded by the body, but substances in the tube are as much outside the body as a finger placed in a doughnut hole is outside the substance of the doughnut.

The processing of food by the digestive tract occurs in the following way. Three times a day, more or less, a bulk cargo of foreign goods—food—enters the alimentary canal through its gateway, the mouth. In the mouth the cargo is

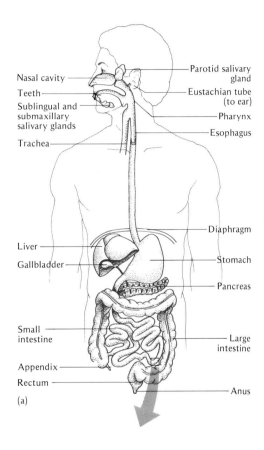

Nasal cavity
Teeth
Sublingual and submaxillary salivary glands
Trachea
Liver
Gallbladder
Small intestine
Appendix
Rectum

Parotid salivary gland
Eustachian tube (to ear)
Pharynx
Esophagus
Diaphragm
Stomach
Pancreas
Large intestine
Anus

(a)

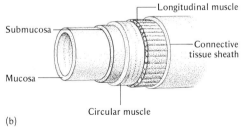

Longitudinal muscle
Submucosa
Connective tissue sheath
Mucosa
Circular muscle

(b)

mechanically degraded by the ripping, grinding, and cutting action of the teeth; the rolling action of the tongue and the secretion of saliva form the food into a pliable ball, or a **bolus.** Food then moves on to the throat (pharynx) and is swallowed. From the throat it is transported to the stomach by way of the esophagus and then on through the intestine; passage of food along the canal is accomplished by rhythmic muscular movements called **peristalsis** (Figure 11–2). Pushed and shoved along by muscles, the bulk cargo is mechanically and chemically changed: the stomach churns the food and mixes it with mucus so that it assumes a pastelike consistency, and here digestion begins. Later, food is moved into the small intestine where "tributaries" carrying digestive enzymes empty; most digestion takes place in the small intestine, and what was bulk cargo becomes a mixture of soluble, fluid materials. The soluble materials are transported across the intestinal wall into the riverways of the circulatory system.

Absorbed and circulated, the molecular constituents of the food enter the body cells, where they are utilized for energy, repair, and building of new cells. The remaining undigested materials (**feces**) pass into the large intestine and leave the digestive tube via the anus.

Now that we have surveyed the general organization of the digestive tube, let us take a more detailed tour and observe its functional specializations. The details of the journey taken by the food we eat are shown in Table 11–1.

11–3 DIGESTION: FOOD DEGRADATION

The digestive tract is relatively indiscriminate. Almost anything and everything digestible that we eat is absorbed without any kind of selectivity. (Other organs, such as the liver and pancreas, actually set the levels of nutrients carried in the blood.) Enzymatic breakdown of

FIGURE 11–2 Peristalsis. (a) The movement of food along the length of the digestive tract can be visualized as a squeezing action, forcing the food along. (b) This mechanism is shown diagrammatically in a section through the intestine. Although in peristalsis, the muscular waves of contraction can move food in both directions, one-way movement is usual. Food distending the tract causes contraction on the headward side and relaxation on the anal side of the distention, producing an easy forward movement.

food within the digestive tract involves one kind of mechanism: **hydrolysis.** Hydrolysis splits molecules by the addition of water. (See Chapter 3 for a discussion of hydrolytic reactions.) In this kind of reaction, the energy change in the molecules themselves is insignificant; that is, little chemical energy of the food is lost. Furthermore, hydrolytic reactions do not require energy (no ATP is consumed). In the stomach and intestine, proteins are digested to polypeptides and ultimately to amino acids, fats are broken down into glycerol and fatty acids, sugars and starches are hydrolyzed to monosaccharides, and nucleic acids are degraded to nucleotides.

Table 11–2 lists the enzymes and other substances secreted by the alimentary canal and its associated organs, and indicates their action on the components of our food.

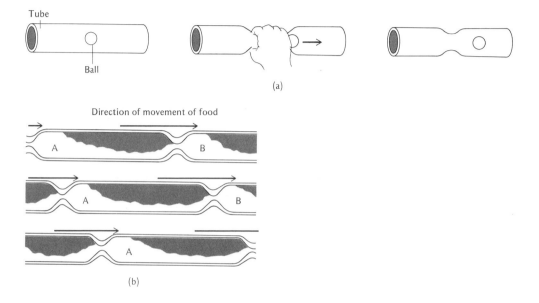

11–4 ABSORPTION: FOOD UPTAKE

Chemical digestion reduces what was once a tasty morsel of food to a solution of fatty acids, glycerol, glucose, fructose, nucleotides, and amino acids. If these small, soluble molecules are to be of any use to the cells of the body, they must be moved out of the intestine into the bloodstream. The process of moving nutrients and water from the intestine to the blood constitutes **absorption.**

The stomach

By the time it reaches the stomach, what was a bolus of finely divided food has been broken down to an acid solution of polypeptides, finely suspended globules of fat, and partially degraded starches.

Only a few substances are absorbed directly across the walls of the stomach. Aspirin, a weak acid, is one of these, and so is ethyl alcohol. Ethyl alcohol is lipid- as well as water-soluble, and it quickly diffuses across the epithelial surface of the stomach to enter the blood. This is why its effects show up so rapidly. Although it is absorbed by the stomach, alcohol is more quickly absorbed across the intestinal walls because of the latter's greater surface area. It is possible to reduce the immediate effects of alcohol, that is, dulling of the higher centers of the brain and the associated euphoria, by diminishing intestinal absorption. Drinking milk before imbibing alcohol or snacking on cheese dips or high-fat hors d'oeuvres delays the emptying of the stomach and consequently slows down alcohol absorption, but it does not prevent it altogether.

The intestine

Approximately 90% of the digested food is absorbed in the 20–25 ft of the small intestine (the duodenum is about 1 ft in length, the jejunum 7–8 ft, and the ileum about 12 ft in length; Figure 11–1). Additionally, about 10% of the body's total water and salt enter the small intestine each day as secretions from the various digestive glands, and approximately 90% of this is resorbed into the bloodstream. Were all the water secreted into the gut lost with the feces, we would have a severe dehydration problem.

Since the small intestine is the place where nutrient and most water absorption takes place, we should expect it to have a large exposed surface area. This is exactly the case. The small intestine is quite long and is packed into the abdominal cavity as a series of coils (Figure 11–1) so that by its length alone a large surface for absorption is provided. However, this is not all, for the surface of the small intestine also

TABLE 11–1 Functional organization of the digestive system

Structure	Description	Activity	Time spent by food
Mouth	Contains teeth and tongue, openings of salivary glands	Teeth grind and break down food, exposing large surface area for action of enzymes. Tongue assists teeth and rolls food into bolus preparatory to swallowing. Saliva dilutes and moistens food. Mucus sticks food together and lubricates bolus for swallowing. From 1 pt to 0.5 gal of saliva is produced each day.	Minutes
Pharynx	Crossroads of digestive and respiratory pathways	Bolus of food is pushed into pharynx voluntarily by tongue. From there, involuntary (reflex) action continues swallowing action.	Seconds
Esophagus	A 10-in. tube connecting pharynx and stomach	Folded walls distend, and peristaltic waves of muscle contraction propel food to stomach. Liquids pass more rapidly along, assisted by gravity.	5–10 sec for solids
Cardiac constrictor	A muscle band closing the stomach entrance	Peristaltic wave pushes against constrictor and causes it to open. Fluids are held until peristaltic wave arrives.	Seconds
Stomach	A J-shaped muscular bag of about 1-qt capacity below ribs, acts as food reservoir; gastric and mucous glands secret about 2 qt of gastric juice each day	Muscular walls churn and subdivide food further. Fine particles are mixed with gastric juice and mucus, producing soupy chyme. Gastric secretion and stomach motility are regulated by nervous and hormonal factors. Storage function eliminates need for many small meals.	2–6 hr
Pyloric sphincter	Muscle closing the exit of the stomach to the small intestine	Gastric chyme is periodically squirted into small intestine. (Emptying of stomach is controlled by various factors acting on sphincter.)	Seconds

Small intestine	A 21-ft coiled muscular tube, 1 in. in diameter, lined by glands and absorptive villi
	Consists of three regions:
	    Duodenum
	    Jejunum
	    Ileum
		Keeps chyme in motion by churning movements. Glands secrete 1.5 gal of mucus and water each day. Disintegrating epithelial cells sloughed off from walls release digestive enzymes.
		
		Hydrolytic digestion continues.
		Absorption of digested materials takes place.
		Absorption of digested materials takes place
			5–6 hr

Pancreas	Large gland opening into the duodenum, producing pancreatic juice*	Pancreatic enzymes continue hydrolytic digestion of nutrients. Pancreatic secretion is regulated by nervous and hormonal factors.	
Liver† and gallbladder	Liver produces dark-green watery bile, which is concentrated and stored in the gallbladder; bile duct connects gallbladder and duodenum	Bile is poured into duodenum, and its secretion is regulated by nerves and hormones. Helps to emulsify fats (break them into small droplets), providing greater surface area for action of the enzyme lipase.	
Ileocolonic sphincter	Valve guarding opening between small and large intestines	Prevents backflow from large to small intestine.	Seconds
Cecum and appendix	Blind sacs at junction of small and large intestines	Nonfunctional (vestigial) in man. In herbivores contain bacteria that are important for the digestion of cellulose.	
Colon and large intestine	Thick-walled, U-shaped tube 2.5 in. in diameter about 5 ft long; secretes mucus; contains large numbers of bacteria	Absorbs water from chyme. Peristalis compacts residue as feces. Stores feces prior to discharge. Mucus lubricates feces and protects against digestive juices from small intestine. Bacteria act on any undigested food and synthesize useful substances, e.g., vitamins and gases such as methane and hydrogen sulfide, which give feces their odor.	12–24 hr
Rectum	Continuation of large intestine guarded by ileocecal valve proximally and anal sphincters distally; storage depot for feces	Some 0.5 lb of compacted, undigestible material plus bile pigments (giving feces color) and bacteria (10%–50% of feces) are eliminated daily as feces. Ileocolonic and anal sphincters regulate emptying of rectum, controlled by defecation reflex.	

* The pancreas also produces the hormones insulin and glucagon, but these are secreted directly into the blood (Chapter 16).
†The liver is an important regulator of dietary substances following digestion and absorption of nutrients.

TABLE 11–2 Digestive enzymes and associated products of the digestive tract and its related organs

Organ	Enzyme	Action	Other comments	
Mouth	Salivary amylase (ptyalin)	Starch → maltose	5%–10% is hydrolyzed in mouth, continued in stomach up to 50%. Neutral pH of saliva is ideal for action of ptyalin.	
Stomach	0.5% hydrochloric acid (pH 1–3)*	Pepsinogen → pepsin	Acidity inactivates ptyalin when bolus is subdivided, sterilizes food, and destroys bacteria.	
	Pepsin	Proteins → polypeptides	Stomach secretes inactive pepsinogen, which is activated by low pH. This prevents self-digestion, since the cells lining the stomach are not exposed to pepsin intracellularly.	
	Mucus*	Protects lining of stomach from pepsin		
	Gastric lipase	Fats → fatty acids		
	Rennin	Aids digestion of milk Casein → milk curd	High in infants and children. Curdles milk, causing it to remain in stomach longer for digestion of proteins by gastric juice. Optimum pH is 5–6, found only in young.	
Small intestine	^{P}Sodium carbonate (pH 8–9)*	Neutralizes acid chyme	Provides proper pH for the action of pancreatic and intestinal enzymes.	
	^{I}Mucus*		Coats intestinal walls with protective effect.	
	^{I}Enterokinase	Trypsinogen → trypsin		⎫
	^{P}Trypsin	Proteins → polypeptides		
	^{P}Chymotrypsin	Proteins → polypeptides	Chymotrypsin is activated by trypsin	Protein digestion
	^{P}Carboxypeptidase	Proteins → polypeptides		
	^{I}Aminopeptidase	Polypeptides → amino acids		⎭
	^{IP}Amylase	Starch → maltose		⎫
	^{IP}Maltase	Maltose → glucose		Carbohydrate digestion
	^{I}Lactase	Lactose → galactose and glucose		
	^{IP}Sucrase	Sucrose → fructose and glucose		⎭
	^{IP}Lipase	Fats → fatty acids and glycerol		⎫
	Bile*	Emulsifies fats	Contains bilirubin (a breakdown product of hemoglobin), various salts, and cholesterol	Fat digestion
	^{IP}Ribonuclease	Nucleic acids → mononucleotides		Nucleic-acid
	^{IP}Deoxyribonuclease			digestion

^{I} Released by the disintegrating epithelial cells of the intestine
^{P} Secreted by the pancreas
* These products are not enzymes, but assist the digestive process as indicated.

contains folds and enormous numbers of fingerlike processes called **villi** (sing. **villus**—shaggy hair; L.) that project into its cavity (Figure 11–3). Both the folds and the villi vastly increase the absorptive surface of the intestinal wall over what it would be if it were entirely smooth. It has been estimated that the total surface area of the small intestine is 2,000 ft² or about 100 times the surface area of the skin. Additionally, each villus is covered with epithelial cells, and the plasma membrane of each cell is thrown into many minute fingerlike projections called **microvilli,** which further enhance the absorptive surface area (Figure 11–4). Thus, the surface of the small intestine is much like a thirsty Turkish towel, with its many fine projections that facilitate the absorption of water; a worn bathtowel, without its nap, absorbs less well.

The villi move back and forth independently of each other, much as a field of wheat sways in the breeze, and these movements are muscular in origin. Motion is greater following a meal, so that absorption proceeds at an even greater rate. The cells at the top surface of the villi are continually sloughed off and replaced as a result of the mitotic activity of cells at the base, and the entire intestinal epithelium is replaced every 36 hours. The continuous discharge of cells into the intestinal space amounts to about 0.5 lb of cells every day!

The passage of nutrient materials across the cellular border of the small intestine is not passive, and the speed with which it takes place cannot be explained by diffusion alone. Rather, it depends on the ceaseless side-to-side bending and shortening of the villi as well as on the active and selective accumulation of materials requiring the expenditure of a considerable amount of metabolic energy. Why is energy required for this selective absorption of nutrients? To move molecules from a region of

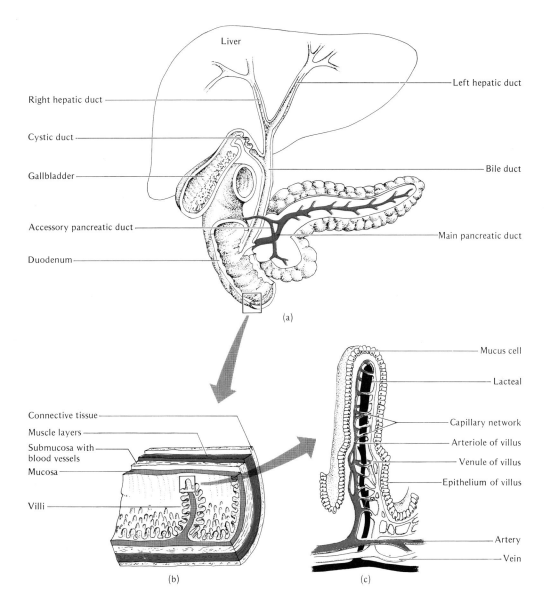

Liver

Right hepatic duct

Cystic duct

Gallbladder

Accessory pancreatic duct

Duodenum

Left hepatic duct

Bile duct

Main pancreatic duct

(a)

Connective tissue

Muscle layers

Submucosa with blood vessels

Mucosa

Villi

(b)

Mucus cell

Lacteal

Capillary network

Arteriole of villus

Venule of villus

Epithelium of villus

Artery

Vein

(c)

lower concentration (in the intestine) to one that has a higher concentration (the bloodstream) requires active transport of the molecules "uphill," so to speak, and the energy for such driving movements is provided by ATP. As you may have anticipated the cells that line the villi of the small intestine are well endowed with mitochondria to generate this "moving power."

Where do the absorbed nutrients go? As the nutrients cross the epithelial lining of the villus, they enter the core region, which contains a capillary bed of blood vessels (Figure 11–3), an arteriole, a venule, and a lymph vessel, the **lacteal.** The amino acids and sugars are pumped across the surface of the villus, enter the bloodstream, and then are carried directly to the liver via mesenteric, gastric, and hepatic portal veins (Figure 11–5). In this way, the liver gets first choice of the nutrients for processing and storage, before they are distributed to the rest of the body. Amino acids will be used for growth, repair, and energy. The liver regulates the level of amino acids and sugar in the blood and acts as a buffer zone between the digestive tract's indiscriminate absorption of nutrients and the rigid requirements of the body's cells. Vitamins, minerals, and water also enter the blood and are transported to the various cells.

Although fatty acids and glycerol do enter the epithelial cells, they recombine to form neutral fats (called **triglycerides**). More than 60% of the fat enters the lacteals in the form of tiny droplets, giving them a milky appearance (*lacteus*—milky; L.). The lacteals drain into the lymphatic vessels that ultimately empty into the blood circulation at the junction of the jugular and subclavian veins in the neck (Chapter 12). A portion of the absorbed fat will be stored in connective-tissue cells called **adipose tissue,** some will be used for the synthesis of membranes, and the rest will be used for energy.

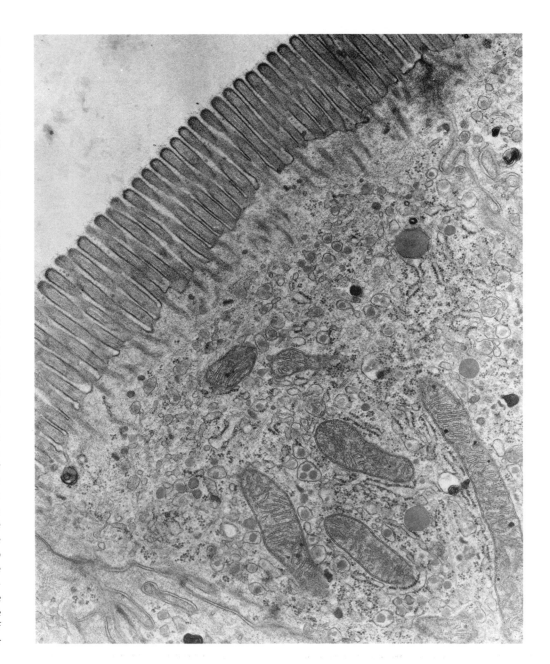

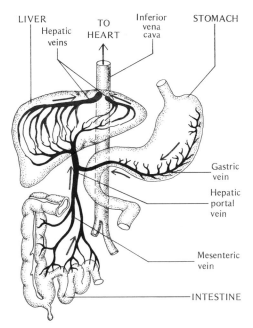

FIGURE 11–5 The hepatic portal system. Blood is carried directly from the stomach and intestines to the liver via the mesenteric, gastric, and hepatic portal veins. Hepatic veins then convey it to the heart by way of the inferior vena cava.

LIVER
Hepatic veins
TO HEART
Inferior vena cava
STOMACH
Gastric vein
Hepatic portal vein
Mesenteric vein
INTESTINE

11–5 DEFECATION: WASTE EXPULSION

The residues of food that cannot be digested or absorbed leave the small intestine and pass through the ileocolonic valve into the large intestine. There, water and vitamins (produced by intestinal bacteria) are absorbed, and waste materials are compacted into feces.

Peristaltic movements in the colon are slow, and 99% of the time there are no contractions, but when the colon becomes overfilled, strong peristaltic waves occur and the fecal material is propelled long distances along the length of the colon. The lower part of the colon (**rectum**) is surrounded by two circular bands of muscle called the **anal sphincters** (literally "that which binds tight"; Gk.). When the sphincters are relaxed, the passageway is opened and material can pass out; when they are contracted the passageway is closed.

When fecal material accumulates in the rectum, it initiates a special defecation reflex

by exciting nerve endings that send their impulses to the spinal cord. Return impulses are transmitted to the colon, rectum, and internal anal sphincter, causing contraction of the muscles in the walls of the colon and rectum and relaxation of the sphincter, which in turn causes emptying of the **bowels** (colon and rectum), provided the external anal sphincter is also relaxed. The external anal sphincter is a voluntary skeletal muscle that guards the outer opening of the anus, and it can be controlled consciously. If the time is not convenient for emptying the bowels, the external anal sphincter prevents defecation in spite of the defecation reflex. Any parent who has lived through bowel training an infant recognizes how long it takes for such conscious control of the external anal sphincter to develop.

Almost all of us are familiar with the natural desire to defecate after a heavy meal or after the first meal of the day. This is caused by the **gastrocolic** and **duodenocolic reflexes.** The filling of the stomach and duodenum results in the stimulation of receptor nerves in the walls of these organs. Impulses travel to the spinal cord and back to the colon, stimulating the muscles in its walls and increasing peristalsis. As a result, fecal material is rapidly moved along the length of the colon, and the defecation reflex is brought into play. The emptying of the rectum after each meal is most evident in infants and the elderly, in whom control of the external anal sphincter is poor.

11–6 NUTRITION: THE USE OF PROCESSED NUTRIENTS

Antoine Laurent Lavoisier (1743–1794) was an elegant aristocrat and a brilliant scientist who met his end beneath the guillotine during the French Revolution. He was executed not because of his scientific pursuits, but because he was a tax collector for Louis XVI. When the

judge was asked at his trial to spare Lavoisier's life, he replied: "The Republic has no need of scientists." After the execution, an old friend of Lavoisier remarked: "It took but a second to cut off his head; a hundred years will not suffice to produce one like it."

Before his untimely demise, Lavoisier clearly stated that food was burned (oxidized) in the body's chemical factories and it was in this way that living organisms derived their energy. Lavoisier also demonstrated that the energy content of any fuel—a lump of coal as well as a cube of sugar or a pound of meat—could be simply determined by burning and then measuring the heat liberated as calories.

Since the days of Lavoisier, we have been calorie-counters. Indeed, we all know that different foods as well as differing amounts of the same food have different caloric values. Generally it can be said that a given amount of fat contains over twice the number of calories as an equal amount of protein or carbohydrate.

All the energy in the food oxidized by the body appears ultimately as heat, which is dissipated into the environment. The rate at which heat is produced is equal to the energy demands of the body, and this is called the **metabolic rate.** The metabolic rate is a measure of energy demands over a certain period of time. It can be measured and expressed in terms of calories. For an average person weighing 150 lb, the bare minimum, that is the amount of energy required to lie in bed and do absolutely nothing, is 1,500–1,800 Calories. This is called the **basal metabolic rate** (BMR) and the energy is used for keeping the heart going and the breathing movements operational, for maintaining body temperature, and in general for keeping the body's cells alive. The basal rate varies with sex and other factors (including age, hormones, nutritional state of the body). An average woman has a somewhat lower metabolic rate (15%–20% lower) than an average man and thus requires fewer calories. Over and above the basal rate, an individual requires calories for work. Caloric requirements vary with different kinds of activities. For example, women require more calories when pregnant or nursing, and a growing child requires more calories per pound of body weight than does an adult; an infant requires 50 Calories per pound of body weight, whereas an adult requires only 18 Calories when sedentary and 35 when active.

Carbohydrates and fats provide energy and are readily interconvertible in the body. The simple exchange of fat for carbohydrate and vice versa explains why fat is deposited when a person eats a carbohydrate-rich diet. Protein, too, may contribute to the energy supply, but the energy demands of the cells are satisfied by carbohydrates and fats as long as these materials are available. This is called **protein sparing.** During starvation the stores of carbohydrate and fat are depleted first, then amino acids serve as fuel. The amount of energy available from amino acids under these conditions can suffice for only a few days, since they must be derived from the structural tissues of the body. Beyond this short time the body begins to consume itself; cells are unable to grow and repair themselves, and ultimately this may lead to death.

Unlike carbohydrates and fats, proteins contain nitrogen. Although proteins, fats, and carbohydrates can all serve as a source of energy, fats and carbohydrates cannot serve as a source of amino-acid building blocks because they lack atoms of nitrogen. Amino acids and proteins cannot be stored (except in eggs and nuts), and excessive amounts of protein are quickly transformed into glycogen and fat and their nitrogen excreted as urea. Thus, although we can exist for some time without fat or carbohydrate, we must have a continuing supply of protein in our diet.

TABLE 11-3   Chemical composition and daily food requirements of a 70-kg (150-lb) human male

Component	Content (gm)	Daily requirement
Water	41,400	
Fat	12,600	Approximately 100 gm ⎫
Protein	12,600	70–100 gm ⎬ Approximately 2,500 Calories*
Carbohydrate	300	Approximately 400 gm ⎭
Na (sodium)	63	3.0 gm
K (potassium)	150	2.5 gm
Ca (calcium)	1,160	1.0 gm
Mg (magnesium)	21	?
Cl (chlorine)	85	2.5 gm
PO_4 (phosphate)	670	1.5 gm
S (sulfur)	112	
Fe (iron)	3	12 mg
I (iodine)	0.014	250 μg
Vitamin A		5,000 IU
Thiamin		1.2
Riboflavin		1.7 mg
Niacin		19 mg
Ascorbic acid		70 mg
Vitamin D		400 IU (10 μg)

*On the average, organic foods (that is, organic, in the chemical sense and as defined on p. 12) contain about 5 Calories per gram; therefore for an adequate caloric supply, the body requires about 500 gm of organic food (1–2 lb) each day. About 14% of the food intake should be protein.

Carbohydrates and lipids from the various species of plants and animals differ only slightly in their composition and in their nutritional value to us. Proteins, however, are unique, and not all proteins contain all 20 different kinds of amino acids. Indeed, certain kinds of proteins are deficient in some amino acids. In general, complete proteins, which are capable of supporting growth when used as the sole dietary supply, are derived from animal sources. Consequently, adequate nutrition solely from plant proteins usually requires a mixed diet of plant foods. Most vegetable matter is low in protein; this means that a vegetable diet ordinarily is supplemented with meat or other protein-rich foods to meet the daily dietary requirements for protein. Roughly it can be said that a human being requires about 1 gm of protein per kilogram of body weight per day. Lack of adequate amounts of protein in the daily diet, or deficiency of amino acids due to imbalance in protein foods, is probably the severest problem facing the impoverished of this world. (See Chapter 26.)

The well-balanced daily diet should consist of protein (15%), carbohydrate (65%), and fats (30%–40%) to supply the energy and structural needs of the body, together with vitamins, minerals, trace elements, and water. These last four substances are important constituents of coenzymes and other metabolic regulators, as well as structural components of the body (Chapter 3). The amounts of the various classes of nutrients included in the diet depend on their role in metabolism. Table 11–3 gives the amounts of food substances required each day and compares them with the amounts found in the body.

11–7 TROUBLE ALONG THE ALIMENTARY CANAL

Under normal circumstances the passage of food along the alimentary canal is regulated by a variety of nervous and hormonal mechanisms, and eating, swallowing, digesting, and defecating all occur without problem. However, the large volume of over-the-counter remedies sold every year for everything from indigestion to constipation testifies to the number of things, real or imagined, that can go wrong.

Tooth decay

Dental **caries,** or cavities, are the most common human ailment. Certain bacteria ferment sugars and secrete acid, and the acid eats away at the teeth and produces cavities. Development of cavities depends on tooth susceptibility, ability of the saliva to neutralize the acids, and the presence of sugar in the mouth. Rinsing or brushing the teeth after meals helps prevent cavities because food particles and the acid secretions of the bacteria are removed and salivary secretion enhanced. Pain (toothache) results when destruction has reached nerve endings in the tooth. Unchecked caries may eventually cause an abcess. It pays to take care of the teeth, not only for reduced dental bills, but because tooth loss causes a marked change in the appearance, as the lips and cheeks sink inward, and in nutrition, since the ability to chew depends on the condition of the teeth.

An old gag has the punch line: "The teeth are all right but the gums have got to go." It bears a grain of truth, for more teeth are lost because of gum disease, especially after age 35, than because of dental caries themselves. The gum tissues surrounding the tooth **(periodontal tissues)** often become inflamed, a condition called **pyorrhea;** the teeth loosen and fall out. Pyorrhea can be caused by the deposition of calculi (bacteria-laden plaques that calcify), improper brushing, trauma from toothpicks and gingivitis (redness or swelling of the gums caused by poor hygiene), deficiencies of vitamins B and C, and bacterial infections such as Vincent's angina (trench mouth).

Heartburn and indigestion

The condition known as **heartburn** actually has nothing to do with the heart except anatomical proximity. It is caused by a backflow into the lower end of the esophagus of highly acidic gastric juice that irritates the lining of the esophagus and causes a burning sensation. In some individuals this condition is chronic, caused by improper closure of the cardiac constrictor muscle at the junction of the stomach and esophagus. Closure of the cardiac constrictor is normally assisted by pressure in the abdominal cavity (the lower part of the esophagus lies below the diaphragm; Figure 11–1a). If the terminal segment of the esophagus is pushed upward through the diaphragm to lie in the thoracic cavity (as occurs in the last 5 months of pregnancy owing to the large size of the fetus), the barrier is not as effective, and there is a tendency for some of the stomach contents to be regurgitated into the esophagus. The stomach acidity leads to the heartburn characteristically present during late pregnancy. It subsides in the last weeks as the fetus moves downward. The tendency for a newborn infant to regurgitate is due to the fact that the lower portion of its esophagus does not lie in its abdominal cavity, so that the constriction

between the esophagus and the stomach is weak. Current advertising notwithstanding, an "acid stomach" is both natural and functional. If the stomach were not acid, protein digestion could not begin. However, heartburn can usually be relieved by drinking a solution of an alkaline substance such as bicarbonate of soda.

Indigestion is common when one is emotionally upset. Secretion of gastric juice by nervous reflex is inhibited by such emotions as anger and fear, so that it is no wonder digestion is impaired when we are under emotional stress.

Peptic ulcers

Peptic ulcers are small craterlike holes eroded in the walls of the stomach (where they are called **gastric ulcers**) or the small intestine **(duodenal ulcers)** and are estimated to affect 1 of 10 adults in the United States; most victims are men. There is some correlation between physiological stress and ulcer development (Box 11B). Ulcers result from insufficient secretion of mucus and oversecretion of gastric juice in the stomach; this erodes the mucosal wall and may perforate the stomach or duodenum itself, causing **peritonitis**[2] and bleeding (hemorrhage). Most commonly a peptic ulcer is found in the duodenum, just beyond the stomach. The pain associated with a duodenal ulcer probably results from irritation of exposed nerve fibers and muscle cells in the region of the ulcer; muscle spasms initiated by acid acting on the muscle may be contributing factors. Ingestion of milk, alkali, and food may temporarily relieve the pain, since these substances tend to buffer the acid in gastric juice and also to delay stomach emptying. The

---

[2] Peritonitis is a bacteria-induced inflammation of the lining of the body cavity; prior to the advent of antibiotics, it was almost always fatal.

Box 11B Ulcers: the price of success

Peptic ulcers occur at almost any age, but most often they develop at about the age of 30 to 35. About 4 times as many men develop ulcers as women. There is a strong relationship between ulcers and personality. The ulcer victim is ordinarily a conflict-ridden, hard-driving, fast-eating, superiority-craving individual; he is more of an action-type than a phlegmatic introspective sort. The highest frequency of ulcers occurs in people under pressure: bus drivers, taxi drivers, executives in the business world.

About 10 years ago some interesting experiments were initiated to investigate the cause of ulcers. Monkeys were trained to avoid an electric shock by depressing a lever, a learning situation that was an easy task for them to master. Although the monkeys learned well, they often died of severe peptic ulcers. Further tests showed that if a monkey was restrained in a chair and shocked at intervals, it developed no ulcers. To clarify the situation, a pair of monkeys were linked to each other by an electrically wired floor and given shocks simultaneously. Thus the treatment administered was exactly the same for each of them. However, one monkey had a lever which would turn off the shock, but the other monkey had a dummy lever; both monkeys received the same number of shocks, but only one of them, the "executive" monkey, could turn off the shock. The executive monkey alone developed ulcers. (Note that the executive monkey received no warning of the shocks to come.)

These studies suggest that physical stress (shocks in this case) is itself not a major factor in ulcer formation, but that psychological or emotional factors are. Of critical importance is the cyclic character of the emotional stress; with 6 hours on and 6 hours off, ulcer development was maximum. Continuous stress, it would appear, produces a kind of stable adjustment to the situation, consequently ulcers tend not to develop.

More recent investigations have attempted to separate the psychological and physical factors involved in the production of stomach ulcers, using rats. Like the monkeys, two rats received electric shocks simultaneously through electrodes placed on their tails. One of the rats heard a beeping tone a few seconds before the shock so that it could predict when the shock would occur, whereas the other received no warning signal. Rats hearing the warning beep developed severe ulceration whereas those that received shocks without warning developed no ulcers. Hence the psychological effect of predictability rather than the shock itself determined ulcer severity. If the animal was provided with a lever by which it could avoid and escape, that is cope with, the shock after the warning beep, ulceration was much less severe.

The latter result was apparently opposed to that obtained with monkeys. This contradiction may be explained by the effects of feedback on coping behavior, that is, if an animal is able to cope with a shock by turning it off, it is obtaining positive reinforcement or so-called relevant feedback from its environment. If the animal attempts to cope with the shock and gets some shock anyway, as did the executive monkeys, it is getting negative reinforcement or poor relevant feedback from its environment; this produces anxiety, the stressor stimulus that ultimately leads to peptic ulcers. If the animal is not offered the opportunity to cope with shock, that is, gets the shock and can do nothing about it, it is not getting positive or negative feedback. The psychological stress is correspondingly reduced, and although the animal is in a stressful situation, ulcer development is reduced because the total amount of stress is reduced.

In a similar situation, people given inescapable shocks but who think they can avoid them by clenching their fist or pressing a button showed less emotional arousal than those given the same shocks, also asked to press a button or clench a fist, but who understand that the shocks are unavoidable. Those who thought they had control over the shock saw their responses as producing relevant feedback—as positively reinforcing—whereas in contrast people who thought they were helpless saw their responses as producing no relevant feedback.

It has further been found that animals receiving positive feedback, that is those able to avoid and escape shock, show an increase in the levels of nervous transmitter substances (norepinephrine) in the brain, whereas their helpless counterparts receiving the same shocks show a decrease. Decreased levels of nervous transmitters in the brain are suspected causes of nervous depression and consequent inability to cope. Thus, initial helplessness in a particular situation may induce biochemical changes in the nervous system; these intensify depressed behavior and the inability to cope could be perpetuated in the nervous system in a vicious cycle.

Once ulcers develop, how can they be treated? Treatment of ulcers involves attempts to reduce the secretion of gastric juice by administration of drugs that block the nervous stimuli for acid secretion. If this does not work, the treatment is to surgically cut the vagus nerve, the principal stimulatory nerve to the secretory portion of the stomach. How can ulcers be controlled without resort to surgery? The simplest method is often the most difficult: avoidance of a stressful environment. Change of job, and the way of life often help as do frequent feedings of a bland diet containing milk and milk products and complete abstinence from stimulants such as tobacco, alcohol, and coffee.

burning abdominal pain in ulcer sufferers generally begins 2–3 hours after eating, when unbuffered acid begins to enter the duodenal region.

Gallstones, bile, and jaundice

Bile is a complex fluid: it contains the pigments **bilirubin** (brown-red) and **biliverdin** (green), which are breakdown products of hemoglobin; a variety of bile salts; and some cholesterol. If the bile duct is obstructed or if the gallbladder, in which bile is stored, has been removed surgically, absorption and digestion of fat are impaired. The bile pigments give the feces their characteristic brown color; thus, bile-duct obstruction, liver damage, or a blood disorder such as anemia (often caused by a lack of hemoglobin) may be reflected by chalky or grayish feces since the bile pigments cannot be poured into the intestine. Occasionally the bile pigments accumulate in the blood because they cannot be excreted and give the skin a yellowish cast—a condition known as **jaundice** (*jaune*—yellow; Fr.). Jaundice has various origins: obstruction of the bile duct, damage to the liver cells caused by cirrhosis, hepatitis, toxic substances such as poisons or alcohol, and excessive red cell destruction due to pernicious anemia or malaria. Since causes are so varied, treatment of jaundice must be adjusted to its mode of production.

Perhaps the most common ailment involving bile is the condition known as **gallstones.** It is normal for bile in the gallbladder to be concentrated 5–10 times by the resorption of water, and under ordinary circumstances the concentrated bile contains no deposits. On occasion, precipitation of crystals of cholesterol occurs, and these are combined with bile salts and bile pigments to produce gallstones. These pretty, yellow-green-tinted gallstones may produce pain and can block the bile duct, thereby

stopping the flow of bile into the intestine and seriously impairing the digestion and absorption of fat. Some gallstones pass out with undigested food, but the obstructive kind must be removed by surgery.

Appendicitis

The blind-ending sac at the tip of the cecum, the **vermiform appendix** (Figure 11–1), is a lymphoid organ that does not perform any digestive function in man. Although it protects against infection, it may become infected itself with bacteria from the rest of the gut, producing an inflamed condition called **appendicitis.** The condition can be treated with antibiotics, or the appendix may have to be removed surgically. The danger of appendicitis is that the infected appendix may burst, spewing its bacterial contents into the body cavity and causing peritonitis.

Terminus: constipation, diarrhea, and colitis

The feces eliminated from the body generally are soft and columnar. If emptying is delayed, there is excessive water absorption, and the fecal matter becomes dry and hard, causing difficulty in evacuation, a condition known as **constipation.** Constipation may also be caused by such emotions as fear, anxiety, or fright, which inhibit the defecation reflex. The headaches and other symptoms that accompany constipation result not from toxins or poisons absorbed from the feces, but from mechanical distension of the rectum. Similar symptoms appear under experimental conditions if the rectum is stuffed with a nontoxic substance such as cotton or an inflated balloon.

The walls of the rectum contain blood vessels. Occasionally, owing to constipation, abdominal pressure (as in pregnancy or severe obesity) or hereditary predisposition, these vessels enlarge, producing **hemorrhoids.** Hemorrhoids may be external or internal and can cause pain and bleeding.

If the contents of the colon are moved too rapidly, there is less time for water absorption, and the fecal materials reaching the rectum are more fluid— a condition known as **diarrhea.** Diarrhea, characterized by loose and frequent evacuations, may result from irritation of the lining of the colon due to the presence of dysentery bacteria, irritants in the food (such as prunes, which stimulate intestinal movements), or toxic substances. Prolonged or severe diarrhea may cause excessive water loss or ulcerative colitis, owing to rapid flow of digestive juices from the small intestine into the colon. This may, in turn, lead to ulcers in the colon wall.

SUMMARY

1. Digestion is the process during which food is broken down from complex insoluble substances into simple soluble ones. It takes place in the digestive tract, which also absorbs the soluble materials and eliminates residues.

2. The digestive tract secretes mucus and enzymes.

3. Peristalsis moves food along the tract from mouth and pharynx via the esophagus to the stomach, small and large intestines, and to the rectum, from which feces are expelled.

4. Carbohydrate digestion begins in the mouth.

5. In the stomach, food is churned and subdivided to chyme in an acidic medium. Hydrolytic digestion of proteins to polypeptides and fats to fatty acids begins.

6. In the small intestine, protein, carbohydrate, and fat digestion continue in an alkaline medium. Products of the liver and pancreas aid in digestion and empty into the duodenum. The duodenum, ileum, and jejunum have a large surface area for absorption owing to their lining of villi and the microvilli of the epithelial cells. Absorption occurs by active transport as well as diffusion.

7. Amino acids and sugars enter the bloodstream via the capillaries of the villus and are carried directly to the liver via the hepatic portal system. Fats, as fatty acids and glycerol, are recombined to neutral triglycerides and enter the lymphatic system via the lacteals of the villi.

8. The gastrocolic and duodenocolic reflexes result in emptying of the bowels (colon and rectum) via the anus.

9. The energy demands of the body are measured in terms of heat (calories) as the metabolic rate. The metabolic rate and caloric requirements vary with sex and daily activity. Energy is usually supplied by fats and carbohydrates, and proteins supply energy as well as building-block materials for structural growth and repair.

10. Unlike carbohydrates and fats, proteins cannot be stored. So-called complete proteins generally come from animal sources whereas plants are lower in protein and usually incomplete in their amino-acid content.

11. We need to eat a well-balanced diet containing fats, carbohydrates, and proteins as well as vitamins, minerals, and water.

12. The digestive process is normally regulated by hormones and nerves.

13. Disorders of the digestive system can include: tooth decay (caries), pyorrhea, heartburn, indigestion, peptic ulcers, peritonitis, jaundice, gallstones, appendicitis, constipation, diarrhea, and hemorrhoids. Some of these, particularly heartburn and ulcers, can be aggravated by psychological factors that produce increased acid secretion by the stomach.

KEY WORDS

digestion
digestive tract
gut
alimentary canal
gastrointestinal (GI) tract
muscle tonus
mucosa
submucosa
bolus
peristalsis
feces
fistula
hydrolysis
absorption
villus (villi)
microvilli
lacteal
triglycerides
adipose tissue
rectum
anal sphincters
bowels
gastrocolic reflex
duodenocolic reflex
metabolic rate
basal metabolic rate (BMR)
protein sparing
caries
periodontal tissue
pyorrhea
heartburn
indigestion
gastric ulcers
duodenal ulcers
peritonitis
bilirubin
biliverdin
jaundice
gallstones
vermiform appendix

appendicitis
constipation
hemorrhoids
diarrhea

TOPICS FOR REVIEW AND DISCUSSION

1. What is digestion? What is its significance to human nutrition?
2. Discuss the structure of the digestive tract and how the structure of specific regions is related to their functions.
3. What is the significance of Beaumont's experiments?
4. How does defecation differ from excretion?
5. Why does the stomach not digest itself?
6. How is the structure of the mucosa related to its absorptive function?
7. Defend the contention that the gut is outside the body.
8. In what ways are digestive enzymes involved in human nutrition?
9. Describe the role of the liver in digestion.
10. How are proteins digested in the human digestive tract?

Bad Heart Would Have Killed Lincoln, Doctor Says.

"President Abraham Lincoln had a bad heart and probably would have died in office even if he had not been assassinated, a doctor says. Although his diagnosis is based on circumstantial evidence, Dr. Harold Schwartz feels certain that Lincoln suffered from a genetic defect known as the Marfan Syndrome. Indications of the syndrome are abnormally long arms and legs, a sunken chest, crossed eyes and a leaking heart valve.

Lincoln had the long arms and legs, the sunken chest, frequent crossing of the eyes, and most likely also had a leaking heart valve when he went to the Ford Theater. In the last two months of his life, Lincoln became easily fatigued, suffered frequent headaches and was bedridden for a time. Schwartz says this is an indication that Lincoln's heart was failing. During this period Lincoln told a friend, "I am very unwell now. My hands and feet of late seem to be always cold, and I ought perhaps to be in bed." This is a sign of bad blood circulation, as would be expected with a failing heart.

Reprinted by permission from the Associated Press.

Today, more than one million Americans die annually of disturbances involving blood vessels and the heart—the cardiovascular diseases. Indeed, more than 55% of all such deaths in the United States this year will be caused by such disorders. To understand these problems we must know something about the circulatory system and how it operates.

INTERNAL TRANSPORT : CIRCULATORY SYSTEM

12–1 WHY DO WE NEED A CIRCULATORY SYSTEM?

Life arose in the sea. The sea not only spawned tiny blobs of living matter, it also nourished them. The early sea, a dilute organic broth, provided the primitive cells with the raw materials necessary to support their metabolism. The cells received food and, if they were aerobes, oxygen from the fluid environment in which they were suspended. Metabolic trash—carbon dioxide and nitrogenous compounds—natural products of cellular activity, were dumped back into the sea. For the earliest creatures, minute in size, exchanges with the environment were a simple affair: across the cell's plasma membrane, the process of diffusion carried in what was needed and carried off what was not. Diffusion worked well because the exchange surface, the loading and unloading platform, was never very far from the innermost recesses of the cell. Even today, all one-celled and some minute many-celled plants and animals rely on similar mechanisms (Figure 12–1a).

As organisms increased in bulk, becoming structurally more complex and consisting of larger and larger communities of cells, problems of transport arose. The "seafront" exchange surface came to be far removed from those cells lying in the interior—the regions of supply and demand came to lie farther and farther apart. How could food and oxygen in the surrounding sea reach the "landlocked" interior cells, and how could their wastes be carried off to be dumped outside the body's surface? One mechanism for providing a "seafront" to "landlocked" cells was to hollow out the body in the form of a tube and move the environmental fluid in and out. This is the situation in saclike organisms, whose modern representatives are sponges, hydras, sea anemones, and jellyfishes (Figure 12–1b). How-

FIGURE 12–1 Exchange of materials between cells and the environment. (a) The one-celled animal *Amoeba.* **(b) The hydra. (c) The flatworm** *Dugesia.*

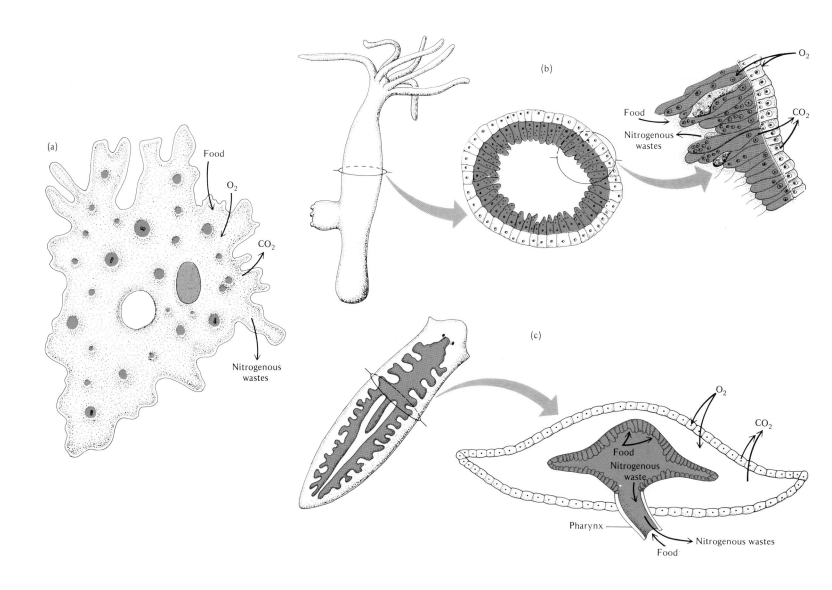

ever, as animal size continued to increase, the simple tube of internal seawater (or fresh water) was no longer enough for efficient exchanges between animal and environment. To allow food and oxygen to reach the cells a considerable distance from the body surface, the tube containing environmental fluid became branched, creating a complicated canal system which assured that exchanges with the fluid environment were possible for most, if not all, cells (Figure 12–1c). Internal cells again obtained a place on the "seafront."

Now other problems developed. Having a canal system filled with a portion of the sea might suffice for small, moderately active creatures such as flatworms, but it certainly would not do for bulky, highly active organisms. Efficiency of exchange and maintenance of a more or less constant internal environment are of paramount importance to such organisms; diffusion is simply not effective over great distances, and ebb and flow in a tubular network fall short of adequacy. In the course of time the problem was solved: the internal sea became enclosed in a tubular network; ebb and flow were replaced by a swift continuous one-way stream; and regulatory mechanisms gradually evolved for maintenance of internal constancy. The fluid was kept moving by a pumping device—a pulsating vessel or heart. As the fluid circulated past the food-and-oxygen-loading platforms at the body's surface, nutrients and oxygen diffused in, and these were then carried to the cells, where they were received; the cells, in turn, gave up their wastes to the circulating fluid, and at the surface unloading platform these were expelled. The circulating fluid became a trade route for effecting rapid exchanges between the outside and the inside of the body.

Thus, in most of the larger active animals, humans included, specialized regions for nu-

trient procurement, gas exchange, and waste expulsion, restricted more or less to the body surface, were linked to the countless deep-lying cells by a system of fluid-filled tubes—the **circulatory system.** Once this had occurred, the cells of the body were bathed from within, and such creatures were to a certain degree independent of the watery environment from which they had come. These organisms could move onto the land, and so some of them did. We are but one of those many land creatures that emerged ages ago from the primeval watery habitat.

12–2 WHAT DOES THE CIRCULATORY SYSTEM DO?

The circulatory system, including the heart, blood vessels, and blood, is the largest organ of the body and in bulk is twice the size of the liver. If the tubular conduits of your circulatory system were laid end to end, they would stretch one fourth of the way from the earth to the moon—60,000 miles.

The muscular **heart** pumps the blood into the arteries, and these branch like a tree to distribute the cargo of the blood to all parts of the body (Figure 12–2). When blood leaves the left side of the heart, it enters the great artery[1] of the body, the **aorta.** The aorta is about 1 in. wide; it curves upward in an arch as it leaves the heart and then runs downward along the backbone into the abdomen. Leading from the aorta are other large arteries that supply blood

---

[1] Note that an **artery** carries blood away from the heart, and a **vein** carries blood to the heart or between other organs (**portal veins,** Figure 11–5). Although the majority of arteries carry oxygenated blood and the majority of veins carry deoxygenated blood, oxygen content cannot be relied upon to distinguish an artery from a vein.

to the head, internal organs, the arms, and the legs. These branch into smaller and smaller arteries, and the smallest branches of the arterial system are the microscopic **arterioles.** The arteries are tubes with relatively thick walls containing elastic and muscle tissues (Figure 12–3a). The strength and elasticity of the walls prevent them from bursting under pressure, and the muscle permits control of their diameter. The finer branches, the arterioles, have their walls composed almost entirely of smooth muscle and are capable of broadening or narrowing their tubular width. The smallest arterioles branch into tiny **capillaries,** composed of a single layer of flat cells (Figure 12–3b). This layer not only forms the walls of the capillaries but also lines the other blood vessels and the heart, so that all the blood in the body is contained within a single continuous envelope. There are thousands of miles of capillaries in an adult person, and they permeate every tissue of the body; indeed, no cell is more than a few thousandths of an inch from a capillary. In the capillary the primary function of the circulatory system takes place: the continuous unloading of cargo for the cells and the carrying away of the products of cellular metabolism. In a sense, the other components of the circulatory system are secondary plumbing for carrying blood to and from the capillaries. From the capillaries blood enters the smallest veins, the **venules,** and these combine to form larger veins. The veins return the blood to the heart. Veins differ from arteries in that their thin walls are not rich in elastic or muscle tissue, but are composed primarily of connective tissue (Figure 12–3c). As a result these vessels are more extensible and less elastic than the arteries; in addition, the veins are provided with one-way valves, so that backflow of blood into the capillaries is prevented. Heart, artery, capillary, and vein form

FIGURE 12–2 A simplified view of the human circulatory system. (From "The Heart" by Carl J. Wiggers. Copyright © 1957 by Scientific American, Inc. All rights reserved.)

a closed system of channels through which the internal river of the blood circulates over and over again.

In summary, the circulatory system is a fluid-filled dynamic system that permits the community of cells to perform their individual but interconnected roles and provides a stable environmental medium for the cells of the body.

12–3 THE HEART: A REMARKABLE PUMP

The double pump

An adult human heart is about the size and shape of a clenched fist (Figure 12–4). It is about 5 in. long and 3.5 in. wide and weighs less than 1 lb. (12–13 oz). The heart is a muscular pump that keeps blood circulating, and without its continued throbbing, life soon ceases. If blood flow to the brain is cut off for more than 5 seconds, we lose consciousness, after 15–20 seconds the muscles twitch convulsively, and after 9 minutes of interrupted circulation the cells of the brain are irreversibly damaged.

The heart lies just behind the breastbone, between the lungs, and above the diaphragm. Surrounding the heart is a double-walled sac, the **pericardium.** The fluid in the pericardium acts as a lubricant, so that with every heartbeat the surfaces glide smoothly over one another. Although centrally located, the heart's axis of symmetry is not along the midline, and the conical portion of the organ slants to the left; this is where the heartbeat is most easily felt and heard (giving rise to the popular but erroneous impression that the heart is on the left side of the body).

At rest the heart pumps 2 oz of blood with every beat, 5 qt per minute, 75 gallons per

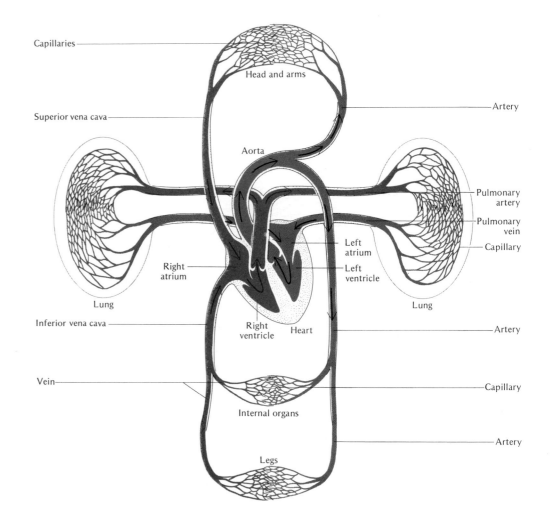

FIGURE 12–3 The structure of blood vessels. (a) Artery. (b) Capillary. (c) Vein.

FIGURE 12–3 The structure of blood vessels. (a) Artery. (b) Capillary. (c) Vein.

FIGURE 12–4 The heart.

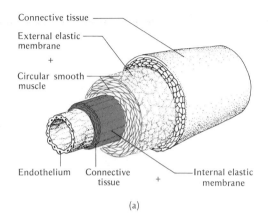

Connective tissue
External elastic membrane
+
Circular smooth muscle
Endothelium
Connective tissue
+
Internal elastic membrane

(a)

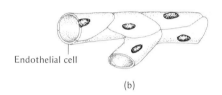

Endothelial cell

(b)

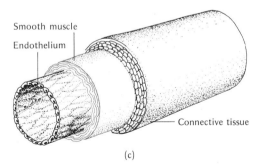

Smooth muscle
Endothelium
Connective tissue

(c)

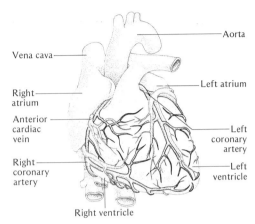

Aorta
Vena cava
Right atrium
Left atrium
Anterior cardiac vein
Left coronary artery
Right coronary artery
Left ventricle
Right ventricle

hour. Each half of the heart has an upper reservoir or collecting chamber, the **atrium** (*atrium—hall; L.*), and a lower pumping chamber, the **ventricle** (Figure 12–5a). Blood returning from the body enters the right atrium and then passes into the right ventricle. Between the right atrium and ventricle are three flaps, or cusps, of tissue that act as a valve (**tricuspid valve**) directing the flow of blood from atrium to ventricle. The flaps of the valve are attached by ligaments ("heartstrings") to small muscles on the inner surface of the ventricle. As blood flows from atrium to ventricle, the valve offers no impedance; but when the ventricle contracts, the flaps of the tricuspid valve move toward one another, sealing the opening and preventing backflow. The ligaments prevent the valve flaps from being pushed backward into the atrium (Figure 12–5b). From the right ventricle the blood travels into the pulmonary artery. The opening is guarded by the **pulmonary,** or **semilunar, valve,** made up of three half-moon-shaped flaps arranged so that backflow from the artery into the ventricle is prevented (Figure 12–5b). The pulmonary artery conducts blood to the lungs. The right side of

the heart is called the **pulmonary pump** since it is involved in a circuit carrying blood to the lungs and back again to the left side of the heart. Oxygenated blood from the lungs returns to the left side of the heart by way of the pulmonary veins. It enters the left atrium, and from there it passes into the left ventricle. The passage from left atrium to ventricle is guarded by a one-way, two-flap valve, the **bicuspid valve.** The bicuspid valve prevents backflow of blood from the left ventricle into the atrium as the thick muscular walls of the ventricle contract to drive blood into the aorta. The opening from the left ventricle into the aorta contains the **aortic valve,** which assures unidirectional flow away from the heart (Figure 12–5b). The left side of the heart forms the **systemic pump,** since it is involved in moving blood from the heart to all the body's organ systems (except the lungs) and back to the right side of the heart.

The opening and closing of the heart valves are passive, depending solely on the way the flaps of tissue are arranged. Because of this it is possible to replace a diseased, distorted, or scarred valve with either a normal one removed from a cadaver or an artificial, ball-and-ring-type valve. The artificial valve closes when a plastic (Silastic) ball seats itself in a ring of steel; the ball itself is prevented from wandering by being enclosed in a meshwork cage on the downstream side (Figure 12–6). Some of these artificial valves last up to 7 years without giving any trouble. However, recent reports indicate that occasionally the valves become infiltrated with fatty materials that cause clot formation; modifications in design are being attempted with a view toward reducing these tendencies.

The wall of the heart consists of three layers: the outer **epicardium,** composed of connective tissue and frequently infiltrated with fat; the inner **endocardium** lining the cavity of the

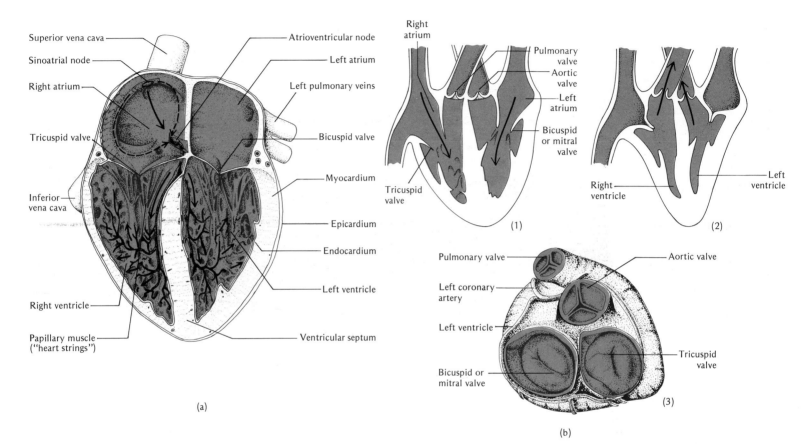

(a)

(b)

FIGURE 12–5 (a) Cutaway view of the heart. (b) Heart valves, their structure and function: (1) filling of ventricles, (2) contraction of ventricles, (3) section through the various valves.

heart, covering the valves, and continuous with the lining of the blood vessels; and the middle layer, or **myocardium,** consisting of a complex meshwork of muscle. The myocardium is responsible for the ability of the heart to contract.

The arrangement of muscle groups in the wall of the heart is such that when the atrial cells are stimulated both atria contract together, and similarly both ventricles contract together when stimulated. The thick ventricular walls have the muscle arranged in rings, whorls, and loops that are connected to each other. This arrangement is very efficient mechanically: ventricular contraction not only pushes blood

out of the ventricles but wrings it out with a twisting motion; the ventricular chambers are the primary elements of the cardiac pump. Before a contraction, the ventricles fill almost completely with blood by elastic recoil from the previous contraction—not, as one might expect, by contraction of the atria themselves. The contraction of the atria completes the transfer of blood, so to speak, by squeezing out the last drops.

Electrical activity of the heart

A heart completely removed from the body but

provided with adequate nutrition will continue to beat rhythmically almost indefinitely. The heartbeat originates in the muscle itself and is therefore said to be **myogenic** (*myo*—muscle, *gennan*—to produce, Gk.). The tempo starts in a region known as the **pacemaker,** or **sinoatrial (SA), node,** which is a group of specialized cells located in the posterior wall of the upper part of the right atrium (Figure 12–5a). These cells generate a brief electrical impulse approximately 72 times a minute, and this impulse spreads quickly over the atria, exciting them almost simultaneously, so that both atria contract at the same time. The impulse also reaches

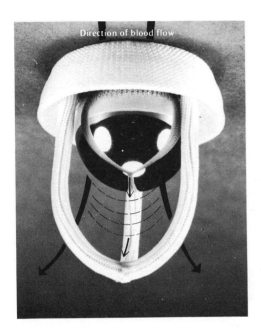

Direction of blood flow

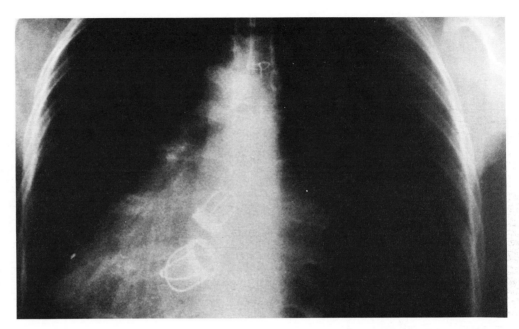

another node of specialized muscle cells situated between the atria and ventricles—the **atrioventricular** (AV) **node.** Here the impulse is delayed briefly (0.07 second) and then is conducted across an insulating ring between the atria and ventricles by a special conducting system called the **AV bundle.** The fibers of this bundle are modified cardiac muscle cells that transmit impulses six times faster than ordinary cardiac muscle. The AV bundle divides into two main branches, which ramify throughout both the ventricles as myriad fine branches called the **Purkinje fibers.** The cardiac impulse spreads via the Purkinje fibers, and as a result the ventricles contract almost simultaneously, making the heart a very effective pump.

In a normal heart, a second impulse cannot be generated or conducted for a period of less than 0.3 second; during this interval the heart

is said to be **refractory.** However, if the conducting pathway is lengthened owing to enlargement of the heart, if there is a decrease in the refractory period owing to disease or drug administration (for example, epinephrine), or if the Purkinje system fails to conduct, then the impulse continues to travel around the heart muscle indefinitely and causes a muscle flutter, or **fibrillation.** Atrial fibrillation is not serious, but if there is ventricular fibrillation, no blood is pumped by the ventricles, and death results. Ventricular fibrillation is easily initiated by an electric shock, especially 60-cycle alternating current, one reason why touching a live household electrical outlet is dangerous.

An isolated heart beats at its own rate without nervous connection. Even when small bits of heart tissue are isolated and placed in a dish containing an artificial nutrient medium, they

continue to beat in rhythmic fashion. As we all know, however, within the body the heartbeat is modified according to work load or state of mind. At rest the heart rate slows; during heavy exercise or because of excitement, its pace is stepped up. When we sleep, the heart pumps 1 gallon of blood per minute, but if we exercise vigorously, it pumps faster, increasing its output up to 5 gallons per minute.

Regulation of the heartbeat occurs via the nervous system. The SA node is supplied by two sets of nerves: **parasympathetic** and **sympathetic** (Figure 12–7). Stimulation of the sympathetic nerve fibers increases the activity of the heart: the heart rate increases, contractions of the heart muscle are stronger, and more blood flows into the coronary arteries, which supply the heart muscle itself. Increased heartbeat from exercise or excitement is due to

FIGURE 12–7 (left) Control of the heartbeat. The parasympathetic (vagus) nerve is inhibitory, and under normal conditions it holds down the beat of the heart by a steady stream of nerve impulses. Excitement, exercise, or emotion triggers impulses from the vasomotor center to accelerate the heartbeat via the sympathetic nerve.

FIGURE 12–8 (below) (a) An electrocardiogram (ECG). (b) The spread of electrical activity across the heart.

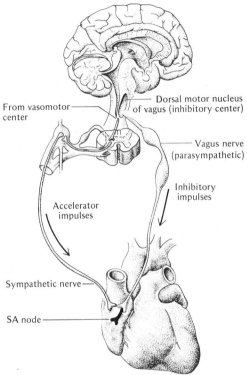

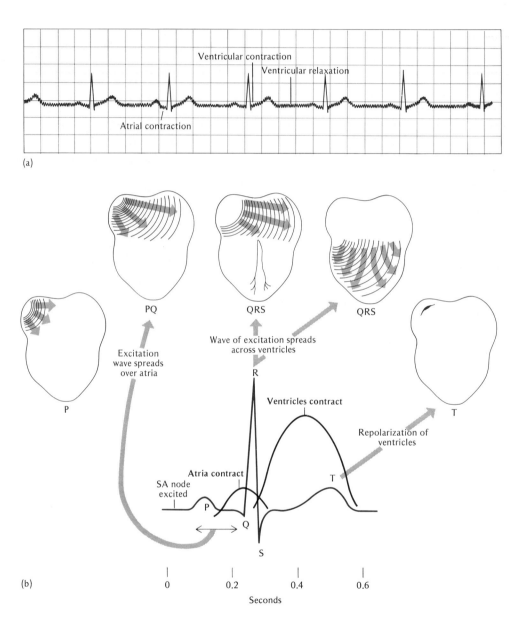

discharge of the sympathetic nerves and clearly is a response to the body's demand (real or anticipated) for rapid transit of blood through the circulatory system. The fibers of the parasympathetic system are inhibitory, and they decrease the activity of the heart.

Although the principal regulatory mechanisms of the heart rate are nervous, other factors modify its action. The hormone epinephrine (adrenaline) acts directly on the heart muscle as well as on the nerve fibers to accelerate the heart; carbon dioxide exerts a dilating effect and increases pumping rate. Mechanical factors such as the pressure of blood within the heart affect the beat.

FIGURE 12–9 Abnormal electrocardiograms.

Cardiac cycle and the electrocardiogram

Physicians listen to sounds of the cardiac cycle by placing a stethoscope on the left side of the chest about 1 in. below the nipple. The sounds heard can be represented as: *lubb, dupp,* pause, *lubb, dupp,* pause. The *lubb* sound is sharp and booming and represents the slamming shut of the atrioventricular valves and the contracting of the ventricles; the *dupp* sound occurs simultaneously with the closing of the semilunar valves. Peculiar wooshes or swishes heard through the stethoscope are called **murmurs,** and many have no clinical significance. However, some murmurs do represent valve malfunction, such as incomplete closure owing to an enlarged orifice or too great a spread in the valve flaps, and in this case there is backflow of blood. In other cases, the murmur indicates a constricted valvular opening due to a hardening of the valvular flap.

The recording of the electrical activity of the heart during a cardiac cycle is known as an **electrocardiogram (ECG).** Since the heart muscle acts as an electrical conducting system, its activity is detected by placing electrodes on the body (a good conductor of electricity). The electrodes lead to an instrument that records the heart's electrical activity. Such a record of activity can be seen in Figure 12–8. As you can see from Figure 12–8b, the first component of the ECG, the **P wave,** represents electrical activity of the atria preceding contraction (**systole**): following this there is a sharp triangular peak, the **QRS wave,** which represents electrical potentials of the ventricles just prior to contraction; then there is a large, slow wave, the **T wave,** which represents electrical activity just prior to ventricular relaxation (**diastole).**

The most important aspects of an electrocardiogram are the time relationships among the various waves. Some abnormal examples are shown in Figure 12–9.

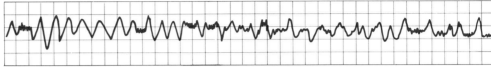
Ventricular fibrillation

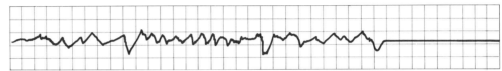
Ventricular fibrillation

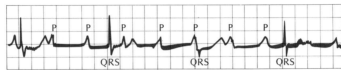
Complete heart block (atrial rate, 107; ventricular rate, 43)

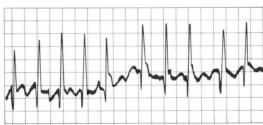
Atrial fibrillation

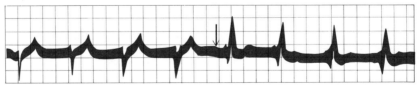
Intermittent right bundle branch block

BOX 12A Artificial pacemakers

Many Heart Pace-makers Malfunction

LOS ANGELES (UPI)—Half of the 250 artificial heart pacemakers recovered from deceased persons in the last year were malfunctioning to some degree, a USC physician said yesterday.

The malfunctions were serious enough to have been life threatening and possibly were a factor in causing death in 16 per cent of the cases, explained Dr. Michael Biltich, assistant professor of medicine at USC.

Biltich said that in 25 to 50 per cent of cases the hospital fails to mail the warranty card to the manufacturer, who then refuses to replace a pacemaker after it malfunctions.

The artificial pacemaker is a mechanical device that supplies to the heart electrical pulses that give the heart muscle a rhythmic beat. The artificial pacemaker is used when the natural pacemaker, the SA node, fails to deliver an electrical signal that can excite the atria and ventricles to activity. For thousands of individuals it is the gift of life, although as the news clipping indicates, it is not without flaws.

In the early 1950s the first artificial pacemakers were introduced. These consisted of an externally worn battery with a connection that ran directly to the heart by wires passed through the skin. Since the heart beats about 40 million times a year, and this would flex the pacemaker wires 80 million times, the problem that most often developed was wire fatigue and breakage. Interruption of the signal by a broken circuit meant heart failure. Recent techniques have gotten around this problem, and at present the wire from the power supply is threaded down the inside of the neck vein, through the vena cava, into the right atrium, through the tricuspid valve, and into contact with the interior wall of the right ventricle (Figure 12–10). The method does not require open-heart surgery and minimizes wire flexion; if the batteries are replaced at regular intervals, a pacemaker can operate for up to 7 years.

The normal heart beats at 70 times each minute; during sleep it drops down to 55 or 60 beats; and during times of stress it may pulsate at 150 beats per minute. Thus for a normal life a person fitted with an artificial pacemaker should have the capacity for varying the electrical impulses between 55 and 150. Such an elaborate device would be not only costly but more failure-prone than one of a simpler design. In practice, therefore, most pacemakers have three switch positions: slow, normal, and fast. These are controlled by the wearer so that the heart beats approximately in accord with its work load.

The most successful modern artificial pacemaker is a transistorized model that picks up impulses from the patient's own SA node; the signals are amplified, delayed for an appropriate interval, and then sent on to the ventricular muscle. In the event the SA node fails to deliver an electrical signal, the implanted pacemaker switches to its own pulse generator and gives 72 pulses per minute. In cases where the heart is blocked irregularly, the artificial pacemaker works on demand and produces a pulse only when the natural pacemaker fails. Since the electrical pulses from the SA node are best picked up on the outside of the heart, the attachment is relatively simple. As one physician put it:

We have had luck. The electrical pulse that stimulates the ventricles happens to be reproducible, using an effect discovered by Luigi Galvani as long ago as 1791. Wrap Galvani's discovery up in some rather commonplace surgical and electrical skill, and you have the present pacemaker. The real achievement would be to match the adaptability of the natural pacemaker with a device linked to the body's own feedback pathways. No one is seriously contemplating that—yet.

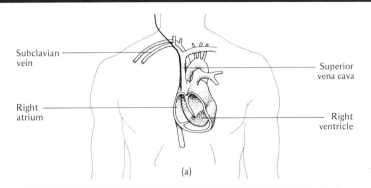

(a)

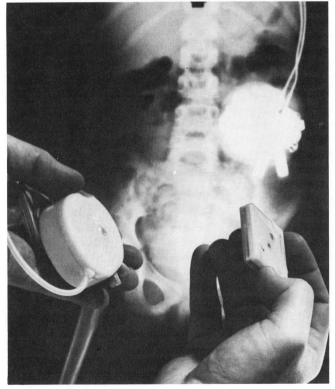

(b)

FIGURE 12–10 The artificial pacemaker. (a) Pacemaker lead passing through a neck vein, vena cava, right atrium, and right ventricle. (b) Two types of pacemakers held in the hand. The one on the right is about the size of a book of matches. In the x-ray film in the background, the larger pacemaker held in the left hand is shown actually implanted. (Courtesy of World Health Organization.)

BOX 12B Heart attack

Each year over 650,000 people in the United States succumb to the nation's number-one killer—heart attack. More than two thirds of the deaths occur within hours (often minutes) after the onset of the attack. Many people believe that a heart attack is an easy way to go—it is quick, simple, painless, and complete; often forgotten are the many thousands who do not die this way, but linger on for years, suffering as heart disease cripples them. What causes heart attack, and why its usual sudden course?

FIGURE 12-11 The progression of atherosclerosis. (a) Normal artery. (b) Partial blockage of artery. (c) Severe blockage of artery. (Courtesy of The American Heart Association.)

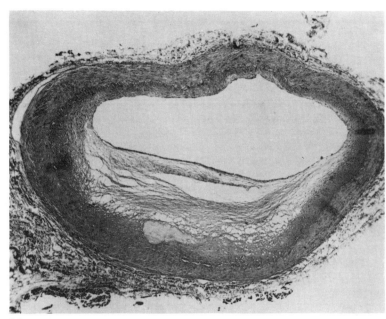

(b)

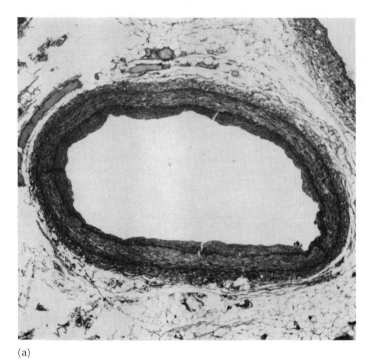

(a)

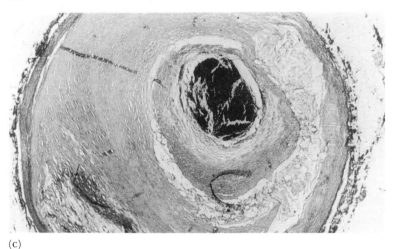

(c)

In an average lifetime, the heart does enough work to raise a weight of 45 tons to a height of 5 miles. To perform such work its demands for fuel and oxygen are great. Unlike skeletal muscle, the muscle of the heart cannot build up an oxygen debt; it must be continuously supplied with oxygen. Whereas other tissues use only 25% of the oxygen carried to them by the blood, the heart uses 80% of its available supply; there is little margin for safety.

The blood passing through the heart chambers cannot be used to supply oxygen to the heart muscle itself; instead blood is piped directly to the muscle cells by two **coronary arteries,** each about the diameter of a soda straw. The coronary arteries originate from the aorta and wrap themselves around the surface of the heart, ramifying to send small branches to all the muscle cells (Figure 12–4). The networks of each coronary artery do not overlap much. Blood returns to the right atrium by a pocket or trough in the wall of the right atrium, the **coronary sinus,** into which the coronary veins empty.

A heart attack is usually caused by a blockage in the coronary arteries, and the muscle and connective tissue fed by the vessels downstream from the blockage soon become starved of oxygen and die. Such blocking may occur quickly or slowly, but in almost all cases it is associated with atherosclerosis. **Atherosclerosis** is a degenerative condition in which the walls of the arteries are invaded with fatty materials, mainly cholesterol. Eventually these deposits become fibrous and impregnated with calcium, forming plates or plaques similar in consistency to bone (Figure 12–11). As the deposits grow, the opening of the blood vessel becomes smaller, until a small blood clot or thrombus, perhaps no bigger than a BB shot, is able to plug it up. If the blockage is in the coronary artery, the individual suffers a "heart attack" or a "coronary," which often causes death. If damage is not extensive because blockage has affected only a small blood vessel, it is possible that the weakened condition may only be temporary.

It will probably surprise you to learn that if you are over 30 years old you are already carrying a few dead areas in your heart or the scar-tissue remains of such local damage. By the age of 40 you will probably experience a few minor coronary blockages, quite tiny, and often mistaken for indigestion pains. Why does such blockage not always produce a coronary attack? If the blockage is gradual, auxiliary blood vessels usually develop, and mild exercise accelerates the development of these accessory blood vessels. As a consequence new blood-supply routes are made available to cells in the region of the blocked artery, and damage is minimal. Yet the coronary occlusion that kills may not always be large. Most deaths are a result of disturbances in the electrical rhythm of the heart. The dying patch of heart muscle downstream from the occlusion may block normal conduction as if it were a tiny electrical signal. An abnormal rhythm is triggered, and ventricular fibrillation results. A fibrillating ventricle pumps no blood, and if the heart is not defibrillated in minutes, death ensues.

12–4 CARDIOVASCULAR PLUMBING AT WORK

Blood pressure

One of the measurements taken during a routine medical examination is arterial blood pressure. The measuring device, a **sphygmomanometer,** consists of an inflatable wrapper (the cuff), a rubber bulb and some tubing, and a glass tube filled with mercury (the manometer) (Figure 12–12). The cuff is wrapped securely around the upper arm and is inflated by means of a rubber bulb. The physician places the bell of the stethoscope on a blood vessel just in front of the elbow, and slowly he releases the pressure in the cuff. In less than a minute the doctor says, "Quite normal. Your blood pressure is 120/80." What does this cryptic, fraction-style notation indicate? How is blood pressure measured? Why are there two pressures indicated?

With each beat of the heart 2 oz of blood are driven from the ventricles into the elastic arteries. The contraction (systole) of the ventricle acting on the blood causes a rise in pressure. While the ventricle fills with blood, it is relaxed (diastole), and the blood pressure in the ventricle is zero. In the artery receiving blood during systole, there is a rise in pressure, and during diastole the pressure falls (but not to zero). Thus, with each beat of the heart there is a rise and fall in the blood pressure of the artery and a corresponding expansion and contraction of the arterial wall.

When the blood-pressure cuff is inflated, the artery in the upper arm collapses when the pressure applied to the artery externally is greater than the pressure exerted by the blood within the vessel. As the pressure in the cuff is gradually released, the first spurts of blood pass through the artery and cause an intermittent sound that can be heard with the stethoscope.

A glance at the manometer at this time shows the maximum pressure in the artery—the **systolic pressure**—and this is recorded as millimeters of mercury. When the cuff pressure is reduced even further, the pressure falls sufficiently low so that blood flows through the artery without any intermittence (the throbbing sounds are no longer heard with the stethoscope). This reading on the manometer corresponds to the **diastolic pressure**—when the heart is resting between contractions and the arteries are contracting by elastic recoil. The diastolic and systolic pressures are recorded fraction-style, with the numerator being the systolic and the denominator being the diastolic pressure. Thus, if the diastolic pressure is 80 and the systolic pressure is 120 mm Hg, the value is written 120/80. There is a wide range of normal pressures: 100–140 for systolic; 60–90 for diastolic.

If the blood pressure recorded for an individual were 200/120, his blood pressure would be abnormally high. The cause of this high blood pressure may be decreased elasticity in the walls of the artery, most frequently due to atherosclerosis. The deposition of calcified plaques prevents the artery from stretching during systole and minimizes elastic recoil during diastole. Low blood pressure can be caused by shock, reduced blood volume, or both.

It is common to measure the pressure in the artery of the upper arm; however, the value for arterial pressure is not the same for the lower arm, lower leg, or other parts of the body. If we look at the blood pressure in the different blood vessels (Figure 12–13), we see a characteristic rise and fall in the arteries, but the average blood pressure decreases continuously as the blood moves farther and farther away from the heart. The pressure is greatest in the aorta, falls off slightly in the arteries, and in the

FIGURE 12–12 Diagrammatic representation of the events recorded by blood-pressure measurement.

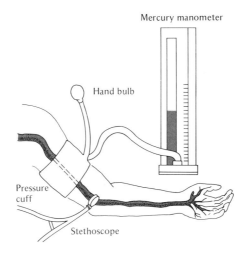

Mercury manometer

Hand bulb

Pressure cuff

Stethoscope

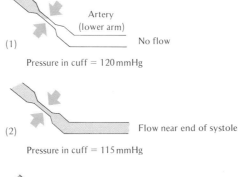

Artery (upper arm)

Artery (lower arm)

(1) No flow

Pressure in cuff = 120 mmHg

(2) Flow near end of systole

Pressure in cuff = 115 mmHg

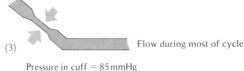

(3) Flow during most of cycle

Pressure in cuff = 85 mmHg

(4) Flow is continuous

Pressure in cuff = under 80 mmHg

arterioles is about one half that in the arteries. By the time blood reaches the capillaries, the pressure is one fifth that found in the large arteries. The pressure reaches its low point in the veins, especially those near the heart. Why is it that the blood pressure drops as blood is moved farther away from the heart?

The frictional resistance of the walls of the blood vessels slows down blood flow, and the narrower their diameter the greater their frictional resistance. Thus, as blood travels away from the pumping force of the heart and passes through narrower and narrower blood vessels, the frictional forces it encounters cause a decline in the flow rate. This accounts for the drop in blood pressure from artery to capillary. Blood-pressure decline is also related to the larger total cross-sectional area of the capillaries (Figure 12–13). The diameter of an individual capillary is so narrow, however, that the blood cells slide sideways in single file. This slow amble of blood through the extensive capillary network allows oxygen, nutrients, and other key materials time to move across the capillary wall into the surrounding fluid-filled tissue spaces. Similarly, wastes move from the cell into the fluid space and into the capillary. (The watery fluid surrounding the cells, which acts as a medium for the loading and unloading, is called **interstitial fluid.**)

Now the blood, laden with cellular wastes, starts its journey back to the heart by moving from the capillaries into the venules. The pressure in the venules and the small veins is quite low because they are even further away from the pump force of the heart and because of the frictional forces in the capillaries. However, the flow rate in the veins is increased greatly because they have a smaller total cross-sectional area than does the extensive capillary network. It is obvious that the pressure in the veins is not sufficient for return to the heart

FIGURE 12–13 Blood pressures and flow rates in various parts of the circulatory system.

FIGURE 12–14 The venous pump: muscles aid in moving blood in the veins toward the heart.

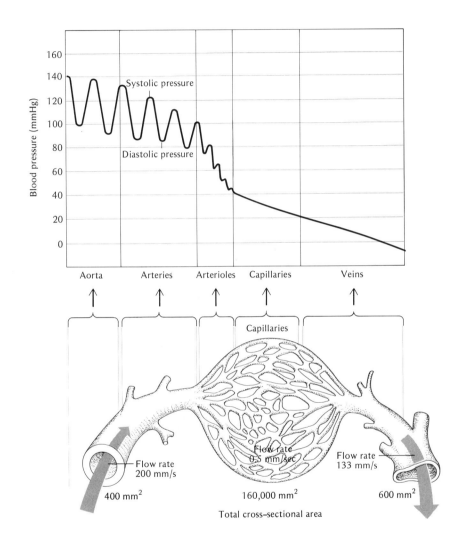

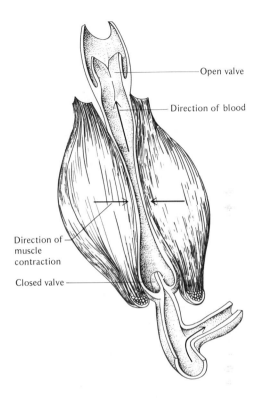

and, since the human subject is in an upright position a good deal of the blood must be driven uphill. Faced with this pressure deficit, how does blood return to the heart?

The movement of blood toward the heart is effected by what is called the **venous pump.** All the peripheral veins contain valves that prevent backflow, allowing the blood to flow only toward the heart (Figure 12–14). In addition, every time a muscle contracts or a limb is moved, the veins are compressed. Since the valves prevent backflow, the muscles are always "milking" the veins and pushing blood in the direction of the heart. Thus, swinging the arms, walking, running, and other body movements all aid in returning blood to the heart. In addition, slight constriction of the muscles in the walls of the veins aids in return flow and prevents pooling of blood in the extremities.

Exchange of materials across the capillary wall

In a 1-in. cube of muscle, it has been estimated, there are over 1.5 million capillaries. The extensive branching, small diameter, thin walls, and great surface area of the capillaries ensure that transfers between the cells and the bloodstream occur fairly rapidly.

The walls of the capillary are freely permeable to water and most small molecules; transport of these materials across the capillary is primarily by diffusion. Exchanges between the blood (contained in the capillaries) and the tissue cells do not occur directly; the tissue cells are surrounded and bathed by interstitial fluid, which acts as the middleman: substances pass from the blood across the capillary wall into the interstitial fluid and into the cell, or the other way around. Because the movement of materials through the capillary wall involves diffusion, a concentration gradient is required to provide the driving force for the net movement of a substance. The maintenance of this gradient is basically dependent on actively metabolizing tissue cells. Let us see why.

The nutrients dissolved in the blood diffuse out of the capillary and into the tissue cells via the interstitial fluid. Since nutrients are continuously being utilized by the tissue cells, there is a rather steady removal of these substances from the interstitial fluid, and as a consequence, a gradient is established in which the concentration of nutrients is greater in the blood than in the interstitial fluid or the cells themselves. Conversely, metabolic wastes are continuously produced by the actively metabolizing tissue

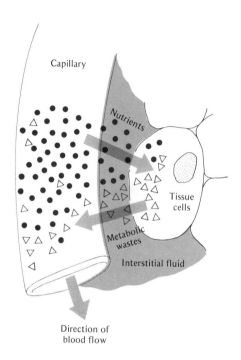

FIGURE 12–15 Diagrammatic representation of nutrient and waste movement between the blood and tissue cells. Substances move across the capillary walls only by diffusion, and thus depend on the maintenance of a concentration gradient between capillary and interstitial fluid.

Capillary

Nutrients

Tissue cells

Metabolic wastes

Interstitial fluid

Direction of blood flow

cells, and as the intracellular concentration increases, these substances move from the cell into the interstitial fluid; this movement, in turn, results in an increased concentration of wastes in the interstitial fluid, and by diffusion these materials enter the capillary. The sequence of events involved in exchanges between capillary and tissue cell is shown in Figure 12–15.

Since fat-soluble materials penetrate the plasma membrane quite easily, these probably pass directly through the capillary wall. Charged molecules (**ions**) and substances that are not fat soluble would have considerable difficulty getting across were it not for the fact that spaces exist between the endothelial cells that form the capillary wall (Figure 12–16). It

is believed that these slow-diffuser molecules move through the intercellular spaces by filtration rather than by direct diffusion across the plasma membrane, and thus they exchange more rapidly than would be anticipated from their diffusion characteristics alone. Additionally, the electron microscope reveals that bulk transport of large and small molecules into and out of the capillaries may take place by pinocytosis (Figure 12–16).

In summary, bulk transport (pinocytosis), filtration, and diffusion are the mechanisms whereby dissolved substances move across the capillary walls, and as a result tissue cells exchange useless wastes for essential nutrients.

Control of fluid movement across the capillary

What determines the direction in which fluids (principally water) move into or out of the capillary? The amount of fluid entering or leaving the capillary is determined by the relative differences in magnitude between the hydrostatic (fluid) and osmotic pressures at each end of the capillary. At the arterial end of the capillary the blood (hydrostatic) pressure is about 30 mm Hg, and that of the surrounding tissue fluid is 10 mm Hg. Thus, the effective hydrostatic pressure tending to drive fluid from the capillary into the tissue spaces is 20 mm Hg. Counterbalancing this force is the difference in the colloidal osmotic pressures of the blood and interstitial fluid. Blood contains relatively large amounts of plasma proteins, to which the capillary walls are not readily permeable; consequently, the interstitial fluid is low in the large-molecular-weight proteins. Because of this, the osmotic pressure of the blood at the arteriole end is higher (25 mm Hg), than that of the interstitial fluid (15 mm Hg). Thus, the effective osmotic pressure, tending to return

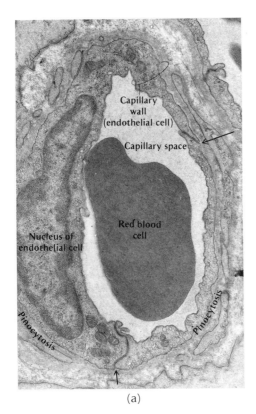

(a)

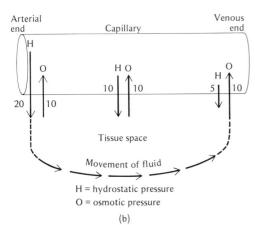

H = hydrostatic pressure
O = osmotic pressure

(b)

fluid to the capillary, is 10 mm Hg. When both the hydrostatic and the osmotic forces are considered, there is an outward filtration pressure of 10 mm Hg (20–10 mm Hg) at the arteriole end (Figure 12–16b). At the venous end the situation is reversed. The blood pressure at this end of the capillary is 15 mm Hg, the tissue fluid pressure is 10 mm Hg, and therefore the effective hydrostatic pressure is 5 mm Hg. The effective osmotic pressure at the venous end of the capillary is the same as at the arterial end—10 mm Hg. Because the osmotic pressure exceeds the hydrostatic pressure at the venous end, tissue fluid is forced back into the capillary. In the middle of the capillary, the hydrostatic and osmotic pressures are about equal, and there is no net movement of fluid (Figure 12–16b).

As long as the osmotic and hydrostatic forces are balanced, there is no net fluid movement (that is, equal quantities of fluid pass in and out of the capillaries); however, any alteration in the balance may allow swelling or shrinking of the tissues. The accumulation of excessive amounts of fluid in the tissue spaces is known as **edema. Shock,** which frequently occurs after surgery, severe burns, injury, or fright, is characterized by increased capillary permeability with escape of blood proteins into the tissue spaces. This reduces the osmotic pressure gradient between the blood and the tissue fluids, return flow is impeded, the blood volume decreases, and, if this continues for an extended time, death can result.

Lymph and the lymphatic system

The body contains an extensively branched tubular network for carrying off excess amounts of interstitial fluid and returning this to the circulatory system. The tubes of this network are thin-walled, blind-ending vessels called **lymphatics,** and they merge to form larger and larger vessels. Eventually, the largest of these drain into two large ducts, the **right lymphatic duct** and the **thoracic duct,** which empty into the blood system in the region of the neck (Figure 12–17). While it is in the lymphatic system, interstitial fluid is called **lymph.** The essential difference between circulatory fluid (plasma) and lymph is that plasma is much richer in proteins.

The lymphatics play an important role in returning protein to the circulatory system. As we have noted, the capillary wall is slightly permeable to protein; as a consequence of this, there is a slow but steady loss of protein from the blood into the interstitial fluid surrounding the cells. However, such proteins cannot return to the circulatory system across the capillary wall because there is insufficient tissue fluid pressure to move them in that direction. The lymphatics themselves provide the mechanism for returning proteins to the blood, since they are quite permeable to these materials (Figure 12–17c). In fact, if such a return did not occur, the osmotic pressure of the blood would fall so low that blood volume would continually be reduced, and death would occur in less than 24 hours. Failure to reduce the buildup of proteins in the interstitial fluid interferes with the mechanism that moves water from the regions surrounding the tissue cells into the venous end of the capillary, there is a net movement of fluid into the tissue spaces, and edema results. The grotesque swelling of limbs seen in elephantiasis (Figure 12–18) is caused by the blockage of the draining lymph channels by a parasitic worm called filaria.

What pushes the lymph along the extensive network of lymphatic ducts to the point where it drains into the blood system in the neck? Lymph flows slowly in the lymph vessels, much more slowly than the blood in its vessels. Two factors determine how fast lymph flows in the lymphatics: (1) tissue pressure and (2) the lymphatic pump. Wherever the tissue fluid pressure rises above normal, the faster is the flow of lymph into the lymphatics. This greater movement of fluid enhances the flow rate of lymph. The **lymphatic pump** is not actually a muscular organ but is the compression exerted during body movement by muscle and other tissues surrounding the lymph vessels. It operates in the following way. Lymph vessels, like veins, have one-way valves that permit flow in only one direction. As each lymph vessel is compressed, fluid is forced toward the neck region and flows into the venous system. Exercise, running, or walking assists in pumping lymph from the tissue spaces back into the bloodstream.

At certain points of juncture of small lymph vessels with large ones (the groin, armpits, and neck, in particular) are located lumpy masses of cells known as **lymph nodes** (Figure 12–19). These play an important role in the body's defense against disease by producing lymphocytes and antibodies and by filtering out bacteria and foreign particles. The **lymphocytes** are specialized ameboid white blood cells that can engulf foreign particles (Chapter 13). All the lymph that drains from the tissue spaces to return to the channels of the circulatory system must pass through the filters of the lymph nodes. In this way lymph is continuously cleansed of foreign debris and bacteria, and the blood is scrubbed clean and pure. It is not unusual that during a severe infection the lymph nodes become inflamed, for they are engaged in protective battle. For example, an infection in the finger or hand may cause a lymph node in the armpit to swell and to become tender and painful. Swollen glands are sometimes in reality enlarged lymph nodes.

Since the lymphatic network is such an ex-

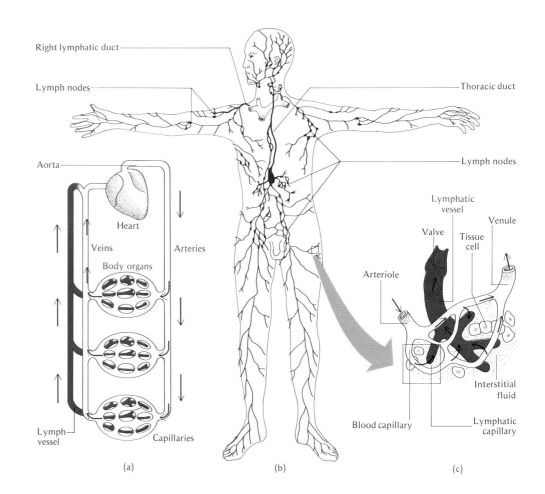

(a) (b) (c)

FIGURE 12–18 A normal leg compared with that of a patient with elephantiasis.

FIGURE 12–19 Lymph nodes and drainage of an infected area.

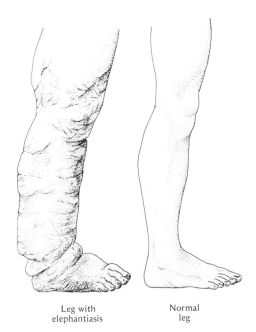

Leg with
elephantiasis

Normal
leg

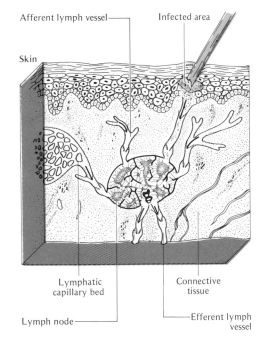

Afferent lymph vessel

Infected area

Skin

Lymphatic
capillary bed

Connective
tissue

Lymph node

Efferent lymph
vessel

tensive drainage system (Figure 12–17), you can readily appreciate the reason why it is often involved in metastatic cancer. Once cancer cells have invaded the lymph vessels, they may become lodged in and proliferate in the nodes. If diagnosed in time, there is the possibility that removal of the downstream lymph ducts and nodes can prevent further spread of such a cancer.

SUMMARY

1. The evolution of circulatory systems was probably linked with increase in body size.

2. Blood flows from heart → aorta → arteries → arterioles → capillaries → venules → veins and back to the heart.

3. Arteries conduct blood away from the heart, veins conduct blood toward the heart or between other organs (portal veins).

4. In the four-chambered heart, one-way blood flow is directed from right atrium to right ventricle by the tricuspid valve, and from the right ventricle to the pulmonary artery by the semilunar valve. From the pulmonary artery blood flows to the lungs.

5. From the lungs, blood returns to the left atrium via the pulmonary vein. One-way blood flow from the left atrium to left ventricle is controlled by the bicuspid valve, while the aortic valve directs flow from left ventricle to aorta.

6. The right side of the heart is the pulmonary pump, the left side is the systemic pump.

7. The heart valves operate passively to prevent backflow of blood when the heart contracts. Diseased valves can be replaced with artificial materials.

8. The heart wall consists of epicardium, endocardium, and the muscular myocardium. Both atria contract together, as do both ventricles.

9. The heart beat is myogenic; that is, it is generated electrically by specialized muscle cells: the pacemaker or sinoatrial node, and the atrioventricular node. Fast-conducting muscle cells, called the AV bundle, radiate from the AV node as Purkinje fibers and help to coordinate ventricular beat. Transistorized artificial pacemakers reproduce or regulate the electrical activity of faulty or failing SA nodes.

10. During a normal heartbeat cycle, there is a refractory period; if this period is decreased, fibrillation ensues. Ventricular fibrillation is especially dangerous, and can be caused by 60-cycle electric current.

11. Nervous control of the heart beat is via parasympathetic (inhibitory) and sympathetic (stimulatory) nerves, which act on the SA node. Hormones (epinephrine or adrenaline), CO_2, and blood pressure may also affect the rate of heart beat.

12. Heart murmurs may represent the malfunctioning of valves.

13. The electrical activity of the heart can be recorded as an ECG. Electrically, each complete heartbeat cycle shows as a P wave (prior to atrial contraction or systole), a QRS wave (prior to ventricular contraction), and a longer slower T wave (reflecting ventricular relaxation, or diastole).

14. Heart attacks are usually caused by blockage of the coronary arteries owing to atherosclerosis.

15. Blood pressure is taken with a sphygmomanometer, and reflects pumping efficiency of the heart. Systolic pressure reflects heart

Box 12C The death of Lenin

It was a bitter cold day in Gorki, a suburb of Moscow. The time was January, 1924. Early that afternoon the body of Lenin, leader of the Russian Revolution, had been brought into the autopsy room. Ten physicians and Joseph Stalin were in attendance. Scalpels and bone scissors cut through the skull and revealed Lenin's brain. The cause of death was clear: the entire left side of the brain was destroyed and a closer examination of the brain showed the reason. A main artery serving an extensive portion of the brain was blocked, and the tissue on the downstream side of the occlusion had died of starvation. As the scalpel blade cut into the blocked artery the knife struck something hard and gritty, as if it were cutting a piece of stone. What was once a resilient, rubberlike artery was now a solid, bonelike rod.

The death of Lenin was not totally unexpected; this was his third stroke. He succumbed to a degenerative disease of the circulatory system called atherosclerosis (commonly referred to as "hardening of the arteries"). Even today over 28 million people in this country have some form of cardiovascular disease. Among your friends, relatives, the people that live in your home town and your county, one out of every seven is afflicted by cardiovascular disease.

contraction; diastolic pressure reflects heart relaxation. High pressures may be caused by atherosclerosis, low pressures by shock, and/ or reduced blood volume. Pressure is greatest in the arteries and least in the veins. The return of venous blood to the heart is assisted by valves in the peripheral veins and contractions of body muscle; together these constitute the venous pump.

16. Exchange of nutrients and wastes between tissue cells and blood occurs by diffusion, filtration, and bulk transport across the capillary and tissue cell walls via the interstitial fluid. Direction of fluid movement across the capillary wall is determined by the relative balance between hydrostatic and osmotic pressures. Alteration of this balance may cause edema.

17. Excess interstitial fluid is returned to the blood via the extensive lymphatic system. When in lymphatic vessels, the fluid is called lymph; it is lower in proteins than blood plasma. Tissue pressure and the lymphatic pump (muscular and tissue compression) as well as valves assist the flow of lymph in the lymphatics.

18. Lymph nodes contain lymphocytes and are important in fighting infection and disease.

KEY WORDS

circulatory system
heart
aorta
artery
arterioles
capillaries
venules
vein
portal vein
pericardium
atrium
ventricle

tricuspid valve
pulmonary or semilunar valve
pulmonary pump
bicuspid valve
epicardium
endocardium
myocardium
myogenic
pacemaker or sinoatrial (SA) node
atrioventricular (AV) node
AV bundle
Purkinje fibers
refractory period
fibrillation
murmurs
ECG
P wave
systole
QRS wave
diastole
coronary arteries
coronary sinus
atherosclerosis
sphygmomanometer
systolic pressure
diastolic pressure
interstitial fluid
venous pump
ions
lymphatics
right lymphatic duct
thoracic duct
lymph
lymphatic pump
lymph nodes
lymphocytes

TOPICS FOR REVIEW AND DISCUSSION

1. List the major functions of the circulatory system.

2. Why do animals of large dimensions require a more complex circulatory system than smaller ones?

3. What are some of the feedback controls involved in the regulation of heartbeat?

4. Why is the heart called a "double pump"?

5. Discuss the relationship of coronary blood vessels to heart attacks.

6. Compare and contrast artificial pacemakers with the heart's natural pacemaker.

7. Describe the relationship of the electrical activity of the heart to the cardiac cycle.

8. How is blood pressure determined and recorded?

9. How is the structure of blood vessels related to their function?

10. What is the lymphatic system? List its major functions.

The chances of saving Professor Benes were slim; a blood clot in the brain would kill him in a matter of hours. To save Benes' life a rather unorthodox plan was developed: miniaturization of a submarine and its occupants, injection into the dying man's carotid artery, transport to the clot by the bloodstream, and destruction of the clot by the piercing rays of a laser gun. Once miniaturization was complete the entire mission had to be completed within 60 minutes for, if not, the submarine and occupants both would return to normal size and kill Benes whether the surgery was successful or not.

In the control room, the television receiver seemed to spring back to life. . . . "They're out of the ear and heading rapidly for the clot." "Twelve minutes. Can they make it. . . ."

"We've entered the brain itself," announced Owens. . . .

He doused the ship's lights again and all of them looked forward in a moment of wonder that put everything else, even the climax of the mission, out of their minds for a moment.

As they drifted through the interstitial fluid along passageways between the cells, they could see the dendrites tangling overhead and for a moment they were passing under what seemed to be the twisted limbs of a row of ancient forest trees. . . .

The submarine came to a halt. For an instant or two, there was silence, then Owens said quietly, "That's our destination, I think. . . ."

Duval nodded, "Yes, the clot."

No human has yet been miniaturized to journey through the body as did the characters in Isaac Asimov's novel *Fantastic Voyage*, but modern techniques in biology have made it possible to blow up and enlarge microscopic structures in such a way that we can gain a "bacterium's eye view" of our own cells. The photograph shows a red blood cell caught in the tangled web of a fibrin clot, a situation much like that encountered by the voyagers in Asimov's science fiction story. (The horizontal black line in the lower right of the photo is about the size of the miniaturized submarine—1 micron!) This chapter will take you on a fantastic voyage through our blood—described by some as "the river of life."

Fantastic Voyage, Isaac Asimov, New York, Bantam, 1966. Reprinted by permission of Otto Klement.

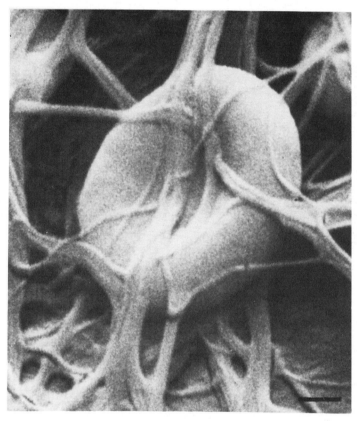

Cover photo by Emil Berstein, from *Science,* 173, August 27, 1971. Copyright © 1971 by The American Association for the Advancement of Science.

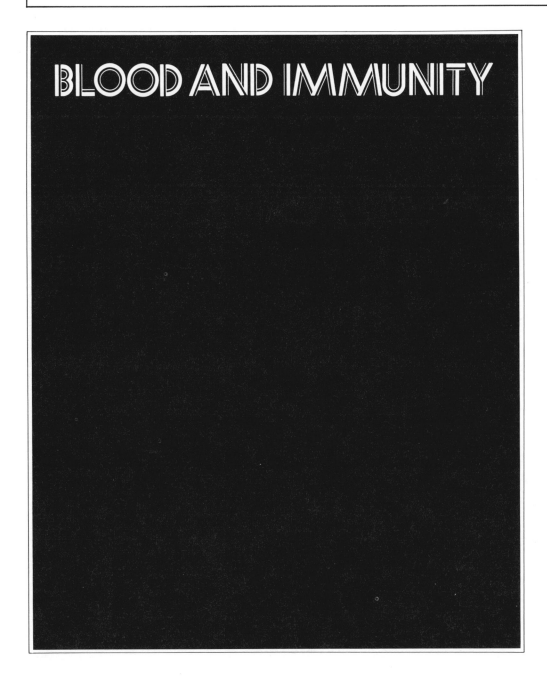

BLOOD AND IMMUNITY

13–1 BLOOD: A TISSUE THAT FLOWS

Blood has always held a special position in the thoughts of man. The ancients recognized that continued loss of blood led to weakness and eventually to death; it was as if a man's life drained out of his body with the outpouring of blood. In days gone by, blood was considered to have many wonderful properties, some of them quite mystical. Even today, we speak metaphorically of the special qualities of blood: virile men are "red-blooded," aristocrats are "blue bloods," undesirable individuals have "bad blood," and close family relations are "blood relatives." Some of these expressions, as we shall see, are biologically unsound, but they emphasize man's long-standing preoccupation with this special fluid.

The quantity of blood in the body is substantial, making up about 7% of the total body weight. On the average we have about 5 qt (almost 10 lb) of blood. In a single day about 5,000 qt of blood (5 qt making 1,000 circuits) course through the body. As a result of this continuous flow, food and oxygen are delivered to the hungry cells, and potentially lethal wastes are carried away; excess heat derived from the metabolic furnaces of our tissue cells is carried off and dissipated in the skin; regulatory substances (hormones) are moved to every part of the body, so that chemical contact between remote organs is maintained; and by virtue of the presence of specific defense agents, foreign invaders are challenged and repelled.

Blood and its derivative, lymph, form a kind of internal sea that continually bathes our cells. Blood is a liquid tissue, and it flows readily because its constituent cells are not joined together in a rigid or semirigid network, but are suspended in a watery medium. The cells float singly and separately and are carried along like flotsam and jetsam in the streaming current. As blood flows through the vessels of the

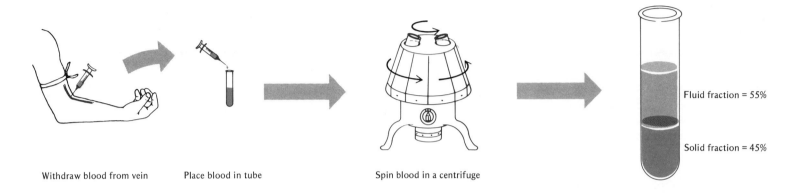

Withdraw blood from vein   Place blood in tube   Spin blood in a centrifuge

Fluid fraction = 55%

Solid fraction = 45%

circulatory system, it consists of a complex, straw-colored fluid **plasma** in which float the suspended **red blood cells, white blood cells, and platelets.** The fluid plasma composes about 55% of the total volume, and the cells (so-called formed elements) make up the remaining 45% of the total volume (Figure 13-1).

Plasma: our internal sea

Blood plasma is basically similar to a dilute solution of seawater and indeed they share the same salts, but in different concentrations (Table 13-1). For some biologists this confirms the speculation that when life arose in the sea those organisms that developed circulatory systems took along with them the seawater characteristic of that ancient geological period, a time when the seas were less salty. (Blood plasma is roughly 0.85% salt and today's sea-water about 3.5% salt.) You know the salty character of plasma if you have ever tasted blood—and who has never sucked on a cut finger?

Since the prime function of the plasma is to act as a medium of transport, it contains substances other than salts. Most of the nonprotein substances are cargo on their way to or from

TABLE 13-1 A comparison of sea water with blood

Component	Seawater (gm/liter)	Blood (gm/liter)
Na (sodium)	10.7	3.2–3.4
K (potassium)	0.39	0.15–0.21
Ca (calcium)	0.42	0.09–0.11
Mg (magnesium)	1.34	0.012–0.036
Cl (chloride)	19.3	3.5–3.8
SO₄ (sulfate)	2.69	0.16–0.34
CO₃ (carbonate)	0.073	1.5–1.9
Protein		70

Relative composition (sodium taken as 100)		
Na (sodium)	100.00	100.00
K (potassium)	2.70	3.99
Ca (calcium)	2.80	1.78
Mg (magnesium)	13.90	0.66
Cl (chloride)	136.80	83.97
SO₄ (sulfate)	7.10	1.73

the tissues. These include nutrients such as sugars (mainly glucose), lipids, and vitamins; hormones; and waste products such as urea (Table 13-2). The concentrations of nutrients and salts, though small, are kept relatively

constant; this is in spite of the fact that blood is continually exchanging materials with the countless numbers of body cells. The constancy of the blood composition is maintained by a series of negative feedback controls—a decline or elevation in a substance initiates responses in a variety of organs, especially the liver and kidneys, until equilibrium is reestablished.

Nonprotein substances make up only a small part of the plasma (about 1%). More than 92% of the plasma is water; the rest consists mainly of proteins—more than 70 different kinds! We need not concern ourselves with all of them, but can focus our attention on three main groups of plasma proteins: **albumin, globulin, and fibrinogen.** Most of the plasma protein is albumin. The globulin fraction consists of a mixture of proteins: the **alpha globulins** are glycoproteins or mucoproteins (carbohydrate plus protein), the **beta globulins** are lipoproteins (lipid plus protein), and the **gamma globulins** are proteins that include most of the circulating antibody. Fibrinogen is the protein that forms the blood clot.

Plasma proteins serve a variety of functions, but their most important role is in the maintenance of the proper osmotic pressure between the circulatory fluid and the fluid in the tissue

TABLE 13–2 Major components of plasma (per 100 ml)

Electrolytes and metals (mg*)	
Bicarbonate	152–190
Calcium	8.5–10.5
Chloride	355–380
Cobalt	0.4–0.7 μg
Copper	88–124 μg
Iodine	5–6 μg
Iron	50–150 μg
Magnesium	1.8–3.6
Manganese	4–6 μg
Phosphate	3.0–4.5
Potassium	15–20 μg
Sodium	315–330
Sulfate	0.5–1.5
Zinc	250–350 μg

Nutrients (mg*)	
Amino acids	30
Glucose	70–100

Vitamins (mg*)	
Ascorbic acid	0.4–1.5
Biotin	1.0–1.7 μg
Carotenoids	100–200 μg
Folic acid	0.6–2.5
Niacin	0.2–0.3 μg
Pantothenic acid	6–35 μg
Riboflavin (B_2)	2.6–3.7
Vitamin A	50–100
Vitamin B_{12}	20–50 μg
Vitamin E	0.6–1.6

TABLE 13–2 (continued)

Metabolic intermediates (mg*)	
Bile acids	0.2–3.0
Choline	26–35
Citric acid	2.0–3.0
Lactic acid	6–16
Lipids	360–820
Cholesterol	150–280
Fatty acids	200–450
Phospholipids	135–170
Pyruvic acid	1.0–2.0

Waste products (mg*)	
Acetone	0.3–2.0
Ammonia	40–70 μg
Bilirubin	0.5
Creatine	0.8–1.0
Creatinine	0.7–1.5
Urea	25–52
Uric acid	3–6

Proteins (gm)	
Albumin	4.6
Antibodies	1.5
Clotting factors	?
Enzymes	?
Fibrinogen	0.38
Globulins	2.6

Hormones (traces)	

Dissolved gases (CO_2, O_2, N_2, etc.)	

*Unless otherwise noted.

spaces, so that exchange of materials between blood and the cells is facilitated. (See Chapter 12 for details of their function in this capacity.) They contribute to the viscosity of the blood and thus play a role in the maintenance of normal blood pressure. Plasma proteins also act to buffer the blood, so that the pH of the plasma is almost always slightly alkaline (pH 7.4).

Where do the plasma proteins come from? The liver is the sole source of fibrinogen, albumin, and most of the alpha and beta globulins, but the gamma globulins originate in lymphoid tissues—from plasma cells and lymph-node cells. Impairment of liver function, such as occurs in chronic liver disease (cirrhosis), is often reflected by a decline of plasma proteins, particularly albumin.

Blood cells: the red and the white

Almost all our blood cells are **erythrocytes** (*erythros*—red; Gk.), or red cells, and these give our blood its characteristic color. The major function of the red blood cells is the transport of oxygen from the lungs to all the cells of the body. They are red because they contain **hemoglobin,** an iron-containing protein that readily combines with oxygen. Oxygen binding to hemoglobin is reversible: in the lungs, where the oxygen concentration is high, hemoglobin loads up and becomes bright red **oxyhemoglobin;** when it arrives at cells where the oxygen concentration is low, it unloads oxygen and becomes maroon hemoglobin once again. (Transport of gases by hemoglobin is discussed in Chapter 14.) About 1 out of every 500 cells in the blood does not contain hemoglobin, and these colorless cells are called **leukocytes** (*leukos*—white; Gk.), or white blood cells. White blood cells have a variety of functions, but the most important of these is to protect the body against foreign invaders. Platelets are quite numerous in the blood; they are fragments of a special kind of bone-marrow cell and play an essential role in the clotting of blood.

Red cells. In the time it takes you to scan this page (about 1 minute) you will produce 140 million new red blood cells to replace the old and the dead. In a single drop of blood there are about 200 million such biconcave discs, each resembling a minute inflated rubber raft, 7–8 μm in diameter and 1–2 μm in thickness (Figure 13–2). Altogether, the number of red blood cells in the body exceeds 25 million million (2.5×10^{13}).

Red cells are unlike any other cells of the body. A mature red cell cannot multiply because it lacks a nucleus. It is in large part a specially designed container for hemoglobin,

FIGURE13–2 The shape of red blood cells as revealed by the scanning electron microscope. (Courtesy of Francois M. Morel.)

and its architecture exposes maximum surface area for exchange of oxygen and carbon dioxide. If it were spherical, we should require much greater numbers of cells to do the job of gaseous transport with the same efficiency.

If mature red cells cannot grow or reproduce, how are new red blood cells made? Erythrocytes in adults are formed exclusively in the bone marrow (located in the central core of the long bones of the body) from nucleated cells called **erythroblasts.** Erythroblasts (*erythros*—red, *blastos*—sprout; Gk.), are actively dividing cells that give rise through a series of stages to erythrocytes. The mature erythrocyte squeezes its way through the walls of the capillaries in the marrow and in this way enters the bloodstream. There it lives for an average of 4 months before disintegrating. The cell fragments are engulfed by phagocytic cells in the liver and spleen—the graveyard of the erythrocyte. The hemoglobin released by the rupture of the aged erythrocyte flows freely in the blood, and in less than an hour it leaks into the tissue spaces. Here it is ingested by phagocytic cells and chemically dismembered to form iron, bilirubin, and amino acids. The amino acids return to the amino-acid pool of the body, while the bilirubin is excreted from the liver in the bile. Iron is recycled: either it is transported to the bone marrow for the formation of more hemoglobin or it goes to the liver for storage in the form of ferritin. When the marrow requires iron, the ferritin is released from the liver, iron is made available, and hemoglobin is synthesized.

About 95% of the protein in a red blood cell is hemoglobin, a thick molasseslike solution enclosed by the plasma membrane. The remaining red-cell proteins are an array of enzymes that enable the cell to survive for about 120 days, though not to grow or reproduce. Why is hemoglobin contained in such a special

cell? Would it not be just as easy to dissolve hemoglobin in the plasma? The answer is no. Hemoglobin in cells rather than in solution is less likely to be lost as the kidneys filter the plasma. Moreover, the hemoglobin in the plasma tends to become oxidized and to be quickly bound to other proteins, both of which occurrences would impair its oxygen-transport capacity. Finally, red cells expedite gaseous transport by causing turbulence in the blood vessels, much as pebbles do in a stream.

White cells. Human white cells (leukocytes) are larger than red cells and have large, often irregularly shaped nuclei. At least five different kinds can be distinguished by their staining properties and the shape of the nucleus. Normally, 65% of the white cells are **granulocytes;** that is, they contain granules that stain readily. About 35% are devoid of such granules and are called **agranulocytes.** The granulocytes are formed in the bone marrow and come in three varieties (Figure 13–3). The agranulocytes come in two varieties, are formed in lymphoid tissue

(lymph nodes, spleen, tonsils, and thymus), and are more abundant in the lymphatic system than in the blood. White blood cells are capable of ameboid movement; they wander by squeezing between the cells that form the walls of the lymph and blood vessels and thus reach the tissues of the body. These wandering cells are important in the body's defense reactions against disease and infection, since they can engulf foreign bodies such as bacteria.

Self-sealing: blood platelets and fibrinogen

If you attach a garden hose to a faucet and turn on the water, it comes out through the nozzle. Puncture the hose and water squirts from the point of puncture—it does not all go where you want it to, and you lose water. Similarly, if one punctures a blood vessel, blood pours out and is wasted. Unimpeded flow of blood (**hemorrhage**) from damaged vessels causes wasting and eventually death. Cuts, scrapes, and tears in the skin with rupture of blood vessels are everyday occurrences; however, in general the flow of blood from a small wound does not continue for long. The blood soon thickens or clots, a scab forms, and the puncture is sealed. In time the scab drops off, and newly formed skin replaces that which was ripped away. Blood contains a self-sealing material that effectively plugs leaks in the plumbing of the circulatory system.

You can demonstrate blood clotting by collecting some fresh blood in a test tube. If you tip the tube upside down immediately, the blood flows out; it is liquid. Wait a few minutes (4–10 minutes is the normal range) and then tip the tube; the blood does not flow. It has solidified and formed a clot. If the reddish clot is allowed to shrink and then is carefully removed from the tube, a clear yellow fluid remains. This is the same material that oozes

	Diameter (μm)	Number/mm³	% Among leukocytes
Erythrocyte	7–8	5–6 million	
Granulocytes			
Neutrophil	9–12	3,000–6,750	60–65
Eosinophil	10–14	100–360	2–4
Basophil	8–10	25–90	0.5–1.0
Agranulocytes			
Lymphocyte	6–12	1,000–2,700	20–35
Monocyte	12–15	150–170	3–8
Platelets	2–3	200,000–300,000	

out of a wound after a clot is formed. The pale-yellow fluid, **serum,** is much like plasma save for one important difference: it lacks a protein called **fibrin.** Fibrin forms the meshwork that traps red blood cells to form the reddish clot.

The sealant is fibrin, but what is the mechanism that leads to clot formation? The patching material of the blood includes a group of soluble proteins and tiny particles called platelets, or **thrombocytes.** Every cubic millimeter of blood contains 5,000,000 red cells and about 7,500 white cells. It also contains about 250,000 exceedingly small platelets. Each platelet is a cytoplasmic fragment of a giant cell known as a **megakaryocyte,** a normal resident of the bone marrow. Platelets are quite stable within the bloodstream, but are fragile outside. When a cut is made in the skin and a blood vessel is torn, the platelets at the wound disintegrate, and the machinery of clot formation begins. Clotting does not occur simply by exposure of blood to air. If blood is carefully drawn from a blood vessel so that no tissue damage results, and this blood is placed in a wax (or other nonwettable) container, it remains liquid. Only when platelets disintegrate, as occurs by contact with a wettable surface such as glass or by contact with damaged tissue, is clot formation initiated.

Clot formation is a complex affair (Figure 13–4). Discoveries since 1947 of the mechanisms involved in clotting have revealed that in addition to the platelets, calcium ions, vitamin K, and at least 10 proteins (called **clotting factors**) found in small amounts in the plasma are required. Deficiency of any one of these can lead to delayed clotting—bleeder's disease, or **hemophilia.** Hemophilic individuals bruise easily, hemorrhages occur in the nose, gums, intestines, and skin—and sometimes even in the brain.

The failure to clot causes one kind of prob-

FIGURE 13–4 Stages in clot formation.

lem, but the reverse situation also causes difficulty. If for some reason platelets gather in large numbers in blood vessels when there is no injury, a thrombus or clot may plug the vessel. If this occurs in an artery, the vital supply of blood to the organ normally fed by the artery could be cut off, with disastrous consequences. For example, a clogged coronary artery can produce a heart attack, a clogged brain artery causes a stroke or worse, and a clogged pulmonary artery may result in death. A clot may form in a leg because of platelet malfunction in an area of sluggish flow. If such a clot breaks loose, it forms an **embolus,** a traveling clot that moves through the bloodstream, and may come to rest and clog a vital artery. The reasons for the spontaneous development of these wayward clots are unknown.

The body has mechanisms for the prevention of clot formation in uninjured vessels. One of these safety factors involves the intact linings of blood vessels which, by virtue of their electrical charges, repel the platelets and prevent their aggregation. Another is the presence in the plasma of **anticlotting factors** such as **antithrombin,** which inactivates thrombin, and **heparin,** which enhances the activity of antithrombin thus preventing platelet breakdown and inhibiting the action of thrombin on fibrinogen. Heparin is often added to samples of blood drawn by syringe to prevent clotting so that plasma and cell samples can be obtained for diagnostic purposes (Box 13A).

13–2 AGAINST THE FOREIGN PERIL: IMMUNITY

Alien agents continually attempt to penetrate the borders of the body. When we bruise or cut our skin, they slip past the skin cells and enter the body; foreign materials creep into the

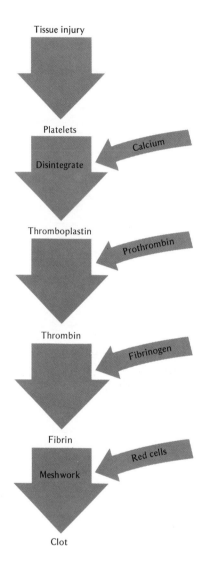

body by way of the nose or any of the other passages from outside the body to the interior; some invaders are carried inside when we swallow our food; dust, microbes, and viruses slip through whatever breach can be found. The continued integrity of the body depends upon its ability to repel the foreign intruder. How is this accomplished?

The body has an elaborate array of defensive weapons geared to prevent foreign agents from invading, establishing themselves, and taking over. The foreign material may be living (such as bacteria) or nonliving (protein or toxin); it may be friend (organ transplant) or foe (filth contaminating a wound). No matter how varied the alien material may be, the defense system depends for its function on a simple and straightforward mechanism: the ability to distinguish between the body tissues (self) and those substances that are dissimilar (not-self). Once a substance is recognized as not-self, the stage is set for battle.

Natural immunity

The close-fitting cells of the skin form a protective wall, providing a natural barrier between the body and the environment. In addition, the body produces watery barriers: for example, mucous secretions in the nose and mouth contain antimicrobial substances to trap and kill bacteria. The salivary glands that open into the mouth produce saliva, which contains bactericidal substances. Coughing repels foreign materials and acts as a watchdog for the lungs. Tears flush the eyes and contain the enzyme lysozyme, which destroys the cell walls of certain bacteria. The ears are protected by waxy secretions, and the acidity of the stomach provides a hazard for bacteria to traverse in their passage from mouth to intestine. The acidity of urine and the regular flushing action of urination contribute to the antimicrobial activity of the urinary system. These systems of defense constitute a part of our **natural immunity** to intruders. We are endowed with them at birth, and they involve anatomical barriers as well as special secretions.

The physical barriers of the body are not impregnable, however. Occasionally invaders sneak past the outer defenses and enter the body itself. (Think of all the small breaks in the skin produced by brushing the teeth, shaving, scratching, and so on.) This could cause problems for the community of cells, especially if the agent penetrates the bloodstream, in which it is likely to be carried to all the internal organs. Moreover, the bloodstream is in many ways a hospitable home for thousands of microscopic organisms, since it provides plenty of oxygen, virtually unlimited food, and usually freedom from competition for the available resources.

Fortunately, there is a second line of natural defenses standing ready to repel those invaders that manage to cross the body's walls and waterways. Manning this line of defense are the cells of the **reticuloendothelial system.** The reticuloendothelial system consists of fixed phagocytic cells lining the sinusoids of the spleen, lymph nodes, liver, and bone marrow. (**Sinusoids** are expanded areas in small blood vessels.) The foreign substances that enter these organs are not pursued, but instead are plucked from the bloodstream by phagocytosis as they pass through the tortuous maze of circulatory channels.

Sometimes a breach in the defense system is counteracted by an active search-and-destroy mission, called **inflammation.** Inflammation is an immediate, localized response to the presence of a foreign substance or, more generally, to tissue injury. Whenever tissue is damaged,

the cells release materials that attract white blood cells (**neutrophils**) from the bloodstream, capillary permeability is increased, and fluid accumulates in the tissue spaces. As a result, the area becomes red, warm, and puffy—in a word, inflamed. The presence of extra fluid dilutes toxic materials, and because some local clotting occurs the damaged area is walled off, thus preventing the spread of the foreign invader.

Neutrophils not only engulf and destroy foreign particles, especially bacteria, but they also clean up the remnants of the damaged cells. When a neutrophil contacts a foreign substance, it quickly surrounds it, and the intruder soon becomes securely enclosed in an membranous sac. Complete engulfment may take less than 0.01 second! The next step is destruction of the alien agent. The lysosomes of the neutrophil move toward the sac, the membranes of the two fuse, and the powerful digestive enzymes of the lysosome are discharged into the sac. The enzymes kill the invader and degrade its molecular constituents into harmless products that are released or are used as a nutrient source by the neutrophil. Some neutrophils die as a result of the phagocytic process, but others continue to ingest, degrade, and inactivate foreign intruders. The mixture of dead cells, bacteria, and neutrophils that accumulates at the site constitutes **pus,** and an extensive accumulation of pus is testimony to the body's battle against invasion.

To summarize: the natural barriers such as the skin, the secretions of cells (tears and mucus), the fixed and mobile phagocytes of the reticuloendothelial and circulatory systems, and the localized inflammatory response are all elements in the natural immune reactions of the body. The system of natural immunity is the first line of defense to clear the body of invaders on initial encounter.

Acquired immunity

Whenever the body encounters an intruder, it attempts to eliminate it by producing an effective countermeasure, one specifically tailored to match the offender. In addition, the body attempts to render itself permanently resistant to that invader and others of its kind. Those mechanisms developed as a consequence of exposure to the foreigner are called **acquired immune reactions.** Acquired immunity may involve specific cells as well as special products of cells, called **antibodies.** An antibody is a specific protein produced in response to the presence in the bloodstream of some material that is foreign to the body. Antibody circulates in the blood, combines with the foreign substance, and destroys it in one way or another.

What enables the body to know that it should produce an antibody and that the invader is a foreigner and not-self? The answer lies in the uniqueness of the molecules of life. Each substance carries its own chemical identification tags, and the molecular configurations of foreign substances that differ from those of the body act as **anti**body **gen**erators, or **antigens.** Thus, antigens are materials (or entire organisms) capable of eliciting an immune response, that is, stimulating antibody production. Antigens can vary: bacteria and viruses that enter the body can be antigens; a skin graft on the body of another person can be an antigen; and proteins, fats, carbohydrates, and nucleic acids can be antigens under the appropriate circumstances. In general, antigens have a common characteristic: they are large molecules, ranging in size from 50,000 to 100,000 molecular weight. For example water, with a molecular weight of 18, and glucose, with a molecular weight of 180, are not antigens, because of their small molecular size.

On occasion the body makes a mistake and

BOX 13A Counting the reds and whites: barometer of disease

The slip shown in Figure 13–5 is a request for blood analysis; such an analysis is often ordered by a physician to determine the state of the patient's health. Most of us are directly involved in just the first stage of blood analysis: donation of a drop or a tubeful of blood. What happens next? If the physician has requested a total blood count, a few drops of blood are diluted with a special fluid. This is well mixed. Then a drop of the diluted blood is placed on a counting chamber (a **hemocytometer**) and covered with a coverglass. The counting chamber is placed under a microscope, and the cells are counted (Figure 13–6). A simple calculation correcting for the dilution gives the cell count per cubic millimeter of undiluted blood. Some normal cell counts of blood are given in Table 13–3.

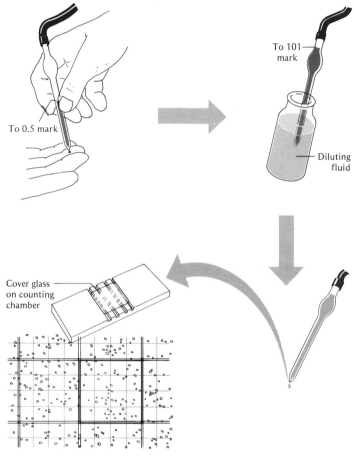

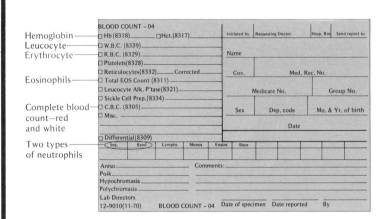

FIGURE 13–5 The lab slip requesting blood analysis.

FIGURE 13–6 Method used in counting blood cells.

TABLE 13–3 Normal cell counts in peripheral blood

Cell	Average count (cells/mm³)	Range of count in 95% of normal population (cells/mm³)
Red cells	5,400,000 (men) 4,800,000 (women) 4,500,000 (infants)	4,500,000–6,500,000 (men) 3,900,000–5,600,000 (women) 4,000,000–5,600,000 (infants)
White cells	7,500 (adults) 12,000 (infants)	4,500–11,000 (adults) 10,000–25,000 (infants)
Platelets	250,000 (all ages)	140,000–440,000 (all ages)

A rising white cell count is often an early clue to a dangerous infection. As the body mobilizes its forces to defend against attack, production of white blood cells increases, and these cells appear in great numbers in the blood. For example, in a person with appendicitis the white cell count may reach 50,000 per cubic millimeter of blood—about seven times the average normal value.

Each red cell contains about 265 million molecules of hemoglobin, and 100 ml of blood normally contains about 15 gm of hemoglobin. It has been shown that iron, vitamin B_{12}, and folic acid, as well as other substances, are necessary for the manufacture of hemoglobin. Thus, a dietary deficiency of any of these substances may produce the condition we know as "tired blood," or anemia, and it may be necessary to supplement the diet with iron, vitamin pills, or both. Anemia is defined in medical terms as a deficiency in the amount of circulating red blood cells or hemoglobin (or both). When for some reason (for example, hemorrhage) the level of red cells declines to one third of normal and hemoglobin concentration falls to one tenth of normal, oxygen is transported with reduced efficiency, and the energy available for cellular work is correspondingly reduced. Consequently, the anemic individual tires easily and often appears pale. The causes of anemia are many, and the disorder should be treated promptly, since a decrease in oxygen-carrying capacity of the blood can result in tissue damage. In addition, decreased viscosity of the blood due to diminished numbers of red cells causes too rapid a flow of blood in the circulatory channels, a condition that promotes excessive strain on the heart and could cause heart failure.

Another aspect of blood analysis involves quality rather than quantity. A drop of blood is smeared out on a glass slide in a thin film. This is dried and stained. The stain shows mature erythrocytes as pink-orange. Abnormalities are indicated if the erythrocytes are too pale or have an abnormal appearance when examined under the microscope. A trained laboratory technician can easily identify the five different kinds of white cells, and by simple count can determine their relative proportions, which may indicate a pathological condition. For example, if a person is infected with parasitic worms such as those that cause trichinosis or hookworm, the eosinophil count rises sharply. In acute appendicitis, neutrophils may rise from 62% to 95%. In infectious mononucleosis, monocytes increase in number and may constitute a major fraction of the white-cell population. Leukemia (literally "white blood": *leukos*—white, *emia*—blood; Gk.) is a cancer of the white-cell-producing tissue, typified by an excessive number of white blood cells in the circulation. The white cells gradually crowd out the red-cell-forming tissue, and there is a corrosive anemia. Ultimately, if the disease is unchecked, death may result.

starts producing antibodies against its own tissues. The disease that develops when this occurs may be quite severe and is called an **autoimmune disorder.** To date there is no good way of undoing the body's immunological confusion, and a progressive wasting away results.

   The primary immune response. The human immune system, weighing about 2 lbs, consists of a number of tissues and organs including the spleen, thymus, tonsils, bone marrow, and lymph nodes. All of these tissues are linked to one another by blood vessels and lymphatics.

   The cells of the immune system that respond to foreign invaders are small, round white cells called **lymphocytes.** Lymphocytes respond to antigens either by the production of sensitized cells or by the formation of antibodies, both of which bind specifically to antigen and inactivate it. Numbering about a trillion in the body, lymphocytes are constantly being formed from precursor or stem cells in the bone marrow. In the time that it has taken you to read the last few paragraphs, your body has produced about 10 million new lymphocytes to replace those that have been damaged or destroyed.

   About half of the lymphocytes, called **T cells,** migrate from the bone marrow to the thymus, a gland that lies in the upper chest (Figure 13–7). Here they remain for a time and multiply; then some of the progeny leave the thymus and, by way of the bloodstream, they either settle down in the spleen, tonsils, lymph nodes, or appendix; or they continue to circulate in the blood. The other half of the lymphocytes, called **B cells,** do not pass through the thymus. (They are called B cells because they were first described in chickens, where their processing takes place in an organ called the **b**ursa of Fabricius; in mammals, including ourselves, the B lymphocytes are presumed to

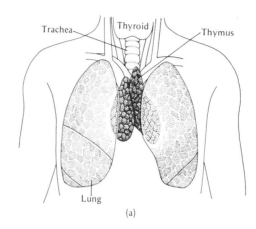
(a)

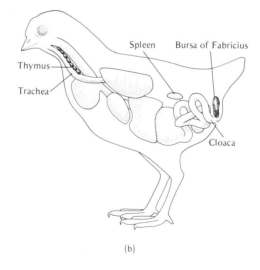
(b)

be processed in the **b**one marrow or the fetal liver.) Once the T and B lymphocytes have been processed, they either enter the circulation or settle down in the lymphoid tissues; now they are capable of reacting with antigen. As we shall see presently, the T and B lymphocytes have entirely different roles in the immune system; however, both are essential for life, and if either the one or the other is lacking, the animal soon dies of infection.

   Resting lymphocytes, prior to antigen activation, produce a number of specific antibody molecules (approximately 100,000), which are strategically placed on the outer surface of their plasma membranes. These molecules are called **antibody receptors,** and in essence each "waits" for an encounter with its specific antigen. Each lymphocyte displays only one kind of receptor. It is believed that by a more-or-less random replacement of amino acids in a particular region of the antibody molecule, millions of different kinds of lymphocytes can be generated, each with different kinds of receptors, and these receptors will fit practically any kind of antigen. If this is indeed so, then how does one's own immune system distinguish between antigens that are part of one's own body and those that are not? The body's immune system avoids reacting with self-antigens not because it is genetically programmed to avoid the manufacture of such receptors, but because in embryonic life the immune system "learned" that certain antigens are "self" and others are "not-self." This property, called **self-tolerance,** is induced when a lymphocyte in the embryo makes contact with self-antigen; lymphocytes with receptors for self-antigen are eliminated or paralyzed. As a consequence, all lymphocytes that display receptors for self-antigen are purged from the immune system during embryonic or neonatal life. What remains is an army of a million or more lymphocytes, each with receptors capable of combining with antigens that are not-self.

T cells are involved in a process called **cell-mediated immunity.** An individual who is exposed to the bacteria that cause the disease tuberculosis (TB), for example, contains circulating T lymphocytes in his blood that will specifically react with the proteins (antigens) of the bacterium. If infection occurs, the appropriate precommitted T cells recognize and react with the antigen, they divide mitotically at the site where the bacteria are present, and they proceed to secrete lymphokines. **Lymphokines** are a grab bag of chemicals that affect the behavior of other cells: some are toxic and kill cells; others inhibit cell division; some halt the migration of phagocytic cells called macrophages so that they accumulate while other lymphokines attract other types of white cells. As a result of lymphokine secretion, the bacteria are ingested and destroyed by the macrophages, and the body is protected. When you are given a tuberculin or Schick test to determine whether you have ever been exposed to TB (and most city dwellers have), the tuberculin antigen is injected under the skin. In 48–72 hours, an inflammation develops at the site of injection if the test is positive. This reaction is called a **delayed hypersensitivity reaction** because the reaction takes days to develop. By contrast, someone who reacts to an antigen within minutes of exposure is said to show **immediate hypersensitivity.** Good examples of such a reaction are the hay fever some people experience after breathing ragweed pollen, and the severe reaction seen in some hypersensitive people who have been sensitized by previous exposure to penicillin or the venom from a bee sting.

The antigen-activated T cell that produces the delayed hypersensitivity reaction does not secrete antibody; however, it does recognize foreign antigens because specific receptors (antibody receptors) are present on its surface. Instead, the activated T lymphocyte divides, releases lymphokines, and participates in a localized mobilization of phagocytic cells at the site of antigen—a response that is called **cell-mediated immunity.** The tuberculin sensitivity test is just one example of cell-mediated immunity. Cell-mediated immunity is involved in contact dermatitis such as that seen when one touches poison ivy, and it is also involved in the rejection of foreign grafts. If a piece of an organ from one animal is grafted to a genetically dissimilar animal, such a graft is rejected. Usually, a skin graft is successful only if the graft is from another part of the individual's own body or comes from an identical twin. If the graft is foreign, wandering T cells enter the area where the graft has been made, lymphokines are produced, and these ultimately kill the cells of the graft; the graft has been rejected. Under some circumstances, activated T lymphocytes will act as aggressor or killer cells that serve to eliminate certain kinds of tumor cells. Indeed, our bodies contain a roving band of T cells that are capable of recognizing and destroying malignant cells if they arise in the body. Through "immune surveillance" by T cells we are continually protected against cancer cells. Some biologists believe that the immune system evolved not to protect an animal against disease agents or to prevent grafts from being made between different organisms, but to police the surfaces of all the body's cells continually so that if antigenically different and potentially deranged cells appeared, they would promptly be destroyed by the mechanisms of cell-mediated immunity.

Certain antigens, especially those of large molecular size with repeating units, as well as bacteria and viruses, do not activate T cells; instead they directly trigger B lymphocytes to activity in a process known as **humoral** or **antibody-mediated immunity.** (Under most circumstances, antigenic stimulation of B cells takes place only when T lymphocytes are stimulated simultaneously; this process is called **immunological cooperation,** but its underlying mechanisms are not completely understood.) The surface of the B lymphocyte, like that of the T cell, bears receptors for a specific antigen, and when the antigen reaches the lymph nodes and the spleen, these B lymphocytes (also called **plasmablasts**) divide and then differentiate into plasma cells. The plasma cells in turn divide rapidly producing a clone[1] or colony of cells (Figure 13–8). This clone of plasma cells includes some that are mature and specialized for the production of antibody and others that will divide at a later time to produce more plasma cells (in the secondary immune response, discussed later). The antigen itself need not enter the plasma cell in order to stimulate division and antibody production. All that is necessary is that antigen be bound to the surface receptor. The plasma cell is a factory for producing antibody molecules (Figure 13–9). By the fifth day after exposure to the antigen, the protein factory is in full production: the antibody genes are transcribed into 20,000 molecules of messenger RNA, which serve 200,000 ribosomes, enabling each plasma cell to produce 2,000 antibody molecules per second. The antibody formed is released from the plasma cells into the circulatory system where it combines with the antigen and neutralizes it. It may take only a few thousand antibody molecules—each specifically tailored to the molecular architecture

---

[1] **clone:** a mitotically produced colony of cells of identical genetic constitution.

of the antigen—to completely inactivate a virus or bacterial cell. The situation has been compared to the state of affairs in which a factory manager (antigen) has only to walk up to the correct building (plasma cell) in order to set off feverish activity within, and there is produced a steady stream of custom-built merchandise (antibody) that is designed to suit him exactly.

When the invader cells or their antigens are neutralized, the primary response to the attack is over. During the ensuing period of 4–6 weeks, the antibodies formed during the primary response tend to disappear from the circulation.

The secondary immune response. Once the primary response to antigen is ended, all is quiet. However, if antigenic challenge again presents itself, then the T- or B-cell response is highly reactive. How is it that the immune system becomes more highly responsive with each additional challenge from antigen?

Not all members of a particular T- or B-cell clone are involved in the primary response of lymphokine production or antibody secretion. Instead, some revert to the resting state and represent the "memory" of the occurrence. Should a particular antigen reappear—be it a graft or a viral infection—immunological memory will produce both an accelerated and an enhanced response. Thus, when an additional antigenic challenge presents itself in the form of a skin graft, the graft may be rejected in a matter of days instead of weeks; if the challenge is in the form of reinfection with a virus, antibody will be formed quickly, it will persist for a greater length of time, and it will be more abundant (Figure 13–10). This is called the **secondary immune response.** As a consequence, the body's defenses are quickly mobilized and the foreign intruder is quickly overwhelmed. The prompt secondary immune response is of great consequence to someone

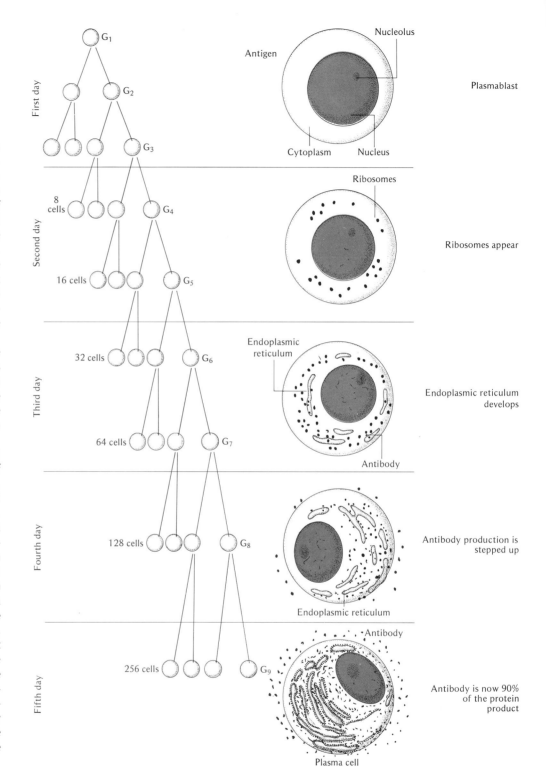

FIGURE 13–9 (below) Plasma cells as revealed by the electron microscope. (Courtesy of K. R. Porter.)

FIGURE 13–10 The formation of antibody with repeated challenges of antigen.

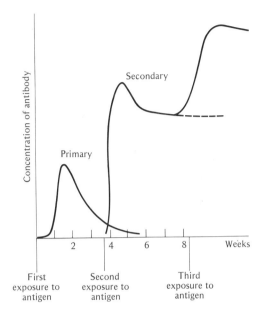

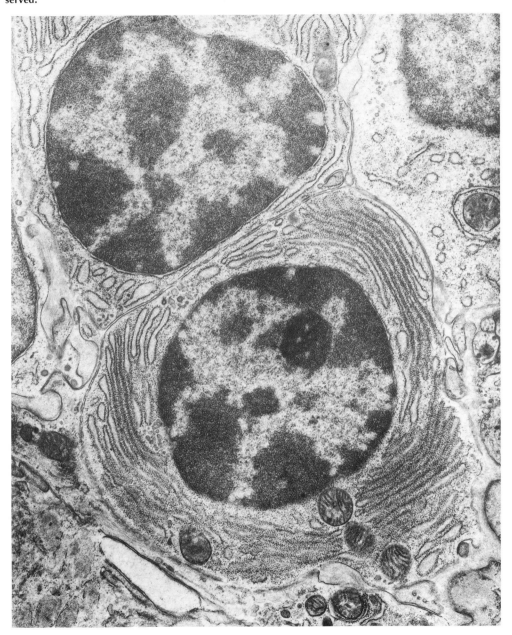

who has been vaccinated or who has recovered from an infection; a second attack by the same antigen stimulates so prompt and massive an outpouring of the correct antibody that the infection is stopped in its tracks. In much the same way that training refines one's physical competence (practice makes perfect), the priming of the immune system through "memory" enhances its capacity to defend the body against alien agents.

Two kinds of acquired immunity: active and passive. The development of antibodies in response to an alien antigen is called **active immunity.** It can come about in either of two ways: (1) infection, that is, by having the disease—the usual occurrence when we recover from such childhood illnesses as measles, mumps, and chickenpox—or (2) immunization, the deliberate inoculation of the disease agent or a close relative (Box 13B).

BOX 13B Putting antibodies to work

There was a time when smallpox was the most feared disease in the world. In eighteenth-century England it killed one out of three children before they reached their third birthday. Those that survived were blind or their skin was pitted with deep, angry scars (pockmarks), bitter reminders of their battle with this "most terrible of all the ministers of death." Smallpox played an important role in the colonization of the Americas. The men of Cortez carried smallpox with them, and shortly after their arrival in Mexico (1520) the disease ravaged the Aztec Indians. More than half the Indians were killed by the epidemic, and this was so demoralizing that their defeat was hastened. The early settlers in the American colonies brought their diseases with them, and smallpox was one. Epidemics broke out frequently; in 1721 more than half of the inhabitants of Boston fell ill with smallpox, and the mortality rate was 15%. This highly infectious disease was also an early kind of germ warfare. It was not uncommon that when the English fought the Indians in North America they contaminated blankets with smallpox, spread them among the highly susceptible Indians, and rejoiced when their enemies died.

Today smallpox is virtually unknown, thanks to the development of the first vaccine by Edward Jenner, an English physician, in 1796. Essentially, Jenner took advantage of local folklore and turned it into a reliable and practical device against the ravages of smallpox. The farmers of Gloucestershire believed that if a person contracted the mild disease cowpox they were assured of immunity to smallpox. Indeed, hardly a milkmaid or a farmer who had contracted cowpox showed any of the disfiguring scars of smallpox. Jenner did an experiment to test this. On May 14, 1796, he took a small drop of fluid from a pustule on the wrist of Sarah Nelms, a milkmaid, who had an active case of cowpox. Jenner smeared the material from the cowpox pustule on the unbroken skin of a small boy, James Phipps. Six weeks later Jenner tested his vaccine (vacca—cow; L.) and its ability to protect against smallpox by deliberately inoculating the boy with material from a smallpox pustule. The boy showed no reaction—he was immune to smallpox. In the years that followed, "poor Phipps" (as Jenner called him) was tested for immunity to smallpox about a dozen times, but he never contracted the disease.

To this day the same treatment works, and in most developed countries smallpox is unknown because of vaccination. Indeed, after Jenner's work it became common to call all immunization shots "vaccinations" after the first practical technique developed by Jenner 150 years ago. But why does cowpox protect against smallpox?

Smallpox is caused by a virus—a tiny bit of nucleic acid enclosed in a protein jacket. Each virus has its own kind of protein coat, conferring specificity upon its owner. Thus, the various kinds of pox viruses—smallpox, cowpox, and chickenpox—are all distinguishable by their protein coats (and, of course, by their genetic codes). The smallpox virus (called **variola**) enters the body through the mucous membranes of the nose and throat when a contaminated article or a droplet containing the viruses comes in direct contact with the body. The viruses spread rapidly via the bloodstream and then localize in the skin and mucous membranes, where they develop into pus-filled sores. In severe cases death may result. (An old curse was: A pox on you and your house!)

Cowpox is a virus disease of cattle, and it is a close relative of the smallpox virus. Cowpox virus **(vaccinia)** and smallpox virus (variola) are antigenically similar; that is, their protein coats have a similar molecular architecture. As a consequence, an individual vaccinated with live cowpox virus develops a mild, single-pustule infection and builds antibodies that will neutralize cowpox virus; such antibodies will also neutralize smallpox virus. Thus, protection is afforded by this vaccination procedure because of the cross reaction of cowpox antibodies with smallpox virus, a condition called **cross immunity.***

The recent conquest of poliomyelitis by vaccination provides an excellent example of how effective vaccination can be. In 1950–1954 the annual incidence of polio in the United States was more than 20,000 cases; in 1961, after immunization had been developed and practiced, this country had only 61 cases! Initially, live poliovirus was treated with chemicals that inactivated its disease-producing capabilities but allowed antigens to remain for induction of antibodies upon injection into a recipient. Later, poliovirus was grown in monkey kidney cells, and from these cultures three mutant lines were isolated that had lost their disease-producing capacity but not their antigenic properties. As a result, effective vaccination with live polio-viruses became practical.

Similarly, live but nonvirulent viruses and bacteria are used in vaccines to provide protection against such maladies as rabies, yellow fever, measles, influenza, diphtheria, whooping cough, and tetanus. The introduced inactivated microbes (or their products) engender immunity without producing disease, and as a consequence sickness and death are virtually unknown from these diseases in the United States and many other countries where vaccination is practiced.

---

*The vaccine for smallpox is prepared today by scratching smallpox or cowpox virus onto the shaved and disinfected skin of a calf. When the pustules are ripe, fluid is collected from them and placed in sterile tubes ready for use.

How does immunization by inoculation or vaccination work? The mechanism is that already described in the discussion of the primary and secondary immune responses. When we are vaccinated against a particular disease agent, we receive a series of graded challenging doses of antigen—small at first and larger later on. The antibody response to the first challenge is usually slow, and the antibodies circulate in the bloodstream for only a short time. Periodic booster shots of antigen subsequent to this initial series serve to keep the immune system primed, and thus immunity remains strong and long. Since the immune reaction is highly specific, immunity gained by vaccination against one disease does not ordinarily protect against another. Thus, commonly we must be vaccinated against each kind of disease with a specific material.

Passive immunity is borrowed immunity. In passive immunity, the antibodies developed in another individual (by infection or immunization) are injected into the body of a person to be protected against that disease. Thus, the recipient of the immune serum containing antibodies does not build his own antibodies against the disease agent (or its product), but simply borrows antibodies that have been made by another animal. Usually these antibodies on loan circulate in the recipient's body for 3–5 weeks and then are destroyed by his own reticuloendothelial system. Passive immunization almost invariably causes the recipient to form antibodies against the proteins in the immune serum and 2–3 weeks later the individual may show signs of serum sickness. Repeated exposure—that is, a second or third dose of immune serum—is often quite dangerous.

Why is passive immunity used when it is such a temporary protective device? Passive immunity is useful because it is immediate, whereas active immunity takes time to develop.

Therefore, passive immunity is used in the treatment of virulent diseases such as acute diphtheria, tetanus (this is the shot of antitoxin), measles, and infectious hepatitis—cases where there is insufficient time for the afflicted individual to produce a strong active immunity. In snakebite, borrowed antibody (if available) can prevent death if administered quickly. A baby is born with antibody borrowed from his mother; the antibody crosses the placenta to enter the child's circulation. Some temporary protection is also transferred to the young via the mother's milk (colostrum), which contains antibodies to a variety of disease agents. In this way a newborn child is protected against poliomyelitis, measles, and mumps. The passive immunity of the infant against poliomyelitis lasts 6 weeks, and measles and mumps immunity for less than a year after birth. To gain lasting acquired immunity, the child must develop his own active immunity.

What is an antibody and what does it do?

Antibodies may be thought of as molecular missiles designed to hit and inactivate a specific alien target, formed in response to the presence of the alien substance (antigen) in the bloodstream. How do antibodies recognize the antigen they are to neutralize, and how do they accomplish their task? These highly specific, tailor-made proteins are of large molecular size (about 150,000 molecular weight), contain several hundred amino acids, and are located in the gamma-globulin fraction of the blood plasma. In spite of the fact that antibodies are giant molecules (**macroglobulins**), the working end—that is the **recognition site** for the antigen—consists of a relatively small region of the entire molecule; similarly antigens, which are also large molecules, have a small region, the **antigenic determinant,** responsible for combining with the antibody at its recognition site. By virtue of their specific geometrical configurations, the recognition site of the antibody molecule is complementary to the antigenic determinant region, and one fits into the other just as a key matches its lock (Figure 13–11). The particular form of the recognition site is produced by slight variations in the sequence of amino acids in the terminal ends of the antibody molecules. As a consequence of such variations in amino-acid sequence, slight differences in the folding of the ends of the protein chains are produced, and these contribute to a different geometric configuration of the recognition site. Each configuration is tailor-made to fit on a specific antigen. Thus, an extra bump or depression in the antibody molecule may completely change its fit with a particular antigen, in much the same way that an extra tooth, a burr or a notch on a key prevents its being used to open a particular lock. It has been estimated that the human body can distinguish at least 10,000 different antigens.

We have thousands of antibody missiles in our immunological defense arsenal, but what exactly does an antibody do, once it recognizes and reacts with its antigen? That depends on the nature of the particular antigen. For example, some disease-producing microorganisms secrete poisons (**toxins**) that destroy host tissue. As a rule these toxins are antigenic, and the antibodies produced against them are called **antitoxins.** Antitoxins can be produced against the toxins of diphtheria and tetanus, thus preventing disease. Active immunity can be engendered by injection of a modified toxin (called **toxoid**), which is nontoxic but antigenic.[2] Some of the snake poisons are enzymes,

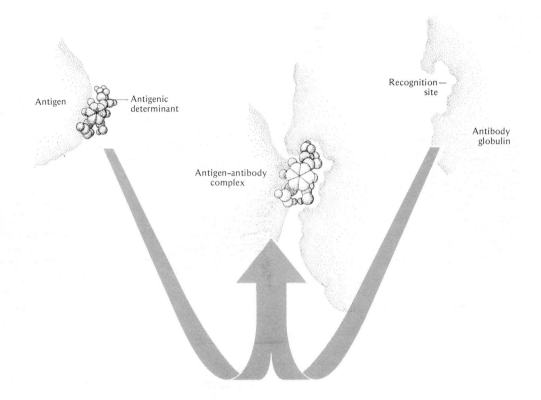

Antigen

Antigenic determinant

Antigen–antibody complex

Recognition site

Antibody globulin

and neutralization of snakebites can be effected by an **antivenin,** which is an antibody (antitoxin) prepared against the venom toxin.

The antibodies produced against viruses may act in a variety of ways. Some antibodies may prevent the adsorption of the virus onto the host cell; other antibodies may prevent discharge of viral nucleic acid into the host cell by the adsorbed virus (this occurs with poliovirus and influenza virus). In the case of vaccinia virus (cowpox), antibody prevents the

escape of the virus from the phagocytic entry vacuole so that viral replication does not occur, and in time the virus is digested within this vacuole. In effect, these varying activities of antibodies neutralize the viral invader.

Certain antibodies react with the lipid layer of the cell walls of bacteria. Antibodies that coat the bacteria in such a way that the bacteria are more easily phagocytized are called **opsonins.** Antibodies that act to clump particles such as bacteria or red blood cells, rendering them more susceptible to removal by phagocytes, are called **agglutinins.** Antibodies that act to dissolve organisms are called **lysins,** and those that react with soluble substances to

[2] It is possible to treat toxins with chemicals or heat so as to destroy their poisonous properties, and then to use these as vaccines. Such an inactivated toxin, called a toxoid, is antigenic and promotes antibody synthesis against the active toxin. The booster shot of tetanus antitoxin is usually a toxoid that results in a buildup of antibody against the toxin.

BOX 13C New organs for old: transplantation

"I am a new Frankenstein," Louis Washkansky told the nurse at his bedside. The macabre humor was appropriate: Louis Washkansky, a sturdy 53-year-old grocer from Capetown, South Africa, had just received the heart of a 25-year-old girl who had died in a traffic accident. Washkansky had literally been brought back to life by a remarkable surgical technique, performed by Dr. Christiaan N. Barnard on December 3, 1967. On that day Barnard made medical history. Other surgeons around the world have performed similar transplants, but of the hundreds of patients who received hearts, only a few dozen are still alive. Most transplant patients survive less than a year with a new heart.

Between 1936 and 1957, many attempts were made to transplant kidneys, but none survived for more than 7 months. By 1977 thousands of kidneys had been transplanted; 55% of the kidneys from related donors (siblings or parents) and 20% of the kidneys from unrelated donors (mostly cadavers) were functioning a year after the transplant operation. There are even reports of kidney transplants functioning for 7 years in related donors and for 4 years in unrelated donors.

The recent success of organ transplants is not due solely to greatly improved surgical skill or to methods of preserving the transplanted organ itself, although these are important. A key factor has been the development of methods to bypass the rejection of the transplant: tissue typing of organs (in much the same way that blood is matched for transfusion purposes) and the use of immunosuppressive agents.

It has been determined that acceptance or rejection of a graft depends on a relatively small number of genes (called **histocompatibility genes**). The closer the genetic constitution of the graft and the host, the better the antigenic match, and the more likely the graft will survive. Before any transplant is attempted, an evaluation of the tissue-transplant antigens is made, using lymphocytes or other white blood cells that carry the histocompatibility antigens. In practice, the genetic constitution of the host is matched as closely as possible with that of the donor. Ideally, grafts should be made between twins or close relatives (brothers and sisters). Once tissue typing is satisfactory the patient who is to receive the donor organ may be surgically drained of lymphocytes to diminish the immune response.

Another means of diminishing the rejection phenomenon is to introduce antibody against the lymphocytes themselves. Thymus glands, lymph nodes, and spleen tissue from human cadavers are extracted and their lymphocytes injected into horses. These are T cells. The horse builds antibodies against the human lymphocytes and later the horses are bled to obtain antilymphocyte serum. When the antilymphocyte serum is introduced into the transplant recipient, the lymphocyte response is diminished, graft rejection is slowed, survival of the transplant is lengthened, and the patient requires lower doses of immunosuppressive drugs. The drawback to antilymphocyte serum is that it can cause serum sickness in man, resulting in damage to blood vessels and the transplant itself.

Administration of immunosuppressive agents also helps to blunt the rejection mechanism. Drugs such as azathioprine, a nucleic-acid antagonist, inhibit the division of lymphocytes and diminish antibody formation. Prednisone, a hormone similar to the adrenal corticosteroids that suppress inflammatory responses is commonly used to delay the rejection response. Another practice is to bombard the lymphocyte centers with ionizing radiations such as x rays. This disrupts the lymphoid cells, and inhibits the rejection of the graft.

Immunosuppression is at best a poor expedient. Upsetting the defense mechanism of the body does more than promote graft acceptance; it promotes the possibility of successful invasion by disease organisms. The graft may be accepted by the host but the host may die of viral or bacterial pneumonia or of a disease produced by some other infectious agent. Thus, an attempt must be made to keep antibody production at such a level that grafts are accepted while an immune response for disease protection is maintained. The balance is obviously very delicate. One hope for overcoming graft rejection lies in the development of blood-soluble tolerance-inducing antigens. These would swamp the immune centers of the host before rejection could be initiated; they would not permit the host to reject the graft and thus would permit acceptance of the graft. Work on these drugs is progressing, and the hope is that they will be available soon so that spare-part surgery will become commonplace.

BOX 13D Wayward antibodies: allergy

Whatever turns your skin to scum
Or turns your blood to glue,
Why that's the what, the special what
That you're allergic to.

OGDEN NASH, 1938

If your eyes tear from pollen, or if you break out in a rash from penicillin, then you know what Ogden Nash was referring to when he penned these lines. We are continually assaulted by foreign substances: pollen, dust, feathers, hair dander, foods, and drugs—to mention a few. To be allergic to these agents, a susceptible individual must be assaulted at least twice. The first time he is sensitized, and the second time he reacts. (This kind of two-time reaction is called an **anaphylactic reaction**.) Foreign substances, such as those named above, may cause a weak immune reaction because they are not very antigenic; they are called **allergens.** When antibody is formed against an allergen, the antibody is incomplete or abnormal in structure; these anaphylactic antibodies (also called **reagins** or **skin-sensitizing antibodies**) have an attraction for receptors on the surfaces of a variety of the host's own tissue cells. The individual shows no outward symptoms. However, on reexposure to the same weak antigen (allergen), the anaphylactic antibody attached to the tissue cell combines with the allergen when it appears in the body, and swelling or destruction of the cell results. This is followed by release of histamine, which passes into the circulation, travels to remote points in the body, and sets up the secondary allergic response—hives, burning in the eyes, shock, and short-ness of breath. Although allergic responses are quite specific, all have one common property: they are annoying. The release of histamine causes capillaries to exude fluid at a rapid rate, blood volume is reduced, venous return to the heart is slowed, cardiac output is diminished; all these symptoms may cause a state of shock to develop, called **anaphylactic shock.** Sometimes histamine causes localized arteriole dilation, there is excessive blood flow in the area, leakage of fluid from the capillary is enhanced, and the skin swells and becomes reddened, causing welts or hives to develop. In hay fever the swelling of tissues due to histamine release occurs in the nose and produces difficulty in breathing; bronchiolar constriction may develop and this may make the individual gasp for breath. Asthma is usually an allergic reaction involving bronchiolar constriction.

There is often a hereditary predisposition toward allergic responses; some individuals seem to produce higher quantities of anaphylactic antibodies than others. The best treatment for allergies is avoidance of the cause. Other tactics include the use of antihistamines or desensitization. In desensitization the individual is repeatedly injected (usually under the skin) with small amounts of allergen so that increased immunity develops slowly, the antibodies formed are complete and act to destroy the allergens before they can reach the anaphylactic antibodies bound to tissue cells. **Antihistamines** neutralize histamine and relieve allergy symptoms such as the sneezy, runny, stuffy nose; the tearing eyes; the dulled hearing, taste, and smell. They do not prevent tissue destruction, however, and their effects are temporary. Without further treatment the allergy could become dangerous. Total treatment such as desensitization should be sought, rather than self-medication with over-the-counter antihistamine remedies.

render them insoluble (and thus much easier to phagocytize) are called **precipitins.**

Although it was once believed that each of the foregoing phenomena was due to a different kind of antibody, it now is considered that these are merely the many faces of the same antigen-antibody reaction. In all cases:

1. Antigen combines with antibody to form a complex.
2. The particular type of complex (agglutination, lysis, or precipitation) depends on the nature of the antigen (soluble, particulate, and so on).
3. The antigen-antibody complex is highly specific.

SUMMARY

1. Blood functions in the transport of food, oxygen, waste materials, and hormones; in the regulation of temperature; and in the control of disease.

2. Blood consists of plasma, red cells (erythrocytes), white cells (leukocytes), and platelets (thrombocytes).

3. The plasma consists of 92% water plus proteins (albumin; alpha, beta, and gamma globulins; and fibrinogen), salts, sugars, lipids, vitamins, hormones, and waste materials. Its composition is controlled by a series of feedback mechanisms. The proteins maintain osmotic pressure, contribute to the maintenance of blood pressure, buffer the blood, and are important in clotting and antibody formation.

4. Erythrocytes are formed in bone marrow. When mature, they are enucleate, biconcave sacs of hemoglobin that can reversibly bind oxygen, forming oxyhemoglobin. Oxygen is picked up at the lungs, where the pressure is high, and released in the tissues, where the pressure is low.

5. Leukocytes are of two basic types: granulocytes (formed in the bone marrow) and agranulocytes (formed in the lymphoid tissue). Both are important in combating disease and infection.

6. A rising white cell count signals disease. Increase in a particular kind of white cell may indicate a specific kind of disease. Anemia may be due to a lowered red cell count, or to lack of iron, vitamin B_{12}, or folic acid which interferes with hemoglobin formation.

7. Platelets, fragments of megakaryocytes, disintegrate when in contact with damaged tissue. Along with calcium ions, vitamin K, and protein (clotting factors), a series of reactions produces a fibrin meshwork that traps red cells and forms a clot, sealing the wound. Lack of clotting factors causes bleeder's disease (hemophilia). Anticlotting factors (antithrombin, heparin) in the plasma prevent platelet breakdown and embolus formation in the absence of tissue damage.

8. Natural physical barriers such as the skin, mucous secretions, saliva, tears, ear wax, stomach acid, urine acidity and flushing action, coughing, and sneezing, all protect the body against foreign substances and disease organisms. Cells of the reticuloendothelial system (fixed phagocytes) that line the sinusoids of the liver and spleen act as a second line of natural defense.

9. Inflammation is a local response to tissue damage. White cells (neutrophils) and fluid accumulate and some local clotting occurs. The neutrophils engulf and destroy the foreign material by means of lysosomes; pus may accumulate.

10. The human immune system consists of the spleen, thymus, tonsils, bone marrow, and lymph nodes, all linked by blood vessels and lymphatics.

11. Acquired immunity involves antibody production by lymphocytes in response to antigens. The reaction is highly specific, and each antigen engenders production of a specific antibody. Autoimmune disorders involve production of antibodies against the body's own tissues.

12. T lymphocytes and B lymphocytes are formed in the bone marrow, but T cells spend some time in the thymus. Each lymphocyte has on its surface about 100,000 molecules, all of one kind: these are antibody receptors specific for one kind of antigen. Receptors specific for the body's own tissues (self-antigens) are normally destroyed early in embryonic life so the immune system is self-tolerant.

13. During the primary immune response to an antigen, specific T lymphocytes divide and secrete lymphokines that inactivate the antigen and allow it to be destroyed by macrophages. Called cell-mediated immunity, this is also responsible for graft rejection and any natural destruction of cancer cells. T cells are also involved in hypersensitivity reactions.

14. B lymphocytes are involved in humoral immunity. In response to antigen, a clone of plasma cells is formed from a specific B lymphocyte. Antigen is bound to the lymphocyte, which produces specific antibody in massive amounts.

15. T and B cells may both cooperate in any response to a particular antigen.

16. A repeated challenge by the same antigen produces a secondary immune response. Members of previously activated T- and B-cell clones retain immunological "memory" of the first antigen, and so an accelerated and enhanced response is induced.

17. Active immunity is produced in response to infection, or immunization. Immunization involves inoculation with graded doses of antigen, attenuated disease agents, or closely related antigens (cross immunity).

18. Passive immunity involves direct injection of antibodies into an infected or poisoned individual who has had insufficient time to develop his own active immunity.

19. Antibodies are giant gamma-globulin molecules (macroglobulins), whose recognition site is complementary in geometric form to the antigenic determinant region; the form of the recognition site is produced by variations in amino-acid sequences. The matching mechanism allows antibody to specifically bind antigen. Inactivation of antigen may be by combination with antibody, forming a complex.

20. Antitoxins are produced in response to toxins. Active immunity is produced by inoculation with toxoid, such as antivenin against snake venom.

21. Viral antibodies may prevent viral adsorption to cells, discharge of viral nucleic acid, or escape of the virus from a cellular vacuole.

22. Bacterial antibodies dissolve bacteria (lysins); or they may clump them (agglutinins), coat them (opsonins), or precipitate them (precipitins), thus making them more easily phagocytized.

23. Allergens cause the formation of incomplete antibodies (reagins or skin-sensitizing antibodies), which are bound to the host's tissue cells. On reexposure to the same allergen, swelling or destruction of the tissue cell with release of histamine causes allergic anaphylactic reactions. Severe reactions may cause a state of anaphylactic shock.

24. Desensitization to allergens involves the induction of active immunity and formation of normal antibodies by inoculation with graded doses of allergen.

25. Individuals who are allergic to certain substances often take antihistamine drugs to relieve the symptoms of allergic reactions. It is important to realize that such drugs simply relieve the symptoms, but do not eliminate the cause of the allergic response.

KEY WORDS

plasma
red blood cells
white blood cells
platelets
albumin
globulin
fibrinogen
alpha globulins
beta globulins
gamma globulins
erythrocytes
hemoglobin
oxyhemoglobin
leukocytes
erythroblasts
granulocytes
agranulocytes
hemorrhage
serum
fibrin
thrombocytes
clotting factors
hemophilia
embolus
anticlotting factors
antithrombin
heparin
hemocytometer
natural immunity
reticuloendothelial system
sinusoids
inflammation
neutrophils
pus
acquired immune reactions
antibodies
antigens
autoimmune disorder
lymphocytes
T cells
B cells

antibody receptors
self-tolerance
lymphokines
delayed hypersensitivity reaction
immediate hypersensitivity
cell-mediated immunity
humoral or antibody-mediated immunity
immunological cooperation
plasmablasts
clone
primary immune response
secondary immune response
active immunity
passive immunity
variola
vaccinia
cross immunity
histocompatibility genes
macroglobulins
recognition site
antigenic determinant
toxin
antitoxin
toxoid
antivenin
opsonins
agglutinins
lysins
precipitins
anaphylactic reaction
allergens
reagins or skin-sensitizing antibodies
anaphylactic shock
antihistamines

TOPICS FOR REVIEW AND DISCUSSION

1. How is the body able to distinguish between self-antigens and foreign antigens?
2. What are antibodies and how do they work?
3. Contrast passive immunity with active immunity.

4. What are the special and distinct roles performed by T and B lymphocytes?
5. Describe the mechanisms involved in the formation of a blood clot.
6. List the components of blood and give a function for each.
7. What is immunization and why does it work?
8. How is antibody specificity related to protein structure?
9. What is a reasonable explanation for the similarity of blood plasma to sea water?
10. Why are transplants from genetically dissimilar animals rejected?
11. List and describe several methods of circumventing graft rejection.

Tragedy under the Sea

It was scheduled to be a routine mission, a dive of about an hour's duration in only a few hundred feet of water off Key West, Fla. The 23-foot long submersible, designed by famed Inventor-Oceanographer Edwin A. Link, seemed more than equal to the task. Since it began operating as an oceanographic research vessel for the Smithsonian Institution two years ago, *Sea-Link* had easily plunged to depths of 1,000 feet. But this time, as the minisub maneuvered in swift currents of the Gulf Stream, routine turned abruptly into tragedy.

Sea-Link was checking and collecting fish traps near the wreckage of an old destroyer scuttled last year to create a man-made reef that would encourage marine growth. Suddenly, *Sea-Link*'s crew heard the harsh, rasping sound of metal rubbing against metal. Apparently pushed off course by an unexpectedly strong current the sub had become ensnarled in cables and other debris. As rescue ships hurried to the scene *Sea-Link*'s crewmen were told to exert themselves as little as possible in order to conserve oxygen. The crew could do little else. At the pressure that exists at a depth of 360 feet (162 lb per square inch) a free swim to the surface was far too risky.

By the time the sub was freed the men had been under water 31 hours. It was a bittersweet success. Two crewmen emerged unharmed from the forward compartment, where the atmosphere had remained at about sea level pressure. But rescuers had to leave 2 others (whose motionless bodies could be seen through portholes) inside the aft compartment while it was slowly depressurized; if the men were still alive, suddenly opening the hatch at sea level would have caused a possibly fatal case of bends. When the hatch was opened, the fears were confirmed: both crewmen had died of carbon dioxide poisoning.

GASEOUS EXCHANGE: RESPIRATION

14-1 WHY DO WE NEED A RESPIRATORY SYSTEM?

We live aboard the spaceship earth. Our precarious existence depends on a self-contained life-support system that provides fuel (food), water, and air. Without these we cannot survive for long. Deprived of food, earthlings like ourselves can live for weeks, without water for a few days; in the absence of air, life is snuffed out in a matter of minutes. Why is this so?

The countless millions of cells of the body all require energy for their operation. They must be supplied with energy-rich fuel molecules, but to release the energy locked within the chemical bonds, oxygen is required. The oxidative combustion of fuel (food), taking place within the metabolic furnace of the cell, liberates energy for the performance of work. The fuel molecules contain carbon atoms, and when the fuel is burned a gaseous ash, carbon dioxide, is produced as a byproduct. Since carbon dioxide is a cellular poison, it must be eliminated directly. Unlike food and water, neither the carbon dioxide that results from metabolism nor the oxygen of the air can be stored within the body. Consequently, oxygen must be continuously supplied from the air around us, and carbon dioxide must be disposed of. How do our cells exchange essential oxygen for useless, poisonous carbon dioxide?

The exchange of gases between the living organism and its environment is called **respiration.** Respiration in man involves three related processes (Figure 14–1). The direct exchange of gases between environment and organism is called **breathing,** or **ventilation;** it primarily involves physical phenomena and is performed by the external respiratory system. Oxygen is transferred from the respiratory system to the blood, by which it is carried to every cell of the body and exchanged for carbon dioxide. The carbon dioxide is carried in the

FIGURE 14–1 (left) The three kinds of respiration.

FIGURE 14–2 (right) A variety of gills. (a) Gills of a fish. (b) Gills of a lobster, with a portion of the gill cover removed. (c) Gills of a clam with one of the shell valves removed.

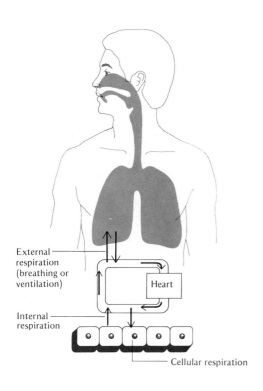

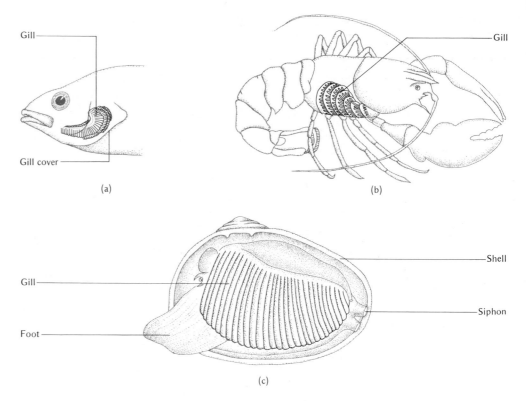

blood back to the respiratory system, where it, in turn, is given up in exchange for more oxygen. The transfer of gases between the blood and other tissues of the body is called **internal respiration** and strictly speaking is a function not of the external respiratory system but of the circulatory system. It completes the essential second stage of the total exchange process between the air and the body's cells. Finally, there is **cellular respiration**—the chemical reaction of oxygen with organic fuel molecules within the cell to release energy, carbon dioxide, and water. This chapter will concern itself with external and internal respiration; cellular respiration is discussed in Chapter 4.

We and all other inhabitants of the surface of the earth are surrounded by an immense ocean of air, the **atmosphere.** The atmosphere presses down on all that live at the surface of this planet of ours, and we live at the bottom of that ocean of air, moving about in it like deep-sea fish. Air is a mixture of gases consisting of 21% oxygen, 78% nitrogen, 0.04% carbon dioxide, and less than 1% argon, helium, neon, and other rare gases. Probably 80% of all the animals that inhabit the earth breathe free air—mammals, birds, reptiles, amphibians, insects, land snails, and earthworms. The remaining 20% "breathe" air dissolved in water—fishes, amphibians, and aquatic inver-

tebrates (such as lobsters, clams, starfish, and sponges).

Gas exchange between an organism and its environment always takes place by diffusion[1] across a moist plasma membrane. Diffusion is a relatively slow process, and in liquids a gas can traverse only very short distances by this mechanism (a maximum distance of 1 mm); thus animals that rely solely on diffusion to obtain sufficient oxygen for metabolic needs and to eliminate carbon dioxide must be small, have a low metabolic rate, or be constructed in such a way that the cells are nowhere very

---

[1] Refer to Chapter 2 for definition of diffusion.

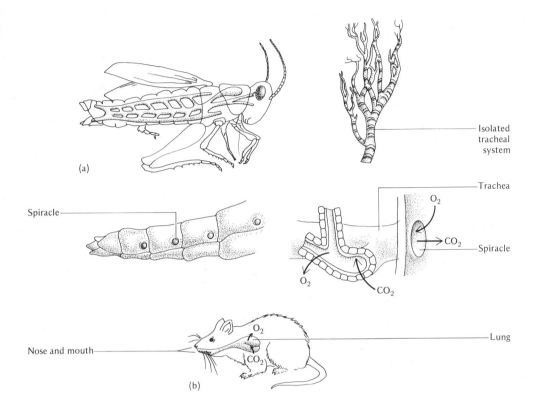

(a)

Spiracle

Isolated tracheal system

Trachea

O_2

CO_2

Spiracle

O_2

CO_2

Lung

Nose and mouth

O_2

CO_2

(b)

distant from the surrounding medium. Larger creatures require the maintenance of a respiratory surface of dimensions large enough to allow sufficient oxygen and carbon dioxide to be exchanged to meet the metabolic requirements of the body; the larger and more active the organism, the more oxygen is required, and the larger the respiratory exchange surface must be. The respiratory exchange surface is usually thin, delicate, and moist and must be not only exposed to the environment for gas transfer but protected against injury. Here is the dilemma: how can a respiratory surface be thin, delicate, large, exposed, protected, and (especially in air) kept moist all at the same time?

One obvious way to expose a large surface area for respiratory exchange would be to utilize the entire body surface; however, such an arrangement would readily suffer mechanical damage and could not be protected easily. Nevertheless, some organisms manage this way—particularly small aquatic animals such as flatworms, sponges, hydras, and frogs. On land, however, the moist and fragile surface would easily dry out and soon be functionless, so that in most organisms the respiratory surface is confined to a specific region of the body where it is protected from damage and desiccation. The respiratory surfaces of water-dwelling animals (such as clams, lobsters, and fishes)

are outgrowths of the body called **gills;** these are kept moist by the water in which the animal lives and are enclosed and protected by a gill cover (Figure 14–2). The land-air dwellers have ingrowths of the body wall called **lungs** (or **tracheal tubes** in insects) that are protected by being tucked inside the body (Figure 14–3); they are kept moist by mucous secretions and communicate with the environment through a chimneylike opening so that water loss and injury are minimized. Since the amount of gas that can be exchanged by diffusion across moist interfaces depends upon the area exposed, the respiratory surface can be increased tremendously by extensive folding. If the environment (air and water) is actively pumped across the respiratory surface (as in ventilation or breathing), efficiency of gas exchange can be further increased. Once the gas has passed across the respiratory surface, efficiency of distribution is increased greatly if the respiratory surface is amply supplied with blood; so that larger animals usually have an efficient circulatory system (Chapter 12). By means of the circulatory fluid, oxygen can be swiftly carried to, and carbon dioxide away from, all the tissues, a system much more effective than transport by diffusion alone.

Gills and lungs have several common properties, including moist surfaces, greatly enlarged surface areas, gas or water pump, protection against mechanical injury, and a circulatory system for transfer of gas between the external environment and the internal body cells.

14–2 THE RESPIRATORY TREE

Air passes from the outside of the body to the respiratory exchange surface, the lungs, via the **respiratory tree,** a complex branching network of passages resembling a tree trunk and its ramifying roots and rootlets (Figure 14–4). As air moves along the respiratory passages, it is both filtered and conditioned.

Air first enters the respiratory tree though the nostrils where projecting hairs filter out dust and debris (Figure 14–5a). The air then enters the nasal cavity, which is divided by a septum and lined by a ciliated epithelium. The cilia distribute a mucus film that traps debris, and in conveyor-belt fashion this mucus sheet is moved toward the throat, where it is swallowed or spat out. Beneath the nasal epithelium is a rich network of blood vessels that helps to warm the air as it eddies about in the nasal cavity. Scroll-like turbinate bones and air spaces in the skull (**sinuses**) form a complex labyrinth of passages that further assist in filtering, warming, and moisturizing the incoming air (Figure 14–5b and c). Even our tears help moisturize the air we breathe, since tear ducts drain the continuously operating tear glands and empty directly into the nasal chamber (Figure 14–5b).

Warmed, humidified, and scoured of dust and small particles, the air passes from the nasal cavity to the throat (**pharynx**), where the air and food passages cross. From the pharynx the air enters the **larynx** (voice box or Adam's apple) through a slitlike opening, the **glottis**, which is guarded by an elastic flap, the **epiglottis.** The glottis remains open at all times except when we swallow. The larynx, made of cartilage, ligaments, membranes, and muscles, includes the **vocal cords,** by which we speak. It leads directly into the windpipe (**trachea),** which is a cylindrical tube, 4–5 in. long supported by regularly spaced rings of cartilage. You can feel these rings by running your finger down the front of your throat (Figure 14–4a). Ciliated and mucus-secreting cells line the trachea and help to trap debris and sweep it upward toward the mouth, where it is either swallowed or spat out (Figure 14–6). The trachea divides into two **bronchi,** which enter the lungs and divide 20–22 times, producing

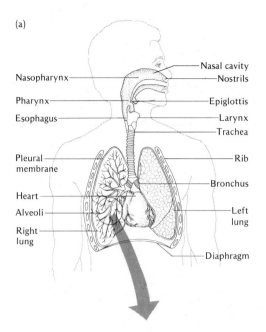

FIGURE 14–4 (a) The respiratory tree of man. (b) Detail of blood supply to the alveolus.

(a)

Nasopharynx

Pharynx

Esophagus

Pleural membrane

Heart

Alveoli

Right lung

Nasal cavity

Nostrils

Epiglottis

Larynx

Trachea

Rib

Bronchus

Left lung

Diaphragm

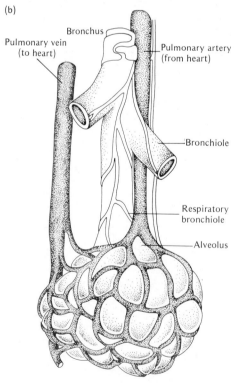

(b)

Bronchus

Pulmonary vein (to heart)

Pulmonary artery (from heart)

Bronchiole

Respiratory bronchiole

Alveolus

smaller and smaller tubes, the **bronchioles** (Figure 14–4b). The smallest have no cilia or cartilage, but an abundance of smooth (involuntary) muscle, and communicate directly with grapelike clusters of thin-walled air sacs (**alveoli).** The bronchioles normally widen and narrow as we allternately inhale and exhale. The alveoli are surrounded by a network of capillaries, and there gas exchange occurs (Figure 14–4b). Oxygen from the incoming air diffuses across the moist alveolar membrane into the blood, and carbon dioxide diffuses in the other direction, from blood to alveolus, from where it can be exhaled.

Our lungs weigh only 2.5 lb. They consist of the millions of tiny alveoli, so numerous that their combined surface area is 70 yd^2, more than 40 times the area of our skin, and equivalent in surface area to about one fourth of a tennis court. The alveoli and associated bronchioles give the lung a soft, spongy texture. In addition, there are elastic tissue, numerous nerves, and blood vessels. The pair of lungs lies free in the chest (thoracic) cavity, attached to each other by the bronchi. The two lungs are not mirror images of each other. In humans, the right lung is the larger of the two and is partially divided into three lobes; the left lung is divided into two lobes (Figure 14–4a). The space provided by the smaller left lung is occupied by the heart.

A thin sheet of epithelial cells, called the **pleural membrane,** snugly surrounds and encloses each of the lungs and is continuous with the lining of the chest cavity. The pleural membranes secret a lubricating fluid that reduces friction and allows the lungs to expand and contract without becoming irritated. The amount of fluid secreted is quite small, and it is normally absorbed by lymph vessels as rapidly as it is produced.

We take 4 to 10 million breaths a year. During the passage of air from the nose to the

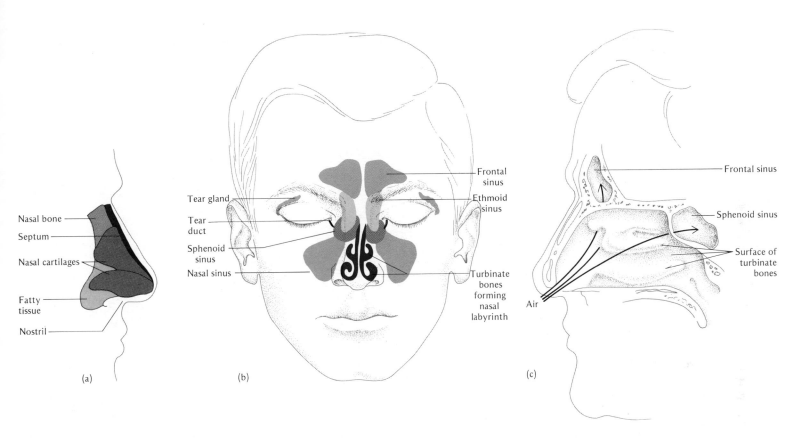

FIGURE 14–5 The nasal cavity, sinuses, and nasal labyrinth. (a) External view of the nose. (b) Frontal view showing the internal arrangement of the sinuses and the turbinate bones. (c) A section through the nasal cavity and two sinuses.

(a)

Nasal bone

Septum

Nasal cartilages

Fatty tissue

Nostril

(b)

Tear gland

Tear duct

Sphenoid sinus

Nasal sinus

Frontal sinus

Ethmoid sinus

Turbinate bones forming nasal labyrinth

(c)

Frontal sinus

Sphenoid sinus

Surface of turbinate bones

Air

lungs, any number of things may go wrong. Table 14–1 catalogs some of the afflictions of the respiratory tree, from coughing to bronchitis.

14–3 ALVEOLUS: RENDEZVOUS OF BLOOD AND AIR

The alveoli of the lungs are moist, diaphanous exchange surfaces with three layers: a thin, fragile epithelium lining the sac; an interstitial layer, also very thin; and an overlay of capillaries arranged in a delicate lacework pattern (Figure 14–4b). The capillary network of the lungs is extensive. Consider the following: in each pound or two of our fatty tissue there are about 6 miles of blood vessels, but if the pulmonary capillaries of our lungs (which weigh a little over 2 lb) were laid end to end, their total length would equal some hundreds of miles.

Gas exchange in the lung occurs in the following way. Air entering the lung has been conditioned and filtered, and it now comes to the blind air sacs, 300 million or more, each enmeshed in a capillary network. Blood enters the capillary network of the alveoli via the pulmonary artery, which in turn comes directly from the heart (Figure 14–7). The meshwork of the microscopic capillaries covers the air sacs as a glove covers a hand, and the interface between air and blood is very thin indeed. Each capillary is only wide enough to permit the blood cells to pass through in single file, so the diffusion path from gas-filled alveolus to blood-filled capillary is extremely short. Oxygen molecules, relatively more abundant on the alveolar side of the interface than in the blood, dissolve in the filmy moisture coating the alveolus, pass across its thin walls, and diffuse into the bloodstream, where they are bound by the hemoglobin of the red blood cells. Oxygen is swept along in the bloodstream, carried back to the heart via the pul-

TABLE 14–1 Trouble along the respiratory tree

Location	Problem	Possible cause	Cure
Nasal cavity	*Bloody nose*	Blowing nose too hard	Time
	Deviated septum: difficulty in breathing	Punch in the nose	Surgery
	Copious secretion from tear glands	Peeling onions or crying	Stop crying, cook onions
	Runny nose: inflammation of nasal lining spreading to sinus lining and causing *sinusitis* and *headaches*	Severe head cold	Antihistamines relieve symptoms by constricting capillaries and reducing secretion of mucus
	Sneezing (a cough in the nose): a reflex action	Irritants in nasal passageway	None, since it is a reflex
Pharynx	*Tonsillitis: sore throat*	Enlarged tonsils	Surgery
	Interference with breathing, *snoring* restless sleep, dull expression from open mouth, which is necessary for breathing	Enlarged adenoids (lymphoid tissues at rear of pharynx)	Surgery
Larynx	*Coughing*	Talking and eating simultaneously; epiglottis remains open too long and food enters larynx	None, since it is a reflex
	Persistent hoarse voice, difficulty in breathing	Cancer of the vocal cords	Surgery
Trachea	*Asphyxia:* choking	Blockage with food, mucus, or blood from an accident	Surgery (cutting an opening and bypassing the trachea; tracheotomy)
Bronchioles	*Asthma:* difficulty in exhaling but not inhaling, for reasons that are not clear	Allergic reaction or nervous stimulation contracts muscles and reduces size of passage	Epinephrine (adrenaline) or other muscle-relaxing drugs
	Bronchitis: inflammation of bronchiolar tree (excess secretion of mucus blocks bronchioles as phlegm, coughing up phlegm weakens linings, and area is susceptible to infection)	Persistent irritation as from colds, infection, cigarette smoke, smog	Wait for cold to get better, stop smoking, reduce air pollution
Lungs	*Emphysema:* impaired lung function, pain (Box 14A)	Chronic expansion and destruction of alveoli	Stop smoking, reduce air pollution
	Pleurisy: pressure, painful breathing, impairment of lung function	Inflamed pleural membranes	
	Pulmonary edema: impaired breathing, pressure, pain	Oozing of body fluid into lungs	Rest, aspiration

FIGURE 14–6 Scanning electron microscope view of the ciliated epithelial cells of the windpipe. A mucus droplet is supported by the cilia. ("Ciliated and Nonciliated Epithelial Cells from Hamster Trachea," Port, C. and Corvin, I. Cover Photo, *Science,* Vol. 177, 22 September 1972.)

FIGURE 14–7 Diagrammatic view of the circulatory pathways leading from the heart to the lung and back to the heart.

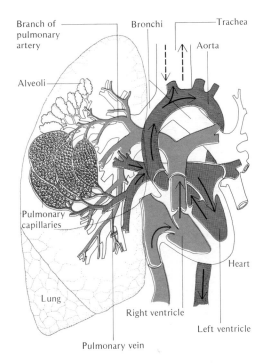

monary vein, and from there distributed by the systemic circuit to the cells of the body.

The lungs do more than deliver oxygen to the blood. Simultaneously they remove the waste product of cellular oxidation, carbon dioxide. Since carbon dioxide is continuously being formed by the oxidative metabolism of the body's cells, it is more concentrated in the blood than it is in the alveolar air. The net movement of carbon dioxide is from the blood into the lung, and in this way, at exhalation, it leaves the body. Although oxygen and carbon dioxide both cross the alveolar interface, they move independently of each other. They are

like total strangers who simultaneously touch the station platform, one boarding and the other leaving the same train.

The delicate meshwork of capillaries surrounding the alveoli of the lungs contains only 2 oz of blood, a volume exactly equal to the amount of blood pumped out of the right side of the heart with each contraction. Thus, with each heartbeat the oxygenated blood in the lung capillaries is pushed back toward the heart, which it enters at the left side, and a new volume of deoxygenated blood is pumped from the heart to refill the pulmonary vessels. When we are at rest, each red blood cell

remains in the pulmonary capillaries for only 0.75 second, but this is time enough for it to unload its carbon dioxide and take on a fresh cargo of oxygen.

At rest the cells of the body require 250 ml of oxygen each minute, and with exercise this requirement is increased 10–20 times—to 5 liters of oxygen each minute (over 1 gallon). Any interruption in the delivery of oxygen to the blood or the tissues, which can occur in any number of ways, results in impairment of cellular function and can cause asphyxia. Plasma in equilibrium with air can take up about 0.25 ml of dissolved oxygen per 100 ml.

BOX 14A Smog, smoking, and emphysema

Each day an average New York City dweller inhales the equivalent of half a pack of cigarettes, even if that individual does not smoke! The dirty air we breathe daily contains about 20 billion foreign particles, and in a lifetime each of us may take in as much as 45 lb of dust. Polluted air, **smog** (*smoke* plus *fog*), blankets most large cities of the world, and our lungs bear scars from breathing this. Smog consists of an aerosol of carbon particles about the size of bacteria, and these settle on the walls of the trachea and bronchioles to be carried out with the mucus sheet. If this were all, we might be in no trouble; however, also present in our filthy air are smaller particles—less than 0.3 μm —and these do reach the alveoli. Their cumulative effects are most clearly seen when the lungs of a newborn child are compared with those of an adult city dweller. The infant's lungs are bright and pink, but those of the adult have a dull pink-gray-mottled coloration.

Smog contains not only particles but noxious and toxic gases such as nitrogen oxides, ozone, sulfur dioxide, and peroxides, and these are not removed by filtration. On a smoggy day in Los Angeles you can smell the ozone and see the brown haze of smog that blankets the basin, choking the inhabitants. With every fire we make, we pollute the air. Coal and low-grade fuel oil send both soot and sulfur dioxide into the air; petroleum refineries, chemical factories, and automobile engines contribute their share of toxic gases. Smog is a by-product of the combustion of hydrocarbons, and the automobile is a prime offender. Every thousand cars that move through a congested city discharge 3.2 tons of carbon monoxide, 400–800 lb of hydrocarbon gases, 100–300 lb of nitrogen oxides, and other lethal fumes. These poison gases, together with cigarette smoke, cause bronchitis, pulmonary edema, and emphysema (Table 14–1). Bronchitis and emphysema were rare diseases 25 years ago; now they constitute a major respiratory ailment afflicting nearly 10 million Americans. How do smog and smoking produce these respiratory disorders?

The smoke from the burning of tobacco or gasoline contains gases that paralyze the cilia of the trachea and the bronchi. The inactivity of the cilia prevents the mucous blanket from moving, and it accumulates in the bronchioles. Inhaled air easily moves past the mucus and swells the bronchioles, but as the air is exhaled the bronchiole walls collapse, the air is trapped and held by the mucus plugging the bronchiole. The trapped air stretches the air sacs, they balloon out forming air blisters, and some burst. In the process the alveolar and elastic tissues of the lung are destroyed. Elastic recoil of the lung tissue is lost. The lungs stretch to compensate for the loss of respiratory surface, they press against the walls of the chest and the diaphragm, the muscles lose efficiency. As a result, the emphysema patient develops a barrel-shaped chest, breathing is difficult, severe oxygen deprivation produces brain damage, and the heart is enlarged and weakened; in severe cases there is a clubbing of the fingers **(pulmonary osteoarthropathy).**

The symptoms begin slowly. Paralysis of the cilia by the toxic fumes, with stagnation of mucus, means that bacterial invaders are no longer swept out, infections develop in the alveoli and spread to the bronchioles—chronic bronchitis develops. Inflammation damages blood vessels, they leak fluid into the lung, and edema results. Respiratory exchange is impaired, and oxygen deprivation ultimately takes its toll. The mortality rate from bronchitis and emphysema due to smog is undetermined, but among cigarette smokers it is 700% higher than among non-smokers. The Geneva Protocol, which the United States sponsored in 1925 and which all major powers have ratified, bans chemical and biological weapons in war. Gas warfare in our cities, however, is at present not only permitted but prevalent.

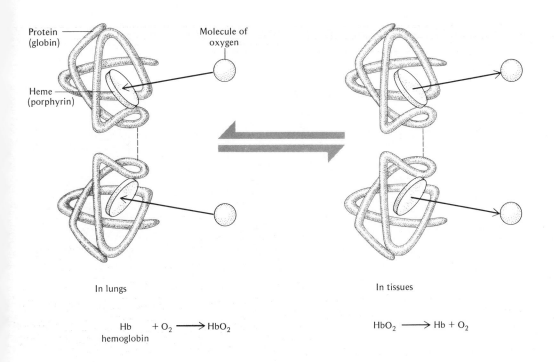

FIGURE 14–8 Hemoglobin. (a) Association of hemoglobin with oxygen. Only two of the four protein chains of a hemoglobin molecule are shown. (b) Detail of the chemical nature of the heme in hemoglobin.

Protein (globin)

Molecule of oxygen

Heme (porphyrin)

In lungs

$$\text{Hb} + O_2 \longrightarrow \text{HbO}_2$$
hemoglobin

In tissues

$$\text{HbO}_2 \longrightarrow \text{Hb} + O_2$$

Oxygen molecule attaches here to heme molecule. Since there are 4 hemes per hemoglobin molecule, 4 molecules of oxygen are carried.

(a)   (b)

This being so, the heart would have to pump 83 liters per minute through the lungs[2] just to meet our needs at rest. This is virtually an impossible task for the heart, and is obviously not the way the body's oxygen demands are met. A critical factor in increasing oxygen uptake by the blood is the ability of hemoglobin, present in the red blood cells, to combine with oxygen. In the presence of plentiful oxygen, hemoglobin forms a complex called **oxyhemoglobin** (Figure 14–8); this is unstable, and under conditions where the oxygen is low

(as in the tissues) the oxygen is released from the complex to form hemoglobin again. As a consequence, hemoglobin loads up with oxygen at the lungs and unloads it at the tissue cells. Because of the presence of hemoglobin, the oxygen-carrying capacity of the blood is increased 65 times, to 20 ml per 100 ml, and this means that at rest the heart need pump only 5.5 liters of blood each minute (which it does). Thus, when oxygen diffuses across the alveolus, less than 2% dissolves in the blood plasma and 98% combines with the hemoglobin, the principal oxygen carrier.

Hemoglobin and its association with oxygen is rather remarkable. Hemoglobin is a conjugated protein. The protein portion, **globin,** is combined with the nonprotein prosthetic

group, **heme,** a flat ringlike structure containing iron in the ferrous state (Fe^{2+}). There are four molecules of heme to each molecule of hemoglobin (Figure 14–8). Oxygen combines with the iron of hemoglobin, and for every molecule of hemoglobin four molecules of oxygen are bound: the loose combination converts the hemoglobin to oxyhemoglobin. The reversibility of oxyhemoglobin to reduced hemoglobin depends on the presence of iron in the ferrous state and intact globin. If the globin is denatured or the iron oxidized to the ferric form (Fe^{3+}), it can no longer carry oxygen.

Two factors influence the carrying capacity of hemoglobin for oxygen: hydrogen ion or carbon dioxide concentration, and temperature. With increased concentrations of carbon

[2] This assumes 100% of the oxygen in the blood is delivered to the tissues, which is not the case. Normally the tissues extract only 20%–25% of the oxygen carried to them.

dioxide, the combining capacity of the hemoglobin for oxygen is less, allowing a greater liberation of oxygen from the blood to the tissues (Figure 14–9); carbon dioxide levels are especially high where cells are most actively respiring, producing larger amounts of carbon dioxide and requiring more oxygen. A rise in temperature also decreases the oxygen-hemoglobin combination. During increased muscular activity, there occur both an elevation in temperature and an accumulation of acid (due to the production of carbon dioxide and lactic acid), which help to liberate oxygen from the hemoglobin to these tissues.

Oxygen is carried by hemoglobin, but how is the carbon dioxide carried from the cells to the lungs? A small amount (5%–10%) of the carbon dioxide is dissolved in the plasma, and a smaller quantity combines with the plasma to form carbonic acid. The remainder (90% of the carbon dioxide) is carried by the red blood cell: approximately 75% in the form of bicarbonate and 20% directly combined to the hemoglobin. Since the reaction of carbon dioxide with hemoglobin does not occur at the same site as the binding with oxygen, both carbon dioxide and oxygen can be carried simultaneously by the red blood cell. Carbon dioxide is 20 times more soluble in water than is oxygen, and because of this it rapidly diffuses from the blood into the alveolus, is unloaded into the lung, and is subsequently exhaled.

14–4 HOW DO WE BREATHE?

We have seen how the respiratory tree is organized and how oxygen and carbon dioxide are exchanged across the alveolar surface. In order to keep a sufficient supply of oxygen in the lungs and to carry away the carbon dioxide, the air in the lungs must constantly be changed. How do we ventilate the respiratory system?

The lungs lie in the chest cavity, bounded

FIGURE 14–9 The loading and unloading of oxygen by hemoglobin. In the lungs (A) hemoglobin loads up with oxygen and is about 100% saturated, but when such fully loaded hemoglobin reaches the tissues (B), where the oxygen concentration is lower, it begins to release oxygen. This release is further accentuated by the presence of carbon dioxide, acid, or high temperature (arrow). For example, under ordinary conditions, when the oxygen concentration

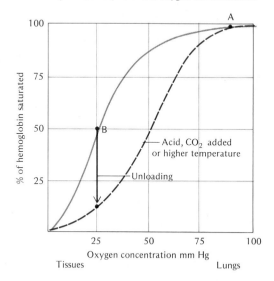

on the top and sides by the rib cage with its associated muscles, and on the bottom by a strong, dome-shaped sheet of skeletal muscle, the **diaphragm** (Figure 14–4). The pleural membranes surrounding the lungs are continuous with the diaphragm. The lungs are virtually in a closed chamber. If you watch your chest during a breathing cycle (inhale slowly and exhale slowly), you will see the chest move up and expand as you inhale and move down and shrink in size as you exhale. If you were carefully to open the chest cavity of an anesthetized animal, the breathing movements of the chest would continue; however, the lungs would not inflate, but would remain collapsed. Thus, it appears that the lungs are acting passively in response to changes in the size of the chest cavity. To keep the air moving in and out of the lungs, a bellowslike air pump is at work, operated by the muscles in the rib cage and the diaphragm. How does this air pump work? If we were to take a pair of balloons and place them in a bell jar (Figure 14–10) with the openings into the balloons connected to the

is about 25 mm Hg, the hemoglobin is only 50% saturated, but in the presence of actively metabolizing cells releasing carbon dioxide, acids, and where the temperature is elevated the saturation curve shifts so that the hemoglobin saturation is below 25%. In other words, it has given up 25% of its oxygen to the cells. The shift of the saturation curve favors oxygen release at sites requiring oxygen, and the loading of oxygen at the lungs, where it is more plentiful.

outside by a hollow glass tube, and the open bottom of the bell jar covered by a rubber sheet, we should then have a setup mechanically similar in its operation to that of the lungs. If we pull down on the rubber sheet, the pressure inside the balloons is reduced, and because the air pressure outside is greater, air enters the balloons until the pressure is equalized: the balloons inflate. Anything tending to decrease the gas pressure inside the balloons causes an intake of air. Similarly, a decrease of gas pressure in the lung causes it to fill with air. During inhalation (also called **inspiration**) the rib muscles contract, drawing the front ends of the ribs upward and outward, an action made possible because the ends of the ribs are attached in a hingelike fashion to the backbone (Figure 14–11). The diaphragm contracts and flattens, and the chest cavity is enlarged. The increased volume of the chest cavity results in a reduced gas pressure in the lungs, and air from the outside rushes in through the windpipe and bronchi to fill the air sacs: the lungs inflate.

The lungs deflate when we exhale (also called **expiration**) by the relaxation of the rib muscles and the diaphragm. The ribs are rotated downward, and the abdominal organs push upward on the diaphragm so it returns to its original dome shape. This causes the volume of the chest cavity to decrease, and the elasticity of the lungs causes them to recoil, contract, and expel the air to the outside. In contrast to inhalation, exhalation is a passive process involving lung elasticity (the lungs themselves are not contractile, having no musculature to speak of).

When we cough or sneeze the abdominal muscles contract and push against the diaphragm, decreasing the thoracic volume suddenly, and there is rapid expulsion of air from the lungs via the nose and mouth. Hiccups are the result of muscular spasms in the diaphragm that cause the chest cavity to enlarge so that

FIGURE 14–10 A mechanical model illustrating the filling and collapse of the lungs during inhalation and exhalation.

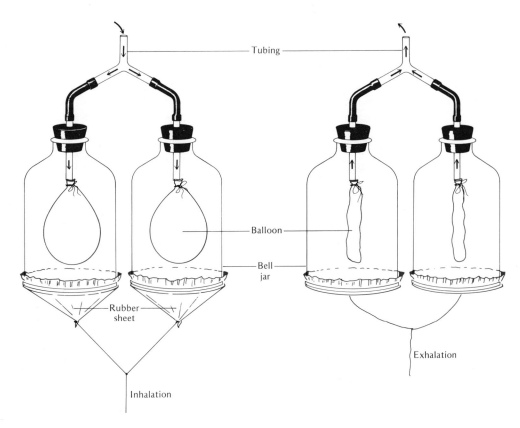

Tubing

Balloon

Bell jar

Rubber sheet

Inhalation

Exhalation

oxygen cannot be exchanged across the alveolar surface. The air in the dead space is the first to be expelled with the next expiration. The last 150 ml of air expelled from the alveoli with each breath also remains in this region, and though rich in carbon dioxide, it is the first to be drawn into the alveoli with the next breath. Thus, with each inhalation only 350 ml of new air reaches the lungs to mix with the 2.5 liters already within.

From an engineering standpoint the dead space in the respiratory system is disadvantageous, since it means that more ventilation is required to compensate for the air in the system that is not exchanged. The system is about 30% inefficient. Theoretically it would be much better if we had a double system of conduits with one-way air traffic; as it is, we have to use the same passage for both intake and exit. Such a design may have certain advantages, however. The use of a single system of conduits for incoming and outgoing air eliminates the need for extra ducts to carry each flow, and thus the lung area for exchange of gases is not reduced by additional tubing. Such a dual conduit system might result in more than a 30% inefficiency.

The air in the dead space, as we have seen, plays no role in gaseous exchange and is richer in carbon dioxide than fresh air. If the dead space of the lungs is increased, as for example by breathing through a long tube such as a garden hose, the air reaching the alveoli is further enriched in carbon dioxide and depleted in oxygen; this could be fatal. At a concentration of 7% carbon dioxide, we gasp for breath, and at 14% we cannot survive.

The compositions of alveolar and atmospheric air are shown in Table 14–2.

14–5 CONTROL OF BREATHING

Try to hold your breath for as long as you can. In a short time you *must* breathe. This simple

air tends to rush inward very quickly. Through a reflex, the epiglottis is quickly closed, the inrushing air sharply strikes the epiglottis, and the characteristic sound of the hiccups is produced. Persistent hiccups can be stopped by inhalation of 5%–7% carbon dioxide (thus breathing into a paper bag often helps, since one is rebreathing the same air, which consequently has a higher carbon dioxide content). If the condition becomes pathological, the phrenic nerve, which runs to the diaphragm and initiates the spasm, may have to be cut.

Men and women are unalike even in their breathing patterns. Men usually breathe with an action involving the diaphragm and abdomen, called **diaphragmatic breathing;** women elevate the ribs with movement of the chest, called **costal breathing.** (The costal muscles are the muscles in the ribs that expand the thorax.) Diaphragmatic breathing is deep and is also seen during sleep. Costal breathing is shallow. Labored breathing, such as seen in runners at the end of a race, involves both diaphragmatic and costal breathing.

With each breath we take in 500 ml of air (Figure 14–12), but only 350 ml of that air actually reaches the alveoli, since the remainder is trapped in the upper respiratory passages (trachea, bronchi, and bronchioles). This 150 ml of trapped air is in dead space, and its

FIGURE 14–11 Breathing movements. (a) Arrangement of ribs and diaphragm produces an enlarged or reduced chest cavity. (b) A mechanical representation of how the ribs are attached to the backbone. Upward movement of the rib cage increases the size of the chest cavity, and downward movement reduces it.

exercise clearly demonstrates that gas exchange (ventilation) is so vital to our being that it is not primarily under voluntary control and there exist safeguards to ensure maximum performance under continual operation.

Nervous control

In 1811 the Frenchman César Legallois found that if the cerebrum, cerebellum, and part of the upper brain stem (medulla) (Figure 17–14)[3] were removed from a rabbit, the animal still continued to breathe in a rhythmic pattern. However, if a small region in the lower medulla was damaged or cut out, breathing ceased. Today, it is still generally accepted that the rhythmic pattern of respiration is controlled by a group of interconnected nerves (neurons) in the lower medulla, the **respiratory center.** There are two self-governing regions in the respiratory center, one expiratory and the other inspiratory, and nerve signals shuttle back and forth between the neurons in these regions. When the inspiratory neurons send impulses to the muscles involved in inspiration (diaphragm and intercostal, or rib, muscles), they simultaneously send inhibitory impulses to the expiratory center to reduce or stop expiratory activity. Then, in a matter of seconds the expiratory center becomes activated, signals are sent to the abdominal muscles, and inhibitory impulses go out to the inspiratory center. Thus, there is mutual inhibition of the two centers and oscillation of neuron activity in each (Figure 14–13).

Although the respiratory center of the medulla is the primary region for control of respiration, it is not the only part of the brain that can influence respiraton. A reflex is involved in preventing overinflation of the lungs. When

---

[3] The nervous system and brain are treated in detail in Chapter 17.

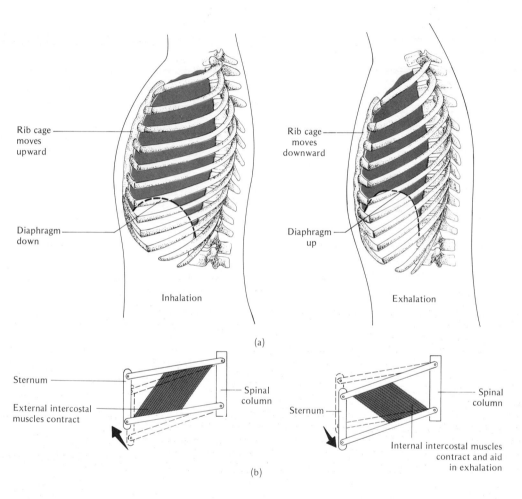

(a)

(b)

TABLE 14–2 Gas composition of the air we inspire, the air we expire, and the air in the alveoli

Gas	Inspired air		Expired air		Alveolar air	
	Partial pressure (mm Hg)	%	Partial pressure (mm Hg)	%	Partial pressure (mm Hg)	%
O_2 (oxygen)	157.0	20.71	110	14.6	105	13.2
CO_2 (carbon dioxide)	0.3	0.04	30	4.0	40	5.3
H_2O (water)	9.5	1.25	45	5.9	45	5.9
N_2 (nitrogen)	593.0	78.00	574	75.5	574	75.6

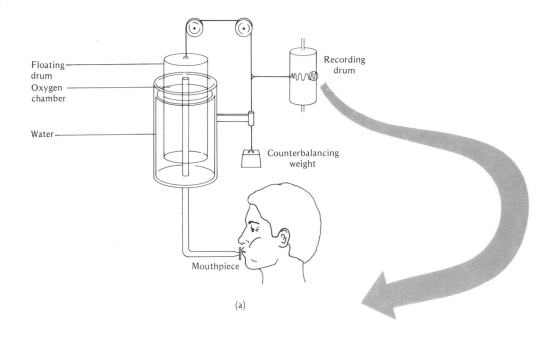

(a)

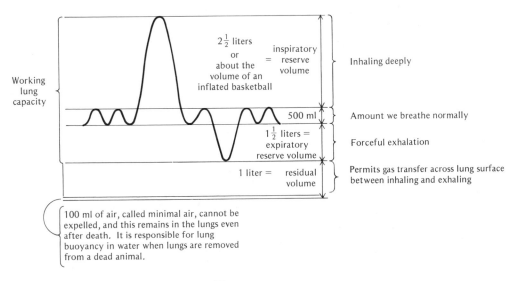

(b)

the lungs are expanded, nerve endings within the walls are stimulated, and impulses pass from the lungs via the vagus nerve to the medulla, where they inhibit inspiration and excite expiration. This prevents excessive lung distension, so that the lungs are not ripped like an overinflated balloon and do not take up too much blood, depriving the left side of the heart of its supply. Thus, the respiratory center has a built-in safety factor with two feedback loops: one loop is intrinsic in the medulla and the other is extrinsic through the reflex system (Figure 14–14). This double control system is characteristic of many important body functions and is essential for a system that must operate without failure for 70 years or more.

The higher centers of the brain (the cerebral cortex) also play a role in respiratory control, and by sending impulses from these centers to the medulla you can voluntarily hold your breath. Such voluntary control of respiration can be maintained for only about 40 seconds, and then the individual breathes even against his own will. The threat of a small child holding his breath until he dies is just a threat; it cannot be done.

The oscillating mechanism of the respiratory center can, on occasion, fail. Most commonly this is due to a cerebral concussion or an abnormal excessive pressure on the medulla. The pressure causes blood vessels that supply the neurons of the respiratory center to collapse; deprived of oxygen, the neurons cease to function and the respiratory rhythm ceases.

The most common way in which the respiratory center becomes inactivated is by the action of drugs that induce sleep—chloroform, ether, morphine, barbiturates, or other central-nervous-system depressants in excessive and toxic amounts. These substances interfere with oxygen utilization by the cells in the respiratory center, impair nerve impulse transmission, and depress blood flow into this region of the brain.

FIGURE 14–13 The respiratory center of man. (a) Location in the brain. (b) Diagrammatic representation of the oscillation between inspiratory center and expiratory center neurons.

FIGURE 14–14 Extrinsic feedback loops acting on the respiratory center.

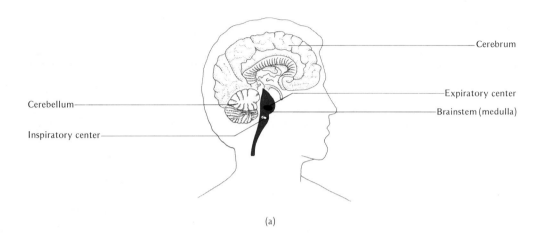

Cerebrum

Expiratory center

Brainstem (medulla)

Cerebellum

Inspiratory center

(a)

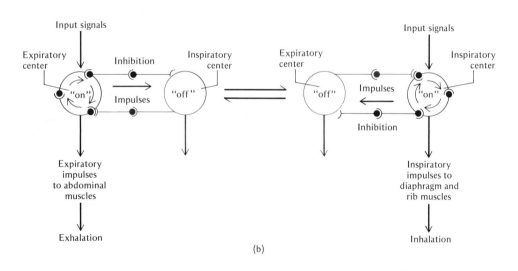

Input signals

Expiratory center

Inhibition

Inspiratory center

"on"

Impulses

"off"

Expiratory impulses to abdominal muscles

Exhalation

Input signals

Expiratory center

Inspiratory center

"off"

Impulses

"on"

Inhibition

Inspiratory impulses to diaphragm and rib muscles

Inhalation

(b)

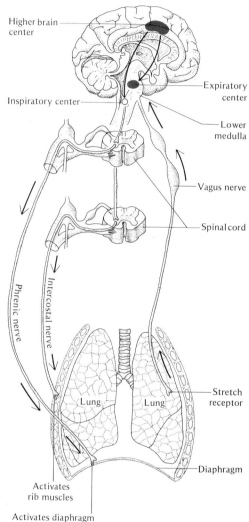

Higher brain center

Inspiratory center

Expiratory center

Lower medulla

Vagus nerve

Spinal cord

Phrenic nerve

Intercostal nerve

Lung

Lung

Stretch receptor

Diaphragm

Activates rib muscles

Activates diaphragm

If such damage to the respiratory center is irreversible, death results. Unfortunately, there are few chemical agents that can excite the respiratory center, so a chemical antidote for an overdose of a sleep preparation is not available; in most cases prolonged artificial respiration is the only means of counteracting severe respiratory depression.

Chemical control

The nervous signals within the respiratory centers of inspiration and expiration oscillate back and forth to produce a resting rhythm of 10–14 breaths per minute, but a short run or sudden fright will convince you that the depth and rate of breathing can be quite variable. During exercise, the breathing rate may be 75 breaths per minute, and the volume of air entering and leaving the lungs goes from about 6 liters per minute to between 80 and 120 liters per minute. The maximum possible rate of breathing is about 300 breaths per minute with a flow of 150-200 liters of air. We obviously have a great reserve capacity for ventilation. When for some reason (for example, increased exercise) the body needs more oxygen or more carbon dioxide must be expelled, the ventilation rate must be increased. How does the respiratory center know what the body's oxygen requirements are?

At first glance it might seem sensible for the amount of oxygen in the blood to influence the respiratory center directly, so that when the blood oxygen level goes down, ventilation increases to compensate; however, this turns out not to be the case. If a person is allowed to breathe and rebreathe air in a closed container, the oxygen in the air gradually decreases, the carbon dioxide content of the air increases, and the respiratory rate increases. If, however, sodium hydroxide or some other chemical is placed in the container to absorb the carbon dioxide so its content does not increase but the oxygen content still decreases, the breathing rate is only slightly increased. Obviously, increased levels of carbon dioxide rather than decreased levels of oxygen stimulate the respiratory center to activity. Convincing proof of this can be obtained by allowing an individual to breathe a gas mixture containing normal amounts of oxygen but increased amounts of carbon dioxide; the respiratory rate is increased even though oxygen levels are normal.

Emotional state, excitement, and exercise increase the metabolic rate, thus increasing carbon dioxide levels in the blood; the rate and depth of breathing are adjusted to meet the body's demand for more oxygen and to prevent accumulation of carbon dioxide. The increased carbon dioxide in the blood travels to the brain and excites the neurons of the inspiratory center; there results an increased volley of impulses to the muscles involved in inspiration, lung expansion becomes greater, and the frequency of the breathing cycle is increased. When the carbon dioxide level returns to normal, the respiratory center returns to its normal oscillatory rhythm.

Carbon dioxide is a poison to most cells, but its complete removal from the blood would be fatal, since it triggers the respiratory center to activity. Rapid, deep breathing **(hyperventilation),** voluntarily undertaken, can so severely deplete the blood of carbon dioxide that the respiratory center may become inactive; dizziness, unconsciousness, and cessation of breathing may result. If the cessation of breathing **(apnea)** lasts long enough, death may ensue.

At birth we take our first breath of fresh air. As the newborn baby is separated from the placenta, its oxygen supply is cut off, the level of carbon dioxide in its blood increases, and this stimulates the respiratory center to send nerve impulses to the muscles of the rib cage and the diaphragm; these contract, and the first breath is taken. If the newborn baby has difficulty in taking its first breath, air containing 10% carbon dioxide may be administered via the mouth, and this usually triggers the respiratory center to become active. Why does the obstetrician slap a newborn baby, so that all of us enter the world crying? Painful stimuli almost anywhere in the body trigger a reflex reaction of the respiratory center, and a good slap on the infant's bottom starts respiration. Other shocks such as a cold shower also trigger receptors to send impulses to the respiratory center and stimulate it; this gives us a breath of fresh air and thus we find it invigorating.

Although the respiratory centers of the brain monitor carbon dioxide levels in the blood and thus regulate ventilation, there is an independent, or backup, system that registers changes in the amount of oxygen in the blood. Receptors chemically sensitive to oxygen are located at the branching of the carotid arteries in the neck and the aortic arch; they are connected to the medulla by nerves (Figure 14–15). If the oxygen level in the blood falls, impulses are sent out from these receptors to stimulate the inspiratory center of the medulla, and ventilation is increased.

Just as there are receptors that activate the ventilation system, so there are receptors that produce a temporary inhibition of breathing. When an irritating gas such as ammonia or acid fumes passes down the respiratory tree, we cannot inhale even if we want to. By stimulating receptors in the larynx, these noxious substances produce a volley of impulses to the respiratory center to inhibit breathing, and we catch our breath. This is a valuable protective device, since it prevents many harmful substances from entering the lungs.

FIGURE 14–15 Chemical receptors acting on the respiratory center.

14–6 LOW PRESSURE: HIGH PRESSURE

Low pressure: high altitude

The traveler who desires to visit Machu Picchu, the lost city of the Incas, must suffer a little to view the prehistoric site of an Andean civilization. This ancient city of the Incas lies nestled high in the Andes mountains, and to reach it one must first fly to Cuzco, Peru, 12,700 ft above sea level. Breathing the air leaves one exhilarated, talkative, and euphoric. These pleasant feelings, however, do not last; one often feels nauseous, a violent headache develops, shortness of breath occurs with the slightest exertion. Some individuals react even more unfavorably: inability to think clearly, irritability, irrationality, speech difficulty, and coldness in the feet. These are the symptoms of mountain sickness, and their only cure is time: in the course of several days these symptoms diminish and disappear.

Why does mountain sickness occur? The basis for such symptoms is a simple fact: gases are compressible. The atmosphere surrounding the earth varies in its air (barometric) pressure with altitude: the higher the altitude, the lower the barometric pressure; in other words the higher the altitude, the fewer molecules of gas there are per unit volume (Figure 14–17). The amount of oxygen in the air is approximately 20% but at higher altitudes, where there is less air, there is correspondingly less oxygen available, and consequently less oxygen is absorbed by the blood. Because of this oxygen deficiency, and since breathing does not occur more quickly, the tissues become starved of oxygen. The brain is particularly susceptible, and symptoms of blurred vision and mental confusion are among the first to appear (up to 15,000 ft). The higher one goes, the greater the problem: at 18,000–24,000 ft the individual gets slaphappy, irrational, and drowsy. The

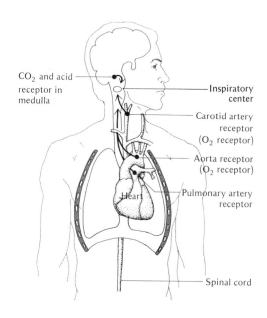

CO₂ and acid receptor in medulla — Inspiratory center — Carotid artery receptor (O₂ receptor) — Aorta receptor (O₂ receptor) — Heart — Pulmonary artery receptor — Spinal cord

time required at various altitudes for oxygen deficiency to cause collapse or coma in a person unaccustomed to high altitudes is shown in Figure 14–18.

Fliers and mountain climbers use oxygen masks to ascend to greater and greater heights. In modern aircraft the cabin is pressurized to about 5,000-ft simulated altitude. The big jets fly at 30,000 ft, yet the passengers and crew have no mountain sickness since the air breathed is at the barometric pressure of 5,000 ft. If the cabin were to leak at that altitude, there would be only a minute of consciousness in which to begin breathing pure oxygen from emergency sources. Most aircraft have automatic devices to provide oxygen in the event cabin pressure is lost.

High pressure: deep-sea diving

There is a beauty beneath the sea that lures

BOX 14B   Why can't we breathe underwater

The body requires about 300 ml of oxygen per minute. Why can't we swim around in the water and obtain oxygen in much the same fashion as we swim in a sea of air? Seawater, or any kind of well-aerated water, contains about 0.3 ml of oxygen per 100 ml of fluid. Each breath of air we take in (about 500 ml) contains about 100 ml of oxygen. Thus in order to process enough water to satisfy our resting need for oxygen for a single minute, we should have to pump 100 liters (about 25 gallons) of water in and out of our lungs every minute. This is impossible for a number of reasons. Our lungs operate on a back-and-forth tidal system, and we move a relatively small mass of respiratory medium (air). If water were to enter our lungs laden with oxygen, it would be mixed with oxygen-deficient water. If the flows were not kept separate we should get very little out of the incoming water. There is another drawback to ventilating our lungs with water as the respiratory medium; water is heavy, and to move it into and out of our lungs (that is, an acceleration-deceleration and reversal of flow of sufficient water) would be very costly in terms of energy expended. To perform such a feat we should require a set of muscles 20 times more effective than the heart (which pumps about 5 liters per minute). Another problem posed in using a lung underwater is that the diffusion path of the gas from

FIGURE 14–16. Fish gills. (a) Diagram of the movement of water through the gill chamber of a fish. (b) Detail of water movement across the gill, showing how the countercurrent flow enriches the oxygen content of the blood.

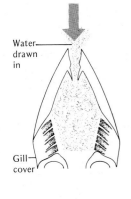

Water drawn in

Gill cover

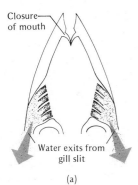

Closure of mouth

Water exits from gill slit

(a)

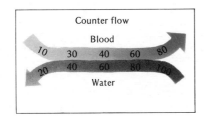

Counter flow

Blood

| 10 | 30 | 40 | 60 | 80 |

| 20 | 40 | 60 | 80 | 100 |

Water

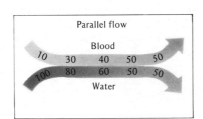

Parallel flow

Blood

| 10 | 30 | 40 | 50 | 50 |

| 100 | 80 | 60 | 50 | 50 |

Water

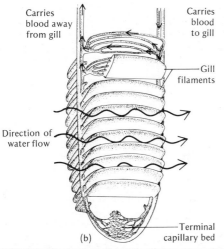

Carries blood away from gill

Carries blood to gill

Gill filaments

Direction of water flow

Terminal capillary bed

(b)

the water to alveolar surface would be very slow because of the slow rate of diffusion of oxygen in water. The rate of diffusion of oxygen in air is 300,000 times faster than it is in water. Clearly, for a warm-blooded animal with a high metabolic rate and a high oxygen demand, air breathing is the preferred mode of respiration. It has one major disadvantage: the danger of desiccation. This is combated by continual production of mucus by the cells lining the lung.

Our lungs are designed to breathe air, not water, and attempts to convert the air-breathing lung to a water-breathing lung have disastrous consequences: drowning. How does a fish breathe underwater? In a "water-breathing" gill the anatomy of the respiratory system is very different from that found in a lung. Gill-bearing animals do not use a tidal, back-and-forth type of respiration, because to move water in, stop it, and then push it out is energetically uneconomical. Fish and other gill-bearing animals have a one-way flow system. Water enters a fish via the mouth, and is drawn into the pharynx with the gill covers (opercula) closed (Figure 14-16a). Then the mouth closes, the opercula move outward, this enlarges the gill cavity, and water is drawn past the gills and exits via the gill slits. A fish does not "rebreathe" its respiratory medium, but constantly takes in new water as it swims along. The direction of water flow over the gill filaments and across the thin-walled gill lamellae is opposite to the flow of blood through the gill, and such a countercurrent flow permits a more efficient and complete oxygenation of the blood than would occur if the currents flowed parallel to one another in the same direction (Figure 14-16b).

If we were to breathe underwater, we should need gills and a redesigned ventilation system, but gills would be useless on land, for they would quickly dry out; the soft gill filaments would collapse and stick to one another, and air could not reach most of the gill surface. A fish out of water or a man underwater cannot respire without some kind of accessory equipment.

men to dive deeper and deeper. There is, however, danger down below, and not only from sharks and other denizens of the deep. As one dives into the ocean, the weight of water pressing down on the body increases with increasing depth, and this causes the barometric pressure of the air in the lungs to increase too—an effect that can be mimicked in a compression chamber. The amount of gas that can pass from the alveolus into the blood, where it may become dissolved in the body fluids, also increases; this can cause serious distrubances in body function.

A diver can breathe compressed air to 200 ft below the sea without any problem of excess absorption, but deeper than this oxygen intoxication results. At pressures encountered below 200 ft (seven times the atmospheric pressure at sea level), oxygen is transported into the blood and body fluids and there it dissolves. The brain is most sensitive to this oxygen excess; metabolism is deranged and nervous twitching, convulsions, and coma result. The danger of oxygen poisoning can be avoided by supplying less and less oxygen in the breathing mixture as the diver descends to greater depths.

Nitrogen, although inert, also dissolves in the body fluids under high pressures. Nitrogen so dissolved may form bubbles as the diver surfaces. At a depth of 200 ft the pressure is four times that at sea level, and if an individual is rapidly brought to the surface, the gases expand in the fluids because the pressure outside the body is now lower than the pressure of the gases dissolved in the fluids. Bubbles of gas form in the cells, blood, and spinal fluid, causing **decompression sickness** (also called the bends, caisson disease, and diver paralysis). The bubbles can plug blood vessels, rupture fiber pathways in the nervous system, and lead to mental disorder or permanent paralysis (see "Tragedy Under the Sea" on page 280). There may be severe pain from nervous-system dam-

FIGURE 14–17 The air (barometric) pressure varies with altitude. There are fewer gas molecules per unit volume at a high altitude than at a lower one. Thus, the air thins out at high altitudes. A barometer is a long glass tube, open at one end, filled with mercury, and placed with the open end in a dish of mercury. At sea level the height of the mercury (Hg) in the tube is 760 mm, and at higher altitudes the air pressure is less and the column height is shorter.

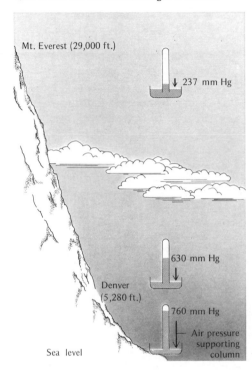

age and bubble formation in the joints. Decompression sickness can be prevented by allowing a diver to come to the surface slowly or by controlling decompression in a decompression chamber so that release of gas is slow.

SUMMARY

1. Respiration is the exchange of gases (CO_2 and oxygen) between an organism and its environment. Breathing or ventilation is the direct physical exchange of gases at the respiratory surfaces (gills or lungs). Internal respiration involves gas exchange between the blood and tissues, and cellular respiration involves the reaction of oxygen with fuel molecules to produce energy, CO_2, and water.

FIGURE 14–18 The time required at different altitudes for oxygen deficiency to cause collapse or coma. (From *Principles and Practice of Aviation Medicine,* Harry G. Armstrong, Baltimore, The Williams and Wilkins Company, 1952.)

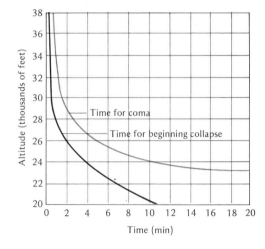

2. Gas exchange between organisms and the environment takes place by diffusion across a large moist surface. In land vertebrates, the exchange surfaces are the extensive lungs.

3. Ventilation of the lungs occurs via the respiratory tree: air passes via the nasal passages and sinuses to the pharynx and larynx and via the glottis (guarded by the epiglottis) to the trachea, bronchi, bronchioles, and the lungs. During its passage, the air is filtered, conditioned, and cleaned.

4. Smoking or breathing polluted air can cause emphysema.

5. The lungs consist of thin-walled alveoli. Oxygen from the incoming air diffuses across the alveoli to the blood where it is lower in concentration. Carbon dioxide diffuses in the opposite direction, from blood where it is high in concentration to the lungs where it is low in concentration.

6. In the blood, oxygen combines with the iron in the ring-shaped heme of hemoglobin to form an unstable complex: oxyhemoglobin. Carried from the lungs to the heart, and then to the tissues where the oxygen pressure is low, the oxyhemoglobin gives up its oxygen to the

tissues. Higher carbon dioxide concentration in the tissues, and rising temperatures in actively respiring cells such as muscle, lower the combining capacity of hemoglobin for oxygen and assist its release.

7. Carbon dioxide is carried by the red blood cell as bicarbonate, combined with hemoglobin, and dissolved in the blood.

8. During inspiration, the rib muscles and diaphragm contract and enlarge the chest cavity; the lungs inflate in response to reduced air pressure. During expiration, the rib muscles and diaphragm relax, the volume of the chest cavity decreases, the lungs recoil and contract, and the inhaled air is expelled.

9. Men exhibit deep diaphragmatic breathing, women show shallower costal breathing.

10. About 350 ml of air is exchanged with each breath, 150 ml is in dead space.

11. Nervous control of breathing is centered in the respiratory center of the lower medulla (brainstem). Stimulatory and inhibitory impulses oscillate between the inspiratory and expiratory centers. Overinflation of the lungs is prevented by a reflex mediated by the the vagus nerve. Voluntary control of breathing is mediated by the cerebral cortex, thus enabling an individual to hold his or her breath. Nervous system depressants can inactivate the respiratory center causing possible death.

12. CO_2 in the blood stimulates the respiratory center. Breathing rate increases when CO_2 blood levels increase. Depletion of CO_2 blood levels by hyperventilation may inactivate the respiratory center, causing apnea and possible death. Increasing CO_2 blood levels in the newborn trigger the respiratory center to activity. Painful stimuli also trigger the respiratory reflex.

13. Depressed blood oxygen levels cause stimulation of the inspiratory center via impulses from carotid receptors that monitor blood oxygen. Laryngeal receptors inhibit breathing when stimulated by irritating gases during inspiration.

14. Mountain sickness is caused by low oxygen levels at high altitude.

15. Oxygen poisoning can result from breathing compressed air at great ocean depths; less oxygen must be supplied as divers descend deeper and are subjected to increasing pressure. Decompression sickness may occur if a diver surfaces too rapidly: excess nitrogen dissolved in the blood while under increased pressure may form bubbles in the blood and tissues if the pressure is released too quickly.

16. We cannot breathe underwater like fish because we cannot move enough water in and out of our respiratory tree to allow complete oxygenation of the blood and efficient gas exchange. Fish have an efficient one-way flow of water with a countercurrent flow of blood in their gills.

KEY WORDS

respiration
breathing or ventilation
internal respiration
cellular respiration
atmosphere
gills
lungs
tracheal tubes
respiratory tree
sinuses
pharynx
larynx
glottis
epiglottis
vocal cords
trachea
bronchi
bronchioles
alveoli
pleural membrane
smog
pulmonary osteoarthropathy
oxyhemoglobin
globin
heme
diaphragm
inspiration
expiration
diaphragmatic breathing
costal breathing
respiratory center
hyperventilation
apnea
decompression sickness

TOPICS FOR REVIEW AND DISCUSSION

1. **Why can't humans breathe underwater?**
2. **What are the three types of respiration?**
3. **How are respiration and breathing related to one another?**
4. **Give the structure and function of the various components of the respiratory tree.**
5. **How do smog and smoking affect lung function?**
6. **What are some of the nervous and chemical controls involved in breathing?**
7. **How is gas exchange accomplished in small and in large animals?**
8. **Draw a mechanical model that can show the manner of expansion and contraction of the lungs. Describe its operation.**
9. **What is the cause of bends or caisson disease?**
10. **Discuss the statement: speech is a by-product of breathing.**

Kidney in a suitcase

For other vacationers, the trip would have been routine. But when Josephine Berman, 43, and her husband toured the Grand Canyon, stopped at Las Vegas and then visited San Francisco last summer, it was something of a medical miracle. For eleven years the pretty Brooklyn housewife has suffered from chronic kidney disease. Like 24,000 other similarly afflicted Americans, she could never go far from the massive dialysis machines that purge her blood of the toxic wastes her kidneys are no longer able to remove. Yet during her 16-day trip, she shunned kidney centers entirely. Her unexpected freedom was the result of a remarkable new device: a portable mechanical kidney so compact it is built into a small metal valise.

"The idea came to me out of simple frustration" says Dr. Eli A. Friedman, inventor of the suitcase kidney. Friedman, director of the renal diseases section at New York's Downstate Medical Center, had planned to take 25 kidney patients on a European holiday, dialyzing them at stopovers en route. But at the last minute, medical authorities in Copenhagen concluded that they did not have enough dialysis machines to handle so many additional patients. Forced to cancel the trip, Friedman resolved to build a dialysis machine that kidney patients could carry on their travels and operate by themselves.

Suitcase kidneys, he feels, will be a boon not only to dialysis patients who want to travel but also to smaller rural hospitals which could use them in emergencies. The little machines could also be used for happier occasions: another of Friedman's patients took a suitcase kidney on his honeymoon.

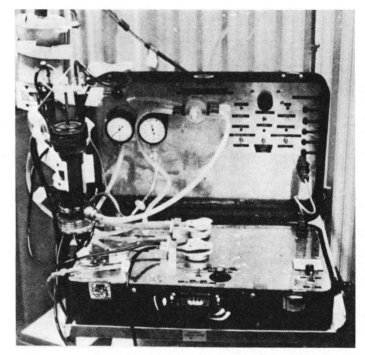

What is a mechanical kidney?
What is a real kidney, and how does it work?

15

CHEMICAL BALANCE: THE KIDNEY

15–1 WHAT DOES THE KIDNEY DO?

The lungs serve to maintain the composition of the extracellular fluid with respect to oxygen and carbon dioxide, and with this their duty ends. The responsibility for maintaining the composition of this fluid in respect to other constituents devolves on the kidneys. It is no exaggeration to say that the composition of the body fluids is determined not by what the mouth takes in but by what the kidneys keep; they are the master chemists of our internal environment, which, so to speak, they manufacture in reverse by working it over completely some fifteen times a day. When, among other duties, they excrete the ashes of our body fires, or remove from the blood the infinite variety of foreign substances that are constantly being absorbed from our indiscriminate gastrointestinal tracts, these excretory operations are incidental to the major task of keeping our internal environment in an ideal, balanced state. Our bones, muscles, glands, even our brains, are called upon to do only one kind of physiological work, but our kidneys are called upon to perform an innumerable variety of operations. Bones can break, muscles can atrophy, glands can loaf, even the brain can go to sleep, and not endanger our survival; but should the kidneys fail in their task neither bone, muscle, gland nor brain could carry on.

Recognizing that we have the kind of internal environment we have because we have the kidneys that we have, we must acknowledge that our kidneys constitute the major foundation of our physiological freedom. Only because they work the way they do has it become possible for us to have bones, muscles, glands, and brains. Superficially, it might be said that the function of the kidneys is to make urine; but in a more considered view one can say that the kidneys make the stuff of philosophy itself.

HOMER W. SMITH, *From Fish to Philosopher*, Boston, Little, Brown and Company, 1953.

Perhaps Professor Smith exaggerated somewhat in stating that "kidneys make the stuff of philosophy itself", but as we shall soon see, his statement is correct in recognizing the central role of the kidney as a regulator of our internal environment.

As the blood flows through the body, its composition is in a state of flux: cellular ashes such as carbon dioxide and urea are dumped into it, and oxygen, regulatory materials, fuels, and building materials are removed. Many organs of the body play a role in this dynamic exchange with blood. The digestive system serves to replenish the supply of raw materials. The respiratory system functions in the removal of carbon dioxide and in replenishing the supply of oxygen. The liver aids in the chemical processing of the blood; it breaks down hemoglobin, regulates the availability of sugar, and forms urea. However, most chemical exchanges with the bloodstream take place in the kidneys. The **kidneys** are primarily concerned with the disposal of liquid sewage and constitute the principal organs of excretion.[1] They are vital to our economy, and although these two organs are only about the size of a clenched fist and account for less than 0.5% of the total body weight, every day they process more than 1,700 qt of blood; with every beat of the heart, 20% of the heart's blood volume flows through the kidneys. The kidneys filter and chemically process this blood and produce urine. The kidney is the body's chemist par excellence, regulating the chemical composition of the body fluids and in effect maintaining the constancy of the internal environment.

What is the internal environment? The human body consists mainly of water; roughly 60% of the body weight is water. Over half is contained in the cells themselves, and the remainder is a dilute salt solution that surrounds and bathes all the cells (Figure 15–1). The liquid portion of the blood (plasma) is part of this bath; as the blood courses through the tiny

---

[1] Excretion involves ridding the body of waste metabolites that it has itself produced, as opposed to defecation, which involves ridding the body of undigested food materials it cannot use.

BOX 15A What is urea?

Much of the food we eat contains proteins. Proteins are nitrogenous compounds, and when their building blocks, amino acids, are used as fuel by cells, the metabolic ashes are carbon dioxide and water plus nitrogen-containing substances. The waste nitrogen must be eliminated from the body, but it cannot be liberated as a gas because too much energy is required to perform such a reaction. Instead, most organisms produce ammonia, urea, or uric acid. Ammonia is a highly toxic gas and is extremely soluble in water. A thousandth of a milligram of ammonia in a liter of blood is sufficient to kill a man. Therefore, ammonia must be eliminated quickly, and aquatic animals flush the gas into the water as quickly as it is formed. Land creatures cannot form ammonia as a waste product because they (and we) must conserve water; instead they convert it to nontoxic urea in the liver. The **renal portal vein** carries urea-laden blood directly from the liver to the kidneys where it is eliminated into the urine. Urea can be tolerated in quantities 100,000 times greater than the lethal dose of ammonia. Therefore, it takes only 1/100,000th the amount of water that would be required for elimination of ammonia to flush out the same amount of urea. Elimination of urea conserves water and in essence permits life on land.

The actual quantity of urea we excrete each day is quite variable and depends on the quantity of protein in our diet, since it is from protein that urea is derived. If we eat a protein-rich diet, we may eliminate about 1 oz of urea each day. The nitrogenous portion of the urea is derived by **deamination** (removal of the amino group) of amino acids. Deamination results in the formation of ammonia:

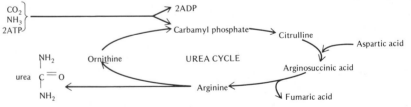

Ammonia formed in the tissues passes to the blood and is carried to the liver. By a series of enzyme-catalyzed reactions (the **urea cycle**), the highly toxic ammonia is combined with carbon dioxide to produce urea, a much less poisonous material.

The urea leaves the liver, enters the blood, is carried to the kidney by the renal portal vein, and eliminated.

FIGURE 15–1 The distribution of fluid in the body.

FIGURE 15–2 The kidney and associated structures.

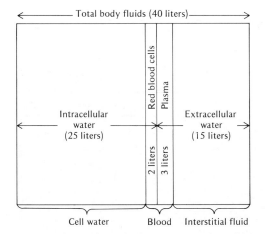

capillaries that pass close to every cell, some of the salty fluid plasma diffuses out, carrying nutrients into the watery solution that bathes the cells (interstitial fluid). Waste products, on the other hand, leave the cells and pass into the blood plasma. In this way, materials essential for cell maintenance, growth, and reproduction are supplied, and poisonous wastes are discarded. The composition of this extracellular solution is quite important, for cells can function properly only if it is chemically balanced; that is, there must be proper proportions of ingredients. Accumulated poisonous wastes such as urea (Box 15A) must be discarded; the ingredients essential to life, such as salts, sugars, amino acids, and other nutrients that are depleted by cell utilization must be retained; acidity-alkalinity (pH) must be balanced; and the volume of the fluid itself must remain fairly constant.

Our kidneys maintain salt balance, excrete nitrogenous wastes such as urea, regulate the acidity of the body fluids, and control the volume of the cellular bathing solution. In short, they critically control the concentrations of most of the substances in the sea within us.

15–2 STRUCTURE AND FUNCTION OF THE KIDNEY

What is a kidney? The human kidney is a dark-red, bean-shaped organ about 4 in. long, 2.5 in. wide, and 1 in. thick. It weighs about 0.5 lb. Each of us normally has a pair of kidneys attached to either side of the backbone above the small of the back (lower abdominal wall) just behind the stomach and liver (Figure 15–2). Attached to each kidney are three tubes: (1) the **renal artery** (*ren*—kidney; L.), which conducts blood from the aorta to the kidney; (2) the **renal vein,** which carries blood from the kidney to a large vein going to the heart; and (3) a **ureter,** which carries urine from the kidney to the bladder (Figure 15–2). The kidneys receive blood, process it, and excrete urine, and the processed blood passes on its way.

How is the kidney organized to process blood and form urine? Each kidney contains about 1 million miniaturized chemical filtration units called **nephrons.** A nephron consists of a twisted hollow tube, closed at one end and open at the other, and a network of associated blood vessels (Figure 15–3). Let us consider the blood supply first. The renal artery that enters the kidney branches to form a series of arterioles, and these distribute blood to each nephron as a spherical cluster of capillaries, the **glomerulus** (little ball, L.) (Figure 15–4). There filtration of the blood takes place. The blood that leaves the glomerulus then flows through another capillary bed surrounding the tubular portion of the nephron. These blood vessels then converge and form a system of veins that merge to form the large renal vein, by which blood leaves the kidney (Figure 15–3).

Now let us look at the tubules of the nephron. The blind end of the kidney tubule is greatly expanded to form a double-walled funnel or capsule, called **Bowman's capsule** (Figure 15–4). Each of the two walls of Bowman's

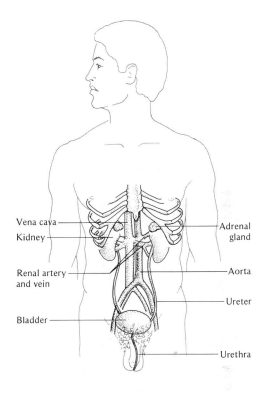

Vena cava
Kidney
Renal artery and vein
Bladder
Adrenal gland
Aorta
Ureter
Urethra

capsule is separated by a fluid-filled space and consists of a single layer of flattened epithelial cells. The invaginated, or cuplike, portion of the capsule almost completely surrounds the coiled ball of glomerular capillaries.

The glomerulus-capsule relationship can be visualized by thinking of a fist that is pushed into an inflated rubber balloon (Figure 15–5). The layer of rubber directly overlying the fist represents the inner surface of the invagination of Bowman's capsule, and the point where the wrist makes contact with the balloon is the opening into the capsule. The blood supply to Bowman's capsule enters at the position of the wrist, and the clenched fist represents the glomerulus that nestles in the capsule. The outer surface of the balloon represents the outer

FIGURE 15–3 (left) The entire nephron with Bowman's capsule cut away.

FIGURE15–4 (right) A scanning electron micrograph of the nephron. (Courtesy of Peter M. Andrews and Keith R. Porter.)

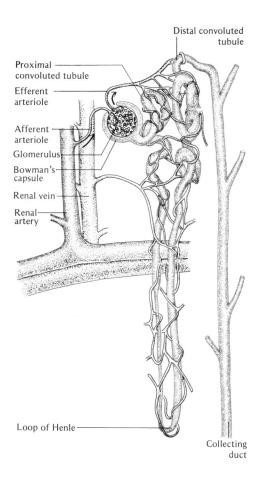

Distal convoluted tubule

Proximal convoluted tubule

Efferent arteriole

Afferent arteriole

Glomerulus

Bowman's capsule

Renal vein

Renal artery

Loop of Henle

Collecting duct

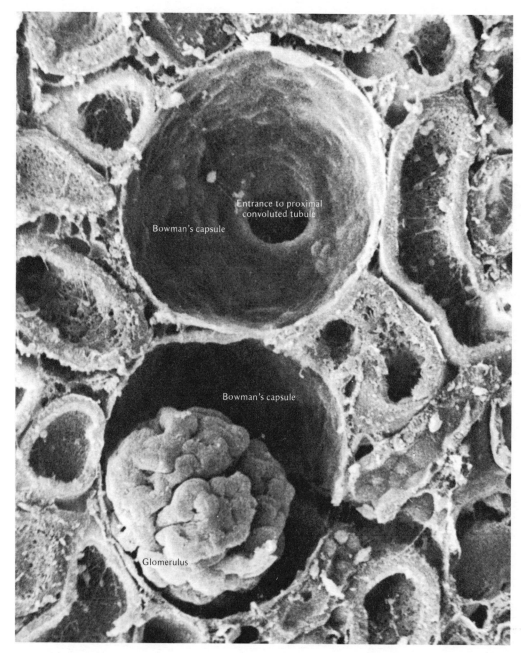

Bowman's capsule

Entrance to proximal convoluted tubule

Bowman's capsule

Glomerulus

surface of the capsule and, in the nephron, is continuous with the tubule leading from it. The tubule leading from the capsule twists in a complicated manner in the vicinity of the glomerulus; then it takes a straight course into the interior of the kidney, where it makes a hairpin turn (the **loop of Henle**); the tubule then runs back to the region of the glomerulus, winds about in a second series of twists, and finally discharges into a **urine-collecting duct** that later connects with the main ureter

FIGURE 15–5 The relationship of the glomerulus to Bowman's capsule visualized using a fist and a rubber balloon.

FIGURE 15–6 The kidney sliced in half to show cortex and medulla as well as the position of the nephron.

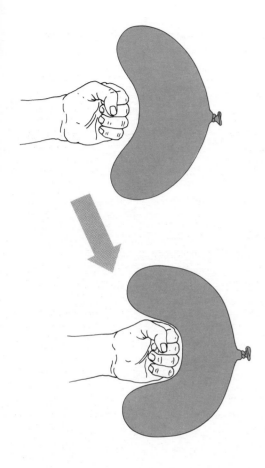

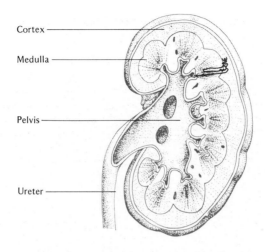

Cortex

Medulla

Pelvis

Ureter

TABLE 15–1 Comparison of the composition of blood, glomerular filtrate, and urine

Component	Blood plasma (gm/liter)	Glomerular filtrate (gm/liter)	Urine (gm/liter)
Glucose	1.0	1.0	0
Sodium	3.0	3.0	3.0
Chloride	3.5	3.5	6.0
Potassium	0.15	0.15	1.5
Uric acid*	0.03	0.03	0.3
Urea	0.25	0.25	20.0
Sulfate	1.0	1.0	1.0
Creatinine†	0.15	0.15	0.7
Phosphate	0.03	0.03	1.0
Amino acids	0.3	0.3	0
Protein	70.0	0.2	0

Blood passing through kidney: 500 gals/day
Glomerular filtrate formed: 50 gals/day
Glomerular filtrate reabsorbed: 49.5–49.75 gals/day
Urine volume excreted: 0.25–0.50 gal/day

*End product of purine metabolism.
†By-product of muscle metabolism.

(*ouron*—urine; Gk.) (Figure 15–3). The meandering of the kidney tubule with its twists and turns serves to compact its length into a small space, microscopic in diameter and only about 1.25 in. long; if the 2 million nephrons in both kidneys were straightened out and laid end to end, they would produce a tunnel about 50 miles long!

The nephrons of the kidney are not arranged in a haphazard manner. If a kidney is sliced longitudinally, two regions can be distinguished: an outer reddish-brown region, the **cortex,** and an inner grayish area, the **medulla** (Figure 15–6). Microscopic examination of the cut surface of the kidney shows that the glomeruli, Bowman's capsules, and the convoluted portions of the kidney tubules are located in the cortex (Figure 15–6), whereas the straight portions of the tubules, Henle's loops, and the urine-collecting ducts are located in the medulla.

Each kidney tubule and its associated capillaries constitute a nephron, and this is the functional unit of the kidney. Because each nephron operates in almost exactly the same way as all the others, we can describe most of the functions of the entire kidney by explaining the function of a single nephron.

The capillaries of the glomerulus have very thin walls, and the electron microscope reveals that they are permeated with many tiny pores (Figure 15–7). The glomeruli of the kidney are mechanical filters, and they permit a cell-free, protein-free fluid to pass from the blood into the tubule. The pores in the capillary wall are of such a size that blood cells and proteins cannot leave, but water and smaller molecules (both useful substances and dissolved wastes)

flow out. Thus, the fluid that passes out of the glomerulus contains water, amino acids, glucose, salts, and urea in exactly the same amounts as are found in blood plasma (Table 15–1); this filtrate of the blood is also nearly identical to the dilute salt solution surrounding all the cells of the body. The total blood flow through the kidneys is normally about 750 ml per minute, or about one fifth of the blood volume pumped by the heart in the same amount of time. Approximately 125 ml per minute of protein-free filtrate is pressed out of the glomerular capillaries and passes into the kidney tubules. The remainder of the blood leaves the glomerulus by way of an (efferent) arteriole.

What supplies the energy for the filtration process? The driving force for moving fluid out

FIGURE 15–7 A small portion of the glomerulus showing the pores (arrows) that act as a filter for the blood. (Courtesy of Marilyn G. Farquhar.)

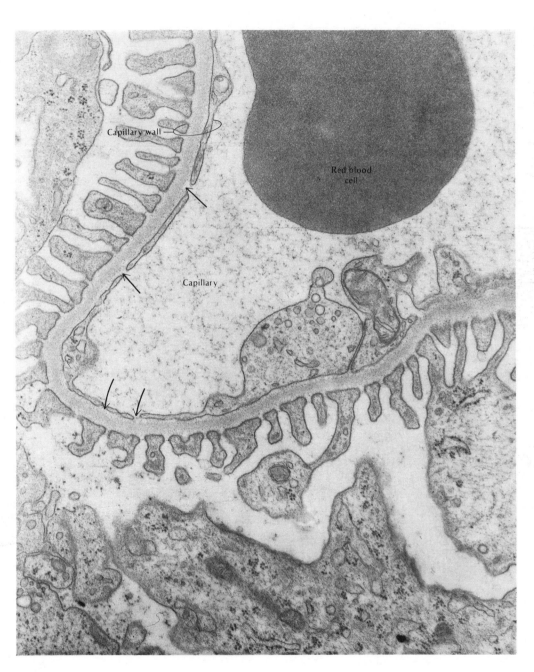

of the glomerulus and into Bowman's capsule is the blood pressure in the capillaries, which is ultimately a result of the work of the beating heart (Figure 15–8). The pressure in the glomerulus is higher than the pressure in Bowman's capsule. Therefore, fluid moves out of the glomerulus via the microscopic pores in the walls of the capillaries and passes into the capsule. After the glomerular filtrate enters Bowman's capsule, it passes into the tubular system of the nephron. Each day about 45 gallons (180 liters) make this journey from the blood to the kidney tubule, but, of course, this is far in excess of what is passed out of the body as urine. Normally we excrete only 1% of the total glomerular filtrate (about 1–2 qt). Where does the remainder go? More than 99% of the glomerular filtrate that enters the kidney tubule is resorbed into the blood by way of the capillary network surrounding the tubule. In this manner the tubules send back to the blood that which is valuable and reusable and leave the wastes inside to be discharged. Additionally much of the fluid lost from the circulatory system is recaptured by the blood. Indeed, if resorption did not take place, we should simply excrete our body fluids into the urine, and the sea within us would soon be depleted.

The cleansing of the blood by the kidney can be analogized to the housewife who wants to clean up a dirty room: instead of merely sweeping up the dust and dirt and disposing of them, she empties the room of all its portable furniture (pictures, tables, chairs, and so on) as well as the waste materials. Once the waste has been disposed of, all the furniture is returned to the room again. At first glance this seems an inefficient way for a kidney to handle purification—throwing out almost everything by filtration and retrieving what the body requires by resorption. Indeed, there are certain drawbacks, but consider the advantage: every half hour our entire blood supply is chemically

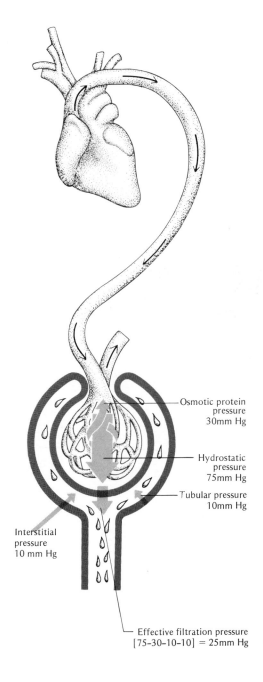

FIGURE 15–8 Production of the effective filtration pressure for producing glomerular filtrate.

Osmotic protein pressure 30mm Hg

Hydrostatic pressure 75mm Hg

Tubular pressure 10mm Hg

Interstitial pressure 10 mm Hg

Effective filtration pressure [75-30-10-10] = 25mm Hg

cleansed in a complete and thorough way, without cutting any corners.

The nondiscriminatory process of filtration is a function of the glomerulus. Selective retrieval of essential materials by resorption and secretion takes place in the kidney tubules; let us examine the retrieval processes.

The glomerular filtrate entering the kidney tubule from Bowman's capsule passes first through the proximal convoluted tubule, then through the loop of Henle, then through the distal convoluted tubule, and finally into a straight collecting duct. Although the initial filtrate is similar in salt composition to the blood plasma (it is **isosmotic**), in its passage through the kidney tubules it will become concentrated, and the urine excreted will be higher in salt concentration (**hyperosmotic**) relative to the blood plasma (Table 15–1). As the glomerular filtrate passes through the proximal convoluted tubule, approximately 85% of the water, sodium ions, and chloride ions; most of the bicarbonate; and all of the glucose, vitamins, and amino acids are resorbed into the surrounding network of capillaries. So powerful and efficient is this resorption process that virtually no glucose, vitamins, or amino acids are found in the exexcreted urine. The proximal convoluted tubule is admirably suited for its resorptive role. The cells lining the proximal tubule have millions of microscopic fingerlike microvilli that serve to increase the resorptive surface area, thus facilitating rapid transfer of water and other essential materials (Figure 15–9). Although water resorption occurs by osmosis, a passive non-energy-requiring process, this is not the case for other essential substances such as glucose, amino acids, and vitamins. These materials are moved against a concentration gradient, and for such metabolic work energy must be expended. The immediate energy source for this "uphill" (active) transport of materials is ATP. The ATP-generating system

of the proximal tubule is found in the large numbers of mitochondria located within the tubule cells themselves. The mitochondria supply the energy for transport, but specificity of active transport is due to the presence of specific **carrier proteins** in the membrane. There are carrier proteins specific for moving certain amino acids across the membrane, other carriers for transporting vitamins, and so on. (The situation is shown diagrammatically in Figure 15–10.)

The tubular fluid, now free of amino acids, sugar, vitamins, and so on, passes from the proximal convoluted tubule to the descending loop of Henle, and from there to the ascending loop. The cells of the ascending limb of the loop of Henle actively pump sodium ions out of the tubule and into the surrounding tissue fluid (chloride passively follows to maintain electrical neutrality). Some but not all of this sodium then diffuses from the tissue fluid into the descending limb so that there is a recirculation of sodium from the ascending limb, into the tissue fluid, to the descending limb, and back again to the ascending limb (Figure 15–11). As a result of the cyclic pumping and resorption of sodium, and by virtue of the organization of the loops of Henle in the medulla of the kidney, a salt (sodium chloride) concentration gradient is maintained in the tissue fluid surrounding the loops of Henle. The concentration is highest in the medulla (at the hairpin turn in the loop of Henle) and lowest in the cortex (at the region of the convoluted tubules). The walls of the ascending limb are impermeable to water, so that only sodium leaves the tubule. As a result, the filtrate in the loop of Henle loses much of its sodium, but little water. Thus, when the filtrate reaches the distal convoluted tubule, it contains less salt than there was in the original glomerular filtrate. Concentration of the filtrate by osmotic resorption of water occurs in the collecting duct. As

FIGURE 15–9 The proximal tubule of the kidney as revealed by the electron microscope. Note the many microvilli and the mitochondria in the cells that make up this tubule aiding in the active transfer of materials from the glomerular filtrate into the blood. (Courtesy of K. R. Porter.)

FIGURE 15–10 Diagram of active transport in the kidney. Specific carriers in the kidney tubule cells pick up molecules and transfer them from the glomerular filtrate to the blood. The energy source for this transport is ATP, generated by the mitochondria of the tubule cells.

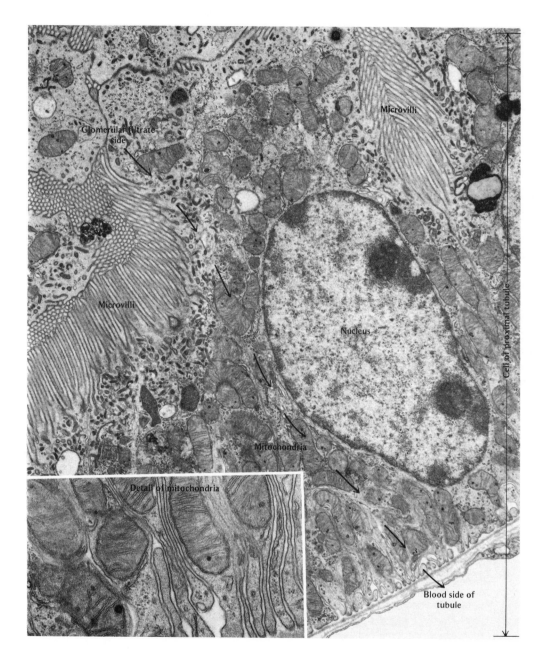

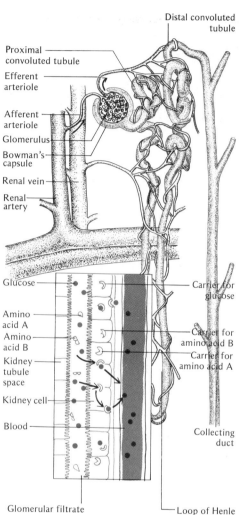

FIGURE 15-11 The mechanism by which urine is concentrated by a nephron in the human kidney.

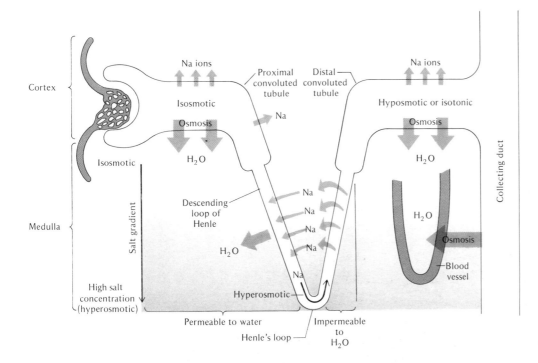

the filtrate moves from the distal convoluted tubule into the collecting duct (which progresses downward from cortex to medulla), it encounters a region of ever-increasing salt concentration in the tissue fluid surrounding the duct. The walls of the collecting duct are permeable only to water, so that water moves passively, by osmosis, from the duct into the sodium-rich (hyperosmotic) fluid surrounding it. As a result the fluid in the duct, now called **urine**, eventually becomes isosmotic with the fluid around the duct, and at the end of the duct the urine is not only much more concentrated than the original glomerular filtrate but is also hyperosmotic to the blood plasma. Thus, of the 180 qt of dilute glomerular filtrate produced in a day, less than 2 qt appear as

concentrated urine. (The concentration of the various salts is shown in Table 15–1.)

The composition of urine depends to a large extent on what is left behind in the tubules, that is, on what the nephrons do not remove from the glomerular filtrate by selective resorption. However, recent evidence has shown that there is also active secretion of certain substances from the capillaries surrounding the tubule into the tubule itself. In the human kidney the amount of such secretion (called **augmentation**) is small and occurs mainly in the convoluted tubules. p-aminohippuric acid, a minor nitrogenous waste product, is secreted into the proximal convoluted tubules, and K^+ and H^+ ions are secreted by the distal convoluted tubules. Secretion of H^+ ions helps to

maintain the pH (acidity) of the blood at a relatively constant level. When the antibiotic drug penicillin is taken, it is secreted into the urine by the distal convoluted tubule. Similarly the antimalarial drug Atabrine is secreted by these tubules. Because these drugs are excreted in the urine, blood levels eventually drop, and for effective therapy administration of the drug must be maintained during the course of the infection.

15-3 REGULATION OF FLUID AND SALT BALANCE

As you undoubtedly realize from personal experience, the output of urine by the kidneys is not constant in amount. A few examples will illustrate how the rate of urine formation varies. If you were to drink about a quart of water over a short span of time (let us say in 15–30 minutes), you would find that your urine production over the next few hours would be greater than your normal output. In fact, if you were to collect the urine produced in the period subsequent to the drinking of the fluid, you would find the excess urine volume to be just about a quart. It is as if the kidneys knew exactly how much liquid was ingested, and to maintain the amount of body fluid at a constant level, they eliminated the excess.

Let us imagine that you spent a day in the desert with only a single canteen of water available. What would be the output and appearance of your urine? Probably you would find that the urine produced was more concentrated, dark yellow, and below normal in amount. Exposed to the hot, dry desert air, you would tend to lose body fluids via sweating, but a considerable fraction of the body fluids would be conserved by the kidneys, as their fluid output was reduced. Again, the kidneys know about the body's fluid balance.

These situations demonstrate that the rate of urine formation is increased or decreased in direct relation to the amount of hydration or dehydration. How do the kidneys know when to excrete water and when to conserve it?

The control mechanism that regulates the body fluids involves the secretion of a hormone called **antidiuretic hormone** (ADH) or **vasopressin.** The system for controlling water conservation or excretion involves the effect of ADH on the resorptive properties of the kidney tubule. When the body fluid supply is low, special sensory receptors located in the hypothalamus of the brain are stimulated. These osmoreceptors send out impulses that are carried by nerve fibers to the posterior lobe of the pituitary gland, and there ADH is released (Figure 15–12). The ADH passes by way of the blood to the kidney, where it affects resorption. The hormone ADH controls the water permeability of the cells of the distal convoluted tubule and the collecting duct. If ADH is present, these tubules become highly permeable to water, so that water leaves the tubules and seeps back into the bloodstream. As a result of this loss of water from the tubules, the urine becomes concentrated. If, on the other hand, the osmoreceptors are not stimulated, as is the case when water in the body fluids is at a normal level, ADH is not released, and the tubules remain impermeable to water. The result of an absence of ADH is the formation of a dilute urine, since salt is resorbed but water is not.

A disease characterized by the production of abnormal quantities of urine is called **diabetes** (to pass through; Gk.), because water that is drunk seems to siphon from the body at such a rapid rate. There are two kinds of diabetes, called **mellitus** and **insipidus.** In the rare disease diabetes insipidus (*insipidus—*tasteless; L.), the urine output may reach 10 gallons a day instead of the more normal

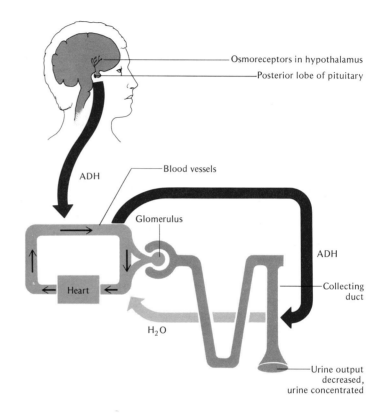

Osmoreceptors in hypothalamus

Posterior lobe of pituitary

ADH

Blood vessels

Glomerulus

ADH

Heart

Collecting duct

H_2O

Urine output decreased, urine concentrated

quantity of 1 qt. Persons so afflicted have an incessant thirst and suffer from a salt imbalance in the body fluids. The disease is caused by a deficiency in the supply of ADH; resorption of water is low, and the urine produced is copious and quite dilute. The more common type is mellitus, or sugar, diabetes, characterized by the presence of sugar in the urine. The primary abnormality of this kind of diabetes is a failure to regulate properly the utilization of glucose due to a deficiency of the hormone insulin (Chapter 16). As a result there is too much glucose in the blood, the glomerular filtrate contains excessive amounts, and only a fraction of this can be resorbed. The excess glucose in

the tubule increases the osmotic pressure of the contents and thus reduces resorption of water. The diabetic thus excretes large quantities of water and sugar and may become dehydrated. Today the disease is easily treated by regular injections of insulin.

As we have have already noted, water resorption in the kidneys is a passive process. The prime factor responsible for water resorption is the active transport of sodium from the tubule into the tissue fluid, as described earlier. Sodium pumping here is regulated by the hormone **aldosterone,** secreted by the cortex of the adrenal glands. The mechanism of action of this hormone is not understood, but by its action sodium is conserved and so is water. When the body fluid volume is decreased owing to dehydration, the amount of sodium in the body fluids increases, and this leads to the production of aldosterone. Aldosterone promotes resorption of sodium (and water), reduces urine output, and conserves water. If, on the other hand, aldosterone secretion is diminished,[2] resorption of sodium is low, and sodium is lost in the urine.

In normal individuals the regulatory mechanisms for retention or excretion of sodium are so precise that the sodium balance in the body does not vary by more than 2% in spite of large variations in salt intake and salt loss (sweating, vomiting, and diarrhea). When the kidneys fail to excrete adequate amounts of sodium, the individual accumulates sodium and water in the body spaces, leading to swelling of the tissues, a condition known as **edema.**

Diuresis

Substances that increase the flow of urine are called **diuretics;** the flow process itself is called **diuresis.** Glucose, salt, and amino acids present in excess cannot be resorbed and water is retained in the tubule to keep these substances in solution. Similarly, meals rich in protein produce diuresis because their byproducts, amino acids and urea, remain in the tubules, water is retained in the tubules to keep these in solution, and urine flow is increased. Caffeine (in coffee, tea, and cola drinks) and alcohol act as diuretics by inhibiting the secretion of ADH. One estimate suggests that 100 ml of water are excreted for every 10 gm of alcohol ingested. The well-known diuretic effect of beer is due largely to its water content, and not its alcohol content; since beer is less

than 4% alcohol, it is less dehydrating than whisky (40%–45% alcohol) and therefore provides a better way of quenching one's thirst.

Emotions may increase or decrease diuresis by affecting epinephrine secretion. Minute quantities of epinephrine cause a constriction of the arterioles leaving the glomerulus, the blood pressure in the glomerulus is elevated, and urine production is increased. Large amounts of epinephrine constrict the arteriole leading to the glomerulus and thus decrease glomerular pressure and diminish the formation of urine.

Diuretic drugs (Diamox, mercurial compounds, thiazides), commonly called water pills, are used to increase the excretion of salt and water in individuals suffering from edema. These diuretic drugs act principally by blocking the specific enzymes involved in tubular resorption, especially those promoting sodium uptake; as a consequence of sodium remaining in the tubules, water resorption is decreased and urine flow is increased. **Dropsy,** a chronic condition of bloating due to excessive resorption of water resulting from heart failure or other causes, may be alleviated by administration of diuretic drugs.

Thirst

Water intake is under voluntary control; we can, of course, drink at will. There is, however, an unconscious control mechanism that tells us when we are thirsty. When water loss has reached 1% of the body weight, the sensation of thirst is experienced. In the hypothalamus there is a localized region of nerve cells called the **drinking center;** when this area is stimulated either electrically or by the presence of high salt concentrations in the body fluids, a reflex is initiated that produces the sensation of thirst. This stimulates us to drink and prompts the kidneys to conserve water. Thirst symptoms

develop because the mouth and pharynx become dry (parched) as a result of declining saliva production. This produces discomfort in swallowing, and relief is obtained only by drinking (wetting the mouth doesn't work). The thirst reflex is much stronger than that of hunger, and the limit of endurance for thirst is a matter of days, whereas hunger can go on for weeks. When water loss reaches 5% of the body weight, a state of collapse ensues; when the water loss is over 10% of an individual's body weight, death is likely to occur.

"Water, water everywhere nor any drop to drink."[3] A man adrift in the ocean with only seawater available cannot satiate his thirst. Drinking salt water will not quench his thirst because the salt content of seawater is three times that of the blood. The excess ingested salt has to be excreted, and this requires more water than was consumed in the seawater. It is a losing battle, and the result is accelerated water loss and a quicker death.

15–4 DISCHARGE OF URINE

Once the filtrate has been processed, all that remains is to carry the urinary wastes out of the body. Exclusive of the nephrons of the kidney, the remainder of the urinary system is simply plumbing for conducting urine to the outside.

Urine passes from the collecting ducts into the pelvis of the kidney, where the ducts unite to form a series of funnels drained by a 1-ft-long tube, the **ureter** (Figure 15–15). Urine is carried along the length of the ureters to the bladder by spontaneous rhythmic peristaltic contractions of the ureter walls, aided by gravity. Ordinarily the urine takes only a minute to go from the kidney to the bladder. Any abnor-

---

[2] This can be accomplished by administration of the diuretic drug spironolactone.

[3] S. T. Coleridge, *The Rime of the Ancient Mariner.*

mality of the ureter may impair its ability to contract, causing urine to collect in the kidney pelvis. This leads to swelling and may promote infection. Urine is stored in the bladder prior to discharge.

Some of the substances contained in the urine are only sparingly soluble in water, and when stagnation of urine occurs, microscopic crystals of calcium phosphate and calcium oxalate may aggregate to form concretions called **kidney stones** or **urinary calculi.** These materials grow in size, and eventually they may fill the kidney pelvis and obstruct the flow of urine in the ureter. These stones cause extreme pain, and passing a stone can be an excruciatingly painful experience. Organic substances such as amino acids (cysteine, for example, is sparingly soluble) and uric acid may also form stones. If it is impossible to redissolve or pass the stones, they must be removed surgically.

The bladder is a muscular bag situated in the pelvic cavity just above the pubic bone. When empty, it resembles a deflated balloon, but as it fills it assumes a pearlike shape and may contain a volume of about 1 qt. Its ability to expand is related to its highly extensible epithelial lining and the stretchability of the layers of smooth muscle on its outer surface. The ureters enter the bladder at an oblique angle to form a flaplike valve (Figure 15–15b). Therefore, any pressure in the bladder tends to close the opening of the ureters, preventing backflow.

Urine leaves the bladder via a single tube, the **urethra** (Figure 15–15), which varies in length according to sex: in males it is about 8 in. long and passes through the penis; and in females it is 1.0–1.5 in. in length. The part of the urethra closest to the bladder is surrounded by two sphincter muscles controlled by the nervous system; one is under voluntary control and the other is not. Ordinarily both sphincters remain closed, so that urine cannot escape

BOX 15B The artificial kidney machine

Perhaps 3 million people in the United States suffer from undiagnosed disorders of the kidney. Sometimes kidney disease reveals itself clearly, and treatment and recovery are possible; but for others there is no warning, and a slow wasting proceeds. No age group, no group of any kind is spared from disorders of the kidneys.

Uremia (urea in the blood) a progressive toxic condition due to kidney failure and related kidney diseases, kills over 100,000 Americans each year. Uremia can occur by accidental ingestion of a poison, shock, physical injury, and disease. Without kidney function, poisonous wastes accumulate in the blood and destroy the body's cells. What would happen if at this moment your kidneys simply stopped working?

Until recently there was little hope for persons suffering from kidney failure, but recent developments in the manufacture of artificial kidney machines provide

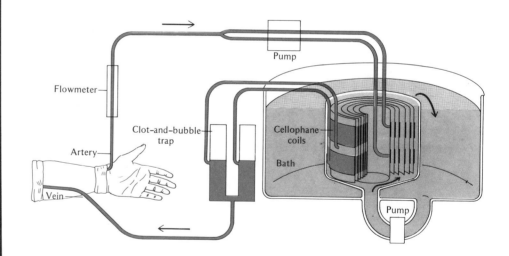

FIGURE 15-13 Diagrammatic representation of the artificial kidney machine. (Modified from *The Artificial Kidney,* **John P. Merrill. Copyright 1961 by Scientific American, Inc. All rights reserved.)**

a mechanism for ridding the body of its poisons. This device permits Josephine Berman (page 300) and countless others to live. How does the artificial kidney machine work?

The natural kidney regulates the body fluid composition by filtration and selective resorption. An artificial kidney does the same things; however, it does not look like a real kidney nor does it operate exactly the same way. The filter of the natural kidney is the glomerulus, and the efficiency per unit surface area of 'these 2 million microscopic filters cannot be matched by any man-made machine at the present time; the natural kidney is a marvel of compact efficiency. The artificial kidney machine (Figure 15-13) consists of a steel tub 2 ft deep and 2 ft wide. In the center of the tub is the filter, a narrow tubular coil of cellophane 25 yd long, providing a large surface area for exchange of materials. Cellophane has pores of the same dimensions as the glomerular capillaries so that the same substances pass across its surface (molecules smaller than 50,000 molecular weight). As blood passes through the cellophane tubing, molecules move across the cellophane according to their concentration in the blood and the surrounding solution (Figure 15–14). For example, if the concentration of substance A is higher in the blood than in the external solution, substance A leaves the blood; if the reverse is true, substance A enters the blood from the external solution. The diffusion of small molecules across a semipermeable membrane is called **dialysis,** and it is the mechanism by which the artificial kidney works. By changing the composition of the solution surrounding the cellophane tubing, the artificial kidney can move materials into the blood or vice versa.

An artificial kidney treatment, or **hemodialysis** (dialysis of the blood), requires that the cellophane tubing of the artificial kidney be primed with an anticoagulant plus blood matched exactly to that of the patient. This assures a continuous flow of unclotted blood from patient to machine and back. The tub surrounding the coils is filled with warm water containing the proper concentration of chemicals (glucose, salt, bicarbonate, amino acids, and other small molecules). The patient is plugged in to the machine by connecting an artery and a vein to the machine (Figure 15-13). Blood, pumped by the patient's heart, flows through the artery into the machine and traverses the long cellophane coil. Purified by dialysis, the blood passes from the cellophane tubing into a clot-and-bubble trap and then back to the body via a tube into the vein. Dialysis takes 6–10 hours, but varies with the individual condition and frequency of treatment.

The artificial kidney cannot secrete or resorb; it can only filter by dialysis, and therefore balance of ingredients in the wash solution is critical. If such balance is not controlled, the blood could eliminate useful materials and take up worthless ones. Clearly hemodialysis is not a procedure that can be readily adapted for home use (see Chapter opener).

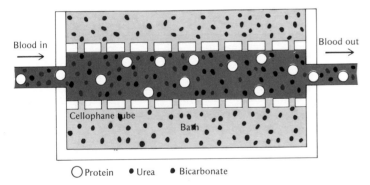

Blood in →

Blood out →

Cellophane tube

Bath

○ Protein   ● Urea   ● Bicarbonate

FIGURE 15-14 The artificial kidney operates by the principle of dialysis. Note how the urea, high in concentration in the blood, moves into the bath, and bicarbonate, high in concentration in the bath, moves into the blood; proteins are retained in the blood because they are larger than the pores of the dialysis (cellophane) tubing. (Modified from *The Artificial Kidney,* **John P. Merrill. Copyright © 1961 by Scientific American, Inc. All rights reserved.**)

FIGURE 15–15 (a) **The structure of the bladder and
its relation to the kidney. (b) Detail of the bladder
wall and the ureter. As the bladder fills, the pressure
(arrow) tends to close the opening from the ureter
into the bladder.**

from the bladder except when there is some to pass and the individual wills it.

Urination (**micturition**) in humans is controlled by a combination of voluntary and involuntary nervous activity. When the bladder fills with about a pint of urine, stretch receptors in the bladder wall are stimulated; these transmit impulses to the spinal cord and initiate a conscious desire to urinate and a subconscious micturition reflex. The subconscious reflex involves motor fibers to the bladder wall and to the internal sphincter muscle of the urethra. The bladder contracts, and the internal sphincter relaxes; then for urine to pass out of the body the external sphincter of the urethra must be relaxed. If the time is suitable for urination, by the conscious activity of the brain (or merely by passage of urine into the urethra), there is relaxation of the external sphincter and urination occurs. If a person wishes to urinate before the micturition reflex occurs naturally, the reflex can be started by contracting muscles in the abdominal wall which push the viscera down on to the bladder; this excites the stretch receptors and the reflex of urination begins. Emotional conditions or excitement also tend to produce the urination reflex.

It is possible to override or inhibit the urination reflex and not succumb to the urge; in this case the desire to urinate passes within a few minutes. However, in a matter of minutes to hours, the reflex returns, and once the bladder fills to capacity the reflex becomes so strong that urination can no longer be prevented by conscious action. A baby has no control over the external sphincter of his urethra, and urination occurs whenever the bladder fills and the urination reflex is initiated. Similarly, a person whose spinal cord has been severed has no bladder control. This loss of voluntary control of micturition is called **incontinence.** Such individuals may require a catheter, a tube

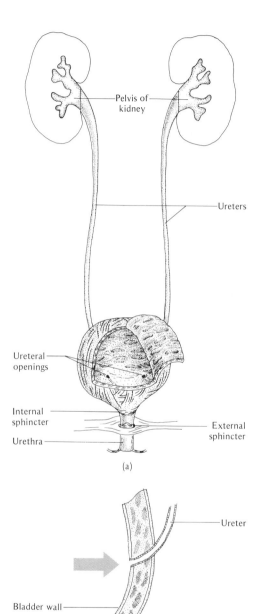

(a)

(b)

inserted into the urethra, so that urine can flow into a convenient receptacle.

Inflammation of the bladder (**cystitis**) may be caused by bacterial infection and occasionally is a result of an inflammation in the kidney itself. Inflammation of the bladder is likely to be more apparent than a kidney inflammation, since it usually causes painful urination. Treatment involves antibiotics that attack the bacteria causing the infection.

SUMMARY

1. The kidneys are the principal organs of excretion, ridding the body of waste and maintaining the constancy of the internal environment.

2. Blood supply to the kidney is via the renal artery and vein. The functional unit of the kidney is the nephron. The renal artery distributes blood to each nephron as a spherical cluster of capillaries called the glomerulus.

3. Owing to the effects of capillary blood pressure, a cell-free, protein-free blood filtrate passes from the glomerulus to Bowman's capsule.

4. From each Bowman's capsule, there is a series of convoluted tubules and an extended loop of Henle. The distal convoluted tubule leads via a urine-collecting duct to the ureter.

5. Henle's loops and the collecting ducts are found in the renal medulla; the rest of each nephron is in the cortex of the kidney.

6. Selective resorption and secretion of 99% of the filtrate occurs in the kidney tubules. The proximal convoluted tubule has a large surface area due to its microvilli. Here, 85% of the water, sodium and chloride ions, bicarbonate, glucose, vitamins, and amino acids are resorbed into the surrounding capillary network. Water enters by osmosis; the other components are actively transported by carrier proteins.

7. Sodium is actively pumped out by cells of the ascending loop of Henle. Chloride passively follows. Sodium then diffuses to the descending limb. Thus, sodium is recirculated and a sodium concentration gradient surrounds the loops of Henle. The concentration is higher in the medulla than the cortex. Since the ascending loop is impermeable to water, sodium is lost here but not water.

8. The dilute filtrate reaches the collecting duct, which is permeable only to water. As the tubule passes through the hyperosmotic renal medulla, the filtrate is concentrated by osmotic resorption of water. The concentrated filtrate is urine.

9. Augmentation of the urine may occur by selective capillary secretion.

10. Urine output depends on a fluid and salt balance that is regulated by antidiuretic hormone (ADH). ADH increases permeability of the distal convoluted tubule and collecting ducts. Hypothalamic osmoreceptors stimulate ADH secretion from the posterior pituitary as necessary.

11. ADH deficiency causes diabetes insipidus, with copious dilute urine. Diabetes mellitus is due to insulin deficiency and increased blood sugar, which spills into the urine, causes osmotic imbalance, and may lead to dehydration.

12. The renal sodium pump is regulated by an adrenal hormone, aldosterone. Failure to excrete adequate sodium causes osmotic imbalances, water retention, and edema.

13. Diuretics increase urine flow and include: high protein diets (which increase amino acids and urea in the tubules); caffeine and alcohol (which inhibit ADH secretion); increased epinephrine secretion in response to emotional stimuli (which increases glomerular pressure and filtration); and diuretic drugs (which block tubular resorption enzymes).

14. The hypothalamic drinking center is stimulated by water loss and initiates a drinking reflex.

15. The collecting ducts unite in the renal pelvis to form a ureter, which carries the urine to the extensible bladder for storage. Urine stagnation may cause kidney stones to form.

16. Urine leaves the bladder via the urethra. Backflow from bladder to ureter is prevented by valves. Discharge of urine is controlled by two urethral sphincters.

17. A full bladder initiates the micturition reflex, which, within limits, can be consciously overridden. Incontinence involves the loss of voluntary control. Cystitis is inflammation of the bladder, which causes painful urination.

18. Ammonia is the nitrogenous waste material of protein metabolism. Soluble and toxic, it is carried to the liver and combined with CO_2 to form urea. Urea passes to the kidneys for excretion via the renal portal vein.

19. Uremia can be treated by hemodialysis using an artificial kidney machine.

KEY WORDS

kidneys
renal artery
renal vein
ureter
nephrons
glomerulus
Bowman's capsule
loop of Henle
urine-collecting duct
cortex
medulla
renal portal vein
deamination
urea cycle
isosmotic
hyperosmotic

carrier proteins
urine
augmentation
antidiuretic hormone (ADH) or vasopressin
diabetes
diabetes mellitus
diabetes insipidus
aldosterone
edema
diuretics
diuresis
dialysis
hemodialysis
dropsy
drinking center
ureter
kidney stones or urinary calculi
urethra
micturition
incontinence
cystitis

TOPICS FOR REVIEW AND DISCUSSION

1. **Why do animals require kidneys?**
2. **What is the nephron and how does it work?**
3. **How is blood pressure involved in kidney function?**
4. **What are the similarities and differences between the natural and artificial kidney?**
5. **How is urea formed? How is urine formed?**
6. **Why is the drinking of seawater dehydrating?**
7. **What is the role of ADH in water balance?**
8. **How does the kidney regulate the fluid volume of the body?**
9. **What is the function of the bladder?**
10. **What is the fate of 99% of the glomerular filtrate?**

BROOM-HILDA

by Russell Myers

TROLLS HAVE VERY UNPREDICTABLE HORMONES!

16

CHEMICAL COORDINATION: HORMONES

16-1 THE NATURE OF HORMONES

Look at yourself. Immediately you recognize what sex you are and so does everyone else. It is a fact of life that in the human species males and females look different. Indeed, our male or female appearance is so important to us that a plaque aboard a United States spacecraft headed for the far reaches of the Milky Way galaxy shows us naked. If it is found by intelligent life outside the earth, even those beings will know what we look like! Our sex differences are present not only externally but in the organs tucked away inside the body (Chapter 9). What mechanisms bring about such differences?

As we saw in Chapter 9, hormones are produced by ovaries and testes, and the action of these chemicals on a variety of tissues determines the external form we recognize as male or female. Indeed, if a boy is castrated before puberty, he grows into a eunuch. The effects of castration make it abundantly clear that for normal development of masculine characteristics there must be a chemical produced by the testes and transported by the bloodstream to all parts of the body. Such chemicals are called **hormones,** a word meaning "to arouse" or "to excite" in Greek. In essence, this is what a hormone does: it activates processes that, prior to its presence, were inactive or operating at a much lower level. A hormone can be defined as a specific chemical substance elaborated by cells in a particular part of an organism (plant or animal) that controls or helps control cellular processes elsewhere in the organism.

Why is it necessary to have a system of hormonal controls in an organism? The division of labor among cells is one of the primary characteristics of the multicellular condition, and it provides for certain advantages. However, division of labor is advantageous only if the component cells are coordinated and integrated by a communications network. In multicellular organisms the communications systems are both hormonal and nervous. The hormonal system generally provides for coordination on a slow-acting basis. Hormonal messages are distributed throughout the body, influencing specific cells in various parts of the body. By contrast, the nervous system is specialized for rapidly conducting electrochemical messages and mediating short-term responses. The signals that traverse the nervous system move along well-defined cables (nerves) and activate a limited number of cells at their terminal ends. Although the hormonal and nervous systems are usually regarded as separate—and in some organisms, such as plants, only one system (hormonal) exists—the two systems have much in common. Both systems influence processes in organs or tissues located some distance from the control center. To a considerable degree both systems operate on negative feedback principles. That is, a message sent out from the control center causes the target organ to increase or decrease its activity, and this response then acts back on the control center; as a consequence the system is self-correcting. The operation of the hormonal and nervous coordinating systems by negative feedback contribute to an organism's functional stability. Because they are error actuated and the controls are built in, such systems are economical and flexible. This is a point to which we shall return in Chapter 21.

16-2 THE HUMAN HORMONAL ORCHESTRA: ENDOCRINE GLANDS

Some glands of the body pour their products into tubes or ducts, and in this way the secretions pass to the outside of the body or to the

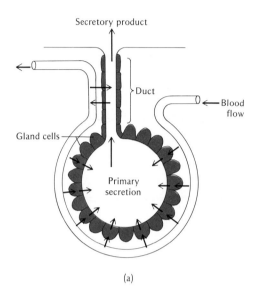

FIGURE 16-1 A comparison of (a) exocrine and (b) endocrine glands. (Modified from *Human Physiology: The Mechanism of Body Function* by A. J. Vander, J. H. Sherman, and D. Luciano. Copyright © 1970, McGraw-Hill Book Company. Used with permission of McGraw-Hill Book Company.)

(a)

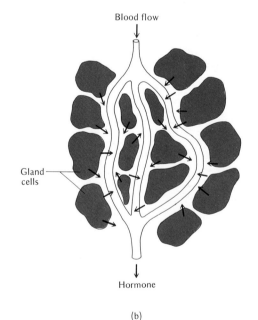

(b)

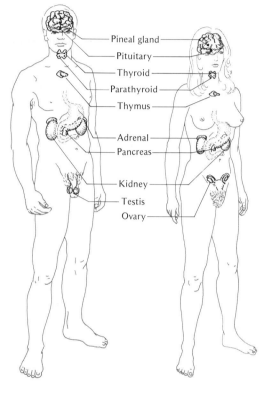

FIGURE 16-2 Human endocrine glands.

cavity of one of the internal organs (Figure 16-1a). Among these **exocrine** (*exo*—outside, *krinein*—to separate; Gk.) **glands** are the mammary glands, sweat glands, salivary glands, gastric glands and the liver. In contrast, hormones are produced by ductless or **endocrine glands** (*endo*—within; Gk.), poured directly into the bloodstream, and swiftly carried throughout the body (Figure 16-1b).

Hormones are effective in minute quantities. The endocrine glands are tucked away in various places in the body—head, neck, abdominal cavity, and gonads (Figure 16-2). They

FIGURE 16–3 The pituitary gland. (a) Location in the head region. (b) Detail of the gland. (c) Structure of the lobes of the pituitary. (d) Feedback control of the pituitary.

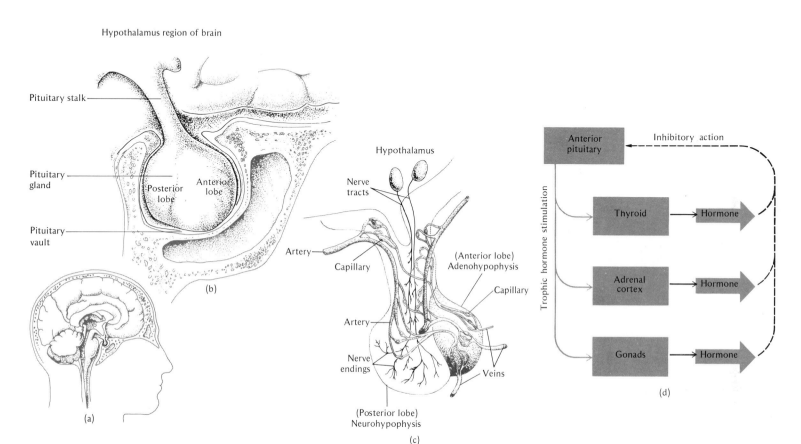

provide an excellent example of the old adage that size is no criterion of function (together they weigh less than 5 oz); the endocrines exercise profound control over such functions as growth and development, regulation of tissue activity, and reproduction. Malfunction of an endocrine gland can lead to serious disturbances, usually owing to either underproduction (hyposecretion) or overproduction (hypersecretion). The causes are manifold: diet, disease, tumors, genetic disorders, and so on.

How is a specific response mediated by the hormonal system? Why does a hormone pro-duce activity in organ A and not in organs B, C, and D? As we might expect, the response of a particular organ to a hormonal stimulus is determined by the specific makeup of that organ. Thus, in coordinating body activity, each hormone has its **target organ** or organs—those it can trigger into activity; the other organs carry on normally. The nature and degree of response of the target organ to a hormone may vary from one time to another, from one individual to another, and from one species to another. This variation in response contributes to an organism's individuality.

Maestro of the orchestra: the pituitary gland

The **pituitary gland,** or **hypophysis,** is a pea-sized organ nestled in a bony cavity below the base of the brain and above the nasal passages (Figure 16–3a). It was first described by the anatomist Vesalius (1514–1564), who mistakenly believed it produced mucus that was discharged into the nose and who therefore called the gland "pituitary" from the Latin word meaning "nasal secretion." The pituitary gland is actually two glands in one, consisting

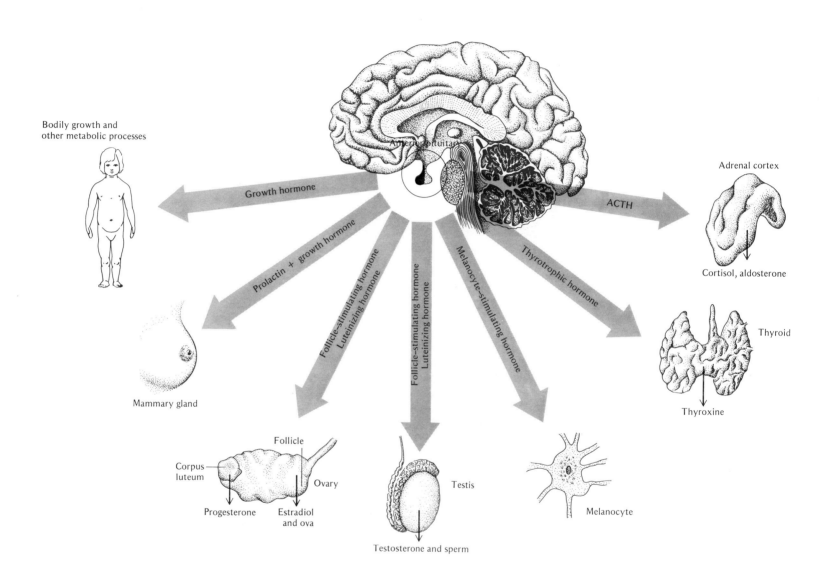

Bodily growth and other metabolic processes

Growth hormone

Prolactin + growth hormone

Follicle-stimulating hormone
Luteinizing hormone

Follicle-stimulating hormone
Luteinizing hormone

Melanocyte-stimulating hormone

Thyrotrophic hormone

ACTH

Anterior pituitary

Adrenal cortex

Cortisol, aldosterone

Thyroid

Thyroxine

Mammary gland

Follicle

Corpus luteum

Ovary

Progesterone

Estradiol and ova

Testis

Testosterone and sperm

Melanocyte

TABLE 16–1 Anterior pituitary (adenohypophyseal) hormones

Hormone	Target organ(s)	Principal effects	Diseases
Follicle-stimulating hormone (FSH)	Ovarian follicle and seminiferous tubules of testis	Stimulates growth of follicles and tubules	Hyposecretion: hypogonadism Hypersecretion: hypergonadism
Luteinizing hormone (LH) Interstitial-cell-stimulating hormone (ICSH)	Corpus luteum and interstitial cells of testis	Stimulates formation of corpus luteum from follicle; stimulates production of progesterone by the ovaries and testosterone by the testes	Hyposecretion: hypogonadism Hypersecretion: hypergonadism
Prolactin (LTH)	Female mammary gland	Stimulates milk production after parturition	
Somatotrophic hormone (STH) Growth hormone (GH)	Body tissues	Increases synthesis of protein	Hyposecretion: dwarfism Hypersecretion: gigantism and acromegaly (Box 16A)
Adrenocorticotrophic hormone (ACTH)	Cortex of adrenal gland, skin, liver, mammary gland	Stimulates production of corticosteroids; increases metabolic rate, glycogen deposition in liver, darkening of skin, milk production	Hyposecretion: Addison's disease (Box 16B)
Thyrotrophic hormone (TH)	Thyroid gland	Stimulates production of thyroxine	Hyposecretion: cretinism, myxedema Hypersecretion: goiter
Melanocyte-stimulating hormone (MSH)	Skin melanocytes	Controls skin pigmentation	

of an anterior lobe and a posterior lobe (Figure 16–3 b and c). Although the anterior and posterior lobes lie side by side, with a portion of the anterior lobe wrapped around the posterior one, the two lobes are quite distinct functionally.

The anterior pituitary (adenohypophysis). The anterior lobe of the pituitary is essential for life. It is the master gland, for many of the hormones it produces control the growth, development, and activity of other endocrine glands (for example, thyroid, adrenals, and gonads). If an animal's pituitary gland is removed or does not function properly, these endocrine organs degenerate or become functionless. In some respects the pituitary functions as the conductor of an orchestra. The maestro calls each orchestral section into activity at the proper time so that a harmony of function is produced. Without a maestro the musicians play when and what they will (as they do while tuning up), and noise instead of music is produced. When the pituitary's stimulatory hormones (called **trophic hormones**) arrive at their particular target organ, they call that gland into activity, and it in turn secretes its own product (usually a hormone). The rate of hormonal secretion by the pituitary and its target glands is controlled by negative feedback mechanisms. They work this way: when the trophic hormone reaches the target organ, the latter produces hormone; when sufficient hormone is produced, it affects the pituitary gland so that output of trophic hormone is diminished. When, on the other hand, the target organ is undersecreting, the pituitary steps up trophic hormone production, and the target organ is triggered to greater activity (Figure 16–3d). Thus, by negative feedback the pituitary and its target organs are integrated in their function, and body activities are coordinated.

The anterior pituitary produces at least seven hormones, all of which are polypeptides, peptides, or proteins: three **gonadotrophic hormones,** which regulate sexual functions in men and women; **adrenocorticotrophic hormone** (ACTH), which controls one function of the adrenals; **thyrotrophic hormone,** which controls the thyroid; **somatotrophic (growth) hormone;** and **melanocyte-stimulating hormone,** which acts on the pigment cells, causing the skin to darken (Figure 16–4). Table 16–1 shows the roles and functions of the hormones of the anterior pituitary.

BOX 16A Giants and dwarfs

The mention of giants and dwarfs conjures up memories of the fairy tales we heard when we were young. Yet these people who live in a make-believe world pressed between the covers of a children's book have their counterparts in the real world (Figure 16–5). In engravings, paintings, and stories of the past we find descriptions of dwarfs, court curiosities believed to be endowed with magical talents. Today the dwarf is frequently regarded as a freak, a part of the circus menagerie. The most famous of all modern dwarfs were Charlie Stratton ("Tom Thumb") and his bride, Lavinia Warren, both less than 3 ft tall; they were displayed on many continents by P. T. Barnum, the master showman. Giants too occur in the real world. Robert Wadlow (1919–1940) of Alton, Illinois, measured 8 ft 10.75 in., weighed 495 lb, wore size-37 shoes, and towered over his 5-ft 11-in. father. He was appropriately called the "Alton Giant."

These extremes in size, the short and the tall, are a result of too little or too much pituitary hormone. The **pituitary dwarf** does not produce an adequate supply of human growth hormone; as a result, the growth rate during childhood is diminished, and in some cases adult size is only twice that at birth. The body is normally proportioned and intelligence is normal, but the physique remains juvenile and sexually immature, and the individual is usually sterile. Treatment of a pituitary dwarf is possible if the condition is diagnosed early and includes the injection of human growth hormone. There are probably about 5,000 pituitary dwarfs in the United States. The treatment period is long (5 years or more) and the hormone availability quite low. In the past the only source of human growth hormone was from the pituitary glands of deceased human donors. To treat a single patient with an underactive pituitary required the processing of hormone from 650 pituitary glands. At present only 50,000 glands are collected each year, hardly enough for all those in need of treatment. In 1971, Professor C. H. Li, an endocrinologist-biochemist at the University of California, and his colleagues announced that after 15 years of painstaking effort they had determined the order of the 188 amino acids in human growth hormone. This breakthrough may enable a production-line syn-

FIGURE 16–5 Too little or too much growth hormone. Insufficient quantities of growth hormone produce a pituitary dwarf; excessive amounts of this hormone result in gigantism. (Courtesy of Syndication International Ltd.)

  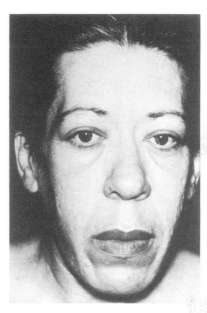

FIGURE 16–6 The progression of acromegaly. (a) At age 20, normal. (b) At age 24, early coarsening of features. (c) At age 40, well-established acromegaly with enlargement of nose and jaw. (Courtesy of Milton G. Crane, Loma Linda University Medical School.)

thesis of growth hormone in the laboratory (probably within 3 years); when that day arrives treatment of children suffering from dwarfism will become commonplace.

Overproduction of growth hormone (usually due to a pituitary tumor) can produce two conditions: if it occurs during preadolescent years it causes **gigantism,** a condition in which the individual is unusually large. If it occurs in an adult, whose long bones have matured so that the shaft has fused with the ends (Chapter 20), there is enlargement of the cartilaginous regions and the bony joints, producing a condition called **acromegaly.** Individuals with acromegaly show disproportionate thickening of the face, the feet, and the hands (Figure 16–6). The chin protrudes; the nose, lips, and extremities enlarge. The face of the individual tends to lack an intelligent expression; the individual suffers from lethargy and severe headaches. Treatment of acromegaly and gigantism is difficult because of the inaccessible position of the pituitary. If oversecretion is due to a pituitary tumor, some benefit may be obtained by surgical removal or bombardment with ionizing radiation (x rays).

TABLE 16–2 Posterior pituitary (neurohypophyseal) hormones*

Hormone	Target organ(s)	Principal effects	Diseases
Vasopressin or anti-diuretic hormone (ADH)	Smooth muscle, especially of arterioles and kidney tubules	Constricts blood vessels, thus raising blood pressure; stimulates resorption of water by kidney (Chapter 15)	Hyposecretion: diabetes insipidus (Chapter 15)
Oxytocin	Uterus and ducts of mammary gland	Stimulates contraction of uterus if primed by ovarian hormones; enables milk-let-down reflex by mammary gland	

*Produced by the hypothalamus and stored in the posterior pituitary.

The posterior pituitary (neurohypophysis). The posterior lobe of the pituitary is both structurally and functionally an extension of the hypothalamic region of the brain (Figure 16–3c). Although at one time it was believed to manufacture the hormones **vasopressin** and **oxytocin,** it is now known that these hormones are produced by the hypothalamus and are merely stored in the posterior lobe. This again shows how closely related the nervous and endocrine systems are, since oxytocin and vasopressin are produced by nerve cells, flow along the nerves in the stalk connecting the hypothalamus and the pituitary, and are later released from the pituitary gland. A summary of the functions of the hormones of the posterior pituitary is given in Table 16–2.

The thyroid gland: controlling the fires of life

The **thyroid gland** is a butterfly-shaped mass of tissue lying in the front of the neck, just below the voice box (Adam's apple). If you place your finger into the notch at the upper-most border of the breast bone the tip of your finger will touch its lower edge. The thyroid produces three hormones: **thyroxine, triio-dothyronine,** and **calcitonin.**

Thyroxine and triiodothyronine. Maria de Medici, wife of Henry IV of France, sat for the painter Peter Paul Rubens in 1625, and the portrait (now hanging in Spain's Prado museum) is today considered a masterpiece. The painting not only attests to Rubens' skill as an artist and his fine eye for detail, but shows quite clearly that Henry's wife suffered from a mild **goiter** (Figure 16–7), an enlarged thyroid gland (*gut-tur*—throat; L.). Many portraits and paintings from this period show the regular and rounded neck, an obvious sign of goiter but considered a mark of beauty in the late Renaissance. Goiter is characteristic of people living in those parts of the world where there is an insufficient iodine supply in the food and water, such as the Alps of Europe and the Great Lakes region of the United States. Although mild goiter might be considered beautiful, extreme enlargement of the gland creates a far from pleasing appearance (Figure 16–8). Goiter is often associated with other symptoms, including dry and puffy skin, loss of hair, lethargy, mental dullness, and a slower-than-normal heartbeat. How is goiter related to thyroid function?

The development of the goiterous condition results from malfunction of the feedback mechanism that operates between the pituitary and its target organs. The thyroid hormones thyrox-

ine and triiodothyronine are iodine-bearing derivatives of the amino acid tyrosine. Both these hormones govern the rate of metabolism of the body cells (although triiodothyronine is 5–10 times as active as thyroxine). High quantities accelerate metabolism, and deficiencies of hormone decrease the metabolic rate. In some respects these thyroid hormones are analogous to the gas pedal in an automobile, which regulates the amount of gasoline fed into the engine and thus controls the rate of the auto's performance (speed and power). Tyrosine can be produced from a wide variety of substances that occur in the body and therefore offers no problems of supply, but iodine must be obtained in the food or water. Production of the thyroid hormones is regulated by the pituitary gland in the following way: Iodine taken by mouth is absorbed by the gut and makes its way via the bloodstream to the thyroid gland. The pituitary secretes thyroid-stimulating hormone **(TSH),** which steps up the trapping of iodine by the thyroid and aids in the synthesis of thyroid hormones (Figure 16–9). Under normal circumstances, the system is self-regulating: an excess of thyroid hormones in the blood suppresses secretion of TSH by the pituitary, and thyroid activity diminishes; when the thyroid hormone level is too low, the pituitary

FIGURE 16–7 Marie de Medici, painted by Peter Paul Rubens in 1625.

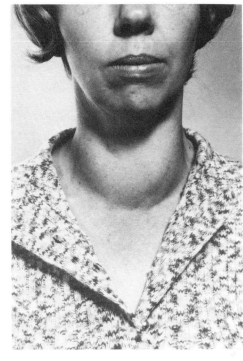

responds by secreting more TSH, thus building up the thyroid's activity (Figure 16–10). If the diet is deficient in iodine, synthesis of thyroid hormones is diminished; this releases the brakes on the pituitary, and production of TSH is increased. As a consequence of the oversecretion of TSH, the thyroid gland enlarges and the goiter forms. Increased size of the thyroid may or may not compensate for the hormonal insufficiency.

Disease and abnormal development may cause a goiter, but most commonly the cause is simply a dietary lack of iodine. Minute quantities of iodine in the diet (as little as 4 gm per year) prevent development of the simple goiter. The introduction of iodized salt has markedly reduced the incidence of goiter in many parts of the world, although as recently as 1960 there were 200 million people afflicted with the condition. Unfortunately, dietary iodine has little effect on an already established goiter.

The symptoms of thyroid deficiency, as seen in individuals with severe goiter, reflect a lowered metabolic rate; the manifestation of other thyroid disorders depends on when the deficiency of hormone sets in. If severe thyroid deficiency starts in early infancy or prior to birth, the result is a mentally defective dwarf, a **cretin** (Figure 16–11a). A cretin never matures sexually, and growth never proceeds beyond that achieved by a 7 or 8 year old. If thyroid hormone administration is begun early enough, there may be some restoration of normal development, but in most cases of arrested development normal growth patterns cannot be initiated again. If thyroid deficiency develops in childhood or adult life, the disease is called **myxedema** (Figure 16–11b). Juvenile myxedema produces a short, squat child with an enlarged head, a short and heavy neck, dry skin, and a face with a dull and stupid expression. In an adult myxedema produces lethargy, obesity, edema, slowed heartbeat, decreased intelligence, and a coarseness of the hair and skin; the individual feels cold—the metabolic fires have been damped. Thyroid hormones administered orally often dramatically reverse these hypothyroid conditions, and the person can lead a normal life by taking daily medication.

FIGURE 16–9 (above) The thyroid hormones and their pathway of synthesis.

FIGURE 16–10 (below) Goiter, the result of impaired negative feedback control in the formation of thyroid hormone.

FIGURE16–11 Too little thyroid hormone: childhood and adult effects. (a) Cretinism in a 2.5 year-old child. (b) Myxedema. (Courtesy of Sanford Schneider, Loma Linda University Medical School.)

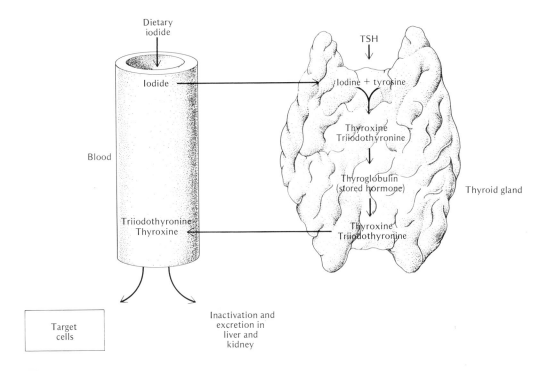

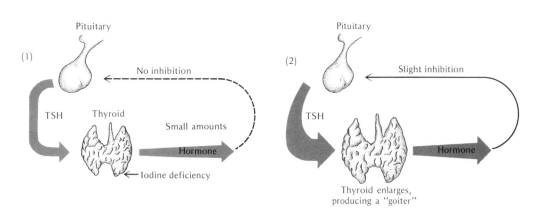

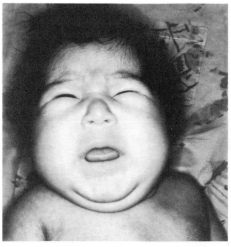

(a)

(b)

FIGURE 16–12 (above) Hyperthyroidism (Graves's disease), with exophthalmos (protruding eyes). (Courtesy of J. Nelson, Loma Linda University Medical School.)

FIGURE 16–13 (below) Scan of the thyroid area after administration of radioactive iodine (^{131}I).

FIGURE 16–14 A summary of the control of calcium metabolism by parathormone, calcitonin, and DHC (vitamin D).

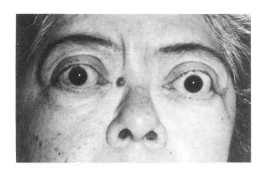

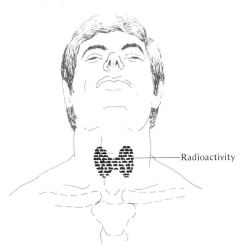

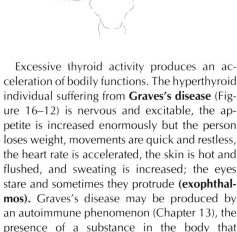

Radioactivity

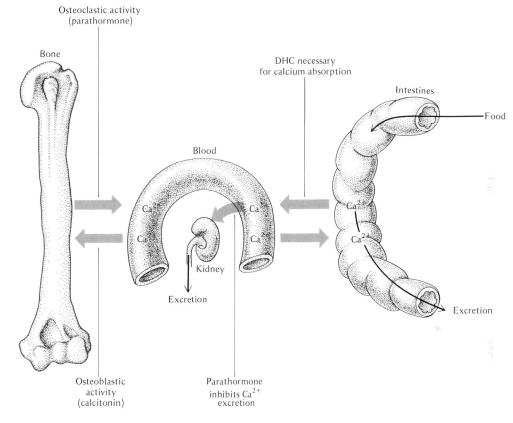

Osteoclastic activity (parathormone)

Bone

DHC necessary for calcium absorption

Intestines

Food

Blood

Ca^{2+}

Ca^{2+}

Ca^{2+}

Ca^{2+}

Ca^{2+}

Ca^{2+}

Kidney

Excretion

Excretion

Osteoblastic activity (calcitonin)

Parathormone inhibits Ca^{2+} excretion

Excessive thyroid activity produces an acceleration of bodily functions. The hyperthyroid individual suffering from **Graves's disease** (Figure 16–12) is nervous and excitable, the appetite is increased enormously but the person loses weight, movements are quick and restless, the heart rate is accelerated, the skin is hot and flushed, and sweating is increased; the eyes stare and sometimes they protrude (**exophthalmos**). Graves's disease may be produced by an autoimmune phenomenon (Chapter 13), the presence of a substance in the body that prolongs thyroid stimulation, or a tumor of the thyroid gland. Treatment involves administration of propylthiouracil (a drug that blocks the synthesis of thyroid hormone), surgical removal of a major portion of the gland (thyroidectomy), or the injection of radioactive iodine (^{131}I). In the last case, advantage is taken of the fact that an overactive thyroid concentrates massive amounts of iodine; 80%–90% of the injected radioactive iodine is localized in the thyroid (Figure 16–13), enough radioactivity to destroy the hyperactive gland cells.

Calcitonin. Bones function as structural supports, but they also play a role in the metabolism of calcium. About 99% of the calcium in the body is deposited in the bones (Chapter 20); the remainder is found in the blood and tissue fluids, where it is important in blood clotting, in neuromuscular function, and in cementing cells together. The levels of calcium in the body fluids are controlled and regulated by the hormones **calcitonin** (produced by the thyroid gland) and parathormone (produced by the parathyroids).

Calcitonin decreases the level of calcium in the body fluids by acting on specialized bone cells called **osteoblasts** (Chapter 20) so that there is increased deposition of calcium salts in the bones (Figure 16–14). Calcitonin secretion is increased when the blood calcium level

BOX 16B Rickets

Rickets is due primarily to insufficient deposition of calcium in growing bones. Although rickets is generally considered to result from a dietary lack of vitamin D, it is in fact not caused by a poor diet but by a deficiency of the sun's ultraviolet radiations. How is ultraviolet light related to vitamin D, and what do both of these have to do with rickets?

Normally, ultraviolet radiation from the sun acts on a prehormone (7-dehydrocholesterol) that occurs naturally in the skin and converts it to a substance called vitamin D_3 (or calciferol). Vitamin D_3 is released from the skin into the bloodstream and reaches the liver, where it undergoes chemical conversion (into

25-hydroxycalciferol) and then travels to the kidney via the blood to become the active material 1,25-**dih**ydroxy**c**alciferol **(DHC).** DHC promotes the uptake of calcium from the intestines and aids in the deposition of calcium salts in the bone. Thus, any factor that reduces the amount of sunlight (and, therefore, of ultraviolet light) reaching the skin reduces the formation of vitamin D_3 and DHC. When DHC is not manufactured, calcium cannot be absorbed from the intestine, and if such a condition persists for many months, the blood calcium level drops and calcium in the bone is automatically removed (under the control of parathormone) to reestablish the normal concentration. The depletion of calcium salts from the bones causes a severe weakening and bowing of the bones: rickets. (Lack of calcium in the diet per se rarely is a cause of rickets.)

The history of rickets is quite interesting in itself. As early as 1650 the disease was described in England; its appearance

FIGURE 16–15 **The feedback loops involved in the regulation of the blood level of calcium. Compare the regulatory process shown here with the feedback circuit in the thermostat (Figure 7–17).**

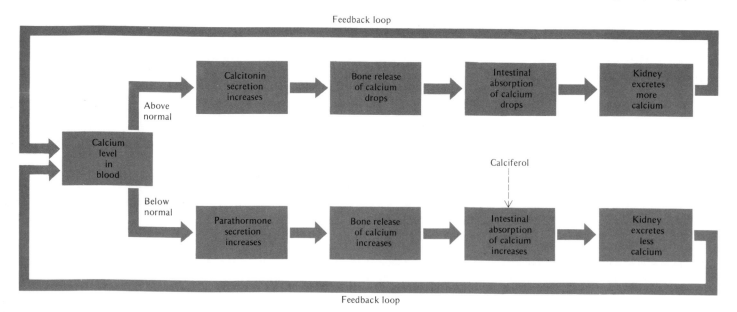

coincided with the Industrial Revolution, which produced widespread use of coal and smoke and increased population densities in the cities with their narrow, dingy, and sunless streets. Rickets was probably the first recognized disease associated with air pollution. Many European physicians noted that rickets was not only a disease of industrialized and urbanized populations, but that it had a seasonal variation. The long, dark winters of Europe and the prolonged confinement of infants during these months powerfully predisposed these children toward rickets. Indeed, it was noted in 1906 that nearly all German children who were born in the fall and died in the spring had rickets, whereas those who were born in the spring and died in the fall were free of rickets. Thus, by the early 1900s sufficient evidence had accumulated to show that rickets was caused by a lack of sunlight. The problem that confronted the physicians was: How can sunlight be provided during the European or North American seasons when the amount of solar radiation is low? In the 1920s it was discovered that ultraviolet irradiation of linseed oil, cottonseed oil, or the plant sterol ergosterol, made such materials capable of curing rickets. In fact, the irradiation of ergosterol is now a routine procedure, and the resultant material, ergocalciferol or vitamin D_2 is added to all milk sold in the United States and Europe (0.01 mg per quart); since vitamin D_2 is a metabolic precursor of DHC, it prevents rickets. If you look at a carton or a bottle of milk you will find the words "400 U.S.P. units of Vitamin D added per quart." The demonstration that dietary materials could cure rickets was further strengthened by the finding that cod-liver oil had a preventive therapeutic effect against rickets. Soon the belief that diet influenced rickets became widespread, and when laboratory animals were fed diets deficient in vitamin D and developed rickets it was "established as fact" that rickets was a vitamin-deficiency disease. Over the years the evidence that rickets is a climate-related disorder was forgotten and the vitamin basis became widely accepted.

Is DHC a vitamin or a hormone? The answer lies in its chemical structure and metabolic role. Calling DHC a vitamin implies that it is a chemical component of a coenzyme, as are all vitamins; but calling DHC a hormone links it to the other hormones (parathormone and calcitonin) that function in the control of calcium and phosphate metabolism in the body (Figure 16–15). Moreover, DHC is a steroid, synthesized in the body as are all other hormones; it is not a dietary substance. Thus vitamin D is a misnomer for a substance that is actually hormonal in its action—a misnomer perpetrated not only in textbooks, but on every carton of milk sold in the United States.

is raised above normal, and this promotes removal of calcium from the blood and its storage in the bones. Insufficient deposition of calcium in the bones during childhood causes **rickets** and is linked to calcium uptake from the intestine and the action upon it of a hormone known as **calciferol** or **vitamin D** (Box 16B).

The parathyroid glands

The parathyroids are four small egg-shaped glands about the size of a raisin, embedded in the back surface of the thyroid. They produce **parathormone,** which plays a vital role in the metabolism of calcium and phosphorus. It regulates the level of calcium in the body fluids by causing an increase in the number and size of certain bone cells, called **osteoclasts** (Chapter 20); these rapidly invade the bone, digesting away large quantities of the bony matrix. Simultaneously calcium is dumped into the extracellular fluids of the body. Since calcium in the bone is bonded to phosphate as $CaPO_4$, phosphate is released along with the calcium. Parathormone compensates for the release of phosphate into the blood by stimulating phosphate excretion by the kidneys. At the same time parathormone inhibits the excretion of calcium by the kidneys, and thus the calcium level in the blood rises (Figure 16–14).

The secretion of parathormone is regulated by environmental factors. For example, when the blood level of calcium is low, owing to an insufficiency of calcium in the diet or a lack of calciferol (or vitamin D, which controls uptake of dietary calcium from the intestine; see Box 16B), the parathyroid glands increase their output of parathormone, and the calcium level of the extracellular fluids increases (derived from the resorption of bone). Ordinarily it takes very little bone resorption to maintain an adequate level of calcium in the body fluids, and since the bones contain a thousand times the content of the fluids, the structure of the bone itself is not impaired. Thus, bone forms a reservoir of calcium, and parathormone regulates the levels of calcium in the body with great sensitivity.

Overactivity of the parathyroids can cause the removal of so much calcium from the bones that they become honeycombed with cavities and so fragile that even walking can induce fractures. Since the parathyroids are embedded in the thyroid and are difficult to distinguish, the most common cause of parathormone deficiency is the inadvertent removal of the parathyroid glands during surgical removal of a portion of the thyroid (for example, in the treatment of Graves's disease). If the calcium level falls to half that required for normal nerve function, the person immediately develops tetany and dies of spasms in the respiratory muscles (Chapter 14). If parathormone is administered in time, however, bone resorption occurs, and the increased level of extracellular calcium restores normal function.

Parathormone thus produces an effect opposite to that of calcitonin, but it differs from calcitonin in another way: it acts slowly, whereas calcitonin acts quickly. It may take hours for the effects of parathormone to become apparent. The secretion of parathormone and the secretion of calcitonin are complementary mechanisms that precisely govern the levels of calcium in the extracellular fluids (Figures 16–14, 16–15).

The pancreas: control of fuel supply

Diabetes mellitus, meaning "sugar siphon," was described by Hippocrates over 2,000 years ago as a condition in which the urine contains excessive amounts of sugar. In seventeenth-century England, the disease was called the pissing evil because there was a marked loss of weight in spite of huge food intake, and it was believed that the substance of the body was being dissolved and poured out through the urinary tract. Before 1913, a child suffering from diabetes would die within 2 years, and if the disease developed at maturity, the afflicted individual could expect to live only 3–4 years. Today, a diabetic individual can expect to live for 60 years or more. The miracle cure for this disease is administration of the hormone **insulin.**

Insulin is produced in the **pancreas,** a gland that lies just beneath the stomach (Figure 16–16). The pancreas is a double gland: it produces digestive enzymes and hormones. In 1869 a German anatomist, Paul Langerhans, observed that the pancreas contains tiny clumps of cells that are sharply separated from the surrounding glandular tissue; these clumps make up only 1%–2% of the total mass of the pancreas and appear as tiny cellular islands or, as Langerhans called them, islets. The **islets of Langerhans** are endocrine glands that produce insulin. It is believed that insulin facilitates the active transport of glucose into the cell across the plasma membrane. In the presence of insulin the entry of glucose is 25 times faster than in its absence. The exact mechanism by which insulin affects the plasma membrane remains uncertain.

How is insulin related to diabetes mellitus? When we eat starch or sugar, enzymes in the digestive juices convert them to glucose; this simple sugar is transferred across the intestinal wall and passes into the blood. When the blood glucose level begins to rise, the islets of Langerhans in the pancreas are triggered to secrete insulin. Part of the absorbed glucose remains in the circulation, but most of it is carried to the liver, where it is transported rapidly across the cell membrane under the influence of insulin and stored as glycogen.

Diabetes may be due to insufficient activity of the cells in the islets of Langerhans or to the presence of circulating antagonists to insulin. Insulin deficiency causes glucose to accumulate in the blood instead of being transported into the cells, and the unused glucose spills over into the urine. The excess amount of glucose in the glomerular filtrate of the kidney diminishes water absorption, and there is excessive output of dilute urine. To compensate for the lack of available fuel (glucose) and energy for cellular activities, fat is mobilized, and fatty acids are released; protein is also used as a source of energy. Repair of injured tissue is slowed, resistance to infection is low, ketone bodies as well as keto acids build up in the blood, and acidosis develops; the urine and breath may smell of acetone. Glucose-deprived cells degenerate, the individual eats voraciously, but remains hungry and loses weight. Acidosis may become severe, resulting in coma and death. Prolonged diabetes causes heart disease, kidney damage, and atherosclerosis. Diabetes is treated simply: daily injections of insulin and a restricted intake of carbohydrate and salt. Most diabetics live active, normal lives when properly treated.

Too little insulin causes cells to be deprived of glucose because the cells are impermeable to the fuel. Too much insulin (because of a pancreatic tumor or injection of too large a dose of insulin) produces the same effect, but for a different reason. Excessive quantities of insulin promote too rapid an uptake of blood glucose; low blood glucose results, and this causes brain cells to be deprived of a constant supply of fuel energy. The brain cells are extremely sensitive to blood glucose concentration and become highly excitable, convulsions occur, then the nerve cells of the brain become depressed, and the individual falls into a coma. Treatment for this may be simply the administration of glucose.

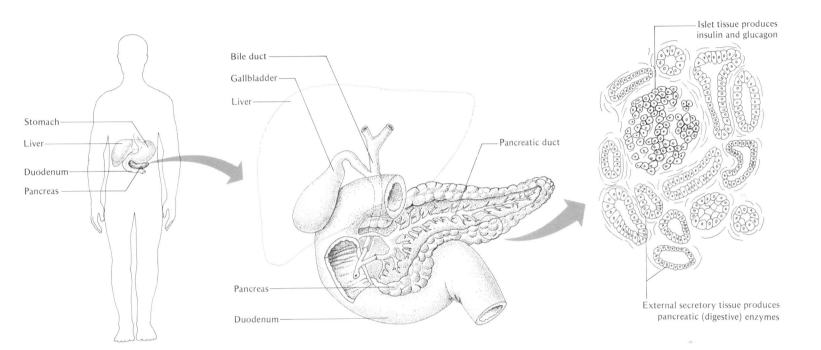

FIGURE 16–16 The pancreas produces both hormones and digestive enzymes. The islets of Langerhans are endocrine organs; they secrete the hormone insulin into the bloodstream.

FIGURE 16–17 (below) Adrenal gland sliced to show internal organization.

Stomach
Liver
Duodenum
Pancreas

Bile duct
Gallbladder
Liver
Pancreatic duct
Pancreas
Duodenum

Islet tissue produces insulin and glucagon

External secretory tissue produces pancreatic (digestive) enzymes

If insulin turns up the set point of the metabolic thermostat, what turns it down? The hormone **glucagon.** Glucagon is an antagonist of insulin, and its secretion by the pancreas[1] results in the breakdown of liver glycogen and release of glucose into the blood. (Glucagon also promotes the release of epinephrine by the adrenal glands.) Secretion of glucagon is elicited by a fall in blood glucose level (below 60–80 mg per 100 ml of blood), and an immediate release of glucose from the liver restores the normal level (90–100 mg per 100 ml).

The adrenal glands

The adrenal glands (ad—on, ren—kidney; L.) sit perched like a three-cornered hat on the upper surface of each kidney (Figure 16–17). Each gland is really a gland within a gland, consisting of a pea-sized central core, the **medulla,** and a thicker outer bark, the **cortex.**

The adrenal medulla. In the developing embryo the medulla arises from nervous tissue; it is really an overgrown sympathetic ganglion whose cell bodies do not send out nerve fibers, but instead release their chemical products directly into the bloodstream, again illustrating the close relationship between the nervous and endocrine systems. The hormone released by the medulla is for the most part **epinephrine** (epi—on, nephros—kidney; Gk.), with smaller

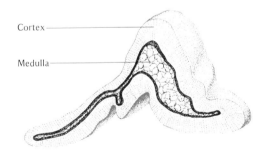

Cortex
Medulla

quantities of a closely related substance **norepinephrine.**[2]

---

[1] The islets of Langerhans contain two kinds of cells, alpha and beta: alpha cells produce glucagon and beta cells produce insulin.

---

[2] Epinephrine is equivalent to the substance known by the trade name Adrenalin; likewise norepinephrine is the same as Noradrenalin.

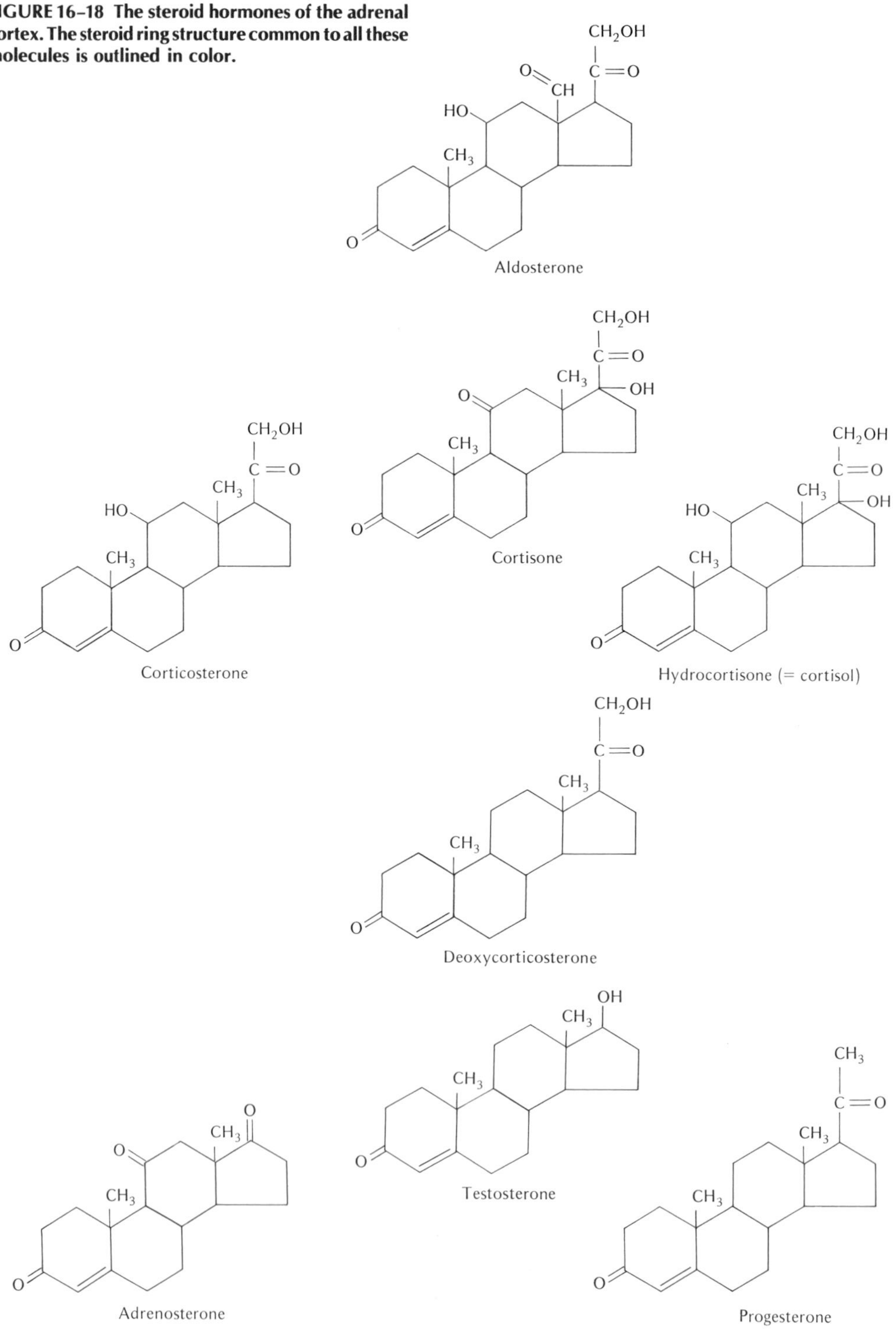

FIGURE 16-18 The steroid hormones of the adrenal cortex. The steroid ring structure common to all these molecules is outlined in color.

Aldosterone

Cortisone

Hydrocortisone (= cortisol)

Corticosterone

Deoxycorticosterone

Adrenosterone

Testosterone

Progesterone

What does epinephrine do? If epinephrine is injected into the body, a complex series of physiological changes occurs: arterioles constrict, and the blood pressure is elevated; liver glycogen is mobilized, and the blood sugar level rises; the pacemaker of the heart is activated, and cardiac output is increased; blood vessels in the skeletal muscles dilate, as do the coronary arteries; the bronchioles dilate, permitting maximal gas flow to the lungs; the spleen contracts, ejecting the stored blood cells into the bloodstream; smooth muscles in the intestine are relaxed, and the sphincters of the anus, urethra, and stomach are constricted. In short, the body is mobilized for unusual exertion. In times of stress, anxiety, or fear the adrenal medulla is activated, and this series of responses is collectively called the **emergency response,** or the **fight-or-flight reaction.** These emergency-response activities of the adrenal medulla are like those produced by the sympathetic nervous system (Chapter 17). The similarity is easily explained by the fact that sympathetic nerves secrete norepinephrine at their ends. Because of this, loss of the adrenal medulla is not fatal; it is backed up by compensatory activity of the nerve fibers of the sympathetic system, and the body can function normally.

The adrenal cortex. For more than a century it has been known that the adrenal cortex (unlike the medulla) is absolutely essential for life. In 1855 Thomas Addison, an English physician, described a rare condition in a patient who had suffered from tuberculosis. The afflicted individual showed a bronzing of the skin, muscle weakness, impairment of kidney function, water retention in the tissue spaces, loss of weight, decreased levels of blood glucose, apathy, and a severe decline of sodium in the blood and tissue fluids. Shortly thereafter the patient died, and upon autopsy

FIGURE 16–19 An abbreviated scheme of the synthesis of the adrenal cortical hormones from cholesterol.

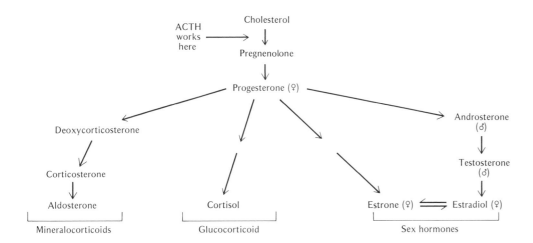

FIGURE 16–19 An abbreviated scheme of the synthesis of the adrenal cortical hormones from cholesterol.

it was found that his adrenal cortex was almost completely functionless. The disease syndrome came to be called Addison's disease; President Kennedy had a mild form of this disease (Box 16C).

Addison's disease involves the failure of the adrenal cortex to secrete its hormones. The cortex of the adrenal gland is literally an endocrine factory secreting many hormonelike materials (about 50 have been identified so far), of which 10 are very active. All the hormones produced by the cortex have the same basic chemical ring structure (Figure 16–18), referred to as the **steroid ring** (Figure 3–10), and they are called **adrenocorticosteroids** or **corticosteroids.** Although the steroid hormones of the adrenal cortex appear to have only minor chemical differences, they do have strikingly different functional properties. The parsimony of chemical synthesis coupled with target-organ specificity produce the economy and variety of control exercised by these molecules. The starting material for the adrenal hormones is cholesterol, and by a complex branching pathway involving addition and deletion of atoms of oxygen, hydrogen, or other small molecules, a group of steroids is produced (Figure 16–19). The steroid hormones fall into three types: (1) **glucocorticoids,** which act primarily in regulating sugar and protein metabolism; (2) **mineralocorticoids** (aldosterone, corticosterone, and deoxycorticosterone), which regulate the amount of sodium and other minerals in the extracellular fluids; and (3) the **sex hormones** (androsterone, testosterone, estrone, and estradiol), which affect the secondary sex characteristics. The synthesis of these hormones is affected by the pituitary gland and by enzymes. ACTH, secreted by the pituitary, affects only the first step (that is, the conversion of cholesterol to pregnenolone); subsequent reactions and conversions are enzyme-catalyzed. If the pituitary gland does not secrete ACTH, the adrenal glands remain inactive; if the cortex is underactive, owing to disease or some other cause, the individual shows the signs of Addison's disease (Box 16C). Table 16–3 summarizes the functions of the most important hormones secreted by the adrenal cortex.

Sex hormones

Development from youth to maturity in humans is marked by dramatic changes in both mind and body, and these changes require coordination. The regulatory processes need not be rapid in action, but they must be continuous, often extending over months and years. The mechanisms for effecting such controls are hormones, especially the sex hormones.

Male hormones (androgens). Removal of the testes **(castration)** clearly reveals that maleness is related to this organ. The male hormones (androgens) are secreted by the testicular interstitial cells situated between the sperm-producing seminiferous tubules. Two androgens are produced, **testosterone** and **androstenedione,** of which testosterone is the predominant one. At puberty the pituitary gland begins to secrete large amounts of ICSH, and this triggers the testis to produce testosterone. Aided by FSH, the testis increases in size, and early spermatogenic activity is induced. Acting in concert, these hormones ensure maturation of sperm, and ejaculation becomes possible. The androgens of the testis are aided by those secreted by the adrenal gland **(androsterone),** but the testicular hormones remain the dominant influence in the production of masculine characteristics (Chapter 9).

Hypogonadism (deficient gonadal function) is present in 0.13% of all males. If it occurs before puberty and involves only pituitary gonadotrophins, testicular activity is impaired, spermatozoa and male hormones are not manufactured, and a eunuchoid individual results. The physique remains adolescent, the pubes and the face remain hairless, the features are delicate, the voice remains high-pitched, there is little sexual desire, and fat tends to accumulate in the hips, breasts, and below the umbilicus. Similar consequences arise from

BOX 16C The critical cortex

The early 1960s were years of optimism and youthful vigor—or so it appeared to some in the United States at the time. John F. Kennedy, 44 years of age, was the president, and his face and mannerisms contributed to the country's feeling of optimism (Figure 16–20). The Peace Corps was formed; the country survived the Cuban missile crisis, but got deeper into war in Vietnam. Although war always loomed on the horizon, the nation seemed to be in good hands. The president had charisma: a flashing smile, sparkling eyes, an easy air, and a deeply tanned skin. However, all in fact was not as it should have been: the smile and relaxed manner covered pain and illness. During the presidential campaign of 1960, the nature of Kennedy's illness came to light. Once or twice a year he required examinations to check on his adrenal insufficiency (Addison's disease). Kennedy had to take regular doses of cortisone from 1947 to 1951 and again from 1955 to 1958 to combat the disease. He said he took oral doses of corticosteroids "frequently when I have worked hard." Full disclosure of John F. Kennedy's adrenal disorder and the effects of the administered hormones on his physical condition has never been made. Yet, it could be said that for a time the health of the chief executive and the fate of the United States depended upon the condition of a pair of adrenal glands.

A patient with Addison's disease cannot cope with stress or resist infection, largely because insufficient amounts of cortisol are produced. Cortisol not only regulates the storage of glycogen in the liver and the amount of glucose formed from protein, but it also reduces the action of insulin on peripheral tissues. Thus, in a person suffering from adrenal insufficiency, the blood sugar level may fall dangerously low. As a consequence the individual suffers from lethargy and muscle weakness.

Cortisol[3] is an especially effective antiinflammatory agent. Presumably it acts by promoting the mobilization of amino-acid reserves so that the extracellular fluids rise in amino-acid content and these are utilized for cellular protein synthesis, thus aiding in the repair and replacement of dead or dying cells.

[3] The common name for cortisol is hydrocortisone.

Cortisol also affects other endocrine glands: it suppresses thyroid function by inhibiting TSH secretion, and it inhibits MSH secretion. Thus, in Addison's disease the skin bronzes (because more MSH is secreted), and the activity of cellular respiration is diminished (owing to underproduction of TSH with consequent decline in the amount of thyroxine circulating in the blood.

The actions of cortisol are so important to the functioning of the body that there exists a special feedback control to ensure that adequate levels are always present in the blood and that these amounts will be increased when necessary. Stress, shock, pain, or any physical damage to the body triggers the hypothalamus to secrete a releasing factor that passes to the pituitary by way of the blood. This, in turn, causes the release of ACTH, which stimulates the adrenal cortex to synthesize and release cortisol into the bloodstream. As the level of cortisol rises, it inhibits further release of ACTH.

FIGURE 16–20 John F. Kennedy. (Courtesy of Wide World Photos, Inc.)

TABLE 16–3 Adrenal cortical hormones

Hormone	Chemical nature	Target organ(s)	Principal effects	Diseases
Aldosterone	Steroid (mineralocorticoid)	Kidney	Controls absorption of sodium by kidney tubules	Hypersecretion: retention of water in tissues (edema) and high blood pressure
Cortisol	Steroid (glucocorticoid)	Body tissues	Regulates storage of glycogen by liver and conversion of protein into glucose; reduces action of insulin on peripheral tissues	Hyposecretion: Addison's disease (Box 16C) and masculinization (adrenal virilism) Hypersecretion: Cushing's syndrome
Androgens	Steroid	Body tissues	Controls development of secondary sex characteristics	Hypersecretion: precocious sexual development Hyposecretion: infantilism

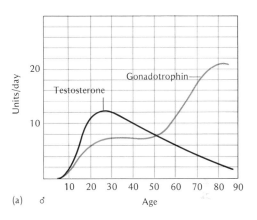

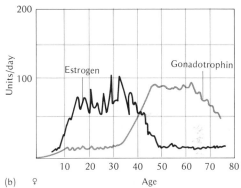

prepubertal castration and **cryptorchidism** (failure of the testes to descend from the pelvic cavity into the scrotum).

Hypogonadism may also occur after puberty, owing to disease (mumps virus causes inflammation of the testes in 20%–30% of affected adult males, but rarely leads to male sterility if the disease is properly treated by a physician), radiation, castration, or chromosomal abnormalities such as Klinefelter's syndrome (Chapter 22). The form of the body remains approximately normal, but there is some enlargement of the breasts and excessive deposition of fat, owing to a lack of testosterone production; such individuals are sterile.

The opposite side of the coin, **hypergonadism** in the male (caused by hereditary factors or by tumor of the pineal, adrenal, pituitary, or testis), leads to excessive development of the genitalia and of the secondary sex characteristics: broad shoulders, hairy muscular body, etc. Prepubertal hypergonadism in males causes precocious sexual development and shortness of stature because fusion of the ends of the long bones with the shaft occurs prematurely.

In their late forties and early fifties many men experience **male climacteric** (male menopause), with a diminution of sexual drive. This is due partly to psychological factors and partly to a decline in secretion of androgens (Figure 16–21) coupled with a rise in secretion of FSH; occasionally there is atrophy of the seminiferous tubules. In some cases of male climacteric the individual benefits from administration of testosterone.

Female hormones. The feminine characteristics of bodily form, menstruation, ovulation, pregnancy, and menopause are all dependent on the female hormones **estrogens** and **progesterone**. The estrogens are produced from cholesterol by the follicle cells of the ovary under the stimulus of the pituitary gonadotrophins (FSH and LH) (Figure 16–19). The pituitary does not secrete these trophic hormones in significant quantities until about age 7–8 years. Then production increases to age 11–13 years, after which time the level remains as in the adult (Figure 16–21). Once the gonadotrophin levels are high enough, the ovaries begin to produce estrogen and progesterone, and body

changes become evident. The breasts enlarge, the pelvis broadens, the uterus increases in size, pubic hair develops, and menstruation and ovulation begin. (The hormonal interactions during the ovarian and uterine cycles, menstruation, menopause, and pregnancy are discussed in Chapter 9.)

Estrogen concentrations in women reach their peak at puberty, and this level tends to inhibit pituitary secretion of growth hormone. Thus at puberty or shortly thereafter in the female, long bones cease to grow and the ends of the bones fuse with the shaft. Male hormone produces no such inhibition of growth hormone, and consequently men tend to be larger in stature than women. Estrogen inhibits the growth of hair on the face, limbs, and trunk, and increases hair growth on the scalp; it causes fat to accumulate in the hips and breasts and promotes the retention of body water. Female hormones, particularly estrogen, tend to reduce the level of cholesterol in the blood, a fact which may have some bearing on the lower incidence of atherosclerosis in women.

If the ovaries of a female are removed prior to puberty, a childlike physique is retained; if they are removed after puberty, the physique becomes masculine, sex drive diminishes, and menstruation ceases. Treatment of a normal female with male hormone produces masculinization. Overactive adrenal glands and ovaries have been shown to produce excessive amounts of androgens and this can lead to baldness and excessive facial hair in women. It is apparent that femaleness depends on suppression of male characteristics.

Other endocrine organs

It has recently been discovered that the **thymus gland,** situated in the neck region (Figure 16–2), secretes a hormone essential for functional maturation of lymphocytes involved in the immune response. The **pineal gland,** a small pea-shaped organ lying on the upper surface of the brain (Figures 16–2 and 17–14c), produces a hormone called **melatonin,** which has recently been implicated in the control of reproduction. Light falling on the retina of the eye increases the synthesis of melatonin by the pineal, and this inhibits ovarian function either directly or indirectly via the pituitary. The onset of puberty is earlier in girls blind from birth than in their normal counterparts, presumably because the blind girls do not produce melatonin and there is less inhibition of ovarian function. Moreover sexual development is de-

layed in children with pineal tumors owing to increased activity of the pineal gland.

The plasma membranes of the cells of a variety of tissues have recently been found to synthesize a family of 20-carbon fatty acids in small amounts. Called **prostaglandins,** these are rapidly degraded, but seem to perform a variety of roles in the body. "Prostaglandins," as one research worker put it, "are hormone-like substances that do anything and everything." He may not be exaggerating. The name prostaglandin is a misnomer based on the erroneous belief that the active substances are derived from the prostate gland alone. Seminal fluid contains 100 times the amount found in any other tissue, and it is probably for this reason alone that prostaglandins were discovered.

One prostaglandin (PGE_2) lowers blood pressure; a closely related one ($PGE_2\alpha$) raises the blood pressure. An intravenous injection of a low dose of either of these stimulates contractions of the uterine muscles. Since prostaglandins are found in the amniotic fluid of women during labor, it very well may be that these substances aid in delivery. If a woman in labor is experiencing difficulty, the injection of a small amount of PGE_2 induces delivery in a matter of hours. Some have suggested that PGE_2 could be a useful agent for inducing abortions. In trials with female monkeys, an injection of PGE_2 shortly after mating produces a sharp reduction in the secretion of progesterone by the corpus luteum, thus preventing implantation of the fertilized egg. This means that PGE_2 could potentially be used as a morning-after, or after-the-fact, contraceptive agent.

PGE_2 reduces arterial pressure and aids patients with hypertension. This same material inhibits gastric secretions in dogs and may be a promising new tool in the prevention of human peptic ulcers. Administered as an aerosol, it relaxes the bronchial passages and aids

asthmatic patients in their breathing. Applied topically to the nose, PGE_2 clears nasal passages by constricting blood vessels. These fatty acids inhibit the release of norepinephrine by sympathetic nerves and thus act as a brake on nerve transmission in the sympathetic nervous system. Prostaglandins may regulate the function of the plasma membrane itself, and in so doing may influence the movement of materials into and out of cells, the formation of cellular metabolites responsible for cell-cell communication, and consequently the transmission of signals from nerve cell to nerve cell.

The future holds great promise for investigation of these substances, and no doubt by the time you read these words we shall know a great deal more about how they act at the molecular level.

16–3 MANAGEMENT OF THE ENDOCRINE ORCHESTRA

The orchestra of endocrine organs is conducted by the maestro, the pituitary gland. The interaction between the pituitary and the other endocrine glands is delicately balanced by negative feedback controls mediated by the hypothalamic region of the brain. The **hypothalamus** is a small mass of brain tissue lying just above the pituitary (Figure 16–3c). The only physical connection between it and the anterior pituitary is a network of capillaries (Figure 16–3c and 16–22). The hypothalamic nerve cells produce substances called **neurohumors** or **neurosecretions,** that are directly conveyed down the length of the nerve cell (by protoplasmic movements called **axoplasmic flow**) and are released locally at the nerve ending (Figure 16–22). The nerve endings come in close contact with the surface of the pituitary; the neurosecretions are picked up by a special capillary network, which moves them on to the pituitary, where they affect pituitary

FIGURE 16–22 The relation of the anterior pituitary
to the hypothalamus. Releasing factors from the
hypothalamus pass to the pituitary via the blood, and
these trigger secretion of the trophic hormones.

function. The hypothalamic neurosecretions
are polypeptides called **releasing factors** and
influence the anterior pituitary to secrete its
trophic hormones. There are five releasing
factors (really hormones by our definition),
each responsible for the secretion of a particular
trophin: FSH-releasing factor, TSH-releasing factor, and so on. A sixth releasing factor inhibits
the release of prolactin from the pituitary.

The hypothalamic releasing factors control
the activity of the anterior pituitary, but what
controls the secretion of releasing factors? The
answer: nerve impulses and hormones. The
hypothalamus, a small area of nervous tissue
is a crossroad of nerve-cell connections, and
it receives nerve inputs from almost all parts
of the body. In addition, it is well supplied with
blood vessels, so that it comes in contact with
the hormones circulating in the bloodstream.
Each specific input to the hypothalamus is
monitored, and the hypothalamus responds by
secreting the appropriate releasing factor. The
nerve cells of the hypothalamus are transducers; that is, they convert one kind of signal
into another—neural messages become hormonal messages, and sensory information is
transferred from the nervous system to the
pituitary. Thus, neural activity may promote
endocrine change, and hormonal secretions
may influence neural activity. Again, the endocrine and nervous systems act as an integrated unit; the anterior lobe of the pituitary is
maestro of the endocrine orchestra, but the
hypothalamus manages the maestro.

16-4 THE MOLECULAR ACTIVITY
OF HORMONES

Hormones provide integrative function between groups of cells, but how do hormones
act at the cellular level? What is the effect of
a hormone on its target cell? In recent years it
has become apparent that hormones exert their

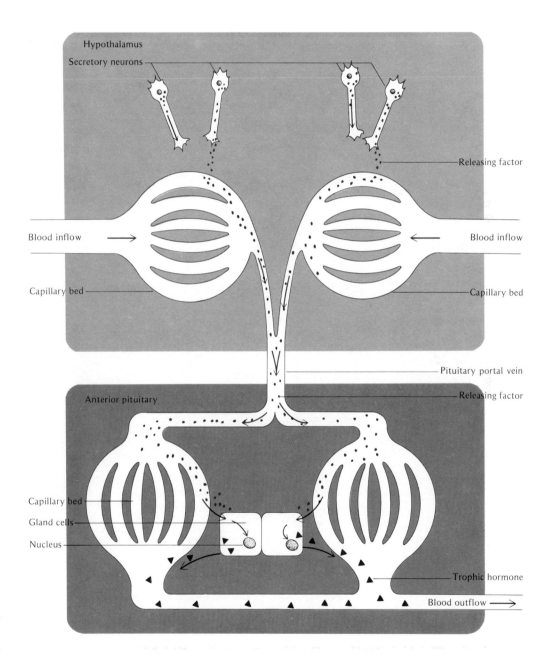

BOX 16D The hypothalamus and history

The hypothalamus, a region of the brain not much larger than a pea, not only directs the pituitary gland but also acts as a command post for the unconscious centers of the nervous system—regulating temperature, sleep, and emotions such as fear and rage.

Consider the effects if a world leader suffered from a pathological disorder of the hypothalamus. How would his sleepiness (narcolepsy), epilepsy, copious production of urine (diabetes insipidus), and alterations of sexual characteristics affect his decisions and the course of history? Napoleon Bonaparte is a case in point (Figure 16-23). As a young man he was very thin, he looked upon sleep as a waste of time (he napped for only 3-4 hours a day), his eyes were alert and piercing, and his manner was imperious. By the time he approached 40 his physical appearance changed radically: his face became soft, fleshy, and feminine-looking; deposits of fat appeared in the hips and chest regions. He needed more sleep and often took naps during the day; he had fits of crying and attacks of fury; he lacked sexual desire and suffered from feelings of grandeur and of baseness. Napoleon himself noted the changes and said to his personal physician upon coming out of his room after an alcohol rub, "Look what lovely arms! What smooth white skin without a single hair! What rounded breasts! Any beauty would be proud of a bosom like mine!" After death the autopsy report confirmed his feminine characteristics. The body was heavily covered with fat. There was scarcely a hair on the body, and that on the head was thin, fine, and silky. Penis and testes were small, the pubis resembled the mons veneris in a woman, the shoulders were narrow and the hips wide. The emperor had changed his secondary sex characteristics. It seems that he suffered from adiposogenital

FIGURE 16–23 (a) Napoleon, first consul. (b) Napoleon, emperor. (Courtesy of the Mansell Collection.)

dystrophy, a disease of the hypothalamus that causes deranged pituitary function and abnormal adrenals, thyroid, and reproductive organs.

With this speculative look at Napoleon can we explain why the great military genius failed to destroy the Russian army at Borodino and then was left with nothing to support his army in Moscow? In 1812 the French grand army retreated from Russia in defeat. Was it because Napoleon was unable to make a decision to attack? Had he delayed too long? A year later (1813) at Dresden his indecision converted victory into defeat once more, when instead of pursuing the enemy he stopped and turned back. At the battle of Leipzig he did not oversee the battle at all, but slept. Waterloo was the final battle: he delayed the attack for 6 hours, and by that time Wellington had received additional support. Much of the time he was fast asleep in a chair! He was physically and mentally worn out at age 45. The fall of the French Empire may well have been caused by the impaired hypothalamic function of Napoleon's brain. What course would European history have taken had the emperor been able to receive treatment for the hypothalamic changes he endured? We can only speculate.

FIGURE 16–24 The second messenger of hormone activity, cyclic AMP.

effects by alteration of enzyme activity, alteration of membrane transport, and gene activation.

Enzyme activity and cyclic AMP

During the 1960s laboratory experiments demonstrated that certain polypeptide hormones (vasopressin, epinephrine, glucagon, trophic hormones) interact with their target cells in such a way that the target cell responds by the formation of **cyclic adenosine monophosphate** (**cyclic AMP** or **cAMP**). Cyclic AMP is made from ATP by the enzyme adenyl cyclase, which is present in the plasma membrane. Hormones activate adenyl cyclase by changing its shape, and the active adenyl cyclase then converts ATP into cAMP. Cells that have cAMP immediately break down glycogen to glucose and thus receive an additional spurt of energy for the performance of metabolic work (Figure 16–24). It is presumed that the specific response of target cells to different hormones is a consequence of different cell types having different hormone-specific receptor sites. Thus, the thyroid gland cells have receptors that react only with the hormone TSH and will not react with other trophic hormones. On the other hand, cells that respond to a variety of hormones presumably have many receptor sites on their plasma membranes. Therefore, liver cells probably have receptors for glucagon, epinephrine, thyroxine, and so on.

In some cases cAMP formation activates enzymes that catalyze reactions important to the function of the cell involved. For example, the cAMP elicited in kidney tubule cells by the hormone vasopressin activates the enzymes that bring about water resorption by these cells. Similarly, cAMP induced in thyroid cells by TSH activates the enzymes in these cells that promote iodine uptake; and the cAMP in liver

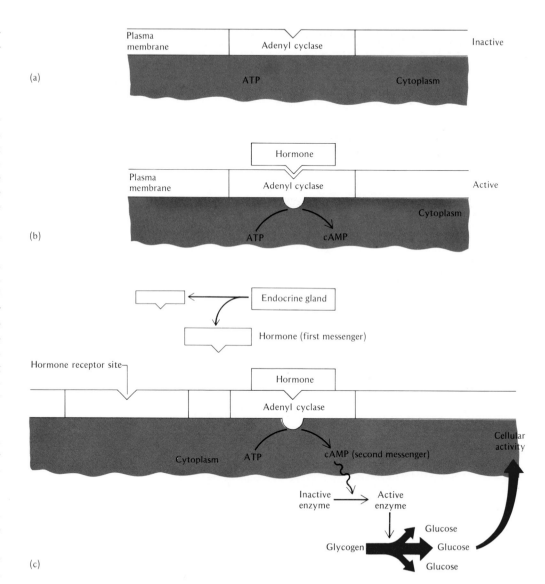

cells, induced by epinephrine and glucagon, activates the enzymes necessary for fat mobilization. Hormone action in these cases may be viewed as a two-messenger system: the first is an extracellular messenger, the hormone itself, which travels from the endocrine gland via the bloodstream to the target cell; the second messenger is an intracellular one, cyclic AMP, which receives the hormonal message in the target cell and initiates molecular (enzymatic) activity.

Membrane transport

Available evidence suggests that the hormone insulin activates the membrane transport of glucose by binding to the plasma membrane and activating a specific carrier for glucose.

Gene activation

There is evidence that some steroid hormones in our bodies work at the level of the gene. In pregnancy and throughout the menstrual cycle, women produce abundant quantities of estrogens and progesterone. One effect of these hormones is the development of secondary sex characteristics, but in addition there is a rapid mobilization of metabolic activities in target cells: the linings of the uterus, cervix, and vagina thicken and become vascularized; their liquid content increases; and new proteins are formed. Evidently under hormonal influence these target organs express bits of their genetic information.

It has been proposed that steroid-hormone molecules such as estrogen penetrate the cytoplasm of the target cell, where they quickly associate with specific receptor proteins (Figure 16–25). The hormone-receptor complex moves from the cytoplasm into the nucleus, where it becomes bound to the chromosomal DNA.

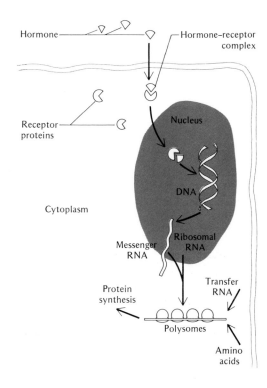

FIGURE 16–25 A model for the mode of action of steroid hormones.

The binding takes place in a matter of minutes, and there is then accelerated synthesis of messenger and ribosomal RNA. RNA moves out of the nucleus and attaches to the polyribosomes of the cytoplasm, and protein synthesis begins. Thus, when the hormone-receptor complex enters the target-cell nucleus, it interacts with the DNA in such a fashion that the activity of specific genes is induced or repressed, as a consequence of which a particular array of genes is turned on or off.

Unfortunately it has not been possible to identify the specific gene products resulting from the activation of target cells by estrogen. Neither do we know whether these products are new or are merely more of the substances ordinarily produced. It is not known whether other steroids act in similar ways to the estrogens. The exact mechanism of steroid-hormone action awaits future research. Solution of this problem holds the key to understanding much of our physiology and development.

SUMMARY

1. Hormones are specific products of one part of an organism that affect cellular processes in another part, and so provide for long-term coordination.

2. In humans, hormones are secreted by endocrine glands and distributed by the bloodstream. Effective in minute amounts, hormones affect growth, development, tissue activity, and reproduction. Hyposecretion or hypersecretion of a particular endocrine gland has serious consequences.

3. Function of endocrine glands and their target organs are integrated by negative feedback control mechanisms.

4. The pituitary gland, located below the brain, is divided into two parts. The anterior pituitary is the master gland; its trophic hormones stimulate many other glands and coordinate their function by negative feedback mechanisms.

5. The secretions of the major glands and their functions are summarized in the following table (Table 16–4).

6. Prostaglandins, hormone-like substances that perform many roles, are synthesized by the plasma membranes of a variety of tissues. They are especially high in seminal fluid. They play a role in the regulation of blood pressure, uterine contractions, and gastric secretions; they relax the bronchial passages and probably regulate plasma membrane function and therefore transport.

TABLE 16-4 The secretions of the major glands

Gland	Location	Products	Target organs or function	Effects of hypersecretion	Effects of hyposecretion
Pituitary	Below the brain				
Anterior lobe (adenohypophysis)		Gonadotrophic hormones	Gonads		
		Adrenocorticotrophic hormone (ACTH)	Adrenals (converts cholesterol to pregnenolone)		
		Thyrotrophic hormone (thyroid stimulating hormone or TSH)	Thyroid (traps iodine, aids in synthesis of thyroid hormone)	Goiter (caused by lack of dietary iodine)	Cretinism, myxedema
		Somatotrophic hormone	Bones and tissues (affects growth)	Gigantism (acromegaly in adults)	Dwarfism (in children)
		Melanocyte-stimulating hormone	Skin melanocytes		
Posterior lobe (neurohypophysis)		Stores vasopressin and oxytocin	Kidneys, blood vessels uterine muscle		Diabetes insipidus
Pineal gland	Upper surface of brain	Melatonin	Reproductive organs, especially ovaries	Delayed sexual development in females	Early onset of puberty in females
Thyroid	In front of and below the voice box (larynx)	Thyroxine Triiodothyronine	Body tissue cells	Accelerate metabolic rate Graves's disease with exophthalmos	Decrease metabolic rate Goiter, cretinism or myxedema
		Calcitonin	Rapidly increases Ca^{++} deposition in bones		
Parathyroids	Embedded in back of thyroid	Parathormone	Slowly increases number and size of osteoclasts and so increases Ca^{++} in blood	Removal of Ca^{++} from bones, bone fractures	Abnormal nerve function, spasms, tetany
Thymus gland	Neck region	Lymphocyte-maturing hormone	Lymphocytes		

TABLE 16-4 (Continued)

Gland	Location	Products	Target organs or function	Effects of hypersecretion	Effects of hyposecretion
Pancreas (islets of Langherhans)	Beneath stomach				
Beta cells		Insulin	Facilitates cellular glucose uptake, especially in liver	Convulsions and coma	Diabetes mellitus
Alpha cells		Glucagon	Facilitates breakdown of glycogen in liver and release of glucose into blood	Diabetes mellitus	
Adrenal gland	Upper surface of kidney				
Medulla	Inner core of gland with nervous origin	Epinephrine Norepinephrine	Generates fight-or-flight reaction (emergency response)	Hypertension	
Cortex	Outer part of gland	About 50 corticosteroids			
		i. Glucocorticoids	Regulate sugar and protein metabolism	Cushing's disease	Addison's disease and masculinization in women
		ii. Mineralocorticoids	Regulate sodium and mineral balance	Edema, high blood pressure	
		iii. Sex hormones	Affect secondary sex hormones	Precocious sexual development	Infantilism, cryptorchidism, male clamacteric, masculinization in women
Skin, kidney, liver	Skin, kidney, liver	Calciferol (vitamin D_3)	Intestines (promotes Ca^{++} uptake) Bones (promotes Ca^{++} deposition)		Rickets

7. The hypothalamus of the brain, lying just above the pituitary, produces neurohumors (releasing factors) which are conveyed to the anterior pituitary by axoplasmic flow. Five hypothalamic releasing factors stimulate release of specific pituitary trophins; the sixth inhibits the release of prolactin.

8. The hypothalamus is the prime regulator of automatic activities such as body temperature control and the basic drives (hunger, thirst, sex, etc.). The hypothalamus receives and integrates nervous and hormonal stimuli, and by secreting appropriate neurohumors to the pituitary, it is able to regulate pituitary activity.

9. Hormones affect cellular activity by altering enzyme activity (by activating adenyl cyclase, which converts ATP to cAMP; cAMP in turn degrades glycogen to glucose or activates other specific cellular enzymes), by affecting membrane transport (specifically of glucose) and by selectively activating genes. The specific response of target cells to different hormones depends on specificity of hormone-receptor sites, either at the level of the membrane or in the cytoplasm.

KEY WORDS

hormones
exocrine glands
endocrine glands
hyposecretion
hypersecretion
target organ
pituitary gland (hypophysis)
adenohypophysis
neurohypophysis
trophic hormones
gonadotrophic hormones
adrenocorticotrophic hormone (ACTH)
thyrothrophic hormone
somatotrophic (growth) hormone

melanocyte-stimulating hormone
vasopressin
oxytocin
pituitary dwarf
gigantism
acromegaly
thyroid gland
thyroxine
triiodothyronine
calcitonin
goiter
thyroid-stimulating hormone (TSH)
cretin
myxedema
Graves's disease
exophthalmos
calcitonin
osteoblasts
rickets
calciferol or vitamin D
parathormone
osteoclasts
dihydroxycalciferol (DHC)
diabetes mellitus
insulin
pancreas
islets of Langerhans
glucagon
adrenal medulla
adrenal cortex
epinephrine
norepinephrine
emergency response or
    fight-or-flight reaction
steroid ring
adrenocorticosteroids or corticosteroids
glucocorticoids
mineralocorticoids
sex hormones
androgens
castration
testosterone

androstenedione
androsterone
hypogonadism
cryptorchidism
male climacteric
estrogen
progesterone
thymus gland
pineal gland
melatonin
prostaglandins
hypothalamus
neurohumors
neurosecretions
axoplasmic flow
releasing factors
cyclic adenosine monophosphate (cyclic
    AMP or cAMP)

TOPICS FOR REVIEW AND DISCUSSION

1. What are hormones?
2. How do hormones control body activities?
3. How do steroid hormones work?
4. Why is vitamin D a hormone?
5. What is the role of cyclic AMP in hormone action?
6. Contrast endocrine glands with exocrine glands.
7. Describe some consequences of hormone deficiencies.
8. How are feedback controls involved in hormone secretion?
9. How do insulin and glucagon regulate the level of glucose in the blood?
10. Describe the relationship of the hypothalamus to hormone secretion.
11. Discuss the statement "Prostaglandins do anything and everything" in terms of the hormonal system.

12. What do we know about gene activation
    and steroid hormones? Why is this im-
    portant to our knowledge of physiology
    and development?

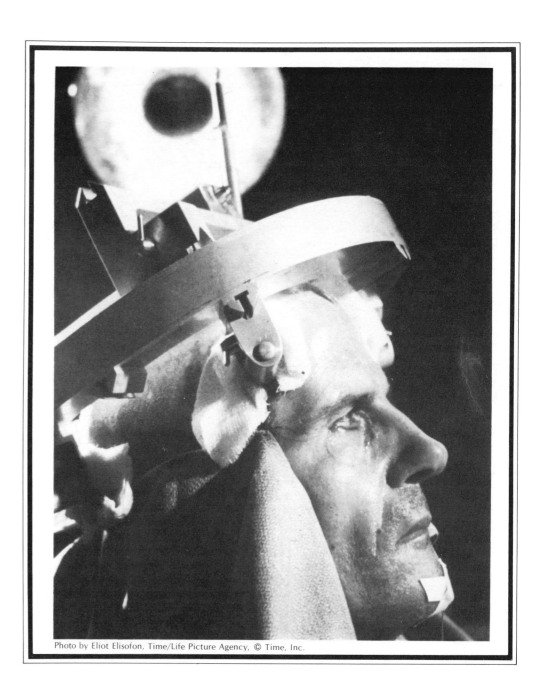

Photo by Eliot Elisofon, Time/Life Picture Agency, © Time, Inc.

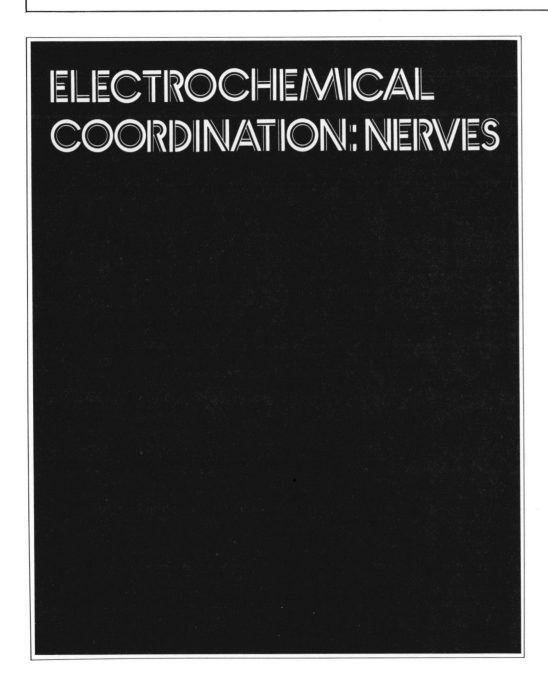

ELECTROCHEMICAL COORDINATION: NERVES

17–1 WHAT DOES THE NERVOUS SYSTEM DO?

The animal is anesthetized and a metal halo is clamped on its head to prevent movement. By means of a high-speed drill, a small shaft is bored through the bony skull and into the coverings of the brain. The shiny, wrinkled surface of the brain is exposed. Using a micro-manipulator, the operator carefully guides a tiny electrode—fine wires insulated with Teflon and scraped bare at the tips—into the soft, jellylike tissue of the brain itself. There is no pain, since the brain has no sensory receptors or free nerve endings in its own tissue. Deeper and deeper within the brain the operator moves the bare stainless-steel wires of the electrode, until he decides that the right position has been reached. Once the electrode has been im-planted, the wires at the scalp end of the electrode are soldered to a small electrical socket, and this is permanently cemented to the skull. The operation is quick, and the instrument implant is so small that in time it will be completely covered by the scalp hairs. After the anesthesia has worn off, plugging into the brain is as simple as putting a lamp plug into a wall socket. When the electrode wires are connected to a power supply, a small pulse of electricity can be sent into the brain and the animal made to perform certain specific move-ments like an electrical toy. When electrodes are connected to the correct region of the brain, the animal gives involuntary but coordinated movements.

The experimental animal we have been de-scribing is not a rat, a cat, or a monkey, but a human—under treatment for psychomotor epilepsy. By delivering electric shocks to the patient's brain, the operator can produce a slow clenching of the hand into a fist. When one such patient was asked to keep his fingers extended through the next stimulation, he re-

BOX 17A

1984 Revisited?

If the late George Orwell were writing a sequel to *1984* today, he would probably reject as archaic the propaganda techniques for controlling people's minds described in his famous anti-Utopian classic. Today, for example, he might envisage a society in which a newborn baby's first experience would be neurosurgery, an operation in which the child's brain was fitted with miniaturized radio devices connected to every major center controlling reason and emotion. Children in such a society might be raised as flesh-and-blood electrical toys, whose ideas and behavior were directed by computer signals. Any aberrant or heretical ideas would be transmitted to the computer, which would be programmed to take appropriate action to restore control.

This may not be as fantastic as it sounds. Some years ago it was discovered that if rats had a stimulating electrode implanted into a particular region of the brain, they could be trained to press a lever that would turn on a stimulating current. Such a rat, when given a choice between feeding and pressing, would prefer to stimulate its median forebrain bundle rather than eat. Indeed, some animals press the lever as many as 8,000 times in one hour. The area came to be known as the reward or pleasure center of the brain. Similar techniques and experiments have been performed on cats, dogs, monkeys, and even man. In one human case where the individual had an electrode implanted in the septal region, the patient declared after electrode stimulation that he felt "good," as if he was building up to a sexual orgasm, although the end point was not achieved. Similarly, unpleasant emotions can be produced in man: stimulation of the amygdala region of the temporal lobe induces fear reactions, and stimulation of particular regions of the hypothalamus incites a variety of emotional responses including rage and terror.

Fortunately, human ignorance about the brain remains so vast that there is no imminent prospect that the techniques being worked on could have Orwellian significance. Nevertheless, the horrifying prospect rises that in the twenty-first century the lexicographers may have to drop the verb "to brainwash" and replace it with "to brainwave."

plied that he could not do so. He commented: "I guess your electricity is stronger than my will."

The wired and electrically manipulated brain represents man's latest method of control over his own behavior. By means of electrodes placed in precisely located spots (more than 400 have been implanted in an experimental monkey for 4 years), the subject can be made to act in a specific way (Box 17A). For example, electrode implantation has been used in the diagnosis and treatment of involuntary movements, intractable pain, and severe epilepsy. When a patient with an implanted electrode feels an epileptic seizure coming on, he can, by using a self-stimulator, induce electrical excitation in a competing area of the brain; this stops the discharge from spreading, and the fit does not develop (Box 17B). Before we can understand how such techniques are possible, it is appropriate to ask two questions: Why do we need a nervous system? and How does the nervous system work?

The body receives messages from its own organs and from the external world. We perceive light, sound, odors, pressure, temperature, chemicals, and the like; we think, we move, we have unconscious thoughts and conscious ones. The normal functioning of the body depends both on receipt of stimuli and on production of integrated responses. For the body to perform its activities in coordinated fashion, there must exist a connecting link between stimulus and response, between receptor organ and effector organ; that link must be a transmitter capable of channeling information from one to the other.

Two systems in our bodies act as coordinating links between stimulus and response: the endocrine and the nervous systems. The endocrine system, as we have seen in Chapter 16, regulates the activities of cells in other parts of the body by means of hormones. When a

FIGURE 17–1 The wiring of the nervous system, as drawn by the Flemish anatomist Vesalius in 1543. (Courtesy of The New York Academy of Medicine.)

hormone is released directly into the blood, some time elapses before it arrives at the target organ. The lag may be seconds, minutes, or hours. Moreover, since the hormone circulates in the blood for a time with little decay in activity, the period during which it can affect the target organ may last for hours or days. Thus, the hormonal system provides for slow communication on a long-term basis. The nervous system, on the other hand, provides for rapid communciation between the various tissues and organs of the body. In large organisms, the organs invoved in the reception of stimuli are on the surface and usually are far removed from the systems involved in reacting to the stimuli (glands and muscles). To link the peripheral listeners with the more internal responders and to effect rapid reactions to stimuli, quick and direct pathways for conducting messages are required, and that function involves nerve cells. The nervous system employs electrochemical messages, **nerve impulses,** that run along specialized nerve pathways receiving and transmitting information to and from various organs. Nerve cells or **neurons** are specialized to conduct electrochemical messages at high speed and when bundled together in cablelike form as **nerves,** are able to regulate the direction in which information flows. Masses of nerve cells produce control centers, such as the brain and spinal cord, which act as central clearinghouses of information, loci of coordination. It can be said that without nervous control, active multicellular animals could never have developed.

17–2 THE NEURON: SIGNAL CELL OF THE NERVOUS SYSTEM

The human nervous system is often likened to a switchboard or a computer, since all three receive signals and send out messages. Electrochemical signals cause the nervous system

to "flash," "blink," and "hum." It is a complex array of "wiring" made up of over 12 billion neurons, interspersed with supportive elements, the **glial cells** (Figure 17–1). In order to arrive at a detailed understanding of how the nervous system works, it is necessary to consider the nature and mode of operation of the structural components themselves, the neurons.

The neuron is the basic unit of the nervous system, and although neurons vary in size and shape (Figure 17–2a), no two being identical, they do show a common functional organization that makes it possible to describe the "typical" neuron (Figure 17–2b). Like other cells, the neuron has a plasma membrane, a nucleus, and various organelles within its cytoplasm. It is a single cell and not structurally continuous with any other cell. The main cell body contains the nucleus, Golgi apparatus, mitochondria, ribosomes, endoplasmic reticulum (Nissl bodies), and a granular cytoplasm. Projecting from this region are many fine rootlike processes called **dendrites;** these receive messages from adjacent cells or directly from the environment. Also extending from the cell body is a single, long, fiberlike extension, the **axon,** which ramifies near its end into fine branches, each terminating in a tiny bulb **(synaptic knob).** The axon passes signals to other neurons or effectors.

The axon is usually covered by a sheath (called the **neurilemma**), formed by special **interstitial cells.** The interstitial cell completely encircles the axon (Figure 17–3a) and may play a role in the nutrition of neurons and in their regeneration following damage. The encircling cells of some axons, especially those in the brain and spinal cord, wind their plasma membranes tightly around the surface of the axon, forming a spiral envelope containing many turns (Figure 17–3b). These tightly wrapped membranes, rich in fatty materials, form an effective insulating layer called the **myelin**

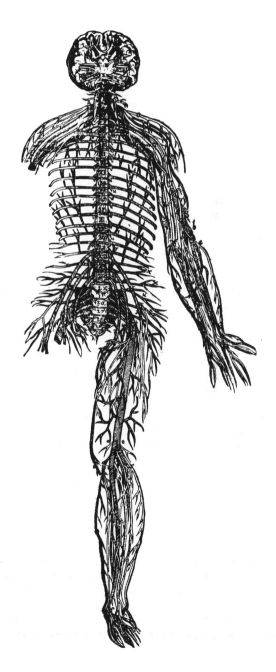

FIGURE 17–2(a) A variety of neurons from different parts of the human nervous system. The axon of each neuron is designated by the letter a, and the dendrites are indicated by the letter d. Dendrites receive excitation from other cells and conduct impulses toward the cell body, whereas axons conduct nerve impulses away from the cell body. (b) The "typical" neuron shows many short dendrites and a single, elongate axon.

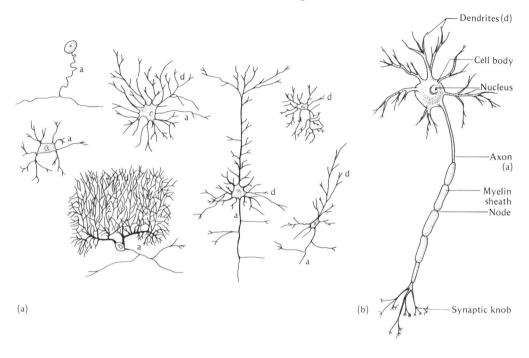

(a)

(b)

— Dendrites (d)

— Cell body

— Nucleus

— Axon (a)

— Myelin sheath
— Node

— Synaptic knob

sheath. The myelin sheath helps to increase the speed of transmission of signals along the axon it surrounds. At intervals along the axon's length the myelin sheath is interrupted; the interruptions (called **nodes**) represent points where one encircling cell ends and another begins. **Multiple sclerosis** (MS), a disease characterized by jerky body and limb movement, double vision, slurred speech, and paralysis, is caused by a patchy destruction of the myelin sheath. Since the myelin sheath is a kind of living electrical tape, much like the insulation around a telephone wire, the destruction of myelin reduces the neuron's ability to conduct impulses. There is considerable cross talk between axons, impulses travel slowly, and there is a tendency for spontaneous firing of neurons. In a sense,

the victim of multiple sclerosis suffers from short-circuiting neurons in the spinal column.

The plasma membrane of the neuron is no different in its structural arrangement from the plasma membranes of other cells, except that it is highly specialized for the transmission of electrochemical impulses. Since the functioning of the entire nervous system depends upon the electrical events occurring in the neurons of which it is composed, understanding that function requires comprehension of some basic principles of electricity.

There are two types of electrical charges in the universe—positive and negative. Like charges repel each other, whereas opposite charges attract each other. When positive and negative charges are separated, an electrical

force tends to draw these charges together. The force that draws these unlike charges together can be measured, and its strength increases when the charges are moved closer together as well as when there is an increasing quantity of charge. The separated electrical charges are capable of doing work if they are allowed to come together. This potential of opposite charges to do work is called **voltage.** The movement of the charges is called **current.**

These principles of electricity apply directly to the neuron. Consider for the moment that the axon of the neuron is a long, thin tube, the walls of which are the plasma membrane. The key to the electrical activity of the neuron lies in the chemical contents of that tube and of the tissue fluid surrounding it. Organic molecules and salts dissolved in the intracellular and extracellular water dissociate to yield electrically charged particles, **ions.** There are important differences between the ionic composition of the tissue fluid and the neuronal fluid. The tissue fluid bathing the neuron contains a high concentration of sodium chloride so that more than 90% of its charged particles are the positive sodium ions (Na^+) and the negative chloride ions (Cl^-). Inside the cell these ions are very low in concentration (less than 10%); the principal positive ion is potassium (K^+), and the negative ions are a variety of organic substances (Org^-) made by the cell itself, such as proteins, amino acids, and citric acid (Figure 17–4a).

The concentration of sodium is 10 times higher outside the neuron than within, and that of potassium is 30 times higher within the neuron than without. The membrane of the neuron is highly permeable to potassium and chloride ions, and only slightly to sodium ions; hence there is a tendency for the potassium ions (K^+) to leak out of the cell at a high rate. As a result of this, as well as of the inability of

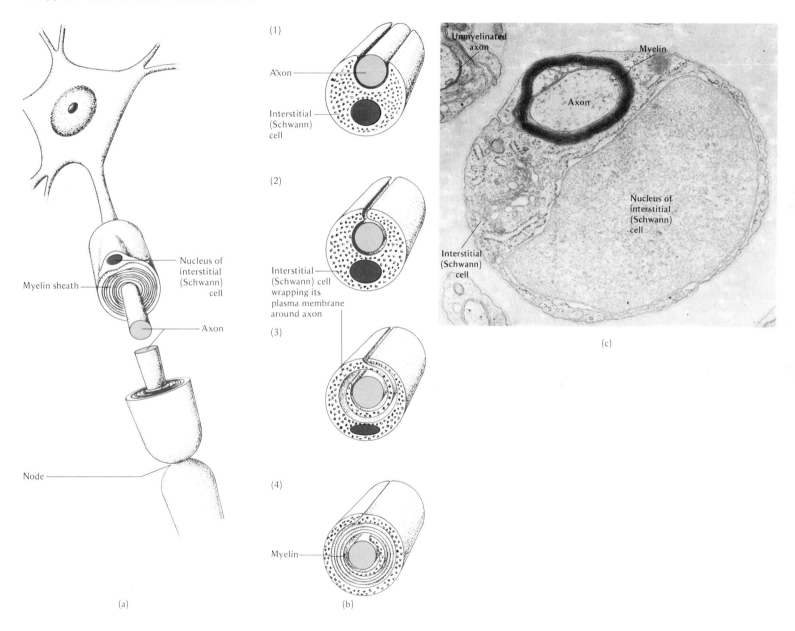

FIGURE 17-3 The relationship of the interstitial (Schwann) cell to the myelin sheath. (a) Diagrammatic reconstruction of the myelin sheath wrapped around an axon. (b) Formation of the myelin sheath, a tightly packed coil of the plasma membrane of the Schwann cell. (1) The Schwann cell partially wraps itself around the axon, and in successive stages (2) and (3) the number of coils is increased until a packed spiral of membranes lying between most of the Schwann cell cytoplasm and the axon is formed (4). (c) Electron micrograph showing the myelin sheath, axon, and nucleus of the interstitial (Schwann) cell. (Courtesy of Mary and Richard Bunge.)

the large, negative organic substances to diffuse across the plasma membrane, the inside of the cell tends to become electrically negative. This tends to attract the positive sodium ions from the cell's exterior to reestablish electrical neutrality. The attraction is counteracted by the cell's lack of permeability to sodium as well as its active extrusion of sodium ions against their own concentration gradient by the so-called **sodium pump** (Figure 17–5). The sodium pump is energy-dependent, but since sodium leaks into the cell slowly, the work expended amounts to only a small fraction of the energy made available by the cell's own metabolism. As a consequence of the outward leakage of potassium, the active extrusion of sodium ions, and the presence of negatively charged organic ions within, the cell maintains a separation of charges; the inside of the neuron is made negative relative to the outside and is thus polarized. The resulting membrane voltage can be measured by inserting into a cell minute electrodes that are connected to a sensitive voltmeter (Figure 17–4b). Measured in this way, the voltage amounts to about 70 millivolts (mV), or approximately 0.1 V. This is called the **membrane resting potential,** and although it may seem small (it is about one fiftieth the voltage of a flashlight battery), in biological terms it is quite substantial.

17–3 THE NERVE IMPULSE: AN ELECTROCHEMICAL MESSAGE

The language of the neuron is an electrochemical one. Sometimes it is difficult to comprehend that thoughts, emotions, learning, and the myriad functions that are a property of the human mind are carried by neurons via electrochemical changes known as nerve impulses, but this is indeed the case. How is a nerve impulse produced?

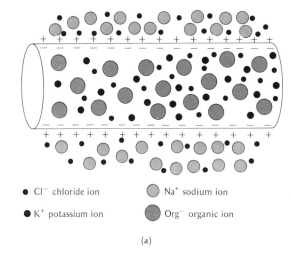

- ● Cl^- chloride ion
- ● K^+ potassium ion
- ○ Na^+ sodium ion
- ● Org^- organic ion

(a)

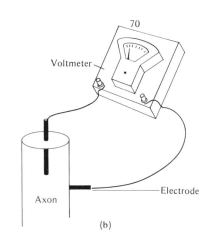

Voltmeter

Axon

Electrode

(b)

When active neurons (and muscle cells) are stimulated, the membrane resting potential changes; hence, these are often called **excitable cells** and the cause of the change is called the **stimulus.** The membrane potential of a stimulated neuron drops from a negative value to zero; that is, the membrane becomes **depolarized.** This is followed by a temporary reversal of polarity so that the outside of the membrane is negative relative to the inside, and then almost as quickly as this occurs the membrane is restored to its original condition **(repolarization)** (Figure 17–6). The rapid depolarization-repolarization of the membrane is called the **action potential** and involves a change in the membrane voltage lasting only 0.001 second. The action potential moves along the length of the neuron, and it is this that constitutes the neuronal message. How does the membrane become depolarized and then repolarized, and how can a wave of depolarization (the action potential) be propagated along the length of the nerve without any decrease in strength so

that a message gets from one point to another without distortion?

The resting potential of the membrane of the neuron is about −70 mV (the voltage is measured relative to the outside, and since the inside of the cell is more negative than is the outside, the voltage is written as a minus value). When the membrane is excited at some point by an appropriate stimulus (such as pressure, electricity, or a chemical agent), the region so affected becomes much more permeable to sodium ions (Na^+) than to potassium (K^+), and the sodium ions rush by diffusion into the cell faster than potassium ions can leave (Figure 17–6b). As a result of this, the voltage drops to zero and then there is a slight overshoot so that the inside of the membrane becomes positively charged in this particular region (Figure 17–6c). The resultant action potential spreads to adjacent areas and is self-reinforcing. How does this occur? The flow of sodium ions through the membrane makes it easier for others to follow, and as the voltage changes,

FIGURE 17–5 Diagram of the sodium pump in a resting nerve membrane.

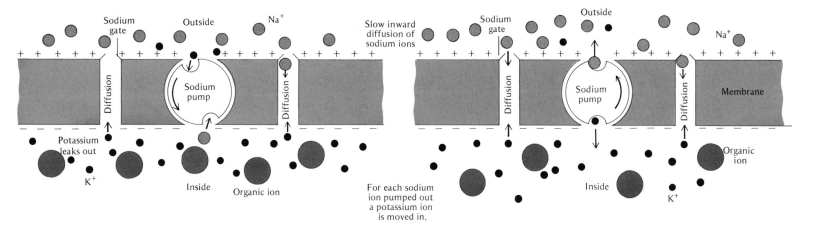

there is a change in membrane permeability immediately adjacent to the region of excitation; this creates a condition favoring sodium flow in this region as well. This process continues, producing an electrical current that flows outward in wavelike fashion from the original region of stimulation along the entire length of the membrane. The moving action potential is the nerve impulse.

Immediately after the peak of the action potential has passed, the process of membrane repolarization begins. The membrane once again becomes impermeable to sodium ions (the sodium gates that were opened are now closed), and the potassium ions, which are more numerous in the cell, easily diffuse outward (the potassium gates are opened wider than during the resting condition). This once again creates a negative charge on the inside of the membrane (repolarization). Repolarization begins at the point of excitation, spreading out as a wave immediately behind the region of depolarization and following a few ten-thousandths of a second after depolarization has occurred. The entire process of depolarization and repolarization takes only a fraction of a second.

For a short time after a particular region has been depolarized, a second action potential cannot be produced, for the membrane must be repolarized, and the membrane is said to be in an inexcitable or **refractory state** (Figure 17–6d). No matter how strong the stimulus, a depolarized nerve will not conduct an impulse. Because of the refractory period, the action potential is unable to reexcite that region of the membrane that it has just passed, and so travels away from its point of origin.

The movement of the action potential along the length of the neuron is often compared to a burning fuse of gunpowder. Once ignited, the heat of the gunpowder burning in one section causes the next section to ignite, and so it burns in a self-propagating manner. Although this analogy has merit, it also has limitations. The gunpowder fuse works but a single time and carries only one impulse, whereas the neuron can carry impulse after impulse over the same pathway. Furthermore, heat traverses the length of a gunpowder fuse, but an electrochemical event produces self-propagating change along a neuron.

In order for the membrane of the neuron to become repolarized to its resting potential, the sodium ions that moved inside and the potassium ions that leaked outside must be returned to their original positions. The restoration is accomplished by the sodium pump (Figure 17–5), an activity that requires energy (ATP is used). However, even when the sodium pump is not operating (as when it is turned off by a metabolic poison such as cyanide; Chapter 4), thousands of action potentials can still be generated in a neuron before it quits functioning altogether. The reason for this is that only a minute amount of sodium passes into the neuron with each impulse, and it takes a considerable amount of sodium in the cell to prevent depolarization completely.

FIGURE 17–6 The nerve impulse. (a) At rest the voltage is −70 mV, the membrane is polarized, and more potassium and organic ions are inside whereas sodium is abundant outside. (b) At the start of depolarization, sodium ions enter the axon and the voltage drops to zero. An action potential has begun. (c) As the action potential develops, the membrane is depolarized; as a consequence of the rapid influx of sodium ions, the interior becomes slightly positive. (d) With extrusion of potassium ions and prevention of sodium entry, the membrane becomes repolarized.

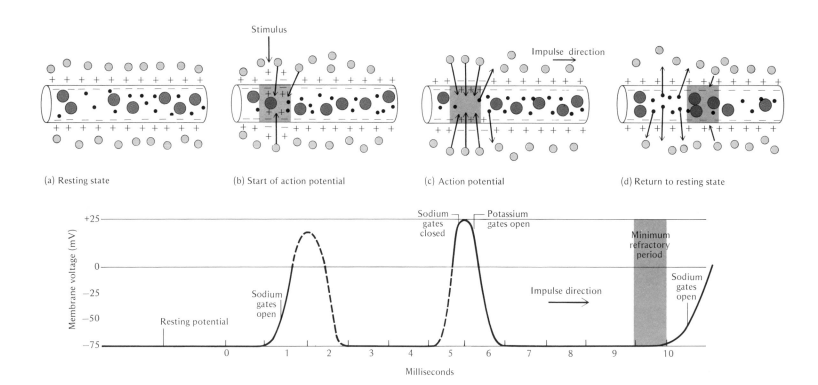

(a) Resting state   (b) Start of action potential   (c) Action potential   (d) Return to resting state

17–4 THE NERVE IMPULSE: PRODUCTION AND CONDUCTION

A nerve impulse can be triggered by a variety of stimuli: chemicals, electricity, and mechanical forces. What determines whether a stimulus will produce a nerve impulse (action potentials) in a neuron?

If a neuron were set up in the laboratory so that the axon could be electrically stimulated under controlled conditions (Figure 17–7a), we should find that depending on the particular axon, there would be a minimum voltage necessary to produce an action potential. This minimal strength is called a **threshold stimulus,** and stimuli weaker than this (subthreshold) would normally produce no action potential (Figure 17–7b). Stimuli greater than threshold (suprathreshold) would elicit an action potential, but this would be no different from a threshold stimulus. One of the most important characteristics of the excitation process in a neuron is that it either leads to an action potential or it does not. If an action potential is produced, its magnitude, character, and duration are independent of the strength or the nature of the stimulus. This relationship is called the **all-or-none law,** and it applies to a single neuron. It means that once the threshold level of stimulation is reached, the response initiated in the neuron does not depend on the kind of initiator or on its strength. It is like setting a mousetrap or cocking the trigger of a gun. Once triggered to activity, the force generated is built in and cannot be increased by stronger stimulation.

Now there develops an important question: If neurons exhibit an all-or-none property with regard to stimulus intensity, how do we detect differing intensities of stimulation—for example, the difference between a sheet of paper or a rock held in the hand or between a light tap and a hard slap on the back? The intensity of stimuli can be determined because different neurons have different thresholds, and a more intense stimulation involves greater numbers of neurons, each of which exhibits the all-or-none phenomenon. Furthermore, although the

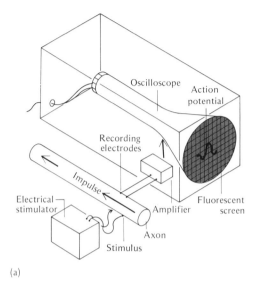

(a)

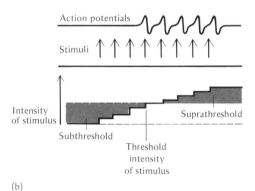

(b)

speed of travel of the action potential along the length of the axon is constant, a more intense stimulus will produce a greater number of action potentials moving down the axon per unit time (greater frequency). We can visualize the situation as follows: the code of the neuron can be thought of as a dot that moves across the page at a fixed rate; if the stimulus intensity is low, the dots are less frequent and few neurons send out dots:

Neuron 1 •   •   •   •

Neuron 2 • • • • • • • •

However, if the stimulus is of high intensity, the dot frequency is high, and many neurons send out signals:

Neuron 1 • • • • • • •

Neuron 2 • • • • • • • • • •

Neuron 3 • • • •   •

Neuron 4 •   •   •   •   •

Neuron 5 • • • • • • • •

Neuron 6 • • • • • •

Thus, to make up for limitations in the coding system, the nervous system employs a large number of channels, each capable of transmitting signals of variable frequency. From the eye alone a million channels run to the brain!

The velocity of transmission varies with the diameter of the axon and whether it is myelinated or not. The myelinated regions along the length of the axon are not continuous, and about every millimeter there is a constriction devoid of myelin, (the **node**). Although depolarization of the axon cannot take place below the myelinated zones, it can take place at these nodes, and impulses in this type of nerve skip along from node to node, producing a jumping (saltatory) conduction of impulses via relay stations. This is a valuable mechanism for transmission of nerve impulses, since it diminishes cross talk between neurons, increases conduction velocity, and minimizes sodium leakage. As a result there is a considerable saving of metabolic energy, and nerve transmission is highly efficient. Next time you pull your hand away from a hot stove, remember that such a reaction is possible because a nerve impulse travels about 300 ft—the length of a football field—in 1 second!

17–5 SIGNALS FROM NEURON TO NEURON: THE SYNAPSE

If the neuron is to function as a signal transmitter and as a part of a communications network, it must be able to send the signal it receives to another neuron or to an effector organ such as a muscle or gland. The junction between one neuron and the next or between neuron and effector is called a **synapse,** and the intervening space is called the **synaptic cleft.** The synapse has the capacity to transmit some signals and not others, and thus it acts as a gate for neuronal transmission. A neuron stimulated in the middle of its axon can conduct impulses spreading out in both directions, but from the axonal end it can transmit its impulses in only one direction. Dendrites can only receive signals and pass them on to the cell body, and the axon carries them from there and transmits them to the next cell. This unidirectional function is imposed upon the neurons by the nature of the synapse.

Synaptic transmission

The dendrites and/or cell body of a typical neuron receive hundreds of small fibers from the ends of the axons of adjacent neurons. These axonal fiber ends terminate in small bulblike swellings called **presynaptic knobs**

BOX 17B Bugging the brain

There are about 12 billion nerve cells in the body, and most of these are always in some state of activity. Even when we are asleep, action potentials race across the network of nerves. Since action potentials are electrical events, they can, in principle at least, be measured with instruments that monitor fluctuations in voltage. We can eavesdrop on these electrical activities by placing a number of electrodes on the various regions of the scalp and recording neuronal signals. To be sure the signals are weak, and not all neurons are active, but a single electrode taped to the scalp can record as little as 0.0005 V. Electrical activity arising in the brain was first reported in 1875 by Richard Caton, an English physician, but it was not until the 1920s that Hans Berger, a German neurologist, devised a system for recording electrical activity on a moving sheet of paper (Figure 17–8). Berger found that there were rhythmic cycles in the brain, and since they appeared as waves on the paper, he called them **brain waves.**

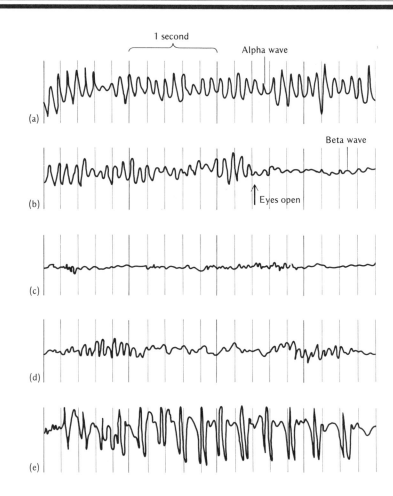

FIGURE 17–9 Typical EEG recordings from the scalp. (a) Normal rhythms characteristic of the relaxed state with eyes closed are called alpha waves. (b) Beta waves appear during wakefulness, at rest with eyes open. (c) Brain waves recorded during rapid-eye-movement (REM) sleep. (d) Brain waves during moderate deep sleep. (e) Petit mal epileptic seizure.

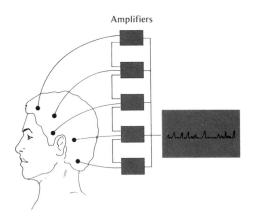

FIGURE 17–8 Recording an electroencephalogram.

The written record of such brain waves, that is, the pattern of electrical activity of the brain, is called an **e**lectro**e**ncephalo**g**ram (*encephalos*—brain; Gk.) or **EEG.** By placing electrodes on various sites on the scalp, we can "bug" different groups of neurons, and although the patterns are extremely complex it is possible to correlate brain waves with various behavioral states.

When we are awake, but at rest with our eyes closed, the EEG shows a regular pattern of activity, **alpha waves** (Figure 17–9a), and this changes markedly when the eyes are opened (Figure 17–9b). When we open our eyes or when there is muscular activity, additional waves, **beta waves,** replace the alpha waves. When we are asleep the electrical activity shows a slow, rolling pattern (Figure 17–9c), and when the sleep becomes deeper the waves become stronger and deeper (Figure 17–9d). If the brain is damaged by hemorrhage, tumor, infection, epilepsy (Figure 17–9e), or injury the EEG can monitor the brain waves and record the location and extent of damage. (Emotional disorders are usually not detected by the EEG.)

Epilepsy is a disorder in the brain's electrical activity. About 1 person out of 200 in the United States suffers from some form of epilepsy. Some famous epileptics were Julius Caesar, Peter the Great, Mohammed, and Dostoevski. It is believed that epileptic seizures are a result of spontaneous, uncontrolled reverberating cycles of electrical activity in the brain's neurons. One portion stimulates another, this activates another, and this in turn stimulates the first, so that a cycle of activity begins and runs its course until the neurons fatigue. While such reverberations are taking place, the individual may have hallucinations and a seizure; the confused circuitry causes a loss of consciousness, and neuronal fatigue promotes sleep. The grand mal, or severe seizure, is uncommon, but milder attacks (petit mal) may occur frequently; the afflicted person believes he has seen flashes of light, there may be sensations of sound and odor and a blackout of consciousness. The majority of epileptics lead near-normal lives owing to treatment with the drug Dilantin (diphenylhydantoin), which suppresses seizures.

(Figure 17–10). The electron microscope has shown that a presynaptic knob has many small sacs, or vesicles, containing molecules of a transmitter substance (Figure 17–11a). When an impulse moves along the axon and reaches the presynaptic knob, it causes a temporary change in the knob's membrane structure, and the vesicles discharge their transmitter substance. The transmitter molecules released into the synaptic cleft (about 200 Å in distance) diffuse across it and alter the membrane potential of the adjacent (postsynaptic) neuron (Figure 17–11b). The electrical change in the postsynaptic cell membrane induced by the transmitter substance is very brief, but if depolarization occurs, an action potential is triggered in the postsynaptic cell.

Let us consider, for example, the case of a particular transmitter chemical, acetylcholine, (ACh). When an action potential arrives at the synaptic terminal, transmitter molecules of ACh are released into the synaptic cleft. The ACh diffuses across the space and attaches to receptor sites on the postsynaptic membrane (Figure 17–11b). The receptor site functions in two capacities: (1) it inactivates ACh because it has enzymatic properties (acetylcholinesterase, which degrades ACh), and (2) it activates the postsynaptic membrane. During activation, there is a rapid influx of sodium ions (the sodium gates are opened), and as a result of the net movement of these positively charged ions into the postsynaptic cell, there is a slight depolarization. When this depolarization, called the **excitatory postsynaptic potential (EPSP)** reaches a threshold level, it initiates an action potential in the postsynaptic cell (Figure 17–12). The EPSP is a local, passively propagated current, and its only function is to help trigger the action potential; it is not all-or-none, but can vary in amplitude. The fact that the EPSP is graded is of critical importance, for it means that a neuron bombarded by signals

FIGURE 17–10 (a) Scanning electron micrograph of the synaptic knobs lying on the cell body of an adjacent neuron. (b) Detail of the synaptic knobs as shown by the scanning electron microscope. (From "Studying neural organization in *Aplysia* with the scanning electron microscope," by E. R. Lewis, T. E. Everhart, and Y. Y. Zeevi, *Science,* Vol. 165, pp. 1140–1143, September 1969. Copyright 1969 by the American Association for the Advancement of Science.)

across many synapses is capable of integrating this information; when the level of chemical transmitter reaches threshold, a distinct action potential is triggered.

The ACh released from the presynaptic knobs does not persist for long, since it is quickly destroyed by acetylcholinesterase. Indeed, such destruction is of considerable functional significance; for if ACh were not rapidly degraded, the stimulatory action would continue indefinitely (the sodium gates would be locked in the open position). Many of the organophosphate insecticides (malathion and parathion) as well as some of the nerve gases are inhibitors of cholinesterase and block the destruction of ACh; as a result the affected animal's nervous system is overstimulated, uncontrolled tremor and spasms result, and death ensues.

The brain and other neuronal centers do not contain solely excitatory synapses; for if they did, neural activity would cause a buildup of excitation, and action potentials would continually be generated, much as occurs during an epileptic seizure (Box 17B). In addition, there are times when inhibition is required for normal action; for example, when you withdraw your hand from a hot stove some of the muscles must be inhibited while others are excited. Obviously there must be some kind of inhibitory transmitter substance to inhibit neurons, the opposite number of excitatory substances. The simplest way of inhibiting synaptic transmission would be to increase the negativity of the resting potential, thus making depolarization more difficult. In a sense, this lessens the likelihood that threshold will be reached. How can such an effect be produced? One of the known inhibitory transmitters is the amino acid gamma aminobutyric acid (GABA). It is believed that this material increases the permeability of the postsynaptic membrane to potassium and chloride ions but not to sodium ions.

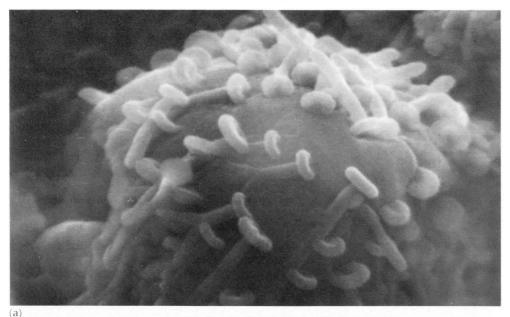

(a)

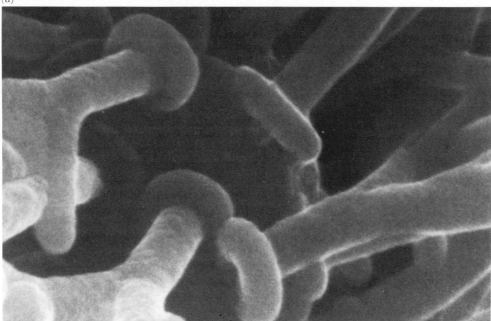

(b)

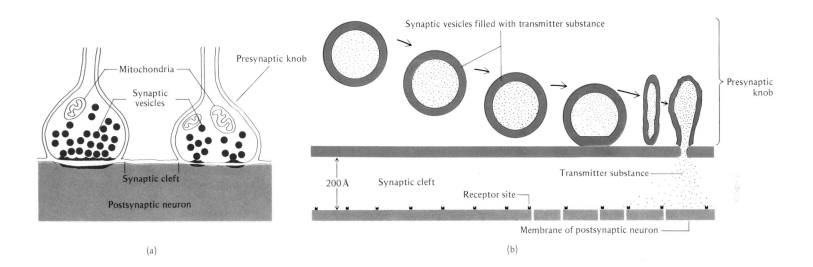

FIGURE 17–11 Structure of the synapse. (a) Diagram of synaptic knobs and synaptic cleft. (b) Schematic representation of the release of transmitter substance from the synaptic vesicles into the synaptic cleft. After discharge, the vesicle must be recharged with transmitter substance.

Labels in (a): Mitochondria; Synaptic vesicles; Presynaptic knob; Synaptic cleft; Postsynaptic neuron

Labels in (b): Synaptic vesicles filled with transmitter substance; Presynaptic knob; 200 Å; Synaptic cleft; Receptor site; Transmitter substance; Membrane of postsynaptic neuron

(a)

(b)

Thus, potassium leaks out of the cell at an accelerated rate, the membrane potential is shifted from -70 mV to -75 mV or -80 mV, and an EPSP of 10 mV which previously would have caused depolarization, now shifts the potential to -65 mV or -70 mV, a value that is below threshold. Thus, inhibition is attained by a brief increase in negativity of the membrane potential. The shift (from -70 mV to -80 mV) is called the **inhibitory postsynaptic potential** (IPSP). The IPSP can grow in graded fashion, as can the EPSP, but with opposite effects on nerve transmission.

Transmission across a synapse depends on diffusion, a relatively slow process; as a consequence, crossing the synaptic cleft is slower than transmission along the length of the neuron. In addition, since synaptic transmission involves release of molecules, which must be resynthesized after depletion, fatigue commonly occurs more quickly at the synapse than in the neuron itself when repetitive stimuli of similar intensity are delivered. At first, neurons

may discharge rapidly, but then, owing to synaptic fatigue, there is a decreased response called **accommodation;** such a mechanism permits nervous reactions to fade away with time.

The human nervous system is a blend of excitation (EPSP) and inhibition (IPSP) between neurons. Every neuron is covered with literally thousands of synaptic knobs (Figure 17–10a) and is subject to a number of excitatory and inhibitory influences. The dynamic balance between these opposing influences produces the events that occur in the nervous system and that we see recorded in EEG wave patterns.

Summation and facilitation

Stimulation of a single presynaptic knob may not cause a postsynaptic neuron to generate an action potential. Oftentimes, for an excitatory postsynaptic cell to generate an impulse, many synaptic knobs must release their transmitter substances simultaneously. As each knob

is recruited toward vesicle discharge, the potential generated on the postsynaptic neuron comes closer and closer to threshold. This progressive recruitment of synaptic knobs is called **summation.** Summation can occur in two ways: time (temporal) and space (spatial). In temporal summation the same presynaptic knob fires several times in rapid succession, so that the transmitter substance is added in amount. For this to be effective, successive firings must occur within less than 0.01 second of each other or else the transmitter substance is destroyed. (Recall acetylcholinesterase is present on the ACh receptor). In spatial summation two or more synaptic knobs fire simultaneously, and their individual effects are additive.

Let us imagine that a number of synaptic knobs secrete their chemical substances into the synaptic cleft, but this is insufficient to produce a potential big enough to generate an action potential; in other words, it is subthreshold. This subthreshold discharge of synaptic knobs may not be completely without effect because it can produce a condition known as **facilitation.** That is, even though no action potential is generated in the postsynaptic neuron, the neuron is primed for excitation and can subsequently be triggered to activity by a subthreshold stimulus. For example, let us say that it takes the discharge of 50 presynaptic knobs to generate an action potential in an adjacent neuron. If only 40 discharge simultaneously, the postsynaptic neuron does not generate an action potential; however, the presence of the chemical transmitter substance in the synaptic cleft facilitates transmission, so that when 10 or more synaptic knobs fire shortly thereafter, an action potential is generated and a nerve impulse begins.

17-6 INTEGRATED CIRCUITS OF NEURONS: THE NERVOUS SYSTEM

Basic design

Pry open the back of a television set or a radio, and you will see a maze of wires and printed circuits. Look inside a computer or a telephone switchboard, and you will find a tangle of interconnecting wires, tubes, and other electronic gadgetry. The wires, tubes, and circuits of these communication devices are the fundamental units of operation, but the sequence and arrangement of the components determine how the machine operates. The fundamental unit of structure in the nervous system is the neuron, and similarly the way neurons are hooked together determines the manner in which the nervous system functions (Figure 17-13).

There are three different functional classes of neurons: **sensory neurons,** which receive stimuli from the environment and transmit information to the central nervous system (brain

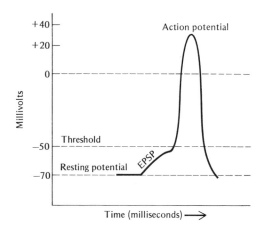

and spinal cord); **motor neurons,** which conduct messages from the brain and spinal cord to the glands and muscles; and **interneurons,** which act in an integrative capacity and shuttle signals back and forth between the neurons of various parts of the brain and spinal cord. Over 99% of the neurons of the body and brain are interneurons.

In a simplified way Figure 17-13 illustrates a schematic wiring diagram of the nervous system. Note that an impulse entering the spinal cord via a sensory neuron has many possible pathways. Rarely does the signal that traverses a sensory neuron directly activate a motor neuron leading to an effector (for example, muscle); typically the signal travels upward via the spinal cord interneurons and through a number of relay centers in the brain before reaching the higher centers. From there the command signals travel down the spinal cord, again via relay centers, and via a motor neuron the effector is triggered to activity. Note that the more neuronal cells in the circuitry, the more flexible can be the response.

The nervous system's circuitry is composed of two basic subdivisions: the **c**entral **n**ervous system **(CNS)** comprising the structures encased within the skull and the vertebral column, and the **p**eripheral **n**ervous **s**ystem **(PNS),** which lies outside the skull and vertebral column but connects up with the CNS via spinal and cranial nerves.

The central nervous system (CNS)

The central nervous system consists of the brain and spinal cord. It serves as a clearing-house for all nerve impulses, controlling, directing, and integrating all messages within the body. The brain and spinal cord embrace a series of hollow spaces (ventricles in the brain and central canal in the spinal cord) that are filled with fluid. This fluid, called **cerebrospinal fluid,** also communciates with and fills spaces between three investing membranes, called **meninges,** that cover and protect the brain and spinal cord.

The brain. The brain, soft gray and furrowed, lies protected from harm within a bony vault, the **cranium** (skull). It is about the size and shape of a head of cauliflower and has been described as a "great raveled knot," a "modest bowl of pinkish jelly," and "a rather messy substance of the consistency of porridge." The exterior of the brain is gray in color, while its inner portions are white. The gray matter consists of the cell bodies of neurons; the white matter is made up of axons. The white appearance is due to the fatty myelin sheaths that surround these axons and act as intercommunicating cables (tracts) between various regions of the brain.

The brain is commonly divided into three main geographical areas: the **forebrain,** the **midbrain,** and the **hindbrain** (Figure 17-14). The forebrain consists of the **cerebrum,** divided

FIGURE 17-13 Simplified circuitry of the nervous system. Neurons are arranged into cables consisting of many axons and dendrites. Axons, bundled together to form a multistranded cable, form the nerve fiber or nerves we commonly see in a dissected specimen. The collections of axons and dendrites in the brain and spinal cord—the information centers—are called tracts.

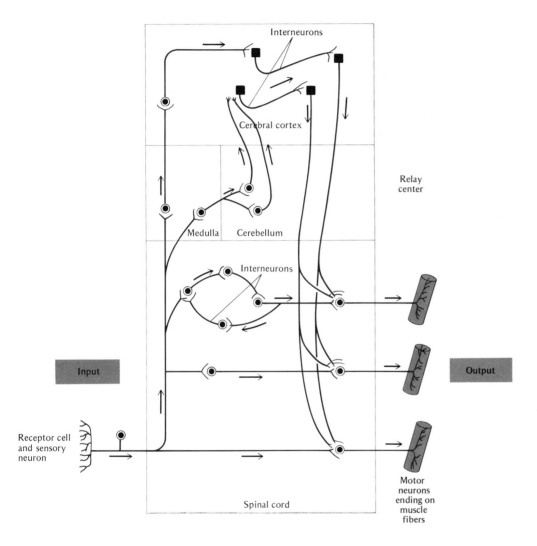

into two lobes, or **cerebral hemispheres,** together with internal structures such as the **thalamus** and **hypothalamus,** and the **limbic system** (Figure 17–14c). The surface layer of the cerebrum is deeply infolded into creases and furrows and constitutes the **cerebral cortex.** The midbrain is an intermediate zone connecting the forebrain to the hindbrain. The hindbrain consists of the **medulla (brain stem), pons,** and **cerebellum.** The brain gives rise to 12 pairs of cranial nerves, running to various regions of the body (Figure 17–14b). Some are sensory, some are motor, and some contain both motor and sensory neurons. Table 17–1 describes the structure and function of the various regions of the brain.

The spinal cord. The brain communicates with the spinal cord through an opening in the base of the skull and is continuous with the spinal cord (Figure 17–14a). The spinal cord is nearly 18 in. long and about as wide as your thumb. It is a hollow tube encased in an elongated, bony tunnel formed by the rings of the vertebrae (the backbone), and consists of millions of neurons arranged for the most part in cablelike fashion. Like the brain, the spinal cord contains gray and white matter, but the arrangement in the cord has the gray matter on the inside and the white matter on the outside (Figure 17–14d). The spinal cord controls many of the involuntary actions **(reflexes)** of the body. From the spinal cord arise 31 pairs of nerves, which pass out through openings between the vertebrae, and these branch and rebranch to innervate the limbs and torso. The spinal nerves and their branches form the PNS, discussed in the next section. All the spinal nerves are mixed; that is, they contain the axons of both sensory and motor neurons. Each spinal nerve emerges from the cord as two strands or roots that unite shortly to form a spinal nerve (Figure 17–14d). Most sensory neurons have their cell

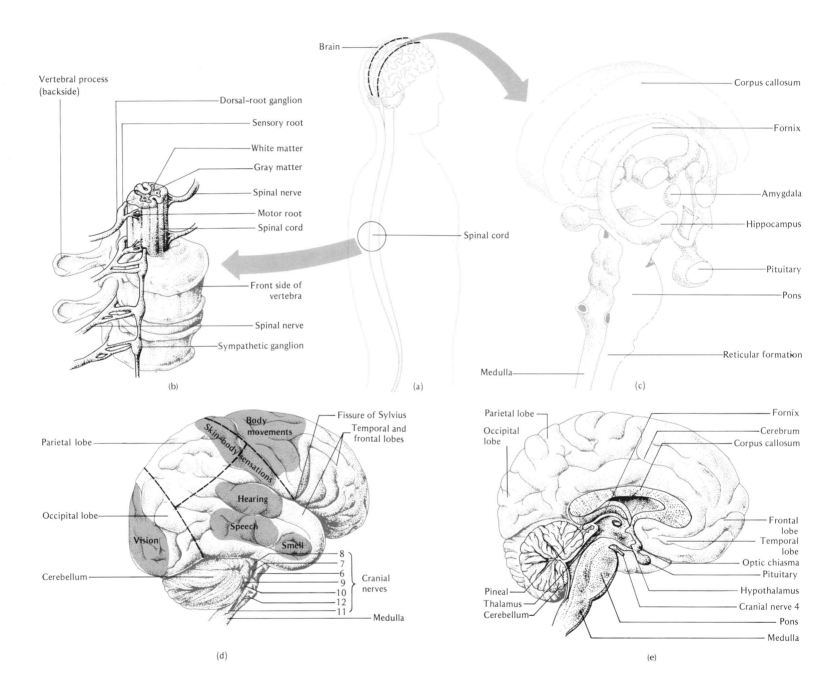

(a)
Brain
Spinal cord

(b)
Vertebral process (backside)
Dorsal-root ganglion
Sensory root
White matter
Gray matter
Spinal nerve
Motor root
Spinal cord
Front side of vertebra
Spinal nerve
Sympathetic ganglion

(c)
Corpus callosum
Fornix
Amygdala
Hippocampus
Pituitary
Pons
Reticular formation
Medulla

(d)
Body movements
Skin-body sensations
Fissure of Sylvius
Temporal and frontal lobes
Parietal lobe
Hearing
Speech
Smell
Occipital lobe
Vision
Cerebellum
8
7
6
9
10
12
11
Cranial nerves
Medulla

(e)
Parietal lobe
Occipital lobe
Fornix
Cerebrum
Corpus callosum
Frontal lobe
Temporal lobe
Optic chiasma
Pituitary
Hypothalamus
Cranial nerve 4
Pons
Medulla
Pineal
Thalamus
Cerebellum

bodies outside the cord (in a spinal ganglion)[1] and enter the cord via the dorsal (back side) root; all motor neurons have their cell bodies within the cord, and their axons leave the cord via the ventral (front side) root. If the motor root is damaged, there is muscle paralysis, but sensation is not lost. In poliomyelitis, there is damage to the cell bodies of the motor neurons within the cord itself, and there results a flaccid paralysis of muscles without sensory impairment. If the dorsal root is damaged, there is a loss of sensation in that part of the body which sends neurons to the cord via that root, but muscle paralysis does not occur. Some forms of syphilis cause dorsal-root degeneration, producing uncoordinated movements and muscle weakness. A complete cut across the cord results in spastic paralysis, loss of all sensation, and loss of urinary and rectal sphincter control. However, spinal reflexes still persist in such paraplegics.

The peripheral nervous system (PNS)

The CNS is concerned with integrating and coordinating all nervous functions, both voluntary and involuntary. All sensory input arises from the environment (external and internal) and the effector organs (muscles and glands) are the ultimate destination of nerve impulses. Lying between the CNS and the environment, and the CNS and the effector organs is a vast network of nerves and ganglia that constitute the PNS. Most of the nerves are mixed and so include both sensory and motor components. No tissue or organ is missed by the complex array of nerves, and thus, through them, the CNS is in continuous contact with every part of the body. We are aware of some of the

---

[1] Ganglia are made up of concentrations of neuronal cell bodies. The singular is ganglion.

TABLE 17–1 The structures of the human brain

Structure	Description	Function
Cerebrum		
Cortex (outer surface)	Thin (1 in.) highly folded sheet with large surface area; most recent evolutionary development; high degree of development	Center of the mind; seat of conscious thought, memory, speech, intelligence, personality, and judgment. Center of sensorimotor coordination.
White matter (inner cerebrum)	Composed of nerve tracts connecting parts of cortex to each other and to rest of brain and spinal cord	Center of sensation, including sight, hearing, taste, etc. (Figure 17-14)
Corpus callosum	Consists of 300 million separate neuronal lines	Connects the halves (hemispheres) of the cortex
Thalamus	Groups of cell bodies arranged in shape of a football	Relays sensory information to cortex (vision, hearing, touch, taste); regulates sleep and wakefulness in conjunction with limbic system

Limbic system
  Hippocampus
  Fornix
  Cingulate gyrus
  Amygdala
	Internal region of forebrain, divided into several regions	Involved in emotion, motivation and reinforcement

Hypothalamus	Linked to sympathetic nervous system and pituitary	Controls basic drives (eating, drinking, sleeping, sexual behavior), regulates temperature, blood pressure, heart rate
Cerebellum ("little brain")	Baseball-sized lump of gray-white tissue lying on either side of the brain stem	Involved in sensory-motor coordination; if damaged, loss of equilibrium and motor coordination results
Brain stem (medulla)	Extension of spinal cord within the skull	Involved in control of respiration, heart rate, and gastrointestinal function
Reticular formation	Consists of fibers and cell bodies running from spinal cord to cerebral hemispheres	Involved in control of arousal and alertness
Pons	Consists mainly of white matter lying anterior to cerebellum and between midbrain and medulla	Acts as bridge linking the various parts of the brain and as relay station from medulla to higher centers

nervous associations: these are the ones that ultimately connect with conscious centers in the cerebral cortex, and we interpret their activities as either sensations (via sensory nerves) or voluntary actions (via motor nerves). But there are other nerve components whose functions ordinarily lie outside our consciousness or volition. These components connect with our visceral organs (heart, lungs, kidneys, blood vessels, intestines), and we are neither aware of what these organs are doing nor have voluntary control over them. This is because the nervous impulses they carry never, or very rarely, reach the cerebral cortex. The motor components of these visceral nerves are extremely interesting and important, and they differ in several respects from all other parts of the PNS. They constitute what is known as the **autonomic nervous system** (ANS). There are two divisions within the ANS—the **sympathetic** and the **parasympathetic.**

Sympathetic nervous system. The sympathetic division is composed of motor nerves and of ganglia that appear as a pair of chains parallel to and on each side of the middle region of the spinal cord (Figures 17–14d and 17–15). The sympathetic nerves exit ventrally from the spinal cord and run to sympathetic ganglia near the cord, where they synapse with motor neurons whose cell bodies are located in the ganglia. The axons of the motor neurons arising from the ganglia innervate the heart, lungs, and sweat glands, and the smooth muscles in the digestive tract, excretory system, reproductive system, and blood vessels. Activation of the sympathetic system causes constriction of certain arteries, acceleration of heartbeat, elevation of blood sugar level, inhibition of gastrointestinal contractions and secretions, dilation of pupils, and many other effects. The sympathetic system mobilizes resources of the body for emergency, effects that

FIGURE 17–15 The peripheral nervous system (PNS) comprises all of the neurons and nerve fibers outside the central nervous system (CNS), and has two subdivisions: (1) the somatic nervous system—nerves running directly from spinal cord to effector organs; (2) the autonomic nervous system (ANS)—nerves leading from spinal cord to effectors synapse first with neurons (ganglia) outside the CNS, and then these send nerve fibers to the effector organs. The ANS, shown below, is composed of two anatomical subdivisions: the sympathetic, where ganglia lie on either side of the spinal cord some distance from the effectors, and the parasympathetic, where the ganglia are in or near the effectors. The sympathetic system mobilizes the body's resources for work or an emergency whereas the parasympathetic is an energy-conserving system.

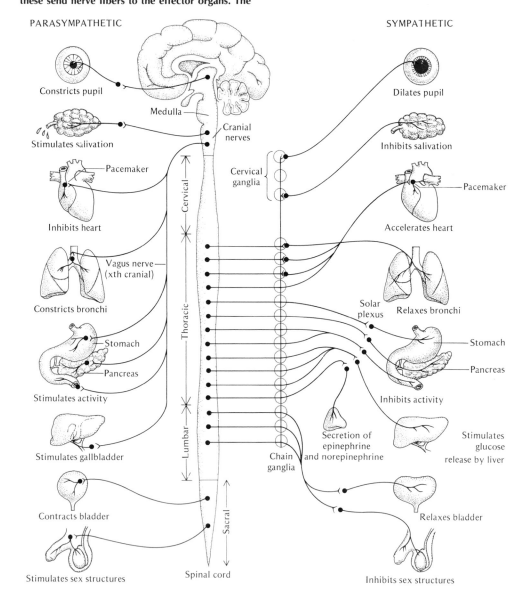

are sometimes called flight-or-fight responses. Thus, in stressful situations (fear, anger, and the like) the heart beats faster, the stomach muscles relax, the pupils widen, and energy is made available. As you may recall these same responses are produced by the adrenal hormone norepinephrine. The reason for the similarity is related to the nature of the transmitter substance in sympathetic neurons: norepinephrine. In actual fact the adrenal medulla is derived from presumptive nervous tissue during development and can be considered a highly elaborate sympathetic motor neuron. Thus, sympathetic neuron and endocrine function are related both chemically and developmentally.

Parasympathetic nervous system. The parasympathetic division is an energy-conserving system; it slows the heart rate, enhances digestive action, promotes muscle relaxation, and causes many arteries to dilate, pupils to constrict, and blood sugar levels to drop. Nerves of the parasympathetic system innervate the same organs as those of the sympathetic system, but originate mainly in the cranial nerves (principally the Xth cranial nerve, the vagus) and in the posterior (sacral) region of the spinal cord (Figure 17–15). The nerves of the parasympathetic system run from their point of origin in the spinal cord directly to their target organs. There they synapse with a motor neuron whose cell body is located in a ganglion close to the organ or even inside it. The parasympathetic neurons have ACh as their transmitter substance, and the nerves are therefore sometimes said to be **cholinergic** (as opposed to sympathetic neurons, which are **adrenergic**). It is obvious that the sympathetic and parasympathetic systems are antagonistic: sympathetic nerves are involved in excitation and emotional stress, whereas parasympathetic nerves are involved in relaxation and maintenance activi-

ties. The condition of a particular organ receiving neurons from the autonomic system is determined by the relative amount of stimulation from the parasympathetic and sympathetic nerves. As a consequence, a dynamic seesaw balance in organ function is attained.

17–7 NEURONAL CIRCUITS IN ACTION: REFLEXES

Perhaps the most familiar of all the neuronal circuits is that involved in the knee-jerk reflex. This is an involuntary stretch reflex—a motor response following sensory stimulation primarily controlled by the spinal cord and not the conscious thinking portion (higher centers) of the brain.

If you sit on the edge of a table and your knee tendon is gently tapped, your leg jerks forward in an involuntary kick. The tapping stretches the extensor muscle, and a sensory neuron relays the electrical signal to the spinal cord. There the signal is transmitted across the synapse to a motor neuron, which transmits the signal from the cord to a muscle. This causes the extensor muscle to contract, and the lower leg raises upward. This is one of the simplest of responses; it involves a neuronal arc (called a **reflex arc**) (Figure 17–16a).

Now let us consider a more complicated situation: pricking one's finger. You are about to pick a rose when your finger hits a thorn. Sensory endings (receptors) are stimulated, an impulse travels to the spinal cord, the motor nerves are triggered to action, and almost instantaneously the entire hand is withdrawn. The stimulus that sent action potentials along sensory neurons not only set off a series of impulses in the motor neurons but also triggered some interneurons within the spinal cord to fire. These signals were transmitted via ascending nerve tracts to the higher centers of

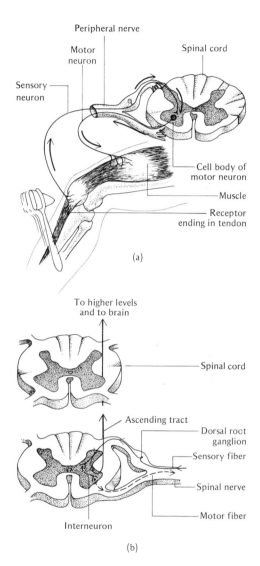

the brain (Figure 17–16b). In the brain, the speech center neurons were activated and you exclaimed, "Ouch!" At the same time the cerebral cortex became aware of pain, and there was initiated a complex series of thoughts: "What happened? Is it bleeding? Let's look at it. You are a fool not to be more careful!" Then there was initiated a series of motor responses carried by a descending nerve tract from the brain to the effectors: "Move the hand in front of the eyes and check on the damage." Again a volley of sensory inputs to the brain: "Looks all right, forget about it." Finally, there was the storage of the event (memory): "Next time you pick a rose be careful to avoid the thorns." During this simple occurrence, thousands of neurons were brought into play: sensation, reflex action, motor response, awareness, speech, observation, evaluation, and storage of information in a memory bank all occurred. With remarkable speed, information was received, sorted, and acted upon.

The reflex involved in withdrawal from injury is quite simple; yet is purposeful and coordinated. The movement of the body is away from the stimulus, and thus further damage to the body is prevented. Thus, the reflex response has purpose, even without the conscious sensation of pain, and most often the sensation of pain occurs after the response has been elicited. Indeed, this withdrawal reflex can be performed even after the spinal cord has been severed from connection with the brain!

Some biologists-psychologists say that the entire function of the nervous system can be explained on the basis of more and more complicated reflexes, that is, more complex circuitry. Another group, however, says that reflexes are the instantaneous, automatic responses to sensory stimuli only, and the more complex patterns of the nervous system are not reflexes but are activities of the higher centers.

These higher centers process information and produce an integrated response; they also store information and play an important role in memory so that appropriate actions will be taken when a similar situation arises. (See Chapter 19.)

17–8 DRUGS AND THE NERVOUS SYSTEM

A chemical agent other than a nutrient that can effect a functional and/or structural change in living tissues is a **drug**. Illegal neurological drugs are called **dope**. Drugs used as a shortcut to happiness, for kicks, are often the same substances employed in medicine for treating disease and alleviating pain; the behavioral difference induced by a drug is often related to the frequency of use, the amount used, and psychic dependence.

How and at what sites do drugs affect the nervous system? While some drugs exert their influence on behavior by affecting the physiology of the nerves themselves, one of the most likely sites of action is at the neural gateway, the synapse. A gate is a control point; locked, it keeps a pathway from being traversed; open, it permits free access. Since the synapse is the neural gate, modifications in neural transmission depend on the status of the gatekeeper, the transmitter substances. Increasing or decreasing the amount of transmitted chemicals, or substitution of molecules with similar action, will profoundly alter the nervous system's activity. Selective action of drugs at the synapse can variously alter thinking, learning, memory and complex patterns of behavior.

Here we treat stimulants, tranquilizers and three other commonly known drugs in some detail. Table 17–2 summarizes available information on a number of classes of drugs that affect behavior.

Stimulants. A stimulant is a chemical substance which "peps" us up, enhances the amount and duration of excitatory synaptic transmitters, and in effect permits more impulses to be transmitted across the synapse. As a consequence the neurons in the nerve network are overactive and most stimuli (even those which are normally subthreshold) trigger the events of excitation. Stimulants such as caffeine, nicotine, amphetamines or "uppers" (dexedrine, benzedrine and methedrine), and strychnine facilitate synaptic transmission. Strychnine in high doses is lethal because neurons tend to discharge spontaneously even in the absence of stimuli. As a result muscles contract spasmodically, ultimately causing death by spastic paralysis of the respiratory muscles. Nicotine acts by mimicking ACh. Amphetamines probably act by inducing excessive release of excitatory transmitter substances such as norepinephrine and serotonin in the brain, particularly the reticular system. They may also block uptake from the synapse so that the transmitter molecules persist, or increase their synthesis, or increase the sensitivity of receptors. Thus, amphetamines lessen depression and fatigue while increasing alertness, motor activity, and verbal activity. However, these elevations of mood are soon followed by severe depression. Often judgment is impaired so that the heart is overtaxed: when exhaustion sets in it is not felt because of the stimulant cover. Taken orally or intravenously, the results are dangerous. Tolerance and psychic dependence (addiction) develop quickly, within weeks. Continued usage ultimately leads to the development of physical and mental problems, and sometimes death ensues. Caffeine appears to be the ideal stimulant since it increases activity, suppresses drowsiness and fatigue, and has few side effects. One cup of coffee (200 mg of caffeine) stimulates the central nervous system sufficiently; moreover, there has never been a death from caffeine overdose!

Tranquilizers and sedatives. Tranquilizers have opposite effects to stimulants. They antagonize synaptic transmitter substances and turn off neural transmission, especially in the higher centers of the brain, and numb one's awareness of the external and internal world—the nerve network "sleeps." The emotions of rage and aggression are centered in the hypothalamus and limbic regions of the brain. The reticular system, the brain's "stay awake" center, is closely linked to these regions. Norepinephrine and serotonin are important neuronal transmitters in these regions of the brain. The tranquilizer reserpine acts by blocking the uptake of norepinephrine into synaptic knobs and its depressant chemical cousin chlorpromazine blocks both the receptors of epinephrine and ACh.

These potent tranquilizers are usually not a part of the drug culture; for the common user and abuser, the barbiturate sedatives ("downers") provide relief from the everyday stresses of life and aid the insomniac. We must be a nation of insomniacs because we consume 200 tons of barbiturates a year, and one out of every two Americans uses barbiturates! Sedatives and barbiturates work in similar ways to tranquilizers, probably through interference with the release of the transmitter substances serotonin and norepinephrine, prevention of their synthesis, decreasing the sensitivity of post-synaptic receptors, or enhancing the decay of the materials once they are released. Barbiturates first affect the reticular system and then directly or indirectly the cortex of the brain. Barbiturates, taken in excessive amounts, can produce oversedation or drunkenness. They are also strongly addictive, and it is not long before there is physical and psychic dependence on these "goofballs." Chronic barbiturate use leads to tolerance within a short period, and excessive doses must be taken to achieve the same state of calm. Abrupt removal from the drug precipitates withdrawal symptoms: "shakes," convulsions, hallucinations. Unable to eat or retain food, panic sets in, fever is present, and if the individual is not healthy, death may result.

Marijuana. Marijuana is derived from the weed, *Cannabis sativa*. It contains a sticky golden resinous substance that is rich in a chemical known as tertrahydrocannabinol, or THC. The THC is responsible for the euphoric effects of marijuana, which tend to be both individualistic and highly variable. Marijuana smokers report feeling detached and free-floating. Visual and auditory perceptions are enhanced, short-term memory is interfered with, speech patterns become disorganized, and thoughts seem to be more lucid, although the train of thought is often lost entirely. THC tends to accumulate in the neurons of specific areas of the brain, in particular the frontal cortex and the hippocampus. The time distortion frequently reported by marijuana users may be related to this, since the hippocampus is involved in the recall of recent events while the frontal cortex is presumed to play a role in evaluating one's actions (Figure 17–14 and Table 17–1).

LSD. The powerful hallucinogen LSD (lysergic acid diethylamide) appears to affect other parts of the brain than does THC. During an LSD "trip" the individual might experience psychoses, disruption of the time sense, strange visions, and behavioral detachment. LSD blocks the effects of serotonin, a neurotransmitter contained in the raphe cells of the medulla, a region that controls slow-wave sleep (Figure 17–14). How does LSD produce its effects? The brainstem cells function in an analogous fashion to a boss's secretary. The secretary is the barrier that must be passed before the boss, that is the cortex, can be influenced. Under normal circumstances the brainstem, like a good secretary, acts to filter, block and scan all the information received from various parts of the body, en route to the brain. When LSD promotes the raphe cells of the brain stem to increase their serotonin content, the filtering action of the secretary is disrupted; no longer can the secretary discriminate among the information—sensory events are both enhanced and distorted, what was once considered insignificant becomes important; sensory inputs are disturbed producing strange feelings, confusion, alarm, and the like. Unpleasant consequences may result.

Alcohol. Alcohol reduces the ability of the nerve cell to conduct messages by its action on the function of the nerve cell membrane. In general, alcohol is a central nervous system depressant that produces a marked decrease in motor performance and mental abilities. Its initial stimulatory effects are probably caused by suppression of the nervous system's inhibitory controls. Alcohol causes increased aggressiveness in many heavy users and may have adverse effects on the liver, stomach, intestines, and even the brain, producing delerium tremens or DTs if the supply is withdrawn. Alcohol is addictive, and treatment of alcoholism is extremely difficult.

SUMMARY

1. The nervous system provides for rapid communication between various tissues and organs, and links the sense organs with effectors.

2. Neurons or nerve cells transmit nerve impulses. Glial cells are supportive in function. The brain, spinal cord, and ganglia are con-

TABLE 17-2 Some drugs that affect behavior.

Drug type	Official name	Slang name	Legitimate medical use	Psychological dependence	Potential for physical dependence
Stimulants	Caffeine No Doz* Coca Cola* Coffee Tea	 Coke Java 	Lessen fatigue and drowsiness; promote alertness	Slight	No
	Nicotine		None	High	No
	Strychnine		None (used in some rat poisons)		
	Amphetamines Benzedrine* Methedrine* Dexedrine*	Uppers Bennies, beans, cartwheels Crystal, speed, meth, crank Dexies, Christmas trees	Lessen fatigue and depression; treatment of obesity; promote alertness, motor activity, verbal activity	High (in weeks)	High
Tranquilizers	Reserpine (Rauwolfia)		Treatment of moderate depression	Minimal	No
	Chlorpromazine Thorazine* Compazine* Stelazine*		Treatment of schizophrenia, depression, delerium	Minimal	No
	Barbiturates Seconal* Amytal* Nembutal* Tuinal*	Downers, goofballs Red devils Blue angels Yellow jackets Rainbows	Treatment of insomnia and tension	High	Yes
Narcotics	Derivatives of opium poppy Opium Morphine Heroin	 Op Horse, H	Treatment of severe pain, diarrhea, cough	High	Yes
	Synthetics Methadone Demerol*		Treatment of persons addicted to heroin, morphine		
Anesthetics	Ether Chloroform		General anesthesia		
	Procaine Novocaine*		Local anesthesia for minor surgery or dental work	No	No
	Cocaine	Coke, snow	Topical anesthesia made systemic by eating coca leaves	Yes	Yes
	Ethyl alcohol (whisky, gin, beer, wine)	Booze, hooch, juice	Sedative, although there is a primary excitatory phase	Moderate	Yes
Hallucinogens	Psilocybin Mescaline (peyote)	 Cactus	Experimental studies on brain function		
	LSD (d-lysergic acid diethylamide)	Acid, sugar		Yes	No
	Marijuana Hashish	Pot, grass, tea, Mary Jane Weed, hash		Moderate	No

*Registered trade name.

Tolerance	Physiological effects (short term)	Long-term effects (psychological and physiological)
Yes	Facilitate synaptic transmission in CNS; therefore, enhance amount and duration of excitatory transmitters	Produce insomnia or restlessness; habituation
Yes	Mimics Ach; excitatory in low doses from smoking	Causes lung cancer, heart and blood-vessel damage, emphysema
	Facilitates transmission by mimicking ACh and depresses IPSP; lethal in high doses because neurons discharge spontaneously in absence of stimuli, resulting in convulsions	Produces death due to spastic paralysis of respiratory muscles
Yes	Stimulate excessive release or synthesis of excitatory transmitters norepinephrine and serotonin in brain; may also block uptake from synapse and increase sensitivity of receptors; structurally resemble epinephrine	Cause malnutrition, paranoid schizophrenia, hallucinations, psychoses; judgment is impaired, heart is overtaxed, exhaustion results from stimulant cover
No	Depletes stores of norepinephrine in sympathetic nerves, thus enhancing activity of the parasympathetic system	Produces drowsiness, drying of mouth, blurring of vision, diarrhea, ulcers
No	Blocks receptors of epinephrine and ACh; CNS depressant	Induces hypothermia, acts as antiemetic, depresses stay-awake reticular network, suppresses hallucinations and delusions
Yes (quickly)	Interfere with synthesis and release of the transmitters norepinephrine and serotonin especially in reticular system; decrease sensitivity of postsynaptic receptors; may enhance breakdown of transmitter substances	Induces sleep, euphoria, drowsiness, impaired judgment, irritability, weight loss; Highly addictive; abrupt withdrawal is severe and unsafe, producing shakes, convulsions, hallucinations
Yes	Act as CNS depressant, inducing muscle relaxation, hypnosis (drowsiness and lethargy), euphoria (well-being), and sleep	Produce constipation, loss of appetite, temporary impotence, addiction with unpleasant symptoms of withdrawal; overdose kills, probably by depressing respiratory center
	Mechanism of action unknown	
	Inhibit axonal transmission of nerve impulses	
No	Prevent synaptic transmission to neurons of CNS	None
Yes	Prevents or reduces uptake of transmitters such as norepinephrine into nerve terminals	Produces euphoria, alertness, addiction
Yes	Acts as CNS depressant, producing loss of motor coordination (high doses) and alertness (low doses); induces euphoria, exact mechanism of action unknown	Causes cirrhosis of liver, physiological dependence, obesity, brain damage, addiction with severe withdrawal symptoms
	May enhance serotonin and norephinephrine levels in the brain because of their chemical similarity to these transmitters; therefore, sensory input enhanced	Precipitate existing psychoses; psycotic symptoms may recur without drug taking
Yes	Rich visual imagery, sensory awareness, "trip," hallucinations	
No	Acts as CNS depressant in high doses, euphoria and stimulant in low doses; exact mode of action unknown	Causes loss of appetite, judgment, and motivation

centrated centers of nerve cells and supporting elements.

3. Neurons conduct messages from dendrites via the cell body to the long axon whose terminal ramifications end in synaptic knobs. The axonal sheath or neurilemma is formed by interstitial cells. Interstitial cells form a tightly wrapped insulating myelin sheath around many neurons, especially in the spinal cord and the brain. The sheath is interrupted by nodes. Multiple sclerosis is due to destruction of the myelin sheath and short-circuiting between neurons.

4. A potential difference is maintained across the neuronal plasma membrane, which is electrically negative inside and positive outside. Na^+ ions are higher outside, whereas Cl^-, Org^-, and K^+ are higher on the inside. The membrane resting potential is -70 mV.

5. When a neuron is stimulated, membrane permeability temporarily changes. Sodium enters and the membrane becomes rapidly depolarized and then just as rapidly repolarized as permeability to sodium is reversed. This is the action potential (impulse), and it passes along the nerve with no decrease in strength.

6. A second action potential cannot be produced until the membrane is completely repolarized and sodium ions are pumped out (the sodium pump). During this period, the neuron is in a refractory state (inexcitable).

7. Stimuli must be stronger than threshold level to produce an action potential. The magnitude, character, and duration of the impulse are independent of the strength or nature of the stimulus (the all-or-none law). Stimulus intensity is determined by variation in neuronal threshold, numbers of neurons stimulated, and frequency of action potentials.

8. Transmission velocity varies with neuronal diameter and myelination. Transmission is saltatory, from node to node.

9. Nerve impulses are transmitted in only one direction because of the synapse. When a neuron is stimulated, presynaptic axonal knobs discharge transmitter substance, which diffuses across the synaptic cleft to the dendrites of the postsynaptic neuron; here, an excitatory postsynaptic potential (EPSP) is triggered, which in turn triggers an action potential when threshold levels are reached.

10. The postsynaptic membrane inactivates the transmitter substance enzymatically. Excitatory transmitters include acetyl choline (ACh).

11. Inhibitory substances such as GABA can lower the resting potential and thus raise the threshold necessary for generation of an impulse (inhibitory postsynaptic potential or IPSP). Nervous activity results from a balance in EPSP and IPSP between neurons.

12. The nerve impulse is subject to accommodation because of synaptic fatigue.

13. Summation involves the progressive recruitment and discharge of presynaptic knobs. It can be temporal and/or spatial. It is subject to facilitation (subthreshold discharge of synaptic knobs).

14. The central nervous system (CNS) comprises the brain and spinal cord. The peripheral nervous system (PNS) comprises all of the neurons and nerve fibers of the CNS.

15. Sensory neurons transmit environmental stimuli to the CNS. Motor neurons transmit messages from the CNS to glands and muscles, and interneurons are integrative.

16. The brain and spinal cord are surrounded and protected by membranes called meninges. Cerebrospinal fluid fills the spaces in the brain, and the central canal of the spinal cord, and the spaces between the meninges.

17. The exterior gray matter of the brain consists of neuronal cell bodies: the internal white matter is made up of axons and their myelin sheaths.

18. The forebrain consists of the bilobed cerebrum, thalamus, hypothalamus, and the limbic system. The exterior cerebral cortex, which is extensively furrowed and highly developed in man, is the center of consciousness. The hindbrain consists of the brain stem or medulla, pons, and cerebellum; and the midbrain connects the fore- and hindbrain. Twelve pairs of cranial nerves arise from the brain.

19. The gray matter is interior to the white matter in the spinal cord. It controls many involuntary actions (reflexes) and gives rise to 31 pairs of mixed nerves. The sensory elements have dorsal roots and ganglia; the motor elements have ventral roots and ganglia.

20. The autonomic nervous system consists of the sympathetic and parasympathetic systems and is concerned with involuntary control.

21. The sympathetic nerves exit ventrally from the spinal cord and synapse with motor neurons in the ventral ganglia. Norepinephrine is the transmitter substance, and adrenergic sympathetic nerves innervate heart, lungs, sweat glands, and involuntary smooth muscles. The system is involved in the emergency response.

22. Parasympathetic nerves originate in the Xth cranial nerve (vagus) and the spinal sacral region. They are cholinergic (ACh is the transmitter) and generally inhibitory. They innervate the same body organs as the sympathetic nerves, and thus a functional balance is maintained between excitation and relaxation and maintenance.

23. Simple reflex arcs involve transmission from a sense organ along a sensory neuron via spinal interneuron and motor neuron to an effector organ. In this way a simple involuntary response is produced. Secondary impulses along the spinal cord to the higher centers of the brain may be generated simultaneously, so that information is processed and a further (perhaps conscious) integrated response may be produced.

24. A chemical other than a nutrient that can effect a functional and/or structural change in a living tissue is a drug. Neurological drugs may affect behavior by acting on nervous physiology or on the synapse, by increasing, decreasing, or substituting for transmitter substances.

KEY WORDS

nerve impulses
neurons
nerves
glial cells
dendrites
axon
synaptic knob
neurilemma
interstitial cells
myelin sheath
nodes
multiple sclerosis
sodium pump
membrane resting potential
excitable cells
stimulus
depolarized
repolarization
action potential
refractory state
threshold stimulus
all-or-none law
brain waves
electroencephalogram (EEG)
alpha waves
beta waves
epilepsy
node
synapse
synaptic cleft

presynaptic knobs
excitatory postsynaptic potential (EPSP)
inhibitory postsynaptic potential (IPSP)
accommodation
summation
facilitation
sensory neurons
motor neurons
interneurons
central nervous system (CNS)
peripheral nervous system (PNS)
cerebrospinal fluid
meninges
cranium
forebrain
midbrain
hindbrain
cerebrum
cerebral hemispheres
thalamus
hypothalamus
limbic system
cerebral cortex
medulla (brain stem)
pons
cerebellum
reflexes
autonomic nervous system
sympathetic system
parasympathetic system
cholinergic
adrenergic
reflex arc
drug
dope

TOPICS FOR REVIEW AND DISCUSSION

1. Draw and label a nerve synapse.
2. What is the function of the synapse? How is drug action related to the synapse?
3. Discuss polarity in the nervous system.
4. What are the differences between the parasympathetic and sympathetic nervous systems?
5. Draw a "typical" neuron and label its parts.
6. Describe a reflex arc.
7. How is the human nervous system organized?
8. In what ways are the nervous and endocrine systems related?
9. Why can't a complete description of the function of a neuron be used to describe the entire operation of the nervous system?
10. Define and describe summation, facilitation, fatigue, and inhibition in the nervous system.

The film *2001: A Space Odyssey* is an adventure concerning deep space exploration. En route to Jupiter the giant antenna system of the spaceship Discovery malfunctions, and Mission Commander Bowman must leave the ship to repair it. The antenna system of the spaceship enables the crew to keep in touch with the Earth almost half a billion miles away, and this receptor system must operate perfectly or else the mission will fail. (Courtesy of Metro-Goldwyn-Mayer, Inc., © 1968.)

BUILT-IN SIGNAL CORPS: SENSE ORGANS

18-1 SENSORY LISTENING POSTS

This morning the alarm clock buzzed, and with eyes closed you pressed the lever that turned it off. With barely opened eyes you checked the weather outside and found the sun shining. Then your eyes opened fully as the lids moved upward. You rolled out of bed, washed your face with cold water to perk yourself up, listened to the early-morning news on the radio, and were completely aroused by the aroma of fresh-brewed coffee. Each day of our lives, mundane or exciting, the world about us is continually being monitored by a battery of listening posts—our sense organs. Our view of the world is imparted by our senses; deprived of all sensation a human being experiences fright, terror, hallucinations, and ultimately madness.

Structurally our communications network consists of signal receivers (receptors), conduction pathways (nerves), and a means for interpreting and/or responding to the signals (brain and muscles or glands). The receptors that monitor the external world are those with which we are most familiar—those of the five senses: sight, hearing, touch, smell, and taste. These are, however, only a few of the many senses with which we are endowed, since, in addition to receiving information from the outside environment, the nervous system receives messages from the internal environment: tension in muscles, tendons, and joints; the feeling of fullness. These sensory receptors are groups of specialized neurons; together they provide us with a wide range of sensory experience and enable us to keep tabs on the workings of the body.

Although sense organs such as the eye, nose, tongue, skin, and ear are responsible for distinct sensations, the sensory receptor itself does not determine what we see, feel, hear, taste, or smell. Indeed, you can prove this to yourself by closing your eyes and lightly tapping on the lids. You see flashes of light that are produced not by light but by touch (pressure). How can pressure signals be received as light? It is clear that what we sense depends not so much on the nature of the stimulus, as on where it is transmitted. Sense organs are essentially **transducers,** designed to collect information of one kind or another and convert it into nerve impulses. Neurons have a limited code of information: the signals are either off or on, and when they are turned on, only the signal frequency can be altered. Only when the signals are transmitted to the cerebral cortex of the brain, where they are decoded and interpreted, do we sense a stimulus. The brain's cortex puts together the coded messages it receives from a variety of sources and integrates, interprets, stores, and acts on them. There is a systematic organization in the cortex, and different regions seem to be responsible for decoding and interpreting different signals (Figure 17–14). Just as tapping on the closed eye produces a sensation of light, so does pressure on the back of the skull. "Seeing stars" as a result of a blow to the head indicates that the light-interpreting region of the brain has received a signal, in this case as a result of a direct pressure stimulus.

One of the frontiers of modern biology involves the mechanism whereby information in the nervous system is coded and decoded to produce a richness of sensation we all recognize. No doubt the future will see this frontier crossed and much advancement in decoding nerve messages and understanding sensation and behavior.

18-2 SMELL: DETECTING CHEMICAL SIGNALS IN THE AIR

The organs responsible for detecting odors, called **olfactory organs,** lie high up in the

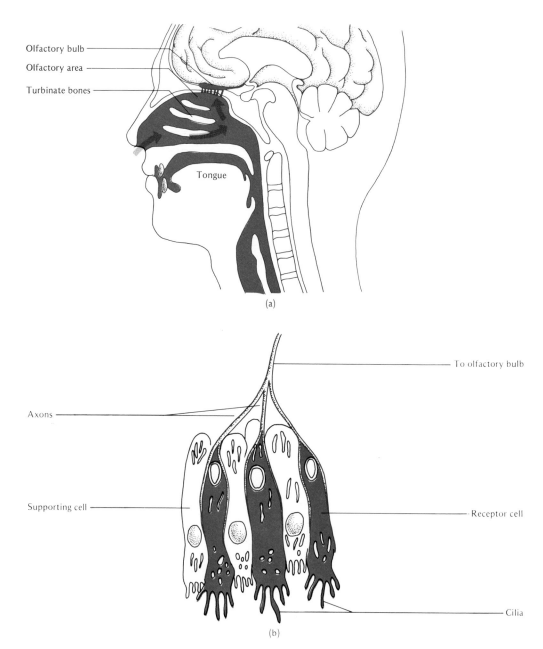

Olfactory bulb
Olfactory area
Turbinate bones
Tongue

(a)

To olfactory bulb
Axons
Supporting cell
Receptor cell
Cilia

(b)

interior of the nose (Figure 18–1a). The receptor cells of the olfactory organs are neurons, each of which bears a cluster of hairlike filaments at its free end (Figure 18–1b). The olfactory cells run toward the brain as the olfactory nerve, where they end in the olfactory bulb. From the olfactory bulb signals are sent for interpretation to the higher centers within the cerebral cortex (Figure 17–14).

Because the olfactory organs are recessed deep within the nasal passages, the material to be smelled must have certain properties. First, it must be volatile, that is, so organized chemically that its molecules evaporate rapidly and can diffuse upward to the receptor or be drawn upward by sniffing. Second, it must be capable of being absorbed and soluble in water and lipids. This enables the molecule to pass through the watery film around the nerve endings and to penetrate the fatty plasma membrane. Third, the substance must be one that is not already present on the surface epithelium of the olfactory organ, so that it elicits a change upon arrival. Only change will cause the sensation of smell. Once the odor molecule has penetrated the receptor cell, it causes depolarization of the membrane and triggers an action potential. Information on the chemistry of the odor recorded as action potentials in the olfactory nerves is then decoded in the brain.

How are we able to discriminate among odors, which number in the thousands? Are there different receptors for each odor? A current theory suggests that discrimination among odors depends upon the presence of a number of olfactory receptors in the nose, each of which has at its tip an absorbent site specific for a particular odor or class of odors. Each receptor terminus will accept only its appropriately shaped odor molecule, just as a lock and key fit exactly. Thus molecules have different odors because they cannot fit into the

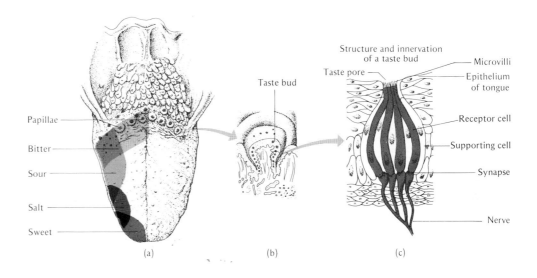

same absorbent site on the olfactory receptor; but if substances of different chemical composition have the same configuration, they can have the same odor because they fit the same site. Differences in orientation of the atoms in a molecule that do not alter surface architecture do not alter odor. Odors may be modified if fatigue has occurred in neighboring receptors. For example, if receptor site A is fatigued, the molecules that would ordinarily fit into A, B, and C can occupy only B and C. This is why, when shopping for perfume, one can smell only a few before they all begin to smell alike—the capacity for discrimination is lost or diminished.

18–3 TASTE: DETECTING CHEMICAL SIGNALS IN SOLUTION

You bite into a thick, juicy steak smothered in mushroom sauce, and your taste buds savor the food. It is really good! Taste is, like smell, a chemical sense. Odorous materials are wafted by air currents into the nose; however, taste

chemicals must be put into the mouth in solution, where they excite taste receptors, the **taste buds** found primarily on raised papillae on the surface of the tongue (Figure 18–2a). Taste is a short-distance sense, for chemicals in solution must be placed directly in contact with the receptors. There are approximately 10,000 taste buds, and they are distributed over the tongue in a definite pattern (Figure 18–2b). The tip of the tongue senses sweet tastes, the sides sense salty and sour tastes, and the back of the tongue detects bitter flavors. Although we have four primary tastes, and all other tastes are combinations of these, all the taste receptors look alike.

Within each taste bud the receptor cells, which are modified neurons, are arranged like the segments of an orange, with the upper tips of the receptor cells extending into a small pore at the tip of the taste bud (Figure 18–2c). The tips contain microvilli, and these provide the contact point between taste molecules and receptor cells. The impulses from the taste buds are transmitted by nerves up the brain stem and then to the cortex. The full detection of taste is related to smell, and often the two cannot be separated. For example, during a stuffy head cold, we lose the sense of taste because we cannot smell!

18–4 SIGHT: RECEIVING LIGHT SIGNALS

Let there be light: and there was light.
Gen. 1:3

The light of the body is the eye.
Matt. 6:22

We and our closest primate relatives are so extremely visual as animals that eyesight is probably our most important environmental sense. Indeed, so reliant are we upon our eyesight that without it we are virtually incapable of surviving unaided.

Eye structure

The human eye is a tender ball about 1 in. in diameter, protected by its location in a socket of the skull. The position of the eyes in front of the head allows for superimposition of the images from each eye, and we see stereoscopically in three dimensions (length, width, and depth).

The optical system for detecting light within the eye is much like that of a camera. The eyeball has a three-layered wall, each concentric layer, or coat, functioning in a specific capacity (Figure 18–3). The outer covering is a tough, fibrous capsule called the **sclera,** the white of the eye; it maintains the shape of the eye and protects it. Muscles responsible for moving the eye in the socket are attached to the outside of the sclera. The front of the sclera, facing the outside world, is perfectly transparent and is called the **cornea.** This is truly the window of the eye, and it is composed of five layers of flat cells arranged like sheets of plate

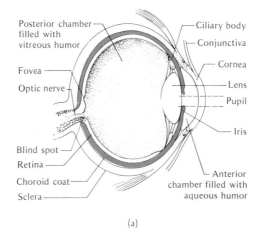

Posterior chamber filled with vitreous humor
Fovea
Optic nerve
Blind spot
Retina
Choroid coat
Sclera
Ciliary body
Conjunctiva
Cornea
Lens
Pupil
Iris
Anterior chamber filled with aqueous humor

(a)

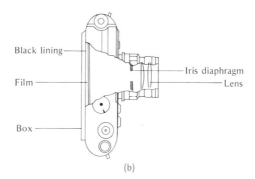

Black lining
Film
Box
Iris diaphragm
Lens

(b)

glass. The layer just inside the sclera, called the **choroid coat,** contains blood vessels and a nonreflecting pigment that acts as a light shield and prevents the light from scattering. Light penetrates the cornea and enters the interior of the eye through the front portion of the choroid coat, which is interrupted to produce a small opening, the **pupil.** The region of the choroid, visible through the cornea and surrounding the pupil as a colored band, is called the **iris.** The iris gives the eye its color, depending on the presence or absence of pigment. If there is a lot of pigment (melanin), the eye color is brown; if pigment is sparse and localized only in the back layer of the choroid, the light is scattered to a greater extent, and the eye appears blue. We are all born with blue eyes because it takes time for melanin production to get under way after birth. In the absence of pigment, as occurs in **albinism,** the eye appears pink because the blood in the blood vessels of the choroid shows through.

To allow for changes in light intensity, most expensive cameras are equipped with a diaphragm (so-called *f* stops). When the light is dim, the diaphragm is wide open; when the light is too bright, it is closed down. The automatic diaphragm of the eye is the iris. The iris can expand and contract, thus enlarging and decreasing the size of the pupil by means of iris muscles that function like the drawstring around a purse. The iris can expand the pupil's diameter sixfold (from $\frac{1}{16}$ to $\frac{1}{3}$ in.), and 40 times as much light can enter the eye when the pupil is expanded as when it is contracted. This dark adaptation takes time, as you know if you have stumbled in a darkened theater after entering from bright daylight.

The elastic **lens** is suspended from the **ciliary body** just below the iris. Composed of concentric layers of fibers and crystal-clear proteins in solution, it serves to focus the light rays onto

FIGURE 18–4 The lacrimal glands, above and on the outer edge of the eye socket, secrete tears that wash away foreign substances and dilute out irritants on the eye's surface. Excess fluid drains through minute openings near the nasal side of each eyelid and enters the nasal cavity through the lacrimal sac and the nasolacrimal duct.

the receptor cells of the film of the eye, the **retina.** The retina is the innermost layer of the eyeball. It is about the size of a postage stamp, not much thicker, and lines the back two thirds of the choroid coat. It contains special nerve cells, called **rods** and **cones,** which convert light rays into nerve impulses.

The coats of the eye are rather soft, and to maintain their position the entire eye is filled with fluid kept under pressure; for the eye to work, the eyeball must be turgid. The space in front of the lens **(anterior chamber)** is filled with a lymphlike watery fluid, **aqueous humor.** It is provided by filtration from capillaries, and is ordinarily drained off into blood vessels supplying the cornea. Behind the lens in the **posterior chamber** is a jellylike but clear fluid, the **vitreous humor.**

The eye is protected in front on its outer surface by a thin transparent layer of epidermal cells, the **conjunctiva.** The conjunctiva is continuous at its periphery with two folds of skin, the **eyelids,** which are aided in their protective function by hairs on their edges, the **eyelashes.** The eyelashes and the **eyebrows** above the eye help to prevent dust, sweat, and so on from reaching the sensitive, moist surface of the eye. The eye is kept moist at all times by the **tears,** a watery secretion of two special **lacrimal glands** (*lacrima*—tear; L.). Excess tears are drained from the eye into the nasal sinuses by the **lacrimal duct,** opening from the inner corner of the lower eyelid (Figure 18–4).

This then is the eye, a living device for photoreception.

Filming the world around us: the retina. The light-sensitive film of the eye, the retina, consists of about 130 million neurons stacked up in three layers. Their function is to convert the electromagnetic radiations of light into electrical signals that can be sent to the brain, where they produce the sensation of seeing.

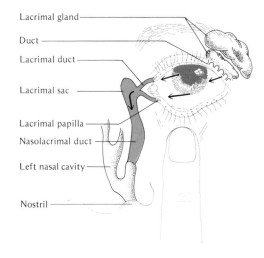

Lacrimal gland
Duct
Lacrimal duct
Lacrimal sac
Lacrimal papilla
Nasolacrimal duct
Left nasal cavity
Nostril

Light that enters the eye first passes through the curved surface of the cornea, continues on through the aqueous humor, and enters the lens by way of the pupil. In the lens the ultraviolet light is filtered out, and the rays of light pass through the vitreous humor and then impinge on the retina. Light hitting the retina first strikes the axons of the optic neurons, next passes through a layer of bipolar cells, and then, at the very back, hits the light-sensitive neuronal elements, the rods and cones (Figure 18–5). It is a curiosity of development that puts the light-sensitive neurons at the back of the retina, directed away from the light, and requires light to pass through two other layers before it strikes the photoreceptor cells. It is for this reason that the retina is referred to as an **inverted retina.**[1]

The rod cells are very sensitive and are more abundant toward the edges of the retina. They

---

[1] Not to be confused with the inverted image formed on the retina.

enable us to see in light too dim to stimulate the cone cells, but they do not give a clear or well-defined image and cannot differentiate color. The cone cells, responsible for detection of color, are more abundant toward the center of the retina and give rise to more detailed and better defined images than the rod cells. They are used for vision in bright light. When light hits the rods and cones, it causes a chemical change in the pigments they contain: the pigments are split, energy is released, ionic changes in the intracellular fluid produce an electrical potential, the neuron is triggered to fire, and this impulse travels to the visual region of the brain, where it is decoded and interpreted as light. The split pigments are reduced in the retina to vitamin A and must be resynthesized if we are to continue perceiving light. If the diet is deficient in vitamin A, pigment synthesis is impaired, and the disease known as **night blindness** results. The rods contain only one kind of pigment, rhodopsin or visual purple, but the protein pigments in the cones come in three different kinds and are sensitive to blue, red, and green light, respectively. We can distinguish 160 different colors, and this ability depends on the amount and kinds of cones stimulated—a sort of neuronal pigment mixing to produce a particular hue, just as the artist mixes colors on his palette. White is perceived if all three cone pigments are equally stimulated by light. Color intensity is determined by the number of nerve impulses produced by the light hitting the cones.

The cornea and lens of the eye focus the light to the **fovea,** a point on the retina no bigger than the head of a pin. The cones are abundant in this region, and there are no rods. The nerve layer above the fovea is quite thin, so that light hits the photoreceptor cells directly. The fovea is the region of keenest vision. On this spot the eye focuses the letters and words

FIGURE 18–5 The human retina is composed of several layers of cells. Incoming light strikes the retinal layers in the following order: nerve fibers, ganglion cells, bipolar nerve cells, rod and cone cells, and pigment cells. The pigment layer acts like the black surface inside a camera and prevents light from being reflected backward; the rods and cones are light-sensitive neurons that transmit their signals to the ganglion cells via the bipolar nerve cells; the axons of the ganglion cells run along the surface of the retina eventually converging to form the optic nerve.

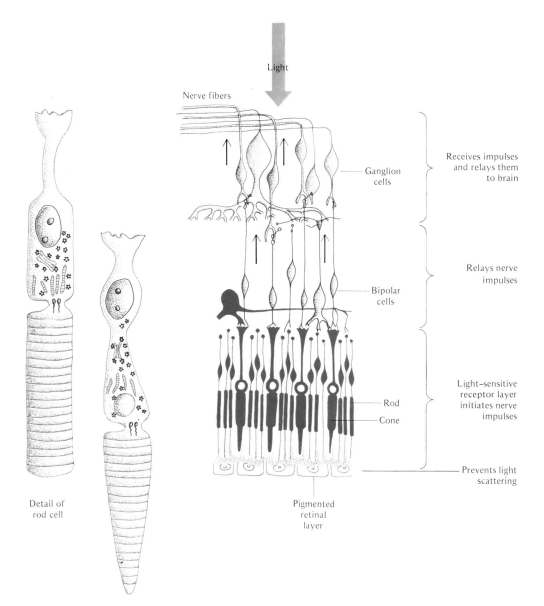

Detail of rod cell

Detail of cone cell

Light

Nerve fibers

Ganglion cells — Receives impulses and relays them to brain

Bipolar cells — Relays nerve impulses

Rod
Cone — Light–sensitive receptor layer initiates nerve impulses

Pigmented retinal layer — Prevents light scattering

you see on this page. Light that strikes the retinal cells outside the region of the fovea gives rise to less well-defined images because there are fewer cones; we cannot make out fine details. This constitutes **peripheral vision.** Peripheral vision is in a sense looking out of the corner of your eye; when it is restricted the person is said to have **tunnel vision.**

If we were to look at the retina through an ophthalmoscope, we should see the fovea as a small depression, and a creamy white disc, the **optic disc,** or **blindspot** (Figure 18–6a). The optic disc contains no rods or cones, but consists only of nerve fibers that are gathered from the retina to form the optic nerve. We are unaware of the blind spot because the eye ordinarily focuses light rays on the fovea. Your own blind spot can be located by following the directions in the caption to Figure 18–6b.

The **optic nerve** passes backward from the retina to the brain. Just in front of the pituitary gland, the optic nerves from each eye partially cross, forming the **optic chiasma.** The nerve fibers originating from the temple sides of the two retinas remain uncrossed, but those from the nasal sides do cross over (Figure 18–7). Thus, the optic nerves that leave the optic chiasma are composed of outside signals of one retina and inside signals of the other. Each optic nerve synapses with cells in a region below the thalamus of the brain and then spreads out to the visual cortex of the brain. The peculiar anatomic distribution of nerves from the two retinas means that damage (such as is caused by an enlarged pituitary or a pituitary tumor) to a single optic nerve affects both retinal images.

Image formation. Just as clear photographs depend upon precise focusing of light on the film of a camera, so clear vision depends upon precise focusing of light rays on the retina. If focusing is poor, the image is blurred.

FIGURE 18–6 (a) The optic disc (blind spot) and the fovea as seen through the ophthalmoscope. (b) Location of the blind spot. Close your left eye and focus your right eye on the cross. Now move the page away from your eye and then toward your eye. At a distance of 6–8 in. the star is not seen.

FIGURE 18–7 (below) The optic chiasma viewed from below.

FIGURE 18–8 (right) (a) Image formation on the retina. Incoming light rays from the object being viewed are bent in such a way that the rays from each point come together at a single point on the retina producing an image. When you look at a candle, all rays reflected from the tip of the flame must be brought together at a point on your retina, all rays from the

base must be brought together at another point, and so on for all parts of the candle. The image formed on the retina is inverted. (b) The divergence of light rays from near and far sources. Light rays from a point source travel outward in all directions. The cornea of a viewer standing near the source will be struck by rays traveling at divergent angles (rays 1–9), but the cornea of a viewer 20 ft or more away, will be struck by fewer rays (rays 4–6).

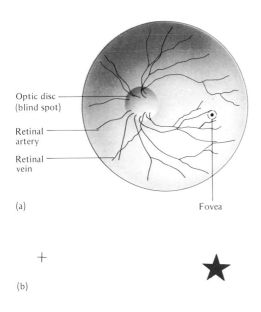

(a)

Optic disc (blind spot)

Retinal artery

Retinal vein

Fovea

(b)

+

★

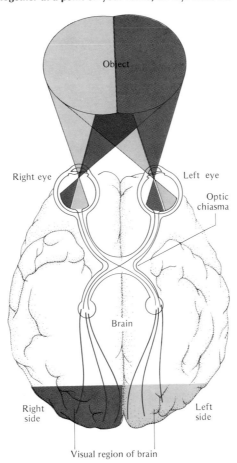

Object

Right eye   Left eye

Optic chiasma

Brain

Right side   Left side

Visual region of brain

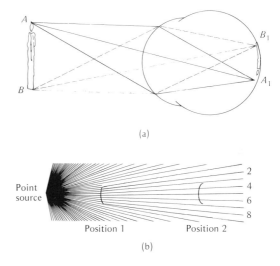

(a)

A B_1

B A_1

Point source

2
4
6
8

Position 1   Position 2

(b)

What is focus? The lens of the eye bends incoming rays of light from objects at which we are looking and brings them together (focuses them) on the retina (Figure 18–8a). If the rays of light are bent incorrectly and come together in front of the retina or behind it, the image is blurred, and we say it is out of focus (Figure 18–9).

As you may know, the image formed on the film of a camera is upside down. The same is true of the image formed on the retina (Figure 18–8a). The world is perceived as right side up because the coded action potentials received by the brain are inverted. This ability to invert retinal images is partly inherited and partly learned.[2] The visual connections of a newborn kitten without visual experience are much the

same as in an adult cat. But a kitten deprived of visual experience for 2–3 months loses much of its ability for visual reflexes and is unable to perceive form. A human wearing prism glasses that invert objects so that the retinal image is right side up is at first disoriented; but in time there is reevaluation of the information on the retina, and proper orientation results. Thus, experience as well as neuronal wiring play a role in our ability to see things as they are.

The cornea of the eye rather than the lens is responsible for most of the bending of incoming light rays. The importance of the lens lies in the fact that it can change its shape by contracting or relaxing the ciliary muscles, thus bending incoming light rays to a greater or lesser degree. This makes it possible to adjust the focus depending on the distance of objects from the eye. (In the camera, we focus on an object by moving the position of the lens rather

[2] This should not be confused with the inverted structure of the retina.

FIGURE 18–9 Focus and the eye. (a) Accommodation. (b) Myopia (nearsightedness). (c) Hyperopia (farsightedness). (d) Astigmatism.

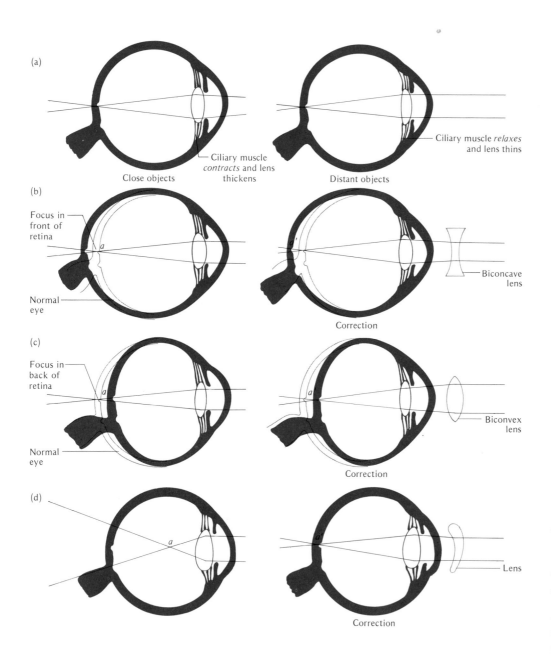

(a)

Ciliary muscle *contracts* and lens thickens

Close objects

Ciliary muscle *relaxes* and lens thins

Distant objects

(b)

Focus in front of retina

a

Normal eye

Biconcave lens

Correction

(c)

Focus in back of retina

a

Normal eye

Biconvex lens

Correction

(d)

a

a

Lens

Correction

than by changing its shape.) Why is it necessary to adjust focus for distance? Light rays reflected from a particular point travel away from that point in all directions (Figure 18–8b). The rays of light entering the eye or the camera from a near object diverge at a much greater angle from each other than rays of light from a distant object, which are almost parallel to one another. Obviously, much less bending is necessary to focus the rays coming from a distant object than is needed to focus the rays coming from a near object. When nearby objects are viewed, the lens thickens; when distant objects are viewed, it flattens out (Figure 18–9a). The change in lens shape to permit focus of close and distant objects is called **accommodation.** Sometimes the cornea and the lens are so constructed that too much or too little bending of light rays occurs, or the eyeball depth is too shallow or too deep. As a result, light does not focus properly on the retina, and this produces conditions of nearsightedness, farsightedness, or astigmatism (Figures 18–9b–d). In **nearsighted** people, light rays from near objects tend to fall on the retina, but rays from distant objects focus in front of it producing a blurred image. This is caused either by a lens that bends the incoming light rays too strongly or by an abnormally long distance between cornea and retina. In **farsighted** people, the distance from cornea to retina may be too short or the capacity of the lens to bend incoming light rays may be weak. Light rays focus behind the retina; the lens may accommodate for this by thickening, but near vision is poor. **Astigmatism** is caused by an irregular curvature of the cornea. As a result, light rays focus at different points behind and in front of the retina, and the image is blurred. These conditions, nearsightedness, farsightedness, and astigmatism, can often be helped by looking through glass lenses constructed according to the defect involved—spectacles.

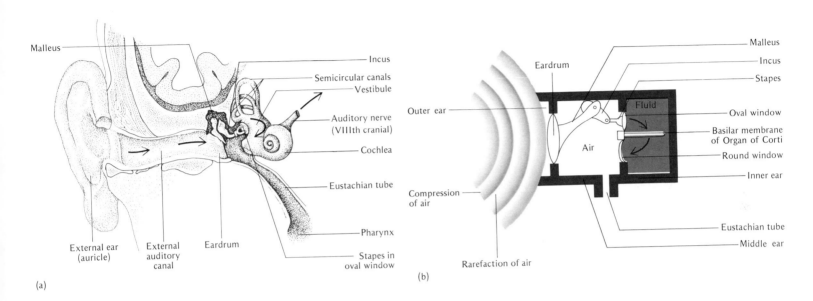

(a)

(b)

18-5 HEARING: RECEIVING SOUND SIGNALS

The buzz of a mosquito, the blended tones of a symphony orchestra, the voice of someone you hold dear, the emotion-laden words of an actor are all sound signals. In order for them to mean anything to us, they must be heard. The ear is a sensitive instrument for detecting air vibrations (sound), and it does this with remarkable acuity.

Before we discuss the reception of sound signals (hearing) let us first ask: What is sound? You are sitting quietly reading when a gust of wind blows against the open door and it slams shut. The kinetic energy of moving air has pushed the door; it moves, but once closed it moves no further. The air rebounds off the door and its frame, sets the adjacent air particles in motion, and causes them to vibrate. The air particles pass the vibrations on to their adjacent neighbors, so that vibrational waves travel from the door to your ear, and you hear the door

close with a slam. The variable compression of the air produces regions of compression and rarefaction, similar to the ripples or waves of water created by a pebble dropped on its surface. Thus, **sound energy** is a disturbance of air molecules transmitted by air.

Our ears can translate air vibrations between extremes of frequency (between 16 and 20,000 cycles per second) into nerve impulses that we interpret as sound. The **frequency** is the number of wave peaks (or cycles) that passes a given point in a second. The faster the vibration, the greater the frequency and the higher the **pitch.** Our best range is above 1,000 cycles per second (to 5,000), and it is this insensitivity to low frequency that allows us to avoid the distraction of noises produced by our own muscles contracting and by everyday bodily movements. (You can hear these low frequencies produced by muscle contraction by sticking a finger in each ear, thus shutting out airborne sound; you will hear a low hum.) Think how unpleasant it would be to hear

every step you took or every beat of your heart as a loud thud.

The **volume** of the sound we hear is a function of the amplitude of the vibrations; the higher the **amplitude** (height of the sound waves) the louder the sound. If a piano, a violin, and a flute all play a note at the same pitch and volume, they will still sound different from each other. The differences are produced by secondary air vibrations, and we call such differences **tone quality.**

The pitch, the volume, and the quality of sound we hear depend upon the interpretation put upon the stimulations received by the sound receptor within the ear. How does the ear collect sound and make us aware of it? The human ear consists of three regions: the **external ear (auricle)** the **middle ear,** and the **inner ear (labyrinth).** (Figure 18–10a). The external ear consists of a flesh-covered, cartilaginous funnel for collecting sound waves. At its narrow end there is stretched a delicate membrane like the head of a drum, the **tympanum (ear drum).**

When air waves strike the eardrum, they set it vibrating (Figure 18–10b); this membrane forms the outer wall of the middle ear, which contains a connected chain of three bones, and these tiny bones are set in motion also. The eardrum is in direct contact with the handle of the outermost of the three middle-ear bones—the **hammer (malleus).** The hammer articulates with the **anvil (incus),** which in turn articulates with the innermost bone, the **stirrup (stapes).** The stirrup's base is shaped like a plunger, and this is attached by a ligament to a membrane called the **oval window** on the inner wall of the middle ear. Two muscles operate the bones of the middle ear; one of these is attached to the handle of the hammer, and its contractions serve to keep the eardrum taut and cone-shaped; the other is attached to the stirrup and makes sure that the bones operate smoothly and do not move apart from one another as they transmit the vibrations from the eardrum to the oval window. The bones of the middle ear are not free to move about, but are anchored in place by elastic ligaments and covered over by mucous membrane. If the middle-ear cavity becomes inflamed, infected, or filled with pus, the bones do not function and hearing is impaired. The oval window and another membrane just below it, the **round window,** separate the middle ear from the inner ear. The round window, like the oval window, is stretched tightly across the entrance to the inner ear.

The inner ear consists of a series of fluid-filled chambers and coiled canals lying in the bones of the skull. The upper group is concerned with the senses of balance and equilibrium. The lower portion of the inner ear contains a long, coiled tube, the **cochlea** (*cochlea*—snail shell; L.). The cochlea is divided into three ducts by membranes. The delicate sound receptor, the **organ of Corti,** lies on one of these membranes (Figure 18–11). Specialized axonal endings, or hair cells, on the

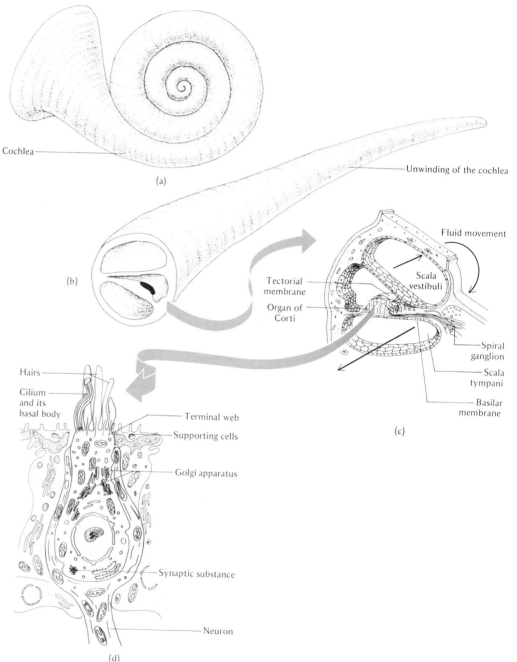

FIGURE 18–11 (a) The cochlea. (b) The cochlea unwound. (Modified from *Biology: Its Principles and Implications,* **2nd Ed., by Garrett Hardin. W. H. Freeman and Co., copyright © 1966.) (c) Cross section through the unwound cochlea showing position of the organ of Corti. (d) Cochlear hair cell, the receptor for sound.**

Cochlea

(a)

(b)

Unwinding of the cochlea

Fluid movement

Tectorial membrane

Organ of Corti

Scala vestibuli

Spiral ganglion

Scala tympani

Basilar membrane

(c)

Hairs

Cilium and its basal body

Terminal web

Supporting cells

Golgi apparatus

Synaptic substance

Neuron

(d)

BOX 18A Unwanted sound: noise pollution

Sound can convey precious information or be totally useless. Noise is useless, unwanted sound, and it can be disturbing and damaging. Today in the United States alone there are over 90 million automobiles, 16 million trucks, and 3 million motorcycles that pollute the air not only with hydrocarbons but with the roar of their engines as well. The clatter of trash collection, the grinding of dishwashers, disposal units, and blenders all contribute to noise in the home. Even the skies rain down noise from jet planes.

The ear, as we have seen, is a sensitive detector of air pressure. The smallest air disturbance that can be heard under ideal conditions in a normal ear is known as the **threshold of hearing.** This forms the baseline for sound measurement in terms of loudness. The unit for measuring loudness of sound is called a **bel,** after Alexander Graham Bell, the inventor of the telephone. A **decibel** (0.1 bel) is used most frequently to describe audible sounds. The bel scale of loudness is logarithmic, just as is the Richter scale for measuring earthquake intensity. This means that a value difference of 1 unit indicates an intensity difference of 10. Thus, 1 decibel, which is the softest sound a human ear can hear, is one tenth as loud as a sound of 10 decibels (or 1 bel) and is one one-hundredth as loud as a 20-decibel (2-bel) sound. A whisper is 30 decibels (Table 18–1) and is 1,000 times louder than the threshold of hearing. As decibels increase, the sound energy increases rapidly; and the greater the energy, the higher the amplitude of the sound waves. Very high-amplitude sound waves (very loud sounds), especially if they are of high frequency, may damage the hair cells of the ear. Sound can be destructive: a sound of 170 decibels focused on a small animal will kill it. Ordinary conversation is about 60 decibels, but the sound in a discotheque may reach 120 decibels. A guinea pig subjected to rock-and-roll music at 122 decibels for periods similar to those to which teenagers subject themselves had widespread and irreversible damage to the hair cells in the organ of Corti.

Since sound travels as waves with areas of compression and rarefaction (Figure 18–10), higher-amplitude waves have greater compression and can produce great pressure. A loud noise, such as the roar of a jet engine, can cause enough pressure within the external auditory canal to rupture the eardrum. Workers around jetports must wear ear defenders to avoid such damage.

Millions of people are going deaf, although they do not realize it. Since we converse at 60 decibels, losses below that level often go unnoticed. For example, a person who has lost 40% of his hearing may not be aware of the loss, since he still can hear other people conversing. Higher levels of speech comprehension may cover up the actual damage. We all tend to tune out extremes of noise. Unfortunately, the insidious damage goes on and the ability to turn off sound in the brain by accommodation may result in an irreversible change and inability to turn on hearing.

TABLE 18–1 Sound level readings

	Decibels
Threshold of pain	130
10 ft away from rock drill	130
Discotheque	120
Motorcycle, power mower	115
2,000 ft from jet taking off	105
Subway station with train coming in	95
50 ft from a moving heavy truck	90
Noisy kitchen in city	85
50 ft from a highway	70
Noisy office	65
Quiet conversation	60
Quiet office	45
Quiet rural night	30
Whisper	30
Threshold of hearing	0

Prolonged exposure to sound between 85 and 115 decibels will begin to cause hearing loss; limited exposure to sound over 115 decibels will cause hearing loss.

Source: Office of Noise Abatement and Control, U.S. Environmental Protection Agency.

organ of Corti are the actual sensory receptors. Vibrations of the oval window are transmitted to the cochlear fluid, the oscillations of the fluid cause the hairs to move, and they rub against an overhanging membrane (the **tectorial membrane**). Their displacement causes the hair cells to produce action potentials that travel along the auditory nerves to the brain for interpretation. The hair cells of the cochlea are arranged in sequence, from those stimulated by low- frequency waves (low-pitched sounds) at the base all the way up to those stimulated by high-frequency sounds at the apex, rather like the keys of a piano. The neurons from each region of the cochlea run to slightly different areas in the brain, and the pitch sensation we experience depends upon the brain area stimulated. High-amplitude vibrations cause similar high-amplitude oscillations of cochlear fluid, which in turn produce more intense stimulation of the hair cells and transmission of more impulses to the brain per unit time, interpreted by the brain as loudness. Very high-amplitude waves (very loud sounds) can damage the sensitive hearing apparatus (Box 18A). Variations in the pattern of hair cells stimulated by secondary sound waves are interpreted by the brain as variation in tone quality.

Why is it necessary to have such a complex setup to stimulate the organ of Corti? If sound waves were to strike the cochlear fluid and the organ of Corti directly, they could not exert sufficient pressure to make it vibrate because fluid is more resistant to flow than air. The eardrum is a relatively large surface (70 mm²) and amplifies the sound waves that strike it. The sound waves are transmitted to the oval window, which has a small surface area (3.2 mm²) by the bones of the middle ear. The total force on the oval window is the same as that on the eardrum, but because the oval window is much smaller, the pressure, that is, the force per unit area, is increased about 22 times (70

= 3.2 × 22). Thus, in transit the sound waves are amplified and the resultant pressure is sufficient to set up a rhythmic vibration in the cochlear fluid. This vibrates back and forth in the channel of the cochlea between the oval and round windows and is further amplified. The amplified vibrations are then sufficient to stimulate the organ of Corti.

If the middle ear were a closed unit between inner and outer ear, there would be great risk of a ruptured eardrum whenever the air pressure on the eardrum changed. For proper function, the air pressure of the middle ear must equal that outside the eardrum. The safety valve of the ear is the **Eustachian tube,** a short passage connecting the middle ear with the pharynx (Figure 18–10). The pharyngeal opening of the Eustachian tube is opened during swallowing, which allows equalization of pressure; normally it is kept closed by pharyngeal muscles. During ascent or descent in an airplane, the pressures on each side of the eardrum are unequal, and you relieve this condition by popping your ears; the procedure allows air to enter the Eustachian tube as you swallow and equalizes the pressures on either side of the ear drum.

18-6 BALANCE AND EQUILIBRIUM

Although it is usually described as a single sense organ, the ear is really concerned with the perception of two totally different kinds of stimuli and fulfills two distinct purposes in our lives. The ear is a sound receptor responsible for the sense of hearing, but it also receives stimuli that make us aware of our movements and enables us to orient ourselves and maintain our equilibrium. The function of orientation is associated with the upper group of fluid-filled **semicircular canals** of the inner ear (Figure 18–10). Three semicircular canals are connected to the cochlea by a **vestibule** containing

two chambers. Each chamber is lined with sensory hairs on which rest crystals of calcium carbonate called **otoliths.** When we move our heads, these crystals pull on some of the hair cells more than others, stimulating them more. These differences in stimulation give rise to action potentials in the axons of the sensory hairs, which are interpreted by the brain and signal the position of the head at any given moment. This gives us our sense of balance.

The three semicircular canals give us our sense of movement and equilibrium. Each canal is oriented in a different plane, and each has a tuft of sensory hairs in a small chamber at its base. When the head is moved, the fluid in the canals lags behind because of its inertia (fluid is resistant to flow). The lagging fluid exerts increased pressure on the hair cells on the lag side and stimulates action potentials. The different stimuli from the three canals are integrated by the brain, and we can determine the direction and movement of our head from these signals.

The semicircular canals are concerned not just with motion itself but with changes in motion. If we move at a constant speed, no motion is detected in the semicircular canals, but stopping or starting motions are both equally effective triggers for activity in the auditory nerve. If we spin on a chair and then stop suddenly, the fluid in the canals keeps moving and continues to stimulate the hair cells so that we detect motion; however, we know from the signals of other sense organs (eyes, touch receptors, and others) that we are stopped and not moving, and so the brain signals that the surroundings must be in motion—the room appears to spin and we may lose our balance.

Impulses in the vestibular apparatus excite the centers of the hindbrain involved in the maintenance of posture and equilibrium, but their activity can stimulate closely related centers and produce vomiting for example. The rocking motion of a ship or a plane stimulates the semicircular canals, and these initiate feelings of motion sickness, nausea, headache, and eyestrain because of the secondary involvement of other centers.

18-7 TOUCH, TEMPERATURE, AND PRESSURE: SKIN SENSES

Perhaps nothing is more important to a small baby or a person in love than a fond caress, a gentle stroking of the skin with the fingers and the palm of the hands. Even the pressure of a handshake feels good—warm and friendly. The body surface is covered with a thin, touch-sensitive wrapper, the **skin.** The skin is composed of two layers: the thin, veneered upper surface, the **epidermis,** and the much thicker and deeper layer, the **dermis** (Figure 18–12a). The dermis forms a mantle containing arrays of nerve endings, and it is these that represent the major sensory receptors of the skin. Touch, pain, heat, and cold are all felt because signals are sent from the skin to the central nervous system. What are these receptors?

Most of our responses to temperature are probably mediated by superimposed nerve endings (Figure 18–12a); when activated by thermal agitation (heat), these alter membrane permeability to produce an action potential. (The mechanism for cold reception is unclear.) The hairs on the body serve as touch receptors, even though they themselves are dead (Figure 18–12b), for around the base of each hair is a collar of sensory nerves that register disturbances in its position. Pressure on the hair is transmitted to the neuron, and from there signals go to the brain. The most highly innervated hairs are found on the face and around the anus and the genitalia.

The most specialized of the skin receptors is the **Pacinian corpuscle,** which is a pressure

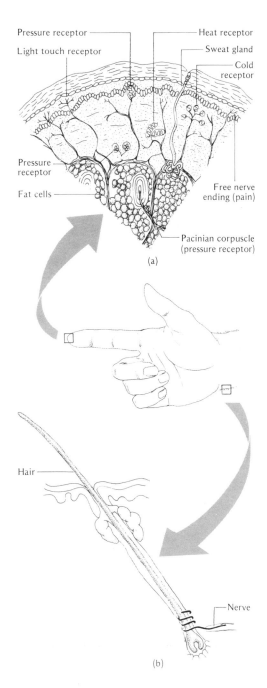

FIGURE 18–12 The sense organs of the skin. (a) Section through the fingertip. (b) The hair and its associated nerve ending.

Pressure receptor
Light touch receptor
Heat receptor
Sweat gland
Cold receptor
Pressure receptor
Fat cells
Free nerve ending (pain)
Pacinian corpuscle (pressure receptor)
(a)

Hair
Nerve
(b)

receptor. The Pacinian corpuscle looks like an onion, with a central nerve ending surrounded by concentric layers of connective tissue or lamellae. The corpuscles are found deep in the dermis, especially on the hands and feet and near joints. They are also numerous in the sheetlike connective-tissue mesenteries that hold the internal organs in place. The lamellae of the corpuscle are tightly wrapped around the ending of the nerve and enclose a small fluid-filled space. When pressure is put on this stack of membranes, a number of "holes" are opened, thus permitting ion flow across the membrane, which sets up the generator potential. The greater the pressure (that is, with an increase in stimulus strength), the greater the number of openings and the more ions flow. Current produced, if sufficient, triggers an action potential. In effect, the Pacinian corpuscle is a biological transducer; it converts pressure stimuli into the electrical energy of a nerve impulse. There is experimental evidence that the determination of pressure change by these corpuscles serves not only for touch sensation but also for the perception of movement.

SUMMARY

1. Sense organs act as transducers, collecting information about the outside world (eyes, ears, nose, touch, and taste receptors) and the internal environment (internal sensory receptors) and converting it into nerve impulses. The cerebral cortex integrates, interprets, stores, and acts on these signals, sending messages to effector organs (muscles and glands).

2. The olfactory organs, high in the nose, detect odors. Volatile, soluble substances trigger action potentials that pass along the olfactory nerve to the olfactory bulb and then to the cerebral cortex. Odor discrimination probably depends on molecular architecture. The olfactory receptors are subject to fatigue.

3. Taste receptors on the tongue are modified neurons with microvilli at their tips. They are stimulated by specific soluble substances and messages are sent to the cortex. The tip of the tongue senses sweet substances, the sides sense salty and sour, the back senses bitter flavors.

4. Superimposition of the images from each eye allows for stereoscopic vision. The outer sclera of the eye is transparent in front, forming the cornea. The choroid coat, with blood vessels and pigment, underlies the sclera and is visible through the cornea as the iris.

5. The amount of light entering the eye via the pupil is controlled by iris muscles. The lens, suspended by the ciliary body, filters out ultraviolet rays and focuses light onto photoreceptor cells (rods and cones) in the retina at the back of the eye. The fovea is the region of keenest vision.

6. The anterior eye chamber is filled with the watery aqueous humor, the posterior chamber with the jellylike vitreous humor.

7. The eyelids, eyelashes, eyebrows, and transparent epidermal conjunctiva protect the eye in front. Tears from the lacrimal glands moisten the eye and drain away via the lacrimal duct.

8. The retina is structurally inverted. Dim-light sensitive rods containing visual purple are abundant at the edges; bright-light sensitive cones containing three different pigments and responsible for color vision are more abundant at the center. Color perceived depends on the number and kinds of cones stimulated.

9. Nerve impulses are generated by light energy that splits the pigments in the rods and cones to vitamin A; energy is released and ionic changes in the cellular fluid produce an action potential that travels to the visual cortex via the optic nerve. Visual pigments must be resynthesized from vitamin A. Dietary deficiency of vitamin A causes night blindness.

10. The optic nerve leaves the eye at the blind spot, which lacks photoreceptors. The nerves from each eye partially cross at the optic chiasma before entering the brain.

11. The retinal image is inverted, and reinverted by the visual cortex.

12. The cornea focuses light onto the retina; the lens makes fine adjustments (accommodation). Nearsightedness, farsightedness, and astigmatism are caused by faulty accommodation, abnormal eye shape, or irregular corneal curvature, and all can be corrected with spectacles.

13. Sound energy, traveling as waves of disturbed air molecules, is perceived by the ear, which can discriminate volume (as a function of sound wave amplitude), tone quality (as a function of secondary air vibrations), and pitch (as a function of sound wave frequency).

14. The external ear (auricle) collects sound waves. The tympanum between the external and middle ear amplifies the vibrations and in turn sets the malleus, incus, and stapes of the middle ear vibrating. The stapes transmits the vibrations to the cochlear fluid in the inner ear (labyrinth).

15. In the inner ear, hair cells of the organ of Corti in the cochlea rub against the tectorial membrane and generate action potentials that are transmitted to the cortex via the auditory nerve. Hair cells of the cochlea are sequentially arranged according to frequency (pitch) perception. Amplitude is reflected by oscillations of cochlear fluid, intensity of hair cell stimulation, and transmission of more or less impulses to the brain per unit time. Variation in the pattern of hair cell stimulation reflects tone quality.

16. Loud sounds produce great pressure in the ear and can cause irreversible damage to the hair cells in the organ of Corti, or rupture an eardrum.

17. The Eustachian tube connects the middle ear and pharynx and equalizes air pressure between middle and outer ear.

18. The semicircular canals and vestibule of the inner ear are responsible for maintaining balance and equilibrium. When the head is moved, otoliths stimulate sensory hairs in the vestibule of the inner ear, and the fluid inertia in the semicircular canals stimulates sensory hairs in a basal chamber.

19. The skin, composed of epidermis and dermis, contains an array of variously distributed touch, pressure, and temperature receptors.

KEY WORDS

transducers
olfactory organs
taste buds
sclera
cornea
choroid coat
pupil
iris
albinism
lens
ciliary body
retina
rods
cones
anterior chamber
aqueous humor
posterior chamber
vitreous humor
conjunctiva
eyelids
eyelashes
eyebrows
tears
lacrimal glands
lacrimal duct
inverted retina
night blindness

fovea
peripheral vision
tunnel vision
optic disc or blindspot
optic nerve
optic chiasma
accommodation
nearsighted
farsighted
astigmatism
sound energy
frequency
pitch
volume
amplitude
tone quality
external ear (auricle)
middle ear
inner ear (labyrinth)
tympanum (ear drum)
hammer (malleus)
anvil (incus)
stirrup (stapes)
oval window
round window
cochlea
organ of Corti
tectorial membrane
Eustachian tube
semicircular canals
vestibule
otoliths
threshold of hearing
bel
decibel
skin
epidermis
dermis
Pacinian corpuscle

TOPICS FOR REVIEW AND DISCUSSION

1. Discuss the statement: the sensory receptor itself does not determine sensation.

2. Describe the structure and function of the human eye.
3. How do we hear? Discuss the relationship of noise pollution to deafness.
4. How are the senses of taste and smell related?
5. What is the fovea?
6. How do rods and cones function in vision?
7. Describe some disorders of the eye and how they may be corrected.
8. What is the relationship of vitamin A to vision?
9. Why are sense organs called biological transducers?
10. Are there only five senses? If not, why not?

C.I.A. Data Show 14-Year Project On Controlling Human Behavior

The Central Intelligence Agency conducted a 14-year program to find ways to "control human behavior" through the use of chemical, biological and radiological material, according to agency documents made public by John Marks, a freelance journalist. "Drugs were part of it," he said, and "so were such other techniques as electric shock, radiation, ultrasonics, psychosurgery, psychology and incapacitating agents, all of which were referred to in documents I have received."

According to Mr. Marks's documents and an earlier Senate investigation, the C.I.A. conducted secret medical experiments from 1949 through 1963 under the code names Bluebird, Artichoke, MK Ultra, and MK Delta. The C.I.A. inspector general report in 1963 described the program as the "research and development of chemical, biological and radiological materials capable of employment in clandestine operations to control human behavior."

Does the above excerpt from the *New York Times* reflect public paranoia about the covert actions of the C.I.A., or does it have basis in fact? Is it possible to modify human behavior by means of drugs, psychosurgery, and the like?

The answer must be a qualified yes. As long ago as the early 1960s, researchers such as Dr. Jose Delgado reported using electrical stimulation to change the brain waves of a chimpanzee, to stop a charging bull, to make a female monkey reject her children, and to perform other similar feats. Dr. Delgado said that his experimental results seemed to "support the distasteful conclusion that motion, emotion and behavior can be directed by electrical forces and that humans can be controlled like robots by pushbuttons." By combining classic concepts about the brain with computer technology and sophisticated engineering principles, scientists are in fact discovering new ways of modifying behavior. To what use are these techniques being put?

In epilepsy, seizure control has traditionally depended on anticonvulsant drugs, but for a small number of so-called intractable epileptics, these drugs are ineffective. For these patients, tiny surgically implanted electrodes act as brain-wave circuit breakers, cutting back on the number of impulses that provoke epileptic seizure. The standard treatment of shaking palsy (Parkinson's disease) involves taking large doses of L-dopa, a chemical that can produce multiple side effects that limit and preclude its use. Experiments with a cannula technique, by which tiny channels can be made in the brain itself, can get around this problem. Through the cannula small amounts of the key treatment chemical, dopamine, can be delivered to the precise spot in the brain where the deficiency lies.

A researcher in California is attempting to restore motion to the arms and legs of monkeys experimentally paralyzed by strokes by using electrodes implanted in their lower brain stems. Cingulotomy, boring holes in the skull and cauterizing bundles of nerve cells that connect various parts of the limbic region of the brain, has produced improvements in alcoholic and schizophrenic patients. Removal of part of the amygdala reduced the tendency toward violent, even homicidal rage in patients at Harvard Medical School. With the help of systems engineering, limited visual and auditory perception may become available to the blind and the deaf. Hyperkinetic children by the thousands are being treated and apparently calmed with drugs such as ritalin.

All of this work is aimed at finding new techniques to help those stricken with brain-associated disorders and not to create some future totalitarian state. It does, however, represent at least a first step down the road toward the nightmare vision of a brain-controlled population.

ANIMAL AND HUMAN BEHAVIOR

19-1 WHY BEHAVE?

It has been so long since you've seen one another. Where has the time gone? How many years ago was it that you went to school together, spent the summer at the beach, and talked over the problems that you'd had the night before? Now the time for reunion is here at last. With great anticipation you approach the house. Nervously you finger the doorbell, almost taken aback by the measured sound of its chimes. Thoughts of the past race through your mind, causing a smile to cross your lips. What fun it will be to relive those days that you shared together! Suddenly, the door opens wide: your uplifted spirits vanish and your facial expression changes from delightful pleasure to stark terror. Lunging forward toward you is a large, ferociously barking dog; its sharp, white teeth gnash and powerful jaws grip your arm. Fear! Your movements become defensive—you raise your arm to protect your face, you move backward attempting to get the dog away from you, and your heart beat quickens. You are frightened, and have good reason to be so.

The response of the dog to your presence, and your reaction to it, are examples of behavior. Succinctly, **behavior** is externally directed activity, often involving the response of an organism to an environmental situation. Responses may include motion or cessation of motion in part or all of the body, or secretion from glands. Responses may be initiated by signals from the environment, such as an attacking dog, the presence of food, a prospective mate, or they may be initiated by internal signals such as hunger, thirst, or inner events such as hormonal changes (Box 19A).

In its many forms, behavior is generally adaptive: that is, an animal responds to appropriate stimuli in order to survive and reproduce itself. Your reaction to an attacking dog was a directed and purposeful action designed to promote your survival. The dog also acted appropriately to promote *his* survival. Accommodations to environmental circumstances such as these tend to promote the well-being of the individual. Other examples of behavioral adjustments could benefit the individual in providing food and shelter, and in activities related to perpetuation of the species: mate selection, care and rearing of the young, and the establishment of a social order.

By and large the machinery that mediates behavioral responses in an animal is the nervous system.[1] The receipt of stimuli by sense organs is usually the first event in behavior, the input to the system. Reaction of an effector such as a muscle or gland is the last event in behavior, the output. Between the two lies a connecting link, the communication or coordinating channels of the nervous system. For example, when the sight and sound of a hostile dog was received by your senses, messages were transmitted via nerves to the appropriate effectors, and action was taken in accordance with these messages.

Since the nervous system and behavior are so intimately related to one another, it is to be expected that what an animal does will vary according to the capacities of its nervous system. Thus, if animal nervous systems are thought of as arrangements of neuronal "wires," then a simple wiring diagram composed of a few neurons, would limit the kinds of behaviors that could be displayed by its

possessor. Conversely, when the wiring arrangement contained many neurons with the possibility for an almost infinite array of complicated nerve-to-nerve hookups, it would be matched by complex and exceedingly intricate behavior patterns. This is exactly what we see in animal behavior. However, circuitry is only one facet of behavioral activity. Behavior is also influenced by the state of development of the nervous system during an animal's lifetime. For example, an undeveloped or immature nervous system is capable of much simpler activities than a mature system, as witnessed by the behavioral changes that take place during the development of an infant into a young child. In addition, nervous function (and thus behavior) is affected by repeated performance and by the environmental changes that have been impressed on it. This component of behavior involves learning, in which an individual's behavior may vary from time to time because previous experiences can modify the response to a particular stimulus. For example, a second visit to your long-lost friend might produce some trepidation on your part, rather than your previous pleasant anticipation. Now you know that there might be a ferocious dog on the other side of the door, and perhaps you will take precautions to avoid a repeat performance of your first visit. You have learned by experience.

Animals with simple or immature nervous systems tend to exhibit fairly fixed or stereotyped behavioral responses, although these may be reversible; animals with more complex or highly developed nervous systems have an increased range of responses and are generally capable of learning; they can respond to their environments with much greater plasticity. As we shall see, through the course of animal evolution, the tendency has been toward the development of nervous systems capable of learning (Figure 19–1). In humans, the fixed or

---

[1] Plants lack both nervous and muscular systems; therefore they do not exhibit the motor activities that are generally referred to as behavior. Plant movements, often called **tropisms,** involve long-term activity performed on a slow time scale. Most of these activities are under the control of hormone-influenced growth patterns.

© Walt Disney Productions

© Walt Disney Productions

BOX 19A Purpose is in the mind of the beholder

People have watched and wondered about animal activities since they first were capable of wonder; however, as a distinct field of study, animal behavior is a newcomer to the scientific scene. The reasons for this are simple enough: behavior is a complex subject that involves the synthesis of several disciplines: neurophysiology, endocrinology, evolutionary biology, ecology, and mathematics. Gaps in our knowledge in these areas limit advances in our understanding of why animals behave the way they do. Furthermore, some observers tend to regard much of animal behavior as purposeful, the result of animal consciousness or awareness, and to assume that human desires and a capacity for reason are shared by other animal species. "The animal does this or that because it *wants...*" is a common way of describing so-called animal "will." This sounds absurd, yet we say a wasp "gets angry," that so-and-so is "clever as a fox," and that birds "love" their babies. Why do we use such phrases? Part of the reason lies in the nature of our language. Human language contains a built-in bias for implying human attitudes, purposes, and motives to animal activities, and this tends to limit the kinds of words we use to describe behavior. The attribution of human conscious attitudes and emotions to animals is called **anthropomorphism:** literally, "human form." Suffice it to say that anthropomorphism is intellectually dangerous and scientifically misleading. Thus, the creations of Walt Disney—Mickey Mouse and Donald Duck to name but two—demonstrate the gamut of human emotions—fear, love, anger, bashfulness—and they delight because they are so human, but in the real world the attribution of such human characteristics to ducks, mice, or other species is totally unwarranted. There is no evidence that conscious awareness—a product of the human brain—is experienced by other organisms. Thus, we should not say that a writhing lobster placed in boiling water feels pain, since the nervous systems of man and the lobster are so very different that such an extrapolation cannot easily be made. Until evidence to the contrary is presented, it is safer not to apply descriptions of pain, fear, love, desire, and so on to animals other than man.

Anthropomorphism has impeded the development of the science of behavior, and it has an ally in teleology. **Teleology** assumes that an animal has an acknowledged goal for its activity. Thus, to say that "wings were developed by birds *in order to fly*," or that "the heart beats faster *in order to* obtain more oxygenated blood" is entirely incorrect. Rather, birds are able to fly *because* they have wings, and the heart obtains more oxygenated blood as a *result* of beating faster.

Both teleology and anthropomorphism are scientifically unproductive and should be avoided.

The study of animal behavior did not become a true science until anthropormorphism and teleology were eliminated. Basically, animal behavior was transformed into a science by one event: the formulation of Morgan's canon. The British biologist C. Lloyd Morgan (1852-1936) rebelled against the use of anthropomorphism and applied an old principle of logic called **Ockham's razor** to behavior. (Ockham's razor was named after the fourteenth-century scholar William of Ockham, the allusion to the razor indicates the cutting out of superfluousness). Ockham's razor states that given a choice of possible explanations for a given phenomenon, the simplest is to be considered the most probable. **Morgan's canon** is a restatement of Ockham's razor, to wit: the actions of an animal should be interpreted in terms of the simplest mechanisms that explain such behavior and permit predictions of future behavior under similar circumstances. (Thus, the writhing of a boiled lobster may be interpreted: the sense organs of the lobster receive temperature stimuli that travel along reflex pathways and result in muscle movements; in the natural environment of a lobster, such movements would be clearly adaptive, contributing to the survival of the individual by enabling it to avoid high water temperatures. No conscious awareness such as pain need be invoked to describe the lobster's response. Note however that the description does not rule out such a possibility.)

Morgan's canon eliminated consciousness in the description of animal behavior and did much to eliminate anthropomorphic interpretations of animal activity.

stereotyped components of behavior play a minor role.

What we call behavior, therefore, is produced by the sum total of all the visible and invisible influences of the past and the present that traverse the neuronal wires, together with the reactions of effectors. Through behavioral adaptations, individuals as well as species can survive the continuing changes that take place in the environment. In addition, through learning and experience animals need not be endowed with fixed programs of genetic instructions designed to meet every situation or contingency, but can adjust to particular events at specific points in time with great plasticity.

19–2 TOWARD UNDERSTANDING BEHAVIOR

The intricate patterns of behavior observed in many animals would not be possible without a muscular system of considerable complexity. Indeed, it is probably true that until a fairly complex muscular system evolved, intricate patterns of behavior were impossible even though the nervous system may have contained the potential for such behavior. Thus, the behaviors displayed by animals depend on a close interplay of both the nervous and muscular systems.

When animal behavior is studied on a comparative basis, however, it becomes evident that evolutionary trends are most clearly reflected in the nature of the nervous system. The most primitive animals, those with simpler nervous systems, tend to show relatively fixed or stereotyped responses to a stimulus or set of stimuli. These responses, called taxes, reflexes, or instincts, are in essence inherited properties of the nervous system. Their occurrence and relative importance in the animal kingdom is illustrated in Figure 19–1. In higher animals

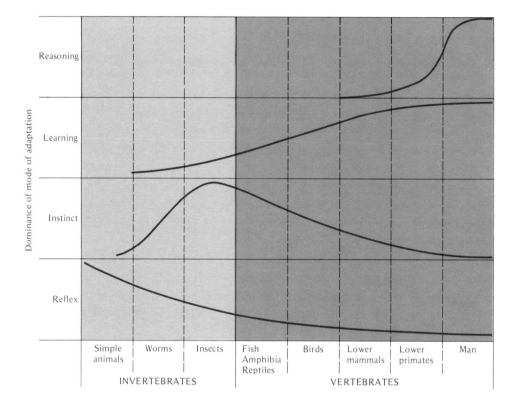

FIGURE 19–1 A schematic portrayal of the changes that occur in the modes of adaptive behavior in the various kinds of animals. (From V. G. Dethier, Eliot Stellar, *Animal Behavior*, 3rd edition, © 1970, p. 91. Reprinted by permission of Prentice-Hall, Inc., Englewood Cliffs, New Jersey.)

such as birds and mammals (including ourselves), stereotyped patterns make up a smaller part of the behavioral repertoire: taxes are almost nonexistent; simple reflexes, though important, make up a small portion of the total; and variable modes of behavior such as learning and reasoning become more prominent. In these animals, patterns of behavior develop or are refined and altered during the lifetime of the individual, and considerable plasticity in behavioral activity is displayed.

By studying the range of animal behavior, it is possible to make a few generalizations:

1. During the evolution of animals, complex patterns of behavior are added to simpler ones and ultimately replace them.
2. In the higher organisms, new behavioral patterns emerge that were not present in lower forms.
3. The mechanisms underlying behavior in simpler organisms often cannot be applied to animals with more complex nervous systems, nor can the processes found in higher forms (such as reasoning) be used to interpret the behavioral activities of simpler forms.

Those who study behavior attempt to understand patterns of activity by looking at their subject in a variety of ways. Most begin by simply observing and then describing behavioral patterns. The dog wags its tail as its master approaches, or the male dog lifts its leg and urinates on a tree. Then they go on to ask the question: How does this activity promote survival of the animal (or the species)? Of what *use* is it in the life of the animal? For example, does the behavior avoid conflict? Is it used to mark a territory? Does it enable mating to take place? And finally, through experimentation, behavioral scientists seek to find answers to the question: What makes the behavior happen? In other words, what causes, or promotes it? As is the case with so many areas of human inquiry, opinions differ not so much with respect to observation and description of behavioral phenomena, but in interpretation of their usefulness in the life of the animal and their causative factors.

During the twentieth century, two schools of thought emerged in the field of animal behavior. One group, the **behaviorists** were primarily experimental psychologists who based their work on the work of an American, J. B. Watson. The behaviorists tried to understand how learning takes place, and they did so by conducting laboratory experiments. They sought to modify behavior in the laboratory by training animals with reinforcement, that is, a positive reward of food or a negative one of electric shock. The Skinner box and the maze were the standard tools of the behaviorists (Box 19B). The white rat became their favorite test animal. The other school of behavior arose in Europe in the early 1900s and consisted of individuals trained in zoology rather than psychology. They called their scientific study of behavior ethology. The **ethologists** studied a variety of animal species both in the laboratory and in their natural environment. They focused

most of their attention on birds, fishes, and insects—animals with relatively fixed, stereotyped behavior patterns. Generally speaking, ethologists do not seek to modify behavior experimentally by training, mazes, and the like, but they examine the role of behavior in an animal's natural environment. The differing attitudes between ethologists and behaviorists, the varied experimental schemes, and the different kinds of experimental subjects promoted strongly divergent opinions about the mechanisms involved in animal behavior. Clashes between the ethologists and behaviorists were a part of the scientific scene more than 30 years ago, and the conflict continues in public to this very day (Box 19C). The ethologists, claimed the behaviorists, tended to explain behavior primarily on the basis of genetically programmed instinct and drive and essentially neglected the importance of learning. On the other hand, the behaviorists were chastised by the ethologists because they restricted their studies to inbred laboratory subjects and highly artificial environments, and because of their perhaps unjustified extrapolations to human behavior. Further, the ethologists said that the behaviorists tended to ignore the genetic basis for behavior. In essence the two schools of thought boil down to this: Is behavior learned or is it innate (hereditary)? Furthermore, which is more important in explaining observed behavior—nature (genes or unlearned behavior) or nurture (experience or learned behavior)? Unfortunately, the nature versus nurture conflict assumes that there are only two distinct and mutually exclusive kinds of behavior, and that each is governed by very different laws. Closer examination of behavioral activity shows that the simple classification of behavior into instinctive (unlearned) versus learned is of limited value. More important is how the behavior develops and how it is adaptive in the

life of an animal. Most animals show an interaction of both kinds of behavior; frequently, one may depend upon the other, as we shall see.

19–3 SIMPLER TYPES OF BEHAVIOR

Taxes

The oriented body movements made by animals toward or away from specific stimuli are called **taxes** (singular: **taxis**). Responses toward or away from light, for example, are called **phototaxes,** responses to gravity are called **geotaxes.** If you open a drawer containing a cockroach, the animal responds to the light by moving away from it (a **negative phototaxis**). A newborn rat, its eyes still unopened, moves upward when placed on an inclined plane resting on the surface of a table, indicating that it has a negative response to gravity (a **negative geotaxis**). Behavioral responses may be the result of interplay between two or more taxes. For example, a goldfish in a bowl of water orients with the belly side down responding to both light and gravity. If one destroys the inner ear of the goldfish where the gravity receptors are located, the fish literally stands on its tail, unable to receive gravitational stimuli, and responds only to the light.

Although these examples of taxes tend to be rather fixed in their response patterns, they are not so rigidly determined that they are without variation. Other factors may modify the basic response. For example, a cockroach in a container that has both a dark and a light side will seek the dark side, but if given an electric shock on the dark side only, it will reverse its normally negative phototaxis. In its simplest form, the animal has learned. The negative geotaxis of a newborn rat changes when its eyes open and it can respond visually to the edges of the inclined plane. Instead of moving directly toward the upper end of the plane, it may turn around and come down on to the table surface on which the inclined plane rests. Clearly, animals respond to more than a single environmental stimulus at one time.

Although lower animals and the immature stages of higher ones will respond to external stimuli in rather simple and fixed ways that we call taxes, the movements need not be "forced" or incapable of modification; rather, they are best described as *relatively* fixed. In addition, taxes may be components in more complex behavioral activity.

Reflexes

Taxes involve responses of the entire body to a stimulus. **Reflexes** involve responses of a part of the body to a stimulus. Both responses tend to be stereotyped and are derived from the wiring pattern of the nervous system. Thus, when you raise your arm to protect your face from an attacking dog, you respond with reflex actions (See also Chapter 17 page 365). Although the difference between a taxis and a reflex is often determined by the amount of the body involved in the response, this is not absolute. The knee jerk is a reflex involving only the lower limb, but if goldfishes or cats are turned on their backs, they right themselves by means of a reflex involving almost all parts of the body. Indeed, the taxes of the cockroaches, newborn mice, and the goldfishes involve a series of reflexes. In this sense, the reflex is a fundamental unit of behavior; to a degree, some of the most complex patterns of behavior may be described as the sum total of an array of reflexes. As a consequence, some students of behavior suggest that even the complex behavior of humans can be reduced to combinations of stereotyped (innate or inborn) reflexes and acquired (learned) or con-

(text continues on p. 398)

BOX 19B Skinner's Utopia: panacea, or path to hell?

"Boy, have I got this guy conditioned! Every time I press the bar down, he drops in a piece of food."

I've had only one idea in my life—a true *idée fixe*. To put it as bluntly as possible—the idea of having my own way. Control expresses it. The control of human behavior. In my early experimental days it was a frenzied, selfish desire to dominate. I remember the rage I used to feel when a prediction went awry. I could have shouted at the subjects of my experiments, 'Behave, damn you! Behave as you ought!'

B. F. Skinner *Walden Two*, 1948

The speaker is T. E. Frazier, a character in *Walden Two* and the fictional founder of the utopian community described in that novel. He is also an alter ego of the author, Burrhus Frederic Skinner, who is both a psychology professor and an institution at Harvard. As leader of the "behavioristic" psychologists, who liken man to a machine, Skinner is vigorously opposed by humanists and by Freudian psychoanalysts.

Skinner's reasoning is that freedom and free will are no more than illusions; like it or not, man is already controlled by external influences. Some are haphazard; some are arranged by careless or evil men whose goals are selfish instead of humanitarian. The problem, then, is to design a culture that can, theoretically, survive; to decide how men must behave in order to ensure its survival in reality; and to plan environmental influences that will guarantee the desired behavior. Thus, in the Skinnerian world, man will refrain from polluting, from overpopulating, from rioting, and from making war, not because he knows that the results will be disastrous, but because he has been conditioned to want what serves group interests.

Is such a world really possible? Skinner believes that it is; he is certain that human behavior can be predicted and shaped exactly as if it were a chemical reaction. The way to do it, he thinks, is through "behavioral technology," a developing science of control that aims to change the environment rather than people, that seeks to alter actions rather

...an feelings, and that shifts the custom-ary psychological emphasis on the world ...side men to the world outside them. ...entral to Skinner's approach is a method ... conditioning that has been used with ...niform success on laboratory animals: ...ving rewards to mold the subject to the ...xperimenter's will. According to Skinner ...nd his followers, the same technique ...an be made to work equally well with ...uman beings.

...Underlying the method is the Skinner-...n conviction that behavior is deter-...ined not from within but from without. ...kinner insists that actions are deter-...ined by the environment; behavior is ...haped and maintained by its conse-...ences." As Skinner sees it, environ-...ents are defective when they fail to ...ake desirable behavior pay off and ...hen they resort to punishment as a ...eans of stopping undesirable behavior. ... short, it is punishment or reward that ...termines whether a particular kind of ...havior becomes habitual. But Skinner ...lieves that punishment is generally an ...effective means of control. "A person ...ho has been punished" he writes, "is ...t less inclined to behave in a given ...ay; at best, he learns how to avoid ...nishment."

...Rats and pigeons were the center of ...boratory experiments in which Skinner ...ntrolled behavior by setting up "con-tingencies of reinforcement"—circum-stances under which a particular bit of desired behavior is "reinforced" or re-warded to make sure it will be repeated. The behavior Skinner demanded of his pigeons was bizarre—for pigeons. He made them walk figure eights, for ex-ample, by reinforcing them with food at crucial moments. By a similar process, Skinner has conditioned pigeons to dance with each other, and even to play ping pong.

All of these conditioning feats were accomplished with the now famous Skin-ner box. It is a soundproof enclosure with a food dispenser that a rat can operate by pressing a lever, and a pigeon by pecking a key. The dispenser does not work unless the animal has first per-formed according to a specially designed "schedule of reinforcement."

Skinner himself admits that "pigeons aren't people," but points out that his ideas have already been put to practical use in schools, mental hospitals, penal institutions, and business firms. Skinner-inspired teaching machines have begun to produce what amounts to an educa-tional revolution. Machines now in use in scores of cities across the country present pupils with a succession of easy learning steps. At each one, a correct answer to a question brings instant re-inforcement, not with the grain of corn that rewarded the pigeon, but with a printed statement—supposedly just as satisfying—that the answer is right.

Skinner is skeptical about democracy. Observing that society is already using such ineffective means of behavioral con-trol as persuasion and conventional ed-ucation, he insists that men of good will must adopt more effective techniques, using them for "good" purposes to keep despots from using them for "bad" ones. In his planned society, he says, control would be balanced by countercontrol, "probably by making the controller a member of the group he controls." This would help to ensure that punishment would never be inflicted, Skinner main-tains, adding that it was the use of "aversive control" (punishment) that doomed Hitler.

The ultimate logical dilemma in Skin-ner's thinking is this: What are the sources of the standards of good and evil in his ideal society? Indeed, who decides even what constitutes pleasure or pain, reward or punishment, when man and his environment can be limitlessly ma-nipulated? To one writer, Skinner's "uto-pian projection is less likely to be a blueprint for the Golden Age than for the theory and practice of hell."

BOX 19C

Genes uber alles?

The concepts are startling—and disturbing. Conflict between parents and children is biologically inevitable. Children are born deceitful. All human acts—even saving a stranger from drowning or donating a million dollars to the poor—may be ultimately selfish. Morality and justice, far from being the triumphant product of human progress, evolved from man's animal past, and are rooted securely in the genes.

These are some of the teachings of sociobiology, a new and highly controversial scientific discipline that seeks to establish that social behavior—human as well as animal—has a biological basis. Its most striking tenet: much of human behavior is genetically based, the result of millions of years of evolution.

Carried to an extreme, sociobiology holds that all forms of life exist solely to serve the purposes of DNA, the coded master molecule that determines the nature of all organisms and is the stuff of genes. As British ethologist Richard Dawkins describes the role of the genes, they "swarm in huge colonies, safe inside gigantic lumbering robots, sealed off from the outside world, manipulating it by remote control. They are in you and me: they created us body and mind: and their preservation is the ultimate rationale for our existence. . . . We are their survival machines."

Angry opponents denounce "so-so biol-ogy" as reactionary political doctrine disguised as science. Their fear: it may be used to show that some races are inferior, that male dominance over women is natural, and that social progress is impossible because of the pull of the genes. Anthropologist Marshall Sahlins dismisses sociobiology as "genetic capitalism"—an attempt to defend the current structures of Western society as natural and inevitable. Harvard evolutionary biologist Richard Lewontin is earthier; he thinks sociobiology is "bullshit."

Sociobiologists call their doctrine "the completion of the Darwinian revolution"—the application of classic evolutionary theory and modern studies of genetics to animal behavior. Darwin's theory, now virtually unchallenged in the world of science, holds that all organisms evolve by natural selection—those that are better adapted to the environment survive and reproduce; the rest die out. Sociobiologists believe the behavior that promotes survival of the winners in the evolutionary game is passed on by their genes. Many recent theorists, such as Nobel-prizewinning ethologist Konrad Lorenz, have focused on the group or species as the primary unit of selection. Darwin wrote that it was the individual organism. But sociobiologists believe it is the genes themselves that conduct the life-or-death evolutionary struggle.

Yet sociobiology did not arise from molecular studies but as an answer to a century-old gap in Darwinian theory: Darwin could not fully explain why some organisms help other members of their species. Since altruistic behavior reduces an organism's chances to survive, evolution should be expected to breed it out of all species. Still, some birds risk their lives for the flock by crying out to warn of the presence of a predator—thus chancing attracting the attention of the enemy and being singled out for attack. Social insects serve the entire community, some going so far as to give their lives to protect the colony from invaders.

Sociobiology tries to resolve the dilemma. Its solution: altruism is actually genetic selfishness. The bird that warns of an approaching hawk is protecting nearby relatives that have many of the same genes it has—thus increasing the chance that some of those genes will survive. Sterile female insects work and give their lives to promote the spread of genes they share with their sisters.

British biologist William Hamilton in 1964 argued that altruism could help an individual spread his genes; the principle helped to explain the social life of insects. In social ants, bees, and wasps, daughters of the queen share an average of three-quarters of their genes (Figure 19–2). Because the daughters are more related to each other than they would be to their own offspring, said Hamilton, it is in their genetic self-interest not to breed but to assist the queen in producing more daughters. Thus the females evolved as sterile workers who cooperate socially for genetically selfish reasons. Some years later, Robert Trivers reasoned that if Hamilton was right, worker ants would spend three times the energy rearing sisters as rearing brothers, because the workers are three times more closely related to their sisters than to their brothers. Trivers and his associates then analyzed thousands of ants of 20 different species and confirmed the 3:1 female dominance—the strongest evidence so far that organisms act as if they understand the underlying genetics.

In the spectrum of current theories about human behavior, sociobiology falls between the thinking of B. F. Skinner, who regards people as pliable beings whose behavior can be almost entirely shaped by their environment (Box 19B), and Lorenz, who believes that man is a prisoner of his aggressive instincts (Box 19D). The strongest argument against sociobiology is that it underrates the

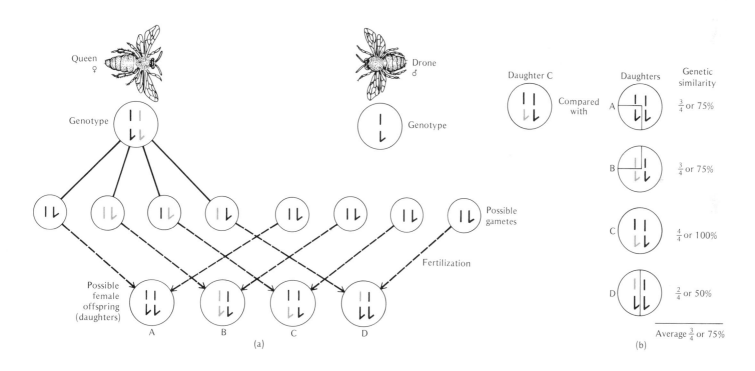

FIGURE 19–2 Female bees: altruistic and sterile. (a) Male bees (drones) are all haploid; the breeding queen and nonbreeding workers (female) are all diploid. Each daughter therefore shares 50% of her chromosomes (and genes) with her mother, father, and potential offspring. (Compare A, B, C and D with queen and drone genotypes.) (b) Pick any daughter genotype (A, B, C or D) and compare it with the possible genotypes of the sisters. Each daughter shares an *average* of 75% of her genes with her sisters. For simplicity, only two chromosomes are shown.

emergence of the human brain, consciousness, and culture. Among humans, learning can be passed on by culture, thus overwhelming the genetic contribution to behavior. Even Edward O. Wilson has stressed his belief that at most 10% or 15% of human behavior is genetically based. The fear of many of sociobiology's opponents is that it will prove nothing but leave a heavy political impact anyway. Sahlins fears it may disappear as a science but go on and on in the popular culture.

Indeed, few academic theories have spread so fast and with so little proof. Apart from the Hamilton-Trivers work on altruism, there has been little to impress the skeptics, and no hard evidence has been presented to show that genes influence human cultural behavior. The power of sociobiology comes from its astonishing promise to link the physical sciences with the human sciences and to bring all behavior from *Drosophila* to *Homo sapiens* under one great discipline. What is more, sociobiology may have come at the right cultural moment. The 1970s have brought with them growing impatience and disillusionment over failed educational and environmental experiments designed to alter social behavior (see facing page to this chapter). The concept of social theorists that man is infinitely malleable and perfectible has fallen into disfavor. At such a time the emergence of a doctrine preaching that man is caught in history, able to exercise free will only within the limits set by his genes, may do very well indeed.

ditioned reflexes. The description of complex behavior in higher organisms by reflexes alone cannot be universally applied, however; consequently more complex behavioral activities are designated by other terms.

19–4 MORE COMPLEX TYPES OF BEHAVIOR

Instinct

The term **instinct** as applied to animals refers to fairly complex, stereotyped behavior patterns that are commonly found among members of a particular species. Most important, instinctive behavior is inherited and unlearned. Generally, instincts are adaptive. Thus, a canary builds a nest just like other canaries even though it is raised in a laboratory and has never seen another canary building a nest. More striking is the fact that the canary builds the nest at the right time of the year so that it is ready to receive the eggs the bird will soon lay. How does the canary "know" how to construct a nest from the proper materials and at such a time that the nest will be ready for the eggs? The answer is an instinct: that is, a genetically programmed behavioral activity.

Man, we are often told, is nothing more than an animal among animals, a naked ape dominated by his killer instincts. Indeed, the novelist Robert Ardrey wrote: "Man is a predator whose natural instinct is to kill with a weapon." The ethologist and Nobel Laureate Konrad Lorenz supports this view by observing that most animal-killing carnivores have instincts inhibiting them from indiscriminately killing their own kind, with two exceptions: rats and men. Is there in fact a human killer instinct? Can it be identified?

Before attempting an answer (Box 19D), let us consider some examples of instinctive behavior so that we may better see how these relate to notions of human instincts.

Instinctive behavior patterns

Reproductive behavior in the three-spined stickleback, a brightly colored freshwater fish, consists of a series of complex behavioral acts: migration, territoriality, aggression, nesting, mating, and parental care (Figure 19–3a). In nature, under the influence of increasing daylength during the spring, there is an increase in gonadal hormones. The male sticklebacks move out of the school in which they ordinarily live, and, directed by temperature, they migrate to warm, shallow water where they establish a territory, build a nest, and defend the territory against other male intruders. A male attracts a female by a zigzag courtship dance, leads the female to the nest, fertilizes the eggs, cares for the developing eggs by fanning them, and once hatched, he protects the young. Such behavior tempts an anthropomorphic interpretation: it almost looks as if the stickleback "knows" exactly what must be done to ensure the perpetuation of the species. However, careful experimentation has shown that these acts are instinctive. The animal follows a preprogrammed, fixed action pattern, which remains blocked until an appropriate stimulus (called a **releaser**) activates the behavior. Releasers are specific stimuli that trigger instinctive behavior patterns. For example, a male stickleback courts a female stickleback only if she has a swollen abdomen. However, a male will also display courtship activity if presented with a wooden model that has a swollen belly or if there are females of other species that are gravid. Thus, the specific releaser for male courtship is the swollen abdomen. The male's aggressive defense of the territory also depends on releasers: the red underbelly of a male or a model of wood that is painted red triggers aggressive displays. Indeed, it has been reported that a male stickleback showed an aggressive display when a red mailtruck passed by the window

BOX 19D Is man an instinctive killer?

Is there, as some ethologists have claimed, a human killer instinct? Does man have an innate, unlearned tendency to kill his fellow man? How is, or was this, adaptive in the life of the human species? Some writers have observed that humans have "natural" tendencies that include the acquisition and protection of property—an instinct for territorial aggression. Taken to its ultimate limits, the implication has been that man instinctively kills his own kind. However, the enormous development of the human cerebral cortex and the associated capacity for learning and reasoning have liberated man from behavior based on simple, stimulus-bound responses. Among humans alone, learning can be passed on by culture, overwhelming to a large extent the genetic contribution to behavior. Because of learning, conscious reasoning, and man's cultural inheritance, would humans be compelled to respond to a killer instinct if indeed it were present? If conscious override of instincts is possible in humans, how are we to determine if man possesses, or indeed ever possessed, such a killer instinct?

One way of understanding and tracing the development of human behavior is by comparing ourselves with our closest relatives, the primates. Investigations of monkeys isolated from birth, with no opportunity for learning from other monkeys, show that aggression seems to be innate. (This is not the same as a "killer instinct," however.) At the period during their lives when aggressive behavior would usually develop, the monkeys showed aggressive tendencies, but since no objects at which the aggression could be directed were present, the monkeys redirected their aggressive tendencies against themselves, biting and even blinding in one case. The same maturational sequences of affection, fear, and aggression observed in primates occur in humans: aggressive tendencies appear in the first 4-5 years of life.

Studies of primates in the wild, however, reveal that these animals tend not to be belligerent unless provoked, and they do not kill their own kind. Although they have sharp canine teeth and show the use of tools, they do not use them for tearing at the flesh of their brethren since they are primarily vegetarians. By extrapolation, this suggests that man as a primate does not have an evolutionary history based on killing. Even evidence that an early human ancestor, *Australopithecus africanus,* was a carnivorous predator does not justify the generalized view that "with his big brain and his stone handaxes, man annihilated a predecessor who fought only with bones . . . . All human history from that date turned on the development of superior weapons . . . for genetic necessity." To some writers, man the tool maker is equivalent to man the weapon user. Is such a conclusion justified? Tools may be used as weapons, but most would agree that the primary use of hammers, scrapers, choppers, and knives was and is as tools rather than as weapons and is culturally inherited and learned rather than behavior passed on in the genes.

Obviously, aggressive behavior is adaptive. Without some aggressive tendencies, organisms would be at the mercy of the nearest predator, and carnivores would never hunt for food; but this is interspecific aggression (between members of different species). Intraspecific aggression (between members of the same species) plays a role in the establishment of social order, mating, territoriality, and other relationships. But for most animals, there are built-in inhibitions against carrying aggressive behavior to the point of killing their own kind. (These are known as appeasement gestures—see page 412.) It is difficult to see the adaptive advantage of a killer instinct that would allow a species to prey on itself. It is more likely that animals possess genetic blocks *against* redirecting interspecific aggression to their own species.

If aggression *is* innate in man, as it appears to be in some primates, it poses interesting questions. What are the stimuli that release human aggressive instincts, for example? What are the behavioral mechanisms that normally turn it off? Is aggression being learned, the instinct for it refined, or is it merely released by watching violent scenes on TV? We have seen in herring gulls and human infants that instinctive behavior patterns can be refined by repetition, practice, and learning. Any system that admits aggression is nonadaptive in current human society, and perhaps seeks to control it, must take these possibilities into account.

The theory of a killer instinct in man is not new; it is merely original sin revisited. If war is indeed in our genes, a result of instincts that may not be denied, then presumably we can do little to prevent it if the appropriate releasers are provided. An unfortunate side effect of this theory is that it removes all responsibility from man for war, murder, and the like. Such a pessimistic and misleading view of human behavior does little to aid investigations of the causes of human aggression, which are probably complex and primarily a result of learned behavior.

FIGURE 19-3 The courtship sequence of the three-spined stickleback. (a) A female with a swollen belly (top right) enters the territory of the male. The male begins a zigzag dance, courts, and leads her to the nest. When she enters the nest (bottom), the trembling movements of the male elicit spawning. Later, the male will enter the nest and fertilize the eggs. (b) Schematic drawing showing the courtship sequence. Each action serves as a releaser for the following action. (From *The Study of Instinct* by N. Tinbergen. Copyright 1951 and reprinted by permission of Oxford University Press.

on which an aquarium was placed. Leading the female to the nest requires the appropriate female response to the male's courtship dance, and the prodding of the female by the male induces her to spawn. If the experimenter prods a female with a glass rod, she also spawns. The presence of eggs in the nest releases parental behavior, but if the eggs are removed the male shows only courtship displays. In other words the programmed behavior of the male will not continued unless and until the appropriate releaser is present (Figure 19-3b). This elaborate behavior in sticklebacks is adaptive in that it assists in bringing together and synchronizing the male and female mating act and tends to prohibit errors in the choice of a mating partner.

What governs the instinctive behavior of an animal such as the stickleback? How is it that out of a wide variety of environmental stimuli presented to an individual, only one specific stimulus triggers an appropriate response? It is believed that the response to a stimulus is related to the internal condition of the animal; this internally initiated condition or degree of readiness is called **drive.** Drives are usually directed toward specific goals, and once the goal has been attained the drive energy is reduced, a condition known as **satiation.** Thus, one may speak of an animal's drive for food (hunger drive), a mate (sex drive), water (thirst), and so on.

In considering drives and releasers as well as the quick execution and rapid termination of instinctive behavior patterns, Konrad Lorenz suggested that the behavioral components could be likened to a hydraulic system similar to that of a flush toilet (Figure 19-4). He proposed that drive energy builds up within an animal—equivalent to the filling of the reservoir in the model. The filled reservoir represents a fixed or stereotyped behavioral pattern and is responsible for drive or motivation. An animal

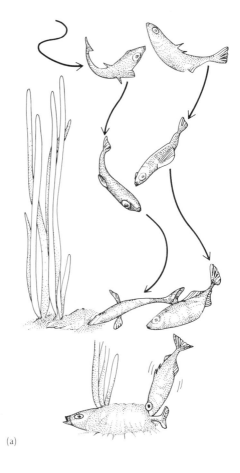

(a)

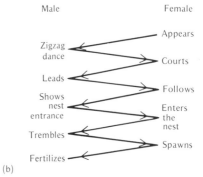

(b)

in this state would tend to appear restless. Ordinarily the release or discharge of activity would be prevented until a specific releaser stimulus was received by a special center in the nervous system, called the **innate releasing mechanism (IRM).** Each IRM would be triggered by its own specific releaser. Upon presentation of the releaser, signified in the model by a weight, a valve is pulled out, the IRM has been triggered; energy in the reservoir is then released into a specific behavioral pathway indicated by the trough, and a specific instinctive response would be produced. Such a model mechanically relates innate driving force, releaser, releasing mechanism, and response in a convenient way, but—and this may be its most serious drawback—the model tells us nothing about the neural or homonal mechanisms involved.

The hydraulic model, by analogy, does suggest an explanation for some kinds of behavioral events that occur in the life of an animal. For example, when caught in a conflict situation between two opposing drives, some animals will respond with apparently irrelevant behavior called **displacement activity.** A nesting bird, when threatened by a predator, is torn between two conflicting drives: to incubate the eggs and to escape the predator. Rather than either of these, it may begin to inappropriately nibble at its feathers. Displacement activity occurs at the point where the two opposing drives are of equal intensity. This was rather beautifully shown by W. C. Dilger in investigating the territorial behavior of male thrushes. Like many male birds, thrushes warn other males away from their territories by loud singing. Dilger placed a stuffed thrush near a loudspeaker and played a recording of a singing male thrush. The bird in whose territory the speaker was located responded as though his territory was being attacked. If the recording was played softly, aggressive behavior was released and

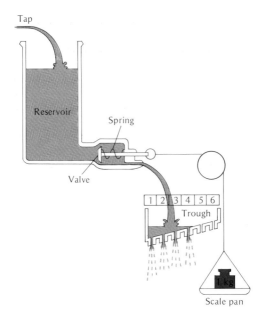

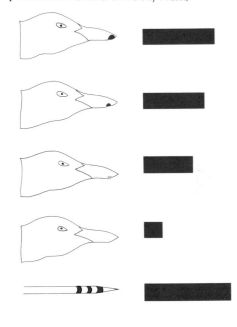

the defending bird attacked the stuffed bird. If the volume was high, escape behavior was released and the defending bird retreated. By alternately raising and lowering the volume, Dilger could make the bird move toward or away from the speaker, even to the extent of teetering on one leg when its conflicting attack and escape drives were perfectly balanced. At this point, however, some birds began to scratch their heads, preen their wings, peck at the ground, or even went to sleep. The thrush, it seems, had to do something with its behavioral energy: if it could not do something relevant to the situation, it did something irrelevant.

To account for displacement, ethologists have suggested that when releasers for two conflicting drives are presented simultaneously, the energy built up in the reservoir is not released through the usual trough; instead, the release valves are blocked, and thus the energy spills over the top of the trough into another kind of activity, one not ordinarily triggered by that releaser. Some authors believe that the thwarting of a single drive may also produce displacement behavior. For example, a thirsty dove prevented from drinking by a sheet of glass in front of its water bowl will peck at the ground nearby. Thwarting or conflict in human behavior is also often resolved by displacement activity; when an individual becomes angered, but for cultural or other reasons is prevented from expression of this rage in a direct way, the individual may resort to pacing, head scratching, yawning, eating, or stretching. On other occasions, the correct behavioral activity may be released but directed at another, more neutral object than normal; this is called **redirected activity.** Thus, a bird that is threatened by an aggressive intruder may attack a branch rather than the intruder. A man who is angry with his secretary may come home and shout at his children.

What is the adaptive significance of such conflict behavior? It has been suggested that displacement or redirection allows the resolution of conflict (that is, the dissipation of built-up energy, according to our model) rather than "freezing" the animal in a state of paralyzing indecision.

How an instinct is learned

The feeding behavior of seagull chicks is thought to be instinctive. When the parent gull lowers its head and points its beak at a week-old chick, it shows a complex but highly coordinated pecking behavior. It grasps the parent's beak and strokes it until the parent regurgitates food for the chick. Analysis of the chick pecking behavior showed that a parent gull was not required; a wooden gull head produced the same pecking behavior in the chick, and the strength of that pecking response was dependent on the presence of a spot on the adult beak (Figure 19–5). Using cardboard cutouts resembling the gull head and with different arrangements of painted spots, it was discovered that the specific releaser for chick pecking was not the parent but the shape of the bill and the presence of a spot. Indeed, the greatest pecking response was elicited by a knitting needle with three white bands at the tip! A classical instinct, or so it seemed: a behavior that was unlearned, and preprogrammed. Recently, however, this instinctive pattern was reinvestigated with a number of different questions in mind. Does the chick require practice to perfect its aim? Why doesn't the chick peck at the parent's red legs or other objects? Using painted pictures of parent gulls, it could be shown that as the chicks matured, their pecking accuracy increased. This increased targeting was directly related to visual experience acquired by the chicks after hatching—in other words, learned behavior. By using vertical rods of different dimensions and by varying the speed of movement of these rods, it was possible to determine the "beak" that would stimulate the greatest response in gulls of various ages. An 8 millimeter rod was preferred no matter what the rate of movement. (This is about the mean beak length seen by the chick.) However, it became apparent that older chicks were much more discriminating—they preferred a live gull parent to a model of a gull. Preliminary experiments also suggested that young chicks could be conditioned in their response by the reward of food—in other words, their innate response could be modified. From this study it was concluded that: a newly hatched chick is clumsily coordinated and has a poorly aimed but instinctive peck that is elicited by a simple stimulus—the shape and movement of the beak. As a result of reward, in the form of food, the chick learns the characteristics of the parent, and by practicing pecking, its aim improves. All this strongly suggests that during the normal development of a so-called instinctive behavior such as pecking, and perhaps of other stereotyped patterns, there is often a learning component. Behavior is thus a combination of innate and learned responses, not simply the one or the other.

Do we humans have instincts? Yes. Without any previous experience, a newborn infant will instinctively suckle the breast and grasp a rod with its hands so firmly that it can support its own weight. Although these patterns tend to fit the term "instinct," that is, they do not seem to depend on learning, they are not refined without experience. The suckling of the infant at the breast becomes more precise and better coordinated with practice. Similarly, the grasping reactions of the infant tend to improve with increasing experience. Again, in the normal development of instinctive behavior patterns, there is a learning component. Thus behavior cannot be clearly separated into two distinct entities that can be labelled "learned" or "unlearned": the development of a behavioral pattern is a complex mosaic produced by the continual interaction of the developing individual with its enviroment. From this, we should be able to conclude that if man has aggressive instincts, he may yet be able to learn to control them (Box 19D).

Learning

"Your mind," said William James, "is the sum of your experience." Indeed, it is as if we begin our lives with minds like clean slates, devoid of any characters, and without any ideas; but in the course of time those slates record all we experience. Consider all that you have experienced since birth, and how your mind has changed as you've matured. You learned to walk, to speak, to read and write, to love and to behave in thousands of other ways. Where would we be without the capacity to learn? Where and how does learning take place?

The experiences of an organism are somehow indelibly recorded in the tangled web of nerve cells. Enduring changes in nervous function that are based on experience permit an organism to alter its future behavior, a condition called learning. When we speak of enduring changes in nervous function and behavior, what do we mean? How long is "enduring"? The nature of the neural changes that take place during the learning process are still incompletely understood. Moreover, as we all know, some of what we learn does not endure—it is forgotten. Is forgetting the permanent loss of what was learned, or is it merely a condition in which the retrieval of information is impossible? Is it a lost or merely a misplaced volume in our library of information? What is the neurological basis for the storage of information? Without memory, learning is impossible. Once we can definitively characterize memory, then the basis of permanence or transience of learning will become clearer. That time is not yet here. For the present, let us hedge by saying that **learning** is a more or less permanent alteration in behavior that occurs as a result of experience.

Habituation. Perhaps the simplest kind of learning is habituation. **Habituation** occurs when a repeated specific stimulus, presented without reward or punishment, results in a decreased response until it may disappear completely. In a sense, habituation is learning *not* to respond to a particular stimulus. It plays an important role in the lives of animals: through habituation, animals avoid wasting time and energy by dropping out responses that are of little significance in their lives. For example, young birds learn not to take flight at the sight of fluttering leaves; crows in a cornfield eventually ignore a scarecrow; when we hear an unexpected sound our attention is aroused, but if the noise recurs repeatedly, we cease to pay attention and eventually may become unaware of it. The "cocktail-party effect" allows us to pay decreasing attention to the group noise around us to the point where we virtually ignore it. Habituation is one of the basic mechanisms involved in training wild animals: if you carefully handle a freshly caught snake, after some hours it may lie completely still in your hands and will no longer strike because it has become habituated to you.

Habituation can be distinguished from the similar phenomena of fatigue and sensory adaptation because it persists even when a stimulus is withheld for some time. A fatigued muscle will fail to respond after repeated stimulation, but this is reversed after a rest period. The sense of smell quickly adapts to odors until they go unnoticed, but again this adaptation is reversed after a period of rest. If you have ever shopped for perfume, you know that after sampling several smells, you can no longer distinguish between them. In sensory adaptation, sensitivity to stimuli within a given sensory modality are temporarily decreased. Following habituation, however, a response can still be elicited by stimuli other than the specific one to which an animal has habituated, even within the same modality. At a noisy cocktail party, you can focus on and respond to a single speaker while disregarding the general noise around you. Habituated to the noise of cars on a highway, you still respond to the screech of brakes.

Classical or Pavlovian conditioning. Perhaps the most familiar of all dogs is not Lassie, but an unnamed dog that lived in the laboratory of the Russian physiologist Ivan Pavlov (1849–1946). Lassie entertains us, but Pavlov's dog instructs us in the mechanisms by which we learn. What Pavlov did was to take an innate reflex, one already programmed into the nervous system, and transfer its response from one stimulus to another; such behavioral modification is called **classical conditioning.** The experimental design was quite simple. When food was presented to a dog, a copious flow of saliva was produced in the dog's mouth by reflex action. However, if a bell was rung regularly at the same time as food was offered to the dog, the ringing of the bell alone would eventually elicit salivation, even when no food was presented. The bell itself is not a specific stimulus since the dog could be conditioned to salivate on a light signal, or the clicking of a metronome, for example. The stimulus of food that normally initiates salivation had been replaced by a **neutral stimulus** (bell, light, clicks, etc.) which prior to the conditioning experience would not produce the response (salivation).

Let us consider another example of classical conditioning involving the reflexive act of limb withdrawal. If we were to administer a mild electric shock to your hand, you would rapidly withdraw your arm. If, however, this shock was regularly preceded by some other stimulus, such as a ringing bell, in time just ringing the bell would cause you to pull away your hand. You have been conditioned! Both examples illustrate how a neutral stimulus can be made to elicit a response which it would not have prior to training.

How does the reflex activity in classical conditioning differ from the innate reflex itself? Innate reflexes persist for life, and given the appropriate stimulus, the response will always occur. However, in classical conditioning the reflex activity will not be produced unless certain factors necessary for its continuance are satisfied. Four factors are a necessary part of classical conditioning: contiguity, repetition, reinforcement, and interference.

1. *Contiguity.* There must be a close association of the neutral stimulus (e.g., a ringing bell) and the unconditioned or **natural stimulus** (e.g., presented food) so that the events occur within a limited space of time, and the order of presentation is specified. For example, the time interval between the neutral stimulus and the natural one cannot be too great, and the natural stimulus (food) cannot be given first. What dog would salivate if you rang a bell after feeding?
2. *Repetition.* The greater the number of pairings made between neutral and natural stimuli, the greater is the response to the neutral stimulus alone.
3. *Reinforcement.* The strengthening of the conditioned response depends on the pairing of a reward (e.g., presented food) with the neutral stimulus (e.g., a ringing bell).
4. *Interference.* If the neutral stimulus is repeatedly presented alone, then the response that it elicits decreases steadily until it is

zero. This is called **extinction.** Pavlov believed extinction was due to new learning that interefered with or inhibited the conditioned response. Thus, the repeated ringing of a bell without any food eventually causes the animal to learn that the bell which once meant ''food'' no longer does; this results in decreased salivary secretion over a period of time, and ultimately the conditioned response no longer occurs.

The following case has been used to support the view of Pavlov that interference is involved in extinction. If during the early period of training an animal is distracted by a loud noise when the bell is rung, there is a reduction in the response, presumably due to external inhibition. However, if the same distracting stimulus is presented during the period when extinction is occurring then the strength of the conditioned response is increased, presumably because this removes the interference of the new learning; this phenomenon is called **disinhibition.**

The Pavlovian concepts of extinction and interference may apply to memory and forgetting. There is the story (probably apocryphal) of a well-known biologist who specialized in the classification of fishes and who later in his career became the president of a large western university. One of his first goals was to learn the names of all the students he met on campus; however, when he discovered that for every student's name he learned he forgot the name of a fish, he gave up the task of trying to learn student names!

Similar situations occur in our own lives. When we learn something and then there is a long interval before we are tested on the material, we tend to forget it. Indeed, the more similar the material learned in the interval between the initial learning experience and the time of testing, the greater is the degree of forgetting. If the material is dissimilar, there is

ANIMAL AND HUMAN BEHAVIOR
19-5 Where and how does learning occur?

404

less loss during the interval, and if sleep occurs between learning and testing, then forgetting is diminished to even a lesser degree. This is why lessons learned just before going to sleep at night are retained with great clarity in the morning.

Operant or instrumental conditioning. A laboratory rat is enclosed in a soundproof chamber with a food dispenser. The rat is rewarded by a pellet of food whenever he inadvertently presses the lever. Eventually the rat learns to press the lever to obtain a pellet of food (Box 19B). How does such instrumental or operant conditioning differ from classical conditioning?

In classical conditioning the learning involved is rather simple and passive. The response elicited by the neutral stimulus is programmed into the nervous system, and there is little variability in the response. The animal hears a bell ring and it salivates. Furthermore, the animal cannot control the stimulus; that is the role of the experimenter. **Operant conditioning** involves a more complex kind of learning, in which the animal exerts some operational control over the stimuli received and its own response; in addition, the behavioral activity that results has some modifying influence on the environment. Consider the caged rat that presses on the lever to obtain a food pellet. Initially, the animal probably jumped around and quite accidentally struck the lever, but after a number of trials it learned to make the proper response. Now it presses the lever as frequently as it requires food; it has modified its feeding situation by its behavioral pattern. This situation is quite different from the bell-food-salivation pattern found in classical conditioning where the animal's behavior does not modify its environment (food received, in this case).

The learning involved in operant conditioning is not restricted to rats; humans do it too.

Las Vegas gamblers, thousands of them standing like robots in front of the slot machines, pour coin after coin into the slot and pull the levers mechanically for hours on end. What makes them behave so when they are fully aware that the odds are against them, that most often they will lose and not win? Why does a gambler continue to pull the lever of a "one-armed bandit" when the rewards seem so remote? The reason is the same as for the rat. The gambler has learned through experience that occasionally there is a jackpot hit, and thus there is a reward for lever pulling (and money invested too).

These cases of operant conditioning reveal that an important element involved in this kind of learning (as in classical conditioning) is reinforcement. Reinforcements are of two kinds: (1) rewards that result in a repetition of the behavior pattern and (2) punishments that tend to suppress a particular behavioral response (Box 19B). Without reinforcement—that is, in the absence of reward or punishment—operant conditioning does not take place. Reinforcement procedures are involved in training dogs to "shake hands," "sit up," "fetch," and so on. They are also used in training young children to be social beings. Behavior which pleases the parent is rewarded, that which displeases may be punished, as when a child is praised for voiding in the toilet and perhaps even spanked for wetting his pants or his bed. The child is being conditioned to deposit his wastes in the same place as the adult members of the family, making life more pleasant for all concerned.

Reinforcement (as well as contiguity, repetition, and interference) apply both to classical and operant conditioning. However, it is important to recognize that although the factors involved are the same, the responses do differ. In classical conditioning the response is a fixed pattern of behavior (a reflex) but in operant

conditioning the animal starts out by expressing a variety of responses that are a part of its behavioral repertoire, and later one or more of them are reinforced through training while other unrewarded responses are either extinguished or subject to habituation. There is another difference: in operant conditioning rewards can be given after each response, or intermittently, without reducing the strength of the response (witness the intermittence of gambling rewards!).

What is the adaptive significance of operant conditioning in the life of an animal? The learning developed through such conditioning enables an organism to gain some control over its environment, and as such it promotes its welfare. Some researchers believe most, if not all, learning involves conditioning with reinforcement and is of overriding importance in human behavior (Box 19B).

Trial and error learning and reasoning. Two dishes are set before you: one is green and the other red. You touch each and rapidly withdraw your hand from the green one because it is hot. Quickly you learn that a green dish means hot. Similarly a rat placed in a simple T maze learns by **trial and error** that one corridor contains food whereas the other does not. In a short time the food-containing corridor is always selected. In these situations a correct response is rewarded (by food) and an incorrect response is discouraged either by punishment (burning) or by an absence of reward (no food). Obviously, operant conditioning and reinforcement are involved in trial and error learning. When you consider your own experiences, you can readily appreciate how much of your learning has been and continues to be based on trial and error. Tastes in food, the ability to play games such as tennis and golf, all involve trial-and-error behavior with reinforcement. In a sense this is a very simple type of problem solving.

Let us now consider a more complex kind of learning situation. You are in a basement room containing two boxes, each two and a half feet in height, and you want to open a window that is ten feet above the ground. How do you do it? Simply by placing the boxes one on top of the other and climbing on the uppermost one. The problem was solved not by trial and error, but by **reasoning** or **insight learning.** Reasoning is the ability of an animal to solve a problem or respond correctly to a situation not previously encountered. Reasoning permits organisms, such as yourself, to apply prior learning to a novel situation and to solve the problem without resorting to trial-and-error behavior. Reasoning is most prevalent in higher primates and man.

Imprinting. One of the most striking things about a mother duck and her ducklings is how closely they follow her. On land they march in single file, and in the water they swim in linear array as if tied to an invisible line. What is behind this kind of behavior?

When Konrad Lorenz examined the strong attachment of young birds toward their mother, he believed it to result from a special kind of learning in which response to objects presented at a critical period in early life, usually quite restricted in time, were "stamped" into the animal's behavior. The special kind of learning involved came to be called **imprinting.** Such attachment does not have to be directed at a natural object, for newly hatched ducklings will form a more or less irreversible bond to the first large moving object they see after hatching, be it a ball or even the experimenter, and will specifically relate to this object from then on. Figure 19–6 shows a clutch of ducklings following Konrad Lorenz, on whom they had imprinted. Here, the natural and adaptive value of imprinting has been lost through experimental interference, and instead of staying close to their mother, the animals follow

another object. Imprinting in the wild is likely to be adaptive, however, because the mother is usually the first large moving object seen after hatching. The newborn animal gets a quick behavioral "fix" on its natural parent, and is bound to her during its early days when its very survival may depend on the attachment.

Imprinting is a form of learning that profoundly affects social life and mating, and it is particularly strong in certain species of large terrestrial and water birds, such as ducks, where early attachment to the mother has great survival value. The early experience, presented at a critical developmental period, produces a lasting and almost irreversible effect. Imprinting, unlike many other types of learning, does not depend on reinforcement or conditioning. Although it is learning, it is apparent that imprinting must be related to special conditions of the nervous system during a restricted period of development, for if the animal is not at this time exposed to the experience, it cannot subsequently be imprinted.

19–5 WHERE AND HOW DOES LEARNING OCCUR?

Almost intuitively one feels that learning must occur within the central nervous system, especially in "the great ravelled knot," the brain. We show we believe this when we say "He hasn't got a thing in his head" in reference to a person of slow learning capacity. How can we determine where learning takes place? Two procedures have been used: neurosurgery, involving removal of a portion of the brain, and electrical stimulation of various loci of the brain. To date, of the large number of studies performed on the brain with an eye toward finding the seat of learning, none have identified a specific site. However, many workers suspect that in humans at least, the center of greatest importance in many kinds of learning, especially reasoning and the development of language, depends chiefly on one area: the layer of gray matter over the cerebral hemispheres, the cerebral cortex (Figure 19–7). It first appears in reptiles, but shows its greatest development in mammals, especially man, in whom the cortex is about as large as all the rest of the brain together.

The early anatomists who mapped the brain's cortex found distinct regions that were either sensory or motor in function. There was however a vast expanse of the cortex that could not be assigned to either of these categories, and this came to be known as the **association area,** providing a link between sensory and motor activity. The neurophysiologists seized upon the association area as the locus of conscious thinking and learning, but later experiments yielded nothing concrete. It is true that destruction of the temporal, frontal, and parietal lobes (Figure 19–7) does produce effects on behavior and on learning, but how does one know that such lesions do not disrupt the normal organization of the brain so that all learning is impaired? No one lesion produces permanent amnesia or inability to learn, but the disruptions do vary in degree. It is as if an animal deprived of a certain region of the brain can compensate and learn with other regions, but at a slower pace. Electrical stimulation of the brain, for example of the temporal lobes, does produce memory related experiences, but this does not occur all the time. In addition, how do we know that the electrically stimulated site is the only region activated? It becomes apparent that learning probably occurs at a variety of loci in the brain and there is no single spot responsible for all learning.

By what mechanisms does learning take place? Are the neurons themselves actually changed? A number of possible leads have been suggested recently, but there is no agreement among scientists that any one of them is the sole—or even partial—mechanism involved.

1. *Facilitation.* It has been shown that if a neuron excites other neurons repeatedly through a particular synaptic connection, the passage of stimuli through this pathway is considerably facilitated. Impulses from the same kind of stimulus are not only channeled through the same pathway in the nerve network of the brain but pass through the synaptic gate with greater ease. This principle has been extended to suggest that if use of a neural pathway facilitates transmission, disuse weakens it. Learning thus could involve the well-worn routes of neural transmission.
2. *Growth of nerve pathways.* Although there is little division of neuronal cells after birth, it is entirely possible that learning involves the development of recurrent nerve circuits in which neurons are activated in sequence so that a closed "memory" loop is formed. As a consequence of such sequential activity, neurons could fire continuously and facilitate synapses that impinge on the loop.

FIGURE 19-7 The parts of the human brain. Many of the major structures of the human brain can be seen on its medial surface. The lines on the cerebral hemisphere represent folds in the cerebral cortex. Note that most of the structures pictured here would be cut in order to prepare a view like this one. The corpus callosum, for example, is made up of fibers running into and out of the plane of the page and connecting the two hemispheres. (From *Elements of Psychology, 2nd edition,* by David Krech, Richard S. Crutchfield and Norman Livson, with the collaboration of William A. Wilson, Jr. Copyright © 1958 by David Krech and Richard Crutchfield. Copyright © 1969 by Alfred A. Knopf, Inc. Reprinted by permission of Alfred A. Knopf, Inc.)

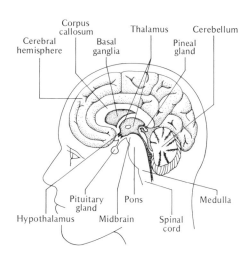

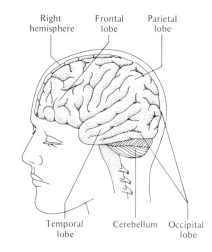

3. *Neuronal size.* This theory suggests that frequent use of certain neurons causes them to enlarge, thus engaging in many more synaptic connections. Conversely, disuse or damage results in decreased neuronal size and impaired levels of function.

4. *Proliferation of cells.* Although the number of neurons in the central nervous system is generally fixed shortly after birth, the supportive elements or glial cells are continually formed during the course of a lifetime. It has been suggested that greater neuronal activity involves greater numbers of glial cells surrounding the neuron and thus learning involves glial cells plus neurons. Supporting this is the finding that in dehydrated rats there is an increase in the division rate of glial cells in the "thirst center" of the hypothalamus.
5. *Dendrite spines.* Electron microscope studies of dendrites of neurons have shown that the surface is not smooth but covered with thousands of tiny spines. Recent evidence suggests that the number of dendritic spines decreases with reduced sensory input and increases with enhanced activity.

At present there is no evidence to exclude any of the above mentioned suggestions, and it is possible that learning involves a combination of some or all of these.

19–6 MEMORY

A creature without memory cannot discover the past; one without expectation cannot conceive a future.
GEORGE SANTAYANA

In the back-to-front world that Alice discovered behind the Looking Glass, the White Queen actually lived in reverse. "There's one great advantage in it," she said, "that one's memory works both ways." Alice remarked "I'm sure mine only works one way. I can't remember things *before* they happen," to which the White Queen replied: "It's a poor sort of memory that only works backwards."

Although our memories don't work forwards as well as backwards, like that of the White Queen, they do allow us to make certain predictions about the future. **Memory** is the information storage and retrieval process that allows us to bring past experience to bear on present activities and thought. Without memory, learning is impossible.

When you look up a telephone number, study a textbook, or hear a conversation, the energy of sound and light are transformed into electrochemical impulses that are carried along the pathways of the nervous system. This pattern of energy transformation, called **transduction,** is the beginning of learning and memory. Once the stimuli are received in the central nervous system, the coded impulses are attended to immediately: that is, there is a conscious awareness of the incoming information: one in which there is a brief storage, what is commonly called **short-term memory (STM);** the other involves more permanent storage, or **long-term memory (LTM).** STM is comparable to instant replay on TV. The capacity for holding information is of brief duration (seconds to minutes) and is usually restricted to about seven items. STM is involved, for example, in remembering a telephone number (up to seven digits), and in the situation where you are asked a question and you reply "What did you say?", but before the question is repeated, you remember what it was. Retrieval of STM is simple and direct, requiring little conscious effort; it stores raw, relatively unprocessed sensory inputs. Unless some further processing occurs, the information held in STM is lost, and the decay is hastened by external interference. Interruption during the time you are looking up a phone number often causes a mistake in dialing, and a return to the phone book for another try.

After a fragment of information is held in STM it is either forgotten or it is transferred to long-term memory (LTM) for more or less permanent storage.

Our understanding of STM is limited indeed. However, it is clear that putting information into the banks of LTM and retrieving it are far more difficult and complicated than from the storage sites of STM. Since associations play an important role in LTM it seems likely that information is stored in an organized fashion. Although the neurological basis for filing information is not clear, there probably does exist some sort of system comparable to the card catalogue in a library. This helps to explain why we can best remember material that we have previously organized systematically, but we have no idea whether remembering is due to methods of storage or ease of retrieval. "Cram" sessions do not promote LTM, rather it is best to space learning at intervals; this seems to facilitate information storage.

The final stage in the processing of information stored in the memory bank is retrieval. How do we pick one piece of information from amongst the billions held in LTM? We really do not know. Retrieval must involve some kind of search through the files, which may or may not be successful. It is easier to recognize material (for example, information presented in a multiple choice or true-false examination) than to recall it without any cues (for example, an objective exam). Perhaps this is true because the former involves a less comprehensive search, since the category has already been given, whereas in the latter case the entire store of LTM must be surveyed for the item of interest. By analogy it is easier to find the weather report in a newspaper if you are given the section where it will be found rather than randomly searching each page of the paper.

What are the mechanisms involved in memory? The early Greeks believed that the fingers of experience wrote upon a blank slate and in this way a mind was filled with information during a lifetime. Strange and even mystical as this concept may seem, there is some support for it as a result of recent experiments. In 1890, William James suggested that every experience leaves a tiny "scar" on the nerve cells and fibers, registering and storing it for the future.

This "scar" of memory has been called the **engram,** and the search for it in molecular and electrochemical terms continues. It has been found that the electrochemical activity of the brain, the EEG patterns, do change as a result of experience. Indeed, the electrical responses of a single neuron can be modified by experience. Although such findings suggest that functional changes occur in the neurons themselves it is not yet clear whether these are the essential components that produce learning, or whether they are merely correlated events in memory-retrieval processes.

In recent years psychologists and neurophysiologists have speculated and found some evidence of a chemical basis for memory. Experiments have shown that training causes an increase in RNA synthesis in the brain which persists for about one hour. It is clear that this is not LTM, but it could be an indication of some of the processes involved in memory storage; on the other hand it could merely be the restoration process of stimulated neurons. Others have suggested that protein synthesis is involved in memory storage; by administration of drugs that inhibit protein synthesis, learning and memory have been impaired. These experiments and others suggest that molecules are synthesized during learning and memory, but it would be presumptuous at this point to conclude that we understand how they function in memory.

The identification of chemical changes in the brain following training has suggested to some that it may be possible to transfer memory from one animal to another. Thus, there has been talk of "memory pills" and the like which will make us "brainier" without recourse to the usual regimens involved in learning. Although there is evidence that injections of RNA and sometimes whole extracts of brains produce learning in animals such as planarian worms and rats, there are an equal number of instances in which the experiments fail. Moreover, it is never quite clear in the seemingly successful trials of memory transfer whether the results are due to a real information transfer or to facilitation of learning. In any case, these experiments provide a fertile ground for exploring the nature of memory. Should information transfer by injection turn out to be feasible, it may be possible to have memory transplants as well as organ transplants. To date, however, no memory molecules are available.

Let us return now to the question of forgetting. Recall that earlier we suggested, in accordance with Pavlov's view, that **forgetting** could be due to interference as a consequence of new learning. It has been suggested that LTM involves both a labile and a stable stage. For information to be stored permanently, it must be consolidated and made stable. If this does not occur, the information is forgotten. For example, if rats are given electroconvulsive shocks each day immediately after training, they are unable to learn a simple task. In humans, a cerebral concussion produces retrograde amnesia, that is the inability to remember not only the injury itself but events immediately preceding the injury. Memory consolidation in LTM thus seems to depend both on timing and electrical activity of the nervous system. It is entirely possible that new learning competes with the consolidation process, and thus we forget. As another cause of forgetting, some have suggested that the early (and perhaps later) neurobiological processes that occur during memory have a fixed lifetime and in the course of time they decay or change. As a consequence, memories fade with time.

Forgetting may in fact be highly adaptive. If we were to retain everything impinging on our consciousness, we should soon drown in a flood of information. Selection of the most relevant and useful information is not only important for survival, but to preserve our

sanity. In this respect, habituation is a form of learned forgetting.

With the advent of computers, tape machines, and other advances of modern technology that allow us to record and preserve information with enormous ease, we are accumulating a flood of material, often with little regard for its worth or relevance. The written word has allowed us to record, accumulate, and pass on the events of our culture, often overriding the genetic components of our behavior; but the recording of culture today has no built-in mechanism for selection, or for forgetting. Faced with such a wealth of information, we may soon be unable to decide what is important and what is not. We will be unable to see the wood for the trees, so to speak. Civilization may not end with a bang but in the chaos and confusion derived from the clutter of our recorded cultural inheritance.

19-7 SOCIAL BEHAVIOR

Man has been described as an aggressive, status-oriented, materialistic, and territorial animal whose behavior is largely determined by his animal heritage. There is considerable appeal in such a description since there are an endless series of examples of people being unpleasant to one another. Hardly a day passes that we are not told of crime in the cities, wars of aggression, class struggles, and the like. Are these products of our social organization? What are the strengths and weaknesses of human society?

Groups and society. It is a fact: no animal is totally solitary. Although individuals may use the mechanisms of instinct, learning, and reasoning to obtain shelter, to find food and to protect themselves from enemies, survival of the species frequently depends on social behavior. What do we mean by social behavior? When animals are grouped together as a result

of their response to environmental factors, we call this **aggregation,** but when the members of a particular species stay together and cooperate because of their responses to one another, we describe this as **social behavior.** For example, if moths circle about a light bulb or houseflies congregate about an open garbage pail, they are behaving as an aggregate or group, but they are not exhibiting social behavior; in these cases, the individuals are behaving exactly as they would if they were alone. Despite the fact that there are many of them, they do not interact with each other to any great extent. Social behavior involves the mutual interactions of two or more individuals so that their behaviors are modified. Such an interaction always involves some communication between the individuals—symbolic signs, auditory signals, chemical agents of language (Box 19E). Mating usually involves a special kind of social interaction, wherein a male and female come together for one purpose. Some authors do not regard this as social behavior, since even solitary species must come together for mating and possibly for rearing of the young. However, the behavioral mechanisms that operate within animal societies are often extensions of those that bring organisms together for mating and operate within a family group. Indeed, insect societies, often considered the pinnacle of social organization, are in fact family groups.

What advantages are there to being social? The formation of social groups—societies if you will—enables a species to go beyond the capacities of its members. In an established society it is possible for individuals to be specialized and to perform different functions so that by virtue of the integration of their special skills that group may display increased complexity and efficiency. Insect societies are an extreme example of this kind of organization (Box 19E). In a sense, by integrating the individual members of a society into a highly organized structure, the whole becomes more efficient and successful than the sum of its individual parts.

The advantages of social life may be many. When a good food source is found by one member, all the others in the group may be attracted to it and share in it (Box 19E). Animals that live in groups tend to be less subject to predation than animals left on their own. For example, when one individual spots an approaching predator it may warn the entire group, or by concerted group action the predator may be driven off. Flocks of birds and schools of fish are so effective in protecting their individual members that predators attempt to isolate an individual before attacking. Further, social behavior may be of critical importance during reproductive periods when territories are established for breeding, for feeding, and for raising of the young. The adaptive significance of social behavior therefore lies in its greater survival value for the species, if not always for the individual (Box 19C).

Territoriality. Territoriality is the exclusive occupancy of a place by an animal or group of animals through aggressive defense or display. It was first described in the 1920s by H. E. Howard for the British warbler. However, since that time the basic concept of territoriality has been extended to animals other than birds. Dragonflies patrol the ponds in which their eggs are laid, crickets will vigorously fight off male trespassers, a lizard will threaten and engage in combat with a male intruder, and salmon and trout occupy territories during the breeding season. Some contend that our closest relatives, the primates, and even we ourselves are territorial animals.

What is the function of territoriality? Territoriality does different things for different species and some animal species show no territorial behavior. However, in many animals,

territoriality plays a role in attracting a mate, in propagating and raising the young, and in ensuring a supply of food. Paradoxically, it may also function in reducing aggression between members of a social group. To maintain a territory a number of behavior patterns are critical. First, the owner of the territory must have the ability to distinguish between those individuals who are intruders, those who are passing through, and those to be attracted to the territory. During the spring the male stickleback fish develops a red belly and establishes a territory. If a strange male enters the territory, the owner assumes a threatening posture, displays its red underside, and warns trespassers off by its threats. The red belly releases male aggressive behavior, for if a female (who lacks a red belly) enters the territory, she does not elicit aggression. Instead the male begins a zigzag courtship dance. Territorial behavior in the stickleback ensures reproduction of the stickleback species. It also increases efficiency in defense of the territory, and the need for combat is reduced. The red belly warns off other males, advertises that the territory is occupied, and fighting is not necessary. Birds achieve the same thing with song. Their repeated calls are advertisements by the males that the territory is already occupied. If notification and threat do not work to repel an intruder, then physical combat may ensue. Generally, the owner of the territory has the advantage, and is virtually undefeatable. Often, territoriality in birds minimizes disruption in large social groups. Many marine birds such as gulls form nesting colonies almost on top of each other, but there is little fighting between pairs and each raises its own family within its own small territory.

Territoriality, generally speaking, increases the efficiency of reproduction and in this way benefits the individual. Territoriality also benefits the species, for under extreme circum-

stances the resources of a territory may be partitioned in such a way that at least some members will reproduce. For example, if a pair of lizards are placed in a large cage they will use the entire territory for feeding and reproducing. If more and more pairs of lizards are introduced into the cage the territories will become smaller and smaller, but the territory is not indefinitely compressible. Ultimately some individuals will not find an area for their needs. Those individuals that do not gain possession of territory do not breed. To this degree territoriality may act to regulate the size of a particular population since the excess males that cannot occupy and defend a territory will leave no offspring. In this way, defended territory may protect a species from violent fluctuations and help to keep populations in line with the available resources.

Social dominance. Many species of animals have established ways of organizing their populations by means other than territories. One of these is by the establishment of rank order within a group. In some ways, ranked order is the equivalent of territoriality but instead of the environment being partitioned into defended areas, the animals participating in the hierarchy partition themselves while occupying a single territory.

For example, chickens are commonly fed by scattering corn or providing an ample supply of food in a series of feeding troughs. If we were to observe a group of 10 hens that had been living together for some time under such feeding arrangement, we would see little evidence of aggressive behavior. If however we modify the feeding situation so that there is a single small trough of grain and the animals are forced to compete with each other for food, then the behavioral pattern changes. One of the hens becomes dominant. It feeds actively, but the others make no effort to share in the food. The dominant hen may scatter some of the grain in its feeding, and others in the group may peck at this, but it is clear that these animals cannot feed directly from the trough. If one of the hens attempts to do so, the intruder is pecked at by the dominant individual, and the intruder shows fear and retires quickly. The dominant individual takes precedence over feeding and the others can feed only when that individual—the alpha—permits. Usually there is a "second-in-command," a beta-level animal that takes precedence at the trough over all but the alpha animal; there may also be a third and fourth level of dominance and so on down the line to the lowest in the order, a hen that gets pecked by everyone and usually is distinguishable by its bedraggled appearance. A **social hierarchy,** or **peck order** has been established within the group—so-called because many early studies on this type of social organization involved chickens. In a peck order, one animal acquiesces to another because of an aggressive relationship. Of course, status within the group may be established not only by pecking but by wing flapping, spur gouging, and other behaviors. The phenomenon of social dominance is quite widespread, and it is most clearly seen in those group-living animals that exhibit a high level of individual aggression. Thus, social wasps, bumblebees, hermit crabs, fishes, lizards, chickens, wolves, baboons, and macaque monkeys often show strong dominance hierarchies.

In stable chicken groups, the order of dominance is clear-cut, and is linear. That is, the alpha individual dominates all the others, whereas the beta individual dominates all but the alpha, and so on. In other cases, there is a two level arrangement in which a single individual is in complete charge and the other members of the group are all of equivalent subordinate rank.

In all of these situations the dominant individual controls the food, water, roosting place,

choice of a mate, and other competitive features. The individuals that rank high in the peck order are easily recognized at sight by their sleek appearance, their confident walk, the way the head is held; in short, by their "high and mighty" display. The hierarchies are formed after encounters among individuals of a group and may involve threats and combat, but once the issue has been settled it is generally semipermanent and mutually accepted. Individuals know their rank and stay in place, and visible fighting is a rare event in the life of the group. Some dominants may posture to "advertise" their alpha status, and in this way through prestige they maintain their dominance. Sometimes it is impossible for an observer to determine a social hierarchy until a crisis situation occurs, when the ranking is displayed very clearly.

While some animals show strong dominance hierarchies, others do not. Thus, in a grazing herd of herbivores the expressions of dominance are very mild and the individuals may feed from restricted bins without any evidence of precedence. Not only that, dominance behavior need not be expressed throughout the life of an animal; it may apply only to competitive situations, especially those involving sexual activities.

Social rank is also well recognized among humans, but here the ranks may depend on the kind of activity involved. Thus, an individual could be an alpha tennis player, but an omega in salesmanship. Often, possessions such as houses, clothing, jewels, and cars are used to advertise the position the owner has, or perhaps would like to have, in the social hierarchy.

What are the advantages of a social hierarchy? The establishment of a social order makes a group a closed society. As a consequence each animal knows every other one, knows its status and is prepared to exclude

BOX 19E The life of the honeybee

Sed inter omniaea principatus apibus, et jure praecipua admiratio.
PLINY. Lib. 11. C5.
(But among all of these species, the chief place belongs to the bees, and rightly are they admired.)

Division of labor in social animals has reached its most complex form in the social insects: the beehive seems to some to provide a model for cooperation in an ideal society. How does it work?

The founder of the hive is the queen bee, an extremely fertile individual who lays two kinds of eggs in cells of the comb: fertilized eggs and unfertilized eggs. The unfertilized eggs develop into haploid males, and the fertilized eggs into diploid females (Figure 19-2. Box 19-C). The males, or drones, have no role in the hive other than to fertilize a future queen; they perform no community tasks—hence the phrase "lazy as a drone." The females, or workers, tend the queen, rear the young, and defend the hive. As a worker matures, she performs a succession of tasks: newly hatched workers prepare cells to receive eggs and food. After a day or two their brood-food glands develop and they feed larvae. Following this, they remove nectar from the field workers, pack pollen loads into cells, and keep the hive clean. At about 2 weeks of age, the wax glands develop and the bees build new cells of the comb. At the end of this stage, some become guards at the entrance to the hive. Lastly, in the third week of life, the bees become foragers outside the hive. Their brood-food and wax glands decline, and they collect nectar, pollen, and water.

The hives are extremely efficient family communities. The bees of a given hive recognize each other by a common odor, and their status and needs are communicated among the members by **trophallaxis**, a passing of food substances from one to the other. Thus, a shortage of nectar, pollen, or water is almost immediately communicated to the foragers. A drop in the number of building bees is communicated by a lack of wax in the passed food, and nurse bees will develop wax glands and become builders. Similarly, a dearth of nurse bees produces a redevelopment of nurse glands among the builders, stimulated by a drop in brood-food substance passed among the community. Alarmed bees secrete a substance that excites the others and makes them aggressive. The queen too produces a substance that is passed around among the workers by the constant mutual feeding and grooming. When this substance is in short supply, the failure of the old queen is communicated, and a new queen is raised. Queens are raised from fertilized eggs laid in special large comb cells. A future queen gets fed more often with higher quality food (called royal jelly) than a future worker, grows larger, and is, of course, not sterile.

Foraging bees are also able to communicate the location of a food source outside the hive to foragers within by means of special dances up and down the hive combs. The dances are very specific; they communicate distance and direction of the food source and even wind velocity. So much information can be communicated by the dance of the bees, it has been referred to as a language.

The division of labor among the three castes in the beehive is extreme. The workers wear themselves out in the service of the community, which benefits at their expense. No matter what, the hive goes on. In a sense, the hive itself is like a superorganism; the members of the hive are the parts of the organism, only a few of which are responsible for reproduction. The queen and the drones are like the reproductive organs of the hive, and the trophallaxis of the members fulfills a similar role to the nervous and hormonal systems in an organism.

Since the genes of the workers die with the workers, the question of how selection operates to maintain these sterile animals has been raised. The answer is that although the workers die, all the female larvae in the hive (offspring of the same queen) have genes similar or identical to each other (Figure 19-2, Box 19C). In this sense, an individual worker's genes are not dying at all, and selection has plenty of similar material on which to operate. Just as an individual organism's somatic cells die with the organism, and this organism depends on its reproductive organs to keep the line going, so a worker depends on the reproductive organs of the hive—the queen and drone(s)—to reproduce its genes. The unit of selection is in fact the entire hive. The genetic odds would be on the maintenance of a system such as a beehive, just as with a whole organism, as long as both continue to be successful in their given environments.

In spite of the fact that hives behave (genetically anyway) much as a single organism, we can compare insect societies with vertebrate and even human societies, at least in terms of the way they function. Both occupy territories, communicate hunger, alarm, hostility, caste, and reproductive status by signals other than language. Members can distinguish each other from non-members, and **kinship**, or degree of relatedness, is important in maintaining group structure and cohesion. In both insect and vertebrate societies, members share the advantages of division of labor. From a human perspective, however, the similarities are perhaps less important than the differences! Genetic variability within a hive is minimal, the members of a hive lack individuality and life is relatively stereotyped. To date, attempts to order human societies so that every member knows his place and contributes equally without argument have always produced internal dissent. Chained as we are to the cycle of individual reproduction, individual freedom is prized over social efficiency, and variety remains the spice of human life.

strangers. If a strange individual attempts to enter such a group, it is usually driven off, or in rare instances, if entry occurs, the individual assumes a subordinate position, thus maintaining the stability of the group. The social hierarchy, once established, is highly stable and this effectively reduces the amount of fighting within the group. As with territoriality, social hierarchies also function to reduce aggression between animals of the same species. The life of the group is made that much more efficient because energy is conserved. For example, when two groups of chickens were studied and one group was permitted to establish a stable hierarchy whereas the other, by frequent introduction of new individuals, was not, the stable group pecked less, ate more, gained more weight, and laid more eggs.

In a peck order it pays to be on top: dominants are able to get more food, they remain healthier, live longer, and tend to reproduce more frequently. Why then should animals subordinate themselves? One suggestion is that it is of benefit to the species. Under conditions of stress, that is when animals have a restricted food supply, dominants gain more food and thus gain more weight while the subordinates starve. As a result dominance leads to survival of some individuals and thus promotes the survival of the group by not permitting the resources to be divided in such a way that none survive. Further, some studies of animal hierarchies show that it is better to be a member of a group, albeit a subordinate, than to be no member at all. Thus, in the European rabbit the dominant individuals enjoy longer lives and produce more offspring than do subordinate members, but the subordinates fare better than the rabbits that are excluded from the group altogether, where they are much more susceptible to predation and starvation. For the subordinate social animal it might be said that half a loaf is better than none.

Courtship, appeasement and ritual. Some of the most interesting behavior patterns observed in animals are those associated with mating. If they occur immediately prior to mating, they may fall into the category of **courtship behavior.** We have already described the elaborate mating behavior of the stickleback (page 398). Courtship behavior such as this serves to establish contact between animals and ensure that mating occurs between the right sex of the right species at the right time.

The sexual partner is often first attracted by special behavior patterns and signals. Birds and insects such as locusts and grasshoppers use courtship songs to attract partners, and many insects use odorous substances. A single female moth can attract males from several miles away. Birds often posture and display, and this is undoubtedly behind the evolution of the magnificent plumage of many male species such as peacocks and birds of paradise. Some of the most interesting courting behavior has been evolved by the Australian bower birds; to attract a mate, the males build large and elaborate courtship bowers and decorate them with twigs, lichens, fruit, flowers, shells, beetles and even bones. Often, a flower or a piece of fruit is presented to the prospective mate.

Many animals, including humans, offer gifts to their prospective mates. Terns offer fish to courted females, and some species of predatory flies will offer the females food before mating; in this way, they avoid the danger of being eaten themselves! Gestures such as this are examples of **appeasement behavior,** where possible aggressive reactions between animals must be overcome before mating can occur. Sometimes, infantile behavior inhibits aggression just as effectively as a gift; females of a species rarely attack the young. Courting male hamsters call like nestling young, and a courting man may speak to his beloved in childlike diminutives. For similar reasons, many animals

appease during courtship by covering or hiding signals that normally release aggression. For example, direct gazing at one another is often an aggressive threat gesture, and during courtship many birds turn their heads away from each other. The black-headed gull presents the back of the head to the prospective mate, so hiding the black face which normally releases aggression (Figure 19–8). "Coy" behavior by human females can be interpreted similarly, and kissing has been regarded by some authorities as a form of appeasement.

Appeasement gestures of various kinds may be used to turn off aggression in other situations, as when a member low in a peck order wishes to turn aside an attack by a dominant individual. Low-order wolves roll on their backs in a defenseless posture; a dominant wolf will not attack another in this position. Is aggression in man inhibited by similar gestures? In human societies, greeting ceremonies such as bowing, raising the hat or a smile of "hello" are thought to be appeasement gestures, which turn aside possible hostility and maintain bonds between individuals. Similarly, attacks on children are normally considered outrageous: in hand-to-hand combat, soldiers will rarely kill a child. Familiarization may act as an inhibitor, for under normal circumstances people are less suspicious of and less likely to attack those they know. The social implications of this are far-reaching. If aggression or even attack is normally inhibited by signals from the attacked, it is not difficult to see how our technology has brought us past the point where such signals can operate effectively. The long-range killing of people unseen by the man with his finger on the button or the trigger removes any possibility for effective appeasement gestures by the victims. Currently, almost the only inhibitor of aggressive behavior on the part of nations is the threat of retaliation. More than 90% of murders in the U.S. are performed with guns of some sort, a quick and effective method. Perhaps if the murderers had to do away with their victims slowly, by hand rather than gun, inhibiting appeasement gestures would have a chance to operate.

Often, social behavior patterns take the form of ritual. In courtship and mating, ritual is very common and appeasement gestures have often become ritualized in this connection. **Ritualization** of a behavior pattern indicates that the signals by which animals communicate are distinct and repeated in a predictable way. Further, the ritualized signals are usually symbolic, in that they may have no obvious connection with the behavior released.

For example, some species of birds show ritual feeding prior to mating, and the premating appeasement feeding of predatory flies also qualifies as ritual behavior (page 412). Ritual preening and grooming prior to mating in many birds and mammals fills the same behavioral function of appeasement. Often, ritual sequences of behavior have become extremely exaggerated and repetitive, as in the elaborate mating dances of many bird species. Courting albatrosses perform a repetitive sequence of behavior, including ritual food begging, appeasement gestures, showing of the nest and preening behavior. The end result is almost a dance: keeping time with each other, the birds wobble from side to side, fence with their bills, make clapping noises by opening and closing their bills sharply, and call to each other. At the end of each courting sequence, the birds bow to each other, point their bills to the ground, sit down, and preen each other. The sequence may be repeated several times before mating actually occurs. These social signals contribute to the mating process, ensuring reproduction.

In other contexts, ritual behavior contributes to the stabilization of animal societies. Ritual fighting may avoid actual combat. Rattlesnakes, for example, never bite each other but push each other from side to side with raised heads: when one is pinned to the ground, it retreats. In most horned mammals, intraspecific fighting is highly ritualized: antelope only spar with the very tips of their sharp horns. Human societies are full of ritual. We have already mentioned the appeasement role of ritual greeting. Organized religion in most of its forms produces cohesion and group efficiency by means of ritual. The pomp and circumstance surrounding kings and queens and parliaments in countries that have them serves to reinforce the top echelons of the dominance hierarchy. Confusion in social interactions, of which courting is one example, are prevented by ritual, and the efficient workings of the community are enhanced.

19-8 PROSPECTS FOR THE FUTURE

Much of man's behavior today is extremely disturbed; until we understand the causes, we cannot hope to find a cure. It has been argued that until we know which specific part of the nervous system is responsible for each behavioral pattern, we cannot truly understand or even compare behavioral activities in animals. One author has said that "our understanding of behavior and its physiological basis is the widest gap between disciplines in science." This may be true. In order to cope with our social problems, however, we must work with what we have.

Some ethologists, particularly the sociobiologists, believe that much of our behavior is innate; the implication has been that we are saddled with an inevitable inheritance, an approach that tempts a take-it-or-leave-it attitude to society. Other workers hold the opposite view, that behavior is determined by the environment: what is inherited is a set of capacities, which can be infinitely molded and shaped by the interactions of their possessor with the surrounding world. The truth is probably somewhere between: we have seen for example that inborn behavior patterns, instincts, can be refined by the environment, but not completely altered. In coping with the future, an eminent ethologist, Irenaus Eibl-Eibesfeldt, cautions us to remember that "man is not easily molded in every direction by the environment. He resists modification in certain areas more than in others."

Our knowledge of how the brain functions is in fact increasing rapidly. The day when thought, memory, learning, reasoning, insight, and the capacity to symbolize can be explained in physicochemical terms is probably closer than we think. Understanding brings the capacity for manipulation (See chapter introduction). The future of man as a species may depend more upon the way in which we use this knowledge than on the possible manufacture of a virulent microbe (Page 106) or the threat of a nuclear holocaust.

SUMMARY

1. Behavior is externally directed activity, often involving the adaptive response of an organism to its environment.

2. Behavior varies according to the capacity and development of the nervous system and effector organs (muscles) and may vary as a result of experience (learning).

3. Anthropomorphism and teleology should be avoided in discussions of animal behavior. Morgan's canon should be applied in interpreting animal behavior.

4. Behaviorists concentrate on learned behavior and its modification in the laboratory. Ethologists focus on the natural behavior of animals often in the wild. Some behaviorists, led by B. F. Skinner, believe human behavior is mostly learned and can be modified by conditioning. Ethologists such as E. O. Wilson and Konrad Lorenz believe there is a large genetic component to human behavior, and thus certain behavior patterns cannot be avoided or modified. As a corollary to this, altruistic behavior in some species can be explained on the basis that survival of the genes is more important than survival of the individual, and such behavior is dictated by the genes.

5. Simple oriented body movements toward or away from specific stimuli are taxes. Reflexes involve responses of part of the body to a stimulus. Some workers have attempted to explain all behavior in terms of innate and acquired reflexes.

6. Instinctive behavior patterns are stereotyped, inherited, and unlearned. Although they can be modified, refined, or even overridden, the potential to perform them cannot be eliminated.

7. Activation of instinctive behaviors often depends on specific releasers. The internal degree of readiness to respond is known as drive. Release of behavioral activity is prevented until the innate releasing mechanism (IRM) is triggered by a specific releaser stimulus. Once the behavioral goal has been reached, the drive is reduced and satiation is achieved.

8. Displacement activity is apparently irrelevant behavior which may be exhibited by animals caught between two conflicting drives, or when thwarted. Released behavior directed at the wrong object is redirected activity.

9. Man probably has innate aggressive tendencies, but they can be modified. The influence of learned behavior and culture is of overriding importance, and humans do not have a definable "killer" instinct.

10. Learning is an alteration of behavior that occurs as a result of experience. Habituation is one of its simplest forms and involves learning not to respond to nonsignificant stimuli.

11. Classical conditioning involves the modification of behavior patterns by transferring a response from one releasing stimulus (the natural stimulus) to another (the neutral stimulus). It involves contiguity of both stimuli, repetition and reinforcement with a reward. It is weakened by interference of other stimuli during the conditioning process. Extinction of classical conditioning occurs in the absence of the neutral stimulus but is subject to disinhibition by the presence of previously interfering stimuli.

12. In operant conditioning, the conditioned animal exerts some operational control over stimuli received and thus exerts controls over its environment. Reward and punishment reinforce operant conditioning; and contiguity, repetition, and interference are also important.

13. Operant conditioning and reinforcement are involved in trial-and-error learning. Reasoning or insight learning is the ability to respond correctly to a novel situation by the application and integration of prior learning.

14. Imprinting is the stereotyped response of animals to objects presented at a critical period early in life. It is subject to interference but does not depend on reinforcement or conditioning.

15. The cerebral cortex is probably the seat of learning and associative abilities, and is highly developed in man. The nervous mechanisms responsible for learning may involve facilitation, growth of nerve pathways, neuronal enlargement and increased numbers of synapses, proliferation of glial cells, or increase in dendritic spines.

16. Memory is the information storage and retrieval process that allows us to profit from experience and to learn. Short-term memory (STM) is brief and quickly lost. Long-term memory (LTM) involves transduction of nerve impulses and storage of learned information and seems to depend on timing and electrical activity of the nervous system. The mechanisms of transduction, storage, and retrieval are not well understood, and the search for the engram in chemical and physical terms continues. Forgetting involves loss or decay of the engram. Some forgetting, such as that involved in habituation, is adaptive.

17. Social behavior involves communication and cooperation between animals, usually with behavioral modification. As such, social groups are distinct from aggregations. Social behavior is important in mating and in rearing of the young; it also enables members of a group to operate more efficiently than they would if they were solitary. Insect societies such as bees and termites are the zenith of social organization, and the members of the group often exhibit structural and physiological as well as behavioral adaptations to group living.

18. Behavior patterns found in social animals include territoriality, social dominance hierarchies, elaborate courtship behavior, appeasement gestures, and ritualization. All of these help to avoid confusion in social interactions; for example, ensuring reproduction, securing a supply of food, raising of the young, limiting aggression, and ensuring the efficient workings and continuation of the community— and hence the species.

KEY WORDS

behavior
tropisms
anthropomorphism
teleology
Ockham's razor
Morgan's canon
behaviorists
ethologists
altruistic acts
taxis (taxes)
phototaxes
geotaxes
negative phototaxis
negative geotaxis
reflexes
instinct
releaser
drive
satiation
innate releasing mechanism (IRM)
displacement activity
redirected activity
learning
habituation
classical conditioning
neutral stimulus
natural stimulus

contiguity
repetition
reinforcement
interference
extinction
disinhibition
operant conditioning
trial-and-error
reasoning or insight learning
imprinting
association area
memory
transduction
short-term memory (STM)
long-term memory (LTM)
engram
forgetting
trophallaxis
kinship
aggregation
social behavior
territoriality
social hierarchy or peck order
courtship behavior
appeasement behavior
ritualization

TOPICS FOR REVIEW AND DISCUSSION

1. What is behavior? Of what biological significance is behavior in the life of an animal?
2. Why are anthropomorphism and teleology not useful in studies of behavior? What is Morgan's canon?
3. In general, how does behavior in higher organisms differ from that in lower organisms? How is this reflected in the structure of the organism?
4. List the differences between behaviorists and ethologists.
5. Discuss the importance of nurture versus nature in animal behavior.

6. What is sociobiology? How does it seek to explain altruistic behavior in animal species?
7. What are taxes? How do they differ from reflexes?
8. Distinguish between instinctive and learned behavior.
9. Discuss the role of innate releasing mechanisms in instinctive behavior. How do the concepts of drives and releasers explain displacement activity?
10. Give examples of how instincts are "learned" by human beings or by herring gulls.
11. Do you believe humans are innately aggressive? Defend your position.
12. How is habituation adaptive?
13. Distinguish between classical and operant conditioning.
14. Of what importance is conditioning in B. F. Skinner's view of the world?
15. What is imprinting? Why is it so important in water birds?
16. Discuss the physical basis of learning in vertebrates.
17. How is STM related to LTM? What is the engram, and how does it relate to memory?
18. Discuss the adaptive value of forgetting to human life.
19. Describe social behavior in honeybees.
20. What is the role of territoriality in social animals? How is it related to the concept of social dominance? Of what major importance are appeasement and ritualization in social animals?

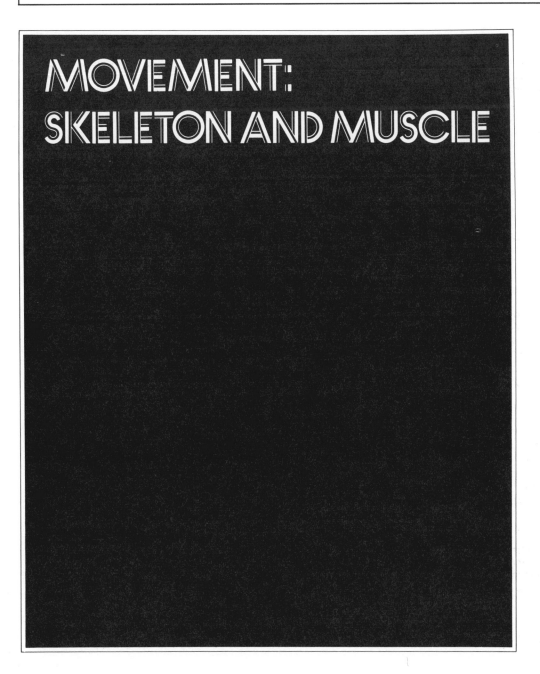

MOVEMENT: SKELETON AND MUSCLE

20-1 THE SKELETON: TOWER OF SUPPORT AND PROTECTION

The Eiffel tower, built by Gustave Eiffel for the Paris exhibition of 1889, was designed to show the potentials of steel as a structural material. Today, this symbol of Paris remains an outstanding example of the strength, flexibility, and endurance of steel and a monument to nineteenth-century engineering skills. The most enduring monument to our own existence—itself a remarkable engineering feat—is the skeleton, for long after the body is dead and the soft tissues are destroyed, this bony tower of strength remains (Figure 20–1). The skeleton consists of 206 individual bones (Table 20–1), some hinged and others fused to one another, and it provides the soft, weak flesh with protection. Together with the muscles that are attached to it, it enables us to stand erect, move, jump, and run. The bony skeleton functions in other ways that often go unnoticed: the bone marrow is the site of formation of blood cells, and the matrix of the bone itself acts as a storage depot for mineral salts such as calcium and phosphorus.

The main axis of the human body is perpendicular to the ground, and the support for this erect position consists of a jointed, flexible column of bones, the **backbone** or **spine.** The individual bones are called **vertebrae** (sing. *vertebra*—joint; L.). This is, of course, the reason why animals with backbones are referred to as **vertebrates.** From the vertebrae of the thorax arise a series of bony struts, the **ribs;** they curve around to the front of the body in barrel-stave fashion and join at the sternum, or breastbone. The rib cage is designed much like a packing crate and shields the heart, lungs, liver, and spleen against damage. Perched atop the backbone is a large bony helmet, the **skull,** with jaws, a prominent chin, and a variety of teeth. The skull and vertebrae enclose and

FIGURE 20-1 A thoughtful skeleton, as described in
a work by Andreas Vesalius in 1543. (Courtesy of
The New York Academy of Medicine.)

TABLE 20-1 The bones of the human skeleton

	No. of bones
Axial skeleton	80
Skull	29
Cranium	8
Face	14
Hyoid	1
Ossicles (ear bones)	6
Malleus	2
Incus	2
Stapes	2
Vertebral column	26
Cervical vertebrae	7 bones
Thoracic vertebrae	12
Lumbar vertebrae	5
Sacrum	1 (5 fused)
Coccyx	1 (3–5 fused)
Thorax	25
Sternum	1
Ribs	24
Appendicular skeleton	126
Shoulder girdle	4
Clavicle	2
Scapula	2
Upper Extremity	60
Humerus	2
Ulna	2
Radius	2
Carpals	16
Metacarpals	10
Phalanges	28
Pelvic girdle	2
Os coxae	2
Lower extremity	60
Femur	2
Fibula	2
Tibia	2
Patella	2
Tarsus	14
Metatarsus	10
Phalanges	28
Total	206

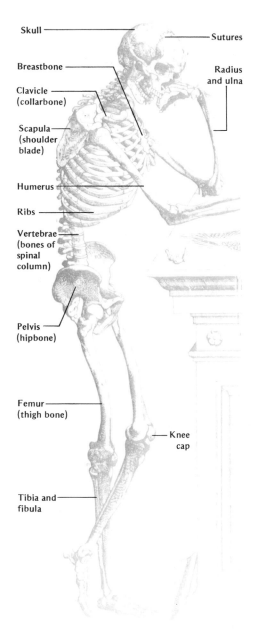

Skull — Sutures
Breastbone — Radius and ulna
Clavicle (collarbone)
Scapula (shoulder blade)
Humerus
Ribs
Vertebrae (bones of spinal column)
Pelvis (hipbone)
Femur (thigh bone)
Knee cap
Tibia and fibula

protect the vulnerable brain and spinal cord
and together with the ribs are known as the
axial skeleton.

When man is upright, almost his entire
weight is carried by the bones of the **pelvis;**
the basketlike pelvis acts as an anchor point
for the legs, supports the soft internal organs
of the body, and marks the lower limit of the
body's trunk or central area. The pelvis is
attached to the lower end of the spine, the
shoulder (pectoral) girdle to the opposite end
below the skull. The shoulder girdle marks the
upper limit of the body's trunk, and from it
stem the bones of the arms. The bones of the
arms and legs, together with the bony pelvic
and pectoral girdles that attach them to the
axial skeleton, are known as the **appendicular
skeleton.**

Bone and cartilage:
living reinforced concrete

Bone often appears rather passive in character
because of its low metabolic rate and great
mineral content, and because the bones we are
used to seeing (such as those from a rib roast
of beef, the carcass of a chicken, fossils or
skeletons in a museum) are dead, dry bones.
However, anyone who has witnessed the
growth of a child or seen how a broken bone
mends itself recognizes the dynamic character
of this tissue.

Bone is an organic cement produced by
specialized living cells (Figure 20–2), **osteocytes** (*osteon*—bone; Gk.). It consists of organic
materials (35%) and mineral salts (65%). The
organic portion consists chiefly of a protein
called **bone collagen,** a fibrous substance
closely resembling the collagen of other connective tissues; this material yields gelatin when
bones are boiled. Between these collagenous
fibers is a gelatinous substance composed pri-

FIGURE 20-2 An osteocyte, as revealed by the electron microscope. (Courtesy of K. R. Porter.)

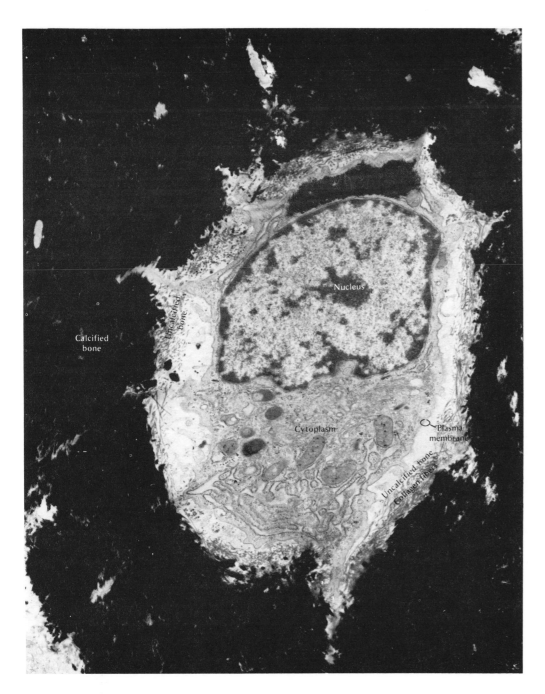

marily of the mucopolysaccharide chondroitin sulfate. The organic materials of bone give it a certain flexibility. The inorganic portion of bone, the material that contributes to its hardness and permanency after death, is a complex crystalline array of calcium salts, calcium phosphate and calcium carbonate. In its organization bone is like steel-reinforced concrete. The protein fibers are analogous to the flexible steel rods, and the calcium salts are like the concrete itself. The flexible elements prevent breakage when tension is applied, and the rigid mineral matrix resists crushing when pressure is applied.

To appreciate the organization of bone, it is necessary to look at it under the microscope. A thin slice of compact bone[1] (such as that of the upper arm) magnified a few hundred times shows a series of concentric layers of intercellular material (**matrix**) and bone cells (Figure 20–3). These are arranged like thousands of miniature solar systems (called **Haversian canal systems**). The hub, or center, of each system contains blood vessels and nerves, and the ''planets'' encircling the hub consist of layers of spiderlike osteocytes separated from each other by a matrix of collagen and calcium salts (the ''outer space'' of the bone). The osteocytes are connected by tiny cytoplasmic canals that permit substances to pass from one cell to another and to and from the blood vessels that lie in the center of the system. In this manner

---

[1] Bone, as seen with the unaided eye, is often described as being either spongy or compact. The spongy type consists of bony crossbars of various thickness and shape (Figure 20–3); such an arrangement gives rigidity and strength. Compact bone appears to consist of a continuous hard mass, and spaces can be seen in it only with the aid of a microscope. Spongy and compact bone are merely different arrangements of the same elements and do not differ chemically.

FIGURE 20-4 The structure of glassy-type cartilage.

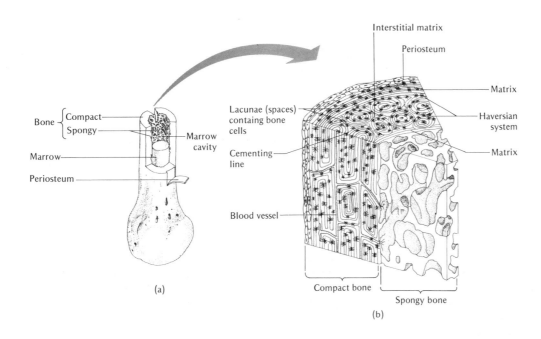

(a)

(b)

Bone { Compact / Spongy }
Marrow
Periosteum
Marrow cavity

Lacunae (spaces) containing bone cells
Cementing line
Blood vessel
Compact bone
Spongy bone

Interstitial matrix
Periosteum
Matrix
Haversian system
Matrix

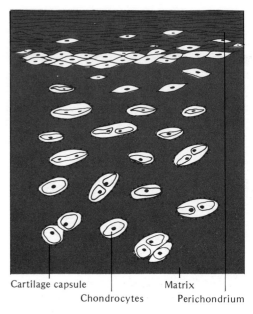

Cartilage capsule
Chondrocytes
Matrix
Perichondrium

the living osteocytes are nourished and their waste products are removed.

All bones are covered with a fibrous connective-tissue sheath called the **periosteum.**

Cartilage, or gristle, is the more flexible structural material of the skeleton, and it contains no mineral salts (Figure 20–4). It is a secretion of specialized cells called **chondrocytes** (*chondros*—cartilage; Gk.). Since cartilage lacks blood vessels, the living cells are nourished by diffusion of materials from the surroundings. Cartilage consists primarily of protein and chondroitin sulfate. Chondroitin sulfate itself is gelatinous, but depending on the nature of its contained fibers, it can be more or less rigid. When chondroitin sulfate is combined with fibers of collagen, it is rigid; this kind of cartilage forms the cushions of the discs between the vertebrae. When combined with elastin, the cartilage is flexible like that found in the outer ear. When there is little fiber content, the cartilage is clear and glassy in appearance; this **hyaline cartilage** is found in the windpipe, the ends of long bones, and the embryonic skeleton. The cartilaginous surfaces are enclosed by a sheath of connective tissue called the **perichondrium,** the source of all new cartilage cells.

Bone growth: chronometer of the skeleton

"The skeleton punches a time clock from birth to death," Dr. W. M. Krogman has said. The sex, approximate age (Figure 20–5), race, and medical history of a deceased individual are accurately recorded in his bones and can be determined from them. For example, it is not unusual for a criminologist, in Holmesian fashion, to determine the sex of a decomposed murder victim from the size of the ridges above the eyes or the shape of the pelvic bones. Indeed, bone is so durable and its developmental patterns so well defined that it provides some of the major clues in unraveling human prehistory. From the structure of fossilized long bones, the size of the skull, and the extent of fusion of the sutures of the brain case, physical anthropologists can often deduce not only the form of members of ancient races, but the kind of work they did, the physical ailments they endured, and the age at which death occurred.

The skeleton of the young embryo initially consists of protein fibers of collagen secreted by the primitive cells called **fibroblasts** (literally

FIGURE 20-5 The age of bone deposition.

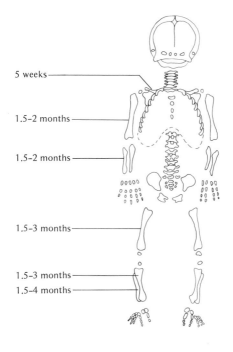

5 weeks

1.5–2 months

1.5–2 months

1.5–3 months

1.5–3 months
1.5–4 months

Fetus

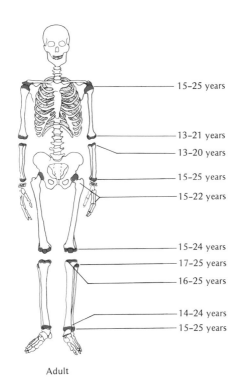

15–25 years

13–21 years
13–20 years

15–25 years

15–22 years

15–24 years
17–25 years

16–25 years

14–24 years
15–25 years

Adult

"fiber makers"). Later, depending on the location in the body, cartilage is deposited between these fibers so that much of the embryo's skeleton consists of fibrous membranes and hyaline (clear) cartilage. By the eighth week, bone deposition, called **ossification,** has begun.

When we examine the growth of bones, we find that no matter what their origin, all bones begin to ossify in the center (Figure 20–6); the process can be most clearly seen in the long bones of the limbs. The bone grows in length in a region just below the tip of the bone and above the shaft itself, called the **growth zone.** The addition of bony material at this region causes the bone to lengthen, and in effect the tip of the bone grows further and further away from the middle of the shaft. This is a practical method of growth because it does not interfere

with the articulation of the head of the bone with the bone next to it. Gradually, the cartilaginous shaft of the bone is replaced by a bony matrix. Increase in girth occurs by additional layering of bone on the outer surface of the shaft. As the bone increases in thickness, bony material is lost from the central part of the shaft, forming an internal cavity (the **marrow cavity**) that gets larger as the bone diameter increases (Figure 20–6b). The removal of material in the marrow cavity results from the activity of special cells, called **osteoclasts.** Osteoclasts are giant cells containing many nuclei; they resorb bone by secreting enzymes that digest the protein matrix and split the bone salts so that they are easily absorbed by the surrounding fluid. The marrow cavity, once occupied by bone and then by osteoclasts,

becomes filled with cells that give rise to red and white cells of the blood. Their hollow structure admirably suits the bones of the limbs to their function. Built on sound engineering principles, they are extremely strong for the amount of material involved in their form. If they were solid, they would break more easily and would be extremely heavy. The hollow units used to scaffold a building under repair or construction employ the same principles—lightness and economy of material combined with great strength.

Just prior to maturity, a long bone consists of two strips of tissue, the growth zones, between the tips of the bone and the shaft. When the growth zones become ossified at about 25 years of age, the bone's length is complete. No further increase in length can occur, but new bone growth can be instituted if there is a break in a bone. When such injury occurs, bone cells in the region of the break become active, proliferate, and secrete large quantities of new bone within a few weeks. Bone healing occurs most efficiently in young people and is optimal when the fractured ends are immobilized, brought in close contact, and properly aligned by a splint or cast or by inserting a bone pin.

The initial shape of a bone during its formation is determined genetically, but mechanical stresses, environmental factors (dietary calcium, "vitamin" D), and muscle tension also influence its form. In large part the architecture of a bone reflects the mechanical stress and strain put upon it, and there are struts, buttresses, crossbars, and so on that act to prevent the structure from collapsing. The marvelous engineering involved in bone structure is a product of and regulated by the strains it receives, since the greater the stress put on the bone the greater the activity of the bone cells. Why this is so is not understood. However, this is the reason why bones that are subjected to continuous and excessive loads ordinarily grow

FIGURE 20-6 The growth of bones. (a) Growth of a long bone from embryo to adult. (b) Diagram showing the surfaces on which bone is deposited and resorbed during growth to account for the remodeling of the ends of long bones with flared extremities. (Modified from *Histology* by Arthur W. Ham. J. P. Lippincott Company, 1969.)

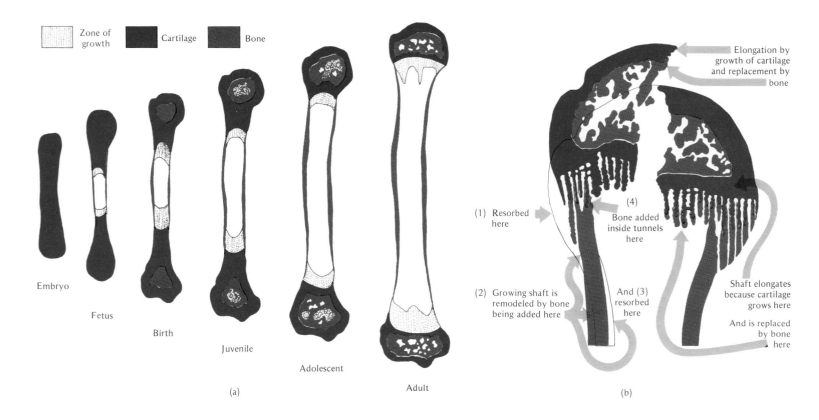

Zone of growth   Cartilage   Bone

Embryo

Fetus

Birth

Juvenile

Adolescent

Adult

(a)

Elongation by growth of cartilage and replacement by bone

(1) Resorbed here

(4) Bone added inside tunnels here

(2) Growing shaft is remodeled by bone being added here

And (3) resorbed here

Shaft elongates because cartilage grows here

And is replaced by bone here

(b)

thick and strong, whereas bones not used for prolonged periods (such as those in a plaster cast or those of a person damaged by poliomyelitis) waste away.

20-2 MOTION: MUSCLE

Poke your dog or cat; touch a frog, fish, lizard, or bird; swat a fly; nudge a grasshopper or an earthworm. They all respond (most of the time) in the same way: they move. Do the same things to a grapefruit, a pinecone, a rose petal, or a kernel of corn, and they remain as they were before. Animals, ourselves included, move in a purposeful fashion, and basic to such activity is the contraction of muscle.

Muscles or flesh are a conspicuous part of the human body, as they are in almost all animals, for approximately one half of our body weight is represented by muscle tissue. Muscles move us from place to place (locomotion), but they are also involved in the less obvious movements of the organs of the body. In the human body there are over 600 different muscles, and when we walk, talk, stand erect, lie down, write, eat, run, dance, creep, crawl, play the piano, and perform a variety of other physical activities, we set these muscles into

operation. In addition to playing a role in motion, muscles give the body its characteristic form. The scaffolding of the skeleton is a prime factor in determining body shape, but the muscles that drape the skeleton produce the contours we regard as graceful and beautiful (Figure 20-7).

The kinds of muscle

Muscles are usually described as **skeletal, smooth,** or **cardiac,** depending upon how they look under the ordinary light microscope (Figure 20-8).

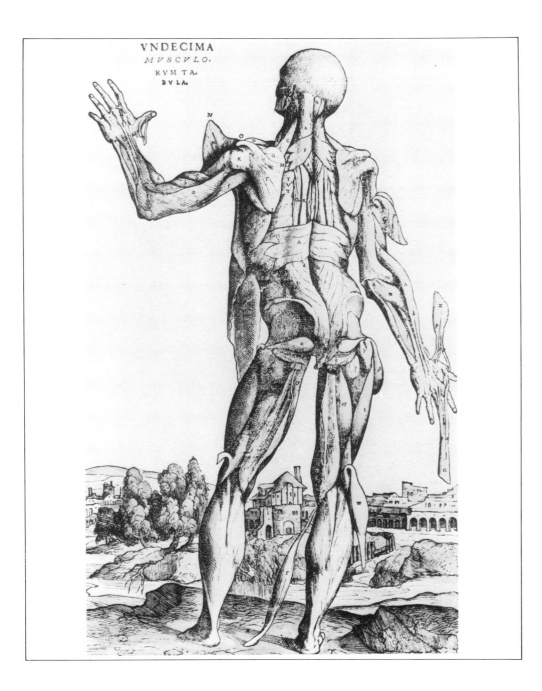

FIGURE 20-7 The muscles of the dissected human body as seen by the Flemish anatomist Vesalius in 1543. (Courtesy of The New York Academy of Medicine.)

VNDECIMA
MVSCVLO.
RVM TA.
BVLA.

Skeletal muscle. Skeletal muscle appears to have cross banding or alternating light and dark stripes running at right angles to the length of the muscle and is sometimes referred to as **striped** or **striated muscle** (Figure 20–8a). It is called skeletal muscle because it is the kind that is commonly attached to the bones of the skeleton. Skeletal muscle consists of bundles of muscle cells; each cell is multinucleate, and sometimes such a cell is called a **syncytium** (*syn*—together, *kytos*—hollow vessel; Gk.). Skeletal muscles make up most of the fleshy parts of the body. They are supplied with nerves that are under voluntary control, and so contraction of various parts of the body can be consciously willed. Skeletal muscles move the limbs, act quickly, fatigue easily, and cannot remain contracted for very long periods of time. A wink, an utterance, the movements of the limbs in walking, dancing, and breathing are all produced by the action of skeletal muscles.

As flesh eaters we recognize two kinds of skeletal muscle; these are the types we commonly call dark meat and light meat, or **red muscle** and **white muscle.** They differ in their color because of the presence of the red pigment **myoglobin,** which is richer in red muscle; it becomes brown upon heating, producing dark meat. Myoglobin is a protein similar to hemoglobin and can reversibly bind molecular oxygen. When oxygen is plentiful, the muscle binds and stores it until there is a demand for it.

Smooth (visceral) muscle. The muscles of the internal organs (viscera), such as those of the gut, uterus, and blood vessels, are distinct from skeletal muscle in that they have no striations. Such muscles are called smooth muscles, and each is made up of groups or sheets of well-defined uninucleate cells (Figure 20–8b). Smooth muscles are not attached to bones; they act slowly, do not fatigue easily, may remain contracted for long periods of time, and are

FIGURE 20-8 The three kinds of muscle of the human body. (a) Skeletal muscle. (b) Smooth muscle. (c) Cardiac muscle.

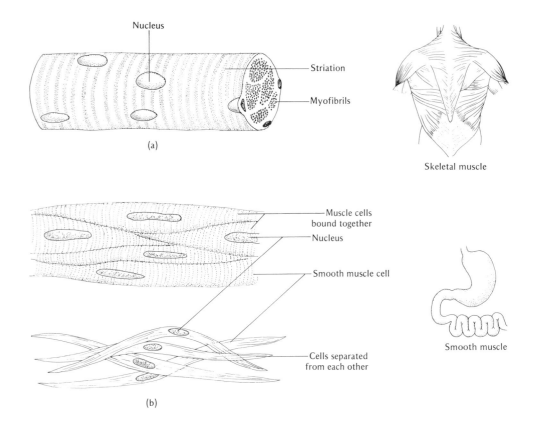

Nucleus

Striation

Myofibrils

(a)

Muscle cells bound together

Nucleus

Smooth muscle cell

Cells separated from each other

(b)

Nucleus

Disc

Striation

Myofibrils

(c)

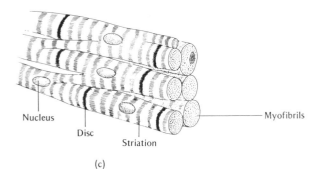

Skeletal muscle

Smooth muscle

Cardiac muscle

not under conscious control. Smooth muscles are also called **visceral** or **involuntary muscles.** When food is moved along the length of the gut, when the uterus contracts to expel the fetus during delivery, or when blood vessels constrict to reduce the flow of blood or elevate the blood pressure, smooth muscles are in operation.

Cardiac muscle. Heart, or cardiac, muscle (*kardia*—heart; Gk.) is located only in the heart. In some respects it embodies the characteristics of both smooth and striated muscles. Cardiac muscle is striated, but it is not under voluntary control (Figure 20–8c). It is not structurally a syncytium (although it functions as one), but it does contract rapidly and is virtually indefatigable—well-adapted for its lifetime job of pumping the blood to every part of the body. Cardiac muscle has an inherent rhythm of 72 beats a minute, and nervous inputs act to modify this, speeding it up or slowing it down (Chapter 12). The metabolism of cardiac muscle is similar to that of red skeletal muscle, but endurance supersedes strength and speed. Cardiac muscle is highly aerobic, and if deprived of oxygen for as little as 30 seconds, it ceases to function.

Characteristics of muscle

All muscle—smooth, skeletal, and cardiac—has the inherent ability to **contract,** a unique characteristic distinguishing it from any other body tissue. Muscles are so organized that when they contract they reduce the distance between the parts they connect or the space they surround. For example, contraction of the muscles that connect two bones brings the attachment points closer together, causing the bone to move. When cardiac muscle contracts, it reduces the space in the chambers of the heart, and this push on the incompressible fluid contained within causes blood to move from the heart into the blood vessels. Smooth muscles are wrapped around the arteries and the intestines like a cuff, so that contraction causes the diameter of these tubes to narrow. Muscles have two properties in addition to contraction: **extensibility** (ability to be stretched) and **elasticity** (ability to return to original length after having been stretched); together, the triad of contractility, extensibility, and elasticity produces a mechanical device capable of performing work, although it is only during contraction that a muscle exerts force. The three kinds of muscles may appear dissimilar, but in the manner by which they perform work, they are alike.

Putting muscles to work

In order for a muscle to produce movement of an object such as an arm or a leg, the muscle must be capable of exerting its force on a movable object; that is, it must have something to pull against. In backboned animals like ourselves muscles are attached to the bones of the skeleton by inelastic **tendons.** The bones, in turn, are connected to each other by **joints,** and each muscle is attached so that it bridges a joint; when a muscle contracts, the skeletal components move. The skeleton thus forms a series of jointed bars and levers with muscles attached—the basic machinery for the movements of limbs, torso, head, and so on.

It is important to realize that muscles are capable of pulling but not of pushing. Reversal of the direction in which a joint is bent must be accomplished by moving a different set of muscles. For this reason, the muscles that move the parts of our body are almost invariably arranged in pairs—one to produce movement in one direction, the other to produce movement in the opposite direction. These paired muscles cooperate to produce body move-

ment since when one is contracted the other must relax to some extent; since they have opposite actions, they are called **antagonists.** By virtue of the ways in which they are arranged, the muscles of the body can move the body parts in innumerable ways and with great precision. For example, the muscles of the upper arm are arranged in antagonistic pairs (Figure 20–9a). One end of the muscle on the front part of the upper arm (biceps) is fixed to the shoulder bone and upper arm bone; these bones do not move when this muscle contracts (Figure 20–9b). The other end of the biceps muscle is attached to one of the bones in the forearm, and it is this bone that moves when the muscle contracts. The muscle on the back part of the upper arm (triceps) is also attached in the shoulder and forearm region. Bend your elbow and with your other hand feel the contraction of the belly (enlarged region) of your biceps muscle; at the same time run your hand around to the triceps and feel its relaxed condition. (There may be some tenseness, but it certainly does not show a contractile force equal to that of the biceps.) Now extend your forearm, and feel the reversal of contraction and relaxation; the biceps relaxes and the triceps contracts. Now bend your arm halfway and contract both biceps and triceps. The arm becomes rigid, because both sets of muscles contract at the same time.

The biceps is a **flexor** and the triceps an **extensor** muscle because the biceps flexes the elbow and the triceps extends it. Similar groups of antagonistic flexors and extensors are found at the wrist, ankle, knee, and other joints. Other antagonistic muscles are **adductors** and **abductors,** which move parts of the body toward or away from the center line of the body; **levators** and **depressors,** which raise and lower body parts such as raising and lowering the lower jaw in eating or speaking; **sphincters**

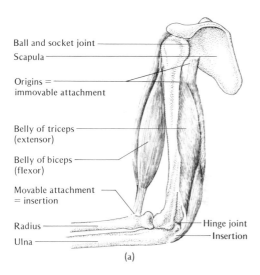

FIGURE 20-9 (a) The arrangement of antagonistic muscles in the upper arm. (b) Flexion and extension of the arm.

Ball and socket joint
Scapula
Origins = immovable attachment
Belly of triceps (extensor)
Belly of biceps (flexor)
Movable attachment = insertion
Radius
Ulna
Hinge joint
Insertion

(a)

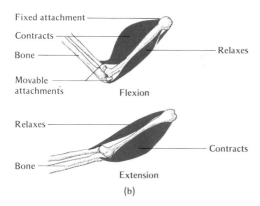

Fixed attachment
Contracts
Bone
Movable attachments
Relaxes
Flexion

Relaxes
Bone
Contracts
Extension

(b)

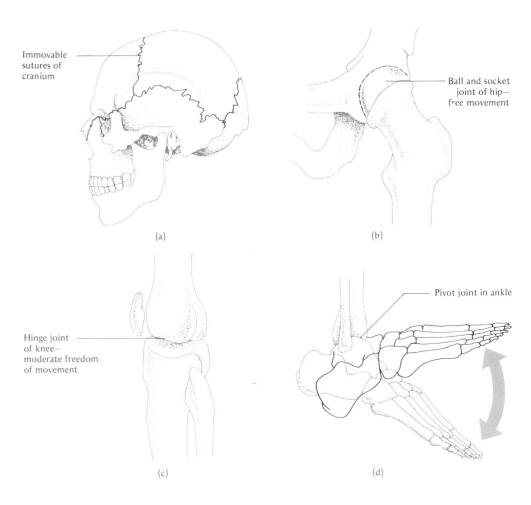

Immovable sutures of cranium

(a)

Ball and socket joint of hip—free movement

(b)

Hinge joint of knee—moderate freedom of movement

(c)

Pivot joint in ankle

(d)

and **dilators,** which decrease or enlarge openings such as the anus and the mouth; **pronators** and **supinators,** which rotate downward-backward and upward-forward, respectively, such as turning the palm down or up.

Joints are the points of junction between bones (Figure 20-10). Some, such as those between the bony plates of the skull, are immovable (Figure 20-10a) because they have become fused. The joints concerned with body

movements are flexible pivot points and vary in form according to their function: a ball-and-socket joint (Figure 20-10b), allowing free movement in several directions, is found in the hip and shoulder; a hinge joint, permitting movement in a single plane, exists in the elbow and knee (Figure 20-10c); and a pivot joint, showing an intermediate range of movement, is found in the wrist and ankle (Figure 20-10d). Whenever two movable bones meet at a joint,

their ends do not contact one another directly; they are covered with a smooth, slippery cap of cartilage (recall the smooth, shiny ends of a drumstick), and the joint (articular) surfaces are encased in a capsule, the **bursa,** made of ligaments and filled with a liquid called **synovial fluid** (Figure 20–11). The synovial fluid is secreted continuously by the membrane lining the cavity (it is a kind of weeping bearing), and its coefficient of friction (a measure of its frictional resistance) decreases with increased movement, a highly advantageous property for a material at a joint surface. As we age, the synovial fluid is not secreted as quickly as needed, the cartilaginous surfaces of the bone ends tend to become ossified, and as a result there is increased stiffness of joints and difficulty in movement. Chronic inflammation of the joints is called **bursitis, arthritis,** or **rheumatism** (Figure 20–11).

Tendons—tough, flexible, and fibrous—are the long cables of connective tissue that anchor muscle to bone. Tendons permit a muscle to act at some distance from the bone it moves. For example, the muscles that move the fingers are located in the forearm, but these muscles are attached by long strands of tendons to the finger bones (Figure 20–12a). This can easily be verified by moving your fingers vigorously and noting the rippling of muscles in the upper part of the forearm; you can also feel the muscles operating with the fingers of your other hand. This is the reason why the muscles in the forearm fatigue and ache when you exercise your fingers for long periods of time, as in typing, writing, or playing the piano. The anatomical arrangement of tendons and muscles in the fingers and forearm permits the fingers to perform delicate and precise movements. If the finger muscles were directly fixed to the finger bones, the hand would be clumsy, indelicate, and muscle-bound (Figure 20–12b).

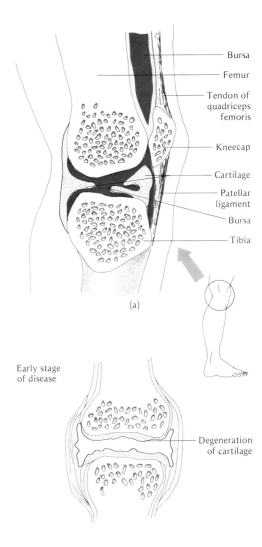

Bursa
Femur
Tendon of quadriceps femoris
Kneecap
Cartilage
Patellar ligament
Bursa
Tibia

(a)

Early stage of disease

Degeneration of cartilage

Late stage

Hardened bone
Bony outgrowth
Cartilage particles
Loss of cartilage

(b)

FIGURE 20-11 (left) (a) Detailed side view of a joint (the knee). (b) Osteoarthritis. In the early stages of the disease the cartilage at the bone surface begins to degenerate. In time, the missing cartilage is replaced by a hardened bony outgrowth that tends to stiffen the joint resulting in considerable pain and difficulty in movement.

The molecular machinery of muscle contraction

When you see an animal move, you are impressed with its agility and grace. Perhaps you also wonder: How do its muscles work to produce such activity? If you have ever pondered such a question, you are not alone, for one of the most perplexing questions concerning muscle has been: How does a muscle convert the chemical energy of fuel into mechanical work? Alternatively, the question could be phrased: What are the mechanisms by which muscles contract? The molecular machinery involved in muscle contraction is probably the same in all types of muscles. What follows here pertains particularly to striated muscles only because, up to the present time, these have been the best studied.

The structure of a skeletal muscle is familiar to all of us, for this is the meat we eat. Let us consider a common example, the thigh (drumstick) of a chicken or a turkey. When the bone is stripped from the drumstick, the bulk of the leg, the fleshy part, consists almost entirely of muscle (Figure 20–13a). The muscles are grouped into bundles, and each is surrounded by a sheath, or wrapper, of connective tissue (Figure 20–13b and c). The connective-tissue sheaths permit the individual muscle bundles to function as discrete units and allow adjacent muscle bundles to slide easily over one another as they contract. The sheaths of connective tissue join each other, and near the ends of the bone they taper and are gathered to form the tendon. If meat is tough, it is usually due to a high proportion of connective tissue relative to the amount of muscle. With advancing age the amount of connective tissue increases, and this is probably the reason for the decreased tenderness of meat from an older animal. No doubt this is also a factor in decreased muscular strength with advancing age (Figure 20–14).

FIGURE 20-12 (a) Tendons in the hand permit delicacy in the fingers, because the muscles that move the fingers are located in the forearm. (b) The bulky appearance of the hand if the muscles were directly attached to the bones in the fingers. (Modified from *Biology, Its Principles and Implications*, 2nd Ed., by Garrett Hardin. W. H. Freeman and Co., copyright © 1966.)

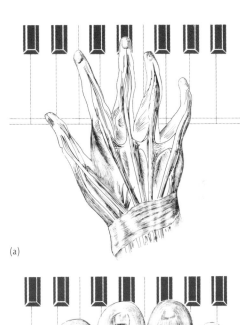

(a)

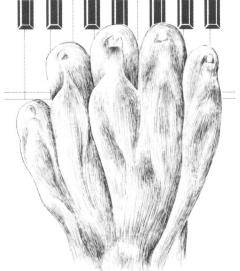

(b)

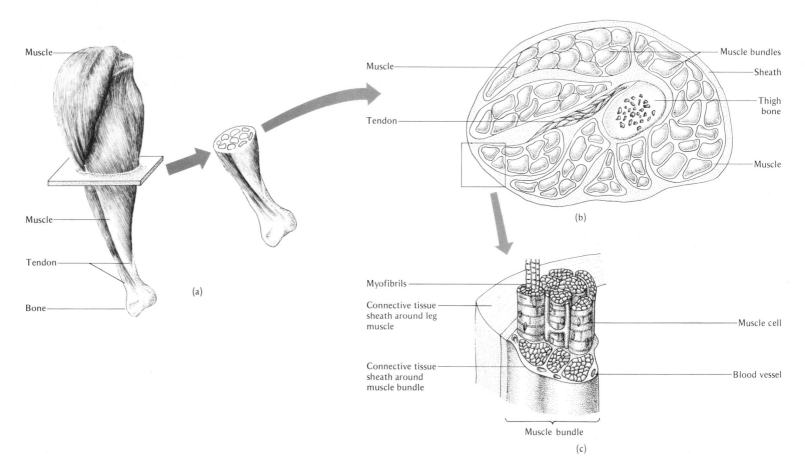

Microscopic examination shows a muscle bundle to consist of muscle cells that run parallel to each other (Figure 20–13c). Each muscle cell is an elongated, multinucleated cylinder of varying length surrounded by a thin, transparent plasma membrane called the **sarcolemma** (*sarcos*—flesh, *lemma*—rind; Gk.) and containing a fluid protoplasm, the **sarcoplasm.**[2] The sarcoplasm contains numerous

[2] Sarco- is the prefix used in words describing organelles of the muscle cell, because flesh is essentially all muscle.

myofibrils (*myo*—muscle; L.) that run the length of the cell and lie parallel to each other. Each myofibril shows alternating light and dark cross bands. Because each myofibril is aligned in register, that is, precisely side-by-side, the muscle cell appears to be striated or cross-striped. Thus, the components that make up the structure of a skeletal muscle can be written as follows:

Myofibrils ⟶ muscle cells ⟶

muscle bundles ⟶ muscle

If a fresh striated muscle is ground up and extracted, the dry chemical components derived are almost entirely protein. (In fresh muscle, 20% is protein and 80% is water.) Approximately 90% of the muscle protein is represented by two specific kinds of proteins: **actin** and **myosin.** When actin and myosin are placed in a watery solution, they spontaneously combine to form a complex called **actomyosin.** If this actomyosin solution is taken up in a medicine dropper and squirted into another salt solution, it will yield jellylike threads of

FIGURE 20-14 Changes in the proportion of muscle and connective tissue in human muscle at various ages. (Modified from *Biology, Its Principles and Implications,* 2nd Ed., by Garrett Hardin. W. H. Freeman and Co., copyright © 1966.)

FIGURE 20-15 (a) Formation of actomyosin thread by squirting a solution of actin and myosin into a salt solution. (b) When ATP is added, the thread contracts. (Modified from *Cell Physiology* by Arthur C. Giese. W. B. Saunders Co., 1962.)

24 years

51 years

Muscle — 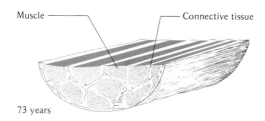 — Connective tissue

73 years

81 years

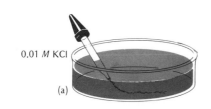

0.01 *M* KCl

(a)

0.01 *M* KCl + ATP

(b)

actomyosin (Figure 20–15a). Such artificial threads contract when ATP is added (Figure 20–15b), but neither myosin nor actin alone will contract in the presence of ATP. It appears that a solution of actin, myosin, and ATP contains all the essential components of the contractile system of muscle, and the actomyosin itself can liberate energy from ATP to provide the driving force for its own contractions.

Careful studies with the electron microscope coupled with the technique of x-ray diffraction suggest that the banding pattern of the myofibril is due to the arrangement of actin and myosin within it. The myofibrils show a pattern of light regions (**I bands**) and dark regions (**A bands**) (Figure 20–16). In the middle of the dark A band is a light stripe called the **H zone,** and in the middle of the light I band is a thin, dark line, the **Z line.** When the myofibril contracts, the I bands and the H zones become narrower, a dense zone appears in the A band, but the A band remains the same width; in effect, the A bands are brought closer together.

To explain such observations H. E. Huxley

and J. Hanson proposed a **sliding-filament model** of the contraction of muscle. According to them, the two types of filaments, thick myosin and thin actin, are interdigitated (Figure 20–17). Each dark A band represents the length of one region of thick myosin filaments; it is darkest at its borders, where the myosin and the actin filaments overlap, and lighter in the middle (H zone), where only the myosin filaments are present. The light I band represents a region of actin filaments, and the Z line is a membranous structure that orients the actin filaments. Movement occurs in the following way. The myosin filaments have fingerlike projections at regular intervals along their length, and these form cross bridges with the actin filaments. The cross bridge is movable, oscillating forward and backward, and the fingerlike myosin projections fit into specialized receptor sites on the actin filament. Attached to the myosin filament is ATP, and when calcium ions appear in the fluids of the muscle fiber (owing to muscle stimulation), the myosin projections form activated myosin, which acts as an enzyme **(ATPase)** and splits the ATP. This provides energy so that the actin filament slides along the myosin filament by a ratchetlike mechanism; the actin filament first locks with one projection, then with one further along, then with the next, and so on (Figure 20–18). According to Huxley, each ratchetlike movement of the muscle filaments—that is, the formation of each new bridge between myosin and actin—requires the energy of one molecule of ATP. When the muscle relaxes, it is presumed that there is no longer a splitting of ATP (calcium ions are removed and ATPase activity is diminished), the myosin bridges no longer combine with the actin filaments, and the muscle spontaneously returns to its original uncontracted state. (The pull of the surrounding elastic connective tissue may also play a role in relaxation.)

FIGURE 20–16 Electron micrograph of striated muscle. (Courtesy of K. R. Porter.)

FIGURE 20–17 The Huxley-Hanson model of muscle contraction. The muscle is uncontracted in (a) and fully contracted in (d).

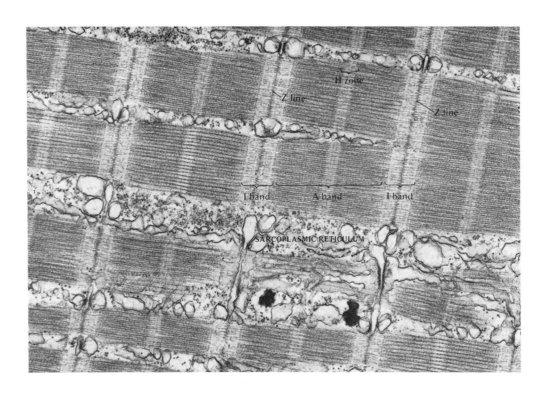

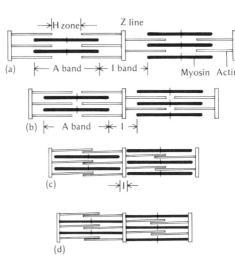

The sliding-filament theory of Huxley and Hanson clearly explains the changes that occur in the banding pattern as the muscle contracts and relaxes. The rigid condition of muscles after death, rigor mortis, is due to the formation of permanent links between actin and myosin in the absence of ATP. In rigor mortis the muscle cell no longer has a supply of ATP, its energy stores are depleted, and the actin-myosin bridges remain permanently fixed as if they had been deprived of their source of lubrication.

Characteristics of muscle contractions

The events taking place during contraction are similar in all muscle cells. Whether you use the muscles in your arm to turn the page of a book or to lift a weight of 35 lb, the molecular mechanisms are identical. However, the force exerted by our arm muscles is not the same in each case: the degree of contractile force is directly related to the amount of work to be performed. Were this not so, muscular contraction would be inefficient and lack precision. How does a muscle regulate its contractile force?

Muscles are supplied with nerves, and each terminal branch of a nerve cell (axon) ends in a special rootlike structure called the **motor end plate** (Figure 20–19a). The highly branched ends of the nerve cell lie in a series of gutters, or troughs, formed by folds in the plasma membrane of the muscle cell (the sarcolemma).

The gap between the ends of the nerve cell and the muscle membrane is a special kind of synapse. As in the case of neuron-neuron interactions, each of the terminal tips of the nerve cell contains small synaptic vesicles containing a chemical transmitter substance (acetylcholine [ACh] is one of the most important). When the nerve cell is appropriately stimulated, the synaptic vesicles discharge their contents into the end-plate gutter, and transmitter substance diffuses across the space between nerve cell and muscle membrane (Figure 20–19b). The transmitter substance acts on the sarcolemma of the adjacent muscle cell and alters its permeability to positively charged ions (particularly sodium ions) present in the surrounding fluid. As the positive ions rush into the muscle cell, the polarity of the sarcolemma becomes first neutral and then reversed; this change in the electrical potential of the membrane at the motor end plate is called the **motor-end-plate potential.** If the magnitude of the motor-end-plate potential reaches a threshold level, the inrushing positive ions cause a

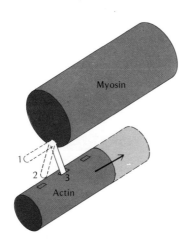

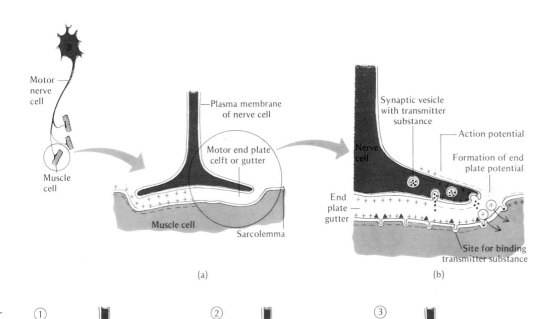

(a)   (b)

①   ②   ③

End plate   Action   Action
potential   potential   potential

(c)

brief decrease in the muscle's membrane (sarcolemma) potential, it becomes depolarized, and the inside of the muscle cell becomes slightly positive relative to the outside. The depolarization of the sarcolemma is propagated as a wave over the entire muscle, and this potential (an action potential) moves outward similar to the way in which waves move outward to the water's edge when a pebble is dropped into a pool (Figure 20–19c). The polarity of the sarcolemma is restored to its original condition just behind the wave of depolarization by the extrusion of positively charged ions (in this case potassium) from inside the muscle cell. As the action potential traverses the length of the muscle cell, it promotes the release of calcium ions from the **sarcoplasmic reticulum** (a modified endoplasmic reticulum that surrounds the packets of myofibrils like a bracelet, Figures 20–16 and 20–20). The calcium ions trigger activation of myosin so that it acts as an ATPase, ATP is split, and the muscle cell contracts as the myofibrils shorten. When the calcium ions are

resorbed (presumably they return to the confines of the sarcoplasmic reticulum) or are bound, there is no longer any activation of ATPase, ATP is no longer split, and the muscle lengthens as the myofibrils relax.

Two points of importance are (1) the motor-end-plate potential and the release of calcium from the sarcoplasmic reticulum are graded responses that depend on the strength of the stimulus, and (2) the action potential is an all-or-none phenomenon—it either is or is not produced—and it does not vary in amplitude with the strength of the stimulus.

What does all this electrochemistry have to do with explaining how a muscle can produce different forces according to its work load? The functional unit of the muscle is a nerve cell and the muscle cells that it innervates, together known as a **motor unit** (Figure 20–19). Usually the muscle cells of a particular motor unit do not lie next to one another, but are spread out over the entire belly of the muscle. In this way stimulation of a motor unit will cause a weak contraction over a broad area of the muscle rather than a single strong contraction at one spot. The number of nerve cells stimulated and

FIGURE 20-20 Diagrammatic reconstruction from electron micrographs of a portion of a muscle cell. The section shows the sarcoplasmic reticulum and its relation to the muscle cell and packets of myofibrils.

FIGURE 20-21 Production of tetanus and fatigue by repetitive threshold stimuli.

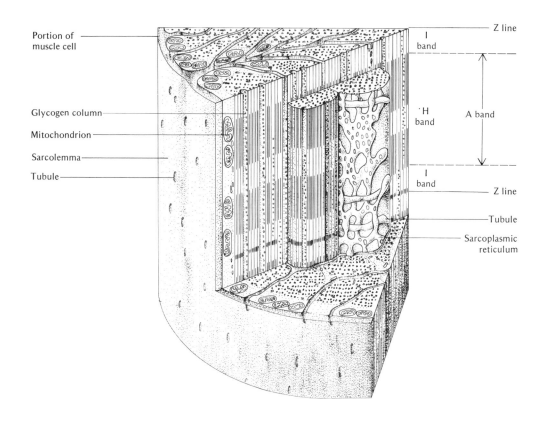

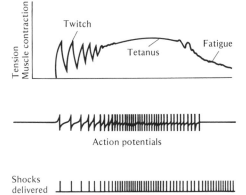

the frequency of the action potentials generated in these nerve cells depend on the amount of stimulation received—for example, the amount of pressure on nerve receptors in the skin. The greater the number of nerve cells carrying action potentials, the more muscle cells there will be contracting. The graded response of a muscle depends also upon how many muscle cells are involved in a particular motor unit as well as the threshold of that motor unit (some will be low, others will be high, and still others will be intermediate). Fine or delicate movements involve only a few muscle cells, and although such a system lacks power it is very precise. The eye muscles are an example of

this, where a single nerve runs to only five or six muscle cells. By contrast, the calf muscle has a single nerve cell innervating 1,900 muscle cells, and when these contract at the same time, the contraction is powerful.

The preceding discussions about the chemical and electrical changes that occur in a muscle during contraction would lead one to expect that muscles cannot respond to frequent and repeated stimuli, since there would not be enough time between stimuli for the muscle to return to its original polarized state. This is indeed the case; muscle cells, like nerve cells, have a short period of time immediately after one stimulus when they cannot generate an

action potential if a second stimulus is administered; this is the **resting,** or **refractory, period.** The refractory period in most muscles usually lasts about 0.02 second.

If stimuli are given with great rapidity (30–350 times a second) and the muscle is unable to relax between stimuli, the contracted state is sustained (Figure 20–21). This condition is called **tetanus** (not to be confused with the disease tetanus, or lockjaw, which produces muscle spasms; Box 20A). Although tetanic contractions do occur normally in our body, if they are continued over a long period of time, the muscle will begin to fatigue and its strength of contraction will diminish even under continued stimulation (Figure 20–21).

What is the reason for muscle fatigue? Why can we not exercise continually and feel no effects? When we use our muscles to exercise, to maintain posture, or to do any other motor activity, muscular contractions are performing work, heat is released, and energy from the breakdown of ATP is utilized. ATP, the driving force of muscle, is formed by the oxidation of fuel materials such as glycogen and glucose that are present in the muscle, with pyruvic acid as the initial end product (Chapter 4). If,

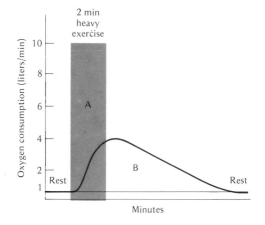

however, the intensity of exercise rises to an extent that blood cannot deliver oxygen in adequate amounts for the complete oxidation of glucose, the muscle contractions occur anaerobically, and lactic acid is formed from pyruvic acid. The lactic acid leaks out of the muscle and passes into the blood. If heavy exercise is continued, the lactic acid content of the blood rises sharply, and eventually so much is present that it no longer leaves the muscle but accumulates there. Excessive amounts of lactic acid depress metabolic activity in the muscle cells and cause decreased irritability. This interferes with the ability of the muscle cell to contract. Muscle fatigue and soreness (cramps) are primarily a result of the accumulation of lactic acid, and this forces cessation of the exercise.

Although the muscle is capable of working in the absence of an oxygen supply for a limited period of time (the exact time being related to the rapidity of lactic acid accumulation), it must eventually have oxygen for oxidizing lactic acid and resynthesizing glycogen and ATP. The amount of oxygen required for these tasks is called the **oxygen debt,** and it is specified as the quantity of oxygen used in excess

BOX 20A Your muscles: use 'em or lose 'em

Muscles become more efficient and capable of performing increased amounts of work as their work load increases, a situation quite unlike the machines we create out of "durable" substances such as aluminum and steel, which become worn with use. The more a muscle is used, the larger its size and strength become. The enlargement (hypertrophy) of muscle resulting from growth or exercise is due to an increase in the diameter of the muscle cells rather than an increase in the number of cells or muscle bundles; it is the number of myofibrils per cell that is increased. This change in girth increases the total force of the contractile process.

A muscle that is not continually used **atrophies;** that is, it shrinks in size and becomes weaker. In poliomyelitis the impaired function of the nerve fibers that transmit impulses to the muscle results in diminished stimulation of the muscle, which causes it to waste (atrophy). Muscles deprived of their nerve supply eventually degenerate; in 6 months to 2 years a muscle may be reduced to 25% of its normal size, and the muscle cells are replaced by noncontractile connective tissues. Often the weakening of muscle can be delayed by direct electrical or mechanical stimulation (massage) of the muscle. Muscle atrophy may also occur when there is an enforced period of inactivity, such as immobilization of a limb in a cast after an operation or during an illness that requires extended periods of bed rest; again skeletal-muscle atrophy can be minimized by massage or limited exercise.

We are born with essentially our full complement of skeletal muscle cells and lack the capacity to produce new ones. Accordingly regeneration of muscle after injury does not occur. However, the surviving cells can increase their size and strength by increasing the synthesis of myofibrils, so that an injured muscle can ultimately perform at its original level with fewer muscle cells. An individual muscle cell may be capable of repair, but any large lesion is replaced by connective tissue, which lacks the contractile properties of the original muscle tissue.

BOX 20B Paralytic poisons

Deep within the Amazon jungle a Jivaro hunter sits quietly behind a tree, blowgun and quiver of darts hanging at his side. Each dart has a needlelike tip that has been carefully coated with a brownish gum containing the substance curare. As a small wild pig emerges from the dense forest and enters the clearing, the hunter levels his blowgun, places a dart into its barrel, and with a silent puff of air the missile is sent toward the prey. The pig jumps as the dart pricks its skin, it lurches forward, sways, falls down paralyzed, and in less than 5 minutes has stopped breathing.

A farmer in the southern United States loads his spray truck with an organophosphate pesticide, and sets out to spray his crops. As he does so he thinks how his crops will flourish, because leaf damage from insects will be minimal. Suddenly, the wind shifts and he is covered with the pesticide. Coming in contact with the pesticide he reacts as would the insect pest: he becomes paralyzed, breathing movements cease, and death occurs shortly thereafter.

These two anecdotes illustrate the use of paralytic poisons to the benefit and the detriment of man. How do curare and organophosphate work?

Nerves coordinate our body movements by regulating the stimuli that reach our muscles. Since nerves produce transmitter substances that carry messages to the muscles, any substance affecting transmission of these messages at the motor end plate will determine whether a muscle contracts or not. The transmitter substance ACh is released by the synaptic vesicles of certain motor nerves and acts to depolarize the muscle membrane (Figure 20-23a). ACh persists for only a short period of time because it is broken down locally by the enzyme acetylcholinesterase, present in the motor-end-plate region. If ACh were to remain indefinitely, the muscle cells would transmit a continuous series of action potentials resulting in muscle spasms and ultimately in death. Certain tribes in South America take advantage of the ACh-cholinesterase system and capture prey or paralyze their enemies by shooting them with darts dipped in the poison curare. When curare enters a wound it is carried to the region of the

FIGURE 20-23 The mode of action of some paralytic poisons. (a) The normal neuromuscular junction. (b) The end plate affected by curare. (c) The end plate affected by organophosphates or by nerve gases.

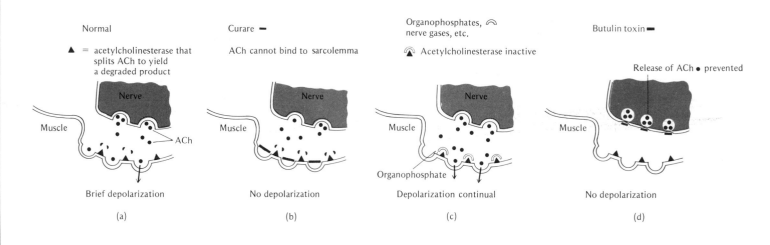

motor end plate, and there it binds to the sarcolemma, preventing ACh from reaching the muscle; thus the muscle cannot be stimulated (Figure 20-23b). Continued blockage of impulses across the motor end plate causes paralysis, and death results from asphyxiation since breathing movements are inhibited.

Neostigmine, physostigmine (eserine), diisopropyl fluorophosphate, and other organophosphates used as nerve gases and pesticides operate by inhibiting the action of acetylcholinesterase, and thus enhance transmission of messages across the motor end plate (Figure 20-23c). As a result these drugs permit the accumulation of ACh (since it is not degraded), there is spontaneous and continued depolarization-polarization of the sarcolemma, the muscle is continually stimulated, and spastic paralysis results. The antidote for these nerve gases is the drug atropine. Atropine blocks the action of ACh, and if used alone would prevent muscle from ever contracting, but since the cause of death from nerve gas is too much ACh, atropine acts as an antidote. (Atropine is also used to dilate the pupils during an eye examination since it relaxes the constrictor muscles of the iris.)

Although curare would seem to have little therapeutic value, its close chemical relative succinylcholine (Anectine), does. During electroshock therapy (EST)* stimulation of the nerves is frequent and so all-encompassing that muscles may contract forcefully enough to cause bones to break. To counteract these muscle seizures, Anectine is administered to relax muscles so that bone breakage is avoided. The dose administered is low, and its effects disappear in a short time.

Botulin toxin, a poison produced by the bacterium *Clostridium botulinum*, is the cause of paralytic food poisoning. This toxin blocks the release of ACh (Figure 20-23d) and prevents muscle stimulation; less than 0.001 mg is sufficient to kill a man. By contrast, tetanus toxin, produced by the bacterium *C. tetani*, acts in the spinal cord apparently by suppressing inhibition of

muscle antagonists (strychnine acts the same way). As a result of the lack of inhibition, antagonistic muscles contract simultaneously and the limbs become locked in a state of spastic paralysis (Figure 20-24).

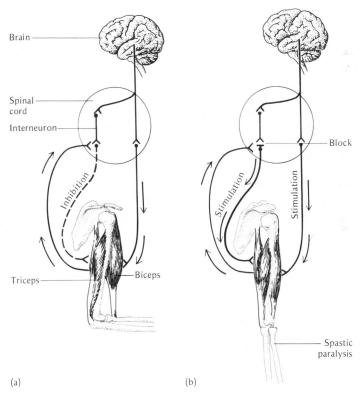

(a) (b)

FIGURE 20-24 The nervous control of muscular action. (a) Normal. When stimulus from the brain reaches the biceps and causes contraction, the triceps is stretched. A stretch receptor sends a signal to an interneuron in the spinal cord so that triceps contraction is inhibited. (b) Tetanus. The inhibitory signal is blocked and both muscles contract simultaneously, causing spastic paralysis.

* EST involves the induction of an epileptic-type seizure by electrical stimulation of the brain and is used to treat severe depressive conditions such as catatonic stupor.

of the resting level during physical activity and ending with complete recovery, that is, a return to the resting level (Figure 20–22). Repayment of the oxygen debt is reflected by our heavy breathing after strenuous activity. Heavy breathing will continue until the oxygen debt is paid in full. During this period, lactic acid is then oxidized to provide energy for the conversion of the remainder to glycogen, and the restoration of ATP from ADP.

SUMMARY

1. The 206 bones of the skeleton provide for protection, muscle attachment, movement and locomotion, storage of calcium salts, and the manufacture of blood cells.

2. The skull and vertebral column compose the axial skeleton. The pelvic and pectoral girdle and the bones of the arms and legs form the appendicular skeleton.

3. Bone consists of 35% organic collagen and chondroitin sulfate and 65% inorganic calcium salts, arranged in Haversian canal systems in which osteocytes embedded in bone matrix surround a central core of blood vessels and nerves. Bones are surrounded by a periosteum.

4. Cartilage, secreted by chondrocytes, contains no mineral salts or blood vessels and is nourished by diffusion. Cartilage may be rigid, flexible, or hyaline. The perichondrium surrounds cartilage and produces chondrocytes.

5. The embryonic skeleton is formed of collagen fibers secreted by fibroblasts, and hyaline cartilage. Ossification begins in the center of embryonic bones. The growth zone of long bones is just below the tip. Bone is also added on the outer surface. An internal marrow cavity is formed by osteoclastic resorption. Mechanical stress, diet, and "vitamin" D influence bone formation.

6. Skeletal, striped, or striated muscles composed of bundles of multinucleate (syncytial) cells are under voluntary control. Red muscle has more myoglobin than white and binds more oxygen.

7. Smooth or visceral muscles, composed of groups or sheets of uninucleate cells, are involuntary, and not attached to bones.

8. Cardiac or heart muscle is involuntary striated muscle. A functional syncytium, it has an inherent (myogenic) contraction rate of 72 times a minute, which can be increased or decreased by nervous activity.

9. Muscles have properties of contractility, extensibility, and elasticity.

10. Skeletal muscles bridge the joints of bones to which they are attached by tendons. Tendons transmit muscular pull and allow muscles to act at a distance from the bones they move. Muscles can pull but not push, so to produce limb movement in both directions, antagonistic muscles must alternately contract and relax.

11. Joints may be fused (as in the skull) or movable. Movable joints may be of a ball-and-socket type (free movement), hinged (movement in one plane), or pivot (intermediate movement). At movable joints, bones are protected by cartilage lubricated by synovial fluid in a bursa.

12. Skeletal muscle cells are elongate, multinucleate cylinders bounded by a sarcolemma and containing sarcoplasm with myofibrils, which are aligned with their cross bands in register.

13. About 90% of skeletal muscle protein is actin and myosin, which forms an actomyosin complex capable of contraction in the presence of ATP.

14. In the sliding filament model of muscle contraction, thick myosin filaments interdigitate with thin actin filaments, producing a banded appearance where they overlap. When muscle is stimulated, Ca^{++} ions appear, activate the myosin (ATP-ase), and split ATP, producing energy. The energy causes actin and myosin to form cross bridges and to slide along each other in ratchet fashion. During relaxation, Ca^{++} is removed, ATP is no longer split, bridges no longer form, and the elasticity of muscle and tissue restore the muscle to its original uncontracted state. In death, permanent cross bridges are formed in the absence of ATP (rigor mortis).

15. A nerve impulse arrives and synapses with a muscle at the motor end plate. Transmitter substance (ACh) diffuses to the muscle sarcolemma, producing a change in membrane permeability. Na^+ ions rush in, the sarcolemma is depolarized, a potential is produced, and if it reaches threshold, it is propagated. Ca^{++} ions are released from the sarcoplasmic reticulum, producing contraction. The sarcolemma is repolarized by the extrusion of K^+ ions. The refractory period is 0.02 seconds. Acetylcholinesterase breaks down the ACh.

16. The muscle action potential is all-or-none. The motor-end-plate potential and the release of Ca^{++} are graded responses.

17. The nerve cell and muscle cells it innervates, which may be spread through the body of a muscle, form a motor unit. The number of nerve cells stimulated and the frequency of action potentials generated depend on the amount of stimulation received. The graded response of a muscle to varying stimulation depends on how many muscle cells are involved in a motor unit, and their thresholds.

18. Tetanus involves a sustained contracted state when muscles are unable to relax between stimuli.

19. Fatigue involves loss of muscle contraction even under continued stimulation. During normal muscle contraction, aerobic respiration produces energy with the formation of pyruvic acid, which ultimately is oxidized to CO_2 and H_2O. Under intense stimulation lactic acid accumulates, which can cause fatigue with muscle cramps and soreness.

20. Muscles hypertrophy with exercise owing to an increase of myofibril numbers and an increase in muscle cell size. Unused muscles atrophy and are replaced by connective tissue.

KEY WORDS

backbone or spine
vertebrae
vertebrates
ribs
skull
axial skeleton
pelvis
shoulder (pectoral) girdle
appendicular skeleton
osteocytes
bone collagen
Haversian canal systems
periosteum
chondrocytes
hyaline cartilage
perichondrium
fibroblasts
ossification
growth zone
marrow cavity
osteoclasts
skeletal, striped, or striated muscle
smooth, visceral, or involuntary muscle
cardiac or heart muscle
syncytium
red muscle
white muscle
myoglobin
extensibility
elasticity

tendons
joints
antagonists
flexor
extensor
adductors
abductors
levators
depressors
sphincters
dilators
pronators
supinators
bursa
synovial fluid
bursitis
arthritis or rheumatism
sarcolemma
sarcoplasm
myofibrils
actin
myosin
actomyosin
sliding filament model
ATPase
motor end plate
motor-end-plate potential
sarcoplasmic reticulum
motor unit
resting or refractory period
tetanus
oxygen debt

TOPICS FOR REVIEW AND DISCUSSION

1. Describe the molecular machinery of muscle contraction.
2. How may paralytic poisons affect the neuromuscular junction?
3. What is the difference between tetanus (lockjaw) and muscle tetanus?
4. What is the cause of muscle cramps?
5. Describe the structure of bone.
6. Name the various types of muscles in the human body and give the characteristics of each.
7. Since muscles can only pull, how are variable kinds of movement made possible?
8. How can graded muscular activity be achieved when a muscle contracts in an all-or-none fashion?
9. How does the skeleton function in support, protection, and movement?
10. What is the role of the sarcoplasmic reticulum in muscle contraction?

Alice looked round her in great surprise. "Why, I do believe we've been under this tree the whole time! Everything's just as it was!"

"Of course it is," said the Queen. "What would you have it?"

"Well, in *our* country," said Alice, still panting a little, "you'd generally get to somewhere else—if you ran very fast for a long time as we've been doing."

"A slow sort of country!" said the Queen. "Now, *here*, you see, it takes all the running *you* can do, to keep in the same place. If you want to get somewhere else, you must run at least twice as fast as that!"

Through the Looking-Glass, Lewis Carroll, 1872. Illustration by John Tenniel.

STAYING IN PLACE: HOMEOSTATIC CONTROLS

21-1 KEEPING THINGS IN ORDER OR MAINTAINING THE STEADY STATE

The desire to understand how a piece of machinery operates provokes one to dissect it and to ask: How do the parts fit together and work? In our previous discussions of the human organism, we have dissected out each system, exposed its parts, and analyzed function in some detail. We tried to reveal clearly both the working parts and their operation. Now we must try to put all the pieces together and to look at the body as it actually exists, an integrated whole.

A living organism is an open thermodynamic system; that is, there is a constant flow of energy and matter between it and the environment, a flow that is in dynamic equilibrium, often called the **steady state.** Steady state does not refer to a static or unchanging condition. When we speak of normal body temperature, blood pressure, or blood sugar level, we are referring to values that fluctuate slightly about a norm. Maintenance of the norm involves the regulation of input and output so that there is little or no *net* change. If our internal and external environments remained constant, there would be no need for regulatory mechanisms; but this is not the case. They both vary. Thus, there must be a system of controls that regulates the rates of exchange between the living system and its external environment. We can think of ourselves as perched on a continually moving treadmill. To survive, we must maintain a position of equilibrium on life's treadmill; that is, we must keep pace with its motion so as to appear motionless. Such an appearance of stability—that is, the maintenance of a steady state—is possible only by continual acquisition of matter and energy. It is achieved because activity of the treadmill is balanced by activity of the organism; backward motion is balanced by forward motion.

As we have seen, every living cell possesses the capacity for self-regulation. Control of enzyme synthesis, product formation, uptake of matter from the environment, and release of energy are all devices employed by cells to cope with the vagaries of the world around them. The mechanisms used are varied: feedback inhibition, enzyme induction, and so on. Such self-regulated control of the internal environment producing a degree of constancy is called **homeostasis** (*homois*—like or same, *stasis*—standing; Gk.). All living organisms are homeostatic; otherwise they could not survive. The more complex an organism is, the more vital the process of homeostasis becomes. When homeostasis fails, the organism first suffers and ultimately may die. Our tissues and organs are functionally integrated by means of homeostatic control systems, each of which is automatic and self-regulating, and operates by negative feedback mechanisms.[1] Such a system consists of three components:

1. A receptor mechanism (a sensor or detector, usually neurons) for monitoring deviations from the steady state
2. A channel or transmission mechanism for processing and transmitting the information received and the appropriate responding information (usually hormonal or nervous)
3. A response mechanism (an effector, usually glandular or muscular) for producing a corrective measure so that there is a return to the steady state

Homeostatic control mechanisms in action have been discussed in previous chapters on reproduction (Chapter 9), the kidney (Chapter 15), the endocrine system (Chapter 16), and the nervous system (Chapter 17). Here we shall examine a single additional example of homeostasis to illustrate how such control enables

[1] A discussion of the principles of negative feedback in the thermostat can be found in Chapter 7.

us to appear motionless in a constantly changing world.

21-2 BODY-TEMPERATURE CONTROL

"Ninety-eight point six." This simple phrase has meant "normal" ever since you can first recall having your body temperature taken. Indeed, it is quite true that although the temperature of the surface of the skin, hands, or feet may vary greatly, the **core temperature,** that is, the temperature of the vital organs—heart, liver, intestines, and blood—is relatively fixed, rarely going below 97°F or rising above 104°F. Why, we might ask, is the body temperature so regulated, and how is such regulation achieved?

We have an elaborate system for controlling the body temperature. The temperature of other animals such as reptiles, amphibians, and fishes—indeed all animals except birds and mammals—fluctuates with that of their environment (Figure 21–1). These animals are called **poikilotherms** (*poikilos*—varied, *therme*—heat; Gk.) or **ectotherms** (*ecto*—outside; Gk.) The disadvantages of ectothermy are many. For example, a lizard on a cool morning is torpid and slow moving, barely able to drag itself into a sunlit spot. All its chemical activities, including the functioning of its muscles, proceed at a slow rate. As the radiant energy of the sunlight warms its body, the temperature of the blood rises, its heart beats faster, its metabolic rate speeds up, and soon it can scurry about in search of food or dart away from an attacking enemy. When the body temperature climbs too high, the animal moves to the shade to cool off, and then it can return to the sun or move onto a sun-baked rock to warm up. During the day a lizard can maintain a constant body temperature by moving in and out of the sunshine, but when the sun sets, it must slow down its activity and become slug-

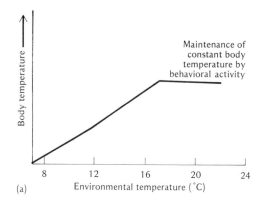

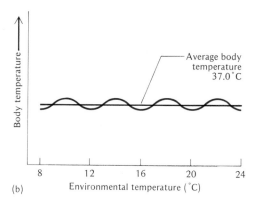

FIGURE 21-1 A comparison between the body temperatures of an ectotherm and an endotherm. (a) A lizard (an ectotherm). (b) A human (an endotherm).

gish. By contrast, our stable body temperature is maintained by the waste heat of our own metabolism, making us largely independent of the temperature of our environment (Figure 21–1). Thus, we humans (and all mammals as well as birds) are **homeotherms** or **endotherms** (*homois*—same, *endo*—within; Gk.). This environmental independence confers great flexibility on homeothermic animals; in a competitive world, homeothermic animals usually assume a dominant position.

A constant internal temperature has other advantages. Each cell of the body is a highly organized system whose chemical reactions

FIGURE 21–2 The role of temperature in enzyme activity. (a) Enzyme 1 shows maximum activity at a lower temperature than enzyme 2, and loses activity at a much lower temperature. At the temperature optimum of enzyme 2, enzyme 1 has lost 50% of its activity. (b) If enzyme 1 were inactivated by a temperature rise, substances B, C, and D would not be formed. If enzymes 2 and 3 were equally efficient at the same temperature, C and D would be formed in equal amounts; but if the temperature optima were different the amount of product from B could also be different. If the rates of reactions catalyzed by enzymes 1, 2, and 3 were increased by a temperature rise, the cell could be depleted of A faster than A could be manufactured.

FIGURE 21–3 Factors involved in heat gain or loss must be balanced to maintain a constant body temperature.

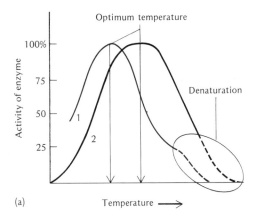

(a)

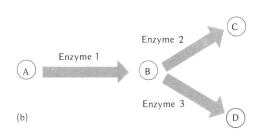

(b)

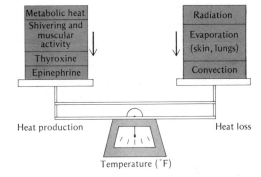

are in large part directed by enzymes, and these are often quite sensitive to shifts in temperature. Each enzyme-catalyzed reaction has its own temperature optimum; that is, there is a temperature at which the reaction proceeds most efficiently (Figure 21–2). Too low a temperature will make metabolic reactions proceed slowly and may prevent the formation of materials needed for other vital processes; conversely, too high a temperature can cause changes in the structure of enzymes (denaturation) so that they are rendered nonfuntional. The temperature limits for man are very narrow—death usually coming above 109.4°F or

below 77°F. Thus, the maintenance of our body temperature may be a life-or-death affair.

How is body temperature of an endotherm such as man achieved? Just as the temperature of a room is regulated by varying the amount of heat produced by the furnace and the rate at which heat leaves the room, so too our body balances heat gain and heat loss (Figure 21–3) in order to keep the body temperature at about 98.6°F. (37°C). As a result of fuel (food) consumption by our cells, some energy is trapped in the form of ATP and is available for cellular work, but the remainder is released as heat. This metabolic heat is produced 24 hours a day, asleep or awake. Additional heat may be produced as a result of physical exercise— muscles doing work when we lift something, or run, or shiver. Conversely, we lose heat by radiation, convection, and evaporation.

Suppose you entered a warm room (for example, 100°F). Would you get warmer or cooler? The answer seems obvious. The heat transferred from one object to another depends on the difference in temperature between them, and the greater the temperature difference the greater is the flow from hot to cool. This heat transfer is called **radiation.** Since the temperature of the room is higher than that of your body, heat is not lost from the body but is gained, and you feel hot. Temperature-sensitive receptors in the skin are stimulated and send coded signals to the hypothalamus via ascending pathways in the spinal cord. The hypothalamus contains a temperature-control center; when its neurons receive impulses from the heat receptors, fewer stimuli are sent to the nerves supplying the muscles of the blood vessels in the skin. Muscles in the walls of the blood vessels relax, the vessels expand (the face becomes ruddy), a large quantity of warm blood flows through the skin, and heat is rapidly lost by radiation. At the same time, there is a decrease in muscular activity (producing the

feeling "it's too hot to move")—the metabolic furnaces are turned down. The air close to the skin will also be heated by the blood flowing near the skin surface, the warmed air will tend to move up and away, to be replaced by cooler air, and so on. This movement of unequally heated portions of a gas (or a liquid) is called **convection,** and by means of convection currents of air the body will tend to lose heat. Finally, impulses from the hypothalamus will be carried by sympathetic nerves to the sweat glands in the skin (Figure 21–4). The sweat glands actively secrete a watery fluid; this is deposited on the skin surface. As the sweat evaporates, it has a cooling effect. Why is evaporation cooling? To evaporate water, that is to change it from a liquid to a gas (water vapor), energy is required. Whenever water is evaporated from the body surface, heat energy is lost, and the temperature of the body surface drops.

An important factor affecting evaporation is the water-vapor content of the air, its **humidity.** When the days are humid, we tend to feel uncomfortably warm because, although the sweat glands are active, the sweat simply does not evaporate, but remains on the skin or drips off. Most mammals lack sweat glands, and they increase evaporative water loss by panting or licking their fur. True panting in man usually

FIGURE 21–4 The structure of the skin.

TABLE 21-1 Average compositions of sweat and plasma

Substance	Concentration in sweat (mg/100 ml)	Concentration in plasma (mg/100 ml)
Sodium	185	325
Potassium	15	15
Calcium	4	10
Magnesium	1	3
Chloride	310	370
Lactate	35	15
Urea nitrogen	20	15
Glucose	2	100
Protein	0	7400

Adapted from Y. Kuno, *Human Perspiration*, Springfield, Ill., Thomas, 1956.

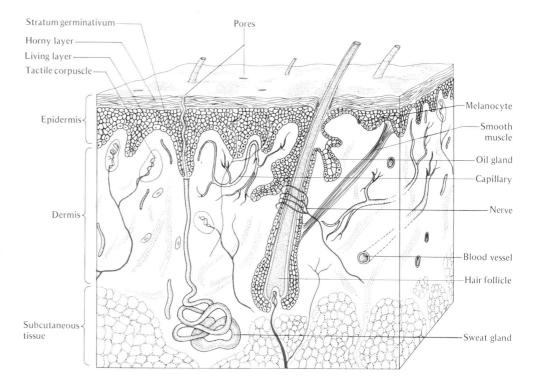

occurs while taking a hot bath when sweating cannot cool the body.

The fact that sweat is not pure water (Table 21–1) is of some importance. Sweat contains some salts, principally sodium chloride, so that excessive sweating can deplete body salt, ultimately causing heatstroke or sunstroke. This condition can be avoided by increasing the intake of salt (usually in the form of salt tablets) as well as water when heavy physical exercise in a hot environment is anticipated.

Let us consider the opposite situation: you enter a walk-in refrigerator set at 32°F (0°C), and you begin to feel cold. During a brief exposure to the cold, epinephrine is released from the adrenal medulla, and this tends to increase the metabolic rate. (If the exposure to the cold lasts a long time such as could occur by living in a cold arctic climate, thyroid hormone is secreted in greater amounts, and there is a long-term elevation in the body's metabolic rate.) Special sensory receptors in the skin send their messages to the hypothalamus, which in turn sends impulses along the de-

scending tracts of the spinal cord to the body muscles, shivering begins, and more heat is generated. This shivering pathway can be suppressed by input from the cerebral cortex, since a cold person ceases to shiver when engaged in voluntary activity such as stomping the feet or running (which also increases heat production). In addition, there is reflex narrowing of the blood vessels in the skin; as arteriole muscles receive nerve impulses, the amount of blood flowing through the skin is decreased, and heat is conserved by reducing the loss from radiation. Hair can help in reducing heat loss from convection: in birds and mammals, when the weather gets cold, feathers and fur puff out to create an insulating blanket of dead air space near the skin surface. The insulation

properties of most clothing result from a layer of warm air trapped in the network of the clothing and not from the material of the fabric itself. Human fur is not much use as insulation since we are virtually hairless, but we attempt to raise up our few remaining short hairs by contracting the muscles attached to each one (Figure 21–4). All that results is a dimpling of the skin—goose pimples—an ineffectual response to the cold. Sweating ceases in the cold, so that evaporative heat loss is minimized. Additional behavioral mechanisms, such as curling into a ball or hunching the shoulders, tend to reduce the surface area of the body exposed to the cold environment and body heat is conserved.

By a balance of heat loss and heat produc-

BOX 21A Fever: resetting the body's thermostat

During the course of an infection such as a common cold or the flu, it is not uncommon to experience **fever**—an elevation of the body temperature. At the beginning the body reacts as if it were subjected to a cold environment—blood vessels in the skin constrict, the skin pales, there is a reduction in heat loss, and you experience chills and shivering. As a consequence of these heat-conserving mechanisms, the body temperature quickly rises; you become feverish, your face is flushed, and you break out in a sweat. Fever is not a breakdown of the temperature-regulating mechanism, but simply a resetting of the hypothalamic thermostat to a higher point. Thus, at the beginning although the body temperature is normal, the body acts as if the core temperature were too low and begins to institute measures both for conserving heat and for increasing heat production (Figure 21-5). As a result the body temperature is deviated, and soon it settles at this higher point. When the fever breaks (called the **crisis**), the set point is automatically lowered; now the core temperature is too high, the skin becomes flushed, sweating begins, and mechanisms for increasing heat loss are set into operation. During the early part of such a bout, blankets are put on to combat the feeling of cold; but when the temperature begins to fall, these are thrown off because the skin becomes flushed and sweating begins.

Fever can be quite harmful. Although the body's tissues can tolerate cooling (temporary body temperatures as low as 85° F do not cause damage), they cannot withstand much heat gain, and chemical reactions may speed up to a dangerous point. Most people suffer convulsions at 106°–107° F, and 110° F is the absolute limit for life. Indeed, when heat cannot be lost from the body fast enough to prevent a rise in body temperature, a vicious positive feedback cycle ensues: metabolic reactions increase their rate as temperature rises; consequently heat production is increased, the temperature of the body climbs, metabolism goes faster, and so on. Unchecked fever (or exposure to extreme heat) can produce fainting and even death. The best course of action: reduce the body temperature by cooling.

What causes a fever? It is believed that chemical agents called **pyrogens** (*pyr*—flame, *gen*—producer; Gk.) are released by white blood cells during the inflammatory response to a foreign intruder. These pyrogens act (in an unexplained manner) directly on the hypothalamus, causing its set point to be pushed abnormally high.

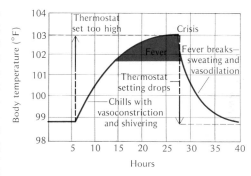

FIGURE 21-5 Fever: a resetting of the body's thermostat.

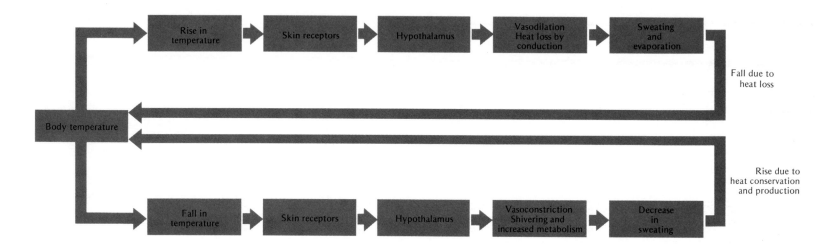

FIGURE 21–6 Feedback loops involved in the maintenance of constant body temperature.

tion, the temperature of the body is regulated; the feedback loops involved are summarized in Figure 21–6. The main center for control of body temperature—our thermostat—is situated in the hypothalamic region of the brain. If the cortex of an endothermic animal's brain is completely removed, such an animal can still regulate its body temperature; but if the hypothalamus is destroyed, this ability is eliminated. The hypothalamus contains two temperature-control centers, one for detecting cooling and one for detecting warming. The anterior (front) part of the hypothalamus contains neurons that are sensitive to warmth, and these monitor the temperature of the blood passing through this region of the brain. When the neurons of the warming center are activated, sympathetic nerves carry impulses to the sweat glands, blood vessels, and adrenal medulla, and there is an increase in heat production and a decrease in heat loss (blood flow is decreased, as is sweating). However, when the cooling center of the hypothalamus is activated, the parasympathetic nerves, which run to the sweat glands and blood vessels, produce essentially the opposite effect.

In the home thermostat, you can fool the system by placing a match or an ice-cold rod under the temperature-sensitive element, thus causing the furnace to go on or off without any previous change in the room temperature. The same can be done with the temperature controls of the hypothalamus. If wires implanted in the hypothalamus are warmed, the animal responds by showing the responses typical of a hot environment—panting, sweating, constriction of skin blood vessels. Conversely, local cooling of the hypothalamus by a chilling of wires implanted in this region of the brain produces responses typical of an animal subjected to the cold.

SUMMARY

1. Living organisms are in dynamic equilibrium with their environments: that is, they maintain a steady state. Although there is input and output, values fluctuate about a norm and there is no net change.

2. Maintenance of the norm involves self-regulation, usually by negative feedback mechanisms. The components of a self-regulatory system are: receptors, transducing and transmitting mechanisms, and response effectors. Self-regulated maintenance of the internal environment is called homeostasis, and when it fails, death may ensue.

3. The body temperature of poikilotherms (ectotherms) fluctuates with their environments, whereas homeotherms (endotherms) have homeostatic mechanisms for maintaining a constant body temperature independent of that of the environment. These mechanisms are under hypothalamic control and include: regulation of metabolic rate; regulation of the blood supply to the skin; regulation of sweat gland secretion; and evaporative cooling, panting, shivering and erection of hairs.

4. Fever involves a resetting of the hypothalamic thermostat to a higher set point under the influence of pyrogens.

KEY WORDS

steady state
homeostasis
core temperature
poikilotherms (ectotherms)

homeotherms (endotherms)
radiation
convection
humidity
fever
crisis
pyrogens

TOPICS FOR REVIEW AND DISCUSSION

1. Define homeostasis.
2. Discuss the meaning of the statement: All the vital mechanisms, however varied they may be, have only one object, that of keeping the conditions of life constant in the inner environment.
3. What is the significance of fever?
4. How is temperature control achieved in the human body?
5. What are some of the mechanisms that conserve heat upon exposure to a cold environment?
6. How do poikilotherms differ from homeotherms?
7. Why could organisms not survive without homeostatic devices?
8. Compare the human thermostat with the mechanical thermostat.
9. Discuss negative feedback controls.
10. What homeostatic mechanisms come into play when the body is subjected to high temperatures?

The past is but the beginning of the beginning, and all that is and has been is but the twilight of the dawn. A day will come when beings who are now latent in our thoughts and hidden in our loins . . . shall laugh and reach out their hands amid the stars.

H. G. WELLS

THE HUMAN POPULATION

We humans have a sense of history. We cope with the present, wonder about our past being and origins, and attempt to forecast the future, hoping that our children will find this planet a habitable place. We are the product of evolution—the ultimate in games of chance. For most creatures—past, present, and future—it is a most unusual game. The participants are unaware that they are playing and know nothing of the game's rules. The prize for winning a round of play is simply the privilege of participating in yet another round of the game; biologically speaking, the winning players perpetuate their kind, and the losers are doomed to extinction. Extinction—leaving the game after losing—is final; there are no second chances. Most of the players will eventually have to pay this penalty. Players entering the next round of the game have a chance of winning that is proportionate to their previous success; that is, the more progeny produced, the greater are the chances that some of these will win too. The winner-take-all game of evolution is further complicated by the fact that the rules for selecting the winners can be changed in a rather unpredictable fashion from one round to the next. And in the final analysis, both environmental surroundings and inherited competence of the participants will determine those individuals selected for another round of play. Nature is both adversary and jury in the evolutionary game. In spite of the scarcity of rules, the players' lack of awareness of any rules, and the hidden goals or purposes of the game, evolution proceeds in an orderly fashion. Only man, himself a successful player of the game, understands its rules and goals. We are aware of our own participation; we know what constitutes success. But this understanding does not allow us to load the dice in our favor; we are still unable to assure the future success of our own kind.

Evolution is to many the unifying concept of biology. The species, a population of freely interbreeding individuals, stands as the fundamental unit of evolution. The individual organism has importance only as a genetic package within the group. Natural selection acts on the hereditary characteristics of the individual, but its prime target is the total collection of genes in the entire population. Thus, fitness (the chance to play) does not apply to an individual gene or its possessor, but includes all the genes of the breeding group and depends upon the way in which these interact with their surroundings. In the chapters that follow we shall study the transmission of genes from parent to offspring (Chapter 22) and the manner in which genes in the population become scarce or frequent (Chapter 23). Thus, we gain insight into how biological evolution—descent with modification—works (Chapter 24).

An appreciation of population biology enables us to discard the notion of a perfect species. There is and there can be no golden mean for any animal or plant. Genetic diversity within a population provides the biological basis for both our individuality and our freedom; it enables a species to hedge its bets in the game of survival. The more different kinds of individuals there are, the less is the chance that a change in the game's rules will eliminate all the players and thus cause a species to become extinct. Species evolve not by deliberate or conscious efforts of their members to meet a particular environment, but because chance variations in the hereditary characteristics of some individuals enable these to survive and reproduce. Thus, the blind and automatic strategy for living creatures playing the game of evolution is expansion of population size and maintenance of genetic individuality.

Species are genetically continuous in both time and space, since the members share a common pool of genes; the individuals of a species form a reproductive community. Species are also a part of an ecological community, since they interact with both the nonliving and the living environment. The world of the next few decades will be determined by an interplay of numbers—numbers of different kinds of people and numbers of environmental resources; since little can be done about the first, it seems critical we examine the second. We must determine the rules that govern our relationship to our natural surroundings (Chapters 25 and 26). Then, and perhaps only then, can we embark on a collective pattern of play to ensure a decent legacy for our progeny.

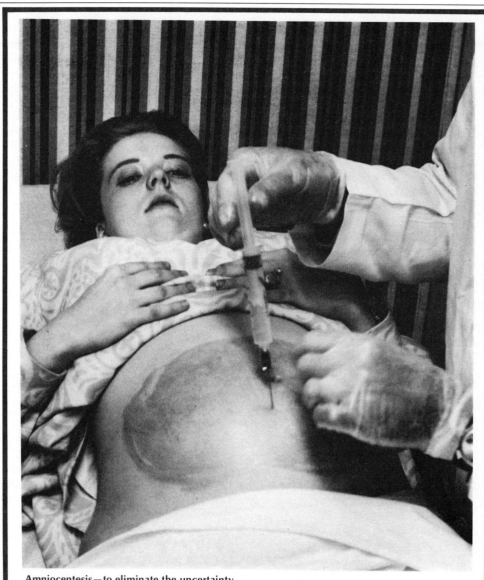

Amniocentesis—to eliminate the uncertainty.
(Photo by Leonard McCombe, Time/Life Picture Agency. © 1972, Time Inc.)

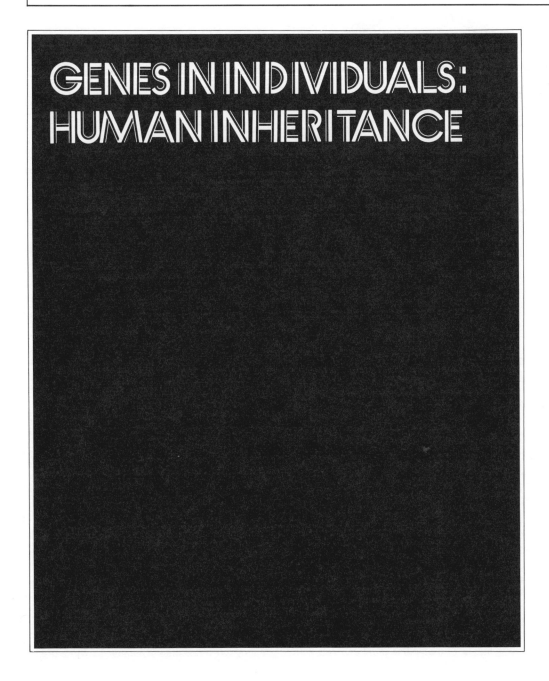

GENES IN INDIVIDUALS: HUMAN INHERITANCE

22–1 COMMUNICATION BETWEEN PARENT AND OFFSPRING

Philip and Mary Sinter are a young, healthy, married couple living in San Francisco. Mary is 4 months pregnant, and the Sinters are elated about the prospects of having their first child. However, a degree of uncertainty tempers their jubilation. Mary's sister, now 4 years old, suffers from an inherited disease called **Down's syndrome (mongolism).**[1] Down's syndrome is characterized by severe mental retardation; the brain never develops beyond the level typical of a 2- or 3-year-old child, although the individual may live 20 years or more (the life expectancy of most of the individuals so afflicted is 12 years). Mary's sister is an imbecile with an intelligence quotient (I.Q.) of 35. She is a cheerful and playful child, although her movements are clumsy and slow; her speech is almost unintelligible. Her hands are short and stumpy, and the little finger is curved abnormally and quite short; even the arrangement of ridges on the fingertips, the fingerprint pattern, is abnormal. The condition of Down's syndrome is most clearly shown in the child's face—a prominent forehead, flattened bridge of the nose, a drooping lower jaw and open mouth, and a skin fold (called the epicanthic fold) at the corners of the eyes (Figure 22–1). What Philip and Mary Sinter want to know is: Will their unborn child also suffer from Down's syndrome?

In previous years the Sinters would have had

---

[1] The first comprehensive description of this disorder was made in 1866 by a London physician, J. Langdon Down (1829–1896). The disease features described by him were quite accurate, but because of the vaguely Oriental appearance of the eyes, he designated the disorder "mongoloid idiocy." The expression is totally unwarranted, since the eyes only superficially resemble those of Oriental peoples.

FIGURE 22-1 (left) Down's syndrome in a 4-year-old girl. (Courtesy of Sanford Schneider, Loma Linda University Medical School.)

FIGURE 22-2 Amniocentesis is accomplished by inserting a hypodermic needle into the amniotic cavity and withdrawing a small amount of fluid and fetal cells. Optimum time for such a procedure is about the sixteenth week of pregnancy.

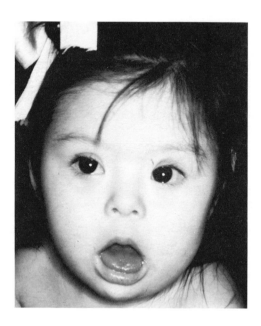

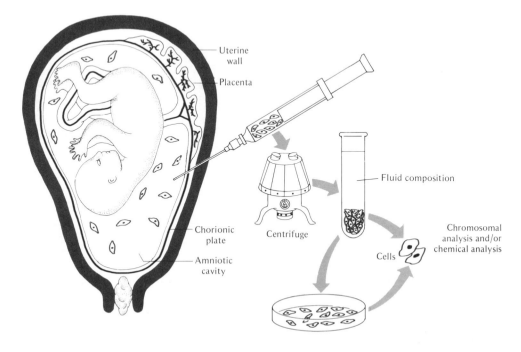

to risk the birth of a child with Down's syndrome. Today, however, it is possible for Philip and Mary to receive genetic counseling, that is, to obtain a good assessment of the chances that their offspring will have a hereditary handicap such as Down's syndrome. Genetic counseling relies heavily on mechanisms involved in inheritance, and the nature of genetic expression. If it seems likely their child will be affected, the Sinters can avail themselves of a new technique called **amniocentesis** and learn with certainty the genetic state of their unborn child. Amniocentesis involves the removal of fluid from the amniotic cavity by needle puncture (see chapter opener and Figure 22-2) and is usually performed between the thirteenth and the eighteenth weeks of pregnancy. Since the amniotic fluid receives cells from the fetus's skin and respiratory tract and from the membranous sac surrounding the fetus, removal of amniotic fluid enables the physician to sample

the cells of the fetus itself. When such cells are grown in an artificial nutrient solution and then studied microscopically or biochemically, it is possible to answer certain questions about the inherited potential of the fetus. If, through amniocentesis, Mary Sinter and her husband learn that she is carrying a child with Down's syndrome, they have two choices: abortion or the delivery of a hopelessly defective infant. Although the genetic counselor will appraise the couple's chances for having a normal or an abnormal child, Philip and Mary must choose their ultimate course of action.

To appreciate the genetic mistake involved in the production of a child with Down's syndrome and the nature of inheritance, we must understand the rules that govern the transfer of information from parent to offspring. **Genetics,** the science of heredity, can be divided into fields of inquiry. One area, called **physiological** or **molecular genetics,** deals with

what genes are and how they work. This aspect of inheritance has been discussed in Chapters 6 and 7. Another area of inheritance, called **transmission genetics** or **classical genetics,** involves the mechanisms by which genes are transmitted from one generation to the next. This chapter will cover the field of transmission genetics, and will not be concerned with the detailed chemical nature of the gene. Here we want to focus attention on a very simple and practical question: What kind of offspring can we expect from a particular set of parents?

22-2 GENES, CHROMOSOMES, AND SEX-CELL FORMATION

Genetic information is carried in the double-stranded molecules of DNA, and that DNA is (at least in eukaryotic cells) associated with protein to form discrete nuclear structures called **chromosomes** (Chapter 7). Most orga-

BOX 22A Preparing chromosomal pictures

The direct observation of human chromosomes was for many years extremely difficult if not impossible. Very few cells could be found in metaphase, when chromosomes are most distinct, and the large number of chromosomes in human cells often caused them to lie on top of each other, making them difficult to count. Indeed, up until 1956 it was believed that humans had 48 chromosomes. In the past 20 years significant technological advances have been made: cells can be grown in glass vessels outside the human body; a chemical agent, colchicine, stops mitosis at metaphase; and cells treated with dilute salt solutions swell and burst, in effect spreading out the chromosomes. Applying these techniques to a drop of blood or to cells removed from the amniotic fluid enables the chromosomes to be analyzed and a **karyotype** (chromosome picture) to be prepared (Figure 22-3). By the use of chromosome typing a number of hereditary defects can be identified, as well as the sex and genetic condition of an unborn child.

FIGURE 22-3 The preparation of a chromosome picture (karyotype) for chromosome analysis. Note that the chromosomes shown are actually pairs of chromatids interrupted at metaphase and held together by their centromere. (Modified from Helena Curtis, *Invitation to Biology,* **Worth Publishers, New York, 1972, page 161.)**

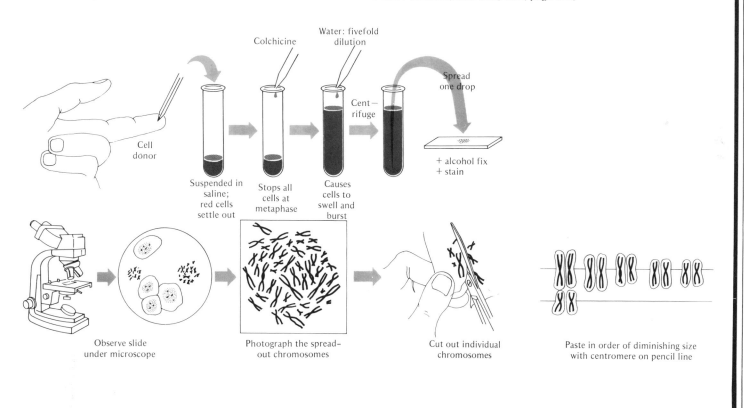

FIGURE 22–4 Human chromosomes. (a) A chromosomal spread from the cell of a human female (top), and (below) the 46 chromosomes arranged in homologous pairs showing the 22 pairs of autosomes and a pair of sex chromosomes (XX). (b) A similar spread from the cell of a human male (top) and (below) arranged in homologous pairs showing 22 pairs of autosomes and a pair of sex chromosomes (XY). (Courtesy of W. R. Centerwall, Loma Linda University Medical School.)

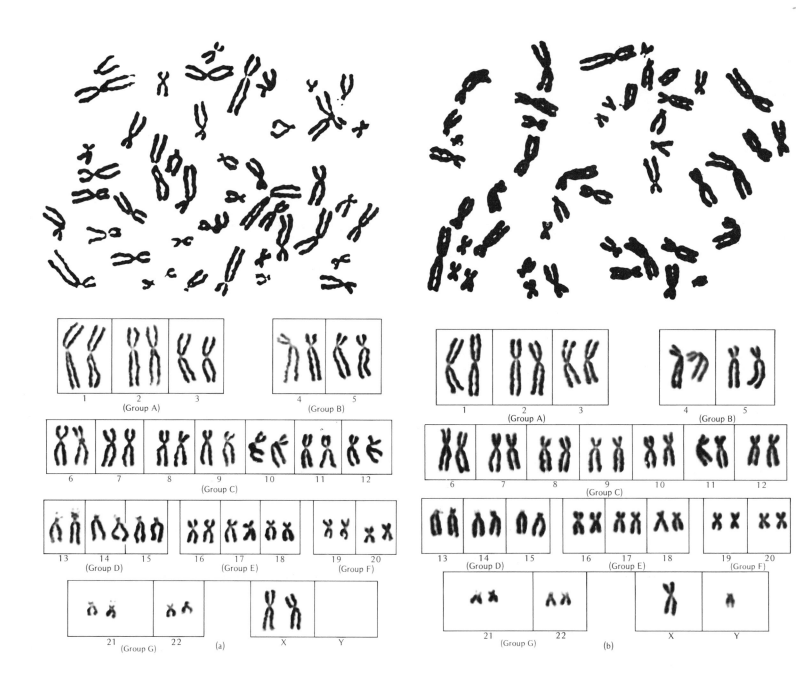

FIGURE 22–5 A schematic illustration of meiosis and fertilization.

nisms carry two sets of chromosomes, and this is called the **diploid number** (2n, where n stands for the number of pairs). The chromosomal picture of a human male and a female shows that each of the body (somatic) cells contains 46 chromosomes, and these are arranged in 23 pairs (Figure 22–4). Each pair of chromosomes has a slightly different appearance and for the sake of convenience can be assigned a number (Figure 22–4). (The method of preparing chromosomal pictures such as the one shown is described in Box 22A.)

Recall that in your body two types of cell division occur: mitosis and meiosis. Mitosis maintains a constant chromosome number, whereas meiosis halves the number of chromosomes.[2] Meiosis occurs only during the course of germ-cell formation and results in the production of **gametes** (sex cells) containing only one member of each chromosomal pair; thus the chromosomal number of a gamete (egg or sperm) is haploid (n). In the human female mature eggs are produced by meiosis and contain 23 chromosomes; by a similar process the human male produces mature sperm containing 23 chromosomes (Figure 22–5). When a sperm fertilizes an egg, the resultant cell, the **zygote,** contains 46 chromosomes (23 pairs). Thus, every cell of the body contains two sets of genes on two sets of chromosomes, except for the gametes, in which one set of genes occurs on one set of chromosomes.

Compare the chromosomes from the male and female; 22 pairs are similar in form. These are the **autosomes.** One additional pair in the female, designated **XX,** is not similar to the

[2] Mitosis and meiosis are discussed in Chapter 9, and the chromosomal movements are illustrated in Figure 9–1. A review of this material will be helpful to the reader.

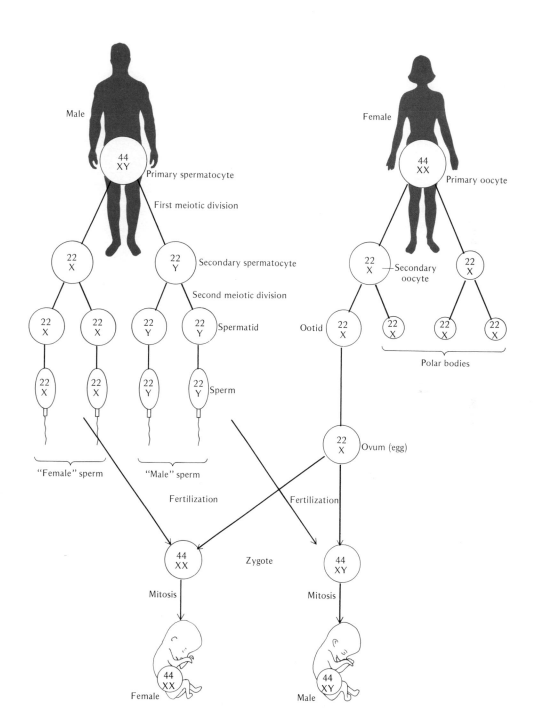

FIGURE 22–6 An albino. (Courtesy of *The Journal of Heredity*, **Vol. 39, No. 5, 1948.**)

equivalent pair in the male, called **XY**. These are the **sex chromosomes.** Meiosis in the human male produces two kinds of sperm (X-bearing and Y-bearing), but the female produces only X-bearing eggs (Figure 22–5). Since the sex chromosomes are responsible for the sex of an individual, the kind of sperm that fertilizes the egg must determine the sex of the offspring. If an X-bearing sperm enters the egg, the genetic constitution is XX or female; if a Y-bearing sperm fertilizes the egg, the offspring will be a male (XY).

22–3 PHENOTYPE AND GENOTYPE

One of the most obvious of all human traits is skin color. Skin color depends principally on the amount of **melanin,** a yellow-black pigment, present in special skin cells called **melanocytes.** If the melanocytes produce a great deal of pigment, the skin is black; if they produce very little, it is white. Melanocytes synthesize melanin from the amino acid tyrosine by a metabolic pathway that can be simply illustrated:

tyrosine $\longrightarrow$ A $\longrightarrow$ B $\longrightarrow$ melanin

Each step in the pathway is catalyzed by a specific enzyme, which in turn is specified by a particular region of the nuclear DNA, a gene:

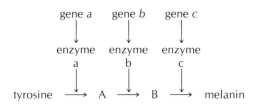

About 1 out of 20,000 humans fails to synthesize melanin because he does not produce one of the enzymes in the path from tyrosine

children produced by the mating. If there are several children from a mating, these are connected by vertical lines to a horizontal one, which in turn links up to the line linking the parents. (b) Matings in generations I and III showing the genotypes of the parents and possible offspring.

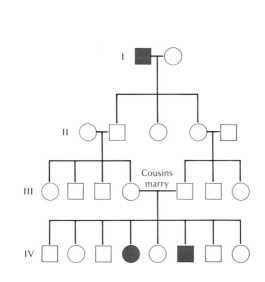

(a)

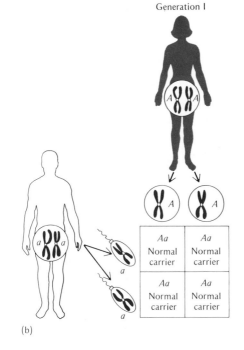

(b)

Generation I

Mating of carrier cousins in generation III

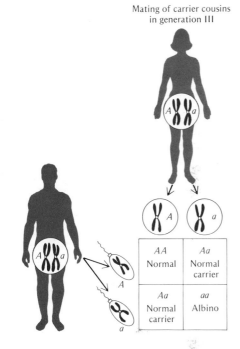

to melanin. The skin of such an individual is white or pink (because the red color of the blood cells in the underlying vessels can easily be seen), the hair is white, and the eyes are pink (Figure 22–6). Such individuals, called **albinos** (*albus*—white; L.), are acutely sensitive to sunlight, are subject to skin cancer, and have poor vision.

The observable trait (appearance) of an albino—whiteness—is called the **phenotype.** Phenotype (*phainein*—to appear; Gk.) refers to the way in which a trait expresses itself, that is, to what can be seen. Blue eyes and brown eyes are different phenotypes; so are red hair, blond hair, brown hair, and black hair. Blood type, nose shape, and many other traits you can think of are also observable traits—phenotypes—and we use these almost every day to

recognize individual persons by their unique array of visible characteristics.

In addition to an individual's phenotype, he also has a **genotype,** that is, a particular set of genetic instructions inherited from his parents. How is phenotype related to genotype? In each of the body cells there exist pairs of genes on paired **(homologous)** chromosomes. Each member of a pair of genes, which may occur in a variety of forms, is called an **allele.** Each allele occupies exactly the same position **(locus)** along the length of its chromosome as its partner allele on the homologous chromosome. Each locus of a single chromosome can carry only one allele. In terms of the primary products of genes, the alleles produce different forms of the same protein. Albinism is caused by a defective gene coding for a defective enzyme

involved in the metabolic pathway of melanin formation from tyrosine. Specifically, the condition arises only when the individual is endowed with two defective genes. (As long as one allele is normal, the individual will produce the enzyme and be normal in appearance.) Tracing the inheritance of a trait such as albinism from one generation to the next is a basic genetic method. Such a hereditary lineage is called a **family tree** or **pedigree.**

The hypothetical pedigree shown in Figure 22–7 illustrates the genotypes of children resulting from the mating of an albino male with a normal female. All the children are normal. What does this tell us? First, the female parent must carry a pair of normal genes; second, the male parent must carry a pair of genes that results in a failure to manufacture melanin. We (text continues on p. 458)

BOX 22B Mendel and his peas

In 1865 Gregor Johann Mendel, an Augustinian monk, described to the Brünn Natural Science Society the inheritance of characters in pea plants after 8 years of painstaking work. The discussion that followed was brief. Sadly, the fundamental importance of Mendel's findings was not understood by anyone in the audience; nevertheless, the society did ask him to publish his lecture in 1866. One copy of the published manuscript was sent to Professor Carl Naegli in Munich, a leading figure in studies of heredity. Letters and views were exchanged between the two men. Naegli, however, did not believe Mendel's explanations and suggested that he work with the hawkweed plant instead of the pea plant. The encounter with Naegli became a disaster for Mendel. After 5 years of intensive effort on the hawkweed, and with his eyesight failing, Mendel wrote to Naegli that the features of inheritance he had discovered with the garden pea did not apply to the hawkweed. (This was in fact not true; the hawkweed rarely reproduces sexually, thus confusing Mendel at the time.)

In 1868 Mendel was elected abbot of the Brünn monastery, an office he thought would allow him ample leisure time to continue his experiments. This was not to be, and by 1871 his productive scientific career came to an end. He died in 1884 at age 62. The mourners who attended the funeral recognized they had lost a dear friend, a dedicated teacher, and a high dignitary, but none at the funeral realized that Mendel had discovered what are at present known as Mendel's Laws of Heredity. The significance of Mendel's work went unrecognized until 1900, when quite by accident it was rediscovered independently by De Vries in Holland, Tschermak in Austria, and Correns in Germany. Since that time the importance of Mendel's contributions to biology has been clear. Today he is generally acknowledged as the father of genetics.

Here is what Mendel discovered about inheritance. At the beginning he asked a simple question: How is it possible to determine the different kinds of offspring that will result when two pure-breeding varieties are crossed (mated)? Then he went on to select the proper material to answer this question. He chose the garden pea *Pisum* because it showed a constancy of certain characters, he could cross-pollinate plants with ease, and the large garden of the monastery would provide enough space for his plants. By 1856 Mendel had selected 22 varieties and arranged their characters in contrasting pairs: smooth versus wrinkled seeds, tall versus dwarf plants, yellow versus green pods, red versus white flowers, and so on. When two plants differing in only one pair of these contrasting characters were crossed, the first generation of offspring, the F_1 **(first filial) generation** as it is called now, were uniform in appearance and showed only one of the two parental characters; this stronger character Mendel called dominant, and the weaker one he called recessive. Thus, when he crossed a plant with wrinkled seeds with another having round seeds, the offspring showed only the round-seed character. He found tall was dominant to dwarf, green pods were dominant to yellow pods, and red flower color was dominant to white.

Next, Mendel allowed the F_1 plants to self-fertilize and the progeny of this **second filial generation**, the F_2, were found to consist of two types of plants, each resembling one or the other of the original grandparents. No plants were intermediate, and thus there was no blending of the hereditary characters. Unlike his predecessors who had also done plant hybridizations, Mendel counted the numbers of individuals with each of the characters in the F_2 generation. He found three fourths of the plants showed the dominant character and one fourth had the recessive character. Thus, when he inbred 929 F_1 plants showing red flower color, he found the F_2 plants to be 705 red-flowered and 224 white-flowered. To explain these results Mendel postulated that each plant contains two hereditary factors—today we call them genes—that affect a particular character, and during the formation of gametes (pollen and egg) these separate. Each gamete contains only a single member of the pair of parental factors (genes), and by combination of different kinds of gametes a hybrid is produced. Mendel showed that the hereditary factors (genes) segregate, pass undiluted and uncontaminated into the gametes, and do not blend when recombined (at fertilization) to form offspring. This is Mendel's **law of segregation**.

Furthermore, when Mendel took the dominant-character plants from the F_2 generation and inbred these to give an F_3 generation, about a third of them bred true. In the offspring of the remaining two thirds, grandparental characters once again appeared in a ratio of 3:1. This could occur only if the F_2 dominant plants were of two kinds—homozygous and heterozygous (hybrid). All this Mendel symbolized in the following way:

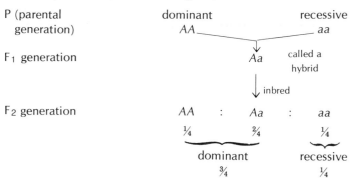

P (parental generation)   dominant AA recessive aa

F_1 generation Aa called a hybrid

↓ inbred

F_2 generation AA : Aa : aa
¼   2/4   ¼
dominant ¾   recessive ¼

If the F_2 Aa were inbred to give an F_3 generation, the progeny would again be:

F_3 generation AA : Aa : aa
¼   2/4   ¼
dominant ¾   recessive ¼

In subsequent experiments Mendel hybridized pea plants that differed not only in one but in two or more hereditary characters. For example, he crossed a pea plant with round and yellow seeds with another having green and wrinkled seeds. The offspring (F_1) were all uniform, showing round and yellow seeds. However, when these were self-pollinated, the F_2 offspring showed all possible combinations of the parental characters in a characteristic ratio: 9/16 round and yellow, 3/16 round and green, 3/16 wrinkled and yellow, and 1/16 wrinkled and green, or a ratio of 9:3:3:1 (Figure 22-8). Mendel concluded that when two or more characters were inherited independently of one another, the offspring produced were the results of random assortment of the hereditary factors in the gametes and subsequent recombination of these during fertilization. This is Mendel's **law of independent assortment.***

To us, unlike Naegli and the people at the Brünn meeting in 1865, Mendel's laws (which apply to peas, flies, mice, humans, and all sexually reproducing organisms) seem obvious, but it should be recalled that all of Mendel's work was done before details of cell division had been worked out, and that he was totally unaware of the existence of chromosomes or genes. Thus, the existence of hereditary factors was deduced only from statistical analysis of experimental results and Mendel's brilliant insight; their physical existence was not discovered for many decades.

*Actually only genes carried on nonhomologous chromosomes are assorted independently, and thus this law is not universal. See page 402.

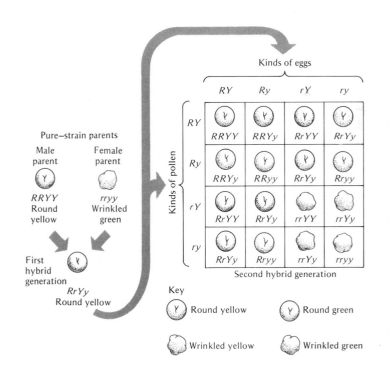

FIGURE 22–9 (left) Meiosis in a heterozygous individual (albinism carrier) illustrates the principle of segregation. Only one pair of chromosomes is shown.

FIGURE 22–10 (below and facing page) Outcomes of nine possible crosses (matings) involving the recessive trait albinism. (Modified from *Human Heredity and Birth defects* by E. Peter Volpe, Copyright © 1971, by the Bobbs-Merrill Company, Inc., reprinted by permission of the publisher.)

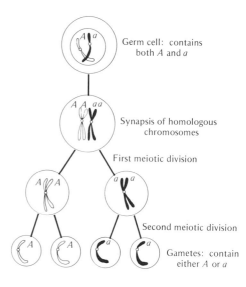

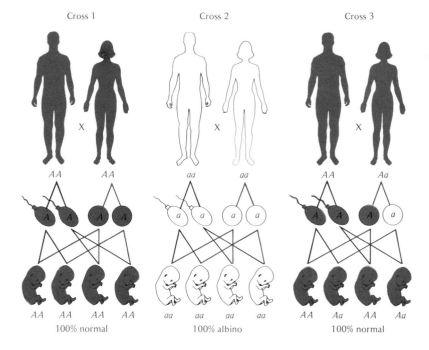

can conclude from this pedigree that the gene for normal pigmentation is **dominant** over its partner allele, the **recessive** gene, because the children, who have one normal and one defective gene, have normal skin color.

The dominant gene is customarily symbolized by a capital letter (in this case *A*); the recessive gene, by a corresponding small letter (*a*). The phenotype of the male parent is albino, and his genotype can be written as *aa*. The female parent is of normal phenotype, and she must be genotype *AA* (proof of this will become apparent shortly). The children of the mating are all of normal phenotype and have the genotype *Aa*. Persons carrying a pair of similar alleles (both parents in this pedigree) are said to be **homozygous;** those carrying unlike alleles (the offspring in this case) are **heterozygous.**

Let us carry the inheritance of albinism one step further and extend the pedigree of this family for two more generations (Figure 22–7); in the third generation two cousins marry, and some of their offspring in the fourth generation show albinism. Therefore, although the phenotypes of the cousins who marry are normal, these individuals must be heterozygous for albinism, that is, of genotype *Aa*; these parents are designated as **carriers** of the trait.

A number of important points emerge from a study of the family pedigree of albinism. First, genes do not blend, and there is no dilution of a genetic trait from generation to generation. Second, genes are discrete entities that are capable of being expressed in subsequent generations. Third, an offspring receives one and only one member of a pair of alleles from each parent (each gamete contains only one member of a pair of alleles). Fourth, albinism results from a change (**mutation**) in the gene coding for at least one of the enzymes involved in melanin synthesis; the presence of one normal allele, the dominant, is necessary for normal pigmentation. Fifth, a trait determined by a dominant gene will not appear in an offspring unless it also appears in one or both parents.

22–4 TRANSMISSION OF GENES

Two principles govern the transmission of genes from parents to offspring: segregation and recombination. Often these principles are called Mendel's first and second laws, after the Austrian monk Gregor Mendel (1822–1884), who worked them out (Box 22B).

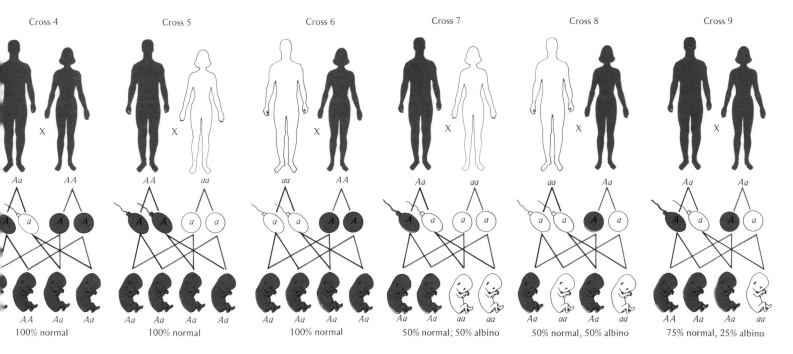

Cross 4	Cross 5	Cross 6	Cross 7	Cross 8	Cross 9
Aa × AA	AA × aa	aa × AA	Aa × aa	aa × Aa	Aa × Aa
AA Aa Aa	Aa Aa Aa Aa	Aa Aa Aa Aa	Aa Aa aa aa	Aa aa Aa aa	AA Aa Aa aa
100% normal	100% normal	100% normal	50% normal; 50% albino	50% normal, 50% albino	75% normal, 25% albino

The **Law of segregation** states the following: (1) each individual starts life as a fertilized egg containing two representatives of each kind of gene; one member of each pair is derived from one parent, and the other member from the other parent; (2) a sexually mature adult (who develops from a fertilized egg) transmits only one member of each pair of genes into each gamete he or she produces. The term segregation refers to the fact that although each pair of genes is associated during the lifetime of an individual, when it comes to gamete formation the two genes clearly separate, or segregate. This is related to the behavior of the chromosomes during meiosis (Figure 22–9). The outcomes of nine possible marriages involving the recessive trait albinism clearly demonstrate the principle of segregation (Figure 22–10).

The Law of independent assortment (also called the **principle of recombination**) states that when an organism matures and produces gametes, the genes may enter into the gametes in a great variety of associations, regardless of their parental origin. Thus, the genes derived from one parent do not necessarily always go together, but they can segregate and form any possible combination; hence the genes are said to recombine (Figure 22–11).

The consequences of the principle of recombination can be examined by following the inheritance of two traits at once. Albinism, as we have already seen, is a recessive trait, and persons of this phenotype have a genotype that can be written aa. Persons who are not albinos carry a dominant gene (A) and can be of two genotypes: Aa or AA (heterozygous or homo-

zygous). Earlobes of humans may be attached or free (Figure 22–12), and the inheritance of this character is much like that of albinism. Those persons having free earlobes carry a dominant gene (F), and such individuals may be of two genotypes: FF or Ff. Attached earlobes are a recessive trait, and individuals of this phenotype are of a single genotype: ff. What kind of offspring would result if an albino man with attached earlobes married a woman with normal pigmentation and free earlobes? The genotype of the albino with attached earlobes is aa ff, and the genotype of his normally pigmented and free-earlobed mate is Aa Ff.

Using the principle of segregation, we expect the gametes produced by the albino to be of a single kind: a. However, the normally pigmented woman can produce two kinds of gametes: *A* or *a*. Since the albino man has attached earlobes and is homozygous for this trait, only one kind of gamete can be produced: *f*. Thus, the gametes of the albino with attached earlobes are of a single kind: *af*. The normally pigmented woman with free earlobes is heterozygous and can produce two kinds of gametes: *F* or *f*. Since the gene pairs (alleles) for albinism and free earlobes segregate and recombine independently of each other, the normally pigmented mate with free earlobes can produce four different kinds of gametes: *AF*, *Af*, *aF*, and *af*.

The kinds of offspring resulting from such a mating can be visualized in diagrammatic fashion by making a square **(Punnett square)**. In such a Punnett square, the genetic constitution of the gametes is written on the margins and that of the offspring (zygotes) is written within the square:

	female gametes			
	AF	*Af*	*aF*	*af*
male gametes **af**	*Aa* *Ff*	*Aa* *ff*	*aa* *Ff*	*aa* *ff*
	normal free	normal attached	albino free	albino attached

There result four different kinds of offspring, each with a different genotype and phenotype, and the offspring occur with about equal frequency. If we look at the pattern of meiosis in the normally pigmented, free-earlobed individual, the principle of recombination is easily visualized (Figure 22–13). What would the offspring look like if both the mother and the father were heterozygous for albinism and free earlobes? (Hint: look at the dihybrid cross made by Mendel shown in Figure 22–8.)

FIGURE 22–11 The principle of recombination, showing that genes derived from one parent do not always go together, but can segregate independently into any possible combination. For *n* genes there would be 2*n* combinations. Thus, in the case shown *n* = 4, and 16 combinations could result; however only 4 of these combinations are shown.

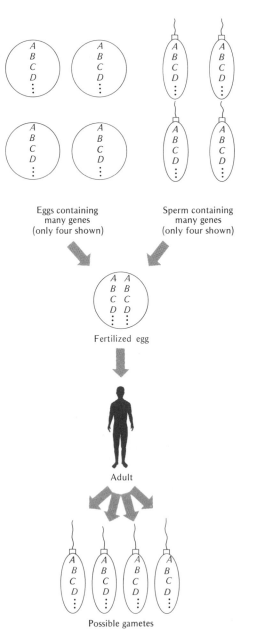

Eggs containing many genes (only four shown)

Sperm containing many genes (only four shown)

Fertilized egg

Adult

Possible gametes

FIGURE 22–12 (a) Attached earlobes. (b) Free earlobes.

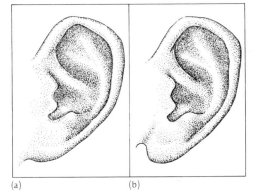

(a)   (b)

Linkage: a case of dependent assortment

Genes are not isolated entities that occur independently of one another in the cell; they are carried on chromosomes. It is the chromosomes that segregate independently of each other at meiosis, during gamete formation, and these same whole chromosomes recombine during gamete fusion (fertilization). It has been estimated that there are about 100,000 genes in man; yet there are only 23 pairs of chromosomes. Thus, many genes occur together on the same chromosome; genes located on the same chromosome cannot separate during meiosis, cannot show independent assortment, and in general must be inherited together as a group. Genes that are on the same chromosome are said to be **linked.** Mendel's law of independent assortment must, therefore, be modified and can apply only to genes that are not linked, that is, to those pairs of genes on different chromosomes.

Crossing over

About 40 years after Mendel's work on peas, W. R. Bateson and R. C. Punnett of Cambridge University in England discovered linkage. They

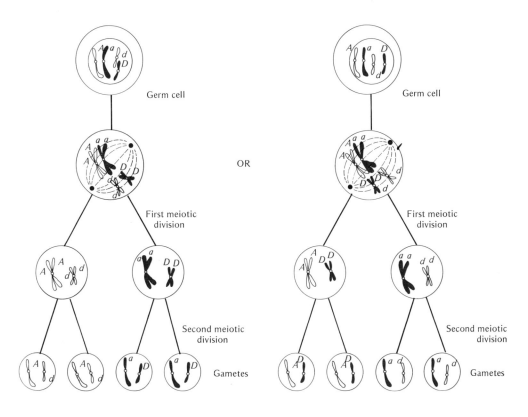

pigmentation and attached earlobes were located on the same chromosome?)

How could such results occur? In 1910 T. H. Morgan, who found similar results while working with the fruit fly *Drosophila,* offered an explanation. Morgan suggested that on occasion the link between two genes on the same chromosome may be broken, and as a consequence new gene combinations could occur. The mechanism for this phenomenon is called **crossing over.** Crossing over is believed to take place when homologous chromosomes synapse during the prophase of the first meiotic division. At that time the chromatids can twist around each other; a break may occur at a corresponding point in each homologous chromatid, and chromatid portions may be exchanged (Figure 22–14).

In the Bateson-Punnett cross of the purple-long heterozygote (*PpLl*) with the round red homozygote (*ppll*), the purple-round and red-long offspring (*Ppll* and ppLl, respectively) must have resulted when the chromatid bearing the *PL* genes crossed over with the *pl* chromatid, giving rise to chromatids carrying *Pl* and *pL*. Upon completion of meiosis some of the gametes would carry these new combinations. Thus, fertilization of a *pl* gamete with these new combinations would produce purple-round and red-long offspring.

When crossing over occurs, the four gametes produced carry four different gene combinations (*PL, pl, Pl,* and *pL*), whereas if it had not taken place only two types of gametes would have occurred (*PL* and *pl*). Crossing over is not an unusual phenomenon restricted to peas or fruit flies; in those organisms where it has been looked for, and when genetic analysis is possible, it has been found to occur. No doubt it occurs in humans as well. Since each chromatid contains many genes, crossing over greatly increases the number and kinds of gene combinations meiosis can produce. (Compare the

showed that in pea plants the genes for flower color and pollen-grain shape were located on the same chromosome. When they crossed pea plants with purple flowers and long pollen grains with plants having red flowers and round pollen grains, the offspring had purple flowers and long pollen grains. Thus, purple was dominant to red, and long pollen was dominant to round. The cross can be written:

$$PPLL \quad \text{X} \quad ppll$$
purple-long   red-round

$$\downarrow$$

$$F_1 \qquad PpLl$$
purple-long

Because the genes for purple-long and red-round are linked, they expected that when they mated their F_1 plants with a red-round plant (*ppll*) they would get only two kinds of offspring, purple-long and red-round. They reasoned this to be so because the *PpLl* plant could produce only two kinds of gametes: *PL* and *pl*. However, when they made such a cross there did occur a few plants that were purple-round and red-long. (Contrast this result with the equal frequencies of offspring found when an albino man with attached earlobes was mated with a normally pigmented woman with free earlobes, as described earlier. What would be the frequency of offspring if the genes for normal

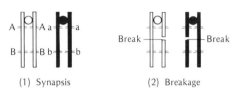

(1) Synapsis   (2) Breakage

(3) Exchange   (4) Separation of homologs

(a)

(1) Prophase I   (2) Prophase I: synapsis of homologous chromosomes

gametes in Figures 9–1 and 22–14.) Therefore, crossing over greatly increases the variation among gametes and offspring in a population. (This is of biological advantage, since it is upon this raw material that natural selection acts to mold evolutionary change.)

Sex linkage

Recall that your sex was determined at the moment of fertilization, when an X- or Y-bearing sperm fertilized an X-bearing egg (Section 22–2). Genes carried on the X chromosome are said to be sex-linked, and they show an interesting pattern of inheritance (Box 22C). A man may show a trait because he carries the gene for that trait, but such a trait may not appear in any of his children. Then the trait may reappear in the son of one of his daughters. (Again we see that genes are not diluted out in succeeding generations, but are distinct bearers of the messages of inheritance.) What mechanism underlies this pattern of inheritance?

The ability to discriminate between the colors red and green is under the control of a gene located on the X chromosome. Inability to distinguish between the two colors is due to a recessive allele of this gene. The pattern of inheritance of red-green color blindness in a hypothetical family pedigree is shown in Figure

22–15a. The children of a color blind man who is married to a normal woman are all normal, but the sons of his daughters show the trait. The inheritance of this trait is shown in Figure 22–15b. The mother is homozygous for the normal allele *(CC)*, but her color-blind husband is **hemizygous**[3] for the recessive allele *(cY)*. Their sons are all normal *(CY)*, since each male offspring receives his X chromosome from his mother; however, since the daughters receive one of their X chromosomes from their father, they have the genotype Cc. The Cc daughter, heterozygous for color blindness, has normal red-green vision but is a carrier for the trait. If she marries a man with normal color vision, their sons will be either color blind or normal, and their daughters will be normal or carriers of color blindness (Figure 22–15c). A heterozygous female *(Cc)* produces two kinds of X-bearing gametes *(C* and *c)*; whereas a male produces only one. Male offspring receive only a Y chromosome from their father but can receive an X bearing either *C* or *c* from their mother; thus the sons can be either one of two

---

[3] Males are described as being hemizygous for sex-linked genes rather than homozygous or heterozygous, since there are no allelic counterparts in the Y chromosome. The Y chromosome apparently carries only those genes responsible for maleness.

genotypes and phenotypes; CY normal, or cY color blind. However, daughters receive an X chromosome from each parent and may be either of two genotypes, Cc or CC, but a single phenotype, normal.

Let us consider the pedigree of a family in which the mother is color blind but the father is normal. In such a case the sons are always color blind, and the daughters are carriers. A girl will be color blind only if she receives the recessive gene from both parents; that is, she must have a color-blind father and her mother must be a carrier or be color blind (Figure 22–15d). If the father is color blind and the mother a carrier, the daughters may be normal or color blind. (See lower right of Figure 22–15a.) Why?

Is XX twice as good as XY? The Barr body

Since human females have two X chromosomes and males have only one, are women twice as well-off genetically as men? The answer appears to be no; let us see why.

In 1949 Murray Barr was studying the nerve cells of cats to determine if physical changes occurred as a result of fatigue. He found no differences in the cells derived from fatigued or from rested cats, but he did notice that in the nuclei of cells of female cats there was

(3) Prophase I: breakage and exchange   (4) Prophase I: crossover completed   (5) Metaphase I   (6) Anaphase I   (7) Telophase I   (8) Prophase II   (9) Anaphase II   (10)

(b)

present a dark-staining body. No such body was evident in the nuclei of cells from male cats. Extending these observations to humans, he found exactly the same situation; a dark nuclear body was present only in cells derived from females (Figure 22–16). The dark nuclear body came to be called the **Barr body,** after its discoverer. What is the significance of the Barr body?

The Barr body occurs only in females, and females have two X chromosomes. It appears that the Barr body represents one of the two X chromosomes, tightly coiled and functionless. Thus, in the female only one X chromosome is uncoiled and functional; in any given cell of a female, the genes on only one of the two X chromosomes are expressed.

Once the Barr body was discovered and its role was understood, it became clear why a female does not have twice as many of the enzymes coded for by the genes on the X chromosome as her male counterpart. Theoretically, in the early embryo half the cells have the paternal X as a Barr body, and the remainder have the maternal X in this inactive state. In a sense the female is a genetic mosaic because it is purely a matter of chance which X chromosome (the one derived from the mother or the one from the father) will become condensed and inactive. The theory that one of the X chromosomes of the female becomes genetically inactive early in embryonic life and remains in this state throughout adult life was first proposed by the British geneticist Mary F. Lyon in the early 1960s. In support of her theory (called the **Lyon hypothesis**), she showed that the coats of female mice heterozygous for sex-linked genes affecting coat color showed splotches of dark and light. This patchy distribution of coat color, Dr. Lyon explained, is due to the early inactivation of one of the X chromosomes (Figure 22–17). A similar condition occurs in calico cats, who have blotches of orange and black on their coats. Calico cats are usually female.

It should be mentioned that the Lyon hypothesis has not yet received universal acceptance, although it does provide a reasonable explanation for the fact that females, with a double dose of the X chromosome, do not produce twice as much gene product as a male with only a single chromosome (carrying the same gene).

22–5 MULTIPLE ALLELES: BLOOD WILL TELL

Blood so dominates man's thinking about inheritance that even today aristocrats are characterized as "bluebloods"; those with behav-ioral disorders are said to have "bad blood"; those with a passionate disposition are "hot-blooded." Indeed it is not uncommon to denote close family relationships (parent and child, brother and sister) by the term "blood relatives." Prior to the work of Mendel, the belief was common that blood was in fact the inherited material—hence the origin of these terms and their link with ancestry. (In the modern context "blood relative" is used in a general way to imply close genetic ties; a more accurate term would be "DNA relative.") Although the ancient belief that blood is the material of inheritance is incorrect, blood type, or group, is inherited. What is the significance of blood type and how is it inherited?

Transfusion of blood into the body's circulatory system can be life-restoring. The first such miraculous cure via transfusion was made in 1665 by Richard Lower, an English anatomist; he transfused blood from one animal to another, and the recipient animal survived and remained healthy. The achievement caused a flurry of excitement in medical circles, and it was only 2 years later that Jean-Baptiste Denis, physician to Louis XIV of France, attempted a similar blood transfusion in a human. By transfusing 8 oz of lamb's blood into the veins of a young son of an important member of the royal court, he temporarily restored the dying

FIGURE 22–15 (a) A hypothetical pedigree for color blindness in a family. (b) The possible offspring produced from the mating of a color-blind male and a normal female (generation I). (c) The possible offspring produced from the mating of a normal male and a normal female who is a carrier for color blindness. (d) The possible offspring produced from the mating of a color blind male with a normal female carrier.

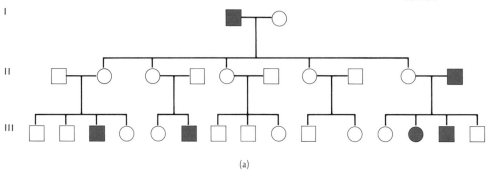

(a)

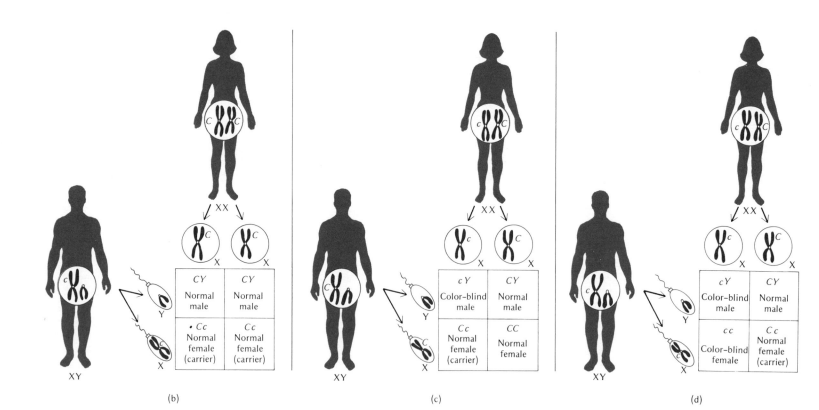

(b)

(c)

(d)

FIGURE 22–16 (a) The nucleus of a cheek cell from a normal male. (b) The nucleus of a cheek cell from a normal female. Arrow indicates the Barr body. (Courtesy of Carolina Biological Supply Company.)

FIGURE 22–17 Diagrammatic representation of the Lyon hypothesis for the formation of the Barr body. (Modified from *Human Heredity and Birth Defects* by E. Peter Volpe, Copyright © 1971, by the Bobbs-Merrill Company, Inc., reprinted by permission of the publisher.)

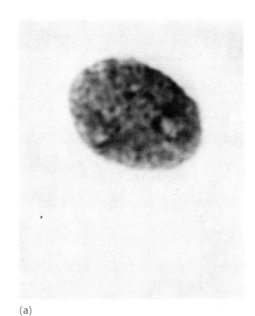

(a)

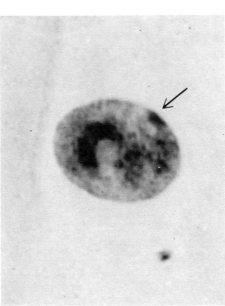

(b)

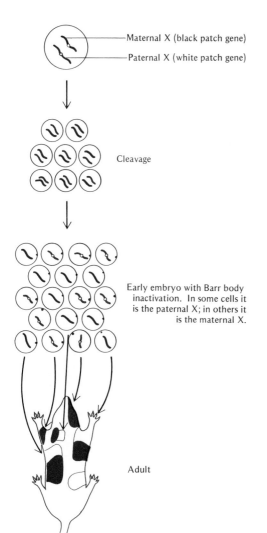

Maternal X (black patch gene)

Paternal X (white patch gene)

Cleavage

Early embryo with Barr body inactivation. In some cells it is the paternal X; in others it is the maternal X.

Adult

child. The momentous event, however, ended in tragedy, and public reaction was so strong that it stopped further experiments with blood transfusion. The practice fell into disrepute and was not employed again until the twentieth century, when enough was learned about blood and about the mechanics of transfusion to make the procedure safe.

The most fundamental of all discoveries concerning blood transfusions came from the work of Karl Landsteiner, a Viennese physician. In the 1900s Landsteiner proposed that the blood of one species of animal is not compatible with that of others; in addition, all humans do not possess the same kind of blood. He found that in human plasma there are antibodies (called **isoagglutinins**) that will clump, or agglutinate, red cells that differ from the person's own. If foreign blood cells are transfused into a person, these cells are clumped and destroyed: blood vessels are plugged, hemoglobin is released into the circulation, the kidney tubules become clogged, and the recipient dies of kidney failure in 7–10 days.

There are two types of antibodies: **alpha** (α) and **beta** (β). The red cells of humans contain specific kinds of antigen, called **A and B agglutinogens.** All human bloods can be classified into four major antigenic groups: A, AB, B, and O. The A and B types contain only the A or only the B agglutinogen, respectively; the AB type contains both A and B agglutinogens; and the O type contains neither A nor B. A person with A-type blood has plasma that contains only β-antibody, one with B-type blood has only α-antibody, one with AB-type blood has neither α nor β antibodies; and one with O-type blood has both α and β antibodies.

When blood is taken from an individual (the **donor**) to be transfused into another individual (the **recipient**), both bloods must first be typed to determine if they are compatible (Figure (text continues on p. 468)

BOX 22C Hemophilia and history

Circumstance often alters the course of history. Disease may be that circumstance. **Hemophilia** (literally "love of blood"; Gk.), a rare disease in which the blood of the afflicted individual clots slowly or not at all, is a case in point. The son of the czar of Russia was a hemophiliac—a bleeder—and because of this the czar and czarina relied heavily on the corrupt and cunning monk Rasputin, who promised a cure. This strange relationship caused a rapid decline in the Russian court. The eldest son of the last king of Spain, Alfonso XIII, died of hemophilia. It could be said that the Russian and the Spanish dynasties both bled to death because they ended with uncrowned successors—the heirs apparent died of hemophilia.

Perhaps the most famous of all hemophilic families is the royal family of Queen Victoria (Figure 22-18). The family tree illustrates that only the male offspring of the queen were bleeders (Figure 22-19). Why is this so? Hemophilia is a hereditary, sex-linked, recessive trait carried on the X chromosome. Since males have only one X chromosome they will show the trait of hemophilia if they carry the defective gene; however, females, having two chromosomes, would have to have a double dose of the gene for hemophilia to be bleeders. This is an unlikely eventuality since the chance of being homozygous for this rare gene is extremely remote. Furthermore, females showing hemophilia do not survive to maturity; the onset of menstruation is fatal. Hemophilic fathers pass the recessive gene to their daughters, but not to their sons (the male child does not receive an X chromosome from his father), and carrier mothers may pass a defective gene to their sons (their surviving daughters will be carriers). Queen Victoria and her two daughters, Alice and Beatrice, were carriers and some of the sons and grandsons of Victoria were hemophiliacs. Since Queen Victoria and Prince Albert were prolific parents and their offspring married into almost all the royal families of Europe, notorious for marrying among themselves, Queen Victoria sowed the seeds that ultimately led to the demise of the European dynasties.

FIGURE 22-18 Victoria and her daughters, shown in a family protrait taken in 1894, illustrate a hereditary dynastic misfortune. Victoria (1) carried the sex-linked mutant gene for hemophilia, as did her daughters Beatrice (2) and Alice (not shown). Alice's daughters, Alexandra (3) and Irene (5), were also carriers. Irene married Prince Henry of Prussia (6), and both their sons were hemophiliacs. Alexandra married Nicholas II of Russia (4), and their son also was a hemophiliac. Also shown are Victoria's sons Edward, the Prince of Wales (7), and Arthur (8), and her daughter Victoria (9), a possible carrier. (Courtesy of The Mansell Collection.)

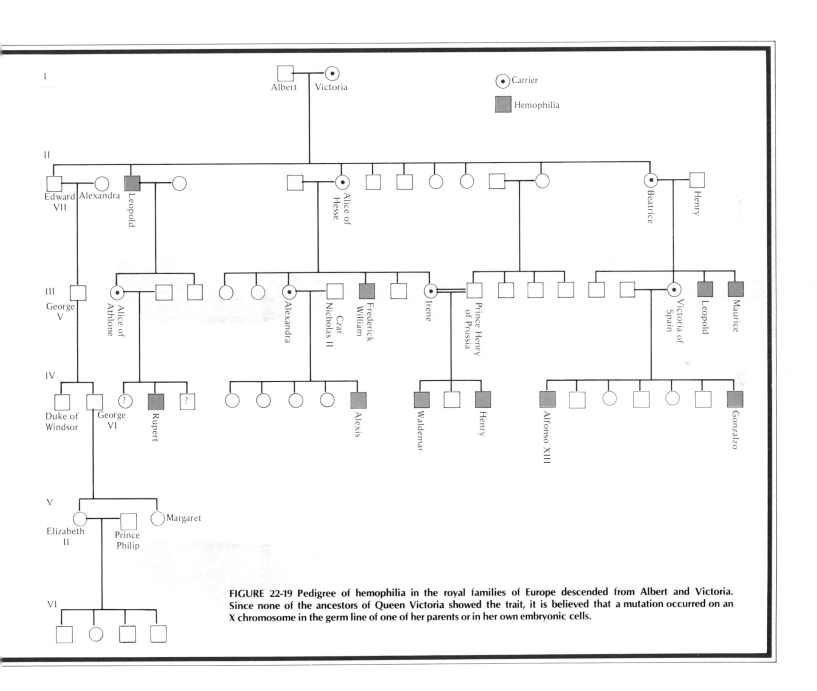

FIGURE 22-19 Pedigree of hemophilia in the royal families of Europe descended from Albert and Victoria. Since none of the ancestors of Queen Victoria showed the trait, it is believed that a mutation occurred on an X chromosome in the germ line of one of her parents or in her own embryonic cells.

FIGURE 22–20 Blood typing. Blood from a donor or a potential recipient is typed by placing two drops of blood on a microscope slide. A drop of α-agglutinin is added to one drop and a drop of β-agglutinin to the other. If clumping occurs in both drops, the blood is type AB; if clumping occurs in neither drop, the blood is type O; if clumping occurs in the drop to which α-agglutinin was added, the blood is type A; and if clumping occurs in the drop to which β-agglutinin was added, the blood is type B.

22–20). If the donor has A, B, or AB blood, his red cells have antigenic properties that could be injurious to the recipient. α-Antibody and A agglutinogen interact to produce clumping of the red cells; similary β-antibody and B agglutinogen cause red cells to clump. Thus, a person with A- or AB-type blood cannot act as donor to a person with α-antibody in his plasma—that is, a person with O- or B-type blood. In the same way, a person with B- or AB-type blood cannot donate blood to an O or A recipient, since the recipient's β-antibodies would cause the donor's red cells to clump (Table 22–1). We might expect that a person with O blood, having both α- and β-antibodies in his plasma, could donate blood only to another O person, since his plasma antibodies would clump the cells of an A, B, or AB recipient. In practice, however, a pint of O-type blood is so quickly diluted in the body that the donor's α- and β-antibodies have no effect on the red cells of the recipient. Thus, a person with O-type blood can donate to anyone, since his cells are not capable of clumping when transfused into another person and his plasma antibodies are diluted and have no effect—he is called a **universal donor.** Because an AB person has no α or β antibodies, he can receive blood from anyone and is called a **universal recipient.**

What is the genetic basis of human blood types? Blood types in human populations are determined by a gene *I* (standing for isoagglutinin) with three different alleles: I^A, I^B, and *i*. Although there are more than two alleles of a single gene in the population, no one person has more than two alleles. (Why?) The I^A and I^B alleles are codominant; that is, if a person receives both an I^A and an I^B allele, he will be of blood type AB. The *i* allele is recessive to the other two, so that $I^A i$ and $I^B i$ genotypes are of A and B blood type, respectively. Persons with a double dose of the recessive gene, that

is genotype *ii*, are blood type O. Table 22–2 shows the relation of genotype to blood type (phenotype).

At the outset of our inquiry into human inheritance we indicated that our main concern would be the simple question: What kinds of offspring can be expected from particular kinds of matings? Blood groups offer a clear system for studies of this kind, and often they play an important role in ascertaining parentage in cases where there is dispute over whether a suspected male is the father of a particular child (Box22D). There are about 20 different kinds of blood groups, and by using some of these rare groups it is possible to determine with great precision that a particular individual could not possibly be the father of the child in question. It should be emphasized that the genetics of blood typing for determination of paternity can only rule out a suspected father, it cannot provide absolute proof of parentage. (Do you think this is so? Why?)

22–6 POLYGENES AND SKIN COLOR

So far we have discussed the expression of traits in discrete, alternative classes (albino or pigmented skin; attached or free earlobes; A, AB,

Type of blood from which test serum was derived	Agglutinin	Type of blood being tested (antigen in red blood cells)			
		A	AB	B	O
A	β				
B	α				

TABLE 22-1 Blood types and transfusion

Donor's blood type	Recipient's blood type			
	A	B	AB	O
A	−	+	−*	+
B	+	−	−*	+
AB	+	+	−	+
O	−*	−*	−*	−

+ Incompatible (agglutination occurs).
− Compatible (no agglutination).
* Theoretically, clumping of the recipient's cells should occur here but because the donor plasma (containing isoagglutinins) is diluted out no clumping of cells occurs.

TABLE 22-2 Genetics of human blood groups

Genotype	Phenotype (blood type)	Agglutinogen	Isoagglutinin (antibody)
$I^A I^A$ or $I^A i$	A	A	β
$I^B I^B$ or $I^B i$	B	B	α
$I^A I^B$	AB	A and B	None
ii	O	None	α and β

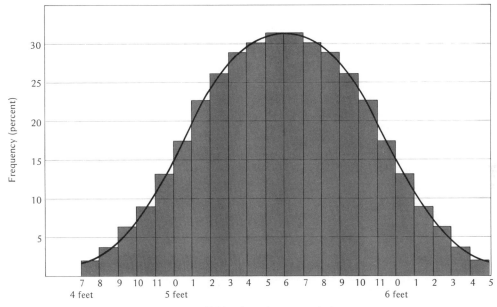

Frequency (percent)

7 8 9 10 11 0 1 2 3 4 5 6 7 8 9 10 11 0 1 2 3 4 5
4 feet   5 feet   6 feet

Height of people to nearest inch

BOX 22D The Charlie Chaplin paternity suit

(Courtesy of Theatre Collection, The New York Public Library.)

In 1944 Charlie Chaplin, the legendary comedian, was involved in a legal battle over the paternity of a child born to Joan Barry, a young starlet. Chaplin described Miss Barry as "a big handsome woman of 22, well built with upper regional domes immensely expansive and made alluring by an extremely low décolleté dress which . . . evoked my libidinous curiosity."

The baby was of blood type B, the mother's A, and Chaplin's O. From what you now know of the inheritance of blood type (consult Table 22-2) could Chaplin possibly have been the father of the child?*

* Blood-group data were not admissible evidence in California at the time of the trial, and since it provided the only evidence that Chaplin was not the father, the court deemed it had no alternative but to declare him responsible for the child's support.

B, or O blood type). There are, of course, a great many traits whose expression appears to be continuous. Some of these traits are height, weight, body proportions, and skin color. For such traits, individuals in a population do not fall into discrete classes, but rather, if arranged in an orderly array, form a continuously graded series (Figure 22–21).

In the early days of genetics it was seriously questioned whether such traits were controlled by genes at all, since genes (according to the principles of Mendel) are particulate entities that segregate and recombine as individual units. There was a strong tendency on the part of early geneticists to ascribe such traits to the simple addition of a series of genes. The logical extension of such thinking is that each individual is merely a collection of unit characters; as we shall see this concept is erroneous.[4]

Subsequent experiments in genetics have shown that quantitative traits are controlled by a large number of genes, called **polygenes,** each producing a small effect. To illustrate polygenic inheritance we shall use a simplified model for the inheritance of human skin color. (There is still incomplete understanding of the inheritance of this trait; the information presented is a model for polygenic inheritance and not a presentation of fact.)

Let us propose that there are just two loci with two alleles governing human skin color

[4] There are actually three kinds of relationships of gene to trait: (1) one-to-one, (2) one gene affecting many traits, as is the case in sickle-cell anemia, and (3) many genes (polygenes) affecting one trait, as is the case in coat color of rats and skin color of man.

through all its ranges. Let the genotype *oopp* represent white; then by substitution of each capital letter, or pigment-producing allele, there results an increase in skin pigmentation. The darkest or blackest skin color will be represented by the genotype *OOPP*. Table 22–3 shows the phenotypes that correspond to the various combinations of these alleles. This scheme directly represents the dosage of pigment-producing alleles. For the intermediate shades, there is more than one possible genotype; the extremes show only a single genotype. This model for inheritance of skin color involves two loci and five trait classes: there is no dominance in this system, and the phenotype reflects gene dosage. Note that such a system features a tendency to continuous gradation of the character; that is, the expressed trait assumes quantitative proportions.

Now let us examine a model in which three loci determine skin color; in such a system the color scale and gene dosage would run from 0→6 (Table 22–4). The result is simply that the color scale is more finely graded between the extremes. Actually, available evidence indicates that inheritance of human skin color is very much as described, that between three and five genes account for the total variation in pigmentation of human skin, and that there is no dominance. (It is not certain that all the genes have equal effects, however, as we have assumed in the illustrated examples.)

The model of polygenic inheritance shown here makes it easy to understand why the mating of the two extremes (that is, black with black or white with white, breed true) whereas mating of those of intermediate grades could produce an entire spectrum of skin color in the offspring. In addition, as one moves in shade toward either extreme the potential for variation among the progeny is reduced. (Contrary to popular belief, two near-whites cannot have throwback black offspring, or vice versa.) Al-

TABLE 22-3 A model for inheritance of human skin color involving two loci with two alleles

	OO	Oo	oo
PP	Black *OOPP*	Dark *OoPP*	Medium *ooPP*
Pp	Dark *OOPp*	Medium *OoPp*	Light *ooPp*
pp	Medium *OOpp*	Light *Oopp*	White *oopp*

Phenotype:	White	Light	Medium	Dark	Black
Color scale:	0	1	2	3	4
Genotype:	*oopp*	*Oopp*	*OOpp*	*OOPp*	*OOPP*
		ooPp	*OoPp*	*OoPP*	
			ooPP		

TABLE 22-4 A model for inheritance of skin color in which there are three loci with three alleles

Color scale:	0	1	2	3	4	5	6
Appearance:							
Gene dosage:	0	1	2	3	4	5	6
Genotype:	*ooppqq*	*Ooppqq*	*OOppqq*	*OOPpqq*	*OOPPqq*	*OOPPQq*	*OOPPQQ*
		ooPpqq	*OoPpqq*	*OOppQq*	*OOppQQ*	*OoPPQQ*	
		ooppQq	*OoppQq*	*oOPPqq*	*OoPPQq*	*OOPpQQ*	
			ooPPqq	*oOPpQq*	*OoPpQQ*		
			ooPpQq	*oOppQQ*	*ooPPQQ*		
			ooppQQ	*ooPPQq*	*OOPpQq*		
				ooPpQQ			
				ooPpQQ			
				oOpPqQ			

though first-generation progeny between the extremes are intermediate, these can in turn produce a variety of shades of skin color.

Note that the pigmentation of human skin, as well as a great many other traits (such as height and weight), is strongly modified by environmental factors. The skin of most individuals darkens upon exposure to ultraviolet light—tanning. Such exposure can often change a lightly pigmented individual into a dark one.[5]

This tends to obscure or blur class boundaries, often making the color of the skin vary greatly between the extremes. The tanning effect shows us that phenotypic skin color is not an absolute index of genotype.

Sometimes the expression of a trait depends on the interaction of genes at two different loci. In the discussion of polygenes it was suggested that skin color depends on the dosage of perhaps as many as five alleles. However, if an individual receives the recessive gene pair for albinism (*aa*), then that individual will be an albino no matter what other gene combination

---

[5] The adaptive significance of this is discussed in Chapter 24.

he has for skin pigmentation. The *A,a* alleles determine whether or not melanin is synthe-sized, and without the presence of the *A* gene for melanin production, the genes governing the intensity of pigment formation are not expressed. Such interaction of the gene for albinism with the genes for pigment intensity is called **epistasis** (literally "standing above") because the *A* locus determines whether or not the alleles for the loci controlling the degree of pigmentation are expressed, and thus it "stands above" them. It is important to distin-guish between epistasis and dominance. Epis-tasis involves interaction between nonallelic genes at two loci, where the phenotypic expres-sion of one gene is masked by an entirely different gene. Dominance refers to the inter-action between alleles, where the phenotype is an expression of one allele at the expense of the other.

22–7 CHROMOSOMAL ABNORMALITIES

Chromosomal abnormalities can be classified in two categories: alterations in the number of normal chromosomes and structural changes in the chromosome itself.

Abnormal numbers of chromosomes

Meiosis is not foolproof. On occasion accidents occur that produce an unequal distribution of chromosomes in the gametes. For example, if the spindle apparatus fails to operate normally, one or more of the chromosome pairs may fail to segregate during cell division, and as a consequence of this (called **nondisjunction**) one of the daughter cells receives extra chro-mosomes and the other daughter cell is lacking in these chromosomes.

Abnormal numbers of sex chromosomes. In 1967 an Olympic gold-medal winner for the women's 100-meter dash was stripped of her

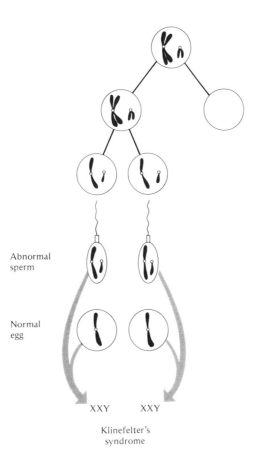

Abnormal sperm

Normal egg

XXY   XXY

Klinefelter's syndrome

XXX   XYY

Triplo-X female   XYY syndrome

records and medals because she failed the "chromosome sex text"; indeed, the Interna-tional Olympic Committee announced that year that all female athletes would be tested for their sex by chromosome examination. How can abnormal numbers of sex chromosomes be produced?

Meiosis in a normal male produces equal numbers of sperm bearing X and Y chromo-somes (Figure 22–5). However, if nondisjunc-tion occurs during the first or the second meiotic division, the sperm may have abnormal numbers of the sex chromosomes (Figure

22–22).[6] If these abnormal sperm fertilize a normal X-bearing egg, the individual suffers from sex-chromosome errors.

If an XY-bearing sperm fertilizes an X-bearing egg, the individual has one sex chromosome too many and the genotype is *XXY*. This error occurs in males with a frequency of 1 in 500 births, and the condition is called **Klinefelter's syndrome,** after the American physician who

[6] We shall assume that in these cases segregation of autosomes is normal.

FIGURE 22–23 (a) Klinefelter's syndrome (*XXY*) in a 24-year-old male. Note the long limbs and horizontal limit of the pubic hair. In some cases there is breast enlargement and a broadening of the pelvis. (b) Karyotype of Klinefelter's syndrome. (Courtesy of W. R. Centerwall, Loma Linda University Medical School.)

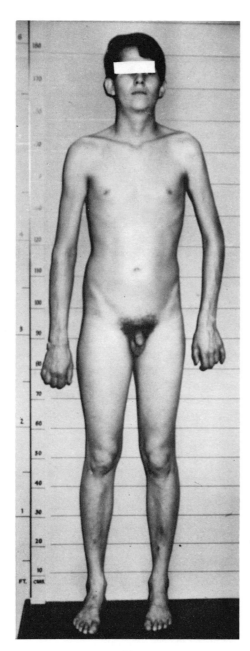

(a)

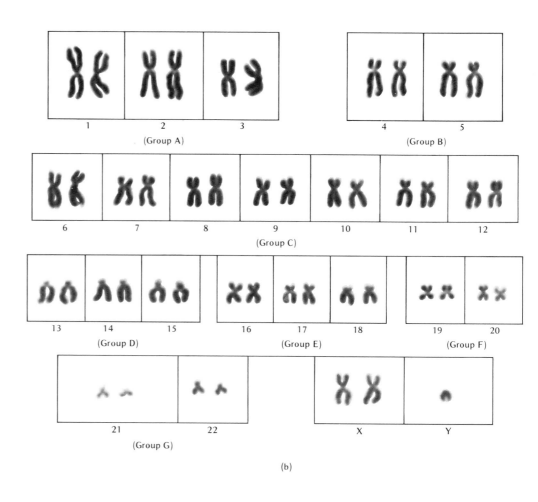

(b)

FIGURE 22–24 Nondisjunction during oogenesis and the kind of offspring that result when these abnormal eggs are fertilized by normal sperm. (Modified from *Human Heredity and Birth Defects* by E. Peter Volpe, Copyright © 1971, by the Bobbs-Merrill Company, Inc., reprinted by permission of the publisher.)

first described it. Individuals with this syndrome have the external genitalia of a male, the testes are small, and sperm are never produced; the breasts are enlarged and body hair is sparse; on occasion the individual is mentally defective (Figure 22–23). These individuals have a single Barr body. Although it has been suggested that Klinefelter's syndrome develops as a result of nondisjunction in the father, it could just as well arise as a consequence of nondisjunction in the mother (Figure 22–24). Indeed, it has been reported that Klinefelter's syndrome occurs more frequently in males born to older women.

Through nondisjunction in the female, eggs may be produced that have no X chromosome; when such an egg is fertilized by an X-bearing sperm, the offspring is of *XO* genotype (*O* signifying the absence of a sex chromosome). Such eggs fertilized by a Y-bearing sperm are not viable and fail to develop. The *XO* individual suffers from **Turner's syndrome** and in external appearance resembles a female (Figure 22–25). However, the breasts are underdeveloped and the ovaries contain only bits of whitish tissue and remain rudimentary. Women with Turner's syndrome do not menstruate or ovulate; they have a thick, webbed neck, are short in stature and below normal in I.Q., and lack a Barr body. The incidence of Turner's syndrome is 1 in 3,500 live female births; only 2% of female embryos carrying the Turner syndrome are born, and 98% are spontaneously aborted.

Thus it can be seen that normal sex determination in the human requires a normal complement of the sex chromosomes, and deficiencies or excess numbers of sex chromosomes result in abnormal phenotypes.

Abnormal numbers of autosomes. Mary Sinter's younger sister suffers from Down's syndrome (page 449). Will Mary's unborn child

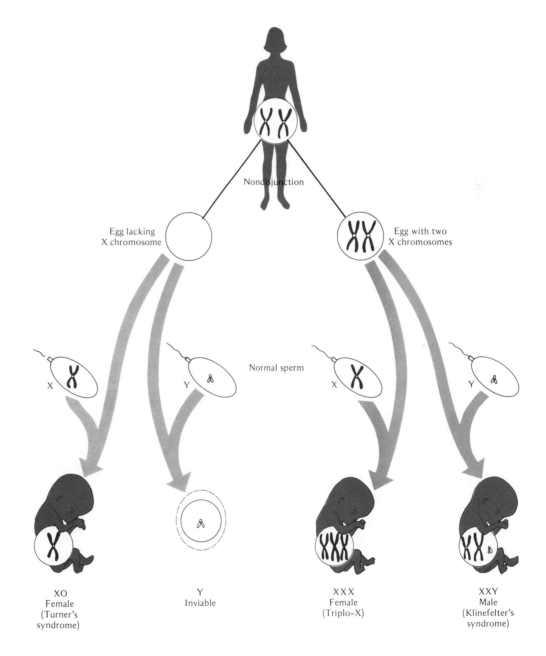

Nondisjunction

Egg lacking X chromosome

Egg with two X chromosomes

X

Y

Normal sperm

X

Y

XO
Female
(Turner's syndrome)

Y
Inviable

XXX
Female
(Triplo-X)

XXY
Male
(Klinefelter's syndrome)

suffer from the same disorder? How does Down's syndrome arise?

Down's syndrome is usually produced by nondisjunction of chromosome 21 during oogenesis (Figure 22–26), although nondisjunction during spermatogenesis could produce it. If we were to examine the chromosomes of Mary's sister we should find that she has 47 chromosomes instead of 46. The extra chromosome is not a sex chromosome, but an autosome (classified by convention as number 21, one of the smallest chromosomes (Figure 22–27).

In almost all cases of Down's syndrome the parents are normal. However, if we examine the relationship between age of the mother and the tendency for offspring to suffer from the syndrome, we find that the older the mother the greater the frequency of such abnormal children (Figure 22–28). Amniocentesis and chromosome analysis will tell Mary Sinter (page 450) whether her unborn child suffers from Down's syndrome. It is highly unlikely that her child will suffer from the disorder, since Mary is young and in good health. However, Mary's own chromosome constitution should be examined (Box 22A, page 451). The reasons for this will become apparent in the next section, when we discuss the genetic cause of a rare kind of Down's syndrome.

Abnormalities in chromosome structure

We have considered alterations in chromosome numbers and their effects. Let us now consider briefly some of the abnormalities that may occur in the chromosome itself. Four major alterations in chromosome structure may occur: (1) **deletion,** where a portion of the chromosome is missing; (2) **duplication,** where a portion of the chromosome is repeated (such effects have not been described for humans); (3) **translocation,** where a portion of one chromosome becomes attached to another chro-

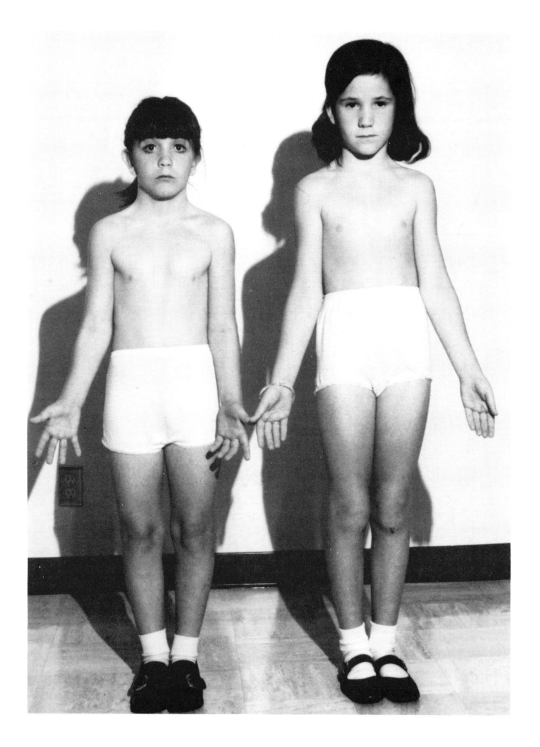

FIGURE 22–25 (a) Turner's syndrome (XO) in an 11-year-old girl. Note the small stature, shield chest, and broad, webbed neck. Standing beside her is her

FIGURE 22–26 The origin of Down's syndrome by nondisjunction of chromosome 21 during oogenesis.

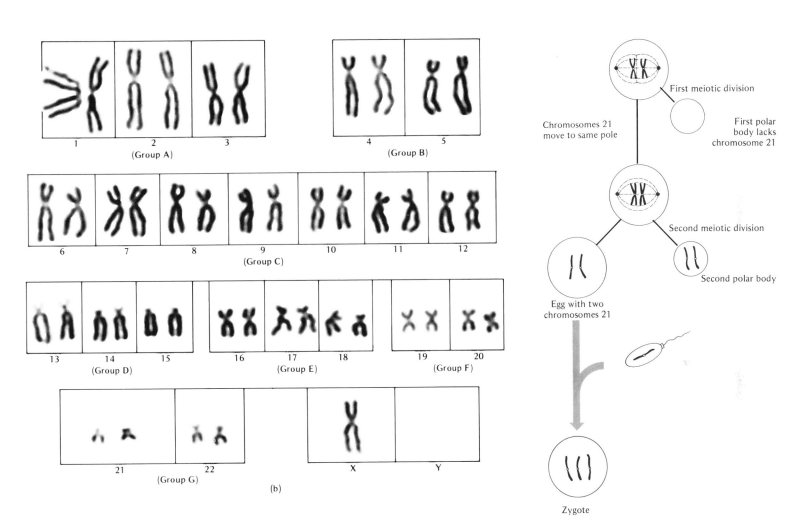

1   2   3
(Group A)

4   5
(Group B)

6   7   8   9   10   11   12
(Group C)

13   14   15
(Group D)

16   17   18
(Group E)

19   20
(Group F)

21   22
(Group G)

X   Y

(b)

First meiotic division

Chromosomes 21 move to same pole

First polar body lacks chromosome 21

Second meiotic division

Second polar body

Egg with two chromosomes 21

Zygote

FIGURE 22–27 Karyotype of Down's syndrome. Note the extra chromosome 21. (Courtesy of W. R. Centerwall, Loma Linda University Medical School.)

FIGURE 22–28 The relationship between Down's syndrome in the offspring and the mother's age. If the mother is over 40, the child's risk is materially increased.

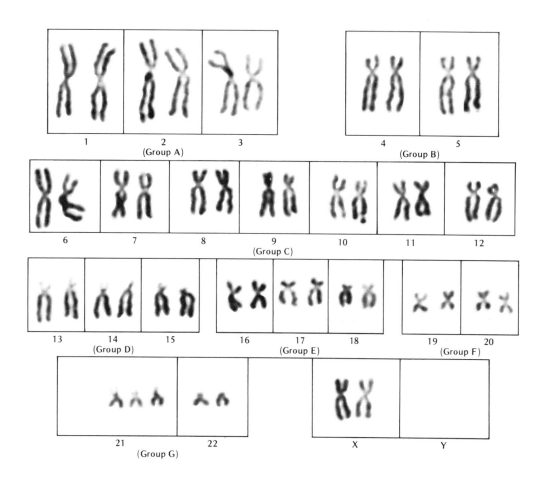

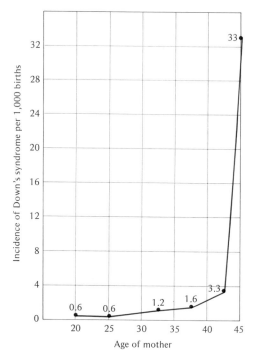

mosome; (4) **inversion,** where the linear array of genes on a chromosome becomes turned around (Figure 22–29).

The consequences of each of these structural changes are quite variable. If a deletion involves the loss of some important genes, it may affect the phenotype or even the survival of the offspring. In 1963 Dr. J. Lejeune described a disorder called **cri-du-chat** (cat-cry) syndrome in infants. The affected child has a moonlike face, utters a cry much like that of a mewing cat, and is retarded both physically and mentally (Figure 22–30). The disorder results from a deletion in chromosome 5. It has been suggested that deletions are really translocations. To lose a piece of a chromosome, two breaks must occur on the same arm and the

FIGURE 22–29 Alterations in chromosome structure.
(a) Base alteration. (b) Translocation. (c) Inversion.
(d) Deletion. (e) Duplication.

FIGURE 22–30 (a) Karyotype of cri-du-chat syndrome. Note the short arm of one of the number 5 chromosomes. (b) Cri-du-chat syndrome. The child is mentally and physically retarded. (Courtesy of W. R. Centerwall, Loma Linda University Medical School.)

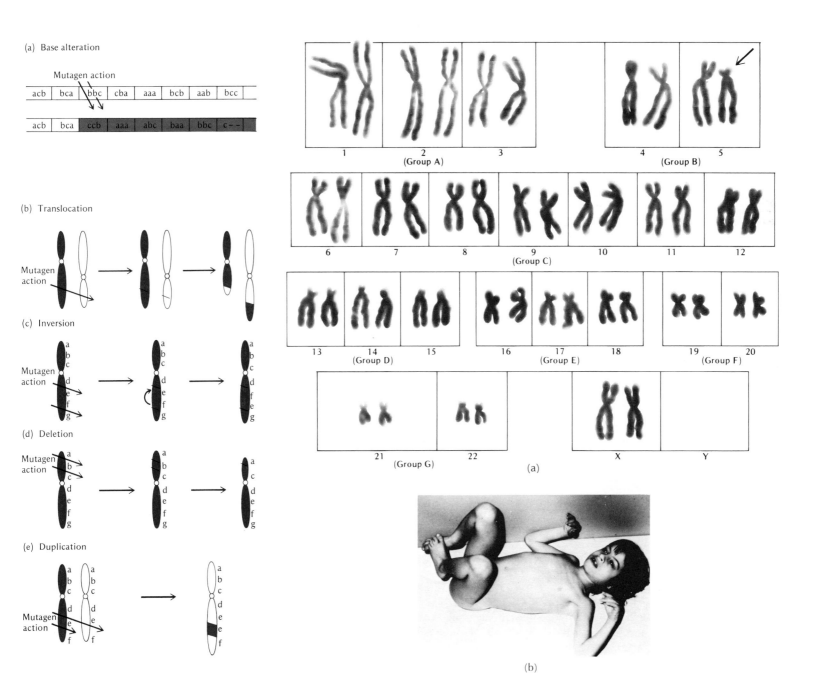

(a) Base alteration

Mutagen action

| acb | bca | bbc | cba | aaa | bcb | aab | bcc |

| acb | bca | ccb | aaa | abc | baa | bbc | c-- |

(b) Translocation

Mutagen action

(c) Inversion

Mutagen action

(d) Deletion

Mutagen action

(e) Duplication

Mutagen action

1   2   3
(Group A)

4   5
(Group B)

6   7   8   9   10   11   12
(Group C)

13   14   15
(Group D)

16   17   18
(Group E)

19   20
(Group F)

21   22
(Group G)

X   Y

(a)

(b)

FIGURE 22–31 Translocation of chromosome 21 produces an elongated chromosome 15. Persons with such a constitution are carriers for translocation Down's syndrome. (Modified from *Human Heredity and Birth Defects* **by E. Peter Volpe, Copyright ©1971, by The Bobbs-Merrill Company, Inc., reprinted by permission of the publisher.**)

ends must rejoin. To produce such breaks, a disruptive agent such as x-rays would have to score two "hits" very close together on a single chromosome, but this is unlikely. It is more reasonable to expect two hits on two different chromosomes. When this occurs, the broken ends remain "sticky" and tend to join up again. If they join with the broken end of another chromosome, a translocation occurs. At meiosis, the two translocated chromosomes may separate to different gametes. Some will have an abnormally short chromosome and some an extra long chromosome. After fertilization, the fetus may suffer an excess or deficiency of chromosomal material—equivalent to duplication or deletion. Duplication effects have not been described for humans. Although one would expect an inversion or a translocation to have little or no effect because no genes are lost or duplicated, this is not so. The order of genes along the length of the chromosome does make a difference, indicating that the position of a gene in relation to other genes affects its function.

We can illustrate the effect of translocation on inheritance by looking once again at Down's syndrome. In 5% of persons with Down's syndrome the chromosome number is 46. It has been found that in these individuals a portion of chromosome 21 is attached to the end of chromosome 15 (Figure 22–31). Thus, the child suffering from Down's syndrome and having 46 chromosomes actually has the substance of three 21 chromosomes; two normal 21 chromosomes and a 21 chromosome attached to chromosome 15. (The only difference evident in the chromosomes as seen under the microscope is that chromosome 15 is lengthened.) A study of the inheritance of this type of Down's syndrome shows that one of the parents (called a **Down's-syndrome carrier**) has 45 chromosomes (the reason being that

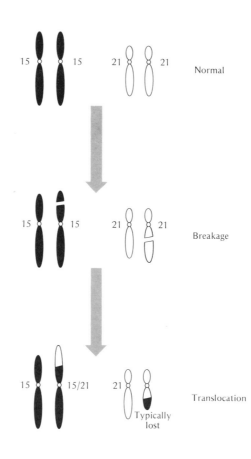

chromosome 21 is attached, or translocated, to chromosome 15). Such a parent may produce four kinds of gametes; when fertilization by a normal gamete occurs, one third of the offspring show Down's syndrome (Figure 22–32). Thus, Mary Sinter's karyotype should be determined before the genetic counselor can assure her it is unlikely she will bear a child with Down's syndrome. Indeed, if she is a carrier, the chances of her child having Down's syndrome will be increased 500 times.

SUMMARY

1. Classical genetics is the study of the mechanisms by which genes are transmitted from one generation to the next.

2. The diploid or 2n chromosome number for humans is 46 (23 homologous pairs). Through amniocentesis and preparation of a karyotype, genetic defects in the unborn can be identified.

3. Haploid gametes are produced by meiosis. At fertilization, the diploid number is restored in the zygote.

4. In humans, there are 22 pairs of autosomes, and one pair of sex chromosomes: XX in females, XY in males. Eggs carry an X, sperm may carry an X or a Y, so sex of the offspring is determined by the kind of fertilizing sperm.

5. The observable appearance of a trait is the phenotype. The genetic constitution of the individual is the genotype. Genes that occur at the same locus on homologous chromosomes are alleles.

6. A gene that exerts its effect on the phenotype when present in only a single dose is said to be dominant; such genes more or less completely mask the effect of an allelic gene, which is said to be recessive. Recessive genes have no effect on the phenotype unless present in a double dose. If both alleles of a gene pair are the same, the individual is said to be homozygous for that trait, whereas if they are dissimilar, the individual is heterozygous for that trait.

7. Gregor Mendel discovered the genetic principles of segregation and independent assortment (recombination) by crossing varieties of peas with contrasting characters. When peas with a dominant trait were crossed with peas with a recessive trait, the F₁ generation all showed the dominant trait. When members of the F₁ generation were crossed with each other,

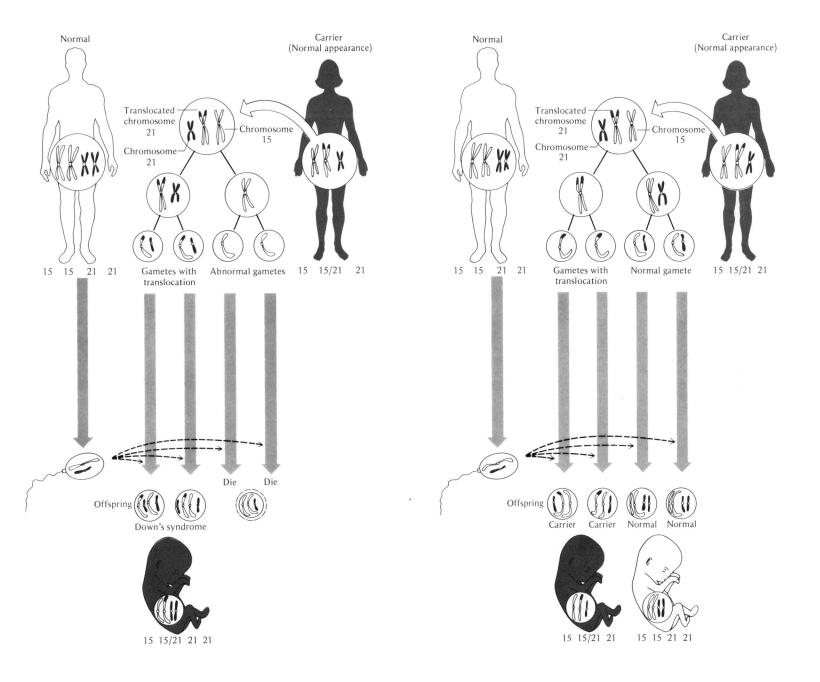

FIGURE 22–32 Transmission of translocation Down's syndrome. There are four possibilities depending on the manner in which the chromosomes are distributed to the gametes during meiosis.

Normal

Carrier (Normal appearance)

Translocated chromosome 21

Chromosome 21

Chromosome 15

15 15 21 21

Gametes with translocation

Abnormal gametes

15 15/21 21

Die Die

Offspring

Down's syndrome

15 15/21 21 21

Normal

Carrier (Normal appearance)

Translocated chromosome 21

Chromosome 21

Chromosome 15

15 15 21 21

Gametes with translocation

Normal gamete

15 15/21 21

Offspring

Carrier Carrier Normal Normal

15 15/21 21   15 15 21 21

the ratio of peas showing the dominant trait was 3 to 1 over those showing the recessive trait. If peas heterozygous for two characters were hybridized, offspring of the F_2 generation showed combinations of the parental characteristics in a 9:3:3:1 ratio.

8. Mendel's first law, the law of segregation, states that genes do not blend, but behave as independent units and are passed intact from one generation to the next. During gamete formation, members of an allelic pair separate (or segregate) so that the gametes contain only a single member of the allelic pair. Mendel's second law, the law of independent assortment or recombination, applies to the inheritance of two or more traits. It states that the inheritance of any gene pair is independent of the inheritance of all other gene pairs, with each trait being expressed independently of any other.

9. In fact, the chromosomes rather than individual genes segregate independently at meiosis, and it is the chromosomes that recombine at fertilization. Genes on the same chromosome are therefore inherited as a group and are said to be linked. Because of this, Mendel's law of independent assortment (recombination) applies to linkage groups of genes on the same chromosome.

10. Linkage groups may be broken by crossing over between homologous chromosomes at synapsis during meiotic prophase, thus increasing variation amongst offspring.

11. Sex-linked genes occur on the sex chromosomes. Since the Y chromosome carries only the genes for maleness and no allelic counterparts for other genes on the X chromosome, males are hemizygous for sex-linked characters. Hence, sex-linked recessive traits such as color blindness and hemophilia occur in males with greater frequency than in females.

12. In females, only one X chromosome is functional in any one cell. The other appears as a Barr body. Females are therefore genetic mosaics, depending on whether the maternal or paternal X chromosome is expressed in a particular cell (the Lyon hypothesis). This explains why Calico cats are usually female.

13. Blood type in humans is determined by a gene with three alleles, I^A, I^B and i. I^A and I^B are codominant, and i is recessive. I^A and I^B homozygotes will be A or B blood type respectively, as will $I^A i$ and $I^B i$ heterozygotes, whereas $I^A I^B$ heterozygotes will be AB; ii homozygotes will be blood type O.

14. A-type blood contains A agglutinogen but β-antibody. B-type blood has B agglutinogen but α-antibody. AB-type blood has A and B agglutinogen but no antibody and O blood has no agglutinogen but both types of antibody. A agglutinogen and α-antibody clump red cells, as do B agglutinogen and β-antibody. Therefore, before transfusion of blood from one individual to another, blood types must be matched so that antibody and agglutinogen of the same type do not mix. People with O-type blood are universal donors, however, since the antibodies are quickly diluted by recipient blood and there are no cells with agglutinogen in the donated blood. People with type AB blood are universal recipients since they have no antibodies to clump transfused cells.

15. There are about 20 kinds of blood groups under genetic control, and blood typing has been used to establish paternity in questionable cases.

16. Quantitative traits are controlled by polygenes. Human skin color is an example in which as many as five alleles may produce a wide range of possible shades of color. Interaction between nonallelic genes may involve epistasis, where the phenotypic expression of one gene or set of genes may be masked by expression of a different gene. The genes for albinism (due to mutation of a gene controlling melanin production) show epistasis over all other skin color genes, for example. Since in albinos melanin is not synthesized, the genes determining intensity of melanin deposition have no effect.

17. Abnormal chromosome numbers may be produced as a result of nondisjunction during meiosis. Abnormal numbers of sex chromosomes may produce Klinefelter's syndrome (XXY feminized males), or Turner's syndrome (XO sterile females). Nondisjunction of autosomes may produce Down's syndrome (extra chromosome 21) with mongoloid physical symptoms, low I.Q., and short life-expectancy; the incidence of Down's syndrome increases with the age of the mother.

18. Deletion, duplication, translocation, or inversion of various chromosomes may produce cri-du-chat (deletion in chromosome 5) or Down's syndrome (translocation of extra chromosome 21 to 15). Chromosome abnormalities such as this may be produced by x rays, which can cause chromosome breaks.

KEY WORDS

Down's syndrome (mongolism)
amniocentesis
genetics
physiological or molecular genetics
transmission or classical genetics
chromosomes
diploid number
gametes
zygote
autosomes
sex chromosomes (XX and XY)
karyotype
melanin
melanocytes
albinos

phenotype
genotype
homologous (chromosomes)
allele
locus
family tree or pedigree
dominant
recessive
homozygous
heterozygous
carriers
F_1 (first filial) generation
F_2 (second filial) generation
law of segregation
law of independent assortment or
   principle of recombination
mutation
Punnett square
linked (genes)
crossing over
hemophilia
sex-linked (genes)
hemizygous
Barr body
Lyon hypothesis
isoagglutinins
α-antibody
β-antibody
A and B agglutinogens
donor
recipient
universal donor
universal recipient
polygenes
epistasis
nondisjunction
Klinefelter's syndrome
Turner's syndrome
deletion
duplication
translocation
inversion

cri-du-chat
Down's syndrome carrier

TOPICS FOR DISCUSSION AND REVIEW

1. What is the relationship of phenotype to genotype?
2. What is a karyotype, how is it prepared, and of what value is it in genetic studies?
3. Discuss the molecular basis for albinism, its mode of inheritance, and how it may have arisen.
4. Why is more known about the inheritance of abnormal human traits than of normal ones? Discuss some structural abnormalities of chromosomes.
5. How are Mendel's laws explained by meiosis?
6. Why is Mendel's principle of independent assortment (or recombination) not universally true?
7. What is polygenic inheritance? Give some examples of polygenic traits and their mode of inheritance.
8. What is sex linkage? What is crossing over and what is its biological significance?
9. Discuss multiple alleles and blood group inheritance.
10. Discuss nondisjunction and some of its consequences.
11. What is the relationship of the Barr body to sex-linked inheritance?
12. Discuss the biological significance of sexual reproduction.
13. Discuss the inheritance of Down's syndrome.
14. Contrast dominance with epistasis.

Detecting an Old Killer

Historically, sickle-cell anemia has been an unheralded killer. It does not occur in dramatic epidemics. Its victims in the U.S. are mostly blacks, and they generally receive less medical attention than whites. The malady affects the red blood cells, which normally are spherical. When the anemia victim is under any stress that reduces the oxygen supply in his blood, his red cells elongate into firm gel-like crescents, called "sickles," that block narrow capillaries and deprive tissues of vital oxygen. The impaired circulation causes great pain in various parts of the body, enlargement of the heart and damage to the brain cells. The tendency of the red blood cells to rupture produces severe anemia.

The estimated 2,000,000 Americans who carry one defective gene sometimes show mild symptoms of the disease, but the one child in 170 who inherits a sickle-cell gene from each parent has barely an even chance of seeing his 20th birthday. If he does survive into middle age, he is likely to be crippled.

Sickle-cell anemia cannot be cured, though treatment can sometimes control its effects. It can be almost completely prevented, provided that partners who both carry the trait avoid having children. The missing element has been a simple test that would allow carriers to be identified and warned. Now this crucial procedure has been perfected.

What is the molecular basis of sickle cell disease? Why is such an obviously debiliating disorder maintained in the population?

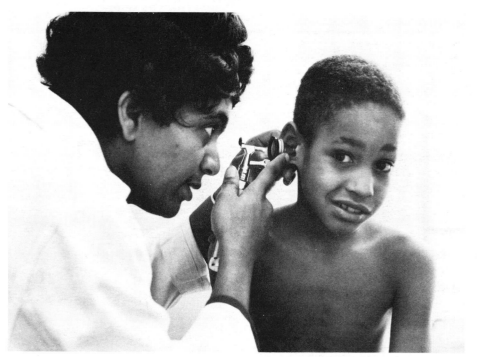

DOCTOR EXAMINING SICKLE-CELL SUSPECT. INCURABLE, BUT AVOIDABLE.
(Photo by Arnold Hinton. Reprinted by permission Monkmeyer Press Photo Service.)

Modified *Time Magazine* © 1971 Time, Inc.

23

GENES IN POPULATIONS: AN INTRODUCTION TO EVOLUTION

23-1 SICKLE CELL: A MOLECULAR DISEASE

Sickle-cell disease was first described in 1910 by Dr. J. B. Herrick in a young West Indian student living in Chicago. When air was excluded from a drop of the patient's blood by placing the drop on a microscope slide, ringing it with petroleum jelly, and covering it with a coverglass, the red blood cells showed a dramatic change in shape from round to a sickle form (Figure 23-1). The patient showed other distressing symptoms—clogged blood vessels, pneumonia, rheumatism, heart disease, inflammation of the soft tissues of the hands and feet, and anemia. What is the cause of sickle-cell disease?

In Chapter 22 we considered disorders caused by abnormal chromosome structure or chromosome numbers. Sickle-cell disease and some 1,500 other human diseases are attributed not to visible changes in the chromosomes, but to alterations in the DNA itself. Some of these disorders are very rare, but others are common. Unfortunately, only about 30 of these conditions can be diagnosed in the unborn child by amniocentesis (Figure 22-2), and sickle-cell disease is not one of them.

Sickle-cell anemia results when the abnormal recessive gene (Figure 23-2) is inherited in a double dose (ss). Heterozygotes, or sickle-cell carriers (Ss), typically show no ill effects; however, when oxygen is limited (for example, at high altitude or on a slide ringed with petroleum jelly), the blood cells of the heterozygote show mild sickling. The heterozygous state is often called **sickle-cell trait.** Individuals who are normal (SS) do not, of course, show the sickling phenomenon.

The molecular basis of sickle-cell disease is a **gene mutation** (change) that occurred centuries ago in Africa. In 1949 Linus Pauling and his colleagues at Cal Tech found that the

FIGURE 23-1 Electron micrographs of (a) normal red blood cells and (b) sickled red blood cells. (Courtesy of Philips Electronic Instruments.)

hemoglobin molecule present in the red cells of persons having sickle-cell anemia was abnormal. The hemoglobin from persons homozygous for sickle-cell anemia had a different electrical charge from that of normal hemoglobin, and heterozygous persons showed about equal amounts of normal and sickle-cell hemoglobin (Figure 23–3). Closer analysis of the sickle-cell hemoglobin revealed that a change in a single amino acid out of the 287 amino acids in each hemoglobin molecule (substitution of valine for glutamic acid, Figure 23–4) produced profound effects on the structure and function of the hemoglobin. The change in one amino acid altered the electrical charge of the hemoglobin, reduced the ability of the molecule to transport oxygen, changed the shape of the red cells, and created a cascade of disease symptoms. Sickle-cell anemia is an example of **pleiotropy,** a situation in which a single gene affects a great many phenotypic characters (Figure 23–5). It is also a classic example of how a change in a single nucleotide in the DNA molecule can produce a distorted message and an inborn molecular error (the DNA code words for glutamic acid are CTT and CTC, whereas for valine the equivalents are CAT and CAC; Figure 6–8).

23–2 OUR CHANGING GENES: MUTATION

What we are and what our children will be depends, in the final analysis, on the nature of the genetic blueprint and how that blueprint is duplicated. Strictly speaking, we did not acquire the genes of our parents, nor will our children receive our genes; rather, parental genes are copied, and the copy is passed on to the progeny. Precision in copying the genetic blueprint and exactness in chromosome duplication ensure that parental traits are passed on intact to the offspring.

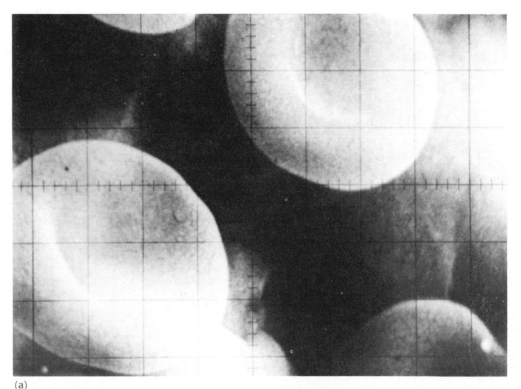

(a)

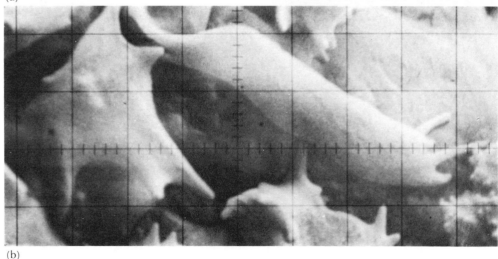

(b)

FIGURE 23-2 The mating of two persons with sickle-cell trait can produce normal offspring, offspring with sickle-cell trait, and offspring with sickle-cell anemia.

FIGURE 23-3 When a solution of hemoglobin is subjected to an electrical field, its pattern of migration in the field depends on its charge. Note the differing patterns of migration for normal and sickle-cell hemoglobin. Note also that the individual with sickle-cell trait has equal proportions of both kinds of hemoglobins.

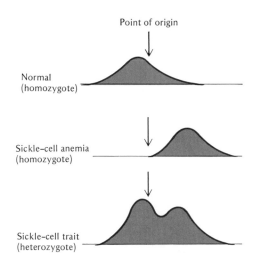

FIGURE 23-2 The mating of two persons with sickle-cell trait can produce normal offspring, offspring with sickle-cell trait, and offspring with sickle-cell anemia.

Sickle-cell trait

Sickle-cell trait

| Ss | |
| S | s |

| SS normal | Ss trait |
| Ss trait | ss anemia |

Point of origin

Normal (homozygote)

Sickle-cell anemia (homozygote)

Sickle-cell trait (heterozygote)

A gene is a linear stretch of the DNA molecule that contains the information for producing a protein chain. The kinds of proteins made depend on how accurately the genetic message is read. The transfer of information from parent to progeny depends on the accuracy with which the DNA molecule duplicates itself. Although the copying process involved in the replication of DNA is quite precise, it is not without error. Once in a while a change occurs in the genetic blueprint; this may involve the simple substitution of one base for another in the DNA molecule. (The molecular basis for such changes is discussed in Chapter 6.) The modified gene is reproduced in its altered form. This heritable alteration in the genetic message, called a gene mutation, has as its basis a change in the base sequence of the DNA.

Mutations are uncontrollable chance phenomena that cause a change in an organism. Such changes may be relatively minor or they may be severe. Indeed, the vast majority of gene mutations observable today in organisms are changes for the worse and may be lethal[1] (the embryo never survives beyond a few hours or days). When we consider that the organisms that populate the earth today are the products of a long evolutionary history, such changes for the worse produced by mutation are not unexpected. The genes we carry today represent the most favorable mutations, selected and accumulated during the course of time; the chance of a new mutant gene arising that confers an advantage over one that has an established historical role is quite unlikely. It would be similar to the replacement of a word or two in this paragraph by another word,

_____

[1] Mutations do occur that are not debilitating; that is, the genes for eye color, blood group, and so on mutate without harm. However, here we are concerned with those diseases that are maintained in the population by mutation.

selected at random, with the retention or improvement of the original meaning. It could happen, but the chances are slim.

If mutation in an organism is often deleterious, what is its significance? Genes arise only from preexisting genes, and mutation is the source of all new genes. Without mutation, evolution could not have progressed beyond the very earliest primitive organisms.

Mutations are chance occurrences. Their appearance has absolutely nothing to do with the needs of an organism or whether the change is advantageous or disadvantageous. On occasion, a mutation may be advantageous, and as such may provide the change necessary for the survival of that population of organisms. Let us consider an example. The bacteria called staphylococci cause a variety of human ailments—sore throats, inflamed ears, boils, and so on. Many staphylococcal infections are easily treated by administration of an antibiotic such as penicillin; penicillin kills the bacteria and the infection disappears. On occasion, the staphylococcal infection does not respond to the antibiotic, and the bacteria grow in its presence; we say that such bacteria have lost their sensitivity to penicillin and that they are now penicillin-resistant. Did the penicillin-resistant bacteria arise by an advantageous mutation because the sensitive staphylococci needed a way of combating the antibiotic? No. A chance mutation in the original bacterial population (which was sensitive to penicillin) produced a small number of individuals capable of growing and reproducing themselves in the presence of penicillin. The mutation was not deleterious to these bacteria in the absence of the antibiotic. However, once antibiotic treatment began, the penicillin killed off the sensitive bacterial cells, and only those bacteria that contained the mutant gene could survive in this new antibiotic-laden environment. Thus, a mutant gene that previously had little or no

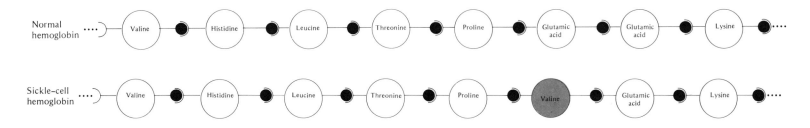

effect proved beneficial to the organism under the appropriate environmental conditions. As we shall see, selection of mutant genes is a continuing occurrence and an important part of the evolutionary process in populations of organisms—human as well as bacteria.

What causes mutations? The cause of natural or spontaneous mutations remains uncertain, but it is well known that many external agents can produce mutations. Among these are background radiation (cosmic rays, radioactive elements in the soil and in the food), temperature, and chemicals (mustard gas, peroxides, and so on). Different genes have different mutation rates, but why this is so remains unexplained.

Some mutant genes are decidedly beneficial; that is, an individual carrying such a gene is better able to adapt to his environment. Others are harmful or even lethal to the individual in whom they are expressed. Since the process of evolution involves changes in the quality as well as the quantity of the genes in all members of a species living at that particular time, the consequences of such mutations for the living population as well as for future generations assume some importance. Harmful mutations carried in the germ cells can be regarded as a **genetic burden** (which is also called the **genetic load**) for the whole population.

The entire genetic variability of all members of an interbreeding population is called the **gene pool.** It may be tempting to believe that lethal genes derived from mutation cannot long

survive in a population because the individuals carrying such genes are unlikely to reproduce or survive, and thus the gene pool will purify itself. This is, of course, true for dominant lethal genes and for individuals homozygous for recessive lethal genes, who will die without reproducing. However, an individual carrying a deleterious gene can function normally if that gene is compensated for by a normal gene; that is, if the individual is heterozygous for the deleterious gene. Thus, some lethal or deleterious genes persist in the gene pool because the mutant gene is masked by the presence of a normal allele or because the effect of the mutant gene, though not completely masked, is expressed to a lesser degree, and the individual with such a **subvital** (harmful but not lethal) **gene** survives at least long enough to reproduce. (Sickle-cell trait is an example of such a condition.) In this way, deleterious mutations may accumulate in the gene pool of generations yet unborn—a grim legacy for our progeny. This raises the following question: Can we reduce the genetic burden we now carry and that of future generations? How can we effect such changes? We shall attempt to answer these questions later in this chapter.

23-3 GENES IN POPULATIONS

In our day-to-day encounters with our fellowmen we tend to use individual differences as a sign of recognition. John has red hair, Philip

has a jutting jaw, Susan has brown eyes and brown hair, and so on. The most casual assessment of the human population convinces us of one thing: we are not all alike. The variety we see around us depends, at least in its fundamental aspects, on two factors: mutation and sexual reproduction. Mutation, as discussed previously in this chapter, is the source of new genes. But how does sexual reproduction contribute to genetic variation?

An organism that reproduces by asexual means, that is, by mitosis, passes on its chromosomes without genetic change, for in this process chromosomes duplicate themselves with great fidelity. As a consequence, the cellular progeny of mitotic division have a genetic endowment identical to that of the parent cell, unless a mutation has occurred. In addition, since the progeny are derived from a single parent cell, the genes in the offspring are in exactly the same combinations as they were in the parent cell. Mitosis, or asexual reproduction, cannot provide a mechanism for recombining the genes of one individual with those of another.

An organism that reproduces by sexual means has a mechanism for shuffling around genes to produce a staggering array of genetic combinations. Let us consider the situation in ourselves. We began life as a zygote, the result of the fusion of two gametes, an egg from the mother and a sperm from the father. Some of the genes of one parent were combined with

FIGURE 23-5 A cascade of consequences connected with sickle-cell anemia.

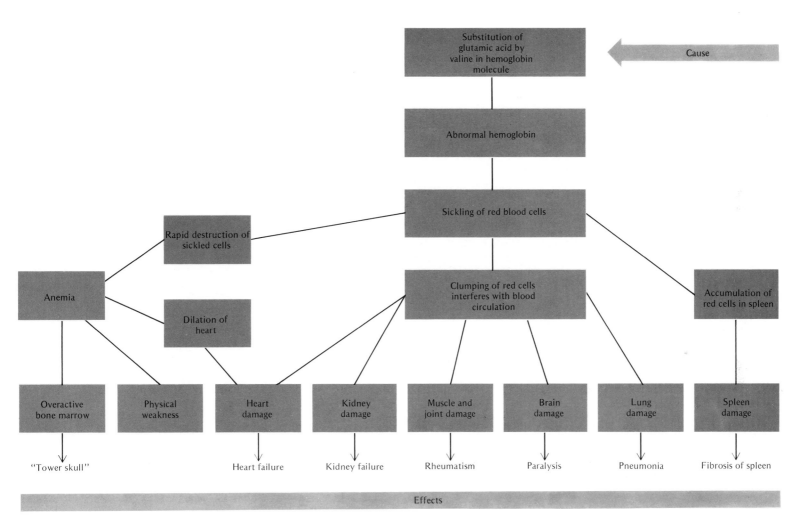

those of the other parent in the formation of the zygote. In addition, through meiosis any particular chromosome may be combined with any other chromosome, thus producing new combinations of genes. The number of possible combinations is given by the formula 2^n, where n equals the haploid number of chromosomes. Recall that the haploid gamete which results from meiosis does not have solely paternal or maternal chromosomes, but a random assortment of these (providing the physical basis of Mendel's principle of recombination (or independent assortment); see Figure 22–13). To gain some impression of the magnitude of such chromosomal combinations, let us look at our own chromosomal constitution. Each of our cells contains 23 pairs of chromosomes, and in theory the number of chromosomal combinations that could be produced during the formation of the gametes is equal to 2^{23}, or about 11 million. The number of possible chromosomal combinations that could arise by the fusion of gametes would be equal to $(2^{23})^2$! This number is further increased by crossing over between the chromosomes (page 461 and Figure 22–14). Thus, the number of chromosomal combinations derived from sexual reproduction becomes almost infinite. It is no wonder that each of us is unique in our different gene combinations.

Genetic variation, a consequence of mutation and sexual reproduction, is the raw material upon which evolutionary change depends. Evolution involves changes in the genetic makeup of a population through successive generations, that is, with the passage of time. An individual cannot evolve, but a population can. At the moment of fertilization the genetic constitution of an individual is fixed for all time, but the genetic makeup of a population can change with time. In the absence of genetic variation no population of organisms can evolve, and in a changing environment such organisms would soon become extinct. Let us look at the genetic variation in a population of organisms and consider the total number of genes in all the eggs and sperm produced by the parents—the gene pool of that population.

When we study the gene pool of the human population, we are interested in the frequency of occurrence of a particular gene in the entire population. To explain the meaning of **gene frequency,** let us consider a population and a single gene locus on a single pair of chromosomes in that population. The two alleles are B and b, and we shall arbitrarily set the frequency of the B allele at 0.5 (that is, 50% of the genes are of this type) and the frequency of b at 0.5. In a more general form we can say that if the frequency of one allele is p and the frequency of the other allele is q, then the collective frequency of alleles at the locus for the entire population is:

$$p + q = 1$$

What would happen to the gene frequencies in such a population if there were completely random mating, if mutations did not occur, and if there were no immigration or emigration? In 1908 G. H. Hardy (a mathematician) in England and W. Weinberg (a physician) in Germany provided a mathematical solution to this question. According to the **Hardy-Weinberg law** the frequencies of the genotypes can be described by expanding the binomial expression $(p + q)^2$, that is:

$$p^2 + 2pq + q^2$$

If the gene pool of the population is composed of 0.5 B and 0.5 b genes in the gametes, we should expect the offspring to occur in three distinct genotypes: BB, Bb, and bb. If p is the frequency of B (0.5) and q is the frequency of b (0.5), then on the basis of the Hardy-Weinberg equation, we expect the offspring of such a population to be:

p^2	+	$2pq$	+	q^2
BB		Bb		bb
(0.5)(0.5)		2(0.5)(0.5)		(0.5)(0.5)
0.25		0.50		0.25

Or, in more familiar terms, 50% of the genotypes in such a population will be either BB (25%) or bb (25%), and 50% of the genotypes will be Bb.

Although in most populations the values of p and q are not equal, the calculations remain exactly the same. That is to say, the more frequent allele in a population will not automatically increase in frequency, nor will the less frequent allele eventually be lost from the population. Let us see why.

Suppose the frequency of the B gene in the gene pool is 0.9 (p) and the frequency of the b gene is 0.1 (q). Using the Hardy-Weinberg equation we find:

p^2	+	$2pq$	+	q^2
(0.9)(0.9)		2(0.9)(0.1)		(0.1)(0.1)
0.81		0.18		0.01

That is, 81% of the population is BB, 18% is Bb, and 1% is bb. If we compute the frequencies for the next generation, we find they are the same ones we started with. Similar calculations for subsequent generations turn up the same result: gene frequencies and genotype ratios are unchanged. The Hardy-Weinberg law states that in a sexually reproducing population under certain conditions of stability, the frequency of genes remains constant from generation to generation. Many aspects of the field of population genetics are based on this law.

The Hardy-Weinberg law describes a population that is unchanged from one generation to the next; that is, it describes a situation where evolution is not occurring. In addition, the Hardy-Weinberg law applies only when the following conditions are met:

1. There must be random mating and no selective advantage of one genotype; that is, all the various genotypes have the same survival value.
2. Mutations do not occur.
3. Migration must not occur.

These conditions are rarely found in nature. Once matings are nonrandom, or mutation or migration occurs, or genotypes differ in their mortality and fertility, the Hardy-Weinberg law does not apply, and gene frequencies will change from one generation to the next. Since the Hardy-Weinberg law depicts a static evolutionary situation that occurs rarely in nature, what is its practical value? The value of the principle is twofold:

1. By assuming the Hardy-Weinberg law does apply to a population and then finding out how gene frequencies deviate from this, we can learn something about the various agents involved in evolutionary change.
2. The formula can be used to estimate the frequency of heterozygotes in a population in which heterozygotes appear normal and thus cannot be distinguished readily. Albino

BOX 23A Ill winds blow over the *Lucky Dragon:*
radiation as a mutagen

The Bravo nuclear test on Bikini atoll took place on March 1, 1954. An unexpected wind carried the radioactive products, fallout, more than 80 miles from the target site and showered the Japanese fishing boat *Lucky Dragon* with at least 15 lb of debris. The 23 crewmen received between 250 and 330 roentgens*, a radiation dose equal to that given off by the entire world supply of radium. One fisherman died, others showed abnormalities in their blood, and most probably suffered genetic damage.

What is radiation and how does it damage the living cell? Atomic radiation, usually referred to as **ionizing radiation,** includes x rays or gamma rays, which are the shortest and hence the most energetic wavelengths of radiant energy. These energetic waves are capable of knocking out electrons from the atoms of materials through which they pass. The loss of an electron creates a charged atom, an **ion,** and this is highly reactive chemically (Figure 23-6a). Through the chemical activity of ions, the molecules of the living cell may be altered—proteins, especially enzymes, may be inactivated, and bases of the DNA may be substituted, deleted, and rearranged (Figure 23-6b). Loss of a large number of bases in the DNA impairs the cell's ability to synthesize proteins and produces errors in coding. The implications of this are far-reaching and the effects could even be lethal.

Ionizing radiations are a part of our ecosystem. Radioactive substances occur in soil, water, air, food, and our own bodies; we are continually being bombarded with cosmic rays from outer space. Together with the other sources of natural radiation, these constitute the **background radiation.** We can do very little about this. Our concern lies with the ever-increasing amounts of radiation that we ourselves have created and that are widely disseminated.

Each of us is exposed to more than twice the radiation received by persons living before the turn of the century. Why is this so? In 1892 Henri Becquerel discovered uranium and the fingerprints of radioactivity—ionization and fogging of photographic plates.

*A roentgen is a measure of radiation. See Table 23-1.

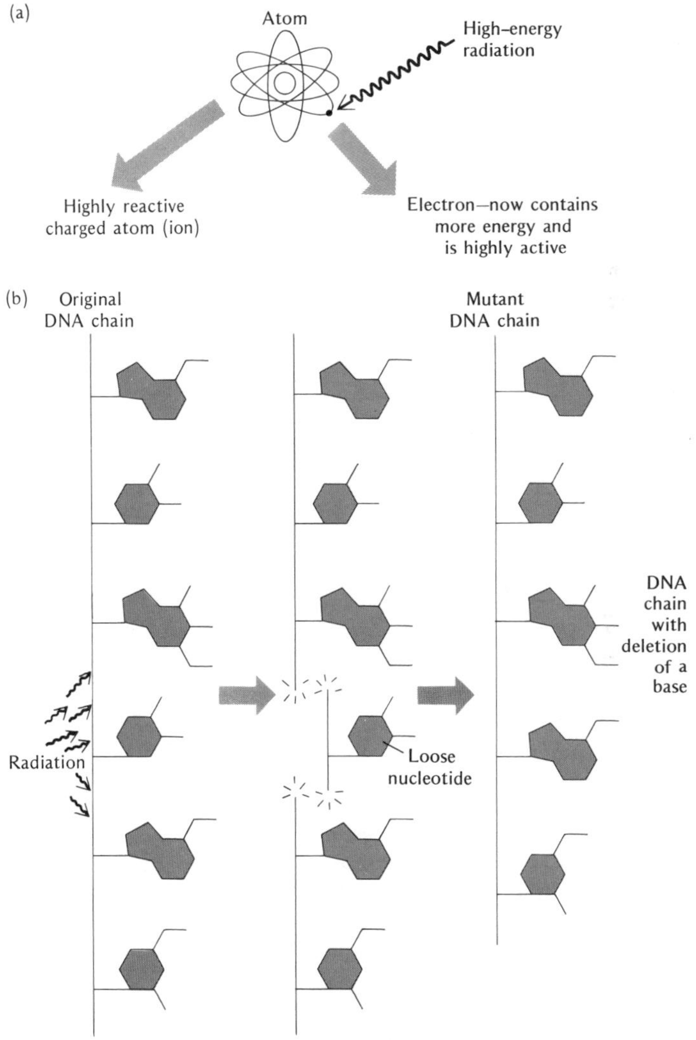

FIGURE 23-6 The effects of radiation on the cell. (a) Ion formation when high-energy radiation strikes an atom. (b) Alteration of DNA base sequence by deletion of a nucleotide (point mutation) when ionizing radiation strikes the DNA chain.

Some years later Pierre and Marie Curie isolated the radioactive element radium. These elements (and many others discovered since) are naturally unstable and undergo spontaneous disintegration. As the radioactive atoms decay, energy in the form of gamma rays and particles is emitted, and this is the source of ionizing radiation. Since 1945 we have been able to split the atom and thus create new and totally artificial elements that also decay to emit deadly gamma rays. Detonation of nuclear weapons creates radioactive substances that are let loose into the atmosphere. This material, falling to the earth, is fallout. In 1895 Wilhelm Röntgen discovered that special high-voltage equipment caused disturbances in the electrons of atoms, producing x rays. X rays could then be produced at will. Gamma rays and x rays are quite similar as sources of ionizing radiation; gamma rays are born in the nucleus of an atom, whereas x rays arise in the electron cloud surrounding the atomic nucleus.

X rays opened up a new era of medicine, for they enabled the physician to see inside the body. Splitting the atom created nuclear weapons and nuclear power plants. New sources of radioactivity could be made available for the treatment of cancer, and radioactive elements provided a means of tracing the metabolism of the cell. These are the benefits of living in the age of the atom. What are the hazards?

Ionizing radiation cannot be seen, but its effects, especially those on living matter, are easily detected. Unfortunately this was not immediately realized by the earliest workers. Pierre Curie placed a small fragment of radium on his arm and burned his skin. The pioneers in radiology took few precautions in their exposure to the x-ray beam and many suffered a life of agony terminated by cancer. The severe injuries caused by the atomic bombs dropped on Hiroshima and Nagasaki are well known. The nuclear reactor successfully employed by Enrico Fermi to produce atomic energy probably contributed to his early death of leukemia at age 54. The signs were all there, but most of the world paid little attention to them. However, in 1927 the subtle genetic effects of ionizing radiation were clearly demonstrated. In that year H. J. Muller bombarded fruit flies *(Drosophila)* with x rays and induced mutations. X rays are **mutagenic** (literally

TABLE 23-1 The radiation we receive each year

Source	Dose (rad)*
Natural background	0.1
Fallout	0.1
Chest x ray	0.001
Pelvic x ray	3–6
Upper gastrointestinal tract x ray	0.2–0.3

*A roentgen is the exposure dose and is defined as the amount of radiation that produces 2×10^9 ion pairs per cubic centimeter of air. (A cubic centimeter of tissue contains 100 billion times this number of atoms.) A rad is a unit of radiation absorption and equals the amount of energy absorbed; 1 rad = 100 erg absorbed per gram of irradiated matter. For our purposes, rad and roentgen are roughly equivalent.

"mutation forming"). Since mutations are random changes in genes, they seldom improve the genetic composition of an organism; most mutations are deleterious, and increasing the mutation rate enhances the probability of disturbing the well-established pattern of inheritance.

Some of the radiation doses we are exposed to are shown in Table 23-1. A dose of 30 rads† is sufficient to double the spontaneous mutation rate. Exposure of a human population of 200 million to 10 rads would cause:

Form of damage	No. of cases in first generation	No. of cases in subsequent generations
Chondrodystrophic dwarfism	1,000	200
Mental and physical defects	80,000	720,000
Infant mortality	160,000	3,840,000
Embryonic mortality	400,000	7,600,000

Such changes occur because the ionizing radiation has interacted with the DNA of the cells undergoing sex-cell formation. The damage to the gametes will show its effect in those yet unborn—a cruel legacy for our children. Exposure to 200 rads causes

†A rad is a unit measure of radiation. See Table 23-1.

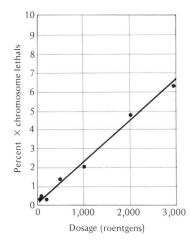

FIGURE 23-7 The relationship of radiation dose to production of mutations in the fruit fly *Drosophila*. It is quite possible that humans are more sensitive to radiation than fruit flies, and therefore lower doses of radiation could produce a similar number of lethal mutations.

chromosome abnormalities in the form of rings, fragments, translocations, and two centromeres on the same chromosome. Larger doses kill most cells. The exact consequences of exposure to ionizing radiation depend on the individual cell and its metabolic state; the most sensitive cells are those undergoing cell division, especially those of the gonads, stomach, intestines, bone marrow, lymph nodes, and spleen.

As Muller showed more than 45 years ago, the production of mutations is directly proportional to the total dose of radiation, and the effects are completely independent of the source (Figure 23-7). We have more to fear from the innocent-looking x-ray machine that may be emitting a higher dose than is necessary and is scattering rays throughout the room than from radioactive fallout! (By mid-1967 less than half of the medical x-ray institutions in the United States met current state regulations or recommendations.) Since there is no threshold dose, no x-ray dose is so small as to produce no mutations. Furthermore, doses are additive; the total amount of radiation received is the important factor, not whether the radiation was given in one concentrated dose or in small doses spread out over a long period of time.

Reduction of total-body exposure to radiation lessens the danger of cancer, especially leukemia. (How x rays induce cancer remains uncertain, but it most likely involves somatic mutations.) The best advice: minimize exposure to radiation, especially that received by the ovaries and testes in persons who have not yet passed the childbearing or child-begetting age.

heterozygotes and sickle-cell heterozygotes are in this category.

23-4 NATURAL SELECTION

We have said that genetic variation is the raw material upon which natural selection works: the term **natural selection** is most frequently associated with the name of Charles Darwin and is the key to his theory of evolution. What exactly is natural selection? Were we really descended from the apes by the forces of natural selection? What is the meaning of a phrase like "survival of the fittest"? Are economic exploitation and genocide forms of natural selection? Let us begin with two examples: industrial melanism and sickle-cell anemia.

Industrial melanism

Traits exhibiting more than one phenotype or morph are called polymorphisms (literally "several forms"). The British populations of the peppered moth *Biston betularia* are **polymorphic** in that there are two color shades: dark black (or melanic) and light gray (Figure 23–8). Collections of these moths in Great Britain prior to 1800 showed a predominance of the light gray form, and the black forms were so rare that they were prized as collector's items. The Industrial Revolution came in the late 1800s, and a striking change occurred in both the British economy and the peppered moth: the cities became filled with factories, and the dark forms of the peppered moth became more numerous while the lighter forms became rare. In 1848 records from Manchester showed the dark form of the moth to be less than 1% of the population, but by 1898, only 50 years later, the dark form predominated and represented 95% of the population. What is the explanation for the rapid color shift in this moth population?

FIGURE 23-8 Dark and light forms of the peppered moth *Biston betularia* (a) on a dark, soot-blackened tree and (b) on a light, lichen-covered tree. (Courtesy of H. B. D. Kettlewell.)

(a)

(b)

During the last 20 years E. B. Ford and H. B. D. Kettlewell have carefully studied these changes in moth coloration, a phenomenon called **industrial melanism.** Their explanation is roughly as follows. Prior to 1800 Britain was agricultural, and the gray color of the moth blended in with the light coloration of the tree trunks on which the moths are often found, protecting the moths from predation by birds. As industrialization proceeded, cities and factories began to dominate the once-rural landscape, and the air was filled with soot and smoke. The replacement of the light-colored moths by the dark moths in regions of heavy industry is attributed to the fact that in these areas the dark forms are protectively colored and more easily escape predation by birds when they alight on the soot-covered trees. Observation of birds feeding on these moths in industrial areas shows that in a single day the birds may reduce the numbers of light-colored moths by 50%, whereas the less-conspicuous dark moths remain relatively free from predation. (Attention to environmental pollution in Britain has caused a decrease in smoke and soot in many industrial areas, and in recent years the light-colored moths have reappeared in ever-increasing numbers.)

The dark and light forms of the peppered moth are controlled by a single gene with two alleles; the dark allele is dominant to the light one. Kettlewell believes that the dark form arose by a mutation in the original population of light moths. At first the dominant mutant gene was disadvantageous—birds could easily see the dark moths on the light tree trunk, and so the dark moths were quickly fed upon. However, industrialization tended to produce an environment favorable for the mutant gene. The mutant gene spread rapidly in the population and soon became the more frequent form. This idea, that the genetic constitution of a population changes because some environmental agent favors some individuals more than others and thus promotes evolutionary change, was first clearly stated by Charles Darwin and Alfred Wallace in 1859. They called the process natural selection.

Natural selection is the mainspring of evolution, but in much of the literature, both scientific and popular, it has been misinterpreted. Some have termed it "survival of the fittest," "nature red in tooth and claw," "the law of the jungle," and a "struggle for life." Such notions are essentially without scientific foundation, for in nature selection rarely involves combat. Plants evolve; yet they never fight with one another. Animals of the same species fight infrequently, and when they do so, killing is the exception rather than the rule. As industrial melanism clearly shows, the dark moths did not take over industrial areas by killing the light forms; they merely survived and reproduced their numbers with greater frequency than did the light forms. As a consequence they became the most numerous in the population. Thus, in the modern view natural selection is defined as **differential reproduction;** that is, those organisms best adapted (or fitted) to their environment are the ones that find food and mates more easily; they tend to produce more offspring, which also survive to reproductive age; and these in turn leave more offspring than others in the population. In the case of industrial melanism, the selective agent is clearly the feeding of predatory birds on the peppered moths; however, in most cases it is impossible to identify the factors or agents that enable one genotype to leave more offspring than another.

Sickle-cell hemoglobin

Natural selection determines whether a gene is rare or common in a population. The process of selection does not create new genes; rather it takes the genes already present in a population and reduces the frequency of a gene that provides less selective advantage. The mechanism whereby this is accomplished is differential reproduction. As a result, and contrary to the belief of some, the most common genes in a population are not always dominants, and rare genes are not always recessives. However, sometimes genes with deleterious effects are maintained at a relatively high frequency in a population. Sickle-cell anemia is an example of such a situation.

Sickle-cell anemia is a severe debilitating disease; the child who is homozygous for this disorder (ss) barely has a chance of reaching maturity, and even if he does survive he is likely to be crippled. Since individuals with sickle-cell anemia usually do not survive to reproductive age, we might expect that the abnormal gene would soon be lost from the population. Indeed, the failure of each homozygote to reproduce would deplete the gene pool of two such genes. Our expectations are not realized. In Africa many populations are polymorphic for the sickle-cell gene. The frequency of sickle-cell trait (Ss) exceeds 20% in many areas, and in some local populations the frequency is as high as 40%. What accounts for the high incidence of the sickle-cell gene, which has obvious deleterious effects when it occurs in a double dose?

At first we might suspect that the sickle-cell gene is maintained at a high frequency because there is an exceptionally high mutation rate from $S \rightarrow s$. However, detailed studies have failed to reveal that the hemoglobin gene is more mutable than others. The geographical distribution of sickle-cell trait shows that it occurs in regions where malignant tertian malaria, produced by the protozoan *Plasmodium falciparum,* is prevalent (Figure 23–9). In 1949 A. C. Allison and his collaborators found that malaria was much less severe in children having sickle-cell trait (Ss) than in children with normal hemoglobin. They postulated that the sickle-cell gene in the heterozygous condition provides some degree of protection against malaria. Thus, in those areas where falciparum malaria is prevalent, persons having sickle-cell trait are more likely to survive to maturity, and these individuals will pass on copies of their genes to the succeeding generation. In malarious areas the heterozygote for sickle-cell hemoglobin is superior in fitness to either of the homozygotes, a condition called **heterosis** or **hybrid vigor.** The S and s genes are both maintained in the population because the loss of deleterious genes by death of the homozygotes (SS and ss), either from malaria or anemia, is balanced by the gain that results from the enhanced survival and reproduction of the heterozygotes (Ss). This phenomenon is known as **balanced polymorphism.**

What happens to the frequency of the sickle-cell gene in areas where malaria has been eradicated? Since the selective advantage of the heterozygote in malaria-free areas would be equal to or less than that of the normal homozygote, we should expect the frequency of the sickle-cell gene to decline. This is exactly what occurs. The frequency of the sickle-cell trait among black Americans living in the United States, where malaria has been eradicated for more than 50 years, is about 14%.

Again it can be seen that natural selection does not simply eliminate the bad genes and keep the good genes. Rather, selection operates within a population by probing the environment with new gene combinations, and in this way the surviving population increases in fitness; in this view, selection is a positive and creative force enabling a population of organisms to survive in a constantly changing and often hostile environment.

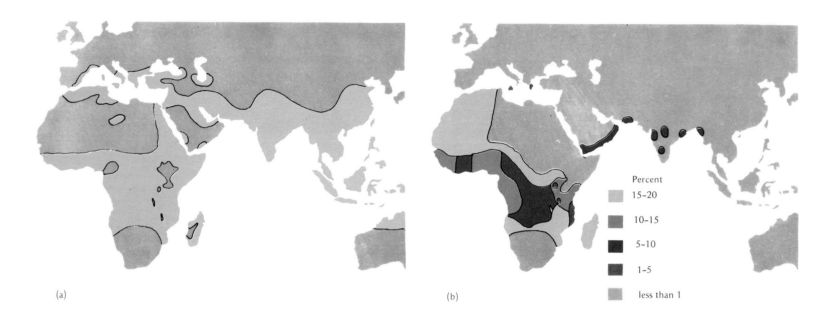

(a)

(b)

Percent

15–20

10–15

5–10

1–5

less than 1

23–5 EUGENICS: IMPROVING OUR GENES BY SELECTION

Negative eugenics

It has been a common practice throughout human history to improve plants and animals by hybridization (controlled mating) and selection. Breeders of livestock have selected for high egg-laying capacity, high meat yield, increased production of milk, high fat content in milk, and so on. The development of high-yield, high-quality strains of wheat, rice, and corn have created "green revolutions" and enhanced food production.

Initially, many such breeding programs were conducted without any genetic knowledge. The general rule applied by the breeder was to cull out the inferior types and retain and inbreed those types carrying the desirable characteristics. The main feature of such a program

is to reduce the variability in the population and to remove the undesirable genes.[2]

In the early 1900s when Mendel's principles were rediscovered, the genetic basis for success in hybridization and selection became apparent. Mendelism, coupled with social-class bias in England and Germany and race-consciousness in the United States, provided the necessary scientific and sociological rationale for fostering programs to "improve" humans by applying the practices of controlled mating and selection. Sir Francis Galton (1822–1911), a cousin of Charles Darwin, publicized the notion of **eugenics** (literally "well born") from 1870 to the time of his death; in 1904 he defined

---

[2] Inbreeding in man—that is, consanguineous marriages—creates social and personal burdens. In such cases, the risk is increased that deleterious genes will be expressed in the offspring, as occurs in infantile amaurotic idiocy.

eugenics as a "science dealing with all the influences that improve the inborn qualities of a race." In 1908 Galton and his followers established the Eugenics Society in England, and a few years later a similar group was founded in Germany. Later, eugenics came to be defined as "the study of agencies under social control that may improve or impair the racial (= hereditary) qualities of future generations either physically or mentally." It was almost inevitable that such a program would attract not only altruists, but many persons with strong racial- and social-group prejudices. In the United States the beliefs of the eugenicists were reflected by restrictive immigration laws and compulsory sterilization programs for certain undesirable phenotypes (based primarily on the overpublicized story of the Jukes and Kallikaks, two families with a large number of mental defectives). Between 1907 and 1931, 30 states passed laws requiring or permitting sterilization for more than 24 different classes of defectives; in other cases individuals were institutionalized to prevent them from reproducing. In this way, it was hoped, the numbers of insane, mentally retarded, and epileptic persons would be reduced in society, and future generations would be free of such burdens. In Nazi Germany, Hitler used eugenics to foster a program of race hygiene and to justify genocide.

Superficially eugenics sounds both scientifically sound and idealistic. If the program were kept free of cranks, would it work? Once methods for predicting the effectiveness of selection became available and the relationship between the frequency of a gene and the efficacy of selection was understood, the hopes of the eugenics movements were diminished greatly. Eugenics wouldn't work. Let us see why.

If **negative selection** is practiced against a dominant gene (that is, if persons of a certain genotype are prevented from reproducing), in one generation the frequency of the gene will be reduced to the level maintained by mutation. The difficulty about applying negative selection to human disorders is that many disorders are not the result of dominant genes. However, the condition known as **Huntington's chorea**[3] is inherited as a simple dominant, and its incidence could be reduced in the population by a program of education and genetic counseling. But there is a disturbing complication: many persons who carry the dominant gene do not know they carry it because the disease symptoms do not develop until middle age, *after* the victim has produced a family and thus passed

---

[3] Huntington's chorea involves a progressive degeneration of the nervous system: it starts with involuntary twitching of the head, limbs, and body; later, there is a loss of mental and physical powers; and finally death occurs.

TABLE 23-2 Effects of complete selection against a recessive trait

Generations	Gene frequency	Recessive homozygotes (%)	Heterozygotes (%)	Dominant homozygotes (%)
1	0.500	25.00	50.00	25.00
2	0.333	1.11	44.44	44.44
3	0.250	6.25	37.50	56.25
4	0.200	4.00	32.00	64.00
5	0.167	2.78	27.78	69.44
9	0.100	1.00	18.00	81.00
10	0.091	0.83	16.53	82.64
20	0.048	0.23	9.07	90.70
30	0.032	0.10	6.24	93.65
40	0.024	0.06	4.76	95.18
50	0.020	0.04	3.84	96.12
100	0.010	0.01	1.96	98.03

From E. W. Sinnott, L. C. Dunn, and T. Dobzhansky, *Principles of Genetics,* 5th edition, New York, McGraw-Hill, 1968.

on the gene for Huntington's chorea. Furthermore, persons with Huntington's chorea are heterozygous for the defective gene, and therefore half the children of a choreic couple would be expected to be normal; prevention of reproduction would disallow both normal and choreic children. (If the condition could be diagnosed in a fetus by means of amniocentesis, parents with a family history of chorea could elect to abort only those fetuses with the disorder; however, it is not possible at present to diagnose a choreic fetus by this technique.)

The difficulty of negative selection against a dominant gene like the one responsible for chorea dims the hopes of a eugenicist; what about negative selection for a gene that is recessive? Again, the problem is recognition. Table 23–2 and Figure 23–10 show that total negative selection against persons homozygous for the recessive gene reduces the frequency of the gene with each succeeding generation. However, recessive genes remain masked in heterozygotes, and the deleterious recessive gene becomes even more difficult to trace. Selection against the homozygous recessive becomes less and less effective as the gene is

reduced in frequency; theoretically, the gene is never eliminated, but by the hundredth generation (in humans that would be 3,000 years) a gene that originally had a frequency of 50% would be reduced to 0.01%. We are drawn, almost inescapably, to the conclusion that negative selection against a recessive gene will not eliminate its presence.

Let us look at a particular case. Persons with sickle-cell anemia (*ss*) are naturally selected against; that is, the condition is so debilitating and death occurs at such an early age that affected individuals are subjected to almost total negative selection. In such a situation, we might ask, how long would it take for the frequency of sickle-cell anemia to be significantly reduced in the United States[4] from its current level of 6 per 1,000 to a level of 6 per 1,000,000? It has been calculated that to reduce the frequency of sickle-cell anemia 1,000-fold would require 395 generations or 10,850 years!

---

[4] In countries such as the United States, where malaria has been eradicated, the gene in its heterozygous condition confers no selective advantage. See page 493.

FIGURE 23-10 The elimination of individuals carrying a recessive gene (homozygous for a particular trait) per generation proceeds at a slower rate as selection pressure decreases. Note that even with complete selection, the percentage never reaches zero.

Thus, active research on sickle-cell hemoglobin and genetic counseling could make sickle-cell anemia a rare disease. If *all* persons with sickle-cell trait (*Ss*) were given genetic counseling and then *all* elected to avoid having children (unlikely possibilities), the frequency of sickle-cell anemia would be reduced to zero in one generation.[5] More importantly, this is not simply theoretical, since persons with sickle-cell trait (*Ss*) can easily be identified by a blood test. Here parental choice alone can make a critical impact on the gene pool.

Each of us carries, in a concealed condition, about four abnormal genes which, if they were to become homozygous, would be deleterious. These deleterious genes constitute our genetic load or genetic burden. The application of modern medical practice, that is, the prolongation of the life of individuals with apparent genetic defects so that they survive to reproductive age, maintains the pool of defective genes. Can we reduce our genetic load by selecting against heterozygotes? First, can we identify such heterozygotes? Second, if we can identify heterozygotes, can we, in a free and democratic society, prevent the marriage of individuals who potentially could produce all normal children (a heterozygote marrying a normal homozygote) or 75% normal children (if both spouses are heterozygotes)? It is clear, for reasons involving recognition of the heterozygote and implementation of eugenics programs, that total selection against heterozygotes cannot be practiced. Finally, we should note that when genes interact in such a way that the heterozygote is of superior phenotype to one or both homozygotes (for example, the person

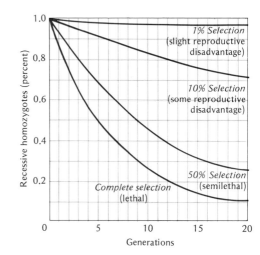

with sickle-cell trait in malarious areas), then complete elimination is theoretically impossible. As Dobzhansky has said: "A genetic good mixer becomes superior to a genetic rugged individualist."

Positive eugenics

Suppose instead of negative eugenics we decided to employ positive eugenics; that is, to encourage those individuals with desirable phenotypes to produce more offspring. Sir Julian Huxley and Nobel laureate H. J. Muller have advocated the establishment of sperm banks in which semen of men judged to be superior would be deposited. The sperm would be used for artificial insemination, and in this manner the frequency of desirable genes in the gene pool would be increased. It works in cattle; why not man?

One difficulty encountered in programs of positive eugenics is the definition of "desirable." Can we really agree on those human traits we would want to see in future generations? Suppose we agree that intelligence,

adaptability, mental stability, courage, perseverance, energy, and compassion are all desirable traits. How can we test for the genetic basis of these traits? As yet, we do not know of a single human gene that produces any of these desirable phenotypes. In addition, we haven't the vaguest notion how such qualities will fare in a human society a few hundred years from now. A society that cannot decide what is desirable today can hardly forecast the needs of society in the future.

Our best hope for the future lies not in positive or negative eugenics, but in genetic counseling, restriction of family size, adoption of children in consanguineous marriages (first cousins), and maximization of individual choice.

Postscript

One of the reasons that man continues and is likely to continue to be the dominant species on the earth lies in the enormous variation in his gene pool, in the large numbers of polymorphisms it maintains and in his control of environmental factors. ... The dangers from restricting human phenotypic variation by eugenic means are trivial, but those from euthenic (environmental) methods and particularly from educational engineering are real. There is little doubt that educational engineering could be used to make similar genotypes into widely different phenotypes by custom tailoring the educational process. But the concern that is felt is for the use of these methods to nullify differences between genotypes by . . . tailoring in the opposite direction to produce uniformity.

I. M. LERNER, *Heredity, Evolution and Society*, San Francisco, Freeman, 1968.

SUMMARY

1. Sickle-cell disease is a genetic abnormality produced by an alteration in the DNA itself. When an individual is homozygous for the recessive gene, sickle-cell anemia results. Such

---

[5] For the purposes of calculation the mutation rate for $S \rightarrow s$ is not considered. Furthermore, mutation would act to delay the reduction of *ss* frequency. The data given are therefore conservative.

individuals show a sickling of their red blood cells with multiple physical side effects. Heterozygotes are sickle-cell carriers; such individuals show a sickling of their red blood cells when placed under oxygen stress and are said to have sickle-cell trait.

2. The molecular basis of sickle-cell disease is a mutant gene that causes a substitution of valine for glutamic acid in the hemoglobin molecule. When a single gene affects a great many phenotypic characters, as in the sickle-cell mutation, it is said to have a pleiotropic effect.

3. Mutations are chance phenomena, and most are deleterious or lethal. Selection of occasional beneficial mutations is of overriding importance in evolution.

4. Subvital genes may be maintained in a population by heterozygotes. Harmful mutations in the gene pool of a population form its genetic burden or load.

5. Ionizing radiation is mutagenic; there is no threshold and its effects are cumulative.

6. Sexual reproduction (as well as mutation) contributes to genetic variation, because of independent assortment and crossing over at meiosis. Genetic variation forms the basis of evolution.

7. The Hardy-Weinberg law states that in a stable population where there is no migration or mutation, where mating is random, and where all genotypes have the same survival value, gene frequency will remain unchanged $(p^2 + 2pq + q^2)$. The law allows prediction of the frequency of heterozygotes with a normal phenotype. Further, measurable deviations in gene frequency indicate agents of evolutionary change are at work.

8. Natural selection can be seen to operate in the situation known as industrial melanism. Changes in the coloration of moths in Britain was due to the selective advantage of certain phenotypes; those phenotypes that were favored became the more frequent form—through differential reproduction.

9. Evolution involves a change in gene frequency in a population. The genes involved may be dominant *or* recessive. Recessive genes such as that for sickle-cell anemia may be maintained in a population in higher than expected frequency because of balanced polymorphism. Here, the normally negative effects of the recessive gene are balanced by positive effects of the heterozygote under certain environmental conditions. (Sickle-cell trait shows balanced polymorphism in areas where malaria is endemic; here the heterozygotes [*Ss*] are protected against malaria by the mild sickling of their red cells; they are therefore at an advantage over both homozygote types [*SS* or *ss*], and are said to exhibit hybrid vigor.)

10. The genetic basis for success in hybridization and breeding livestock and crops became apparent when Mendel's principles were rediscovered. Eugenicists then sought to apply the same principles to human populations.

11. Negative selection against deleterious human genes is extremely difficult. If the gene in question is recessive, eradication from the gene pool is all but impossible, even if heterozygotes can be easily identified and prevented from breeding. The heterozygote may show heterosis (hybrid vigor), tending to maintain rather than decrease the frequency of the deleterious gene.

12. Drawbacks of positive eugenics programs include the difficulty of defining desirable traits and their genetic basis as well as that of predicting their survival value in a changing world. Man's best hope lies in his enormous and variable gene pool.

KEY WORDS

sickle-cell trait
gene mutation
pleiotropy
genetic burden or genetic load
gene pool
subvital gene
ionizing radiation
ion
background radiation
mutagenic
gene frequency
Hardy-Weinberg law
natural selection
polymorphic
industrial melanism
differential reproduction
heterosis or hybrid vigor
balanced polymorphism
eugenics
negative selection
Huntington's chorea
positive eugenics

TOPICS FOR REVIEW AND DISCUSSION

1. **What is pleiotropy? Give an example.**
2. **How do x-rays act as mutagens?**
3. **What assumptions have been made in the derivation of the Hardy-Weinberg law, and why does it represent a static evolutionary situation?**
4. **What is natural selection?**
5. **Discuss industrial melanism and its significance to evolutionary thinking.**
6. **Discuss why and how the sickle-cell gene is maintained in certain populations.**
7. **Discuss the causes of genetic diversity in populations and the evolutionary significance of such diversity.**
8. **Why is evolution essentially an irreversible process?**
9. **Discuss why natural selection and mutation are necessary prerequisites for evolution.**
10. **Discuss genetic load and its causes.**

Hollywood's view of Australopithecus (the man-ape) as seen in the movie *2001: A Space Odyssey.* (Courtesy of Metro-Goldwyn-Mayer, Inc. c 1968.)

THE EVOLUTION OF MANKIND

24-1 MAN'S PLACE IN NATURE

In 1838 Charles Darwin, 26 years of age, returned from a worldwide scientific expedition aboard H.M.S. *Beagle* (Figure 24–1). This voyage marks the starting point of his theory of evolution, a theory that began a revolution in biological thinking. On his trip Darwin observed varieties of animals he had never encountered before, and he wondered how such diverse creatures came into being. Nine years later, thinking on this question, he wrote to a friend: "At last gleams of light have come, I am almost convinced (quite contrary to the opinion I started with) that species are not (it is like confessing a murder) immutable." Among the places visited by Darwin on his trip were the islands of the Galapagos. Using patient observation and keen insight (he knew nothing of Mendel or the principles of genetics), Darwin recognized that the organisms isolated for thousands of generations on the various islands of the Galapagos group had gradually evolved in distinctive ways, producing separate species. What is a species?

Life has a common structural and functional basis. There is a unity in biology, and all living creatures are endowed with certain similar properties. However, there is another aspect to life: diversity. Each living organism is distinct and separate from every other one. So impressive is the diversity of life that down through the ages man has been preoccupied with attempts to make some order out of it. To do this he has created systems of classification. Aristotle (384–322 B.C.), the ancient Greek naturalist, believed that all living things could be arranged into two great classes or kingdoms: plants and animals. Within each of these two kingdoms further refinements were possible, so that animals and plants could be arranged in a hierarchy. This hierarchy, or ladder, of nature had the simplest creatures on the lowest rungs

FIGURE 24-1 (a) The voyage of H.M.S. *Beagle.* (b) Portrait of Charles Darwin age 31, 5 years after he returned from his voyage. Painted by George Richmond, the original hangs in Down House, Downe, Kent. (Photo Derrick Witty. Copyright Gorge Rainbird Ltd. Reprinted by courtesy of The Royal College of Surgeons of England.)

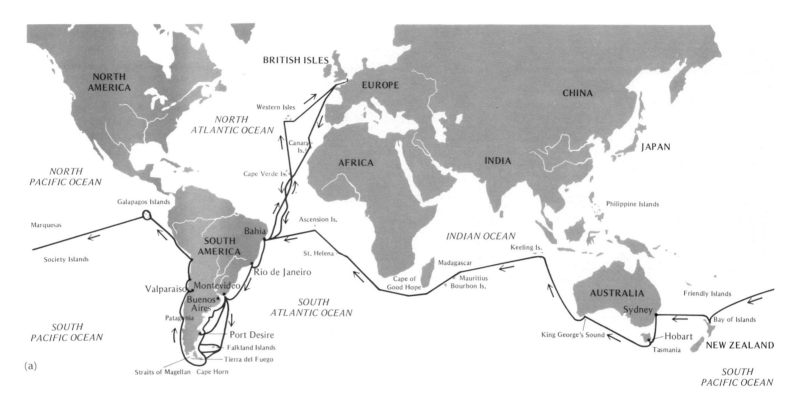

(a)

and the more complex ones at the top. Those animals having a backbone, or spine, were placed on the upper levels; man, of course, placed himself at the top, and the lower levels were occupied by spineless creatures. Similar arrangements were made for plants: lower plants were represented by algae and ferns, and the higher plants included those having flowers, seeds, and fruits. The presumption in all this was that the kinds of animals and plants were immutable, or fixed, and each was the product of divine creation. Carl von Linné (1707–1778) (Figure 24-2), believing in original creation and lack of change in organisms, developed an improved system of classification for plants and animals. In 1737 he published *Systema Naturae*, in which he established the

convention of using two Latin[1] names for each organism, the **genus** and the **species** (implying simply "kind"). Thus, according to Linnaeus's system all humans now living are classified as *Homo* (meaning "man") *sapiens* (meaning "wise"). Linnaeus applied the term species to plants and animals that were significantly similar to each other, descended from significantly similar forms that had themselves arisen by special creation and that did not interbreed with other species. Although Linnaeus's view that species remain unchanged through time,

---

[1] In the eighteenth century Latin was the language of scholars. The scholars even Latinized their own names; for example, Carl Von Linné became Carolus Linnaeus.

(b)

FIGURE 24-2 Carolus Linnaeus (1707–1778), who devised a classification scheme for plants and animals. (Courtesy of The Bettmann Archive, Inc.)

FIGURE 24-3 The hierarchy of classification.

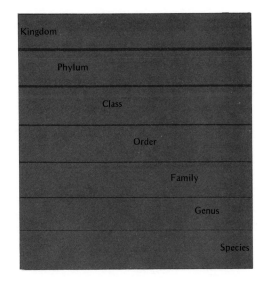

In classification, a biologist uses past experience and a number of criteria to narrow the possibilities so that a particular species is placed closest to its relatives. Similarity of **structure, function,** and **development** all enter into the classification of organisms. Structure refers to body form or morphology; function, to the manner in which the organ systems operate; and development, to sequences in the life history of the individual organism. Table 24–1 illustrates the classification for the relatives of the domestic cat and of man. According to this table, the species within a genus are quite similar, but genera differ more markedly in their characteristics. Our cats have more in common with lions and dogs than they do with us; we, on the other hand, are more closely related to apes and monkeys than to our dogs and cats. Indeed, we are the sole living members of the genus *Homo;* there is no species except *H.sapiens.*

24-2 WHAT IS A MAN?

The resemblance in structure between one organism and another indicates the degree to which they are related, and this is the basis of classification. Who are man's closest living relatives? When man's body structure is compared with that of other organisms, we see that in common with fish, amphibians, reptiles, birds, and a variety of furry creatures he possesses a vertebral column, and so he is classified as a vertebrate (subphylum: Vertebrata). Warm-blooded and air-breathing, with a hairy skin and young that are born alive and nursed at the breast (*mamma*—breast; L.), he is entitled to membership in the class of mammals (class: Mammalia). Birds, reptiles (lizards and snakes), amphibians (salamanders, toads, and frogs), and bony fishes neither nurse their young nor have hair, and they belong in different classes: Aves, Reptilia, Amphibia, and Osteichthyes,

a concept called the **fixity of species,** was entirely incorrect (as Darwin clearly recognized), his classification scheme proved so useful in categorizing the various plants and animals that it is used to this day. Thus, the form of the classification system of Linnaeus remains, but the reasoning behind it has changed. Today the classification of organisms represents a summary of physical resemblances as well as evolutionary relatedness. This involves two important ideas· (1) two or more organisms having a large number of significant similarities in function and form must be closely related, and (2) close relationship means common ancestry.

According to the modern view, biological species are defined as "groups of interbreeding natural populations that are reproductively isolated from other such groups."[2] By this defi-

nition, members of the same species are potentially capable of mating and producing fertile offspring, and these in turn can produce fertile progeny. This definition of a species says nothing whatever about how the population looks, and members of the same species—males, females, and young—may differ markedly or slightly in their form **(morphology).**

The Linnaean system of classification as used today recognizes both the kingdom and the species; between these levels are five other categories. Thus, species are grouped into **genera** (sing., genus), genera are grouped into **families,** families into **orders,** and orders into **phyla** (sing. phylum). The various phyla make up a **kingdom** (Figure 24–3). The levels between genus and kingdom are strictly artificial groupings of organisms having similar qualities; there is no biological definition for these categories.

A biologist who attempts to classify a new plant or animal must observe the structure of the specimen and then determine the proper pigeonhole into which that creature belongs.

[2] E. Mayr, *Populations, Species and Evolution* Cambridge, Mass., Harvard University Press, 1970, p. 12.

FIGURE 24-4 A gallery of primates.
(a) Chimpanzee, an anthropoid ape. (Courtesy of the
American Museum of Natural History.)

TABLE 24-1 Classification of the domestic cat and man

Species	Genus	Family	Order	Class	Phylum
(F. domestica) Cat	(Felis) Cat Lynx	(Felidae) Cat Lynx Lion	(Carnivora) Cat Lynx Lion Dog	(Mammalia) Cat Lynx Lion Dog Monkey Horse Man	(Chordata) Cat Lynx Lion Dog Monkey Horse Man Bird Lizard Frog Fish
(H. sapiens) Man	(Homo) Man	(Hominidae) Man	(Primates) Man Monkey Ape Lemur Tarsier	(Mammalia) Man Monkey Lynx Cat Dog Horse	(Chordata) Man Monkey Lynx Cat Dog Horse Bird Lizard Frog Fish

respectively. Among the mammals, man has features that place him in the order Primates, along with the tree shrews, lemurs, tarsiers, monkeys, and apes. What are the characteristics of the primates?

A glance at Table 24–2 and the photographs in Figure 24–4 will show you that man shares many physical features with the modern primates. This alone is enough to suggest that man and the primates are closely related and probably descended from a common ancestor. There are, however, additional lines of evidence that, taken together with shared physical characteristics, are very convincing. For example, the chemical composition of blood of humans, apes, and monkeys is quite similar. A precipitating antiserum prepared against hu-

TABLE 24-2 Characteristics shared by man and anthropoid apes

1. Face replaces muzzle.
2. Sense of smell is reduced; emphasis is on vision—stereoscopic and color.
3. Tactile hairs on nose are lost (whiskers are reduced or absent).
4. Menstrual cycle occurs.
5. Pectoral mammary glands are present.
6. Single uterus is present; only one or two offspring are born at a time.
7. Breeding groups are formed, with elaborate or prolonged maternal care of the young.
8. Hands have five fingers with an opposable thumb.
9. Clavicle (collarbone) braces the shoulder against the rib cage, giving the arm strength for use in swinging (brachiating) movements.

(b) Orangutan, an anthropoid ape. (Courtesy of the American Museum of Natural History.)

(c) Colobus, an old-world monkey. (Courtesy of the American Museum of Natural History.)

(d) Gorilla, an anthropoid ape. (San Diego Zoo Photo.)

(e) Gray gibbon, an anthropoid ape. (San Diego Zoo Photo.)

(f) Macaque, an old-world monkey. (San Diego Zoo Photo.)

(g) Langur, an old-world monkey. (San Diego Zoo Photo.)

(h) Lemur, a prosimian. (San Diego Zoo Photo.)

(i) Tarsier, a prosimian. (San Diego Zoo Photo.)

man blood gives a strong precipitation reaction when mixed with the blood of a gorilla, a somewhat weaker reaction with orangutan blood, and a very weak reaction with baboon blood. When the precipitating antibody to human blood is mixed with ox blood, there is virtually no precipitation. Man and other primates, particularly the apes, share similarities in blood types such as ABO, MN, and Rh. We and our primate relatives are subject to the same diseases; syphilis attacks man and the chimpanzee with great intensity, but the disease is less severe in the orangutan and quite mild in the baboon. The chromosomes of man and the apes are also quite similar; chimpanzee and gorilla chromosomes are much more like our own than those of the orangutan and the gibbon. All evidence supports the contention of T. H. Huxley that "whatever system of organs be studied, the structural differences which separate man from the gorilla and chimpanzee are not as great as those which separate the gorilla from the lower apes."

The hierarchy of categories for the primate order is shown in Figure 24–5, and the position of man is clearly indicated. Each step in the classification scheme tells us how closely man shares characteristics with the other primates. The lower the category in the hierarchy, the more specific are the features of those members of the group, and the more similar they are.

Thus far we have considered man as a primate. But how does man differ from the other anthropoid primates? What are the unique features of man that enable us to say: "This specimen before us is a human being and not a gorilla or a chimpanzee"?

Man is bipedal and walks erect upon the soles of his feet; other primates are four-footed (quadrupedal). Man's legs are long, greatly strengthened, and well-provided with powerful muscles used in walking upright. The adoption

FIGURE 24-5 Classification of the primates.

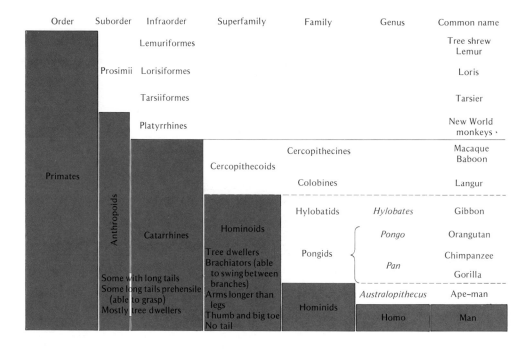

of an erect stance frees the hands and arms from a role in locomotion. The hand of man is fashioned more effectively into a device for making and using tools. The human cranium (skull) is dome-shaped and increased in volume over that of man's closest primate relatives in order to accommodate his greatly expanded cerebrum. The human brow is steep and high in contrast to the retreating forehead and the ridged brow of the apes. The skull is poised directly above the spinal column, and the pelvis is expanded to form a basket that provides support for the internal organs. Other distinguishing human features include the distinct chin, the female breasts (which are permanently enlarged mammary glands), and, when compared with the primates, the enormous penis.

Man's closest living relatives, the apes, are born naked, but in time they grow a coat of hair. Man is a naked ape; he is born so and remains so.

Finally, man's ability for thought and language, attributes that shape much of his day-to-day and historic behavior (that is, culture), are developed to a level of complexity that is unique to the human species. No other species within the entire animal kingdom can begin to approach the technology of human civilizations, with all its vast implications. Indeed, it is probably true that most of the significant evolutionary trends for our species will be cultural. The differences between man and the primates are presented in Table 24–3 and Figure 24–6.

FIGURE 24-6 A comparison of the body structure of three primates. (a) Macaque monkey. (b) Gorilla. (c) Man.

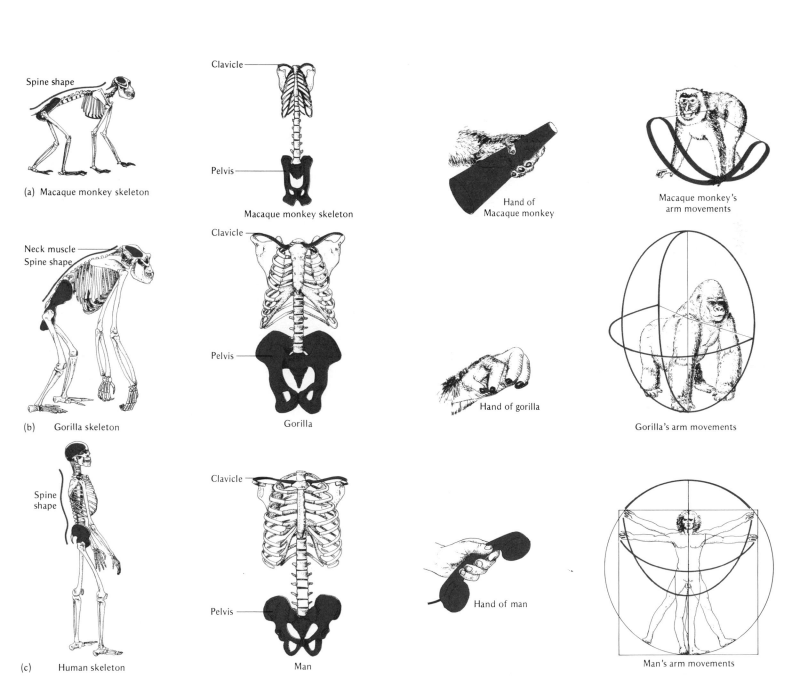

Spine shape

(a) Macaque monkey skeleton

Clavicle

Pelvis

Macaque monkey skeleton

Hand of Macaque monkey

Macaque monkey's arm movements

Neck muscle
Spine shape

(b) Gorilla skeleton

Clavicle

Pelvis

Gorilla

Hand of gorilla

Gorilla's arm movements

Spine shape

(c) Human skeleton

Clavicle

Pelvis

Man

Hand of man

Man's arm movements

TABLE 24-3 Distinguishing features of anthropoids and man

Characteristic	Anthropoids	Man
Locomotion	Brachiating	Bipedal
Shoulder	Very mobile	Very mobile
Stretching arms	To side	To side
Thumb	Small, but used	Large, much used
Growth rate	Slow	Slower
Sexual maturity	8 years	14 years
Full growth in males	12 years	20 years
Infant dependency	2 years	6–8 years
Female receptivity to sexual intercourse	At ovulation	Continuous
Long bones of lower extremities	Shorter than those of upper extremity	Longer than those of upper extremity
Tarsal bones	Rather short, toes rather long	Rather long, toes rather short
Trunk	Long compared with lower extremities	Short compared with lower extremities
Vertebral column	Straight or curved uniformly backward	Alternately curved backward and forward
Leg	Curved, knees turned outward	In upright posture, straight in knee and hip joint
Joint between skull and vertebral column	At back of skull	Almost in center of base of skull
Canines	Large fangs	No larger than premolars
Crown of first lower premolar	Has bladelike cutting edge	Unspecialized (bicuspid)
Dental arch	Laterally compressed with side rows of teeth almost parallel	Rounded without sharp angles
Jaws	Long and large	Short
Face	Long, in front of brain, protruding	Short, steep, under brain
Brain	Average size one third that of human	Large

24-3 DARWINISM: FROM FINCH TO MAN

Man is the dominant species of vertebrate on this planet. The question of how man came to be has preoccupied human thought through all of recorded history and, no doubt, even before that period. Each culture has produced its own story of creation, often steeped in mysticism, magic, and symbolism. The Bible describes the creation of the world in 6 days; on the final day God, we are told, made man in his own image and gave him dominion over the fish of the sea, the fowl of the air, over the cattle and over the earth, and over every creeping thing that crawls upon the earth. The notion of the special creation of man in a single day was so dominant in man's thinking that in 1650 Archbishop James Usher of Armagh, Ireland, calculated the date of creation to be 4004 B.C., on October 23, at 9 o'clock in the morning! However, by the late 1700s there appeared reason to doubt the theory that man was created in a single dramatic event. At that time ancient human bones and odd-shaped stone tools were found in France, Germany, England, and Belgium—many associated with the bones of extinct animals. Speculation arose that there was a connection between extinct animals, ancient tools, and extinct humans. Critics were quick to point out that such associations could have occurred by chance and that these findings did not constitute proof that all these things existed at the same time.

A turning point in understanding man's origins came not from studies of human bones themselves, but from establishment of the age of the earth. James Hutton (1726–1797), a Scottish physician, and Sir Charles Lyell (1797–1875), a geologist, broke with tradition and suggested that the age of the earth could be determined by studying the successive layers of the rocks (strata). The present, they contended, is the key to understanding the past, and only the geological processes that can be observed in action today could have been involved in forming the rock strata. To Hutton and Lyell, the changes in the earth's crust were not the product of repeated catastrophes or some special event, but rather of an orderly process that goes on in nature all the time. Through the passage of long periods of time, the surface of the earth was changing. Strata were produced slowly, taking many thousand years to form. Hutton and Lyell also contended that the extinction of many species and the formation of new ones must have occurred slowly and continually throughout past time. Support for this contention came from discoveries in the strata of the preserved remains (fossils) of species that do not exist today.

While geologists made progress in estimating the age of the earth, naturalists began to recognize similarities between men and apes. In 1809, Jean Baptiste de Lamarck (1744–1829) stated that men were descended from apes; naturally the view met with opposition. The idea of man's relation to primates was not really accepted until after publication of the writings of Charles Darwin (1809–1882) (Figure 24–7). The idea is refuted in some quarters even today.

In 1859 Darwin's work, *On the Origin of Species by Means of Natural Selection, or the Preservation of Favored Races in the Struggle for Life,* appeared. Although he was not the first to propose a theory of evolution, his work was presented in a sober and subdued manner; the intellectual climate was right, and his scholarly ideas were amply documented with supporting evidence; the idea took root. It could be said that Darwin made evolution respectable. Not only did he provide evidence for the process of evolution and a reasonable explanation for the diversity of organisms, but he proposed a mechanism whereby evolution could work. The mechanism proposed by Darwin was **natural selection.[3]**

What was Darwin's evidence for evolution? Charles Darwin was 22 when he set sail on H.M.S. *Beagle* from Plymouth, England, on

---

[3] The idea of natural selection was independently developed by Alfred Russel Wallace (1823–1913); however, Wallace's work was never as well documented as that of Darwin, and thus his contribution is often minimized. In truth, he should share in some of the credit today, as he did at the time.

December 17, 1831. In 1835, after the ship reached the Galapagos Islands, he became convinced that evolution must have occurred and that species arose from preexisting species. The Galapagos is a group of 15 or more rocky islands off the coast of Ecuador; it is believed that the islands rose from the ocean floor by volcanic action a million or more years ago (Figure 24–8). The islands were never part of the mainland of South America, nor were they ever connected to one another. In surveying the animal life on the islands, Darwin found very few vertebrates; however, a group of small birds—finches—were quite abundant, and they showed extreme diversity in the form of their beaks as well as in their feeding habits. There were 14 different species of finches; these have been found nowhere else in the world. How, Darwin asked, did these species of finches arise?

Darwin surmised that the ancestors of the present finch populations were blown to the islands by high winds from the mainland. In time some of these original colonizers spread from one island to the others. Since the finches do not fly readily over large stretches of open water, these populations tended to remain pretty much on their own rocky islands. The original immigrants were split up into separate island populations and came to be geographically isolated. Hence, they could no longer interbreed. In time, chance variations arose among the finches on the separate islands (Figure 24–9a). There was, however, a problem; most of the islands did not contain just a single species of finch, and on some there were up to 10 different species. How did Darwin account for this? He visualized the process as follows. The original population of finches found no predators or other bird competitors on the islands, and so they multiplied their kind. However, in time, food for the finches (primarily ground seeds) came to be a limiting

FIGURE 24-8 The Galapagos Islands, where Darwin began to formulate his theory of the evolution of animal species. (a) Migration path of the ancestral finch from Ecuador. (b) The exploration pathway of the *Beagle* during its time at the Galapagos Islands.

factor. Only those individuals capable of adapting to new environments could survive and reproduce themselves. Darwin believed that the original finch population contained individuals with many differences. If any of these variations permitted individuals to exploit environments not yet occupied by other finches, these individuals would have a greater chance of surviving and leaving offspring (survival of the fittest). Thus, the successful colonization of previously unoccupied habitats would lead to a diversification of species, and each of these species would be better fitted to survive and reproduce under the new conditions than in the ancestral habitat. This spread of a population into a new environment accompanied by divergent adaptive changes from the original one is called **adaptive radiation.**

Adaptive radiation is clearly seen among Darwin's finches. Six of the ground-finch species have heavy, parrotlike beaks for crushing seeds, others have flower-probing beaks for feeding on the flowers of the prickly-pear cactus. One of the tree finches is woodpeckerlike; it has a stout beak for boring into wood, but instead of using its tongue it searches for insect larvae by using a cactus spine. Another finch is warblerlike, with a beak adapted for picking off small insects from the bushes (Figure 24-9b).

According to Darwin, the original population of finches came to be divided into species that were localized in their habitats; offspring of the better-adapted forms survived in greater numbers than offspring of less well-adapted finches. Aware of the way in which man could control plant and animal traits by artificial selection, Darwin used the term natural selection to indicate that some external agent, rather than conscious choice of the organism, was responsible for favoring certain individual differences and eliminating those that were injurious.

Darwin's theory of evolution and the role of

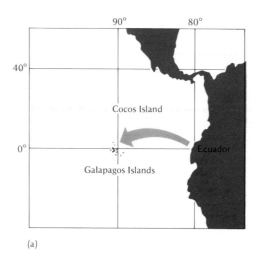

(a)

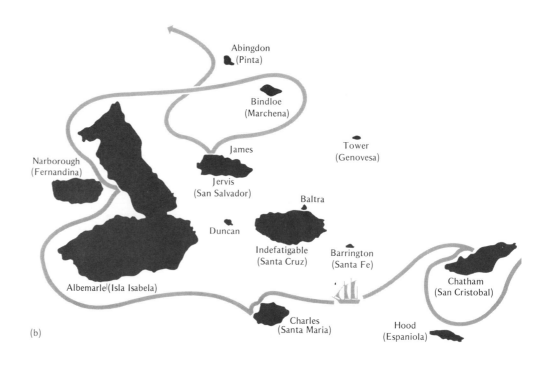

(b)

FIGURE 24-9 Darwin's finches. (a) How the finches on the Galapagos Islands may have speciated. (b) Adaptive radiation in Darwin's finches.

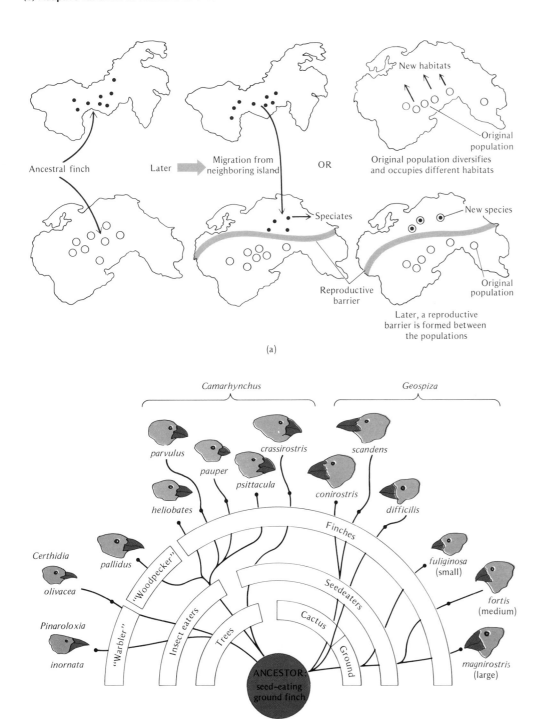

(a)

(b)

natural selection was based on the premise that there exist natural checks on increases in population size; some of the limiting factors are predators, food availability, competitors, climate, and disease. As a consequence, he reasoned, more individuals of each species will be born than can possibly survive, and thus there ensues a struggle for existence; those individuals within a population having a particular endowment that gives them better survival chances will be naturally selected. These selected varieties will in time propagate themselves. Thus, **Darwinian fitness** is reproductive fitness, and not, as some have implied, a survival of those capable of destroying their own kind. Natural selection is a blind and opportunistic force that shapes the changes in a population producing new species; it acts as a sieve to cull out the unfit and permits only the fittest to pass through and reproduce their kind; it enables organisms to fit their surroundings, but it cannot anticipate the changes necessary for survival in future environmental situations.

Darwin recognized the importance of variation in a population and its relationship to the evolutionary process. However, he had no knowledge of heredity, and so he ascribed the source of variation to natural forces; to him this was simply an act of faith. Today it is recognized that mutation and sexual recombination are the sources of variation in a population, and these provide the raw materials for evolution. Darwin's mechanism for evolutionary change, natural selection, by itself cannot account for evolution. Another force is necessary for a single species to evolve into two or more species: this force is **reproductive isolation,** which prevents interbreeding between species and populations. In the modern view, **evolution** is a change in the gene pool with the passage of successive generations. Evolution of new species involves (1) hereditary

FIGURE 24-10 Cartoon lampooning Darwin's suggestion that man had an apelike ancestor. (Courtesy of the Granger Collection.)

variation due to mutation and sexual recombination; (2) natural selection, that is, differential reproduction of certain genotypes; and (3) genetic isolation.

Darwinism came to be a focal point in human thought, and although Darwin never dealt with human evolution in *On the Origin of Species,* it was inevitable that his theory would be applied to man. Debate on the topic of man's origin became lively, and the notion that men were descended from monkeys was the talk of fashionable salons of the day. In 1871 Darwin himself entered the discussion of human evolution with his publication, *The Descent of Man,* in which he presented evidence for his theory of evolution and explained how it shed light on the apelike origins of man. For this he was frequently lampooned in the popular press (Figure 24–10). Despite this, in the hundred years since the publication of *On the Origin of Species* his theory has, in a somewhat modified form, survived attacks from various social, religious, scientific, and political groups. Evolution is a biological fact. Man is a product of evolution, as is every organism that inhabits earth.

24-4 THE ROAD TO MAN

A visit to a zoological garden, a view of monkeys and apes, and we are immediately struck with similarities of appearance, facial expressions, manner of movement, and behavioral patterns between these primates and ourselves. In the face of this sort of circumstantial evidence, it is difficult to believe that not until the nineteenth century were the similarities between men and apes recognized. Hindsight is perfect vision, so let us not be too harsh in our judgment of the past. One of the main stumbling blocks in understanding human evolution was the strong current belief in the

MR. BERGH TO THE RESCUE.

THE DEFRAUDED GORILLA. "That *Man* wants to claim my Pedigree. He says he is one of my Descendants."

Mr. BERGH. "Now, Mr. DARWIN, how could you insult him so?"

divine creation of animals. Indeed, the Linnaean pattern of classification was considered to be the Creator's plan, and the interrelationships of organisms a part of supernatural design. Then, in 1859 with the publication of *On the Origin of Species*, the idea that man could have evolved from a primatelike ancestor began to take hold. At first, it was contended that man was descended from the apes, but later it was clearly recognized that man could not have evolved from apes like those living today, since they, like us, are recent products of evolution. Rather, modern man and apes must both be descended from a common ancestral form, now extinct.

In *The Descent of Man* (1871) Charles Darwin built his case for human evolution on evidence from living men and primates, but he believed that remains of the missing link between apes and men—the extinct common ancestral form—would be found fossilized in the rocks. Indeed, we clearly recognize now that the emergence of man can best be appreciated by arranging the different known fossil forms according to their positions in the geological time scale. Before we begin the story of man's ancestors, we should recognize that the fossil record of man and the primates is far from complete. Yet despite this limitation there is sufficient fossil evidence to suggest the sequence of bodily changes that must have occurred during the process of hominization (emergence of *H. sapiens*). Unhappily, speech and social organization are not fossilized, and thus our reconstructions of these attributes of ancestral man remain strictly hypothetical.

The family tree

Who are the nearest relatives of man? The various primates of today are man's closest living relatives, and these are at the ends of

TABLE 24-4 A calendar of mankind (from Ernst Mayr)

Origin of	Date*	Million years ago
Earth	January 1	4,500
Life	April 15	3,200
Vertebrates	November 27	420
Primates	December 24	75
Hominoids (anthropoid apes)	December 29, 1:00 P.M.	30
Hominids (ape-man)	December 30, 9:45 P.M.	14
Australopithecus (man-ape)	December 31, 4:15 P.M.	4
Homo (man)	December 31, 9:30 P.M.	1.3
H. sapiens (modern man)	December 31, 11:27 P.M.	0.3

*Based on the fact that a year has 525,600 minutes. Therefore 4,500 million years equals 525,600 minutes and 1 million years, 116.8 minutes.

divergent lines of evolution. By comparing them with one another biochemically, or by comparing their anatomy with that of fossil forms, we can picture evolutionary relationships in the form of a family tree (Figure 24–11). To provide some idea of how recently modern man has emerged, Table 24–4 shows the history of the earth and dates the events leading to modern man; the time scale is shown in terms of millions of years as well as on the basis of a single year.

The insectivores. The primates were derived from small insect-eating animals whose fossil remains are found in rocks more than 75 million years old. The tree shrews of today are closely related to these fossil animals and are descended in a relatively unchanged form from the earliest types of primates. Tree shrews, together with lemurs and tarsiers, which are also derived from the earliest types, are grouped

as prosimians (*pro*—before, *simia*—ape; L.). Prosimians were abundant worldwide 55 million years ago; they disappeared from North America 25 million years ago, and today are found only in tropical Asia, Africa, and the island of Madagascar. Prosimians have poorly developed brains, but they do have grasping hands (and sometimes feet) and opposable thumbs (and big toes), and some have nails instead of claws. Monkeys, apes, and man (anthropoids) probably all arose from prosimians.

Dryopithecus. The point at which anthropoids diverged from prosimians is uncertain, since the fossil record is quite poor, but in the 1930s L. S. B. Leakey found on an island in Lake Victoria an apelike ancestor, which he called *Proconsul*. *Proconsul* (now called *Dryopithecus*) lived more than 20 million years ago and has the traits of both a monkey (hand, skull, and brain) and an ape (face, jaw, and teeth). What makes it so certain that *Dryopithecus* is really anthropoid and not prosimian? The pattern of teeth provides the best proof; prosimians have 34 teeth, but anthropoids have only 32. It is believed that *Dryopithecus* or a similar apelike form gave rise to the modern anthropoids; it may even be possible that he was an ancestor of man—the "missing link," or common ancestor.

Ramapithecus. In the 1930s in India and in Africa, jaws were found in which the teeth were distinctly human in character (incisors and canines uniform in size), rather than apelike (with projecting incisors and canines) (Figures 24–12 and 24–13). This fossil, called *Ramapithecus*, is 14 million years old and is the best example of an apelike form ancestral to man. *Ramapithecus* is thus on the direct evolutionary line to man, and not a common ancestor of apes and monkeys. The hominid line (apelike men) leading to modern man diverged from the

FIGURE 24-11 The family tree of man and his primate relatives.

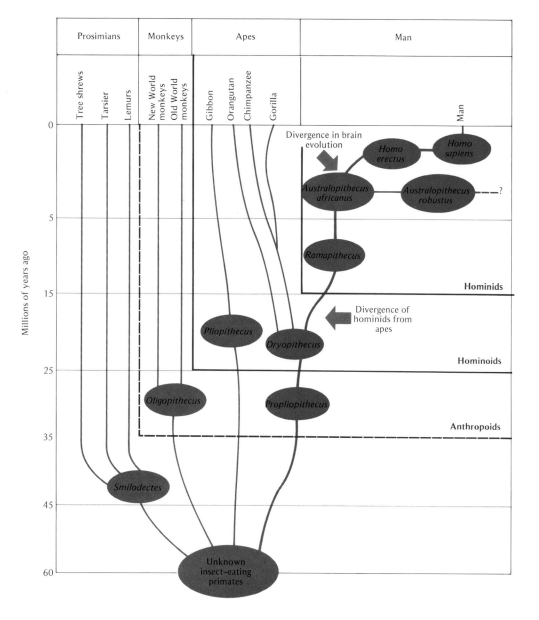

hominoids (manlike apes such as the gibbon, orangutan, chimpanzee, and gorilla) about 15 million years ago.

Australopithecus: the man-ape. A remarkable discovery was reported by Professor Raymond Dart of Johannesburg, South Africa, in 1925. Dart found among a student's fossil collection a skull that resembled that of a baboon, but a baboon skull unlike any seen before. Eventually, it turned out to be the skull of a 5- or 6-year-old child, and Dart called the specimen *Australopithecus africanus*, the South African ape. Several years later, other skulls and jaws were found in Africa. *Australopithecus* is between 2 and 5 million years old and is probably a direct ancestor of *H. sapiens*. What do we know at present about *Australopithecus*?

The general features of *Australopithecus* reveal that he was a small (about 4-ft tall) individual, weighing 50–90 lb, who stood and ran on his hind legs. The brain was about the size of modern ape brains, or about half the size of the brain of modern man (Figure 24–14). The jaws were massive and projecting and the molars very large. The incisor teeth were relatively small, and the canines were not of the sharp, projecting kind common to apes (Figure 24–13). One of the factors indicating that *Australopithecus* was a bipedal creature is the position of the foramen magnum, the hole through which the spinal cord passes through the skull to connect to the brain; it faces almost downward. Thus, this preman's head was perched on top of his spine, indicating an erect, upright posture. In quadrupeds (four-footed animals) the spinal cord leaves the cranium from the rear (Figures 24–6 and 24–12). The hipbones provide the most impressive evidence that *Australopithecus* was more manlike than apelike. In the apes the hipbone is narrow and long and prevents the animal from standing erect; whereas in man it is flat and broad, with

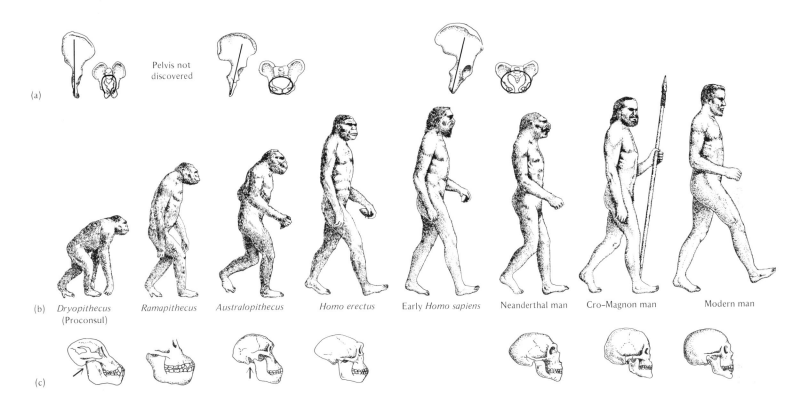

FIGURE 24-12 An artistic reconstruction of the road to man, based on fossil evidence. (a) Pelvic bones. (b) Reconstruction from skeletal remains. (c) Skull.

(a)

Pelvis not discovered

(b) *Dryopithecus* (Proconsul)   *Ramapithecus*   *Australopithecus*   *Homo erectus*   Early *Homo sapiens*   Neanderthal man   Cro-Magnon man   Modern man

(c)

a rear-projecting flange, an adaptation to the human erect posture (Figure 24–12). The flange acts to anchor the large buttock (gluteal) muscles that are used for balancing the trunk on the lower limbs in standing or walking; it forms a kind of basket to cradle the internal organs; and it gives attachment to the abdominal muscles that help maintain abdominal tone during standing. The abdominal muscles form a strong barrel that prevents the internal organs from sagging due to the pull of gravity. In quadrupeds, the internal organs are suspended from the back (Figure 24–6). Thus, the anatomical evidence indicates that *Australopithecus* stood and walked almost as we do ourselves.

There persist many questions about these preman creatures. Some of the most perplexing are: Why did they get up on their hind legs? Why did the canine teeth become reduced in size? The answers are, of course, conjectural, since there is no fossil evidence for behavior. There is, however, good evidence that during the period in which *Australopithecus* lived there existed considerable expanses of lush savannah with scattered shrubs, trees, and grasses. There were berries and roots in abundance, and because such areas were suitable for grazing, these savannahs were well stocked with game. These areas provided new habitats, abundant in food, and so we surmise the

australopithecines came down from the trees in which their own apelike ancestors lived in order to avail themselves of these new sources of food. Initially, they foraged and returned to the trees, continuing to move from tree limb to tree limb by brachiating movements of their arms; but later they spent more and more time on the ground. Although descent from the trees does not always result in evolution of upright posture in primates (witness the gorilla), through a lucky combination of anatomy and habits, these ape-men became bipedal. Being bipedal meant that the hands were freed from locomotor function and could be employed in manipulative skills such as carrying and dragging objects, fashioning of tools and weapons, and so on (Figure 24–15). Although some biologists suggest that upright posture preceded the use of the hand in fashioning and in using tools, there are compelling arguments that the reverse is true—that toolmaking ability preceded the evolution of upright posture. Apes and monkeys, although tree dwellers, do use their hands and are tool users. They throw stones, they poke with sticks, and they produce sticks by stripping the leaves from vine stems. Tool users get along best by walking on their hind limbs and having their hands free for holding objects; thus, where tool using is important for survival, selective pressure would be toward an upright posture. With the advent of toolmaking, hunting for big game became a possibility, and the brain and the hand were now subject to the molding force of natural selection. Those individuals with better manipulative skills and learning ability would be better adapted and more likely to survive in the new habitat. The destruction of food outside the mouth by the use of hands in ripping and tearing, and the ability to fashion tools as weapons of offense and defense meant that the teeth (the canines especially) were no longer of such importance for survival, and selective forces began to modify the structure of the skull. In time, the use of tools as weapons of offense and defense became the dominant habit, and with this sort of expertise *Australopithecus* could wander further away from the trees.

There is good evidence that *A. africanus* used the thighbone of an antelope as a bludgeon for slaying baboons. He was agile, quick-footed, and like ourselves, a meat eater. He became a creature of the open ground, and by means of natural selection his offspring became better adapted for a predatory life on the ground. Once hunting game became a part of

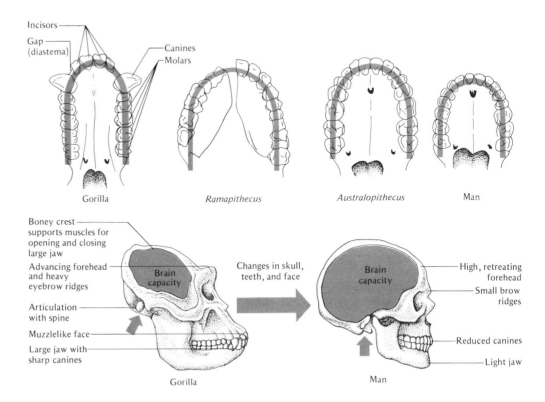

the life of *Australopithecus,* there was selective pressure toward the establishment of base camps and hunting groups, the division of labor, the accumulation of information, and the fabrication of improved weapons. All these required an increase in brain size and some form of communication. In this way, stored information could be passed on and used to advantage. The best evidence we have for this is the progressive development of pebble tools (Figure 24–15). Thus, speech and language became decisive events in the process of hominization. Language is not only an old adaptive pattern in humans, but it is apparently unique to man. Though many animals make sounds that are signals transmitting information, these constitute a closed system of calls that are all-

or-none. As far as we know, only the language of man can convey many bits of information at the same time.

Once man moved out of the forest he was faced with new dangers of predation; banding together to hunt and acquiring some kind of aggressive behavior pattern would have enhanced the chances for survival. In addition, the earliest humans must have established close-knit social groups for breeding. Much instinctive behavior came to be replaced by learning during a prolonged childhood. Complex learned behavior, based on acquired information, was passed from generation to generation and roughly can be called culture. Behavior became the strongest of all selective pressures for human evolution. Thus, we are

FIGURE 24-14 The estimated brain size of anthropoid apes, man-apes, and man.

FIGURE 24-15 The evolution of tool use.

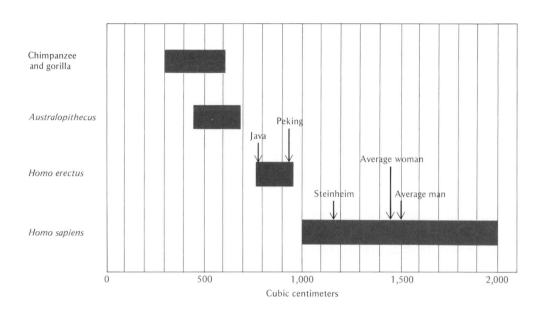

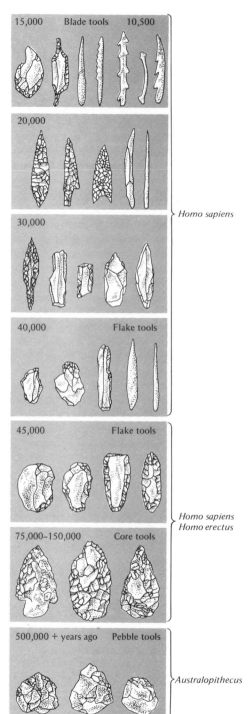

drawn to the conclusion that human evolution and especially the foundations of human intelligence, had their roots in a mosaic of adaptive features: bipedalism, hand and tool use, brain development, speech, and breeding pattern. Evolution did not occur to the same degree in all these characteristics at the same time, and apparently the first may have been bipedalism. All evidence seems to suggest that *Australopithecus* was the earliest human.

Homo erectus: the first true man. The next stage of human evolution shows a phenomenal increase in brain size and stretches backward to a time slightly over 1 million years ago. The discovery of the fossil remains of these "true men" (*Homo*) came in 1887. In that year, E. Dubois, a Dutch anthropologist, discovered in

Central Java a skull cap, a lower-jaw fragment, and several thighbones that aroused considerable excitement among students of human evolution. The skull was obviously apelike but the legbones resembled those of a human (Figure 24–16). This anthropoid ancestor of man, apelike in skull and human in upright posture, Dubois called *Pithecanthropus* (*pithekos*—ape, *anthropos*—man; Gk.) *erectus*. It is more commonly known as Java Man. For nearly 50 years there was skepticism about Dubois's find because the skull and thighbones seemed so dissimilar in character, but in the late 1930s other fossils of Java Man were discovered. During the latter part of the twentieth century, the status of Java Man was reconsidered, and as a result of these studies he is now termed

FIGURE 24-16 Java Man, skull and reconstruction. (Courtesy of the American Museum of Natural History.)

H. erectus (erect man). The skull of Java Man is unlike that of *Australopithecus*. Its capacity is intermediate between the average skull capacity of the largest living apes and the skull capacity of modern man, but it is twice that of *Australopithecus* (Figure 24–14). Java Man has a flat, receding forehead, the eyebrow prominences are enormous; the foramen magnum is far forward under the skull, much as it is in ourselves; and the thighbones in shape and proportion are like those of a modern human. All this indicates that although *H. erectus* did not exhibit great intellect and possessed an apelike face, he was too brainy to be classified as an ape. He walked upright and had heavy neck and jowl musculature to support his powerful chewing jaw (Figure 24–12).

Finds related to Java Man were made in the late 1920s by a scientific expedition in the village of Choukoutien about 40 miles southwest of Peking, China. There, in a vast cavern, were unearthed the remains of 40 or more individuals. The diggings also disclosed ample evidence of the use of fire and a plentiful supply of stone tools. This fossil man, first called *Sinanthropus pekinensis* (Chinese man of Peking), had a larger cranial capacity than Java Man. However, the jaws were massive and projected forward beyond the nose and the rest of the face, and the limb bones were distinctly characteristic of upright posture, although the standing height was only about 5 ft. Because of his similarities to Java Man, Peking Man too is now classified as *H. erectus*.

It is obvious from the fossil record of man that upright posture and other bodily changes came in advance of cranial development, and we are led to the conclusion that evolution of the postcranial (below the skull) skeleton of man was completed over 1 million years ago.

Let us try to recreate a picture of the life of *H. erectus*. He was an expert hunter of game and dined mostly on meat. He used fire and made quartz tools and a cradle type of hand ax (Figure 24–15). He was cannibalistic and occupied cave sites for about 200,000–300,000 years. The earliest signs of fire date back 400,000 years. *H. erectus* was probably a fire stealer rather than a fire maker; no fire-producing implements have been retrieved from any of the investigated sites. Use of fire enabled the cooking of food and made caves habitable even in the colder seasons. Bones of *H. erectus* have been found in Africa, Java, and China, and present thinking is that these men originated in Africa. How they spread through the continents is unknown, but it is obvious that such migration took place over the centuries and over many generations. Whether *H. erectus* wore clothes or built permanent habitations in addition to caves, we have no idea. We picture him as a hunter in small bands, but we know nothing of his social structure, religion, art, or any other cultural practices. The characteristics of such a culture have left no fossil remains.

We do not know with any certainty what became of *H. erectus*, but perhaps it is not too far from the truth to surmise that he changed with time to eventually become as we ourselves.

Homo sapiens: modern man. About 500,000 years ago the mental development of man had expanded to such a point that the fossils can be considered representatives of *H. sapiens* (Figure 24–14). The earliest find of this period was a skull unearthed in a cave on Gibraltar in 1848; it caused little stir. Then in 1856 near Düsseldorf, Germany, in a valley called Neander, a skull cap and some bones were dug out of a limestone cave. The skull cap was thick, there were massive eyebrow ridges, and the retreating forehead was much flatter on top and bulging in the back than the skull of any present-day human (Figure 24–17). The remarkable discovery, coming just 3 years before the publication of Darwin's *On the Origin of Species*, caused much controversy. T. H. Huxley

FIGURE 24-17 A comparison of the skull of Neanderthal Man (left) and Cro-Magnon (right). (Courtesy of Carolina Biological Supply Company.)

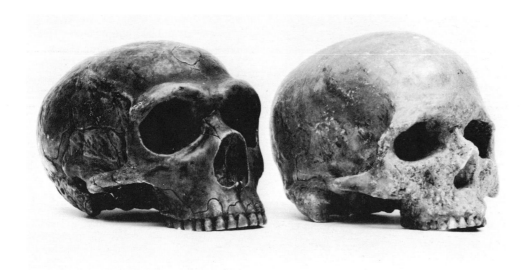

recognized some very peculiar things about this creature, who came to be called Neanderthal Man. The brain case was of the same volume as those of some primitive tribes, and the bony configuration marked it as the most apelike of all human skulls. Later diggings in Belgium produced skeletons similar to that of Neanderthal Man, and these occurred with the remains of a mammoth, a wooly rhinoceros, and other extinct animals. Neanderthal remains have also been found in Italy, France, Palestine, Siberia, southern Russia, and North Africa. Neanderthal was obviously a man who roamed Europe about 100,000 years ago, but a man quite unlike those who unearthed his ancient bones. The classic Neanderthal type disappeared about 30,000 years ago. Neanderthal has become the classic concept of the caveman. His name has been synonymous with brutishness: a short thick body with limbs of ungainly proportions, a lumbering gait, and a stooped posture. His features—heavy brow ridges, re-

treating forehead, large jaws, and shape of the limb bones—all suggest that he was more apelike than *H. sapiens,* yet as we have seen, *H. erectus* had limbs that were already of the human type. Was Neanderthal a distinctive and specialized form of *H. sapiens,* a sideline in the evolution of modern man? Recent examinations of skeletons suggest that his stooped appearance may have been the result of disease rather than low evolutionary status. One theory is that he suffered from a deficiency of calciferol, aggravated by the diminished sunlight during the Ice Age, and developed severe rickets. New archeological finds in Iraq indicate that he was a skilled hunter and toolmaker, practiced crude surgery, and buried the dead with formal rites. The coarse features of Neanderthal Man are undeniable, but his cultural behavior patterns indicate he was not as apish as early writings suggest. Even today, the exact status of Neanderthal Man remains controversial, but the best available evidence supports

his being a member of an extinct race of *H. sapiens.*

If Neanderthal Man was not in the direct ancestral line of modern man, then who were our ancestors? In 1933 a fossil skull and face were discovered near Steinheim, a town not far from Stuttgart, Germany, and their owner was given the name Steinheim Man. The age of the fossil is about 300,000 years, and the cause of death is clearly visible in the depression of the left temple: *H. sapiens steinheimensis* was killed by a blow to the head. The forehead is only moderately recessive, the brow ridges are pronounced, the nose is broad, but the facial and dental configuration is that of man (Figure 24–18). The cranial capacity is at the lower end of that of modern man (Figure 24–14). Other fossils of the Steinheim type have been unearthed at Swanscombe, England, and included with the human bones were hand axes. The cranium of the Swanscombe find housed a brain just about the size of a modern human female's brain.

It is apparent that although populations of *H. erectus* still existed during the Lower Pleistocene Era (700,000–1,000,000 years ago), individuals of the *H. sapiens* type were also present. Gradually, as the selective advantages of the *H. sapiens* traits began to take effect, there would be a greater proportion of *H. sapiens* in the population.

The final phase in the story of man's emergence came about 20,000 years ago, when the Neanderthal inhabitants of Europe were abruptly displaced by a people of a completely modern European type. The remains of these people were found at Cro-Magnon in southern France, and the fossils have therefore been called Cro-Magnon Man. In reality Cro-Magnon Man is just one manifestation of *H. sapiens.* This group of humans was tall, and muscular, with a large cranial capacity and fine facial features—in all respects indistinguishable from modern man. Cro-Magnon Man was the last of the Old Stone Age men. He lived in rock shelters and was an accomplished artist; his paintings and sculptures are some of the finest examples of human creativity.

It is unlikely that man originated in a restricted geographical locality. The Garden of Eden exists only as a metaphor. The progenitors of the hominoids and the earliest representatives of the family of man were probably distributed as widespread breeding communities in India, Europe, Asia, and Africa. The separation of modern man into geographical races probably took place as long as 500,000 years ago. The Asian forms migrated to Polynesia and Australia. Some 20,000 years ago Asian man crossed the Bering Strait and reached the American continent (Figure 24–19).

24–5 THE RACES OF MAN

The biology of race

All men have been created equal; most certainly they are not all alike. The idea of equality derives from ethics; similarity and dissimilarity are observable facts. Human equality is not predicated on biological identity, not even on identical ability. People need not be identical twins to be equal before the law, or to be entitled to an equality of opportunity.

And yet, equality is often confused with identity, and diversity with inequality. They are confused so chronically, persistently and obstinately that one cannot help suspecting that people have deep seated wishes to confuse them.

The race problem as described by Professor T. Dobzhansky is a real one; there is no doubt that the linked concepts of race and prejudice are deeply rooted in man's inhumanity to man. A clear understanding of the biology of race should demonstrate the fallacious reasoning behind racial prejudice.

Humans, you and I, make up a single species. By this we mean we are a group composed of interbreeding populations that are reproductively isolated from other such groups. However, our being of a single species does not reveal anything about the appearance of the members of that species. Indeed, at the species level, it is reproductive compatibility and not appearance that is of importance. Thus, although black bears and grizzly bears occupy the same territory, they do not interbreed, and the same is true for the other species that share our territory; a species remains as a distinct biological entity because its gene pool is isolated from other gene pools.

Species are made up of populations of individuals, and within these populations there exists two kinds of variations: individual and geographical. Individuals in a particular locality, be they bears or humans, show differences in their sex, age, physique, coloration, and so on. This individual variation within a population—**intrapopulation diversity**—depends on

genetic variability in a single trait or many traits and is called **polymorphism** (*poly*—many, *morphe*—form; Gk.). It is polymorphism that enables you to tell Jack from Jim and Ellen from Mary. Polymorphic traits may be continuous (weight, height, age) or discontinuous (sex, blood type, eye color) in the population.

Now let us imagine that we expand our examination of a particular species and survey populations in different geographical localities. If we were to study the local populations of humans in Canada, Sweden, China, Ghana, and India, we should find them to be polymorphic. However, if we compared one geographical population with another we should soon see that the polymorphisms of one population were not identical with the polymorphisms of another; this is **geographical variation,** and because of it, it is sometimes simple to identify individuals coming from one area or another. It is interesting to note that the closer the populations are geographically, the more similar they are, and the further apart they are geographically, the more different they appear. Sometimes the variations from one locale to the next are not sharply defined, but in other cases the distinctions are quite clear. A species showing geographical variation, that is, **interpopulation diversity,** is called **polytypic,** and each population within that species can be shown to differ in the incidence of certain genes. Such differentiated populations occurring within a species, and differing in their gene frequencies, are termed **subspecies** or **races.** Many species of plants and animals, including man, are polytypic and consist of races or subspecies; under the appropriate conditions, that is, genetic or reproductive isolation, subspecies may eventually develop into full species.

The human species is polymorphic and polytypic. Why is there polymorphism? Because of gene mutation and sexual recombi-

FIGURE 24-19 The spread of man across the face
of the earth.

nation. Why are there races? There are two principal reasons: (1) populations that make up a race or a subspecies are or have been partially isolated geographically from other populations of the same species; if such isolation is continued, gene flow from one population to the next is restricted, and the gene pool takes on a distinctive character; (2) geographically separate populations are subject to different selective pressures, and as a consequence the populations tend to differentiate from one another. Although in most cases it is impossible to identify what is being selected for, it is well established that isolation and selection promote race formation. In this respect it is easy to see that a race is an incipient species—a species in the making.

The term race, as defined biologically, is equivalent to a subspecies and involves populations within a species that are characterized by differences in gene frequency. There is no agreement on the classification of the races of man. Boyd recognizes 6 races, Coon describes 30 races, and Garn classifies human populations into 9 categories and 34 local races (Tables 24–5 and 24–6). Although it is easy to describe differences among human populations in various geographical localities in terms of physique, skin color, hair shape, and so on, many of the traits used in describing human races have only a partial genetic basis and are strongly influenced by environmental factors such as exposure to the sun and the adequacy of nutrition. Such traits are difficult to measure, and at present there is virtually no information on their genetic basis. So far no clear-cut racial difference can be shown to exist for these traits. Furthermore, physique and skin color, the two characters most commonly used in race designations, are not critical characteristics of *H. sapiens* from an evolutionary point of view. As far as survival and selection go, it is difficult to

Dryopithecus and *Pliopithecus*
Apes
8–18 million years ago

Australopithecus
Man-apes
2–5 million years ago

Homo erectus
Early man
400,000–700,000 years ago

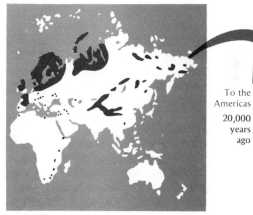

Homo sapiens
Modern man
300,000 years ago

To the
Americas
20,000
years
ago

TABLE 24-5 Boyd's classification of the races of man by blood-group frequencies

European (caucasoid)
High frequencies of Rh, cde, and CDe; moderate frequencies of the other blood-group genes; M usually slightly above and N below 50%

African (negroid)
Very high frequency of Rh, cDe; moderate frequencies of the other blood-group genes

Asiatic (mongoloid)
High frequency of B; little if any cde

American Indian
Most homozygous O, but sometimes high frequencies of A; absence of B; little if any cde; high M

Australoid
Moderate to high A; little or no B or cde; high N

Early European
High frequency of Rh, cde; no B (Basques are survivors of this group)

Modified from William C. Boyd *Science*, **140**, 1057–1064, 1963.

---

pinpoint advantageous or disadvantageous characteristics in an age where man creates his own environment. It could well be argued that the important features of our species are intelligence, compassion, imagination, courage, perseverance, adaptability, and so on. Biologically, the racial differences described by anthropologists remain quite minor and of a secondary quality. The biological unity of modern man—that is, his being a single species—is much more significant. The gene pools of man exhibit a genetic and biological continuity in both time and space. Furthermore, no genetic isolating mechanisms exist for separating the races of man, and even social barriers are inefficient for separating the races.

The significance of race and its origins

"He is a pure-blooded Indian." "The Aryan race was one of superior intellectual and phys-

TABLE 24-6 Garn's classification of human races

European
1. Northwest European: population of Scandinavia, northern France and Germany, the Low Countries, the United Kingdom, and Ireland
2. Northeast European: population of Eastern Baltic, Russia, and modern Siberia
3. Alpine: population of central France, southern Germany, Switzerland, and northern Italy, to the Black Sea
4. Mediterranean: population surrounding the Mediterranean, eastward through Asia Minor

Indian
1. Indic: population of India, Pakistan, and Ceylon (Note: Garn calls this race Hindu, but inasmuch as Hindu properly identifies a religion, we feel that Indic is more appropriate.)
2. Dravidian: aboriginal population of southern India

Asian
1. Classic Mongoloid: population of Siberia, Mongolia, Korea, and Japan
2. North Chinese: population of northern China and Manchuria
3. Turkic: population of western China and Turkestan
4. Tibetan: population of Tibet
5. Southeast Asian: population of South China through Thailand, Burma, Malaya, the Philippines, and Indonesia
6. Ainu: aboriginal population of Japan
7. Eskimo: population of northern maritime fringe of North America and ice-free fringes of Greenland
8. Lapp: population of arctic Scandinavia and Finland

Micronesian (no local races distinguished)

Melanesian
1. Papuan: population of mountain highlands of New Guinea
2. Melanesian: population of coastal area of New Guinea and most of the other islands in the Melanesian archipelago

Polynesian
1. Polynesian: aboriginal population

2. Neo-Hawaiian: nineteenth- to twentieth-century blend of Polynesian, European, and Asiatic

American
1. North American (Indian): aboriginal population of Canada and the continental United States
2. Central American (Indian): population of the Southwestern United States, Mexico, and Central America to Brazil
3. South American (Indian): population of all South America, except Tierra del Fuego
4. Fuegian: population around the Straits of Magellan
5. Ladino: new Latin-American population resulting from blending of Mediterranean, Central and South American (Indians), Forest Negroes, and Bantu
6. North American: Colored eighteenth to twentieth-century population blended of Northwest European and Africans

African
1. East African: population of the East African Horn, Ethiopia, and Nilotic Sudan
2. Sudanese: population of the Sudan, except for Nilotics
3. Forest Negro: population of West Africa and most of the Congo
4. Bantu: population of South Africa and adjacent parts of East Africa
5. Bushman-Hottentot: surviving post-Pleistocene population in South Africa
6. Pygmy: small-statured population living in the equatorial rain forest
7. South African Colored: population of South Africa produced by a blend of Northwest European and Bantu, plus some Bushman-Hottentot

Australian
1. Murrayian: aboriginal population of southeastern Australia
2. Carpentarian: aboriginal population of central and northern Australia

FIGURE 24-20 Distribution of skin pigmentation in the world.

Lightest

Medium-light

Medium

Medium-dark

Darkest

ical capabilities." "The Jewish race . . ." "Anglo-Saxons belong to the white race." Are these statements correct? Of course not. Pure races do not exist; races consist of populations showing genetic variability from one another, and therefore a race is defined statistically, not individually. No individual can be "typical" of a race. A few misguided individuals have applied a typological definition of race to man and have classified him according to artificial criteria based primarily on skin color. This is contrary to the biological concept of race. Racism takes as its basis the erroneous belief in a type specimen for each species or race. Inherent in such thinking is belief in the special creation of each species that remains unchanged throughout the course of time and belief that each kind of creature can be described in an ideal form. Such notions are without scientific foundation. Race, defined in biological terms, is not equivalent to religious groups, language groups, cultural groups, or social status. The classification of human races is quite arbitrary, and often it is used for rather trivial scientific reasons.

Recognizing the arbitrary nature of race designation in humans, let us consider two of the traits that have been used traditionally—skin color and physique—and ask the questions: What is the adaptive significance of race? and What are its origins?

Skin color. If we survey the distribution of skin pigmentation throughout the world before European exploration and colonization blurred population characteristics, we find that dark pigmentation occurred principally among people living around the equator (Figure 24–20). Lighter skin pigmentation occurred in populations living further away from the equator. What is the adaptive significance of this? There seems little doubt that degree of skin pigmentation is related to environmental conditions. W. F. Loomis suggests that the degree of

pigmentation regulates the amount of calciferol (vitamin D) synthesized by the action of ultraviolet light on the skin. Calciferol regulates calcium metabolism, and too little results in inadequate deposition of calcium in the bones; such a deficiency disease, called rickets, produces twisted legs and spine (Chapter 16, Box 16B). In a dark-skinned individual living in the tropics where the ultraviolet radiation is abundant, many of these rays would be screened out. However, as man moved out of the tropics he received less ultraviolet light; he would be faced with a deficiency of calciferol if dark skin prevailed. Those mutations that resulted in a reduction in skin pigmentation would be favored in the temperate zones, since light skin would promote the reception of adequate amounts of ultraviolet light by the skin, and permit the synthesis of calciferol in normal quantities. Loomis's hypothesis is attractive, but it requires further documentation.

Physique. Body form may be related to cold and heat adaptation. A thin, lean individual has a high surface-volume ratio and is well adapted to radiating heat: in warm climates this could offer some selective advantage. Features such as a flat nose, narrow eyelids, padding of the eyelids, and a stocky body with short limbs may all be adaptive for survival in cold climates, since body heat is conserved. Selection for these alternate physiques doubtless entered into the evolution of the Watusi, on the one hand, and the Eskimo, on the other; these are examples of extreme types, however, and this selective pressure has had little effect on other geographically isolated groups.

24–6 WHAT IS MAN'S EVOLUTIONARY FUTURE?

It is difficult for most of us to conceive of how long we were in becoming what we are today.

The development of *H. sapiens* from *Australopithecus* covers a span of 1 million years—equivalent to 15,000 human life spans or about 40,000 generations. A single lifetime is but a fleeting moment in the time of man, and it seems to each of us that the evolution of our species stands still. However, change is part of the nature of things. Since change will continue as long as the universe exists, what does the future hold for man? The prophets of our future speak with assurance and certainty; we shall satisfy ourselves with speculation.

Will man speciate in the future? Probably not. *H. sapiens,* the species to which we belong, has failed to speciate and will probably continue in this way for two reasons: (1) man shows great ecological diversity, that is, he occupies a wide variety of habitats, and thus a burst of adaptive radiation into new habitats (such as occurred with Darwin's finches on the Galapagos Islands) is unlikely; (2) reproductive isolating mechanisms develop slowly and never have persisted long enough in humans to prevent gene flow between the races of man. Indeed, the greater mobility of man and his relative independence of the natural environment make geographical isolation and species formation near impossibilities.

Mutation, sexual recombination, and natural selection led to the emergence of man. Man is the first, and at present the only, species that can control its environment to any large extent and plot its evolutionary course. Owing to advances in the medical sciences, many countries have largely succeeded in controlling early death from disease. Many individuals with genetic disorders, who in previous years would not have survived to produce offspring, are now surviving and producing families. Will maintenance of such deleterious genetic factors impair our future? Provided medical care facilities are available, our future will probably be affected to an insignificant degree.

Increased intelligence has been of adaptive value to man's survival in a changing world, but since the time of Neanderthal Man there has not been much change in the size of the brain. Will man's brain increase in size? There is no genetic reason why this could not happen if natural selection favored the survival of such a trend. However, it is possible that brain size has reached a plateau, since enlarged size of the brain of a baby would be perilous during childbirth unless there were (1) alterations in the size of the female pelvis or (2) reduction in the duration of pregnancy; greater brain growth after birth than *in utero* would be another alternative. Such changes have not been selected for, and it has been suggested that increase in the size of the brain ceased not because of selection against increase in size but because larger brain size did not carry selective advantages.

Are we free of natural selection today? When we consider infant mortality and differential fertility in those countries where medical facilities and food are in short supply, it is obvious that natural selection still operates in human populations. However there has been a reduction in selective pressure for adaptations to the environment such as the ability to resist cold and heat, natural resistance to various diseases, and so on. Our future will depend not solely on our ability to adapt to new habitats, but on our ability to devise a mode of life (culture) that is compatible with the lives of members of our own species and of all other creatures on this planet.

SUMMARY

1. Species are groups of interbreeding natural populations that are reproductively isolated from other such groups.

2. The Linnaean system of binomial nomenclature, in which living organisms are known by both genus and species name, is used today. The Linnaean concept of the fixity of species has been superseded by the Darwinian view that organisms of similar structure, function, and development are closely related and have a common ancestry.

3. Species are classified into a hierarchy of genera, families, orders, phyla, and kingdoms, reflecting both their similarities and differences.

4. Man belongs to the animal kingdom, the subphylum Vertebrata, class Mammalia, order Primates, family Hominidae, genus *Homo* and species *sapiens*. Humans are the only living species in the genus *Homo*.

5. Man shares many features with other primates. Distinctly human features include bipedalism, structure of the hand, cranium, brow and pelvis, distinct chin, female breasts, enlarged penis, reduced body hair, enlarged brain, capacity for thought and language, and highly developed culture and technology.

6. Establishment of the geological age of the earth eroded theories of the special creation of living organisms. With the publication of Charles Darwin's *On the Origin of Species* and its evidence for natural selection, the theory of evolution began to be accepted.

7. Adaptive radiation and speciation among finches of the Galapagos Islands were one of Darwin's case studies in evolution.

8. Mutation and sexual recombination are the sources of genetic variation in a population, providing the raw materials on which natural selection can work. Reproductive isolation prevents interbreeding between populations, and frequently produces changes in the gene pool. Evolution thus involves hereditary variation, natural selection of the best-adapted (fittest) organisms, differential reproduction (survival of the fittest), and genetic isolation.

9. Evolutionary theory also applies to man, whose closest living relatives are the primates. Today's primates were descended from an ancient insectivore prosimian stock, whose modern representatives are tree shrews, lemurs, and tarsiers.

10. Fossil ancestors of man include *Dryopithecus (Proconsul)*, with both monkey and apelike traits; *Ramapithecus,* an apelike man with humanoid incisor and canine teeth; *Australopithecus,* a bipedal erect humanoid with reduced canines, toolmaking ability, a savannahlike habitat and a brain about half the size of modern humans; *Homo erectus (Pithecanthropus* or Java man and *Sinanthropus),* the first true men with greatly enlarged brains and upright posture showed cannibalism, used fire, and lived in caves.

11. The change from tree-dwelling habitat to life on the plains was probably important in the selection of upright bipedalism. Increased brain size and mental capacity followed since selective pressure favored social development, cooperation, and communication abilities.

12. *Homo sapiens* probably appeared about 500,000 years ago. Neanderthal man, an apelike cave dweller who disappeared about 30,000 years ago, was a skilled hunter and toolmaker with a developed culture; Neanderthal was probably a race of *H. sapiens.* Steinheim man and Cro-Magnon man were other Stone Age races of *H. sapiens.*

13. Modern man probably became separated into geographical races about 500,000 years ago. Geographic isolation is responsible for the production of polytypic species with different gene frequencies forming incipient species (subspecies or races).

14. Isolation of geographic groups restricts gene flow, and such isolated groups are subject to different selection pressures. Thus, isolation and selection promote race formation. Races are subspecies characterized by differences in gene frequency.

15. Modern man is a single species with no true subspecies. Genetic differences among various races of man are insignificant from an evolutionary viewpoint.

16. Differences in human skin pigmentation probably arose as a result of selection for less pigment in areas of less sunlight, since pigmentation regulates calciferol production in the presence of ultraviolet light. Differences in body form may be related to cold and heat adaptation.

17. *Homo sapiens* is unlikely to speciate in the near future. Given modern mobility, it is unlikely that reproductive or geographical isolation or adaptive radiation into new habitats will occur. Cultural adaptation to a changing world will probably be of more importance than selection and genetic adaptation.

KEY WORDS

genus
species
fixity of species
morphology
families
genera
orders
phyla
kingdom
structure
function
development
strata
natural selection
adaptive radiation
Darwinian fitness
reproductive isolation
evolution
insectivores
Dryopithecus
Ramapithecus
Australopithecus
Homo erectus
Homo sapiens

intrapopulation diversity
polymorphism
geographical variation
interpopulation diversity
polytypic (species)
subspecies or races

TOPICS FOR DISCUSSION AND REVIEW

1. Why can't natural selection alone account for speciation?
2. List the similarities and differences between man and other primates.
3. Discuss the statement: *Australopithecus* was a man-ape and not an ape.
4. How can evolution be a blind force and yet lead to species that are better adapted than their predecessors?
5. Define a biological species. What are the weaknesses of such a definition? Strengths?
6. What is the significance of Darwin's finches to our thinking about evolution?
7. Discuss the biological basis for race formation and why discussions of human racial superiority are without biological meaning.
8. Describe the significant stages in human evolution and the fossil evidence for each stage.
9. Distinguish between polytypic and polymorphic.
10. John D. Rockefeller, Sr., once told a Sunday School class: "The growth of a large business is merely survival of the fittest. . . . The American Beauty Rose can be produced in the splendor and fragrance which bring cheer to its beholder only by sacrificing the early buds which grow around it. This is not an evil tendency in business. It is merely the working out of a law of nature and a law of God." Discuss the biological error implicit in such thinking.

MAN AND THE ENVIRONMENT

25-1 WHAT'S ECOLOGY?

You are, for the moment, an astronaut about to leave for a mission to the moon. To provide you with the necessities of life, the small space craft is well stocked with food, water, and oxygen; receptacles will be provided to store your liquid and solid wastes, and the carbon dioxide you produce by respiration will be trapped chemically. Prior to lift-off, all the life-support systems as well as the rocket hardware must be in "go" condition, or your mission will surely end in disaster.

Stocking a spacecraft with the necessities for life is an obvious prerequisite for manned spaceflight. Life-support checklists are routinely employed for a space vehicle not much larger than an automobile, but for another kind of spaceship, the planet earth, environmental checklists and a balanced environment are often ignored. Months, perhaps years, of planning go into a week-long manned voyage to the moon, but a trip in space lasting 70 years receives little thoughtful consideration. Indeed, as we shall soon see, the spaceship earth, carrying over 4 billion human inhabitants and a variety of other living creatures, is as fragile a craft as any yet fabricated by man. Our survival depends upon maintenance of our life-support systems and an understanding of our own ecology, tasks more easily identified than accomplished.

What is ecology and what relationship does it bear to the spaceship earth? Perhaps the best way of defining ecology is to look at biology in terms of its levels of organization (Figure 25-1). The molecules of life are organized in specific ways to form cells, cells are grouped into tissues, and tissues are arranged to produce functional organs. The body organs are integrated to produce organ systems, and the entire array of these systems constitutes an organism. Organisms exist not just as single individuals,

FIGURE 25-1 The hierarchy, or levels of organization, in biology.

but in groups called **populations.** The various populations of organisms that interact with one another form a **community;** interdependent communities of organisms interact with the physical environment to compose an **ecosystem.** (The prefix eco- is derived from the Greek oikos, meaning "house," and in its broadest sense "house" is taken to mean environment, both living and nonliving.) Finally, all the ecosystems of the planet are combined to produce a level of organization known as the biosphere (Figure 25–2). The **biosphere**—that film of air, water, and soil extending to a depth of no more than 10 miles—is the relatively thin rind surrounding the surface of the earth where we and all other living creatures exist. **Ecology** (literally, "a study of the environment") is the field of biology concerning itself with the levels of organization beyond that of the individual organism—population, community, ecosystem, and biosphere (Figure 25–1).

The biosphere consists of a variety of ecosystems—pond, lake, forest, field, ocean, and so on. However, no ecosystem is an island unto itself. A pond or a lake may be surrounded by a field or a forest, and these in turn may be enclosed by rivers, oceans, and mountains. The existence of such continuity means there are no discrete boundaries between ecosystems, and each ecosystem is partially dependent on what is happening in other systems. Therefore, when a study is made of a particular ecosystem, factors outside the system must be considered as well as those within. As a consequence of this, ecological studies are often complex. (Some of these external factors can be eliminated in the laboratory or in special field experiments when specific ecological principles are studied and tested.)

All living things, from men to the smallest microbe, live in association with other living things . . . an equilibrium is established which permits the different

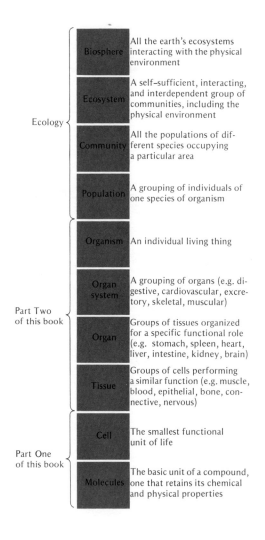

Ecology
- Biosphere: All the earth's ecosystems interacting with the physical environment
- Ecosystem: A self–sufficient, interacting, and interdependent group of communities, including the physical environment
- Community: All the populations of different species occupying a particular area
- Population: A grouping of individuals of one species of organism

Part Two of this book
- Organism: An individual living thing
- Organ system: A grouping of organs (e.g. digestive, cardiovascular, excretory, skeletal, muscular)
- Organ: Groups of tissues organized for a specific functional role (e.g. stomach, spleen, heart, liver, intestine, kidney, brain)
- Tissue: Groups of cells performing a similar function (e.g. muscle, blood, epithelial, bone, connective, nervous)

Part One of this book
- Cell: The smallest functional unit of life
- Molecules: The basic unit of a compound, one that retains its chemical and physical properties

components of biological systems to live at peace together, indeed often to help one another. Whenever the equilibrium is disturbed by any means whatever, either internal or external, one of the components of the system is favored at the expense of the other.[1]

If the ecosystem becomes imbalanced, changes occur that tend to restore the equilibrium, but on occasion the changes are so overwhelming that the compensatory processes could exclude many forms of life. As we shall soon see, this concept is at the core of man's current environmental problems.

25–2 ENERGY FLOW

All actively metabolizing organisms consist of a complex organization of organic molecules dispersed in a watery medium. To maintain such organization, and to have it reproduce itself, there must be available a continuing supply of energy. Energy drives the matter of life.

As the earth travels through space, it exchanges little in the way of matter with the rest of the universe. Indeed, it may be said that life started on this planet with a fixed supply of raw materials, and these are continually being recycled. By contrast, the earth is continuously receiving energy from the sun, and it will continue to do so for an almost infinite period of time. Sunlight is the power of life. This is true whether the living form is an enormous blue whale or a tiny hummingbird, a small child on a tricycle or a bulky football player; sunlight powers the pen that writes these words and provides the energy for you to comprehend their meaning. Although all of life depends on the sun, only photosynthesizing creatures are

[1] R. Dubos, The Germ Theory Revisited. Quoted in S. Wolf and H. Goodell (eds.), H. G. Wolff's Stress and Disease, 2nd ed., Springfield, Ill., Thomas, 1968.

FIGURE 25-2 The thin film of life—the biosphere.

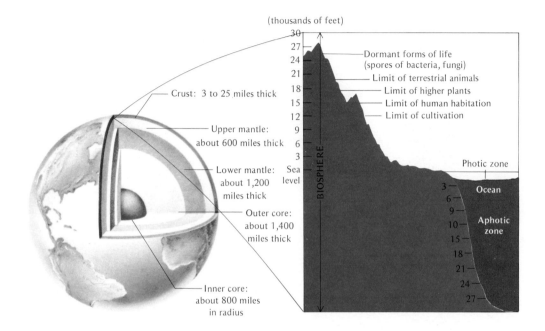

(thousands of feet)

Crust: 3 to 25 miles thick

Dormant forms of life
(spores of bacteria, fungi)

Limit of terrestrial animals
Limit of higher plants
Limit of human habitation
Limit of cultivation

Upper mantle:
about 600 miles thick

Lower mantle:
about 1,200
miles thick

Photic zone

Sea
level

Ocean

Outer core:
about 1,400
miles thick

Aphotic
zone

Inner core:
about 800 miles
in radius

BIOSPHERE

capable of trapping light energy directly and, by using simple raw materials, synthesizing organic compounds; these organisms are called **producers.** In discussions of energy, the term producer may sometimes be misleading. Producers do not, indeed they cannot, generate energy out of nothing; rather, they convert light energy into chemical energy. Producers are, in fact, converters or transducers of energy, taking the radiant energy of the sun and chemical elements from the earth and transforming these into useful, energy-rich organic molecules. Because photosynthesizing organisms synthesize their own food (organic molecules), these producers are also called **autotrophs** ("self-feeders").

Since photosynthesizing organisms capture light, they are obviously confined to that portion of the biosphere that receives solar radiation

during the day. This region, where fixation of light energy occurs, is restricted to the clearest waters of the oceans to a depth of 300 ft (Figure 25–2). In turbid lakes, rivers, and streams the light may penetrate only 1 in., and thus the availability of light energy for photosynthesizing plants occupying this habitat may reduce their activity to a narrow zone just below the surface of the water. On land, photosynthesis is restricted to an altitude below 18,000 feet, not by the availability of light but by low temperature and lack of oxygen.

The energy flow from the sun to the photosynthesizing autotroph is not the end of our ecological story; it is merely the beginning. Some organisms meet their nutritional needs by feeding on other organisms rather than by manufacturing their own food; these are called **heterotrophs** ("other feeders"). The hetero-

troph is a consumer; heterotrophs that feed directly on plants are called **herbivores,** or **primary consumers.** Not all animals feed on plants; some feed on other animals. These are also heterotrophs, but they are **secondary consumers,** or **carnivores.** A secondary consumer derives its energy indirectly from the producer (the autotrophic plant) by way of the primary consumer (the herbivore). The hamburger you ate for lunch makes you a secondary consumer. The steer that supplied the hamburger meat was the primary consumer, and the grass upon which that steer fed to provide the meat was the producer.

It is important to recognize that the flow of energy through the ecosystem is unidirectional, not cyclic. There is no recycling of energy in the universe. The chemical constituents of life—nitrogen, carbon, oxygen, water, and the other mineral elements—circulate in the biosphere, but energy does not. This means that the flow of energy is always:

producer $\longrightarrow$ herbivore $\longrightarrow$ carnivore
autotroph $\longrightarrow$ heterotroph

Let us consider the consequences of this unidirectional flow of energy through the ecosystem. The total amount of solar energy fixed on the earth sets an upper limit on the total amount of life the biosphere can support. Another limit is set by the pattern of energy flow through the biosphere. For example, the total supply of solar energy reaching the earth amounts to 13×10^{23} gram-calories per year, or 2.5 billion horsepower. This amounts to about 25% of the total incoming solar radiation, the remainder being absorbed by clouds, dust, water, and atmospheric gases. The absorbed radiation serves during its temporary stay on the earth to melt ice, to evaporate water, to generate winds and wave currents, and to warm the waters and land masses of the planet.

FIGURE 25-3 The average solar radiation received by the earth.

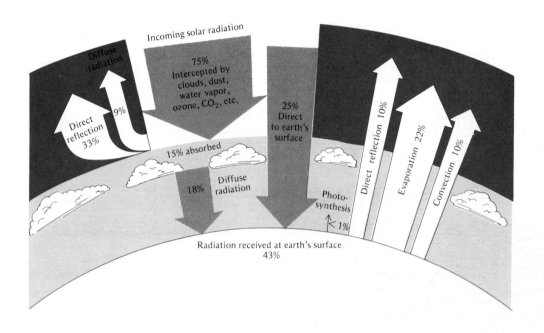

Incoming solar radiation

Diffuse radiation

75% Intercepted by clouds, dust, water vapor, ozone, CO_2, etc.

9%

Direct reflection 33%

15% absorbed

18% Diffuse radiation

25% Direct to earth's surface

Photo-synthesis ↑ < 1%

Direct reflection 10%

Evaporation 22%

Convection 10%

Radiation received at earth's surface 43%

25-3 FOOD CHAINS AND ECOLOGICAL PYRAMIDS

The transfer of food energy from plants to animals and then to other animals by successive stages of feeding is called a **food chain.** In some cases, the relationships between the organisms involved are so complex that the chain is in the form of a highly complicated and branching network called a **food web.**

Two food chains, one land-based and the other aquatic, are illustrated in Figure 25–4. A more complicated food web is illustrated in Figure 25–5. It should be noted that there are four components in these food chains: plants, herbivores, carnivores, and decay organisms. The products of decay are returned to the plant in the form of utilizable nutrients, and thus the matter in a food chain is continually being recycled.

If a food chain is arranged in levels, the first level is occupied by the producers: in a land-based ecosystem these may be grasses and trees; in an aquatic ecosystem, algae, seaweed, and the microscopic phytoplankton. The second level is occupied by the herbivores: on land hoofed animals, rodents, birds, worms, insects, and fungi are herbivores, or primary consumers; in water-based ecosystems molluscs, worms, herbivorous fishes, and microscopic zooplankton occupy this level. The third level is occupied by the carnivores, or secondary consumers: on land these are mountain lions, hawks, foxes, and so on; in the water the carnivores are fishes, seals, and birds. On the fourth level are the carnivores that themselves feed on other carnivores: otters that eat trout, trout that eat insects, and foxes that eat owls. Many organisms, including man, eat both plants and animals and are called **omnivores.** Thus, we and they enter the food chain at a number of levels.

Much of this absorbed radiation is subsequently lost into space: 10% by direct reflection, 22% by evaporation, and 10% by convection (Figure 25-3). The actual amount of available solar energy trapped by photosynthesizing organisms is quite small—less than 1% of the total incoming solar radiation. As small as this value is, it provides sufficient energy to synthesize more than 200 billion tons per year of dry organic matter for the world, including our own food.

When an animal feeds on a plant, it succeeds in extracting only 50% of the calories stored in the plant's organic molecules;[2] the remainder of the calories is dissipated in the form of heat. Of the extracted calories, the animal uses 70%–80% for powering the living system, and only 20%–30% are actually built into organic substance. Thus, the efficiency of conversion of plant matter into animal matter is about 10%. A secondary consumer, one that feeds on animal tissue, does a little better (because animal matter has less indigestible material than a plant), but still the efficiency of conversion never exceeds 20%.

The unidirectional flow and efficiency of energy utilization account for the need for a continual source of energy to prevent the collapse of an ecosystem. Indeed, no ecosystem can maintain itself for an extended period if it is deprived of an energy supply.

---

[2] The conversion of energy from one form to another is not 100% efficient—a situation described by the Second Law of Thermodynamics. This law states that when a process involving energy transformation occurs, a fraction of the available energy is dispersed in the form of heat and is unavailable for work. (See Chapter 4.)

**FIGURE 25–4 Two food chains. (a) Land-based.
(b) Water-based (aquatic).**

Sun

Plants

Herbivores

To air and soil reservoirs

Primary carnivores

Decay organisms

Secondary carnivores

(a)

Sun

Phytoplankton
(producers)

Seaweed
(producers)

Zooplankton
(herbivores
primary consumers)

Decay organisms

Browsers
(herbivores
primary consumers)

Plankton predators
(secondary consumers)

Predators
(secondary consumers)

Large carnivores
(tertiary consumers)

(b)

FIGURE 25-5 A highly simplified food web.

FIGURE 25-6 The pyramid of biomass in a land-based food chain.

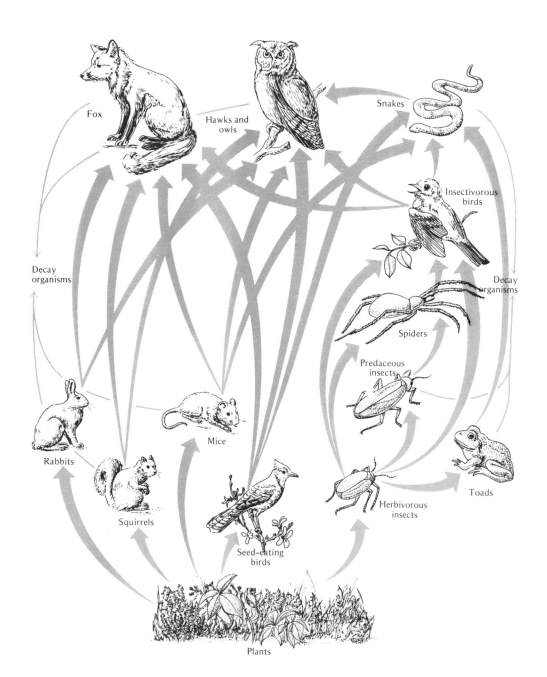

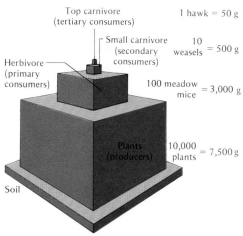

In an ecosystem such as a pond or a forest, the energy base is provided by the green plants (producers); the next level above this is occupied by primary consumers (herbivores); and succeeding levels above this are made up of secondary, tertiary, and quaternary consumers (carnivores). If we represent this situation graphically by taking the total amount of dry weight of the organisms in each of these levels, a pyramid of **biomass** is obtained (Figure 25–6). Another way of looking at the composition of an ecosystem is to represent the energy content at the various levels. The lower levels have a greater energy content than do the higher levels (Figure 25–7). Pyramids of biomass and energy for an ecosystem are based on the Second Law of Thermodynamics.

No one would dare construct a building without an adequate foundation, or base, upon which to build; the larger the building, the greater the base must be. Similarly, the biological upper limit of biomass is set by the primary

FIGURE 25-7 The energy pyramid in a land-based food chain.

FIGURE 25-8 The relationship between length of the food chain and energy efficiency in feeding. Eating further down on the food chain supplies a larger amount of energy.

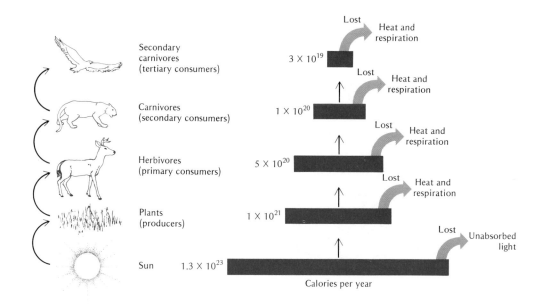

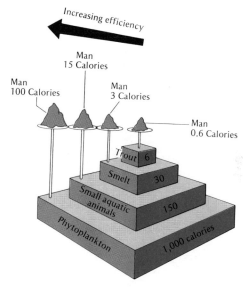

producers, the photosynthesizing organisms responsible for trapping the energy of sunlight. The biomass at each succeeding level is in part determined by the biomass (or energy) available at the level upon which it rests. This has great practical significance: the shorter the food chain, the greater will be the available food energy. The poorer underdeveloped nations of the world take full advantage of this in their utilization of vegetables and grain foods rather than meat, since energy losses are kept to a minimum when man acts as a herbivore. In countries where meat is a part of the diet and the economy is impoverished, animals used as the source of meat are scavengers, consuming plant wastes not utilizable by man. The pig is a prime example; it eats the garbage man does not or cannot eat. Only the affluent countries of the world can afford to use grain, itself useful as human food, for feeding cattle, since these are energetically expensive to produce. Of particular importance in supporting an ever-

increasing human population will be attempts to shorten the length of food chains. Thus, more people could be supported on a diet of soyburger than on hamburger or on trout (Figure 25-8). If the world population grows and outstrips the capacity of the earth for meat production, we may indeed be forced to a grazing diet.

25-4 THE DANGER OF ACCUMULATED CHEMICALS

We have considered the flow of energy in a food chain and seen that only about 10% of the energy obtained at one level is stored in the living tissues and available for transfer to the next level; typically 90% of the energy is lost as heat to the environment at each step and is, therefore, not usable by the organism. Another facet of food transfer is the use to which the food itself is put. Usually less than 20% of the food obtained by an animal is used

for the construction of the animal's tissues; the remainder is used for work or eliminated by excretion. Small though this amount is, it provides a mechanism for concentrating materials: a substance that is not eliminated is retained, and when passed from one feeding level to the next, it may be concentrated in the tissues two or more times. Some substances, such as mercury, lead, cadmium, and asbestos, are neither used nor eliminated but are stored in the body, where they may exert a cumulative and damaging effect on a variety of physiological processes. Organic pesticides of the chlorinated hydrocarbon variety (DDT, for example) are another class of long-lasting cumulative poisons stored in living tissues. Other substances may be used by the body even though they are damaging. For example, radioactive elements such as strontium 90, cesium 137, and iodine 131 may be concentrated by transfer from one feeding level to the next; cells receiving high doses of ionizing radiations from such elements

'Tomorrow, the world!'

MERCURY

PESTICIDES

MAULDIN

can be severely damaged and gross abnormalities induced. Persistent pollutants such as these pose a serious environmental hazard to the "top dogs" in an ecological pyramid, because by the time they reach the top level they have been concentrated many times over. We, as "top dogs" in our particular food chain, are very susceptible; furthermore, in the more affluent developed nations of the world the food chains are longer, and so the concentration of pollutants may be severe in the members of such societies. Let us look at some of these pollutants that endanger us with almost every meal.

Mercury

Madness or mental derangement afflicted the Hatter in Lewis Carroll's *Alice in Wonderland*. Minamata disease—characterized by numbed muscles in the arms and legs, impaired speech, and blurred vision—appeared in 1953 in Japanese fishermen who ate seafood from Minamata Bay; of the 134 victims, 47 died. The cause? Mercury poisoning.

The element mercury (quicksilver) is not a poison (except when inhaled), and a person could swallow up to a pound with no adverse effects; neither are inorganic mercury compounds such as calomel and mercuric oxide important environmental hazards. However, organic mercurials such as ethyl mercury and methyl mercury are deadly. In large amounts methyl mercury can cause brain damage, induce chromosome breaks, and intoxicate the unborn fetus so that it develops cerebral palsy. The chemical basis of such effects seems to be the strong affinity of the organic mercury compounds for the SH (sulfhydryl) groups in proteins, especially those in cellular membranes, causing changes in electrical potential, ion distribution, and movement of fluid. Methyl mercury and ethyl mercury are resistant to degradation and persist in the environment for weeks or months; thus their effects are most damaging and essentially irreversible.

Mercury poisoning can occur by occupational exposure or environmental accumulation. The Mad Hatter's mental derangement was probably caused by chronic exposure to mercury fumes from the mercuric chloride commonly used in processing beaver pelts into men's hats. The fishermen of Minamata Bay developed symptoms of mercury poisoning as a result of accidental ingestion of methyl mercury. The Chisso Chemical Corporation, a plastics manufacturer, used mercuric chloride in one of its processes, and for each gram of plastic produced, 60 gm of mercuric chloride were dumped into Minamata Bay. Once this had settled to the muddy bottom, it was converted by bacterial action into methyl mercury. The mud's methyl mercury content was concentrated to 50 parts per million (ppm) in the fish, and since the fishermen and their families lived almost exclusively on bay fish the concentration in the humans reached over 100 ppm.

Evidence of abnormal amounts of mercury in water, fish, and game birds has turned up in at least 33 states. The concentrations of mercury in unpolluted waters is about 1 part per billion (ppb), but in the waters around plants producing caustic soda the concentration was found to be 1,800 ppm. Tuna fish have been found to contain 0.13–0.25 ppm, and in some catches of swordfish the concentration is even higher. Symptoms of methyl mercury poisoning have been associated with blood concentrations of 0.2 ppm, but in order to maintain this level a 154–lb man would have to ingest 9 lb of fish containing 0.5 ppm each week. The U.S. Food and Drug Administration (FDA) has estabished a concentration of 0.5 ppm as the allowable limit in fishes used for human food.

The principal sources of mercury are fungicides; plants that manufacture chlorine and caustic soda, where mercury is used as a catalyst and is washed away in the residual brine; lumber and paper mills; and paint-manufacturing factories. The mercury deposited in fresh waters may remain for centuries after mercury dumping has been stopped. Although there is a natural background of mercury, this is no reason to discount the danger. Mercury is a cumulative poison, and the more mercury released into the environment the greater are the chances that the tolerable level will be exceeded. By restricting entry of mercury into the environment from industrial sources, by controlling its use as a fungicide, and by reducing the ingestion of heavily contaminated foods, we may avoid future tragedies from mercury poisoning.

DDT and other chlorinated hydrocarbons

About 100 years ago an obscure German chemist, Othmar Zeidler, synthesized di-chloro-diphenyl-trichloroethane, or DDT. It lay on the shelf until 1930, when a Swiss entomologist, Paul Muller, tested it for its killing properties on insects and found it most effective. In 1942 DDT came to the United States, where its usefulness as a potent insecticide was quickly recognized. In many countries around the world spraying DDT eliminated the diseases typhus, malaria, yellow fever, and plague—all of which are transmitted by insects. Since World War II the use of DDT has saved many millions of lives and prevented untold human suffering. So widespread was the use of DDT in controlling pest-borne diseases that now the compound is everywhere—the land, the sea, the fatty tissues of animals and humans, and even breast milk. Nobody, no thing is free of DDT. Why does DDT, one of the best and cheapest of the pesticides, now contaminate our environment,

not only killing pests but destroying helpful birds and insects and sickening farm workers?

DDT, a chlorinated hydrocarbon, is an exceedingly stable material; it breaks down in the environment very slowly (its half-life is 15 years).[3] When DDT is sprayed, it is carried into the air, and depending on the weather conditions, it is rapidly and widely spread from ecosystem to ecosystem. Since it is not very soluble in water (1 ppb), it tends to remain in the environment. In a Long Island salt marsh that had regularly been sprayed for mosquito control over a 20-year period, the mud in the marsh contained 32 lb of DDT per acre! Although the amount in the water may be infinitesimally small, aquatic organisms tend to accumulate DDT because it is fat-soluble. As a consequence, long after the spraying, organisms are continually exposed to DDT, which they absorb and retain in their fatty tissues. If the organism is a plant, little DDT is lost by respiration; when the plant is eaten, the DDT is transferred virtually intact and concentrated in the fatty tissues of the herbivore. At each feeding transfer, the concentration of DDT in the flesh of the predator is increased (Figure 25–9). Thus, in the Long Island marsh sprayed with DDT, the plankton (microscopic plants and animals) contained 0.04 ppm; the fish (minnows), 1 ppm; and the birds that fed on the minnows, 75 ppm. Similar high concentrations of DDT are found in carnivorous birds, mammals, and man. Human fat on the average contains 12 ppm, and in human breast milk the concentration may be as high as 5 ppm. According to the FDA, the permissible level of DDT in food is 0.05 ppm; thus, if human breast

---

[3] Half-life is the time required for half of a substance to be degraded. Thus, if the half-life of a substance is 100 years, only half of the starting material would present after that period; in 200 years, only 25%; in 300 years, only 12.5%; and so on.

milk were sold with the same restrictions as cow's milk, a nursing mother's milk would be declared unfit for human consumption.

DDT is a poison, and it kills not only insects but myriad other creatures. Unfortunately, we have no clear understanding of the exact mechanism whereby it exerts its killing effect, but it appears likely that its site of action is in the nervous system, since typical DDT poisoning involves tremors and convulsions. One neurophysiologist claims that DDT is an irreversible nerve poison (even in man) and is especially damaging to the central nervous system. DDT has been implicated in human disorders such as cerebral hemorrhage, hypertension, cirrhosis of the liver, and cancer. Dieldrin, a related chlorinated hydrocarbon with a shorter half-life (7 years), is four times as toxic as DDT; it also has been implicated as a causative agent in human disorders. Chlordane and lindane, related pesticides, fall into the same category.

While DDT toxicity has been difficult to demonstrate unequivocally for man, its role in the decline of several species of birds seems much clearer. Birds such as the falcon, pelican, and eagle are particularly vulnerable because they are "top dogs" in the food chain and may receive very high concentrations of DDT. In 1940 it was found that the shells of the eggs of these birds were reduced in thickness.[4] The weakened shells crack or break quite easily in the nest, and the embryos die. The accumulated pesticide also depresses the bird's estrogen levels, thus delaying the time of breeding and impairing reproduction. DDT is a problem for both birds and men.

The widespread use of DDT not only almost eliminated many insect-borne diseases, but by

---

[4] Probably by DDT inhibition of the enzyme carbonic anhydrase, which makes the calcium of the blood available to the oviduct for shell deposition.

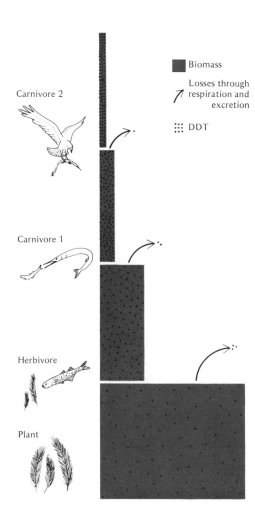

Biomass

Losses through respiration and excretion

DDT

Carnivore 2

Carnivore 1

Herbivore

Plant

protecting crops from insect damage and destruction, increased yields by 30%–40% in underdeveloped nations and by 10% in developed countries. DDT was soon hailed as a miracle insecticide, a cure-all for insect-borne plagues, and a powerful tool in the war against starvation. Then these hopes were dashed: by 1948, 12 insect species had developed resist-

ance to DDT; in 1954, 25 species were resistant; by 1960, the number had increased to 137 species; and by 1965, 165 species showed DDT resistance. The more we sprayed, the greater were the numbers of resistant species. This is because resistance is a genetic trait; some individuals in an insect population survived exposure to DDT, and their progeny were able to thrive, mate, and lay eggs with complete immunity to DDT. Consequently, the numbers of resistant individuals in that species increased dramatically in only a few generations.

Resistance to DDT was only a part of the problem. DDT and its chemical relatives are not specific. DDT and related compounds constitute an "elixir of death" for birds, fishes, mammals, insects, and perhaps even man. A single aerial spraying against mosquitoes killed 60,000 milkfish in a Philippine farm, and 700,000 hatching salmon were poisoned by the DDT in their own eggs!

Yet, not all the dangers of DDT stem from resistance, biological magnification, and persistence. A few parts per billion of DDT in the water depresses the photosynthetic activity of algae, and thus the primary productivity of a lake is diminished, and ecological disruption begins. Sometimes the spraying of an insecticide destroys the wrong organism at the wrong time and promotes ecological imbalance. For example, the red-spider mite is a destructive pest of spruce trees, but normally it is kept in check by a predatory mite. When spruce forests were sprayed to control the spruce beetle (another insect that damages the spruce needles), the predatory mites and the spruce beetles were both destroyed, but the red-spider mites living on the undersides of the needles survived. The year following the spraying saw a wave of red-spider-mite damage in excess of that ever done by the combined efforts of spruce beetles and red-spider mites.

On December 31, 1972, 10 years after Rachel Carson wrote *Silent Spring* and warned of the threat posed by indiscriminate and widespread use of pesticides, the Environmental Protection Agency banned almost all domestic uses of DDT. The growers of cotton, peanuts, and soybeans—three crops that account for almost all of the 14,000,000 lb of DDT used in 1970—must use methyl parathion, a toxic but highly degradable pesticide. In spite of the ban, 26,000,000 lb of DDT will still be exported for use in malaria control.

The limited ban on DDT in specific areas seems to be in the public interest, and it will force us to think of insect-control measures less hazardous than the indiscriminate aerial spraying of pesticides. Integrated control, that is, the rational use of ecological factors and the moderate use of chemicals, seems to provide a reasonable alternative to DDT and other toxic and persistent pesticides. The future of pest control appears to lie not in the development of more potent chemical insecticides but in the combined use of safe and specific predators and parasites for eliminating crop pests, the trapping of pests by sex attractants, the release of sterile males, and the avoidance of single-crop planting.

Atomic wastes

On the morning of July 16, 1945, at a desert site near Alamogordo, New Mexico, the first atomic bomb was exploded. Since that day we have lived with the atom bomb and its pollutants. Each atom bomb, by a chain reaction, converts a small amount of radioactive matter into blast (50%), heat (35%), and radioactivity (15%). The radioactive materials produced in an atomic blast and let loose into the atmosphere ultimately settle to the earth by gravity and are called **fallout.**

The most important fact about the radioactive materials **(isotopes)** in fallout is that for all practical purposes they have the same chemical properties as the nonradioactive forms. Thus, radioactive iodine(^{131}I) can be incorporated into thyroxine, the thyroid hormone, as easily as nonradioactive iodine (^{127}I). When radioactive isotopes decay, they emit ionizing radiations, and these can damage living tissue or induce mutations (Chapter 23). The exact effects of these radiations depend on the intensity of the radiations, the rate of decay of the isotope, and the place in the body where it is absorbed and stored.

Strontium 90 is a fission product released as a result of nuclear-bomb tests. It is easily spread in the air and can make a complete circuit of the earth in 16–25 days—the exact amount and distribution depending on climatic conditions. Strontium 90 is particularly damaging to the blood-forming centers in the bone marrow, since it is chemically very similar to calcium and thus tends to be accumulated in the bones and other tissues rich in calcium. Strontium 90 is relatively long-lived, with a half-life of 27 years, so that its ionizing radiations remain hazardous for a person's entire lifetime.

The route of entry of strontium 90 into human tissues is quite direct. The fallout containing the isotope enters plant tissues from the soil or the air; these plants are eaten by herbivores such as cows, and the isotope accumulates in their bones and milk. The biological concentration is quite efficient, and soon there is much more strontium 90 in a quart of cow's milk than in a pound of grass. Other contaminated foods serve as additional sources. Lactating mothers also concentrate strontium 90 in their breast milk and pass it on, further concentrated, when breast-feeding.

Environmental pollution with radioactive waste increases in direct proportion to the number of nuclear weapons tested above ground.[5] "Clean" bombs, if such exist, are preferable to "dirty" ones, but abolishing above ground testing would be the best course of action. As the demand for energy increases over the next few decades, we shall have to depend more and more on the use of nuclear reactors. The gravest disadvantage of these energy generators is that the residues of fission are radioactive and must be disposed of. Already hundreds of millions of gallons of radioactive wastes are in storage, and since they will remain radioactive for hundreds of years, temporary storage underground is only a stopgap. Furthermore, the radioactive atomic power plants themselves must be dismantled and buried after a short working life of only 20 to 30 years. If these wastes were ever released into the environment the consequences could be disastrous. Accidents in nuclear power plants or plutonium-processing plants could provide additional environmental contamination—a danger that increases with the number of facilities. Is there a fail-safe device that can prevent a radioactive leak from a nuclear power plant located in an earthquake-prone area such as coastal California?

25–5 RECYCLING MATTER

As long as the sun shines and organisms are capable of photosynthesis, there remains good reason to believe that, barring an all-out atomic holocaust or some other widespread disaster, life can continue on this planet. However, the biosphere and its contained ecosystems have no extraterrestrial sources of the materials of life—substances such as carbon, nitrogen, oxygen, potassium, and water. These essential materials must be continually recycled through the ecosystem if the communities that are a

---

[5] As of this writing France and China are the only nations indulging in this practice.

part of it are to persist. To make a cycle of matter operate, there must be more than producers and consumers: there must exist a group of organisms known as **decomposers** or **decay organisms.**

Decomposers are chiefly species of bacteria and fungi. They do not consume food in the same way that an herbivore or a carnivore does; instead they release enzymes into the body of a dead plant or an animal, degrading and digesting the organic constituents of these creatures, and then they absorb the soluble nutrients directly. Much of the organic matter degraded by the decomposers does not go into their own bodies, but is released into the environment as minerals, and these materials are made available to the producers. (Recall that autotrophs require simple inorganic substances for the synthesis of their cellular constituents.) Thus, producer, consumer, and decomposer form a linked chain for the recycling of matter in the biosphere (Figure 25–10). Now let us consider the recycling of four essential materials: carbon, nitrogen, phosphorus, and water.

The carbon cycle

Life as we know it is based on the chemical element carbon. In order for a green plant to manufacture organic materials, carbon must be "fixed"; that is:

$$CO_2 + H_2O \xrightarrow[\text{from sun}]{\text{radiant energy}} \text{"CH}_2\text{O"} + O_2$$

carbon   water   carbohydrate   oxygen
dioxide

When a plant is eaten by a primary consumer, the consumer obtains carbon atoms in its food; when the primary consumer is fed upon by a secondary consumer, there is another transfer

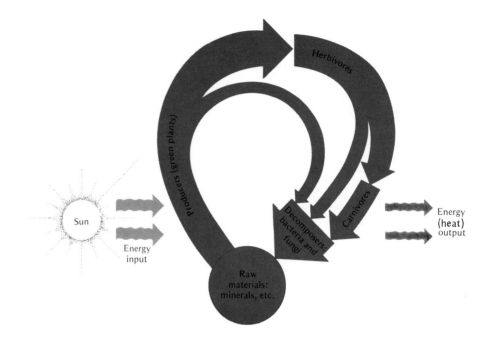

of carbon atoms. By the process of respiration, organic (carbon-containing) materials provide the living organism with energy, and one of the waste products of such cellular metabolism is carbon dioxide. The decomposers also enter into the cycle of carbon, for they degrade the organic matter of dead plants and animals into carbon dioxide.

Although some carbon is tied up in the organic molecules of living organisms, most carbon is quickly returned to the atmospheric pool in the form of carbon dioxide[6] by the processes of respiration and decomposition. Some carbon-containing materials do not enter

[6] It is estimated that the oceans contain 50 times more carbon dioxide than the amosphere: this oceanic reservoir regulates the atmospheric content, which is about 0.03%.

this rapid recycling route, and incompletely decomposed organic matter may be transformed and stored as fossil fuels such as coal, oil, peat, and natural gas. Carbon may also be removed from the cycle for longer or shorter periods of time by the formation of limestone $(Ca + CO_2 \rightarrow CaCO_3)$, produced by the activities of organisms such as foraminifera and corals or by geological forces. These deposits of carbon are returned to the cycle when fossil fuels are burned and limestone weathers. The carbon cycle is summarized in Figure 25–11.

The nitrogen cycle

Nitrogen gas makes up 78% of the atmosphere, but very few organisms can utilize nitrogen gas directly. Nitrogen is a chemical element required by all organisms, since it is an essential

FIGURE 25-11 The carbon cycle.

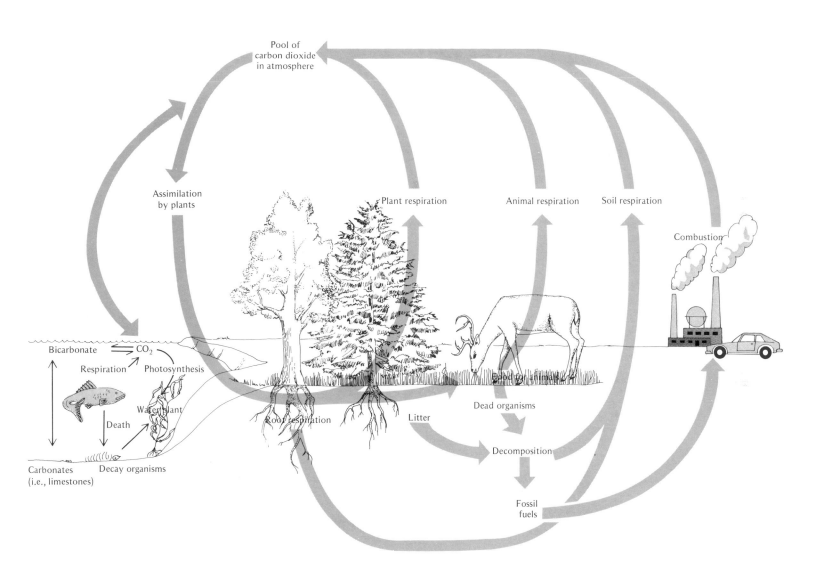

Pool of
carbon dioxide
in atmosphere

Assimilation
by plants

Plant respiration

Animal respiration

Soil respiration

Combustion

Bicarbonate ⇌ CO_2

Respiration   Photosynthesis

Water plant

Death

Root respiration

Litter

Dead organisms

Carbonates
(i.e., limestones)

Decay organisms

Decomposition

Fossil
fuels

FIGURE 25-12 The nitrogen cycle.

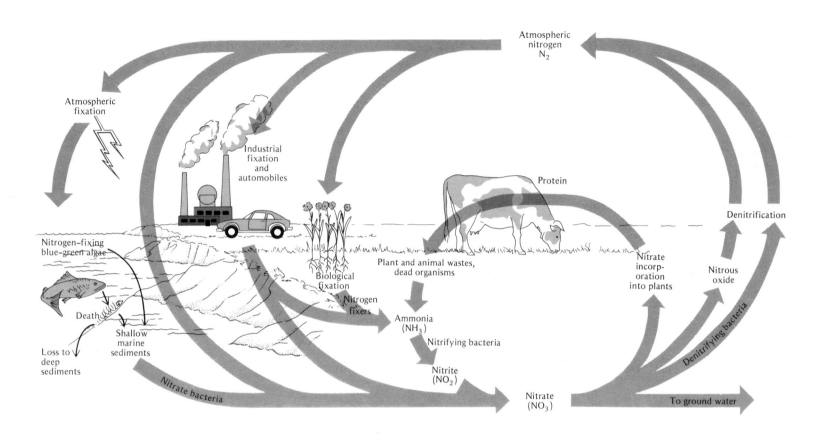

ingredient of both amino and nucleic acids. How does nitrogen get from the atmosphere into the biosphere?

Certain bacteria and blue-green algae can convert gaseous nitrogen into nitrates: $N_2 \rightarrow NO_3$. These nitrates can be directly utilized by plants when the bacteria and algae die, and animals eventually obtain their nitrogen from plants. The conversion of nitrogen into nitrate is called **nitrogen fixation,** and organisms that can do this are called **nitrogen fixers.** One of the most important of these nitrogen fixers is a group of bacteria (*Rhizobium*) that lives in the root nodules of legum-

inous plants—beans, clover, peas, and so on. Other soil bacteria (notably *Azotobacter* and *Clostridium*) also fix nitrogen. In addition, nitrogen is fixed by lightning and the internal-combustion engine, and man supplements the available soil nitrogen for crops by adding nitrate fertilizers. The decomposition of animals and plants by the action of bacteria and fungi releases ammonia gas (NH_3). Ammonia is highly toxic to most organisms, but a special group of bacteria, the nitrifying bacteria, is not susceptible to its lethal action and instead degrades the ammonia to nitrite ($NH_3 \rightarrow NO_2$); subsequently the nitrite is acted upon by nitrate

bacteria to produce nitrate ($NO_2 \rightarrow NO_3$). In this manner, by the action of nitrogen fixers, nitrifiers, and nitrate bacteria, nitrate—the essential nutrient for plant growth—is made available in the soil, and direct fixation of atmospheric nitrogen by green plants is not necessary.

The completion of the nitrogen cycle, that is a return of nitrogen to the atmosphere from the soil and organisms, involves another group of bacteria, the **denitrifying bacteria.** These bacteria degrade nitrates, nitrites, and ammonia to liberate gaseous nitrogen into the atmosphere. The nitrogen cycle is shown in Figure 25-12.

FIGURE 25-13 The phosphorus cycle.

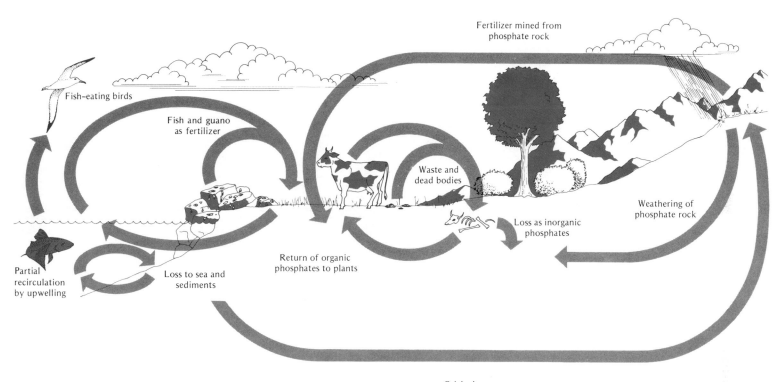

Fertilizer mined from
phosphate rock

Fish-eating birds

Fish and guano
as fertilizer

Waste and
dead bodies

Weathering of
phosphate rock

Partial
recirculation
by upwelling

Loss to sea and
sediments

Return of organic
phosphates to plants

Loss as inorganic
phosphates

Original source:
uplift by geologic processes

The phosphorus cycle

Phosphorylated sugars, phospholipids, ATP, DNA, and RNA are all key materials in the living cell; all contain phosphorus. Life as we know it is based on the element phosphorus as well as on carbon, and we know of no organism that can exist without a supply of phosphates. Phosphorus tends to be accumulated in the bodies of organisms, it cycles through the biosphere slowly, and it is ecologically significant because of its limited availability. In some instances, it is the amount of phosphorus alone that determines ecosystem productivity.

The principal reservoirs of phosphorus are phosphate rocks, guano (seabird excreta), and fossilized animals. Through leaching and erosion, mining, and application of fertilizers, phosphorus is released to become available as phosphate to plants. In both plants and animals the phosphate becomes dispersed into a variety of organic constituents; these ultimately return to the soil by decomposition. Soil continually loses phosphate by leaching, and this ends up in the ocean. There it may be used in the marine ecosystem or it may be lost for long periods of time in deep ocean deposits. The phosphorus cycle is shown in Figure 25–13.

The water cycle

Most of the stuff of life is water. Without water life cannot exist. Unmanned spaceflights to Mars and Venus have shown that on these planets there is virtually no water, and manned flights to the moon indicate that it is devoid of water. Mars, Venus, and the moon are all lifeless celestial bodies. By contrast, about 70% of the surface of the planet earth is covered by water. Water is a renewable resource that is recycled again and again in a complex fashion.

The principal water reservoirs are the oceans. Through evaporation, precipitation, runoff, and

FIGURE 25–14 The water cycle.

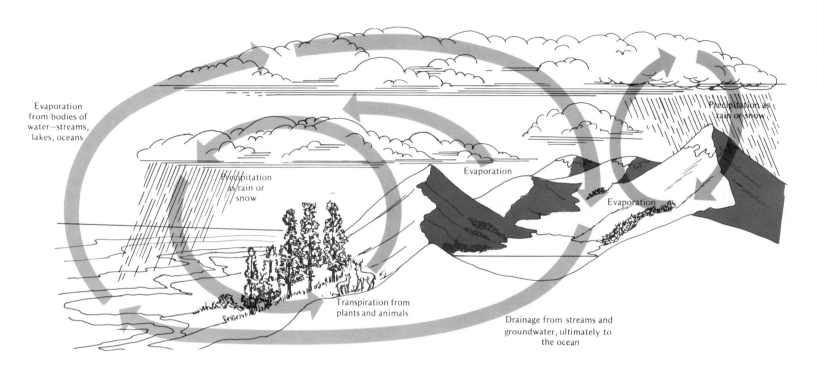

Evaporation
from bodies of
water—streams,
lakes, oceans

Precipitation
as rain or
snow

Evaporation

Precipitation as
rain or snow

Evaporation

Transpiration from
plants and animals

Drainage from streams and
groundwater, ultimately to
the ocean

groundwater seepage, water is constantly circulating, and generally the total water cycle remains in balance (Figure 25–14). However, it is important to recognize that the relative as well as the absolute amounts of water in an area determine the nature of the ecosystem. The pattern of water distribution on the land via precipitation varies greatly and depends on atmospheric circulation as well as on the topography of the land masses (Figure 25–15). Consequently, some regions of the world are dense forests, while other areas are barren deserts.

It takes vast quantities of water to produce food (recall that in photosynthesis six molecules of water are used in the formation of one molecule of sugar). In more concrete terms, it takes 60 gallons of water to produce 1 lb of dry wheat, 200–250 gallons of water to produce

1 lb of rice, 2,500–6,000 gallons of water to produce 1 lb of meat, and 1,000 gallons of water to produce 1 qt of milk. Industrial processes utilize staggering amounts of water: it is estimated that 100,000 gallons of water are used in the production of a single automobile, and the manufacture of a single copy of a Sunday newspaper uses 150 gallons! The amounts of water we waste are phenomenal: shaving, washing, and brushing teeth use 5 gallons; a quick shower or a load of wash takes 30 gallons; and flushing the toilet may use as much as 7 gallons. In 1900 each American disposed of 525 gallons of water daily, but by 1960 the per capita use had risen to 1,500 gallons. It is estimated that by 1980 the daily consumption will be 2,500 gallons.

About 97% of the earth's water is salty, and the remainder is mostly in the form of ice in

Antartica and Greenland. Only a minute fraction (0.06%) of the total fresh water is available for use by living organisms. If man removes more fresh water from the land masses than is replaced via the water cycle, he puts himself and all other living things in jeopardy. Witness the water shortages that frequently occur during the summer months in areas of the United States such as New York, Massachusetts, and California. Groundwater reserves act as "water capital," but these are being reduced by withdrawals at an alarming rate; once withdrawals exceed capital reserve, the water bank will fail. One authority on the subject contends that Europeans extract three times more water than the cycle returns for human use, and Americans take out twice the amount returned. Clearly the trend cannot continue unabated. If we are to avoid running out of water, we shall have to

FIGURE 25-15 The pattern of annual precipitation in the world.

Over 80 inches annual mean rainfall

From 40 to 80 inches

From 20 to 40 inches

From 10 to 20 inches

Under 10 inches

be more efficient in our use of it. Once water becomes very short in supply,[7] agriculture will fail, food supplies will diminish, and mass starvation will result. Then it will truly be as Coleridge's Ancient Mariner complained: "Water, water everywhere, nor any drop to drink."

25–6 HOMEOSTASIS AND SUCCESSION IN THE ECOSYSTEM

Homeostasis

An organism maintains its internal environment in a rather constant state by a series of negative feedback loops, a condition known as **homeostasis.** When an organism is out of balance, function becomes abnormal, and disease or death may be the consequence of such an unhealthy situation. The same principle applies to the natural ecosystem. An unbalanced ecosystem leads to instability; survival of the occupants is jeopardized; and if balance is not restored, the system may undergo irreversible damage.

What is a balanced ecosystem, and how does it come about? Balance in any system implies an equality of opposing forces: a seesaw is balanced when the weights of the persons on each end are equal. In an ecosystem such as a grassland, a forest, or a desert, there are many opposing forces: organisms are born while others die, water is lost from the soil by evaporation and returned via precipitation, nutrients are leached out of the soil and utilized by producers in the ocean. All these events tend to balance each other out.

Ecosystem homeostasis can also be seen in the relationship between a predator and its

---

[7] It is estimated that by 1980 the American requirement will be 700 billion gallons, but according to technological estimates only 650 billion gallons will be available.

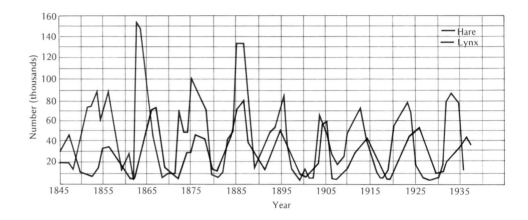

FIGURE 25–16 Fluctuations in populations of the Canada lynx (predator) and the snowshoe hare (prey). (After MacLulich, University of Toronto Studies, 1937, Biological Series, 43.)

prey. When the prey is abundant, predation is heavy because the predators can find and capture the prey quite easily, but as the number of prey organisms diminishes (from predation as well as other factors), the food available to the predators declines and their number drops. Such an oscillating cycle of population density for the Canadian lynx, a predator, and the snowshoe hare, its normal prey, is shown in Figure 25–16.

The balance in nature can also be seen in the cycling of matter. The regulation of carbon dioxide levels in the atmosphere provides an example of this. Carbon dioxide is quite soluble in water, and the oceans constitute a vast reservoir of this material. The greater the amount of carbon dioxide in the atmosphere, the warmer the earth becomes because of the **greenhouse effect.** The greenhouse effect occurs because carbon dioxide in the atmosphere acts as do the glass walls of a greenhouse. The glass walls transmit sunlight into the greenhouse, where it is absorbed and reradiated as heat that does not readily exit through the glass; as a consequence the temperature inside the greenhouse becomes higher than that outside. (A similar situation occurs with an auto-

mobile that is closed and sitting out in the sun.) If the carbon dioxide levels in the atmosphere were to increase (owing, for example, to excessive burning of fossil fuels), the atmospheric temperature would increase; this would tend to melt the polar ice caps, and the release of water would cause the ocean levels to rise. The increase in ocean levels would absorb the excess carbon dioxide, the carbon dioxide in the atmosphere would be diminished, the temperature would become colder, the ice caps would once again be formed, and the ocean level would decline. Thus, the system is buffered against severe fluctuations by a homeostatic control system.

Homeostatic controls help to maintain balance in an ecosystem; however, such control measures are not always able to cope with imbalance. If the limits of the control system are exceeded, the survival of the ecosystem may be jeopardized. Let us consider some instances where the homeostatic control system becomes overloaded.

Prior to 1907 the Kaibab plateau in Arizona supported a deer herd of about 4,000, and this level was maintained by a variety of predators such as coyotes, mountain lions, and wolves

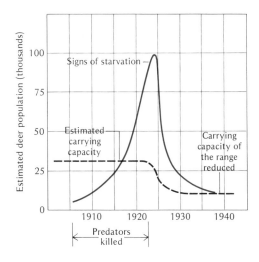

FIGURE 25-17 Population explosion and subsequent crash in the deer herd on the Kaibab plateau of the Grand Canyon. (After Leopold, 1943. Wisconsin Conservation Dept. Pub. 321.)

(Figure 25–17). To "help" the deer out, man decided to eliminate the predators. Between 1907 and 1923, 3000 coyotes, 670 mountain lions, and 11 wolves were killed by hunters, and the deer population "exploded" to 100,000. In the winter of 1924–1925 more than 60,000 deer starved to death, thus reducing the population. The wholesale winter starvation and damage to the habitat by overgrazing continued to reduce the size of the deer herds below the original number (Figure 25–17).[8] Starvation due to man's "help" killed more deer than predators would have taken. This illustrates the tragedy that follows destruction of balance in population size by a simple short-term alteration in species competition.

As the world population continues to rise, it will be necessary to provide technological benefits to many underdeveloped nations. It has been suggested that damming the Amazon River would provide an economic resource that could be of considerable value to the entire continent of South America. However, the destruction of the Amazonian forests by flooding behind the dam could have disastrous consequences. Flooding the Amazonian jungle would reduce the amount of carbon dioxide trapped by plants in photosynthesis, and this would cause a rise in the carbon dioxide level in the atmosphere. As South America became industrialized, burning of fossil fuel would increase, and excessive amounts of carbon dioxide would be liberated into the air, perhaps overtaxing the buffering capacity of the oceans. As the carbon dioxide level continued to rise, the greenhouse effect would result in elevated temperatures, and the polar ice caps would

---

[8] A recent study (G. Caughley, *Ecology* 51:53, 1970) suggests that the decrease in the deer population was primarily a result of habitat alteration by fire and overgrazing by sheep and cattle, and not of predator reduction per se.

melt. A temperature rise of only a few degrees would raise the sea level 30 ft. and most of the coastal land areas (where most of the human population lives) would be flooded. A simple Amazon River dam could drown the world's continents!

The complexity of the homeostatic devices in the ecosystem, about which we have incomplete knowledge, leads us to conclude that man should be cautious in his attempts to modify the balance of nature.

Succession

If you have ever started a small vegetable garden and then left it unattended for weeks or months, you have been witness to the changes in a mini-ecosystem that occur with the passage of time: crops of vegetables are succeeded by crop after crop of weeds. In the same way, plant and animal populations undergo a series of progressive changes. The sequence of alterations through which an ecosystem passes as time progresses is called **ecological succession.** For any particular en-

vironment these changes follow a rather predictable pattern, with particular communities of organisms paving the way for others that will come, until a final stage, or **climax** is reached. A climax is final or unchanging only in a relative sense, because through the passage of geological time even a climax ecosystem undergoes changes with alterations in temperature, altitude, precipitation, sunlight, and so on.

If one follows ecological succession in an abandoned farmland that changes into a forest, or a shallow lake that becomes a marsh and then a forest (Figures 25–18 and 25–19), a number of characteristics become clear: (1) the change in species composition from the early, or pioneering, community to the climax community involves increasing plant and animal diversity; (2) biomass, the total amount of organic materials present, increases at every stage until it reaches a maximum at the climax; (3) food chains become more elaborate at every stage until they reach maximum complexity at the climax; and (4) the climax community is relatively stable and dominated by longer-lived plants.

A climax community is called a **biome.** The nature of the climax community in a given area is determined largely by its climate; the plants and animals that make up a biome share common environmental problems. North American biomes include tundra, forest, grassland (prairie), and desert (Figure 25–20).

Why does succession occur? Could a climax community not be established immediately if soil, climate, rainfall, and seeds of the right plants were present? Probably not. Let us see why. As we follow succession in communities in a given location, we observe that it takes place in a regular and predictable fashion. The pioneering invaders are the grasses that exploit habitats newly opened up by catastrophe—fire, flood, and manmade clearings. Grasses grow fast on bare ground when they receive adequate sunlight and rainfall and relatively small amounts of nutrients. Shrubs and tree seedlings, the next group of immigrants into the area, develop more slowly and cannot at first compete with grasses. However, most grasses are annuals and must reseed themselves and start all over again each year; thus, they come to be at a competitive disadvantage with the perennials—shrubs and young trees—and in time the pioneering grasses give way to these intermediate species. Shrubs build sizable aboveground (stems and branches) and underground (roots and stems) storage structures; their roots use soil nutrients and water more efficiently than grasses, and their branches begin to exclude sunlight from the grasses beneath them. Perennial shrubs do not die back each year, and the shrub-held habitat slowly becomes less suitable for the yearly invasion of grass seeds. Thus the shrubs begin to take over. Shrubs provide a conducive environment both for themselves and for young tree seedlings— shade and a favorably altered soil. The final invaders are equilibrium species (usually, though not exclusively, trees)[9] that produce massive structures and storage organs; because such organisms have adequate reserves of energy, they can often withstand long stretches of environmental bad times. Trees provide leaf litter, and this rich source of organic material, when acted upon by decomposers, yields humus. Humus alters the soil itself, promoting water retention and aeration; consequently nutrient cycling becomes more efficient. The increasing shade provided by the canopy of trees tends to exclude the grasses, and these

[9] Equilibrium species have relatively stable population sizes and are adapted to living in more or less predictable environments.

Time (years): Community type:	1–10 Grassland	10–25 Shrubs	25–100 Pine forest	100 + Hardwood forest

Birds
- Grasshopper sparrow
- Meadowlark
- Field sparrow
- Yellowthroat
- Yellow-breasted chat
- Cardinal
- Towhee
- Bachman's sparrow
- Prairie warbler
- White-eyed vireo
- Pine warbler
- Summer tanager
- Carolina wren
- Carolina chickadee
- Blue-gray gnatcatcher
- Brown-headed nuthatch
- Wood pewee
- Hummingbird
- Tufted titmouse
- Yellow-throated vireo
- Hooded warbler
- Red-eyed vireo
- Hairy woodpecker
- Downy woodpecker
- Crested flycatcher
- Wood thrush
- Yellow-billed cuckoo
- Black and white warbler
- Kentucky warbler
- Acadian flycatcher

	1–10	10–25	25–100	100 +
Number of common species[a]	2	8	15	19
Density (pairs per 100 acres)	27	123	113	233

[a] A common species is arbitrarily designated as one with a density of 5 pairs per 100 acres or greater in one or more of the four community types.

FIGURE 25-19 Succession in a shallow lake with conversion into a dry marsh.

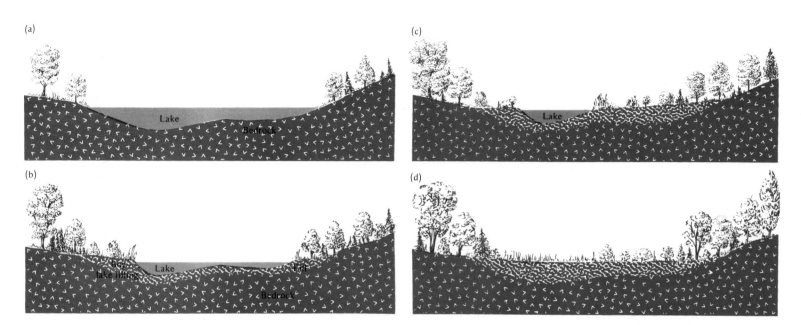

(a)

(b)

(c)

(d)

disappear. Shade-loving plants (ground cover or underbrush) move in. The flowers and fruits of these and of the trees act as sources of nourishment for a variety of animals—species diversity increases.

The significant features of succession now become clearer. Pioneering species put most of their energy budget into seed production—indeed, that is how they are dispersed from one habitat to another. Intermediate species have restricted capacities for dispersal, but they tend to occupy a habitat longer and more tenaciously by building woody structures that can resist environmental changes and by storing nutrients underground. Equilibrium species build woody roots, trunks, and branches, and are adapted to living in a predictable environment; in addition they themselves modify environmental fluctuations by acting as windbreaks, resisting flooding, and retarding soil

erosion. Because the climax community is dominated by equilibrium species that have massive woody structures (think of a redwood), it has increased biomass. In time, the opportunistic pioneers must be displaced by the more stable equilibrium species; in effect each community in the successional continuum sows the seeds of its own demise.

Is there a grand design to succession so that communities must inexorably proceed to the climax—that "single perfect community" for a given geographical area? No. Successions occur because of the forces of natural selection:

In the . . . Darwinian view of ecology there is no organizing deterministic principle behind succession. . . . Complex ecosystems are the product of crowding many species into restricted spaces and forcing them to live together and to adapt to each other's presence.
P. COLINVAUX, *Introduction to Ecology*, New York, Wiley, 1973.

25-7 AGRICULTURE AND UNBALANCED ECOSYSTEMS

The most fundamental principle of ecology is that everything is connected to everything else. In the words of the English poet Francis Thompson: ". . . thou canst not stir a flower, without troubling of a star." All of us resident on the spaceship earth depend on chemical recycling, and any break in the cycle or alteration in the rate of cycling by removing too much or too little chemical matter may spell ecological disaster. Agriculture may be the agent to disturb the natural cycling processes.

Agriculture is the source of unprecedented supplies of human food and has contributed significantly to our population growth, urbanization, and industrialization. That's the good news; now for the bad. Agriculture by its very nature is an artificial and unbalanced ecosys-

FIGURE 25-20 The biomes of North America.

tem. Farmers plant opportunistic pioneering plants because these grow quickly on bare soil and yield edible reproductive structures such as cereal grains and storage stems (sugar beets and potatoes) instead of inedible woody branches and roots. By agricultural practices the trend of succession is reversed, and ecological changes are artificially kept in check. For example, when forests become crop fields, equilibrium species are displaced and often become extinct; massive inputs of nutrients via fertilizers are required because mineral cycling is low. Crop fields are maintained perenially "young" by the constant attention of human labor. It is necessary to prevent insect pests, fungal parasites, and drought from destroying the precious harvest; to buffer against weather extremes; and to restrict the spread of disease agents or predators. Ecological simplicity and uniformity replace complexity and diversity. Consequently, any society dependent on a single crop is threatened more by drought or plague than another society working on a diversified ecosystem.

As the agricultural revolution increased food production, the population grew and people moved from the farms to the cities. The growing demand for food by the ever-increasing population created a drain on the ecosystem; with a greater drain and a slower return, the ecosystem came to be disrupted. Once-fertile lands became barren. The farmer's choice was clear: refertilize or move on to another region where the soil fertility is high. In some cases, by proper management, the soil has been enriched; but in other cases it has been destroyed.

In 1924 a cloud of dust appeared over New York City and blew into the Atlantic Ocean. It was an ecologically important dust cloud, since it was a significant part of the once-rich Oklahoma topsoil. What caused this? In 1889 Oklahoma was opened to homesteaders. The population rose from 60,000 to 390,000 in a single

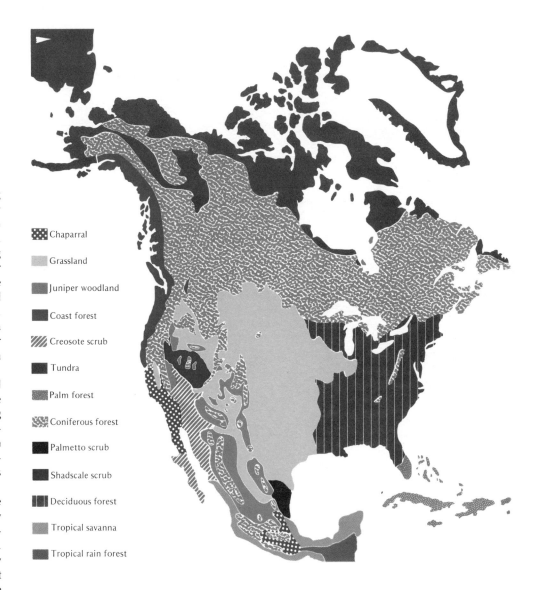

Chaparral

Grassland

Juniper woodland

Coast forest

Creosote scrub

Tundra

Palm forest

Coniferous forest

Palmetto scrub

Shadscale scrub

Deciduous forest

Tropical savanna

Tropical rain forest

BOX 25A

Nuclear Waste Object of Study

The federal government has embarked on a costly program that may show the American public whether the nuclear industry can clean up after itself.

The Department of Energy has contracted with three western firms to prepare a master plan for decontaminating and dismantling unneeded government-owned nuclear reactors. Estimates of overall cost range as high as $4 billion.

Atomics International of Canoga Park, along with two firms located at the federal Hanford Atomic Reservation near Richland—United Nuclear and Battelle Northwest—have been given the task of gathering and evaluating data on about 350 outdated reactors, laboratories and related facilities throughout the nation.

Joe Nemec's role in the program involves developing an overall plan for dismantling four Hanford reactors once used to produce plutonium for the government's nuclear weapons program. The four reactors, none of which has been in operation since 1967, are all included in a preliminary "top 10" list for the entire nation.

The reactors at the Hanford complex have been responsible for tens of millions of gallons of highly radioactive liquid waste, which must be isolated from man for several thousand years. The cores of dormant nuclear reactors contain equally radioactive substances.

More than a half-million gallons of liquid waste has leaked from Hanford's underground storage tanks since 1958, and while such leaks dramatize the problems involved in safekeeping such dangerous substances, government officials maintain that no health hazard has yet resulted.

In January, Nemec supervised dismantling of a small test reactor at Hanford. The project took eight months and cost $230,000.

By comparison, Nemec estimates it will cost between $15 million and $25 million to dismantle each of the four production reactors, taking at least two years, depending on the availability of federal funds, to complete each job.

But Nemec thinks it's worth the cost to prove to the industry's critics that "we can clean up our own mess," not to mention the cost in maintaining these facilities in their present state.

Reprinted by permission of United Press International.

"Maybe we could package it as a soil conditioner."

year, and the settlers lived by working the land. The flat, rockless expanses of rich topsoil easily yielded to the plow, and at first the harvests were good. To make way for crops and cattle, the settlers killed off the natural herbivore, the bison, and permitted domestic cattle to overgraze, thus preventing the grasslands from reseeding themselves. The land soon became susceptible to erosion by heavy rains, and as the rainwater ran off instead of percolating downward, the water table came to be lower and lower. Depletion of water meant death for perennial plants and the prairie grasses. Soil destruction was further accelerated as the hooves of the vast numbers of cattle compacted the soil and prevented the natural seepage of air and water through the soil. The digging and turning over of the Oklahoma prairie soil rendered it vulnerable to the weather, since there were no longer prairie grasses capable of holding the soil in place during periods of drought. The common agricultural practice of pulling weeds between the rows of crops left the soil susceptible to erosion when heavy rains came. The practice of dust mulching—chopping the top few inches of soil into a fine dust—to bring water to the surface by capillary action during short droughts tended to decrease the concentration of organic matter in the soil (it is brought to the surface and quickly oxidized). Gradually the soil lost its fertility. In less than 35 years the soil was severely depleted of its nutrients, refertilization was incomplete, and the winds and water eroded away the once-rich topsoil. Finally, in 1924 a severe drought struck, the deficient and unprotected soil simply blew away in the winds that swept across the landscape. Rich prairie became a dust-bowl because of man's agricultural mismanagement. Ironically, the prairies that were not managed by man easily survived the prolonged drought and remain fertile to this day. Some say it will take several thousand years to restore the dust-

bowl to its full fertility—fertility lost in a single generation by poor agricultural planning.

Man's agricultural practices and the consequent disruptions in the ecosystem continue nonetheless. In 1971 the Aswan High Dam in Egypt was completed. It was designed to provide hydroelectric power and to increase crop yields. By damming the Nile, it was argued, excess water and fresh soil from the Abyssinian mountains could be stored, and controlled flooding would increase soil fertility. Indeed, higher yields of cotton, grain, fruits, and vegetables have occurred during the dry summer months, and regions of land once barren have been reclaimed. Yet the situation is not without its disadvantages. Natural flooding ordinarily washed away the salts in the soil, but artificial flooding tends to leave the salts. Thus, the salt content is increasing in the soil year by year, and ultimately crop yields will decline unless salt accumulation is checked. The sediment carried by the natural flooding of the Nile prevented seepage, strengthened the dikes, and minimized coastal erosion by the Mediterranean currents. Since the advent of the dam, these protective features of the Nile delta's sediment have been lost. Now the sediment settles out behind the dam, and silt-free waters race toward the sea, eroding river banks and damaging bridges. In addition, since the water flowing out of the dam contains less silt, its nutrient content is lowered so that the agricultural land must be fertilized by chemical additions. Paradoxically, the dam has created a water shortage in Egypt. At present, the reservoir is not even half filled because water is continually lost by seepage through porous rock and rapid evaporation produced by the hot winds in Upper Egypt. The nutrients that once flowed into the Mediterranean nourished a food chain leading to the sardine, and since these nutrients are now blocked by the dam, the yield of sardines has severely declined (annual loss of

18,000 tons). Fortunately, the fish catch in Lake Nasser (12,000 tons in 1970), behind the dam, has almost balanced these losses. However, what will the fishing prospects be when the lake becomes clogged with silt? Most seriously, the Aswan Dam has created a massive health problem. The year-round reservoir of standing water provides a suitable breeding ground for snails that tranmsit the human blood fluke. About 80% of the population living along the irrigation canals served by the dam are infested and seriously damaged by this parasite. Higher crop yields are hardly balanced out by the increase in human misery from the abundance of disease from blood flukes.

The lessons of the past—Oklahoma's dust bowl and the destruction of human resources by the Aswan Dam—are clear: if technology is to save us from agricultural disruptions, it must recognize that in ecology there is no such thing as independence.

SUMMARY

1. Ecology is the field of biology that concerns itself with studies of populations, communities, and ecosystems as they interact in the biosphere.

2. Energy is not recycled through the ecosystem; its flow is unidirectional, from producers (autotrophs) to herbivores, carnivores, and omnivores (heterotrophs).

3. According to the Second Law of Thermodynamics, energy conversion is always less than 100% efficient; therefore a continual supply of energy is needed to maintain an ecosystem.

4. The transfer of food energy between the plants and animals of successive feeding stages is called a food chain or food web. Food webs usually involved plants, herbivores, carnivores, and decay organisms that recycle the matter in a food chain or web.

5. Because of the Second Law of Thermodynamics, ecosystems can be arranged in pyramids representing biomass and energy. At the base are the producers (plants), followed in ascending levels by herbivores (primary consumers), carnivores (secondary consumers), and secondary carnivores (tertiary consumers). Omnivores may enter the pyramid at a number of levels.

6. The biomass (or energy) at each successive feeding level of an energy pyramid is determined by the biomass (or energy) available at the level on which it rests. Approximately 10%–20% of the energy at one level is available to the next level.

7. Plants are at the base of all food chains. By shortening the food chain, the number of energy conversions is reduced and available food energy is increased.

8. When materials that are not used or eliminated are passed from one food level to the next, they may be concentrated two or more times over in the tissues of the consumer. The "top feeders" of a food chain therefore can be the most endangered by environmental pollutants. Examples of such cumulative and damaging substances are: mercury, DDT (and other chlorinated hydrocarbons), and radioactive elements.

9. Matter, including substances that contain carbon, nitrogen, phosphorus, is recycled through ecosystems by decay organisms, chiefly bacteria and fungi. Thus, producers, consumers and decay organisms continually recycle the matter of the biosphere.

10. Ecosystem homeostasis depends on factors such as the recycling of matter and predator-prey relations. Upsetting the homeostatic balance can destroy an ecosystem.

11. Ecological succession is the sequence of alterations through which an ecosystem passes with time, from an early pioneering community until a relatively stable climax community or biome is reached. During succession, the forces of natural selection lead to increasing organism diversity, biomass and food-chain complexity.

12. Biomes are climax communities whose nature is largely determined by climate.

13. Agricultural practices produce unnatural and unbalanced ecosystems. Although it is the source of unprecedented food supplies, agriculture reverses normal succession: soil must be artificially enriched to maintain a perpetually "young" community. Examples of ecological disasters produced by poor agricultural practices include the Oklahoma dustbowl of the 1920s and the Aswan Dam of the 1970s.

KEY WORDS

populations
community
ecosystem
biosphere
ecology
producers
autotrophs
herbivores or primary consumers
carnivores or secondary consumers
food chain
food web
omnivores
biomass
fallout
isotopes
decomposers or decay organisms
carbon cycle
nitrogen cycle
nitrogen fixation
nitrogen fixers
denitrifying bacteria
phosphorus cycle
water cycle
homeostasis
greenhouse effect
ecological succession
climax
biome

TOPICS FOR REVIEW AND DISCUSSION

1. **Discuss the meaning of the term ecology.**
2. **Why does succession take place in an ecosystem?**
3. **Discuss the effects of DDT on an ecosystem.**
4. **How do agricultural practices serve to disrupt ecosystems?**
5. **Discuss the statement: Energy does not cycle in the biosphere.**
6. **What are food chains and food webs?**
7. **How is matter recycled in the biosphere?**
8. **What are some of the dangers of atomic wastes?**
9. **What role do decomposers play in the recycling of matter in the biosphere?**
10. **What is the relationship of the Laws of Thermodynamics to ecological pyramids?**
11. **Why is a pound of meat more expensive than an equivalent weight of carbohydrates?**
12. **Why is it essential to direct urban growth away from prime land?**
13. **Discuss the mechanisms by which chemicals accumulate in the environment.**
14. **Discuss homeostasis in the ecosystem.**

"Are you kidding?"

ABOARD THE SPACESHIP EARTH

26–1 THE INHABITANTS OF THE SPACESHIP EARTH AND HOW THEY GREW

In April 1970, the world watched in helpless awe as the three astronauts aboard the Apollo 13 spacecraft radioed back to the earth that several systems were malfunctioning and there was little chance they could return to earth. As we followed the perilous journey of Apollo 13 and its occupants on radio, on TV, and in the press, another dangerous situation received little or no publicity. During the same 4 days that Apollo 13 was in space, aboard the earth 116,000 human beings died of starvation or malnutrition, and half of these were children under 5 years old. Luckily, a corrective solution was found for Apollo 13, and the astronauts did return safely home. No universally acceptable solution has yet been found to save the diseased and dying occupants of the spaceship earth.

The spacecraft we now inhabit, one of finite size and resources, seems to be coming apart at its seams. Material wealth increases, and yet the environment grows poorer; leisure time is more abundant, but there is less space and beauty to enjoy; technological improvements create environmental hazards. Air and water in many places exhibit signs of complete overload, and our overcrowded cities impose intolerable psychological stress. What has gone wrong? The human population today is 4.2 billion, and it is on the increase. Every minute on board our fragile craft more than 200 babies are born. The 70 million new passengers that arrive on earth each year must be fed, clothed, and housed. Each inhabitant makes demands on the finite resources available on the earth, and in turn each contributes to global pollution. There are more hungry and malnourished people on earth today than there were human inhabitants in 1850. Poverty, starvation, destruction, depletion of natural resources, and pollution all ride the coattails of unchecked population growth. Simply stated: our well-being is threatened because there are too many of us.

Around the bend

Examination of the growth curve describing the increase in man's numbers during the millennia shows that not only has the total population increased markedly over the past 500,000 years, but the rate of increase has also steadily risen (Figure 26–1). The simplest way of appreciating the rate of increase is to look at the **doubling time,** that is, the time it takes for the population to double its size. When the human population went from 5 million in 8000 B.C. to 100 million in A.D. 1, the population increased itself 20-fold; to achieve this required about 4 doublings: 5 → 10 → 20 → 40 → 80. This took place over a period of 8000 years, so that the calculated doubling time was 2000 years (8000/4). From A.D. 1 to A.D. 1650 the population swelled from 100 million to 500 million, a 5-fold increase requiring 2.5 doublings, or a doubling time of about 600 years (1600/2.5). The next doubling—from 500 million to 1000 million—took only 200 years (1650—1850), and the human population doubled again in 80 years (1850—1930). By 1975, only 45 years from our last calculation the world population doubled again. At the present rate of increase, by 2005, a time most of us may live to see (and perhaps regret), the world population will be about 8 billion! Why does the growth rate (or doubling time) change? A population of humans, or any other organism for that matter, tends to increase because the birth rate and the death rate do not keep pace—additions (births) are greater than subtractions (deaths). Reproduction is clearly a positive feedback system, and an unchecked population

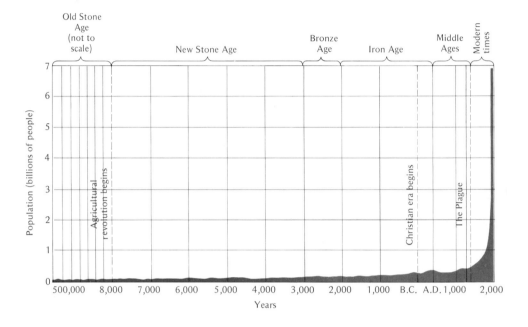

increases faster and faster, approaching infinity. To understand the nature of this growth phenomenon, let us compare **geometric** (also called **exponential**) and **arithmetic** growth. In arithmetic growth the increments are stepwise and occur by a fixed amount: 2 → 4 → 6 → 8 → 10 is an example of arithmetic growth with an increment of 2. In exponential growth the increaments are doubled in each successive step, and a J-shaped growth curve results (Figure 26–2). (Examples of exponential growth can be seen in a rapidly dividing population of mitotic organisms or when a sheet of paper is folded on itself. 1 → 2 → 4 → 8 → 16 and so on. In quantitative terms, a single sheet of paper folded on itself 38 times and doubling its

thickness with each folding would reach from New York to Los Angeles—about 3000 miles. Doubling the paper 45 times would produce a thickness equivalent to the distance from the earth to the moon—240,000 miles; doubling it only 8 more times would give a thickness of 93 million miles—from the earth to the sun!

Recognition of the nature of population growth and its application to man is not new. In 1755 Benjamin Franklin wrote:

There is, in short, no bound to the prolific nature of plants and animals. . . . Thus, there are supposed to be now upwards of one million English souls in North America (though it is thought scarcely 80,000 had been brought over sea) and yet there is not one the fewer in Britain, but rather more . . .

FIGURE 26-2 A comparison between arithmetic and
exponential growth.

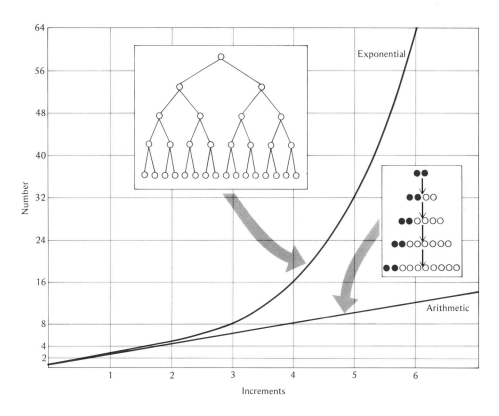

"THEIR BREEDING RATE IS INCREDIBLE."

The population bomb

Growth rates. Two people meet on a corner.

First person: We are all indebted to the male alligator for our very lives.

Second person: The male alligator? Why?

First person: During a single year the female alligator lays over a hundred eggs, but the male alligator eats 98 of these, allowing only a few to survive. So you see, if it were not for the male alligator, we would be up to here (pointing to his nose) in alligators!

This old joke has real meaning in human terms. The present explosive rise in the world's population is due to the fact that the birth rate exceeds the death rate. During the last 50 years, and especially since the end of World War II, the death rate has been declining because of improved sanitation and medical care (Figure 26–3). Consider the case of Sri Lanka: from 1871 to 1950 the annual growth rate[1] was

---

[1] To calculate the growth rate of a population, we simply subtract the death rate from the birth rate (disregarding immigration and emigration): thus, population growth equals birth rate minus death rate. Let us assume that in 1 year the birth rate for the world was 34 per 1,000 persons in the population, and the death rate was 14 per 1,000. The population growth is therefore 20, and the yearly rate of growth is 2% (20/1000). Using prepared tables it is possible to calculate the doubling time of a population for a given growth rate (Table 26–1). For example, if the growth rate is 2% per year, the population (irrespective of its original size) will double itself in 35 years.

1.7%, but in 1950 DDT was introduced to eliminate the malaria-carrying mosquitoes; the birth rate remained relatively constant, but the death rate declined; as a result the yearly growth rate today is 2.8%. Since 1650 the human population has not had a negative growth rate. Between 1650 and 1750 the world growth rate was 0.3% per year, between 1750 and 1850 it was 0.5%, and between 1850 and 1950 it was 0.8%. By the 1960s the growth rate was close to 2%.

Probably more shocking than this is the fact that even if the birth rate were to decline sharply, the population would continue to rise for many years (unless, of course, death rates were also to rise). For example, if every couple

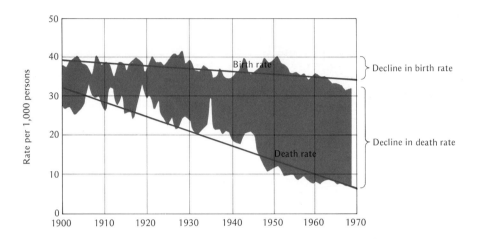

FIGURE 26-3 Changes in the birth and death rates over the past 70 years in Sri Lanka. Note that the decline in the death rate greatly exceeds that in the birth rate. (After Population Reference Bureau.)

TABLE 26-1 Doubling times for various rates of annual increase

Percentage rate of increase	Doubling time (yr)
0.6	116
0.8	84
1.0	70
1.5	47
2.0	35
2.5	28
2.8	25
3.0	23
3.5	20
4.0	18

in the United States had only 2 children (a 1:1 replacement of individuals) and there was no immigration, the United States population, would not level off until the year 2037, and at that time the population would be 275 million. Thus, 60 million people would be added before the U.S. population stabilized. It is obvious that a rational approach toward **zero population growth (ZPG)** requires a lowering of the birth rate and long-range planning for at least 50 years. Shall we have that much time? As the astronauts on Apollo 13 were stranded in space, we are stranded on our planet—we do not have the option of leaving the spaceship earth and beginning anew. All we can do at this point on the bend in the exponential growth curve is try do diminish the steepness of the upward climb.

Perhaps more disturbing, and not revealed by world growth rate figures, is the country-by-country variation in growth rate. At present the growth rate in the United States is 0.6% and at this rate our population will double in 116 years (Table 26–1). Africa and Latin America have current growth rates of 2.7%, and their population will double in 26 years; Asia currently has a rate of 2.0% with a doubling time of 35 years. Within each of these conti-

nents the growth rate also varies (Table 26–2). What the political, social, and economic results of such diverging growth patterns among the nations of the world will be is highly speculative; one possibility is that the have-nots will be increasingly agitated to partake of the holdings of the haves.

Health and nutrition must, of necessity, decline in those countries that do not roll back their population growth rate or increase the capacity of their environments to support their population. If malnutrition, starvation, and disease exist, the death rate will exceed the birth rate, the growth rate will become negative, and the population will tend to decline or level off. Pestilence, war, and famine, three of the Four Horsemen of the Apocalypse, provide natural checks on the growth of a population. Decline in the human population from such causes is not unknown. Unfortunately, such catastrophes do not occur by human design, and they strike unexpectedly in supposedly safe areas. In 1348–1350 the Plague probably killed about 25% of the European population. In this century, famine has contributed to between 5 and 10 million deaths in Russia (1918–1922,

1932–1934), 4 million deaths in China (1920–1921), and 2 to 4 million deaths in Bengal (1943).

Natural measures must ultimately check unbridled population growth. However, this is not a very pleasant way to achieve stability. The question presently facing us is whether we can forestall these catastrophes by instituting voluntary measures today that will stem the tide of population growth. Can we do it in time?

Population structure. The size of a population and whether it will grow or decline depend on the difference between the birth and death rates, which in turn depend on the age and sex ratios of that population. In particular, the number of women of childbearing age in a population is crucial in evaluating its growth rate. Our present world population is young; the young of reproductive age constitute the gunpowder of the population explosion. These young people are already on board the spaceship earth, and when such a combustible fuel is ignited it can only have disastrous consequences. Compare the age structure of the United Kingdom (Britain) in 1970 with that of

TABLE 26-2 Population growth rate, doubling time, and population size for selected countries of the world

	Annual growth rate	No. years to double population	1977 population (millions)	2000 population estimate (millions)
Mexico	3.5	20	64.4	134.6
Honduras	3.5	20	3.3	6.9
El Salvador	3.2	22	4.3	8.6
Ecuador	3.2	22	6.4	12.2
Algeria	3.2	22	17.8	36.5
Morocco	3.2	22	18.3	35.5
Iraq	3.2	22	11.8	24.3
Guatamala	3.1	22	6.4	12.2
Peru	2.9	24	16.6	31.2
Pakistan	2.9	24	74.5	145.5
Iran	2.9	24	34.8	66.1
Liberia	2.9	24	1.7	3.0
Brazil	2.8	25	112.0	205.0
Malaysia	2.8	25	12.6	21.6
Bolivia	2.6	27	4.8	8.7
Ethiopia	2.5	28	29.4	53.8
Tanzania	2.5	28	16.0	33.1
Burma	2.4	29	31.8	53.3
Thailand	2.4	29	44.4	84.6
India	2.1	33	622.7	1,023.7
Haiti	2.0	35	5.3	7.9
World	1.8	38	4,083.0	6,182.0
Puerto Rico	1.7	41	3.2	4.1
China, People's Rep.	1.7	41	850.0	1,126.0
Chile	1.6	43	11.0	15.8
Argentina	1.3	53	26.1	32.9
Japan	1.1	63	114.0	133.0
Spain	1.0	69	36.5	45.4
Australia	0.9	77	13.9	19.6
U.S.S.R.	0.9	77	259.0	314.0
Canada	0.8	87	23.5	31.6
United States	0.6	116	216.5	262.5
France	0.4	173	53.4	61.7
Denmark	0.4	173	5.1	5.4
Sweden	0.2	347	8.2	9.2
United Kingdom	0.1	693	56.0	61.9
West Germany	-0.2	—	16.7	17.7

From *Population Reference Bureau 1977 World Population Data Sheet.* Washington, D.C., Population Reference Bureau, 1977.

India (Figure 26–4). Since the graphs show ages on a percentage basis, the two countries can be directly compared even though the total population sizes differ. It is obvious that India had more young people than Britain; this is the condition found in countries showing rapid population growth. In 1970 more than 44% of the population in India was under 15, but in Britain only 23% of the population was in this age group.

What does the preadolescent bulge in the population of India mean? First, birth rates are high and infant mortality is low. Second, a society so constituted is burdened with a relatively large number of nonproductive individuals. Although in these societies large families are considered an economic advantage—children work on the land and provide security for the parents in their old age—this is a myth. Children in vast numbers put an extreme burden on the educational system, they lead to shortages of teachers and teaching time, and in general, they tend to leave school early. In the economy of such a country, children surviving into adulthood strain the funds available for medical care and housing. It is the young who are punished for parental sexual indulgence and the desire for economic security. A preadolescent population bulge not only strains the present economy but has a continuing negative effect. As the children of the "bulge" reach reproductive age, an astronomical rise in population size will occur if the birth rate remains the same, and this, in turn, will once more inflate the preadolescent group. Thus, the population bulge tends to be self-perpetuating.

Although we have observed that death rates may have a marked effect on the growth of a population, it must be recognized that the age at which death occurs is a critical factor. If the victims are in the postreproductive period, the future growth of the population will not be changed at all. In many underdeveloped countries the death rate is high for infants and those of postreproductive age. In these same societies the average age of marriage is very young, and there is a tendency to have a great many children during the reproductive period (perhaps to compensate for the high degree of infant mortality), so the population continues to soar.

Prognostication of population size is a complicated affair. Predicting population changes must take into account the proportion of women of reproductive age, the size of the reproductive group, and fertility (likelihood of producing a child) patterns. These, of course, are influenced by biological, sociological, religious, and political attitudes, many of which cannot be adequately accounted for in statistical evaluations of a population. Social factors and religious and political attitudes may encourage or discourage births: so may the changing role of women in society and the changing attitude toward the child as a supporter of aged parents. Immigration or emigration may affect population structure. War and military service also modify the reproductive potential of a nation, but since these are rather unusual situations, they tend to confound the predictions. To illustrate this consider the distribution of the female population in Sweden in 1910, 1930, and 1960 (Figure 26–5). The shift from a rapidly expanding population to a relatively stabilized population was influenced by the social and economic effects of World War I and World War II. No study of the characteristics of human populations with regard to size, density, death rate, birth rate, migration, sex ratio, and so on could forecast such a dramatic shift. Predicting where the human population will be in the next few years requires knowledge from many fields and involves great uncertainty.

In a developed country such as the United States or Britain, the number of dependent

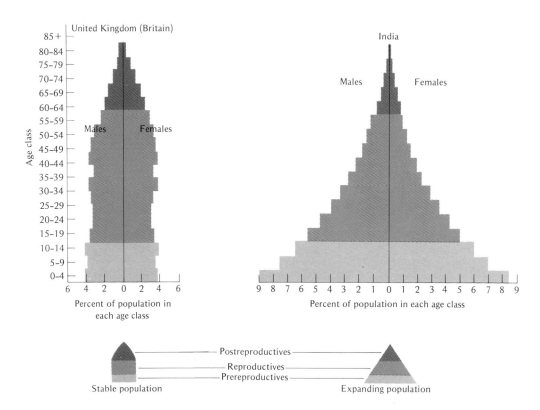

young is lower than in an underdeveloped country, the work force is proportionately larger, and economic productivity is often greater (Figure 26–6). In the United States and Britain infant mortality is low, and so is the death rate of the elderly. Despite the fact that the birth rate and the fertility rate (number of births per 1,000 women aged 15–44) in the United States has been dropping to the level necessary for ZPG (Figures 26–7 and 26–8). the United States will probably not attain ZPG in this century. It cannot be stated with any certainty what path we as a nation will follow, but if we want to maintain a nongrowing population of 300 million, now is the time to

plan for it. If we are to maintain legal immigration at its present level, 1.7 should be the average number of children per woman so that a stable population size can be achieved before the middle of the twenty-first century.

Distribution of human populations. The most significant features of the current population problem are the finite size of the earth, human reproductive potential, and time. *Homo sapiens* appeared on this planet about 1.5 million years ago. During the Stone Age (1 million years ago), the human population, numbering in the vicinity of 100,000 was restricted to the continent of Africa (Figure 26–9). Stone-Age man was a nomadic hunter,

FIGURE 26-5 Population age distribution of females in Sweden in 1910, 1930, and 1960.

FIGURE 26-6 Population structure of the United States, 1975. (After Population Reference Bureau, 1975.)

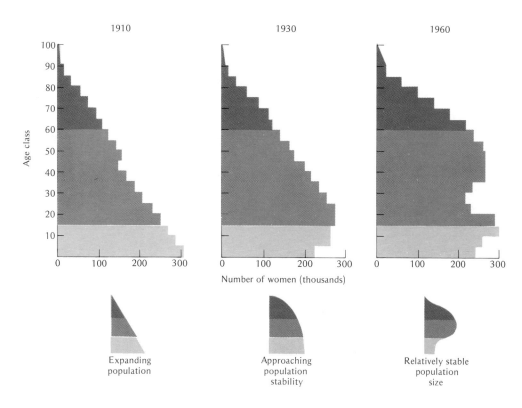

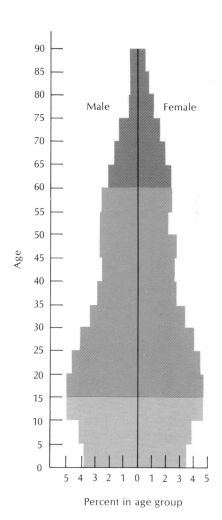

and his numbers were limited by the availability and abundance of his prey as well as by what he could capture. In those days, 2 miles of good hunting territory supported a man. About 8000 B.C. the agricultural revolution took place; man learned how to raise crops and domesticate animals. Once man ceased to hunt and began to till the soil, his wandering stopped. He settled on the land, formed agricultural communities, and improved methods for producing food.

Since there was increased availability of food plus virtually limitless land (more than 20 million square miles of good land is present on the earth's surface), the population increased; humans spread across the continents of Africa, Europe, and Asia, and their density was 0.06 person per square mile. At the time of the birth of Christ, village and urban farming was commonplace, methods of storing and transporting food were improved, and since a farmer could produce more than enough food for his own family, people were freed from the land and could pursue other occupations. Villages contained larger and larger numbers of people, many engaged in a variety of occupations indirectly connected with agriculture, and soon these centers came to be large enough to qualify as cities. It has been estimated that by about 1 A.D. the human population density

FIGURE 26–7 (above) Birth rate in the United States, 1910–1975. (After Population Reference Bureau, 1975.)

FIGURE 26–8 (below) Fertility rate in the United States, 1910–1974. (After Population Reference Bureau, 1975.)

was 2 persons per square mile. By 1650, the development of science, technology, and industrialization brought about new changes in the environment and in human populations. New land, the Americas, opened up and the human population grew; the density was then 5 persons per square mile. Two hundred years later (1850), the population reached a density of 12 persons per square mile, and by 1950 it had reached a density of 20 persons per square mile. It has been estimated that by the year 2000, when the human population reaches 6 billion, there will be 60 persons per square mile; given the present rate of growth, by 2600 there could be 1 person per square foot of the earth.

Cities first arose in Mesopotamia only 5,000–6,000 years ago, but at the present rate of emigration and growth, by 1980 more than half the world's population will live in cities. Not only that, by 2020 half of the human beings in the world will live in a city having 1 million persons, and by 2044 everyone will live in a city of that size. By 2044 we shall exist in wall-to-wall megacities, with the largest containing 1.4 billion people (the projected world population being 15 billion). It is a perplexing and strange situation to us, the inhabitants of cities, to consider that in 1776 not a single city of 1 million existed.

The trend toward city dwelling, or urbanization, in the United States reflects a worldwide change. Accelerated movement from the agricultural regions toward the cities began in 1800. At that time 6% of the United States' population was urban; by 1850 the proportion living in cities was 15%, by 1900 it was 40% and today more than 70% of our fellow beings in this country live in the cities or the suburbs. Underdeveloped countries show a similar trend: in South America more than 55% of the population lives in cities; Nairobi, the capital

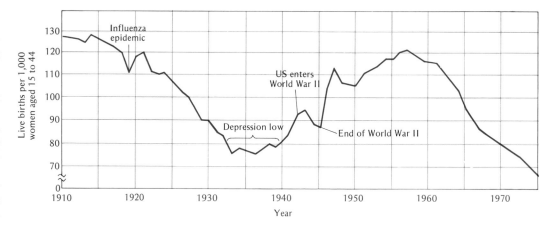

Date	Area populated	Assumed density per square mile	Total population (millions)
600,000 B.C.		0.006	0.125
200,000 B.C.		0.02	1
20,000 B.C.		0.06	3.34
8,000 B.C.		0.06	5.32
4000 B.C.		1.0	86.5
A.D. 1		2	133
1650		5	545
1750		7	728
1800		10	906
1900		15	1,610
1950		20	2,400
2000		60	6,270

Agricultural revolution → (between 8,000 B.C. and 4000 B.C.)

Americas opened up and industrial revolution → (between 1650 and 1750)

FIGURE 26-9 Increase in population density through the ages.

of Kenya, grows larger in population by 7% each year; Lagos, the capital of Nigeria, grows at a rate of 14%; while Los Angeles grows at an annual rate of less than 3%.

Cities have good museums, libraries, theaters, universities, shopping centers, symphony orchestras, and transportation. They also produce many problems that rural areas do not: overcrowding, loss of identity, alienation, anonymity, slums, pollution, and economic dislocations. One continually wonders: Are the advantages of urban life balanced by the disadvantages?

26-2 IMPLICATIONS OF POPULATION GROWTH ON SPACESHIP LIFE

The human population of the earth is increasing exponentially, but planetary resources are not. Obviously, at some point, as the songwriter said, "Somethin's gotta give." Put another way, if you have a normal pulse beat, it will not quite keep up with the growth in world population. Every time your pulse throbs, the human population of the world will have increased by one. Let us look at some of the implications of unlimited population growth.

Our hungry planet

In 1967 it was estimated that 20% of the people in the world were not receiving enough calories per day; that is, they were undernourished. Deprived of calories, the individual cannot meet the body's energy demands, tissue reserves are drawn upon, and there is a gradual wasting away, with death as the final consequence. In addition, 60% of the world's population was malnourished; that is, they were lacking in essential food materials, mostly pro-

FIGURE 26-10 The geography of hunger. (Data from *FAO Production Yearbook,* 1968.)

High calorie, high protein

High calorie, minimum protein

Low calorie, minimum protein

Low calorie, low protein

teins and vitamins. The hungry people of this planet (Figure 26–10), undernourished and malnourished, number between 1 and 2 million. Tonight when you go to sleep, remember that more than half the children of the world will go to bed hungry.

In the underdeveloped nations malnutrition takes two forms: **marasmus** and **kwashiorkor.** Marasmus develops when both calories and protein are in short supply; most victims are children less than 1 year old who have been weaned too early or have been given a poor substitute for mother's milk. The child with marasmus is emaciated, his skin is wrinkled, and his eyes are large and sunken (Figure 26–11a). Kwashiorkor in African means "the rejected one" and is derived from the fact that children with such symptoms often are displaced from the mother's breast by the arrival of a new baby. The oldest child, about 1 year old, still requires protein, but since the mother can suckle only one child, the displaced child is fed a carbohydrate diet and develops a protein-deficiency disease. The hair and skin are discolored, the belly is bloated, due to fluid imbalance and edema, physical growth is impaired, appetite is poor, and in general the child is lethargic (Figure 26–11b). Over half of infant deaths (ages 1–4 years) in underdeveloped countries result from kwashiorkor. Those children who survive are damaged in other ways: brain development is retarded (the I.Q. of such children is about 13 points lower than the I.Q. of those raised on adequate protein diets), and the body is dwarfed. Protein deficiency during infancy is essentially irreversible, and provision of a normal diet later does not compensate for the early slowing of growth. This is probably because during the first 3 years the brain grows to 80% of its adult size (with more than 50% of its dry weight being protein), and the entire body grows to 20% of its size at maturity.

We, the affluent, drink half of the world's milk and eat three-quarters of the world's meat.

FIGURE 26-11 (a) Marasmus. (Photo by Terence Spencer, Courtesy of Time-Life Picture Agency, copyright©Time, Inc.) (b) Kwashiorkor. (Courtesy of Sanford Schneider.)

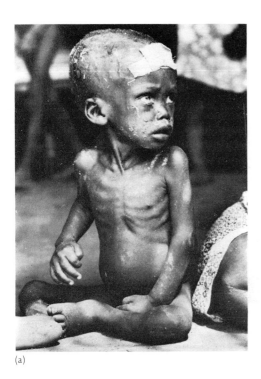

(a)

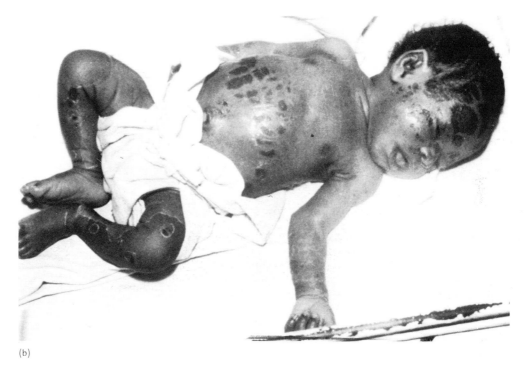

(b)

Dogs in this country eat enough protein to satisfy the requirements of 4 million human beings! Because it is more economical, people living in underdeveloped countries—and even the poor within our own nation—depend on a diet of grain rather than meat. In Asia, 97% of the diet is grain, mainly rice, but among well-fed North Americans only 75% of the diet is in the form of grains such as rice, corn, and wheat. The dispartiy between the nutritionally well-off developed countries and the hungry underdeveloped countries is illustrated by the use of grain in the United States: 20 million tons of plant protein are used for feeding livestock, and from this only 2 million tons of meat protein are obtained. The lost protein—18 million tons—could provide 25% of the protein for the world's population!

The way to feed the hungry, some say, is to cultivate more land, to develop new high-yield strains of grain plants, and to farm the sea. On the surface some of these suggestions have merit, but closer examination shows their deficiencies in solving the hunger problem.

The **carrying capacity** of an ecosystem—the amount of biomass produced—depends on the primary productivity of its photosynthesizing plants. This varies from ecosystem to ecosystem (Figure 26–12). Maximum plant productivity is achieved in efficient agricultural operations such as those found in the developed countries of the world, and the ideal maximum production capacity based on photosynthetic efficiency and nutrient availability is about 20 gm per square meter per day. In the United States there used to be about 9 in. of topsoil that

could support good harvests. However, in the last 300 years about one third of that has been destroyed. Farm lands are now converted to suburbs, highways, garbage dumps, and factories at a fantastic rate: 2 acres per minute or 1 million acres per year. Every day in California 300 acres of agricultural land are lost to urban sprawl, and this state currently produces 29% of the nation's food (Box 26A).

Some have suggested that by utilizing land that could be farmed, but at present is not, we could support a larger population. Theoretically, it is possible to increase productivity in arid (dry) and semiarid regions by irrigation (witness southern California and Israel), but to do so requires massive irrigation and vast nutrient additions to improve the soil's fertility. For example, growing crops requires great

BOX 26A Last chance to keep off the grass?

The admonition to "Go west, young man!" was taken literally by thousands of Americans during the 1950s and 1960s. The slow-down in migration during the early 1970s shows signs of reversing itself as winter-weary Easterners seek to escape to the sunshine state. What effect is this having on California?

California, the nation's number one agricultural state, contributes 9% of the total value of agricultural products in the U.S., with 36 million acres of land generating $44.5 billion worth of agricultural industry receipts in 1976. Without question, agriculture is an extremely important industry in California, and the quality and quantity of the three basic natural resources that are required to support agriculture (soil, air, and water) need to be maintained. Urban development has been diminishing California's commercial agricultural lands by about 20,000 acres annually over the past two decades. However, 50,000 acres of previously nonirrigated lands are said to have been coming into production annually for the first time, resulting in a net annual increase of 30,000 acres of agriculture production in California each year.

The major drawback to this equation is the fact that it takes far more energy to convert raw land to producing land than it takes to farm the best soil. Poor soils require more heavy equipment, more fertilizer, and hence more energy in order to start producing than do good soils (Table 26-3). One way to conserve our resources is therefore to use our best soils for farming and poorer soils for urban use.

We are not doing this. Since 1965, 23% of California's prime agricultural land and 14% of lesser quality land has become urbanized. Furthermore, since the best land is more intensively cultivated than land of poorer quality, the current pattern of urban expansion will have a more dramatic impact on agricultural production than if the impact were more evenly distributed. If urban projections made for the period between 1965 and 1980 prove to be forerunners of urban expansion, more than two-thirds of the prime land will be in urban use by the year 2100, and none will be left for agriculture by the year 2200, in the absence of four factors. These factors are: (1) a decline in population growth; (2) technological advancements increasing agricultural yield on low-quality lands; (3) widespread introduction of a totally new concept of food and fiber production; and (4) external controls that direct urban expansion to less productive land.

Since the first 3 factors seem at present unlikely to occur, what if anything can be done about redirecting urban expansion?

Currently, state and federal government agencies are examining the problem, but at the local level the pressures for urban development are often enormous. These can reach the dimensions of polarized conflicts, with builders lining up on one side and environmentalists on the other. Some cities in California have put an upper limit on the number of building permits that can be issued annually; others have attempted to plan and direct growth by rating the merits of proposed developments, by imposing agricultural zoning in prime areas, or by attempting to establish agricultural preserves in threatened locations. However, some have questioned the right of government to limit the rights of property owners and the freedom of Americans to move about and live where they please. Legislation at the state and federal level will probably be required. In the meantime, the future of California's agriculture industry hangs in the balance.

TABLE 26-3 The agricultural demand for energy

Source	(%)
Natural gas (used in the production of fertilizer)	53
Electric power	16
Diesel fuels	17.5
Gasoline	10
Liquid propane gas	2.25
Aviation fuel	.75
Total	100%

LAST CHANCE TO KEEP OFF THE GRASS

Drawing by Richter;
© 1976 The New Yorker Magazine, Inc.

FIGURE 26–12 Primary production in grams of dry matter per square meter per day, in the major ecosystems. (Modified from *Fundamentals of Ecology,* Second Edition, by Eugene P. Odum, copyright© 1959 by W. B. Saunders Co.)

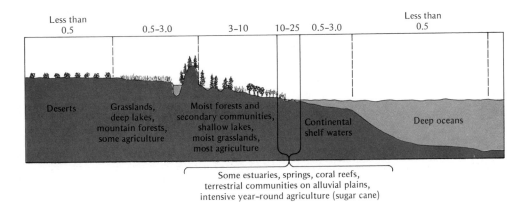

Some estuaries, springs, coral reefs, terrestrial communities on alluvial plains, intensive year–round agriculture (sugar cane)

quantities of water—200 gallons to produce the 2,500 calories required by one person each day—yet 97% of the earth's water is salty. Desalting oceanwater for irrigation is costly and not yet technologically feasible. By 1984 desalination could supply only 3% of our water needs, and by 2010 it is estimated that 270,000 desalination plants would be required to meet the need for water. What would we do with the salt that is left after desalination? What kinds of pipelines would be needed to transport the fresh water inland? The expense and environmental dislocation produced by mass desalination could be too costly to make the procedure practical.

Enhanced food yields can be obtained by using new strains of wheat and rice—called the **green revolution.** Employing such strains Mexico has increased its wheat yield sixfold in 20 years. However, the new high-yield varieties require large amounts of water, fertilizer, and pesticides. The ecological consequences of heavy fertilizer and pesticide use are not completely understood and could be dangerous. And, in certain parts of the world the soil is so poor and the climatic interactions so severe that even the addition of fertilizer, water, and agricultural know-how cannot produce crop

diversification or increase yields. Paul and Anne Ehrlich have said:

"In many ways the problem of revolutionizing underdeveloped countries' (UDC) agriculture is inextricably tied up with the problem of UDC development. Shortages of capital, demand, resources and trained technicians, lack of effective planning, and the absence of adequate transport and marketing systems all tend to combine with extremely high rates of population growth, malnutrition and disease to make any kind of development extremely difficult, and thus retard agricultural development. It is a vicious cycle—one that the Green Revolution may not be able to break.[2]

If not the land, how about the sea? Why are the oceans less productive than the land (Figure 26–12)? Large bodies of water, such as open oceans and large lakes, are at a production disadvantage because a large portion of the light energy is absorbed by the water before it reaches the site of maximal mineral supply. In addition, the oceans and larger lakes tend to lack adequate amounts of nutrients. The myth of a food-rich sea is just that—an unsubstantiated illusion. According to J. H. Ryther of the

---

[2] P. Ehrlich and A. Ehrlich, *Population, Resources, Environment,* San Francisco, Freeman, 1970.

Woods Hole Oceanographic Institution, the maximum fish yield of the oceans is 100 million metric tons, and that is only 1.5 times the record fish harvest of 1968. To increase yields it might be possible to harvest lower down the food chain, but this does not seem profitable or practical at present or in the near future. More human calories would be spent on harvesting a thin surface film of plankton than would be gained in consuming the catch!

What about farming the sea by culturing fish and other marine organisms? Could this not be achieved by fertilizing and harvesting the oceans as we do the land? Upwellings of nutrient-laden waters occur naturally in the coastal zones, and such areas have a significantly higher productivity than the open ocean (Figure 26–12). For example, off the coast of Peru, nutrient upwellings permit large blooms of phytoplankton, fish and fish-eating birds are abundant, the seabirds produce guano in large quantities, and this can be harvested for the production of fertilizer. However, enrichment of the oceans by inducing artificial upwellings would be expensive in terms of energy input. Furthermore, in the offshore areas where such a project might be feasible water pollution poses a serious threat.

Finally, even if we could grow seaweed and algae at reasonable cost, how easy would it be to get people to eat them? How tasty and nutritious will seaweed- or algae-burgers be? We are drawn, almost inescapably, to the conclusion that the seas are essentially biological deserts and likely to remain so.

In spite of the gains made by the green revolution there are more hungry people today than ever before. The more people we feed, the more there are to reproduce. As we feed more people, the population rises; pollution becomes more severe, species composition is adversely affected, and consequently food productivity declines. The land bank is bankrupt.

We cannot farm the sea; we cannot produce sufficient quantities of synthetic food cheaply; and we cannot reduce our caloric requirements significantly. Even if we could grow enough food to feed our present population, this would only buy time; it will not curtail population growth. Feeding the world's population treats a symptom, but not the disease. To survive, we must control the disease itself, and this means curbing population growth.

Depletion of resources

Every person on the earth depletes its resources. Paradoxically, it is the developed and affluent countries, whose growth rate is decidedly less than that of the underdeveloped nations, that constitute the most serious threat to our environment. The United States, with 6% of the world's population, uses 50% of the world's resources of fossil fuels, metals, radioactive substances, and the like. The Western nations together, including the United States, Russia, and Japan, account for 25% of the world's population but use 90% of the earth's natural resources. As the consumption of these resources increases, it is automatic that pollution will also increase.

The resources available to man are classified as renewable or nonrenewable. Renewable resources can, with proper management, replace themselves in time; plants and animals are renewable resources. If these are mismanaged, as in the despoliation of forests or the excessive killing of deer, the ecosystem becomes severely imbalanced and renewal is virtually impossible. Nonrenewable resources cannot be replaced; although these materials are not actually lost, they are changed by man into forms that can no longer be used. Thus, fossil fuels are burned to carbon dioxide and water, radioactive elements decay, and the fertility of the soil may be lost. The mineral

resources of our planet have been here since the earth was formed. They are being depleted at an ever-increasing rate because there are more of us using them and our present technology consumes enormous quantities of these substances.

How long will the nonrenewable resources last? Estimates vary with the particular resource. For example, aluminum supplies may last for 570 years, coal and oil for 300 years, but tin, copper, lead, uranium, and zinc may be depleted in one generation (30 years). The rate of depletion is staggering when we consider that it took 1 billion years to produce the earth's coal deposits that we are using up in only a few hundred years. In the last 50 years we have used up more fuel and minerals than in the entire history of man. Most of these materials end up in the sea, too diluted for reclamation.

The consequences of depletion of nonrenewable resources are obvious: substitute materials must be developed and costs must rise for those substances that become less abundant. Increasingly metals will have to be replaced by plastics, but the substitution is of limited value since plastics are derived from fossil fuels. By switching to new sources of structural materials, we shall deplete other resources.

We are also using up our energy resources at a prodigious rate. Early man used 2,000 Calories per day, but Americans now use 200,000 Calories each day. The developed countries use 77% of the world's coal, 81% of its oil, 95% of its natural gas, and 80% of the nuclear energy. Nuclear fission, it is estimated, can supply about half our energy needs by the year 2000, but supplies of uranium (on which the process depends) are expected to last only 30 years more. Discovery of new radioactive fuels and the development of controlled nuclear-fusion reactors could enable future energy demands to be met, but the problems of thermal water pollution and accumulation of radioac-

tive wastes would remain. Nuclear power plants require cooling water in massive amounts, and raising the temperature of the water would soon disrupt the aquatic ecosystem (Box 26B).

The decline and fall of clean air

People were crawling about, people were whimpering, people were screaming, gasping for breath, touching each other, vanishing in the dark, and ever and anon being pushed off the platform on to the live rail.[3]

These words tell of the collapse of a totally automated society; a society completely subterranean, globally uniform, in which people were fed on synthetic foods and breathed fetid air. The people, if they could be called that, became used to the fetid air and the tasteless food and the dimming illumination until the power source decayed. Fiction or our future? Will the world end from a nuclear holocaust or with a cough, a wheeze, and a gasp from emphysema-scarred lungs?

The blight of light. Brown air and gray air are two types of **air pollution**. Brown air covers Los Angeles; it is a pollutant of young cities, and its prime source is the automobile's internal-combustion engine (Figure 26–13). When air containing nitrogen and oxygen is heated to 3,000°F, as it is in the automobile engine, nitric oxide (NO) is produced; this reacts with ozone (O_3) in the atmosphere and yields the highly toxic yellow-brown nitrogen dioxide (NO_2). Nitrogen dioxide is harmful to the delicate lining of the lungs, contributing to the development of bronchitis, cancer, emphysema, and cardiovascular damage.

The unburned hydrocarbons of gasoline,

---

[3] E. M. Forster, *The Machine Stops.*

FIGURE 26-13 Photochemical air pollution. The disappearing landscape during a smoggy day in California. (Courtesy of Air Pollution Research Center, Riverside, Calfiornia.)

released into the atmosphere in the presence of nitric oxide and sunlight, undergo chemical reaction to produce **p**eroxy-**a**cetyl-**n**itrate (PAN), formaldehyde, and other oxidants that cause the eyes to burn and tear, damage lungs, kill vegetation, and rot rubber and nylon. The abundant sunshine in southern California contributes to the production of ozone by the reaction $NO_2 \rightarrow NO + O$, the oxygen reacts with unburned hydrocarbons and atmospheric oxygen to yield ozone. Automobiles, sunlight, and ozone are a self-perpetuating and dangerous combination that produce photochemical air pollution. Ozone at 1 ppm disrupts photosynthesis (Figure 26–14) and at 0.1 ppm adversely affects the lungs. On a clear day the ozone content is about 0.02 ppm, but on summer air-polluted days in Los Angeles and environs it may be as high as 0.5 ppm!

Incomplete combustion of coal, oil, and gasoline results in carbon monoxide formation. Carbon monoxide inactivates hemoglobin and cytochromes—respiratory pigments. Living for 8 hours in an atmosphere polluted with 80 ppm of carbon monoxide is equivalent to the loss of 1 pint of blood through hemorrhage. In Tokyo the carbon monoxide content of the air often reaches 40–50 ppm. On an automobile-jammed highway a motorist may be exposed to 400 ppm! The symptoms of mild carbon monoxide poisoning are headache, nausea, dizziness, blurred vision, and muscle pain. High levels of carbon monoxide cause convulsions, unconsciousness, and death.

Climate also contributes to air pollution. During a thermal or temperature inversion, a layer of warm air sits over a mass of cool air and prevents vertical mixing of the air. The air stagnates and pollutants accumulate (Figure 26–15). Life in such a city becomes nightmarish. High levels of photochemical air pollution have been recorded in Sydney, Australia; Tokyo, Japan; Boston, New York, and Atlanta.

However, in the Los Angeles basin air pollutants are a way of life.

The breath of death. To medieval man the odor of sulfur dioxide meant the Devil was close by. Medieval man was not far wrong, for the gray air of older industrialized cities (New York, Chicago, London, Manchester) bedevils the inhabitants with this same gas. Gray air (smog) is encountered in cities where fossil fuels are burned; in addition to sulfur dioxide, it contains soot and dust (particulates) plus toxic metals such as lead and beryllium. Automobiles emit little sulfur dioxide because when gasoline is refined from fuel oil the sulfur is removed. However, coal and heavy oils do contain sulfur, and when these are burned, the sulfur combines with oxygen to form sulfur dioxide. Exposure to 0.2 ppm of sulfur dioxide for 24 hours constitutes a serious health hazard; in New York in August 1970 the range was 0.04–0.11 ppm. Sulfur dioxide reacts with water vapor to produce droplets of sulfuric acid. The Parthenon and other monuments of human civilization are disintegrating because of exposure to sulfuric acid. Smog may destroy not only man's monuments but man himself. In 1901 over 1,000 deaths in Glasgow and Edinburgh were related to smog, and in December 1952 the London smog caused up to 4,000 deaths.

What can be done about air pollution?

1. Spend more local and federal money to clean the air.
2. Reduce the sulfur content of fossil fuels before burning.
3. Decrease hydrocarbon and nitric oxide emissions from automobiles by installing smog devices on the exhaust.
4. Use cleaner energy sources (solar and hydroelectric) instead of fossil fuels.
5. Improve mass transit in cities, reducing the automobile load.

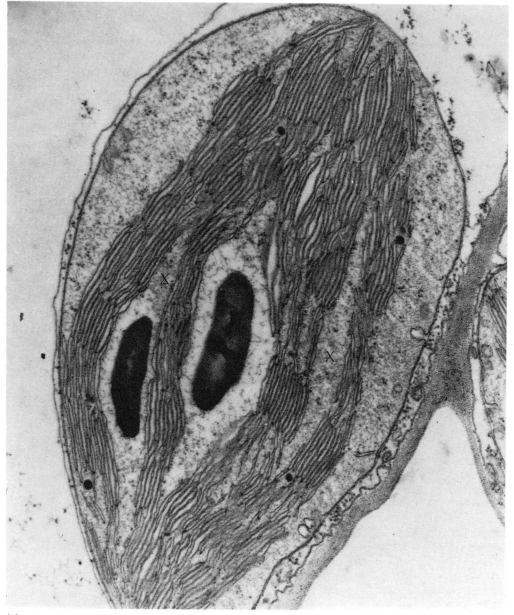

(a)

FIGURE 26-14 Electron micrographs of the leaves of an 8-day-old bean plant exposed to ozone at a concentration of 1 ppm for 30 minutes, a level comparable to that found in the Los Angeles basin. (Courtesy of W. W. Thomson.) (a) (facing page) Normal chloroplast. (b) (below left) Beginning of smog damage; granulation appears between stacks of membranes (X). At this stage (3 hours after gassing) the plant would appear normal to the unaided eye. (c) (below right) Extensive chloroplast damage with extensive granulation and disruption of membranes 24 hours after exposure. The leaves of this plant would have a reduced photosynthetic efficiency.

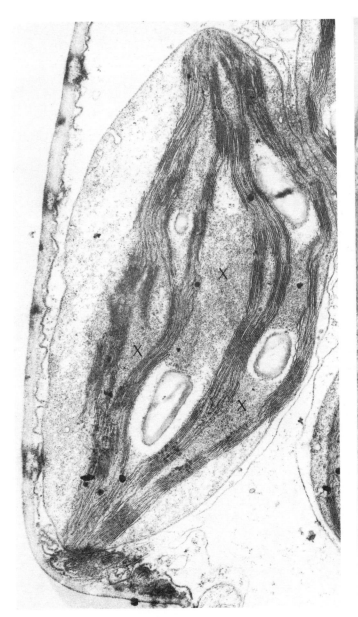

(b)

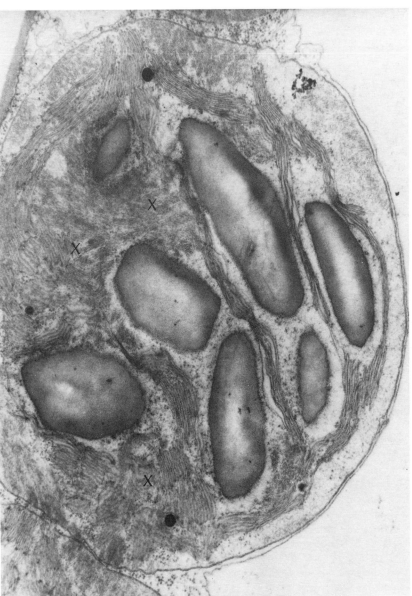

(c)

FIGURE 26–15 Thermal inversion. (a) Normally pollutants are carried away from the ground level because warm air rises. (b) During a thermal inversion, a layer of warm air at higher altitudes traps cooler air beneath and this tends to accumulate pollutants close to the ground level. (c) (facing page) Aerial view of the inversion. (Courtesy of Air Pollution Research Center, Riverside, California.)

6. Employ electrostatic precipitators to remove soot and other particulates from the air.
7. Curb sources of pollution—automobile and industrial. (Each day in Los Angeles 4 million cars burn 8 million gallons of gasoline producing 3 million tons of carbon monoxide, 200,000 lb of nitrogen dioxide, 400,000 lb of hydrocarbons and other noxious substances.)

In 1542 Juan Rodriques Cabrillo, the Portuguese explorer, reached Los Angeles. He saw the region blanketed by smoke from Indian fires. He called the basin *La Bahia de los Fumos*—Bay of Smoke—and sailed away convinced nobody could live there. Cabrillo may be proved right if air pollution is not checked.

"Water, water everywhere Nor any drop to drink"

As human populations increase in size, waste output also rises. There are two major categories of waste: liquid and solid. **Liquid waste** includes materials that, treated or untreated, are eventually flushed into the waters of the earth—the rivers, lakes, and oceans. Here they are diluted, and agents of decomposition such as bacteria break them down and recycle their materials. Until quite recently it has been assumed that the capacity of our natural waters to accept, to dilute, and eventually to purify our liquid wastes is virtually limitless. However, the volume of waste has reached such proportions that dilution no longer suffices, and the decomposers cannot work fast enough and may be overcome. The natural process of self-purification is not enough. This has serious implications because these same rivers and lakes serve as our sources of drinking water. Sometimes the wastes themselves are discharged into a river and inadvertently recycled before they are broken down—carried downstream to a neighboring town and added to the

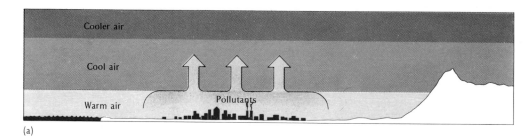

(a)

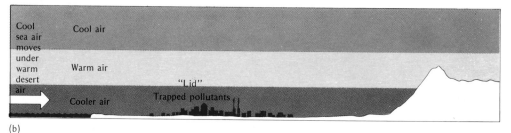

(b)

(c)

BOX 26B Thermal pollution: we can't break even

According to Webster's dictionary, **pollute** means to make or render unclean or impure. Heat energy—waste heat—discharged into the environment is called **thermal pollution.** It is a consequence of the Second Law of Thermodynamics: every energy conversion must add heat to the environment. Thus, the ultimate and inevitable pollutant is heat.

As our demands for energy increase along with population growth more heat will be generated. It is estimated that by the year 2000 the total energy used in the world will have increased by 500%, and in the United States it will rise by 800%. Our power needs double every 10 years. (Think of all the electrical appliances in use right now!) If the insatiable demand for energy were to continue unrestricted, by 2175 all the available land in the United States could be occupied by power plants. Since all our fossil fuels and uranium will be gone long before that time, nuclear fission in breeder reactors (ones that regenerate nuclear fuel) will assume prime importance in supplying our energy needs. Although the fuel in such power plants can be recycled, the heat energy cannot, and it must flow downhill (from warm to cool) into the environment.

Although some of this heat can be used for raising the temperature of greenhouses, heating buildings, and so on, the cost of piping warm water is high, and utilization methods are imperfect. The irony of thermal pollution is: the warm water that results from cooling of power plants is at too low a temperature for use as an energy reservoir, but is warm enough to threaten the aquatic ecosystem. Most aquatic animals cannot exist at temperatures greater than 35°C. As the temperature rises, the solubility of gases decreases; thus oxygen levels in the water will decline, and organisms with a high oxygen demand will be damaged or killed. By contrast, higher water temperatures plus increased nutrient load will promote excessive growth of algae—eutrophication—and the ecosystem will be further disrupted. Thermal pollution can make many of our rivers and streams uninhabitable for fishes by the year 2000. Furthermore, as the temperature of the seas increases, the polar ice caps will begin to melt and the level of the oceans will rise. Cities such as London, Los Angeles, and New York will disappear beneath the waters along with a large part of the earth's land mass.

The "energy crunch" of the winter of 1973–1974 gave people in industrialized countries a foretaste of life in a world where energy is permanently in short supply. The fact that the world's sources of energy are dwindling rapidly was brought home to many of us in the form of lowered thermostats, long lines at gasoline stations, and shortages of many commodities. This confirmed some sufferers in their opinion that more energy is needed. But did they consider the results of increased energy use? Are we caught in a thermodynamic race we cannot win? Will this race destroy all earthly order? If we do nothing, yes! However, by restricting population growth, reducing energy consumption and above all, using what we have more wisely, we may yet postpone (almost indefinitely) such a thermodynamic destiny. Ideally, we should ration energy consumption, reduce electrical demands and raise the cost of power to discourage its reckless use. We can do it if we want to: there is still time left.

'My! How I envy your dazzling white wash!'

TABLE 26-4 Treatment of sewage to recycle water

Sewage treatment	What's done	Cost cents/ 1,000 gal	What's left
Primary	Screening and grinding remove particles	3.5	Bacteria, pesticides, phosphates, metals
Secondary	Activated sludge or trickling filters containing bacteria and protozoa remove nutrients	8.3	Pesticides, metals
Tertiary	Chlorination and flocculation remove organic matter and soften water; activated charcoal removes odors	13–34	Fresh, drinkable water

water supply. As Tom Lehrer tells us: "The things we throw into the [San Francisco] Bay they eat for lunch in San Jose." If measures are not taken to improve treatment of sewage, the pathological bacterial and viral content of our water could provide a source of widespread disease. In a recent survey, 144 out of 155 United States cities with populations over 250,000 had measurable traces of sewage in their drinking water.

Agriculture uses water and pollutes it. An orange grove covering 1 acre requires 800,000 gallons of water. Spraying of herbicides and pesticides and intensive cultivation to increase crop yields also contaminate our waters. Farm animals produce 10 times the amount of sewage humans do! All this ends up in the water.

Additionally, industrial plants contribute over 500,000 different chemicals to our waterways. What can be done?

1. Recycle sewage water (Table 26–4).
2. Restrict the introduction of agricultural and industrial wastes to the water supply.
3. Conserve water, using only the minimum necessary.
4. Curb population growth.

The problem of solid wastes

We live in a throwaway age. Hundreds of products are disposable, and many items are packaged in no-deposit, no-return containers. In truth, however, matter—whether it be called disposable or throwaway—cannot be gotten rid of. We can move it, dump it, bury it, burn it, and transform it, but it can never be totally eliminated. Thus, much of our **solid waste**—rubbish, refuse, garbage—remains to litter the landscape.

As a human community grows, so does its waste output. Today's urban dweller discards 4–5 lb of solid waste daily—twice the amount his parents discarded 20 years ago. Currently, Americans produce 73.5 million tons of gar-

bage a year, and in New York City more than 5 million tons of garbage are collected annually. The bill for collection and disposal is billions of dollars each year, and it is easy to see why when we learn it costs more to dispose of the *New York Sunday Times* than to buy it.

The age-old attitudes toward disposal of solid wastes are proving unsatisfactory. Dumping garbage provides a breeding ground for disease-carrying rats, flies, and mosquitoes and a source of contamination of the water supply. Land for garbage dumping is limited, and because the dump-sites are far removed from the urban source, transportation costs are often quite high. Burying wastes requires large tracts of land and locations that are not directly leached into the waterways. Sanitary landfills—compacting garbage and covering this with gravel, soil, or clay—eliminate pollution and unsightliness, but it is estimated that by the year 2000 we shall have run out of landfill sites and the oceans will be too polluted to receive additional garbage. Only combustible garbage can be burned, and this leads to air pollution.

What can be done to improve the disposal of the mountains of solid waste we produce? Recycle it. The aluminum cans that now clutter the environment from sea to shining sea can

FIGURE 26-16 The growth of a bacterial culture.

be remelted, and the 6 million cars junked yearly can be "mined" and melted down. By providing something equivalent to a mining-depletion allowance to car manufacturers, salvage could be made as economical as refining the earth's metallic ores. Glass bottles can be used again and again; a large monetary deposit might supply an incentive for returning these to the store. We recycle only 20% of the 60 million tons of paper products used annually—why not more? The recycling of a ton of paper would not only cut down on litter but could avoid chopping down 17 trees. As Lamont Cole has pointed out:

"Why can't a product be kept off the market until a satisfactory method for its disposal is discovered? And shouldn't the price of a product reflect the cost to society and its ultimate disposal?"

Self-destruct cartons and bottles or Mount Trashmores (a ski and toboggan run south of Chicago made from a mountain of trash) scattered across the country hardly seem the perfect remedy for disposal of solid wastes. A more reasonable approach is to make the recycling of wastes profitable, thus balancing economic and ecologic forces.

26-3 LIMITS ON POPULATION GROWTH: STRATEGY FOR SURVIVAL

Two hundred years ago Thomas Malthus, an English clergyman, wrote *An Essay on the Principle of Population* (1798) in which he stated that a population which is unchecked increases in a geometric fashion, but the food supply for such a population increases only arithmetically. The consequences of overpopulation were, to use Malthus's words, "misery and vice" (starvation, disease, and war). These, he believed, would act as natural checks on population growth and would tend to restore the population to its optimal density. According

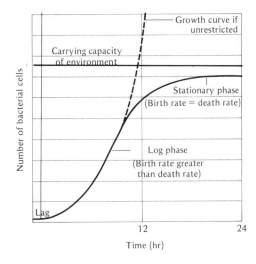

to the Malthusian theory, a natural population has an optimal density, and this density remains relatively constant for a long time.

We shall examine the validity of Malthus's theory for man, but before we do that, a few preliminaries are necessary. If we were to fill a jar with a nutrient solution and then add some bacteria, it would not be long before the nutrient solution became turbid from a large bacterial population. Closer inspection would show that growth began slowly, then became rapid, and finally slowed down (Figure 26–16). The bacterial growth curve is **S-shaped** (or **sigmoid**). In more descriptive terms the growth of the population shows a **lag phase,** then accelerates exponentially **(log phase),** and finally ceases and the population remains relatively fixed in size **(stationary phase).** The question is this: What determines what the population size will be?

Under ideal conditions the growth rate of a population is exponential, and the organisms reproduce themselves without impedance. Opposing this growth rate **(biotic potential)** and

keeping the population in check is the **environmental resistance,** the sum total of all the environmental factors that act to limit the population. Some of the limiting factors are space, temperature, food availability, oxygen supply, and so on. Stabilization of the size of the bacterial population probably occurs because of the depletion of food and oxygen, the accumulation of wastes, and other environmental changes that tend to bring the population into a state of dynamic equilibrium—when the death rate equals the birth rate. This equilibrium level is the carrying capacity of the environment, and it represents the maximum population size that can be supported by that particular environment. Although this part of the growth curve is shown as a smooth line, the population actually tends alternately to overshoot and undershoot this level. These fluctuations reflect changes in environmental resistance as well as differences in the responses of the organisms. If the environment changes (for example, by removal of toxic wastes or availability of new sources of food or new places to live), the carrying capacity will also be shifted.

We can visualize the Malthusian theory in terms of a feedback control system: each population has its own set point around which there may be size fluctuations. To maintain the set point there are a number of feedback controls—when environmental resistance increases, it curbs overpopulation; when it decreases, the population increases in size.

Is there self-regulation in the growth of the human population?

According to Malthus, there is. Malthus concluded that the population of Britain was about optimum in 1800; that is, it was in dynamic equilibrium and if it were to grow (he assumed a doubling time of 25 years), there would be

widespread misery in Britain. Malthus was severely criticized for his theory, and Dickens and Hazlitt excoriated him publicly. Indeed, during the 36 years that Malthus lived after publication of *An Essay on the Principle of Population,* both population and prosperity increased in Britain. As a matter of fact, population growth and affluence have been most evident for the last 150 years in the developed countries of the world. The human population has grown exponentially since the dawn of man, and there is no evidence of a deceleration in growth rate. Does the Malthusian theory apply to some organisms, but not to man?

If the data on the growth of the human population (Figure 26–17) are plotted on a logarithmic scale, it is possible to indicate both time and population size in some detail and to smooth out year-to-year fluctuations. The population growth curve shows not a single surge, but three surges and two set points: exponential growth occurred 600,000 years ago, corresponding to the cultural revolution; another surge took place 8,000 years ago, marking the onset of the agricultural revolution; and the last one began 200 years ago, establishing the beginning of the industrial revolution. Between the cultural and the agricultural revolutions and between the agricultural and the industrial revolutions, the human populations remained relatively stable and fluctuated around an equilibrium level. The reasons for this abrupt exponential growth of the human population during these revolutions were enhanced food production, opening up of new territories and continents, mechanization, and urbanization. Thus, with each improvement in man's ability to obtain food and space, there was established a new carrying capacity, and equilibrium occurred around a new set point. Note that no set point has been reached following the last surge—the curve is still rising. It should be noted also that the carrying capacity can rarely

FIGURE 26-17 A log-log plot of the human population over the last million years.

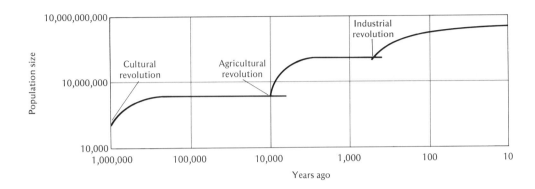

TABLE 26-5 Maximum world population potentially capable of being supported at present United States levels of consumption of existing resources.

Resource	Population (billions)
Maintaining or increasing the overall quality of life	0.5–2
Preserving the quality of our life support system	1–5
Heat buildup	10
Food	30
Oxygen	100
Space*	100
Energy*	100
Water*	100

*See J. A. Campbell, *Chemical Systems*, San Francisco, Freeman, 1970.

be predicted in advance, but is determined empirically by the present time and conditions.

Thus, although for some time Malthus was considered to be wrong and his critics right, in fact the reverse is true. The reason for the paradox was circumstance: Malthus lived during the time when the industrial revolution was in progress, and a new equilibrium level was being established. It is in the long view of human existence that Malthus's theory proves correct.

What is the equilibrium level for the human population and how can it be achieved?

What is the carrying capacity of our planet? No one is certain whether it is 3 billion or 100 billion; what is certain is that there is a limit. The limiting factor could be air, water, energy, food, heat buildup, space, or the accumulation of pollutants. J. A. Campbell has estimated the world population that could be supported on existing resources provided consumption remains at the level in effect at present in the United States (Table 26–5); these estimates assume that we shall employ new and more efficient methods for locating and using resources and that the usage will not disturb the balance in the biosphere. These are big assumptions indeed. For example, according to one estimate, the largest world population that could be supported in a fashion similar to that of the average United States citizen today is around 1 billion, and that is less than one fourth of the earth's present population. Today most humans, including more than 20 million Americans, live in impoverished and unhealthy conditions. There is yet another defect in these optimistic estimates: if all the world's people were to consume nonrenewable resources at the level of present-day Americans, the metal ore in the earth would be depleted in 24 hours!

We must establish spaceship rules for life on this planet. No longer can we live by what economist Kenneth Boulding calls cowboy or frontier rules, where everything is available in abundance and life is free and easy. One of the first of these spaceship rules must be a decision to establish a set point for our population, and then to curb population growth by governmental, religious, and individual action. Acceptable solutions for limiting population size will vary from person to person and country to country, but the goal for all must be the same: ZPG. We must lower the birth rate, for in this finite world a system involving positive feedback (such as reproduction) cannot be endured for long. Ultimately, negative feedback controls will emerge to achieve a measure of stability. The natural checks on the human population when it begins to exceed the carrying capacity of the planet will be widespread disease, starvation, warfare, social disorganization, and depletion of the life-support systems.

What can be done to achieve ZPG and maintain a high quality of life as soon as possible?[4] The following recommendations can be made:

1. Reduce the number of spaceship passengers by
   a. Having birth control methods freely applied and readily available
   b. Providing tax incentives to encourage smaller families
   c. Developing educational programs on family planning

[4] Another plan is propounded by P. R. Ehrlich and R. L. Harriman, *How To Be a Survivor*, New York, Ballantine Books, 1971.

d. Making adoptions easier
e. Encouraging late marriages

2. Reduce the level of consumption and recycle materials by
   a. Cutting down on use of power and nonrenewable resources
   b. Providing tax incentives for recycling and reducing pollution
   c. Improving quality and durability of products

The only question with relation to the future of world population is whether the inevitable decline in rate of growth will come about through a tremendous upsurge in the death rate or through a drastic fall in the birth rate. Over the long haul these two solutions represent our only choice. The former will mean unprecedented human misery; the latter, unprecedented human wisdom.

From *Man . . . An Endangered Species,* U.S. Department of the Interior, Conservation Yearbook No. 4, Washington, D.C., GPO, 1968.

As the theologian H. Cox has said: "Not to decide, is to decide."

SUMMARY

1. The human world population today is 4.2 billion and growing annually by 70 million. Because population is increasing geometrically, doubling time is currently only about 30 years.

2. Because of improvements in medical care and living standards, the death rate continues to decline drastically, but the birth rate remains constant. Only a decline in the birth rate or an unpleasant increase in the death rate can produce zero population growth (ZPG).

3. In predicting growth rates, population structure is critical. Size and sexual composition of the reproductive age group; average age of marriage; age at which death occurs; fertility patterns; and sociological, political, and religious attitudes are all important.

4. With improved agricultural technology and the opening up of industrial job markets, trends in population distribution have been toward concentration in urban centers and away from rural agriculture regions. Advantages of urban life may be offset by the problems of overcrowding, however.

5. People in overcrowded, underdeveloped nations generally suffer from marasmus (caloric and protein deficiency) and from kwashiorkor (protein deficiency), both of which hit hardest at the young.

6. Carrying capacity of an ecosystem depends on its primary productivity. Productivity of the earth can be increased by enhancing agricultural efficiency through the use of fertilizer, advanced technology, and the development of high-yield plant strains (the so-called green revolution). However, the degree of increase has limits. Suggestions for farming arid lands, irrigation through desalination, and farming the oceans all have major drawbacks. Furthermore, the more people we feed, the more there are to reproduce. The best solution to overpopulation is to curb population growth.

7. As resources are depleted, pollution increases. Air pollution from automobile exhaust and industrial discharge is a major problem in many areas of the world. Liquid waste, especially sewage, is increasingly difficult to treat and dispose of as its volume increases with the population. Some solid waste can be recycled, but more needs to be done in this area.

8. Thomas Malthus in 1798 pointed out that while food supply increases arithmetically, population increases geometrically. The sigmoid growth curve applied to the human population shows that there have been three phases of exponential growth corresponding to the cultural, agricultural, and industrial revolutions. In each case, new carrying capacities were established as food, water, and other environmental factors ceased to be limiting.

We are still in the log phase of the last growth period, and a new carrying capacity for the earth is yet to be established.

9. Once the Malthusian population limit is reached, natural checks on population growth are likely to include uncontrollable disease, starvation, warfare, social disruption and depletion of life-support systems. If this is to be avoided, we must implement measures to control population growth, to reduce levels of consumption, and to recycle materials.

KEY WORDS

doubling time
geometric or exponential growth
arithmetic growth
zero population growth (ZPG)
marasmus
kwashiorkor
carrying capacity
pollute
thermal pollution
air pollution
liquid waste
solid waste
sigmoid growth curve
lag phase
log phase
stationary phase
biotic potential
environmental resistance

TOPICS FOR REVIEW AND DISCUSSION

1. **What are the ecological consequences of a declining death rate and a rising birth rate?**
2. **Discuss why malnutrition is so prevalent in the world today.**
3. **How are pollution and population size related to one another?**

4. Describe some methods for curbing population growth.
5. Why is farming the sea considered to be a myth?
6. What is the Malthusian theory of population growth?
7. What is smog and what are its causes?
8. What is carrying capacity and how does it influence the size of a population? Relate your answer to the human population.
9. What are the difficulties inherent in predicting the population size of a particular country 20 years hence?
10. How can we prevent the depletion of the earth's natural resources?
11. Why is the Green Revolution a stopgap measure in controlling worldwide hunger?
12. Why do the developed countries of the world have a greater impact on the world's natural resources than do the underdeveloped countries?
13. Why is thermal pollution the ultimate pollutant?

EPILOGUE

When we first started down this path (and it has been somewhat long and circuitous), we were looking for an answer to the question: What is life? We have not answered this seemingly simple question altogether satisfactorily because the question is not simple at all. In attempting to find an adequate response to this query, we have explored much of what is known and have defined areas that remain for future investigation and discovery. The three sections of this book form a hierarchy of human biology. First we took an excursion into the microscopic world of the cell—its structure, function, and regulatory controls. Then we moved on to the larger world of the human organism—in health and sometimes in disease. Finally, we traveled to the broader realms of inheritance, population genetics, human evolution, and ecology. We glanced at some of the dismal consequences of overpopulation and took an overview of the effects of environmental tampering by human hands.

And now the path ends. What is the message of all these pages? What have we learned from these explorations into the meaning of life?

Clearly, there are solutions to many of our contemporary problems. Mankind is possessed of energy, imagination, reason, determination, and compassion: with these we can solve our most pressing difficulties, and our future can be bright. There is no reason why generations yet unborn cannot reap the rich rewards of a productive and meaningful life here on earth.

We travel together, passengers on a little space ship, dependent on its vulnerable supplies of air and soil...preserved from annihilation only by the care, the work, and I will say the love, we give our fragile craft.

ADLAI E. STEVENSON

APPENDIX

SIMPLE CHEMISTRY FOR BIOLOGY

ELEMENTS AND ATOMS

1. The universe consists of matter and energy.
2. **Matter** has mass and occupies space; that is, it has weight, volume, and density. **Energy** neither occupies space nor has weight; energy is measured by its capacity to do work.
3. Matter is composed of basic substances called **elements,** which cannot be decomposed into simpler constituents by chemical means.
4. At present 105 elements are known (new ones are continually being discovered), and each is designated by a chemical symbol of one or two letters. Thus, hydrogen is represented by H, oxygen by O, carbon by C, nitrogen by N, chlorine by Cl, and so on.
5. The smallest unit of an element, one that retains the chemical characteristics of the particular element, is called an **atom.**
6. An atom, the basic structure of matter, consists of a positively charged center, or **nucleus,** surrounded by a swirling cloud of negatively charged particles called **electrons:**

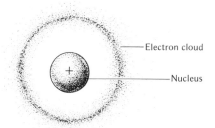

7. The nucleus of an atom contains **protons** that carry a charge of +1 and **neutrons** that have no electrical charge.[1] Both pro-

---

[1] Hydrogen, the lightest element, has no neutrons.

tons and neutrons have a mass of 1 (based on an arbitrary unit of atomic mass).

8. The **atomic number** of a chemical element represents the number of protons in the atomic nucleus; the **mass number** of an element is the sum of the masses of the protons and neutrons. Thus, the nucleus of a carbon atom and an oxygen atom would look as follows:

Carbon

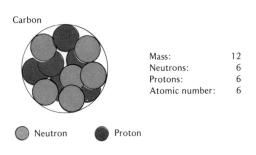

Mass:	12
Neutrons:	6
Protons:	6
Atomic number:	6

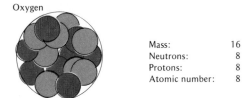

Neutron   Proton

Oxygen

Mass:	16
Neutrons:	8
Protons:	8
Atomic number:	8

9. **Isotopes** are atoms of the same element having the same chemical properties and the same atomic number, but differing in atomic mass. The difference in mass is due to varying numbers of neutrons in the nucleus, the number of protons remaining the same for all isotopes of a given element. Thus, carbon, with an atomic number of 6, has three isotopes with masses of 12, 13, and 14:

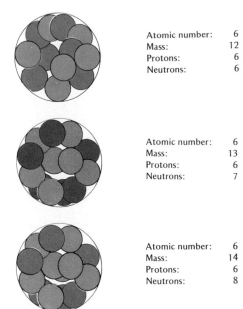

Atomic number:	6
Mass:	12
Protons:	6
Neutrons:	6

Atomic number:	6
Mass:	13
Protons:	6
Neutrons:	7

Atomic number:	6
Mass:	14
Protons:	6
Neutrons:	8

Oxygen, atomic number 8, has three isotopes with masses of 16, 17, and 20. Some isotopes (those designated as **radioactive**) are unstable and undergo nuclear breakdown, liberating radiations and smaller particles. The radioactive isotope of carbon, named carbon 14, is frequently used to follow the chemical reactions of carbon-containing substances in an organism; this and similar radioactive elements are called **tracers.**

10. The electrons surrounding the atomic nucleus have a charge of −1 and have virtually no mass (1/1850 the mass of a proton). Since the electron cloud has a negligible mass, it is the protons and the neutrons in the nucleus that give an atom almost all its weight. The electrons surrounding the positively charged nucleus occur in discrete shells or levels called

orbitals some distance removed from the nucleus. It is impossible to determine the exact location of a particular electron in its orbital; however, for convenience the electrons of an atom can be shown schematically as a planetary system with discrete orbits for the electrons. For example, an atom of hydrogen can be represented as:

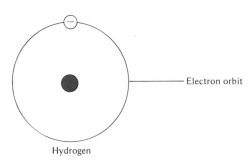

Electron orbit

Hydrogen

Carbon can be represented as:

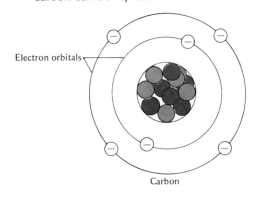

Electron orbitals

Carbon

11. In a neutral atom the number of negatively charged electrons that orbit the nucleus is balanced exactly by the same number of positively charged protons within the nucleus. A neutral atom has no net charge.

12. The chemical properties of an atom are generally determined by the number of electrons in the outermost orbital(s).

COMPOUNDS, CHEMICAL BONDS, AND pH

13. The maximum number of electrons each orbital can contain at any one time is 2 for the orbit closest to the nucleus and 8 for the second closest. (Although some atoms contain other orbitals that can hold more than 8 electrons, these atoms behave as if they were stable when the outermost orbital contains 8 electrons.)

14. An atom having the full complement of electrons in its outermost orbital is chemically unreactive or inert.

15. An atom having fewer than 8 electrons in the outermost orbital tends to gain or to lose electrons so that when this orbital becomes filled or emptied the atom attains a stable configuration. (The hydrogen atom, with only a single orbital, requires 1 additional electron to achieve stability.)

16. When the atoms of two or more elements combine chemically, the resulting product is called a **compound.** Thus, atoms of oxygen and hydrogen may react chemically to form the compound water. Note that the chemical properties of a compound are different from those of its constituent atoms.

17. The basic unit of a compound, one that retains the chemical and physical properties of that compound, is called a **molecule.** For example, two atoms of the element hydrogen and one atom of the element oxygen are chemically bonded together to form a molecule of the compound called water, and the combination of two atoms of oxygen forms molecular oxygen, a gas.

18. The attractive force between two or more atoms in a molecule is called a **chemical bond.** Each bond represents a certain amount of potential chemical energy.

19. An atom that gains an electron becomes negatively charged, whereas an atom that has more protons than electrons is posi-

tively charged. Such electrically charged atoms are called **ions.** Ionization may occur when an atom of one element donates one of its electrons to an atom of another element.

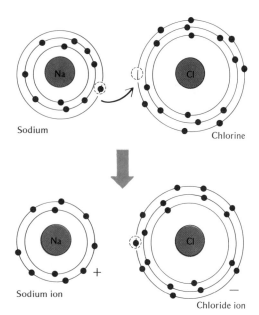

Sodium   Chlorine

Sodium ion

Chloride ion

20. If two ions have opposite electrical charges, they are electrostatically attracted to one another, and the resultant compound is held together by a chemical bond called an **ionic bond.** For example, the attraction of a positive sodium ion to a negative chloride ion produces sodium chloride, or table salt, by ionic bonding. Crystalline sodium chloride is an ionic compound since its constituents are ionized.

21. The ions of ionic compounds have a tendency to break their ionic bonds **(dissociate)** when placed in solution (usually water). This is because the water tends to keep the ions apart. Thus, when table salt, sodium chloride (NaCl) is dissolved in water, free

sodium (Na^+) and chloride (Cl^-) ions are formed.

22. Solutions containing ions can conduct an electrical current. Because of this property, ionic compounds are often called **electrolytes.** By contrast, substances such as table sugar (sucrose), which do not dissociate into ions when placed in water, are **nonelectrolytes.**

23. Acids and bases are important ionic compounds. An **acid** is a substance that dissociates to yield hydrogen ions (H^+), also called **protons** because they are hydrogens with only a bare proton and no electrons. A **base** is a substance that dissociates to yield hydroxyl ions (OH^-) or that can accept H^+ ions.

24. The acidic or basic (alkaline) nature of a solution can be measured and described on a scale called **pH.** The pH scale runs from 0 to 14. The midpoint or neutral value in the scale is 7, the pH of pure water. At pH 7 the number of H^+ and OH^- ions is equal. pH values greater than 7 are found in basic solutions; numbers less than 7 refer to acid solutions.

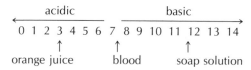

acidic   basic

0 1 2 3 4 5 6 7 8 9 10 11 12 13 14

orange juice   blood   soap solution

25. pH is defined as the negative logarithm of the molar concentration of H^+ ions in a 1-liter solution. A liter of pure water has a hydrogen ion concentration that is 1×10^{-7} molar (or 1/10,000,000), and when written as a negative logarithm gives a pH value of 7. A 0.1 molar solution of hydrochloric acid (about 3.6 grams of HCl per liter) would dissociate to give a H^+ ion concentration of 1×10^{-1} molar, and its pH would be 1 (highly acidic). This is a millionfold increase over the hydrogen ion concentra-

tion of pure water (pH 7). Since the pH scale is logarithmic, a change of 1 pH unit involves a 10-fold change in H^+ concentration.

26. Ionic bonds involve a complete transfer of electrons from one atom to another. However, in some cases chemical bonding involves electron sharing rather than transfer. This is called a **covalent bond.**

27. Carbon combines with the atoms of many other elements by forming covalent bonds; water is made up of two atoms of hydrogen covalently bonded to one oxygen atom. Hydrogen gas is an example of a molecule held together by a covalent bond:

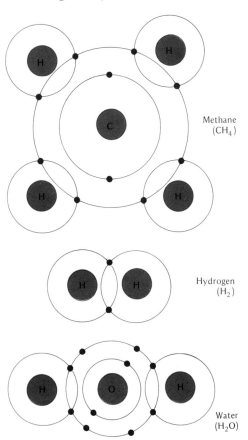

Methane
(CH_4)

Hydrogen
(H_2)

Water
(H_2O)

28. In addition to covalent and ionic bonds, there is another weak chemical bond, the **hydrogen bond.** Hydrogen bonding occurs where hydrogen atoms in molecules are electrically attracted to unbonded electrons of nitrogen and oxygen—the hydrogen atom acts as a bridge between two other atoms. Water molecules are bonded together by hydrogen bonding, and the three-dimensional structure of proteins and nucleic acids is maintained by hydrogen bonding.

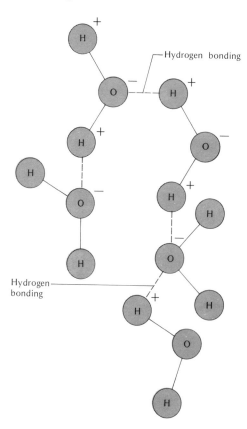

Hydrogen bonding

Hydrogen bonding

WATER

29. Life as we know it is impossible without water. All cells contain water, and it is only in a watery environment that cells can function. A water molecule is composed of two hydrogen atoms covalently bonded to one oxygen atom. The water molecule is polarized; that is, the molecule has a positively charged end and a negatively charged end. The polarity is a consequence of the nucleus of the oxygen atom attracting more electrons than do the hydrogens, and thus the oxygen end tends to be more electronegative. Also, the geometric configuration of the molecule is such that both hydrogens are placed at one end, creating two positively charged regions and a single negatively charged region:

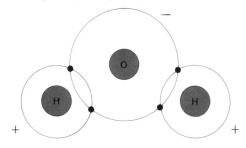

30. Water is a universal solvent, dissolving many substances, and thus permitting chemical reactions to take place.

31. Water has a tendency to favor ionization of substances in solution.

32. Water plays an active role in certain chemical reactions, for example, hydrolysis.

33. Water maintains temperature constancy, permitting an organism to resist sudden changes in temperature. This is accomplished because water has a high heat capacity; that is, it can lose and gain large amounts of heat with little change in its own temperature.

B.C.   **by Johnny Hart**

© Field Enterprises Inc. 1972. Reprinted by permission.

34. Water is more dense at 4° C than at 0° C; for this reason, ice floats. This also means that a body of water such as a lake freezes from the top down and not from the bottom up, allowing organisms to be protected in deeper waters.

CARBON

35. Although water is important in living systems, the chemistry of life is centered on the chemical properties of the carbon atom.

36. Carbon, atomic number 6, has 6 protons, 6 neutrons, and 6 electrons. Since the outermost orbital of carbon contains 4 electrons, it must borrow, donate, or share 4 more electrons to become energetically stable.

37. The 4 electrons in the outer orbital of the carbon atom which can engage in chemical bonding can be represented:

38. Carbon atoms can share electrons with hydrogen, nitrogen, oxygen, and other carbon atoms to form stable compounds. The bonding in such compounds is covalent.

$$\cdot \overset{\cdot}{\underset{\cdot}{C}} : \overset{\cdot}{\underset{\cdot}{C}} \cdot \qquad \text{or} \qquad -\overset{|}{\underset{|}{C}} - \overset{|}{\underset{|}{C}} -$$

39. The covalent sharing of a pair of electrons between two carbon atoms—the most common kind of bonding in organic compounds—is called a **single bond.** However, carbon atoms may also share two or three pairs of electrons to form **double** or **triple bonds.** For example, a molecule of ethane has a carbon-carbon single bond, ethylene has a carbon-carbon double bond, and acetylene has a triple bond between carbons:

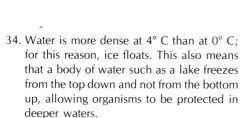

acetylene

Oxygen, with 6 electrons in its outer shell, needs only 2 more electrons to become energetically stable. If two atoms of oxygen share two pairs of electrons with a single carbon atom, a stable gas, carbon dioxide, results; it contains two carbon-oxygen double bonds:

$$\overset{\cdot\cdot}{\underset{\cdot\cdot}{O}} : : \overset{\cdot\cdot}{\underset{\cdot\cdot}{C}} : : \overset{\cdot\cdot}{\underset{\cdot\cdot}{O}} \qquad \text{or} \qquad O = C = O$$

double bonds

40. Carbon-carbon bonding can produce molecules that may be long or short, branched or unbranched, straight or circular. The variety of carbon compounds is virtually endless.

CHEMICAL REACTIONS

41. In principle, all chemical reactions are reversible. That is:

$$A + B \rightleftharpoons C + D$$

42. All chemical reactions involve changes in energy content.

43. Atoms or molecules interact during a chemical reaction in such a way that electron sharing or exchange takes place.

44. The probability of a chemical reaction occurring is increased when the likelihood of electron-electron interactions is increased.

45. Three ways of increasing the likelihood of two substances interacting and thus enhancing the rate of a chemical reaction are:

a. Increase the concentration of the reactants (A + B) or decrease the concentration of the products (C + D)
b. Increase the temperature or pressure
c. Arrange the molecules or atoms so that collisions are likely to occur; that is, subdivide or dissolve the substances or provide a large surface area.

46. Chemical reactions in which the products contain less energy than the reacting materials release energy (usually in the form of heat) and are called **exergonic reactions.** Exergonic reactions are downhill reactions:

$$A + B \longrightarrow C + D + energy$$

47. Reactions in which the products contain more energy than the reacting materials absorb energy and are called **endergonic reactions.** Endergonic reactions are uphill reactions:

$$energy + A + B \longrightarrow C + D$$
$$or$$
$$A + B \longleftarrow C + D + energy$$

48. In the living system endergonic and exergonic reactions are coupled to one another in such a way that the energy released from an exergonic reaction is used to drive the endergonic one.

49. If the chemical reaction of substances A + B to give products C + D is left to itself, chemical equilibrium will result; that is:

$$A + B \rightleftharpoons C + D$$

At equilibrium the concentrations need not be the same for reactants and products, but the rates of forward and reverse reactions must be the same.

50. A living system cannot exist in a state of chemical equilibrium. In the chemistry of the cell there must be a directionality to the flow of carbon compounds and/or energy. In the reaction of substances A + B to yield the products C + D, the direction will be A + B → C + D and not the reverse if:

a. The concentration of products (C + D) is less than that of the equilibrium concentrations.
b. The products C + D are removed or used in subsequent reactions
c. Energy is released, thus giving C + D less energy content than A or B

51. Whether a reaction will occur (even if it is exergonic, or "downhill") depends on whether the reactants receive the proper amount of activation energy. The activation energy is the "push" necessary to start the downhill reaction, and its magnitude varies for different reactants.

52. A driving force for chemical reactions can be provided only by a transition from a highly organized state of matter to one that is less organized (unless energy is supplied).

53. Since living organisms obey the same chemical and physical laws as nonliving systems, understanding the mechanisms of chemical reactions enables us to comprehend the way in which living systems—we—work.

MEASURING UP OR OUR SYSTEM IS DYING INCH BY CENTIMETER

"I'd walk 1.67 kilometers for a Camel."

© CAMEL

"A miss is as good as a kilometer"? Or would you believe: "A gram of prevention is worth a kilogram of cure"? So may speak the next generation of Americans.

The United States is moving toward the conversion of its present system of measurements to the metric system, long in use in all the major countries of the world except those in the British Commonwealth. And the British have already begun walking the last kilometer.

The basic reason for the changeover (whose cost should run into the millions) would be to improve the American trading position. At present a large percentage of American appliances and other products are not marketed abroad because they are unsuited to the international scheme of things. Repairs, for example, cannot be made locally because screw threads are different or other standard components do not fit.

How deep will the conversion go? Will it penetrate our literature and songs? Many would be aghast at the thought of converting, "I love you, a bushel and a peck" to liters and hectoliters. Will angry men cry: "I'll beat you to within a centimeter of your life?" Or, "Give him a millimeter and he'll take a kilometer?"

The National Bureau of Standards firmly says no. "The rule of reason" will apply, although sports fans may have to make an adjustment, for example, when they hear that the Redskins are on the 30-meter line with 3 meters to go.

But the drift to conversion seems inexorable despite the opposition from some industries concerned about the resultant confusion—and the cost.

The American chemical and pharmaceutical industries are now largely on the metric system. So are such Government agencies as the National Aeronautics and Space Administration and important parts of the weapons and aircraft industries. Last month the General Electric Corporation, whose products for foreign consumption are already made to metric dimensions, came out in favor of conversion, but only in a carefully timed manner in which each industry would follow a pace suited to its needs and problems. The trouble with too slow a changeover, however, is that it prolongs the period of confusion. Conversion, in that sense, is like pulling teeth. It's an awful experience, but the sooner it is done, the better.

It used to be said, "A pint's a pound the world around"—that is, a pint of water weighs one pound everywhere. Today it would be more accurate to say: "Wherever in the world I am, a cubic centimeter is a gram."

WALTER SULLIVAN.*The New York Times*, May 14, 1972.

LENGTH

Metric to English
1 km (kilometer) = 0.62 mi (mile)
1 m (meter)   = 3.28 ft (feet)
1 cm (centimeter) = 0.394 in. (inch)
1 μm (micron)   = 1/25,400 in.

English to Metric
1 mi (mile) = 1.609 km (kilometer)
1 yd (yard) = 0.9144 m (meter)
1 ft (foot) = 30.48 cm (centimeters)
1 in. (inch) = 2.54 cm = 25.4 mm
(millimeters)

Metric system
1000 meters = 1 kilometer (km)
0.01 meters = 1 centimeter (cm)
0.001 meters = 1 millimeter (mm)
0.000001 meters = 1 micron (μm)
0.000000001 meters = 1 nanometer or millimicron (nm or m μ)
0.0000000001 meters = 1 Angstrom (Å)

WEIGHT

Metric to English
1 kg (kilogram) = 2.2 lb (pounds)
1 gm (gram)   = 0.035 oz. (ounce)

English to Metric
1 oz (ounce) = 28.35 gm (grams)
1 lb (pound) = 16 oz (ounce)
             = 0.4536 kg (kilogram)

Metric system
1000 gm = 1 kilogram (kg)
0.001 gm = 1 milligram (mg)
0.000001 gm = 1 microgram (μg)

VOLUME

Metric to English
1 liter = 0.2643 gallon
1 ml (milliliter) = 0.03 oz (ounce)

English to Metric
1 gal (gallon) = 3.7854 liters
1 qt (quart)   = 0.946 liter = 946 ml (milliliters)
1 pt (pint)   = 0.473 liter = 473 ml (milliliters)
1 oz (ounce) = 0.029 liter = 29 ml (milliliters)

Metric system
1 liter = 1,000 ml (milliliters)
0.001 liter = 1 ml (milliliter)
0.000001 liter = 1 μl (microliter)

TEMPERATURE

Metric to English
Degrees Fahrenheit (°F) = (°C × 9/5) + 32
Note: 9/5 = 1.8

English to Metric:
Degree Celsius (°C) = (°F − 32) × 5/9
Note: 5/9 = 0.5556

INDEX

Page numbers in boldface type refer to definitions of words; these words are also in boldface type on that page of the book. Page numbers in italic type refer to pages where a figure and/or table appears.

A

A band, **429**, *430*
Abcess, dental, 231
Acetylene, 587
Abdominal striae, 187
Abductor muscle, **425**
Abiotic synthesis, **18,** 19
Abortion **216,** 336
Absorption, intestinal, **224,** 227, 228
Accommodation, **359, 380**
Acetabularia, 44–45
Acetone, 330
Acetyl coenzyme A, **90**
Acetylcholine. *See* ACh
Acetyl group, **90**
Acetylcholinesterase, 357–359
ACh, 357, 358, 365
    accumulation, 435
    mimics, 366, *368–369*
    in muscle contraction, 430
Acid, **585**
Acidity. *See* pH
Acidosis, 330
Acquired immunity, 265, **271,** 273
Acromegaly, **323**
ACTH, **321,** 333
Actin, **428,** 429
Action potential, **352–360**
    frequency, 355
    initiation, *360*
    muscle, 431
    trigger, 357, 358
Activation energy, *72*
Active acquired immunity, **271,** 273
Active site, **73**
Active transport, **32–34,** 93
    in digestion, 228
    of glucose, 330
    in kidney, 307, *308*
    sodium, 310
Actomyosin, **428,** *429*
Adam's apple, **284,** 286
Adaptation
    behavioral, 389, 409
    and conditioning, 404
    in early man, 517

and instinct, 398
    sensory, 402–403
Adaptive radiation, **510,** *511*
Addison, Thomas, 332
Addison's disease, 333, 334
Adductor muscle, **425**
Adenine, 14, **109,** *110,* 111, *114, 115,* 116
Adenohypophysis, 178, **321**
Adenosine diphosphate, **85**
Adenosine triphosphate. *See* ATP
Adenyl cyclase, 340
ADH, **310,** 311. *See also* Vasopressin
Adipose tissue, 228
Adiposogenital dystrophy, 338–339
Adolescent
    female, 178
    male, 170
ADP, **85**
Adrenal cortex, 310, **331,** 332–335
Adrenal glands, 187, *331–333,* 336
Adrenal hormone, 365. *See also* Norepinephrine
Adrenal medulla, **331–**332, 365, 442
Adrenal tumor, 335
Adrenaline. *See* Epinephrine
Adrenergic neurons, **365**
Adrenocorticosteroids, 333
Adrenocorticotrophic hormone. *See* ACTH
Adsorption, 14
Adsorption site, 72, *73*
Aerobic respiration, **86,** *87,* 89–92
Afterbirth, 207, *209*
Age
    and Down's syndrome, 476
    population distribution, 559
    and weight, 63
Agglutinins, **274**
Agglutinogen, **465,** 466, 468
Aggregation, **408**
Aggression, 398, 399
    and alcohol, 367
    inhibition of, 412–413
    and instinct, 398, 399
    in man, 399
    reduction of, 409, 412
    ritual, 413
Aging
    muscle, 429
    and sunburn, 156
Agranulocytes, **262**
Agricultural revolution, 548, 574
Agriculture
    California, 564

development of, 559
    irrigation, 563
    and unbalanced ecosystems, 547–549
Air
    alveolar, *292*
    composition of, 282, *292*
    inspired, *292*
    expired, *292*
    pollution, 154, 288, **566–**570, *567*
    pressure, 296, 297
    respiratory passageway, 284
Albinism, **376,** 459
    inheritance of, *458*
    pedigree, *455*
Albino, *454,* **455**
Albumin, **260,** 261
    and pinocytosis, 34
Alcohol, **367,** *368, 369*
    absorption, 224
    diuretic effect, 311
Alcoholism, treatment, 388
Aldosterone, **310,** 311, 333
Alfonso XII, 466
Algae, classification, 48
Alimentary canal, **221.** *See also* Digestive system
Allantoic arteries, **206**
Allantoic veins, **206**
Allele(s), **455**
    blood type, 468
    multiple, 463–468
Allergens, **276**
Allergy, 276
Allison, A. C., 493
All or none law, **354,** *355*
Alpha antibodies, 465, 468
Alpha globulin, **260,** 261
Alpha helix, **68**
Alpha waves, *356,* **357**
Altitude, 296, *297*
Alton giant, 322
Altruism, 396
Altruistic behavior, 396
Alveolus (alveoli), **284,** 285, 287, 289, 290
Amazon ecology, 545
Ameba. See also *Amoeba*
    movement, 28
    nuclear transplants, 43, *44, 45*
Ameboid cells, 254. *See also* White blood cell
Amenorrhea, **187**
Ames, Bruce, 157
Ames test, 157
Amino acid(s), 13, **66,** 67–70

Amino acids(s) (continued)
  activated, 118, 119
  combinations, 67
  essential, **113**
  laboratory synthesis, 13, 14
  membrane penetration, 30
  R-group, **66**
  stereoisomers of, 61
  uses, 228
Aminopterin, 75–76
Ammonia
  breathing, 295
  elimination, 302
  in nitrogen cycle, 540
Amnesia, 406, 408
Amniocentesis, *448,* **450,** 474, 483, 495
Amnion, **206,** 208
Amniotic cavity, and amniocentesis, 450
Amniotic fluid, 208
  and prostaglandins, 336
Amniotic sac, *194, 208*
Amoeba, 238
Amphetamines, 366, *368–369*
Amplitude, sound, **381,** 383
Amygdala, 348, *363,* 388
Amylase, 74
Anaerobes, **86**
Anaerobic respiration, **86**–89
Analog, **74**
Anal sphincter, **229**
Anaphase, *130,* **133,** 137
Anaphylactic reaction, **276**
Ancestor
  ape-like, *512, 513*
  common, 513
Androgens, **170,** 172
  climacteric, 335
  in women, 336
Androstenedione, **333**
Androsterone, **333**
Anectine, 435
Anemia, 233, 267
  chalky feces, 233
  sickle cell, 482–486, *494–496*
Anesthetics, *368–369*
Animal, **48**
Animal cell, *24, 135*
Animal classification, 499, 500, 501
Animalia, 48
Ankle joint, *426*
Annuals, 546
ANS, **364,** 365

Antagonists, muscle, **425**
Ant behavior, 396
Anterior chamber, eye, **377**
Anterior pituitary, 178, **321**
Anthropoids, *502, 503, 507, 508*
Anthropomorphism, **391**
Antibody, **265,** 268, 273–277
  alpha, 465, 468
  anaphylactic, 276
  beta, 465, 468
  blood, 465, 468
  circulating, 260
  formation, *271*
  and immunization, 272–273
  -mediated immunity, **269**–273
  molecular weight, 273
  polyoma virus, **159**
  production, 269–273, *271*
  receptors, **268**
  response, 273
  skin-sensitizing, 276
  and viruses, 274
Anticlotting factors, **264**
Anticodon, **118,** 119
Antidiuretic hormone. *See* ADH
Antigen(s), **265,** 268–270
  -antibody combination, *274*
  blood, 465, 468
  booster, 273
  recognition, 273
  tissue transplant, 275
  tolerance-inducing, 275
  tuberculin, 269
Antigenic determinant, **273,** 274
Antihistamines, **276**
Anti-inflammatory agents, 39
Antilymphocyte serum, 275
Antimetabolite, **74**
Antithrombin, **264**
Antitoxin, 273, **274**
Antivenin, **274**
Anus, 229
Anvil, ear, **382**
Aorta, **239**
Aortic valve, **241**
Apelike ancestor, *512, 513*
Apnea, **295**
Apollo 13, 553
Appeasement behavior, **412–413**
Appendicitis, **233**
Appendicular skeleton, **418**
Appendix, 225, **233**

Aqueous humor, **377**
Ardrey, Robert, 398
Areola, **211**
Aristotle, 48, 499
Arithmetic growth, **554,** *555*
Arm buds, *199*
Arsenic and cancer, 154
Arteriole, dilation, 276
Artery, **239,** *241*
  allantoic, **206**
  atherosclerotic, *65*
  brain, 264
  blockage, *248*
  carotid receptor, 295
  clogged, 264
  coronary, **249,** 264
  plaque, 250
  pulmonary, 264
  renal, **303**
  umbilical, 207
Arthritis, **426,** *427*
Artificial heart valve, *243*
Artificial insemination, 176
Artificial kidney, 300, *312–313*
Asbestos
  and cancer, 154
  and environment, 533
Asexual reproduction. *See* Mitosis
Asphyxia, 286, 287, 435
Aspirin, absorption, 224
Association area, **405**
Asters, 132
Asthma, 276, 286
  and prostaglandins, 336
Astigmatism, **380**
Astral rays, 132
Aswan high dam, 550
Atabrine, 309
Atherosclerosis, 64, *65,* *248,* **249**
  and cholesterol, 64, 66
  in diabetes, 330
  in women, 336
Atmosphere
  CO_2 regulation, 544, 545
  composition, **282**
  early earth, 10, 12, 13
  oxidizing, **12,** 18, 19
  pollution of, 154
  pressure, 296, 297
  reducing, 12, 18, 84
  temperature regulation, 544
Atom, **583,** 584

Atomic bomb, 537
Atomic energy, 490
Atomic number, **584**
Atomic waste, 537, 549
ATP, 84–**85**
　in aerobic respiration, 91–92
　in anaerobic respiration, 89
　calories in, 89
　and cyclic AMP, 340
　in excretion, 307
　in Krebs cycle, 91
　and muscle contraction, 429, 430, 432
　and sodium pump, 353
　structure, *85*
　uses, 93
ATP-ase, **429**
Atrioventricular node, **243**
Atrium
　contraction, **241,** 242. *See also* Systole
　electrical activity, 245
　fibrillation, 243
Atropine, 435
A-type blood, 465, 468
Augmentation, 309
Auricle, **381**
Australopithecus, 399, *498,* 514–518, *521*
Autoimmune disorder, **268,** 327
Autonomic nervous system. *See* ANS
Autonomous growth
Autosomes, *451,* **453**
　abnormal number, 473–474
Autotrophs, 17, **18**–20, 48, **529**
AV bundle, **243**
Avery, O. T., 108
Axial skeleton, **418**
Axon, **349,** *350,* 352–360
Axoplasmic flow, **336**
Azathioprine, 215

Ⓑ

Backbone, 361. *See also* Spine
Bacteria
　antigenic properties, 265
　chromosome, 136
　classification, 48
　and decay organisms, 538
　denitrifying, **540**
　and disease, 9
　division, 135
　and DNA, 136, *137*
　dysentery, 234

　and genetic engineering, 124–125
　growth curve, *573*
　infection, 47
　lysis, **53**
　nitrifying, 540
　nitrogen fixing, 540
　penicillin resistant, 485, 486
　phagocytosis, 39
　soil, 540
　tetanus, 86
　tooth decay, 231
Bacteriophage, **53.** *See also* Virus
　information capacity, 116
　life cycle, *52*
Balance, sense of, 382, 384–385
Baldness in women, 336
Ball and ring valve, 241
Ball and socket joint, *426*
Balsa wood, 47
Barnard, Christiaan N., 275
Barnum, P. T., 322
Barr, Murray, 462
Barr body, 462, **463,** *465,* 473
Bartholin's glands, **182**
Basal metabolic rate, **229.** *See also* Metabolic rate
Base, **585**
Base-pairing, **111,** 117, 119
Base-pair substitution, 161
Basophils, *263*
Bateson, W.R., 460
Battery, *84*
B cells, **268,** 269, 270
Beadle, G. W., 112, 121
Beagle, H.M.S., 499, *500, 509*
Beatson, G., 158
Beaumont, William, 222
Becquerel, Henri, 489
Bee
　behavior, 396
　dance, 411
　drone, 410
　genetics, 397
　nurse, 411
　queen, 391, 410
　social life, 410–411
　venom, 269
　worker, 410
Beer, diuretic effect, 311
Behavior, 388–415, **389**
　aggressive, 398–399
　altruistic, 396
　appeasement, **412**–413

　birds, 396
　children, 404
　classical conditioning, **403**
　complexity, 390
　conflicting, 401
　and consciousness, 391
　control, 348, 388
　courtship, **412**–413
　development, 390, 399
　disturbed, 413
　dogs, 404
　and drugs, 366, *368–369*
　evolution of, 390, *392,* 396, 516
　and future of man, 413–414
　genetic basis, 393, 396, 398
　herring gull chick, 401–402
　imprinting, **405**
　infantile, 412
　instinctive, 398–402
　learned, 393, 402–408
　manipulation, 413
　mating, 408, 412–413
　monkey, 399
　operant conditioning, **404**
　pigeons, 395
　plant, 390
　primate, 399
　problem solving, 404–405
　psychohydraulic model, *401*
　purpose, 391
　rats, 395
　reasoning, **405**
　reproductive, 398
　ritual, 412–**413**
　social, 408–414
　stereotyped, 390, 392, 393
　technology, 394–395
　and temperature control, 440–443
　territorial, **409**–413
　unlearned, 393
Behaviorists, 393, 394–395
Bel, **383**
Bell, Alexander Graham, 383
Bends, 298
Benign tumors, **152**
Benzedrine, 366, *368–369*
3:4 Benzypyrene, 154
Berger, Hans, 356
Beriberi, 74
Beta antibodies, 465, 468
Beta globulin, **260,** 261
Beta waves, *356,* **357**

Biceps, *425*
Bicuspid valve, **241**
Bile, 233, 262
Bilirubin, **233,** 262
Biliverdin, **233**
Biology, organization of, 528
Biomass, 545, 547
  pyramid of, **532,** 533
Biome, **545,** *548*
Bioplasts, 35
Biosphere, **528,** *529*
Biotic potential, **573**
Bipedalism, 514–517
Birds
  behavior, 396, 405
  courtship behavior, 412–413
  habituation in, 402
  immune system, *268*
  imprinting, 405
  nest-building instinct, 398
  ritual behavior, 413
  song, 409
  territoriality, 400–401, 409
Birth, 208–210
  canal, *208*
  date calculation, 187
  defects, 213–215, 217
  and lactation, 211
  multiple, 181
  rate, 555–558, *560,* 575
Biston betularia, 491, *492*
Bladder, 172, 181, 311, 312, *314*
  cancer, 154, 161
Blastocyst, *197*
Blastoderm, *197*
Blind spot, **378,** *379*
Blood, *51,* 258–264
  analysis, 266–267
  antibodies, 465, 468
  antigen, 465, 468
  calcium levels, 327–329, *328*
  cells. *See* Red blood cell; White blood cell
  cholesterol levels, 65
  clot, 189, 260, 262–264. *See also* Thrombus;
    Embolus
  composition, *260, 302*
  count, *267*
  dialysis, *312–***313**
  donor, **465,** 468
  filtration, 303–309
  flow rate, 250, *251*
  fluke, 550

formed elements, **260**
  -forming tissue, *51*
  functions, 259
  glucose levels, 39, 59, 330, 331
  group, 502, *522*
  menstrual, 181
  nitrogen bubbles, 298
  nutrients, 260
  osmotic pressure, 252, 253
  oxygen-carrying capacity, 289
  pH regulation, 309
  plasma. *See* Plasma
  platelets, *51,* 261–264
  pressure, 252
  quantity, 259
  recipient, *465,* 468
  salts, 260
  separation, 260
  sugar level, 228
  transfusion, 463–465, 468
  types, 463–465, *468,* 506
  vessels, 364, 419. *See also* Artery; Vein
  viscosity, 261
  volume, 253
Blood pressure, 249–252
  in glomerulus, 311
  and heartbeat, 244
  measurement, *250*
  and prostaglandins, 336
Blue baby, 210
Blue-green algae, 48
Bokanovsky's process, 168
Bomb calorimeter, **89**
Bonaparte, Napoleon, *338–339*
Bond(s)
  chemical, **585**–587
  covalent, **586**
  double, **587**
  energy, 83
  high energy, **85**
  hydrogen, **586**
  ionic, **585,** 586
  single, **587**
  triple, **587**
Bone(s), *50,* 416–422
  calcium, 328–330
  cells. *See* Osteocyte
  collagen, **418,** 420
  fetal, 205
  fossil, 420
  growth, 420–*422*
  growth zone, **421**

marrow, 261, 268
  matrix, **419**
  metabolism, 327–330
  penis, 174
  salts, 419
  of skeleton, *418*
  structure, 419–*420*
Booster, 273
Borlaug, Norman E., 104
Botulin toxin, 435
Boulding, Kenneth, 575
Bowels, **229**
Bowman's capsule, **303,** *305,* 306, 307
Brain, *51,* 347, 349, 356–357, **360,** 361. *See also*
    CNS; *specific region in question*
  and alcohol, 367
  artery, 264
  cortex. *See* Cerebral cortex
  damage, 357
  embolism, 189
  embryonic, *199*
  electrical stimulation. *See* ESB
  electrode implantation, 34
  evolution, *517*
  fetal, 205
  human, 398
  hypothalamus, 310
  and learning, 405–408
  limbic system, *362*
  neural activity, 358
  pleasure center, 348
  reflex arc, 366
  relay centers, 360
  size, *517*
  stroke, 264
  structure, *362, 363*
  ventricles, 360
  stem. *See* Medulla
  waves, **356,** 357
Braun, A., 161
Bread baking, 88
Breaking of the waters, 208
Breast(s), 177, 211–213, *212,* 506
Breastbone, 417
Breast cancer, 154, 156, 158, 161
Breathing, 290–296, **281.** *See also* Respiration
  control of, 291–296
  costal, **291**
  cycle frequency, 295
  diaphragmatic, **291**
  inhibition, 295
  movements, 290, *291, 292*

Breathing (continued)
  nervous control of, 292–295
  rate, 295
  reflex, 293
  underwater, 296–297
Britain, population structure, 557, *558*
Broad ligament, **178**
Bronchi, **284**
Bronchioles, **284,** 286
  collapse, 288
  constriction, 276
Bronchitis, 286, 288
  and air pollution, 566
  and smoking, 155
Brood-food glands, 410
Brow, human, 506
Brown, Robert, 42
B-type blood, 465, 468
Buerger's disease, 155
Bulbourethral gland, **172**
Burkitt's lymphoma, 158
Bursa of Fabricius, *268*
Bursa joint, **426**
Bursitis, **426**
Butter yellow, 154

C

Cabrillo, Juan Rodriques, 570
Cadmium, 533
Caesar, Julius, 357
Caffeine, 311, 366, *368–369*
Caisson disease, 298
Calciferol, 328–**329**
  in Neanderthal man, 519
  and skin color, 523
Calcitonin, **324, 327,** 330
Calcium
  and blood clotting, 263
  bone salts, 419–421
  diet, 77
  ions, 431
  metabolism, *327,* 328–330, 523
  in muscle contraction, 431
  oxalate, 312
  phosphate, 312, 419
  regulation, *328–330*
Calculi
  dental, 231
  urinary, 312
Calf muscle, 432
Calico cats, 463

Calorie(s), 83
  annual solar energy, 103
  in ATP, 89
  deprivation, 561, 562
  food, 229
  in glucose, 89
  work requirement, 230
Camera, *376*
cAMP, **340,** 341
Campbell, J. A., 575
Canada lynx, 544
Cancer, 150–163, **152.** *See also* Tumor
  and air pollution, 566
  and analog inhibitors, 75–76
  and atmospheric pollution, 154
  benign, 152
  bladder, 154, 161
  breast, 154, 156, 158, 161
  causes, 159–162
  cervical, 181
  and cigarettes, 154–155
  and DDT, 536
  death rate, 151
  dietary, 154
  and folic acid, 75–76
  gene, 161
  and lymphatic system, 253–255
  lung, 154–155
  metastatic, 154, 255
  mortality rate, *156*
  mouth, 154
  and mutations, **160,** 161
  occupational, 153–154
  prostate, 161
  and radiation, 491
  and recombinant DNA, 122
  screen, 157
  scrotal, 153–154
  and selective gene activation, 161–162
  skin, 155–156
  transmission of, 158
  uterine, 154, 181
  and viruses, 53, 157–160, 162
Cannabis sativa, 367
Capillary, **239,** *241, 253*
  alveolar, 285, 287
  exchange across, 250, 252–*253*
  glomerulus, 305
  hydrostatic pressure, 252
  nephron, 303
Caponized fowl, 154
Carbohydrates, **58–**62

  as antigens, 265
  digestion, 224, *226*
  energy, 230
  fixation, 37. *See also* Photosynthesis
  as fuel, 84
  production in photosynthesis, 102
Carbon
  atom, 583, 584, 587
  cycle, 538, *539*
  in primitive atmosphere, 10
Carbon dioxide
  atmospheric regulation, 544, 545
  atmospheric screen, 18
  blood level, 290, 295
  exchange, 262, 287
  in respiration, 87
  and respiratory rate, 295
  solubility, 290
  structure, 587
Carbon monoxide, 567
Carboxylase, **88**
Carboxyl group, 67
Carcinogens, **153–**158
  Ames test for, 157
  chemical, 153–155
  dietary, 154
  environmental, 153–155
  hormonal, 154
  occupational, 153–154
  physical, 155–157
  x-rays, 156–157
Cardiac constrictor muscle, 231, 225
Cardiac cycle, 245
Cardiac muscle, 51, **422,** *424*
Caries, **231**
Carnivores, *531, 533*
Carotenoids, 99
Carrier
  albinism, **458**
  color blindness, 462
  Down's syndrome, **478**
  genetic, **458**
  for glucose, 341
  hemophilia, 466
  molecules, 33
  proteins, **307**
Carrying capacity, **563,** 573–575
Carson, Rachel, 537
Cartilage, *50,* 418–422, *420*
  composition, 46
  hyaline, **420,** 421
  joint, 426

Castration, 317, **333,** 335
Catalyst, **72**–75. *See also* Enzymes
Catheter, 314
Caton, Richard, 356
Cavities, tooth, 231
Cecum, 225
Cell(s), 23. *See also specific cell type in question*
  anaphase, *133*
  animal, *24, 135*
  blood. *See* Red blood cell; White blood cell
  bone. *See* Osteocyte
  bone marrow, 261
  cancer, *150*–153
  cleavage, **195,** *196*
  contact inhibition, 159, **160**
  and cyclic AMP, 340
  differentiation, 136
  diploid, **169**
  drinking. *See* Pinocytosis
  eating. *See* Phagocytosis
  "engine", *59*
  equator, **132**
  excitable, **352**
  Fallopian tubes, 41
  germ, **170,** 177
  glial, **349**
  glycolysis in, 94
  haploid, **170**
  as heat engines, 83
  hormone receptors, 340
  interphase, *130, 131*
  interstitial, 172
  intestinal, 129
  isothermal, 83
  metabolism, 56–79
  metaphase, *132*
  movement, 40
  muscle, *428, 432*
  nerve, 93, 129, 135. *See also* Neuron
  oxygen requirement, 287
  pacemaker, 43
  plant. *See* Plant cell
  plasma, 269, 270, *271*
  prokaryotic, 136
  prophase, *130, 131*
  respiration, 80–94
  resting, **130**
  retinal, *376*–378
  sap, **39**
  Schwann, *351*
  Sertoli, **170**
  somatic, **170**
  specialization, 48–51, 136
  stained, *29*
  structure, 23–55, *24, 25*
  telophase, *134*
  tracheal, 41
  turgid, 40
  wall, *47*
Cell division, 128–136, **130,** 453. *See also* Mitosis; Meiosis
  and cancer, 151
  implications of, 135–136
  initiation of, 146
  plant cell, 134–135
Cell-mediated immunity, 269
Cellulose, **47, 62,** *63*
Central nervous system, *51,* **360**–363, *362*
Centriole, **40,** *41,* 42, **130,** 136
Centromere, **132,** 133
Cephalin, 65
Cerebellum, **361,** *363*
Cerebral concussion, 293, 408
Cerebral cortex, **361,** *363,* 364, 366, 374–376
  and drugs, 367
  and learning, 405
  respiratory control, 293
  and shivering, 442, 443
  visual, 378
Cerebral hemispheres, **361**
Cerebral palsy, 534
Cerebrospinal fluid, **360**
Cerebrum, **360,** *363*
Cervix, **181,** 182, 208
Cesarian section, 191, *208,* **216**
Ceylon, growth rate, 555, *556*
Chain terminator codons, **119**
Chaplin, Charlie, 469
Chargaff, Erwin, 109
Chemical(s)
  body, 57–58
  bonds, **585**–587
  carcinogens, 153–155
  energy, 85–86
  and environmental accumulation, 533–537
  food, 57
  reactions, 587–588
Chickens, peck order, 409–**410**
Chimney sweeps, cancer in, 153–154
Chimpanzee, *502*
Chin, 506
Chloasma, 187
Chlordane, 536
Chloride ions, 350, 352, 358
Chlorine atom, 583, 584
p chlorophenyl dimethyl urea. *See* CMU
Chlorophyll, **36, 99,** *100*–103
  a and b, 99–101
Chloroplast, 99, 102–103, **36**–38
  development, 37
  grana, 103
  in eukaryotes, 48
  smog damage, *568, 569*
  stroma, **36,** 103
  structure, *38*
Chlorpromazine, 366, 368–369
Cholesterol, 64, **66,** *332*
  and bile, 233
  blood levels, 65
  structure, 67
Cholinergic neurons, **365**
Cholinesterase inhibitors, 358
Chondrocytes, **420**
Chondroitin sulfate, **45**–46, 419, 420
Chorion, **206**
Chorion frondosum, **206**
Chorionic gonadotrophin, **187**
Chorionic villi, **206,** 207
Choroid coat, **376,** 377
Chromates, 154
Chromatid(s), **132**
  arm length, 133
  metaphase, *139*
  structure, 136, *137*
Chromatin, **130,** 131, 134
Chromoplasts, **37**
Chromosome(s), **42, 130, 169, 450**–454
  abnormal arrangement, 214
  abnormal segregation, 471–478
  abnormal structure, 471–478, 491
  activity, 138–140
  analysis, *451,* 474
  anaphase, **133,** 137
  coiling, *131*
  daughter, **133**
  deletion, **474,** 476
  diploid number, **453**
  duplication, *139,* **474,** 478
  egg, 453
  in eukaryotes, 48
  female, *133*
  fruit fly, *137*
  gene order on, 478
  giant, **138,** 140, 141
  homologous, **169,** 170, **455**
  hormone binding, 341

Chromosome(s) (continued)
  human, *452*
  independent segregation of, 460
  inversion, **476,** 478
  lag, 136
  lampbrush, **141**
  metaphase, 133, 137
  puffs, **140**
  segregation, 138
  sex, *452,* 453, **454,** 462, 463
  sex test, 471
  sperm, 453
  sticky ends, 478
  structural alterations, *477*
  structure, 131, *132,* 136–138
  synapsis, **170**
  telophase, 137
  translocation, **474,** 476, 478
  uncoiling, 134
CIA, 388
Cigarettes and cancer, 154–155
Cilia, **40,** 41
  in eukaryotes, 48
  and mucus, 47
  paralysis, 288
  structure, 41, *42*
Ciliary body, **376**
Ciliary muscles, 379
Cingulate gyrus, *363*
Cingulotomy, 388
Circulation
  fetal, *207, 210*
  in newborn, *210*
Circulatory system, 236–257, **239,** *240*
  and respiration, 282
  functions, 239–240
  pulmonary, *287*
Circumcision, **174**
Cirrhosis, 233, 261, 267
Cisternae, Golgi, 34
Citric acid cycle, *90*–91
City, development of, 560, 561
Classical conditioning, **403**–404
Classification
  animal, 498–501
  cat, *502*
  man, 500–*502*
  of organisms, 48, 501
  plant, 498–501
  primates, *506*
  races, *522*
  tissues, *49–51*

Cleavage, **134,** 195, *196, 197*
Climacteric, male, **335**
Climax
  community, 547
  ecological, **545**
  male, 172
Clitoris, **182**
Clone, 128, **269**
Clostridium botulinum, 435
Clot, 189, 260, 262–264
Clubfoot, 213
Clyman, Martin, 191
CMU, 103
CNS, *51,* **360**–363, *362*
Coacervate(s), **14**
  droplets, 15, 17
  formation, *15*
  history of, *18*
  of proteins, **68**–69
Coal
  and cancer, 154
  depletion, 566
  tar, 154
Cobalt, in diet, 77
Cocarboxylase, **88**
Cochlea, **382,** 384
Cockroach, taxes, 393
Cocktail party effect, 402
Cod liver oil, 329
Codons, **117,** 119
  chain terminator, **119**
Coenzyme(s), **75**
Coenzyme A, **90**
Cofactors, 74–77
Coitus, *175,* **182**–184
Coitus interruptus, **185**
Colchicine, 132, 133, 451
Cold reception, 385
Colitis, 233–234
Collagen, **46,** 66
  bone, **418,** 420
  fibers, *46*
Collecting duct
  kidney, 307
  permeability, 310
Colloid(s), 14, **27,** 28
  sol-gel state, *28*
Colloidal droplets, *27*
Colloidal osmotic pressure, 252
Colobus, 503
Colon, 225
  peristalsis, 229

  ulcers, 234
Color, inheritance of, 454–455, 468–471
Color blindness, 462, *464*
Color perception, 377
Colostrum, **211,** 273
Columnar epithelium, 49
Communication, 408
  hormonal, 349. *See also* Hormone
  nervous, 349. *See also* Nervous system
Community, 528
  pioneering, 545
Complementary base pairing, **111,** 117, 119
Compounds, **585**–587
Concentration gradient, 252
Condensation reaction, **60**
Conditioned reflex, 394, 398
Conditioning
  behavioral, 395
  classical, **403**–404
  operant, **404**
Condom, **185**
Cones, retinal, **377,** *378*
Confinement, date, 187
Conflict behavior, 401
Congenital aberrations, **214**
Conjugated protein, **69**–70
Conjugation, hemoglobin, 69
Conjunctiva, **377**
Connective tissue, *50*
  capsule, 152
  mitosis in, 135
  muscle, 427, *428, 429*
Consciousness and behavior, 391
Conservation of energy, Laws of, 82
Constipation, **233**
Consumers, 97, **529,** 530, *532, 533,* 538
  primary, **529,** *532, 533*
  secondary, **529,** 530, *532, 533*
Contact inhibition, **160**
Contiguity, 403, 404
Contraception, **185**–189
  during lactation, 213
Contraceptive, prostaglandins as, 336
Contraceptive pill, 154, **188**–189
Contractility, muscle, 425
Contraction(s)
  muscle, **424,** 427–433
  uterine, 208
Control system, feedback, 142
Convection, **441,** 530
Convoluted tubule
  kidney, 305, 307, *308, 309*

Convoluted tubule (continued)
  permeability, 310
Copper, in diet, 77
Copulation. See Coitus
Cornea, **376,** 377–380
Corona radiata, **179,** 184
Coronary artery, **249**
  clogged, 264
Coronary sinus, **249**
Corpora cavernosa, **174**
Corpus albicans, **180**
Corpus callosum, 363
Corpus luteum, **180,** 181
Corpus spongiosum, **174**
Cortex
  adrenal, 310, **331–335**
  brain. See Cerebral cortex
  cerebral. See Cerebral cortex
  kidney, **305,** 307
Corti, organ of, **382–384**
Corticosteroids, **333**
Corticosterone, 333
Cortisol, 334
Cortisone, 39
Cosmologists, **10**
Costal breathing, **291**
Coughing, 264, 286
Counseling, genetic, 450, 495
Courtship behavior, **412–413**
Courtship dance, stickleback, 398, *400, 409*
Covalent bond, **586**
Cowper's glands, 172
Cowpox virus, 19, 272, 274
Cramps, muscle, 433
Cranial capacity, early man, 514–520
Cranial flexure, *199*
Cranial nerves, 360, 361, 365
Cranium, **360,** 506. See also Skull
Creation, 508
Cretin, **325,** *326*
Crick, Francis, 110–111
Crickets, territoriality, 409
Cri-du-chat
  karyotype, *477*
  syndrome, **476,** *477*
Crisis, temperature, 443
Cristae, mitochondrial, **36**
Cromagnon man, *518, 520*
Cross bridge, 429
Cross immunity, 272–273
Crossing over, 460–**461,** *462–463*
Cryptorchidism, **174,** 335

Crystal(s)
  calcium, 312
  DNA-diffraction pattern, **110**
  growth of, 6
  suspension, 30
Cuboidal epithelium, *49*
Cumulus oophorus, **179**
Cultural revolution, 574
Culture, 506
  evolution of, 516, 524
  human, 398
  inheritance, 399, 408
Curare, 434
Curie, Pierre and Marie, 490
Current, **350**
Cuzco, 296
Cyanide, 353
Cyclamates, 154
Cyclic adenosine monophosphate. See Cyclic AMP
Cyclic AMP, **340–**341
Cystitis, **314**
Cytochrome(s), 70, 77, **91,** 101
Cytokinesis, **134**
Cytoplasm, 26–30, **27,** 43
  blood cells, *263*
  cleavage, **134, 195,** *196*
  colloidal properties of, 28
  division of, 130, **134**
  furrowing, **134**
Cytosine, **109,** *110,* 111
  genetic code, *114, 115*
  RNA, 116

D

D and C, 216
Dance, bee, 411
Dark phase, photosynthesis, **102**
Dart, Raymond, 514
Darwin, Charles, 491, 493, 494, 499, *500,* 508, *509*
Darwin's finches, 509–512
Darwinism, 508–512
Daughter chromosome, **133**
Dawkins, Richard, 396
DCMU, 103
DDT, 535–537
  and death rate, 555
  resistance, 536
Deafness, 383
Deamination, **302**
Death rate, 555–558
Decay, 9

  organisms, 530, **538**
  primitive earth, 14
Decibel, **383**
Decidua, **206**
Decomposers, **538.** See also Decay, organisms
Decompression sickness, **298**
Deer, population regulation, 544, 545
Defecation, 229
Defense reaction, 262
Degenerate code, **114**
deGraaf, Regnier, 179
Dehydration, 310, 311
Dehydrogenation, 84
Delayed hypersensitivity reaction, **269**
Delerium tremens, 367
Deletion
  DNA, 489
  genetic, **474,** 476
Delgado, Jose, 388
Democracy, 395
Denaturation of proteins, **68–**70
Dendrites, **349,** *350*
  and learning, 406
  transmission direction, 355
Denis, Jean Baptiste, 463
Denitrifying bacteria, **540**
Dentition, primate, *516*
Deoxycorticosterone, 333
Deoxyribonucleic acid. See DNA
Deoxyribonucleotide, **108,** *109, 116*
Deoxyribose, **108,** *109*
Depolarization, muscle, 431
Depression and drugs, 366, *368–369*
Depressor muscle, **425**
Dermatitis, contact, 269
Dermis, **385**
Desalination, 565
Descent of Man, 512, 513
Desensitization, 276
Development, **501**
  and behavior, 399
  chloroplast, 37
  human, 194–219
  nervous system, 405
Deviated septum, 286
Dexedrine, 366, *368–369*
Dextrose, **59, 61.** See also Glucose
DHC, **328**
Diabetes, **310**
  and genetic engineering, 123
  insipidus, **310**
  mellitus, **310, 330**

Dialysis, kidney, 300, **313**
Diaphragm, **290**
  camera, 376
  contraceptive, **186**
Diaphragmatic breathing, **291**
Diarrhea, **234**
Diastole, **245,** 249
Diastolic pressure, **250**
Diatoms, 48
Dichlorophenyl dimethyl urea, 103
Dichlorodiphenyl trichloroethane. *See* DDT
Dieldrin, 536
Diet
  balanced, 230
  and cancer, 154
  carbohydrate, 562, 563
  essential amino acids, 113
  folic acid in, 76
  minerals in, 76, 77
  vegetable, 230
  vitamins in, 76
Diethylstilbestrol, 154
Differential reproduction, **493,** 512
Differentiation, cell, 136
Diffraction pattern, **110**
Diffusion, **30,** 282–283
  in bone, 420
  capillary, *252*
  and circulation, 237
Digestion, 86, 220–235, **221**
Digestive system, **221,** *223,* 225, 302
Dihybrid cross, 460
1,25 dihydroxycalciferol, 328
Dilantin, 357
Dilation and curettage, **216**
Dilator muscle, **426**
Dilger, W. C., 400
Dimethylbenzanthracene, 154
Diphenylhydantoin, 357
Diptheria, 273
Diplococcus pneumoniae, 108
Diploid, **169**
  bees, 397, 410
  chromosome number, **453**
Disaccharide, **59**–60
  synthesis, 60, *62*
Disease(s). *See also specific disease in question*
  autoimmune, 327
  bacterial, 9, 47
  and birth defects, 215
  blood fluke, 550
  defense reaction, 262

genetic, 106, 123, 483
gum, 231
heart, 155
and immunity, 271–273
liver, 233, 261, 367
maternal, *217*
and microorganisms, 9
Minamata, 534
Parkinson's, 388
protein-deficiency, 562
and protozoa, 9
sickle cell, 483–486, 493
vaccine, 272–273
viral, 47, 53
yeast, 9
Disinhibition, **403**
Displacement activity, **400,** 401
Dissociation, **585**
Disulfide bridges, **71**
Diuresis, **311**
Diuretics, **311**
Diver paralysis, 298
Diversity, 499
  genetic, 447
  interpopulation, **520**
  intrapopulation, **520**
  species, 547
DMBA, 154
DNA, **108**–127
  bacterial, 136, *137*
  and behavior, 396
  chain length, 136
  deletion, 489
  distinction from RNA, 116
  double helix, 110–*111,* 112
  duplication, 142
  frog, 138
  fruit fly, *137*
  hormone binding, 341
  human, 136, 138
  ligase, 125
  mutation, 485
  nitrogenous bases, 108, *109, 110,* 114
  nucleotides, **108,** *109,* 110
  organization, 111
  packaging, 136–138
  periodicities, **110,** 111
  polymerase, **111**
  polyoma virus, 159
  prokaryotic cells, 136
  recombinant, 106, 122–125
  redundant, 138

replication, **112,** *139*
staining, 42
sticky ends, 125
sugar. *See* Deoxyribose
transcription, 115–119, 117
uncoiling, 136
virus, 136, *137*
Watson-Clark model, 110–111, 122, 136, 137
x-ray diffraction pattern, **110**
Dobzhansky, T., 496, 520
Dominance
  and epistasis, 471
  hierarchy, 410–413
Dominant gene, **458**
Donor, **465,** 468
Dope, **366**–367, *368–369*
Dostoevski, 357
Double bond, **587**
Double helix, 110–*111,* 136, 137
Doubling time, 103–104, **554,** *555, 556, 557*
Douche, vaginal, **186**
Down, J. Langdon, 449
Down's syndrome, 136, **449,** 450, 473–*475*
  carrier, **478**
  karyotype, *476,* 478
  transmission of, 478, *479*
Dragonflies, territoriality, 409
Drinking center, **311**
Drive, 393, **400,** 401
  conflicting, 401
  thwarting, 401
Droplets, colloidal, *27*
Dropsy, **311**
Drosophila
  chromosomal puffs, *140*
  mutations, 490
  nucleotides, 138
Drugs, **366**–367. *See also specific drugs*
  addiction, 366, *368–369*
  and behavior, *368–369*
  and birth defects, 214, *215*
  diuretic, 311
  immunosuppressive, 275
  kidney secretion, 309
  and lysosomes, 39
  and memory, 407
  and pinocytosis, 34
  psychic dependency, 366, *369*
  and respiratory center, 293
  sleep-inducing, 293
  withdrawal, 366, *369*
Drumstick, muscle, 427, *428*

Dryopithecus, **513,** 521
Dubois, E., 517
Dugesia, 238
Duodenocolic reflex, **229**
Duodenum, 224, *225*
   microvilli, 228
   ulcers, **231,** 232
Duplication, genetic, **474,** 478
Dustbowl, 548, 550
Dwarf, 322–323
Dysentery, 234

E

Ear, *381–384*
   defenders, 383
   drum, **381–384**
   embryonic, *201*
   and natural immunity, 264
Earlobes, *459–460,* 461
Early man, 514–520
Earth
   age of, 12, 508
   history of, 10–21, *12, 13, 18, 20*
   spaceship, 3, 527, 552–577
ECG, *244,* **245**
Eclipse period, **53**
E. coli. See Escherichia coli
Ecological succession, **545–**547, *546*
Ecology, 526–581, **527**
Ecosystem(s), **528**
   and agriculture, 547–549
   balance, 544–549
   carrying capacity, **563**
   climax, **545**
   energy flow, 528–531
   unbalanced, 547–549
Ectoderm, *198*
Ectotherms, **440**
Eddy, Bernice E., 159
Edema, **253, 311**
   in pregnancy, 187
   pulmonary, 286
EEG, *356,* **357,** 359
Effector, 440
Egg(s), 184
   abnormal, 473
   chromosomes in, 453
   fertilization, 184, 195, *196*
   fertilized, 178, 185. *See also* Zygote
   haploid, 453
   human, 46

implantation, 178
   production. *See* Oogenesis
   sperm penetration, 46
   yolk, 206
Ehrlich, Paul and Anne, 565
Eibl-Eibesfeldt, I., 413
Eiffel, Gustave, 417
Ejaculation, 172, 174, *175,* 183, 333
EKG. *See* ECG
Elasser, W., 7
Elasticity, muscle, **425**
Elastin, **46**
Elbow joint, 426
Electricity, principles of, 350
Electrocardiogram, 244, **245**
Electrode implant, 348
Electroencephalogram, 356, 357, 359
Electrolytes, **585**
Electron(s), **583,** 584
   acceptors, **100,** 101
   flow, 101, 103
   microscope, 29
   orbitals, **584,** 585
   in oxidation, 84
   in reduction, 84
Electroshock therapy, 435
Elements, **583,** 584
   abundance in universe, *10*
Elephantiasis, 253, *255*
Elimination, 233–234
Embolism, 189
Embolus, **189, 264.** *See also* Clot
Embryo, **195**
   2.5 weeks, *198*
   3 weeks, *198*
   4–5 weeks, *199*
   6 weeks, *200*
   7 weeks, *201, 206*
   abnormal, 214
Embryogenesis, **185**
Embryonic disc, *197*
Emergency response, **332**
Emission, **183**
Emphysema, 155, 286, 288, 566
Emulsion, **28**
Endergonic reactions, **588**
Endocardium, **241**
Endocrine gland, *50, 318–336. See also specific
   gland in question*
Endocrine system, functions, 348–349
Endoderm, embryonic, *198*
Endometrium, 181, 187, *197*

Endoplasmic reticulum, **34,** *35,* 134
   in muscle, 431
   and nuclear membrane, 45
   in prokaryotes, 48
   rough, 34
Endotherm, 443
Energy, **81, 583**
   and absorption, 227
   activation, *72*
   and active transport, 33
   agricultural demand, 564
   atomic, 490
   barrier, 72
   behavioral, *400–401*
   bond, 83
   and carbohydrates, 58
   cascade, **86,** *87, 99, 100*
   chemical, 85–86
   crunch, 571
   and cyclic AMP, 340
   daily requirement, *80*
   depletion, 566
   ecosystem, 528–531
   of fats, 62
   and feedback control, 146
   food, 86, 229
   and food chains, *532, 533*
   in glucose, 83
   and kidney filtration, 305–306
   kinetic, **82,** 86
   light, 99–101
   nuclear, 566
   photosynthetic yield, 98
   and pinocytosis, 33
   population demands, 571
   potential, **82**
   on primitive earth, 13, 18, 19
   and respiration, 80–94
   and seed production, 547
   solar, 99
   sound, **381**
   and steady state, 439
Engram, **407**
Entelechy, 6
Entropy, **82**
Environment, 526–581
   atomic wastes in, 537
   and behavior, 393
   carcinogens in, 153, 155–157
   carrying capacity, 573
   DDT in, 535–537
   internal, 301, 302

Environment (continued)
mutagens, 161
pollution, 533–537
Environmental resistance, **573**
Enzyme(s), **6,** 17, 70–74, **73.** *See also* Catalysts
active site, **73**
activity, 340–341
constitutive, 144
and cyclic AMP, 340–341
denaturation, 441
digestive, *226*
and energy, 86
function, 112
induction, 143–146, *144, 145,* 148
inhibition, *74, 75*
lysosomal, 35
poisoning, **74,** 75
and proteins, 66
red blood cell, 262
restriction, 125
of seminal fluid, 172
specificity, 73
substrate-interaction, *73*
synthesis, 114, 144–146
and temperature, *441*
Eosinophils, *263,* 267
Epicardium, **241**
Epidermis, **385**
Epididymis, **172**
Epiglottis, **284,** 291
Epilepsy, *356,* **357,** 358
control, 388
ESB treatment, 348
Epinephrine, **331,** 332
and diuresis, 311
effects of, 332
and heart rate, 244
and temperature control, 442
Epistasis, **471**
Epithelial cells, 287
Epithelial tissues, *49*
Epithelium, nasal, 284
EPSP, **357,** 359, *360*
Equilibrium
population, 575
sense of, 382, 384–385
species, 546–548
Erection, penis, 174
Ergosterol, 329
Erythroblasts, **262**
Erythrocyte, **261.** *See also* Red blood cell
ESB, 347–348, 356–357, 405, 406

Escherichia coli
crippled, 123
enzyme induction in, 144
genes, 138
and genetic engineering, 124–125
isoleucine in, 146, 147
and recombinant DNA, 106, 122–123
Esophagus, 225
EST, 435
Ester linkage, **63**
Estradiol, 189, 333
Estrogen(s), **179**–181, **335–336**
and breast enlargement, 211
and cancer, 154
and gene activation, 341
level, 189
structure, 67
synthetic, 189
Estrone, 333
Ethane, 587
Ethologists, 393, 413
Ethology, **393**
Ethyl alcohol, in respiration, 87
Ethylene, 587
Eugenics, 108, **494–496**
Eukaryotes, **48**
Eunuch, 317, 333
Eustachian tube, **384**
Evaporation, 530
Evolution, **6,** 369, 390, *392,* 447, 491, 493, **511,** 516
cultural, 516, 524
finch, 509–512
future, 523–524
and genetics, 482–497
Homo sapiens, 518–520
human, 498–525
of language, 516
penicillin-resistance, 486
physique, 523
and recombinant DNA, 123
of respiration, 94
skeleton, *515,* 518
skin color, 523
of tool use, *517*
Exchange
food, *238, 239*
gas, *238*–239
Excitable cells, **352**
Excitatory postsynaptic potential, 357, 359, 360
Excretion, 300–315, **302**
Exercise and respiratory rate, 295
Exergonic reactions, **588**

Exhalation, 290, *291*
Exocrine glands, *49,* **318.** *See also specific gland in question*
Exophthalmos, **327**
Experiment, **78**
Expiration, **290,** *291*
Expiratory center, 292, *294, 295, 296*
Exponential growth rate, 573, 574
Extensibility, muscle, **425**
Extensor muscle, **425**
External ear, **381**
Extinction, **403,** 447, 508
Extracellular substances, 45–47
Eye, anterior chamber, **377**
embryonic, *199, 200, 201*
fetal, 205
muscles, 432
structure, *376–377*
Eyebrows, **377**
Eyelashes, **377**
Eyelids, **377**

F

F_1 generation, **456–457**
F_2 generation, **456–457**
Face, embryonic, *201*
Facilitation, 359–**360**
in learning, 406, 408
FAD, 91
Fallopian tubes, *177,* **178,** 180, 184
blockage, 181
cells, 41
ligation, 191
Fallopius, Gabriel, 178
Fallout, **537**
Family, **501**
bee, 410
groups, 408
tree. *See* Pedigree
Farming the sea, 565
Farsightedness, **380**
Fatigue
muscle, *432*
odor, 375
sensory, 402–403
synaptic, 359
Fat(s), **62–66**
absorption, 233
as antigens, 265
digestion, 224, *226,* 233
energy, 230

Fats(s) (continued)
  hydrolysis, *64*
  membrane penetration, 30
  metabolism, 330
  neutral, *64*
  polyunsaturated, 65
  saturated, **64**
  unsaturated, **64**
Fatty acids, **63,** *330*
  absorption, 228
  membrane penetration, 30
Feces, 229, 233
Feedback, **143**
  circuit, *142*
  control system, **142**
  inhibition, **146,** *147*
  negative. *See* Negative feedback
  in population control, 573
  positive, **143**
  respiratory loops, *294*
  and temperature regulation, 444
  and thyroid malfunction, 324–325
Feeding behavior, chicken, 409–410
Female
  adolescent, 178
  age distribution, 559
  breathing pattern, 291
  chromosomes, *451*
  facial hair, 336
  genitalia, *182*
  gonadotrophic hormone levels, 335
  hormones, 177, 335–336. *See also specific hormone in question*
  infertility, 181
  masculinization, 336
  puberty, 178, 335
  reproductive system, 174, *177, 178*–182
Fermentation, **88**
Fermi, Enrico, 490
Fern test, **186**
Ferredoxin, 101
Ferritin, 262
Fertility
  patterns, 558, *559*
  rate, U.S., *560*
Fertilization, 168, 169, 174, 183–185, 195, *196, 453*
  multiple, 181
Fertilizers, 104, 548
Fetal membranes, *207*
Fetus, **202**–205
  8 weeks, *202*
  12 weeks, *194*

3 months, *203*
4–5 months, *204*
6–9 months, *205*
  circulation, *210*
Fever, **443**
Fibrillation, **243**
Fibrin, 66, **263**
Fibrinogen, **260**–264
  and menstrual blood, 181
Fibroblasts, **420**
Fibrous connective tissue, *50*
Fight or flight reaction, **332,** 365
Fighting, ritual, 413
Filaria, 253
Finger muscles, 426, *427*
Filtration
  blood, 303–309, *307*
  capillary, 252
Finch, evolution, 509–512
Fingers, embryonic, *201*
Fire, use of, 518
First filial generation, **456**–457
First Law of Thermodynamics, 82
Fish yield, 565
Fitness, 447, 493, **511**
Fixed control system, **142**
Fixing, **29**
Flagella, 40, **41**
  in eukaryotes, 48
  in prokaryotes, 48
  structure, 41, *42*
Flavin adenine dinucleotide. *See* FAD
Flexor muscle, **425**
Fluid
  balance, 309–311
  distribution in body, *303*
Fluoride, as enzyme poison, 74
Focus, 378, 379, *380*
Folic acid
  analogs, 75–76
  deficiency, 75–76
  in diet, 76
Follicle, Graafian, **178,** *179*
Follicle-stimulating hormone. *See* FSH
Food
  chain(s), 97, **530,** *531*–537, 545, 550
  chemicals in, 57
  daily requirement, *230*
  derivation, 97
  digestion of, 220–235
  and energy, 86
  glucose source, 62

  oxidation, 229
  poisoning, 435
  production increase, 104
  pyramid, 97, *98*
  web, **530,** *532*, 533
Foramen magnum, 514, 518
Foramen ovale, 210
Ford, E. B., 492
Forebrain, **360**
Foreplay, 182
Foreskin, **174**
Forgetting, 402, 403, **408**
Fornix, *363*
Fossil(s), 513
  bones, 420
  man, 512–520
  phosphate content, 541
Fovea, **377,** *378, 379*
Fox, Sidney W., 14, 16
Frequency, sound, **381,** 383, 384
Fructose, *59*, 60
Fruit fly
  chromosomes, *137*
  DNA, *137*
Fruit ripening, 47
FSH, **170,** 178–180, 189, 333
  at climacteric, 335
  releasing factor, 337
Function, **501**
Fungi
  classification, 48
  and decay, 538
Furrowing, **134**
Furylfuramide, 157

G

Galactose, *59*
β galactosidase, 144–146
Galapagos islands, 499, 509, *510*
Gallbladder, 225, 233
Gallstones, **233**
Galton, Sir Francis, 494
Galvani, Luigi, 247
Gambling, 404
Gametes, 450–454, *453. See also* Egg; Sperm
Gamma aminobutyric acid. *See* GABA
Gamma globulin, 66, 67, **260,** 261, 273
  structure, 67
Gamma rays, 490
Ganglia, *364*
Ganglion, spinal, 363

Gas
  and breathing inhibition, 295
  diffusion in, 30
  exchange, 280–299
  natural supply, 566
  nerve. *See* Nerve gas
Gasoline and cancer, 154
Gastric glands, 318
Gastric juice, 231
Gastric ulcers, **231,** 232
Gastrocolic reflex, **229**
Gastrointestinal tract, **221.** *See also* Digestive tract
Gel, chondroitin sulfate, 46
Gelatin, 26–31, 418
Gene(s), 114, 116, 119, 137, **138, 169,** 450–454
  action, 146, 148
  activation, 162, 341
  activity, 138–148
  allele, **455**
  bee, 411
  and behavior, 396, 399
  cancer, 161
  defective, 214
  deleterious, 496
  dominant, **458**
  duplicated, 138
  E. coli, 138
  flow, 521
  frequency, **488,** 493, 495, 496, 521
  function, 122
  histocompatibility, 275
  lethal, 486
  linked, **460,** 461
  locus, **455**
  mutation, **483**–486. *See also* Mutation
  operator, **145**
  order on chromosome, 478
  polyoma virus, 159
  pool, **486,** 488, 511, 522
  in populations, 486–492
  recessive, **458,** *496*
  regulator, **144,** 145
  selective activation, 161–162
  segregation of, 459
  sex linked, 462
  structural, **144,** 145
  structure, 136
  subvital, 486
  transmission, 458–463
  transplantation, 124–125
  viral, 159
Genetic burden, **486,** 496

Genetic code, **112**–123, **114,** *115,* **117,** 136
  transcription, 115–119, **117**
  translation, 115–120
Genetic counseling, 450, 495
Genetic diseases, 106
Genetic diversity, 447
Genetic engineering, 106, 108, 122–125
Genetic load, **486,** 496
Genetic mosaic, 463
Genetic selfishness, 396
Genetic switching, 141, 146
Genetic variation, 486–488
Genetics, 106–127, **108, 450**
  bee, 397, 410–411
  and behavior, 393, 398
  blood groups, 468
  classical, **450,** 448–480
  hawkweed, 456
  Mendelian, 456–457
  molecular, **450**
  peas, 456–457, 461
  physiological, **450**
  population, 486–492
  skin color, 468–471
  transmission, **450,** 448–480
  virus, 116
Genitalia
  female, *182*
  male, *171,* 173–174
Genotype, 454–458, **455,** 470, 488
Genus, **500,** 501
Geographic isolation, 521
Geographic variation, **520**
Geology, 508
Geometric growth, **554,** *555*
Geotaxes, **393**
Germ cells, **170,** 177, *200*
Ghosts, **26**
Giant chromosomes, **138,** 140, 141
Giants, 322
Gibbon, *503*
Gigantism, 322–**323**
Gill clefts, embryonic, *199, 200*
Gills, *282,* **283,** 296–*297*
Gingivitis, 231
Gland(s). *See also specific gland in question*
  adrenal, *331*–333
  brood-food, 410
  endocrine, *50,* 318–336
  exocrine, *49,* **318**
  lacrimal, **377**
  male accessory, 170, *171,* 174

  parathyroid, 187, 329–330
  pineal, **336**
  pituitary, 170, 187, *318,* **319**–324
  sweat, 441
  thymus, **336**
  thyroid, **324**–329
  wax, 411
Glandular epithelium, *49*
Glans penis, **174**
Glial cells, **349,** 406
Globin, 69, **289**
Globulin, **260**
Glomerular filtrate, 306, **307**
Glomerulus, **303**–*307, 305, 306,* 311
Glottis, **284**
Glucocorticoids, **333**
Glucose, *59,* 60
  active transport, 330
  aerobic respiration of, 86, *87,* 89–92
  anaerobic respiration of, 86–89
  blood level, 39, 59, 330, 331
  calorific value, 89
  carrier, 341
  and cyclic AMP, 340
  energy content, 83
  food source, 62
  membrane penetration, 30
  molecular weight, 265
  oxidation in muscle, 432
  phosphorylation, **87**
  regulation of, 310
  respiration of, 86–95
Glutamic acid, 122, 484
Glycerol, **63,** 228
Glycogen, **62**
  and cyclic AMP, 340
  metabolism, 330
  in muscle, 88
  structure, *63*
  uterine, 180, 206
Glycolysis, **88,** 94
Glycoprotein, 70, 260
Goiter, 324, *325, 326*
Goldberger, Joseph, 75
Goldfish, taxes, 393
Golgi, Camillo, 34
Golgi apparatus, **34**–35, *36,* 48
Gonad, **170.** *See also* Ovary; Testis
Gonadotrophic hormones, **321,** 335
Gonorrhea, 181
Gorilla, *503, 507*
Graafian follicle, **178,** 179

Graft rejection, 269, 275
Grana, **37,** 103
Grand mal, 357
Granulocytes, **262**
Grasping reflex, 402
Grasses, 545, 546
Grave's disease, **327,** 330
Gray matter
 brain, 360
 spinal cord, 361
Greenhouse effect, **544**
Green revolution, 494, **565**
Griffith, F., 108
Gristle, 46. *See also* Cartilage
Grooming, 413
Groups, 408
Growth, 5–6
 arithmetic, **554,** *555*
 autonomous, 152, 161
 bone, *420–422*
 cancer cells, *160*
 crystals, 6
 curve, *573, 574*
 exponential, **554,** *555*
 hormone, **321**–323, 335–336
 population, 554–577, *556*
GTP, 119
Guanine, **109,** *110,* 111, *114, 115,* 116
Guano, 541
Guanosine, 119
Guanosine triphosphate. *See* GTP
Gull behavior, 409, *412*
Gum disease, 231
Gurdon, J. B., 128
Gut, **221.** *See also* Digestive system

H

Habituation, **402**–403, 408
Hair
 curling, 336
 facial, 336
 pigment synthesis, 112, 114
 and temperature regulation, 442
 touch reception, 385
Hallucinogens, *368–369*
Hamilton, William, 396
Hammer, ear, **382**
Hand
 fetal, *204–205*
 human, 506
 tendons, *427*
Hanson, 429, 430

Haploid, **170**
 bee, 397, 410
 chromosome number, 453
Hardy, 488
Hardy-Weinberg Law, **488**
Harelip, 213
Haversian canal system, **419**
Hawkweed genetics, 456
Hay fever, 269
Hearing, 381–384
 threshold, **383**
Heart, 81, **239,** *241,* 240–248, *242*
 attack, 248–249, 264
 defects, 214
 disease, 155, 330
 electrical activity, 242–247
 embryonic, *199*
 fetal, 210
 fibrillation, 243
 innervation, 364
 and kidneys, 302
 ligaments, 241
 Lincoln, Abraham, 236
 murmurs, **245**
 muscle contraction, 242
 myogenic rhythm, **242**
 pacemaker, **242,** 246–247
 transplants, 275
 valve, **241,** *242, 243*
Heartbeat
 regulation of, 243–244
 refractory, **243**
Heartburn, **231**
Heat
 metabolic, 84
 pollution, 571
 reception, 385
Heat engine, 92
Height
 distribution, *469*
 normal range, 213
Helium, in primitive atmosphere, 11, 12
Heme, 69, 70, **289**
Hemizygosity, **462**
Hemocytometer, **266**
Hemodialysis, 313
Hemoglobin, 66, 77, **261,** 262, *289*
 and air pollution, 567
 blood concentration, 267
 breakdown products, 233
 genetic expression, 124
 manufacture, 267
 molecular weight, 67

 oxygen carrying capacity, 289–290
 oxygen loading, *290*
 sickle cell, 67–68, **121,** 122, 484, *485, 486,* 493,
  496
 structure, 67–69, *70*
Hemophilia, **263, 466,** 467
 and genetic engineering, 123
Hemorrhage, **262,** 263
Hemorrhoids, 187, **234**
Heparin, **264**
Hepatic portal system, *229*
Hepatitis, 273
Herbicides, 572
Herbivores, 410, **529,** *531*
Heredity
 laws of, 456–459
 library of, 116
 origin of, 17
Herrick, J. B., 483
Heterosis, **493**
Heterotrophs, **17**–20, 48, **529**
Heterozygote, **458**
 frequency, 488
 in sickle cell, 483, 493
Hexose, **58–59**
 diphosphate, 102
Hiccups, 290, 291
High energy bond, **85**
Hindbrain, **360,** 384
Hip joint, *426*
Hinge joint, *426*
Hippocampus, 363, 367
Hippocrates, 330
Hiroshima, 103, 490
 and birth defects, 217
 and leukemia, 156
Histamine, 276
Histidine, 157
Histocompatibility genes, 275
Histone, **136,** 140, 148
Hitler, 395, 495
Hives, 276
Hodgkin's disease, 158
Homeostasis, 438–445, **440**
 ecosystem, **544**–549
Hominid line, 513, 514–520
Homo erectus, 517–518, *521*
Homo sapiens
 classification, 500–*502*
 distribution, 521
 evolution of, 518–520
Homogenized milk, 28
Homologous chromosomes, **169,** 170, **455**

Homozygote, **458**
  sickle cell, 484
Honeybee. *See* Bee
Hookworm, 267
Hormone(s), 316–345, **317.** *See also specific hormone in question*
  adrenal, 331, *332, 333,* 365
  adrenal cortical, 334, *335*
  adrenocorticotrophic, **321**
  and behavior, 398
  and breast development, *212*
  carcinogenic, 154
  DNA binding properties, 341
  female, 177, 212, 335–336
  and gene activation, 341
  gonadotrophic, **321**
  growth, **321**–323, 335–336
  hypothalamic, 336–337
  male, 170, **333.** *See also* Androgens
  melanocyte-stimulating, **321**
  molecular activity, 337, 340, 341
  neurohypophyseal, *324*
  pancreatic, 330, 331
  parathyroid, 327, 329–330
  pituitary, 170, 212, 318–324, *320, 321*
  receptors, 340, 341
  regulation of production, 148
  steroid, *341*
  synthesis, 341
  thyroid, 324–329
  thyrotrophic, **321**
  trophic, **321,** *337*
Howard, H. E., 409
Humidity, **441**
Humoral immunity, **269**–273
Humus, 546
Hunger, 561–563, *562*
  drive, 400
Huntington's chorea, 495
Hutton, James, 508
Huxley, Aldous, 168
Huxley, H. E., 429, 430
Huxley, Sir Julian, 496
Huxley, T. H., 506, 518
Huxley-Hanson model, *430, 431*
Hyaluronic acid, **45**–46
  in corona radiata, 184
Hyaluronidase, 46
  in sperm, 184
Hybrid, 457
  vigor, **493**
Hybridization, 494
Hydra, 238

Hydration, 310, 311
Hydrocarbons
  and cancer, 154
  and smog, 567
Hydrocephalics, 213
Hydrogen
  atom, 583, 584
  bonds, 122, **586**
  ions, 309, 585. *See also* pH
  peroxide, 72
  in primitive atmosphere, 10–12
Hydrogenation, of oils, **64**
Hydrolysis, **224**
  fat, *64*
  starch, 62, *63*
  sucrose, 59–60
Hydroxyl ion, 585
Hymen, *182*
Hypergonadism, **335**
Hyperkinetic children, 388
Hyperopia, *380*
Hyperosmotic medium, **31**
Hyperosmotic urine, **307**
Hypersecretion, endocrine, **319**
Hypersensitivity
  delayed, **269**
  immediate, **269**
Hypertension, and prostaglandins, 336
Hyperthyroidism, *327*
Hyperventilation, **295**
Hypogonadism, **333,** 335
Hypoosmotic medium, **31**
Hypophysis, **319**–324. *See also* Pituitary
Hyposecretion, endocrine, **319**
Hypothalamus, **336,** 337, 348, **361,** *363*
  drinking center, **311**
  hormones, 324
  and lactation, 212
  and Napoleon, 338–339
  osmoreceptors, 310
  releasing factors, **337**
  and temperature regulation, 441–442
Hypothesis, **8**
  one gene–one enzyme, **112**
Hysterectomy, **216**
H zone, **429,** *430*

I band, **429,** *430*
ICSH, **170,** 172, 179, 333
Idiocy
  mongoloid, 449. *See also* Down's syndrome

phenylketonuric, 113
Ileocolonic sphincter, 225
Ileocolonic valve, 229
Ileum, 224, *225*
Image formation, **378,** *379*
Immediate hypersensitivity, **269**
Immune response
  to polyoma virus, **159**
  primary, 268–270
  secondary, 270–271
  suppression, 275
Immune system, 268
Immunity, 264–277
  acquired, **265, 271,** 273
  active, **271,** 273
  cell-mediated, **269**–273
  fetal, 205
  humoral, **269**–273
  natural, 264–**265**
  passive, **271,** 273
Immunization, 271–273
Immunological cooperation, **269**
Immunosuppression, 275
Implantation, blastocyst, *197*
Imprinting, **405**
Inbreeding, 494
Incontinence, **314**
Incus, **382**
India, population structure, 557
Indigestion, **231**
Induction, enzyme, 143–146, *144–145*
Industrial melanism, 491, **492,** 493
Industrial revolution, 574
Infantile behavior, 412
Infection, 47. *See also* Disease
  and immunity, 271–273
Infectious mononucleosis, 267
Infertility, 181
Inflammation, **265**
Influenza
  vaccine, 273
  virus, *53,* 274
Inguinal canal, **172,** 174
Inguinal hernia, 172, **174**
Inhalation, 290, *291*
Inheritance, 107–112
  of albinism, *458*
  of color blindness, 462, *464*
  cultural, 399
  human, 448–480
  polygenic, 124, 468–471
  sex linked, 462
  of skin color, 468–471

Inhibitory postsynaptic potential, **359**
Innate releasing mechanism, **400**
Inner ear, **381,** 384
Inoculation, 271–273
Insect
    courtship behavior, 412–413
    respiratory system, *283*
    society, 408
Insecticides, 104, 358
Insectivores, 513
Insemination, artificial, 176
Insight learning, **405**
Inspiration, **290,** *291*
Inspiratory center, 292, *294, 295, 296*
Instinct, 393, **398**–402
    aggressive, 398, 399
    human, 402
    and learning, 401–402
    refinement of, 401–402
Instrumental conditioning, 404
Insulin, 66, 310, **330**
    disulfide bridges in, *71*
    E coli production, 106
    genetic basis, 123–124
    and membrane transport, 341
Intercourse, sexual, **182**–183
Interference, 403, 404
Interneurons, **360,** 365
Interphase, 130, *131,* 132, 135, 138
Interpopulation diversity, **520**
Interspecific aggression, 399
Interstitial cell(s)
    nerve, **349,** *351*
    testis, 172, 333
Interstitial cell stimulating hormone. *See* ICSH
Interstitial fluid, 250, 252, 253, 303
Intervillous space, **207**
Intestine
    absorption, 224, 227, 228
    and alcohol, 367
    cell lifespan, 129
    embryonic, *200*
    innervation, 364
    fetal, 205
Intrapopulation diversity, **520**
Intraspecific aggression, 399
Intrauterine device, **188**
Invasive tumor, **152**
Inversion, **476,** 478
Inverted image, 379
Inverted retina, **377**
Involuntary muscle, **424.** *See also* Smooth muscle

Iodine in diet, 77
    radioactive, 327
    and thyroid function, 324–325
Ions, **93, 252, 350, 489, 585.** *See also specific ion*
    in question
Ionic bonds, **585,** 586
Ionization, 585
Ionizing radiation, **489**
IPSP, **359**
Iris, **376**
IRM, **400**
Iron, dietary, 76–77
    in hemoglobin, 289
    in primitive atmosphere, 10
    filings, catalytic properties, 72
Irrigation, 563
Islets of Langerhans, 330, *331*
Isoagglutinins, **465**
Isolation, geographic, 521
    reproductive **511**
Isoleucine, 146, *147*
Isomers, structural, *59, 60*
Isosmotic kidney filtrate, **307**
Isosmotic medium, **31**
Isothermal cells, 83
Isotopes, **537, 584**
IUD, **188**

J

Jacob, F., 145
Jacob-Monod model, 145–146
James, William, 407
Jaundice, **233**
Java man, 517, *518*
Jaw, embryonic, *200*
Jejunum, 224, *225*
Jenner, Edward, 272
Jet noise, 383
Joint, **425**–*426*
    inflammation, 426
    stiffness, 426
Jugular vein, 228
Jukes, 494

K

Kaibab plateau, 544, *545*
Kallikaks, 494
Karyotype, **451**
    cri-du-chat, *477*
    Down's syndrome, *476, 478*

Turner's syndrome, *475*
Kennedy, John F., 333, *334*
Keratin, 66, 71
Ketone bodies, 330
Kettlewell, H. B. D., 492
Kidney, 300–315, **302,** *303*
    artificial, 300, *312*–313
    and ATP, 93
    cortex, **305,** 307
    dialysis, 300, **313**
    function, 303–309
    innervation, 364
    medulla, **305**
    resorption by, 306, 307, 309–311
    secretion, 307, 309
    stones, **312**
    structure, 303–309, *305*
    transplants, 275
    tubule, 304–310, *308*
Kinetic energy, **82,** 86
Kingdom, 499, **501**
Kinship, **411**
Kissing, 412
Klinefelter's syndrome, 335, **471,** *472, 473*
Knee jerk reflex, 394
Knee joint, *426*
Koch's postulates, **158**
Krebs cycle, **90,** 91, 92, 94
Krogman, W. M., 420
Kwashiorkor, **562,** *563*

L

Labia majora, **182**
Labia minora, **182**
Labor, **208**
    and prostaglandins, 336
Labyrinth, **381**
Lacrimal duct, **377**
Lacrimal glands, **377**
Lactation, **211**–213
    fetal, 205
Lacteal, 228
Lactic acid
    in muscle, 432–433
    in respiration, 87
Lactiferous duct, **211**
Lactiferous sinuses, **211**
Lactose, **60,** 144, 146
Lacunae, endometrial, **206**
Lag phase, **573**
Lake Nasser, 550

Lamarck, Jean Baptiste de, 508
Lamellae, chloroplast, **37**
Lampbrush chromosome, **141**
Landsteiner, Karl, 465
Langerhans, Paul, 330
Language, 408
   bee, 411
   evolution of, 516
   human, 506
Langur, *503*
Lanugo, **205**
Large intestine, 225
Larvae, bee, 410
Larynx, **284,** 286
Lavoisier, Antoine Laurent, 229
Law of independent assortment, 457, 459, 460. *See*
   also Principle of recombination
Law of segregation, **456, 459**
Laws of thermodynamics, **82,** 86, 532, 571
Lead
   air content, 568
   and environment, 533
Leakey, L. S. B., 513
Learning, 390, 393, 398, **402–408**
   and aggression, 399
   and DNA, 124
   and drugs, 366, *369*
   impairment, 405–406
   imprinting, 405
   insight, **405**
   location of, 405–408
   mechanisms of, 406
   and memory, 406–408
   and operant conditioning, 404
   trial and error, 404–405
Lecithin, 65, *66*
L-dopa, 388
Lederberg, Joshua, 106
Leeuwenhoek, Anton van, 7, *9*
Legallois, Cesar, 292
Leg buds, *199*
Leguminous plants, 540
Lejeune, J., 476
Lemur, *504,* 513
Lens, **376,** 377, 379, 380
Letdown reflex, 211, *213*
Lethal genes, 486
Leucoplasts, **37**
Leukemia
   and analog inhibitors, 75–76
   and blood count, 267
   frequency, 156

and radiation, 491
and viruses, 158
Leukocytes, **261.** *See also* White blood cell
Levator muscle, **425**
Lewis, W. H., 34
Lewontin, Richard, 396
LH, **179,** 180, 189, 335
Li, C. H., 322
Life, **5, 6**
   characteristics of, 5–6
   Mars, 4
   origin of, 4–21, 237, 260, 508
Ligaments, heart, 241
Light energy, 99, 100, 101
Light phase, photosynthesis, 99–102
Light reception, 376–380
Lignin, **47**
Limbic system, *363*
Limeys, 74
Lincoln, Abraham, 236
Lindane, 536
Linkage, 460, 461
Linnaeus, Carolus, 500, *501*
Lipase, 63, 74
Lipids, 26, **62,** 63–66
Lipoprotein, 70, 260
Liquid, diffusion in, 30
Liver, 225, 228, 318
   and alcohol, 367
   cirrhosis, 233, 261, 367
   damage, 233
   embryonic, *200*
   functions, 302
Living things, characteristics of, 5–6
Lizard
   temperature control in, 440
   territoriality, 409
Locus, gene, **455**
Log phase, **573**
Long term memory, **407**
Loomis, W. F., 523
Loop of Henle, **304,** 305, 307
Lorenz, Konrad, 396, 398, 400–*405*
Loudness, 384
Lower, Richard, 463
LSD, **367,** 368–369
LTM, **407**
Lubricating fluid, 182
Lungs, **283,** 284, 286
   breathing movements, 290, *291*
   cancer, 154–155
   capacity, 291, *293*

gas exchange, 285
innervation, 364
water breathing, 297
Luteinizing hormone, **179,** 180, 189, 335
Lyell, Sir Charles, 508
Lymph, **253**–255, 259
   nodes, **254,** *255,* 268
   valves, 254
Lymphatic duct, right, **253**
Lymphatic pump, **254**
Lymphatics, 228, **253**–255
Lymphocyte(s), **254, 268.** *See also* B-cells; T-cells
   maturation, 336
   staining properties, *263*
Lymphoid tissue, 262
Lymphokines, **269**
Lyon hypothesis, **463,** 465
Lyon, Mary F., 463
Lysergic acid diethylamide, **367,** *368–369*
Lysins, **274**
Lysosome(s), **38,** *39,* 40
   enzymes, 35
   of neutrophil, 265
   structure, *39*
Lysozyme, 264

M

Macaque, *503, 504, 507*
Machu Picchu, 296
Macleod, C. M., 108
Macroglobulin, **273**
Macrophages, 269
Magnesium, dietary, 76–77
Magnification, **29**
Maintenance, 5–6
Malaria, 493
Malathion, 358
Male
   breathing pattern, 291
   chromosomes, *452*
   climacteric, **335**
   genitalia, *171,* 173–174
   gonadotrophic hormone levels, **335**
   hormones, 170, **333.** *See also* Androgens
   hypergonadism, 335
   hypogonadism, 335
   menopause, 335
   reproductive system, 170, *171*–174
Malignant tumors, **152,** 153
Malleus, **382**
Malnutrition, 104

Malthus, Thomas, 573, 574
Maltose, **60**
Mammary glands, **211,** 318. *See also* Breast
Mammary tumors, 158. *See also* Breast cancer
Man, *507, 508.* See also *Homo sapiens*
 brain size, *517*
 characteristics of, 501, 502, 506
 DNA, 136
 early, 514–520
 killer instinct, 398–399
 races of, 520–523
Manganese, dietary, 77
Marasmus, **562,** *563*
Marfan syndrome, 236
Marijuana, 367, *368–369*
Marrow cavity, **421**
Mars, life on, 4
Marsh, *547*
Masculinization, 336
Mass number, **584**
Mating, controlled, 494
Mating behavior, 408, 412–413
 sticklebacks, 398, *400, 409*
Matter, **81, 583**
Maze, 393, 404
McCarty, M., 108
Measles vaccine, 273
Mechanists, 7
Meconium, **205**
Medici, Maria de 324, *325*
Medulla
 adrenal, **331,** 365, 442
 brain, 361, *363*
 kidney, **305**
 respiratory center, **292,** 293, *294, 295, 296*
Megakaryocyte, **263**
Meiosis, 167–170, *168, 169, 453*
 abnormal, 471–478
 in oogenesis, 179, *185*
 in spermatogenesis, *172, 173*
 and translocation, 478
 and variation, 486–488
Melanin, **454,** 455
 eye, 376
 synthesis, 470–471
Melanism, industrial, 491, **492,** 493
Melanocyte-stimulating hormone, **321**
Melatonin, **336**
Membrane(s)
 depolarization, **352,** 353, *354,* 357
 embryonic, *197*
 fetal, *207*

functions, 47
 lysosomal, 38
 and mercury affects, 534
 muscle. *See* Sarcolemma
 nuclear, 134
 plasma. *See* Plasma membrane
 pleural, **284,** 290
 postsynaptic, 358
 potential, 352, 357
 refractory state, **353**
 repolarization, **352,** 353, *354*
 resting potential, **352**
 semipermeable, **31**
 tectorial, **384**
 transport, 341
 unit, **26**
 voltage, 352
Memory, 366, 402, **406**–408
 cells, 270
 chemical basis, 407
 consolidation, 408
 decay, 408
 and drugs, 366, *369*
 electrochemical basis, 407
 long term, **407**
 loop, 406
 molecular basis of, 407
 short term, 367 **407**
 transfer, 407–408
Menarche, **178**
Mendel, Gregor Johann, 456, 458
Mendel's Laws of Heredity, 456–459, 494
Meninges, **360**
Menopause, **178**
 and breasts, 211
 male, 335
Menstrual cycle, **181**
 and lactation, 213
Menstruation, **178,** 181
Mental retardation
 and PKU, 113
 and Down's syndrome, 449
Mercury
 and environment, 533, 534, 535
 poisoning, 534–535
Mesoderm, embryonic, *198*
Messenger RNA, 116, *119, 120, 140,* 341
Metabolic rate, **229**
 and respiration, 295
 thyroid regulation, 325
Metabolism, **6**
 bone, 327–329

 calcium, 327–329, 330, 523
 cellular, 56–79
 and cyclic AMP, 340
 development of, *17*
 and diet, 231
 fat, 330
 heat of, 84
 melanin, 454, 455
 muscle, 424
 origin of, 17
 phenylalanine, *113*
 protein, 330
 sugar, 330–331
 and temperature regulation, 441, 442
Metaphase, *130,* **132**
 arrest, 451
 cell, *132*
 chromatid, *139*
 chromosome, 133, 137
Metastatic cancer, 255
Metastatic tumor, 153
Methane, in primitive atmosphere, 12
Methedrine, 366, *368–369*
Metric system, 589–590
Microfilaments, **40**
Micro-organisms and disease, 9
Microscope, 29
Microtubules, **40,** 42
Microvilli
 intestinal, **227,** *228*
 taste bud, 375
Micturition, **314**
Midbrain, **360**
Middle ear, **381,** 382, 384
Middle lamella, 47
Migration, 398
Milk
 and alcohol absorption, 224
 composition, *213*
 homogenized, 28
 lactose content, 60
 production volume, 213
 secretion. *See* Lactation
 strontium 90, 537
 sugar. *See* Lactose
 witch's, 205
 and vitamin D, 329
Milky Way, 11
Miller, Stanley L., 13, *14*
Minamata disease, 534
Mineralocorticoids, **333**
Minerals

Minerals (continued)
  absorption, 228
  ecosystem cycling, 548
Mitochondria, **35**–36, *37*
  cristae, **36**
  in eukaryotes, 48
  in kidney tubules, 307
  in prokaryotes, 48
  and respiration, 94
Mitosis 128–136, *130,* **134,** *168–169,* 453
  animal cell, *135*
  implications of, 135, 136
  metaphase arrest, 452
  timing of, 135
  variation, 486
  zygote, 195
Mittelschmerz, **186**
Mixed nerves, 361, 363
Mohammed, 357
Molecular action, *27*
  of hormones, 337, 340, 341
Molecular weight, 273
  and antigens, 265
  glucose, 265
  hemoglobin, 67
  sucrose, 67
  virus protein, 67
  water, 67, 265
Molecule(s), **585**
  carrier, 33
  cellular, 56–79
  charged, 65
  membrane penetration, 30
  odor, 375
  promoters, **146**
  in solution, 30
  symmetry, 60
  water, 586
Monera, **48**
Mongolism. *See* Down's syndrome
Monkey, *504, 505, 507*
  behavior, 399
Monocyte, *263*
  count, 267
Monod, J., 145
Monosaccharides, **58**–*59,* 60
Mons veneris, **182**
Mora, P., 7
Morgan, C. Lloyd, 391
Morgan, T. H., 461
Morgan's Canon, **391**
Morning sickness, **187**

Morphology, 501
Morula formation, *196*
Mosaic, genetic, 463
Motion detection, 384
Motion sickness, 385
Motor end plate, **430,** *431*
  potential, **430,** 431
Motor nerves, 361, 363, 364
Motor neurons, **360**
  damage, 363
  in reflex arc, 365
Motor responses, 366
Motor unit, *431*
Mountain sickness, 296
Mouse
  bladder cancer, 154
  respiratory system, *283*
Mouth, 225
  cancer, 154
Movement, 6, 416–437
  cell, 40
  fetal, **205**
  perception, 384–385
mRNA. *See* Messenger RNA
MSH, 334
Mucopolysaccharide, bone, 419
Mucoprotein, 260
Mucosa, **223**
Mucus, 47
Muller, Hermann, 160
Muller, H. J., 490, 496
Muller, Paul, 535
Multicellular organisms, 48
Multicellularity, **26**
  origin of, 20
Multiple alleles, 463–468
Multiple sclerosis, **350**
Mumps
  immunity in newborn, 273
  virus, 335
Muscle(s), *51,* 422–437, *423*
  abductor, **425**
  adductor, **425**
  aging, *429*
  antagonists, **425**
  atrophy, **433**
  bundles, 427
  cardiac, *51,* **422,** *424*
  cardiac constrictor, 231, 255
  cell, 135, *428, 432,* 435
  ciliary, 379
  components of, 428

contraction, **424,** 427–433
  contractility, **425**
  depolarized, 431
  depressor, **425**
  dilator, **426**
  elasticity, **425**
  extensibility, **425**
  extensor, **425**
  fatigue, *432*
  finger, 426, *427*
  flexor, **425**
  graded response, 432
  heart, 242
  kinds, 422–424
  levator, **425**
  motor end plate, **430,** *431*
  myoglobin, **423**
  paralysis, 363, 366
  poison, 434–435
  pronator, **426**
  red, **423**
  refractory period, **432**
  respiration in, 88
  skeletal, **422.** *See also* Striated muscle
  smooth, *51,* **422,** *423, 424*
  sphincter, 312, 314, **425**
  striated, 422, 423, *424, 428, 430*
  supinator, **426**
  tension and bone growth, 421
  tetanus, **432**
  tonus, **223**
  white, **423**
Muscular system and behavior, 392
Mutagens
  environmental, 161
  nitrous acid, 122
  radiation, 122, 488, 490–491
Mutation, **6, 122, 458,** 484–486, 490
  base-pair substitution, 161
  and cancer, **160,** 161
  Drosophila, 490
  frameshift, 161
  lethal, 122
  myelin sheath, **349**–**350,** 351, 355,
    360
Myocardium, **242**
Myofibrils, **428**–*432*
Myogenic heartbeat, **242**
Myoglobin, **423**
Myopia, *380*
Myosin, **428,** 429, 431
Myxedema, **325,** *326*

N

NAD, **87,** 91
NADP, 101
Naegli, Carl, 456
Nagasaki, 156, 217, 490
β-napthylamine, 154
Narcotics, *368–369*
Nasal cavity, 284, *285,* 286
Nasal epithelium, 284
Nasal labyrinth, 284, *285*
Natural selection, **15,** 17, **491**–493, **509**–512, 521
Nausea in pregnancy, **187**
N-dimethylaminobenzene, 154
Neanderthal man, *519*
Nearsightedness, **380**
Nectar, 410
Negative feedback, **143,** 144, 146
    and hormones, 318, *319,* 321
    and population growth, 575
Negative geotaxis, **393**
Negative phototaxis, **393**
Nelms, Sarah, 272
Neoplastic state, **152,** 161
Nephron, **303,** *304,* 306, 307
Nerve(s), 346–371, **349**
    cells, 93, 129, 135. *See also* Neurons
    conduction velocity, 355
    cranial, 360, 361, 365
    gas, 74, 93, 358, 434–435
    and heartbeat regulation, 243, *244*
    impulse, **349,** 352–360. *See also* Action potential
    and learning, 406
    motor, 361, 363, 364
    mixed, 361, 363
    olfactory, 375
    optic, **378**
    poison, 536
    root, 361, 363
    sensory, 361, 363, 364
    spinal, 360, 361
    transmission, 336, 354, 355, 358–360
    vagus, 293
Nervous system, 346–371, *349*
    autonomic, **364,** 365
    and behavior, 390, 392
    central, *51,* **360**–363, *362*
    circuits, *361*
    design, 360–365
    development, 405
    and drugs, 366–367, *368–369*
    cf. hormonal, 318, 348–349

    parasympathetic, **365**
    peripheral, **360,** 361, 363–365, *364*
    sympathetic, 332, **364,** 365
Nervous tissue, *51*
Nesting
    birds, 398
    sticklebacks, 399, *400*
Neural tube, *198, 199*
Neurilemma, **349**
Neuromuscular transmission, *431*
Neuron(s), *51,* **349**–352
    activity, 356
    adrenergic, **365**
    cell bodies, 360
    cholinergic, **365**
    circuits, *361*
    expiratory, 292, *294*
    fluid, 350
    inspiratory, 292, *294*
    insulation, 350
    and ion concentration, 350, *352*
    kinds, *350,* 360
    motor, 363, 365
    nutrition, 349
    olfactory, *374,* 375
    optic, 377
    polarization, 352
    postsynaptic, 357, 359
    refractory state, **353**
    regeneration, 349
    resting potential, **352**
    retinal, 377
    sensory, 365
    signal frequency, 355
    size and learning, 406
    taste, 375
Neurosecretion, **336**–337
Neurosurgery, 405
Neutral fat, 64
Neutral stimulus, **403**
Neutrons, **583**
Neutrophils, *263,* **265**
Newborn
    circulation in, *210*
    immunity in, 273
    respiratory reflex, 295
Niacin deficiency, 87
Nickel and cancer, 154
Nicotinamide adenine dinucleotide, **87,** 91
Nicotinamide adenine dinucleotide phosphate, 101
Nicotine, 366, *368–369*
Night blindness, **377**

Nine-plus-two arrangement, 42
Nipple erection, 182
Nitrates, 540
Nitric oxide, 566
Nitrites, 540
Nitrogen
    atom, 583, 584
    bubbles in blood, 298
    cycle, 538, *540*
    dioxide, 566
    elimination, 302, 303
    fixation, **540**
    fixers, **540**
    in primitive atmosphere, 10
Nitrogenous base
    DNA, 108, *109, 110,* 114
    RNA, 115
Nitrous acid, 122
Node
    lymphatic, 254, *255,* 265
    nerve cell, **350, 355**
    sinoatrial, **242**
Noise pollution, 383
Noncyclic photophosphorylation, **101**
Nondisjunction, **471,** 473
    chromosome 21, 474, *475*
Nonelectrolytes, **585**
Noradrenalin, 331
Norepinephrine, **331,** 365, 366
    and ulcer formation, 232–233
Normality, **213**
Nose, 264, *285,* 286
Nostrils, 284
Nuclear breeder reactors, 571
Nuclear energy, 566
Nuclear envelope, 42
Nuclear fusion reactors, 566
Nuclear membrane
    disintegration, 130, 132
    and endoplasmic reticulum, 45
    prokaryotic, 48
    reconstruction, 134
Nuclear pores, *43*
Nuclear reactors, 549
Nuclear waste, 549
Nucleic acids, 77, 107–127
    as antigens, 265
    antagonist, 275
    prokaryote, 48
    synthesis, 14, 130
Nucleolar organizer, **43, 134**
Nucleolus, **43,** 130, 134

Nucleoplasm, 26, **43**
Nucleoproteins, 70
Nucleotides, **108**
  DNA, **108,** *109,* 110
  pairs, 136
  RNA, 116, 119
  sequence, 138
  structure, *109*
  substitutions, 122, 161
Nucleus, 26, 42, *43–45*
  Acetabularia, 45
  atomic, **583,** 584
  function, 44–45
  interphase, **130,** *131*
  pores, 45
  prokaryotic, 48
  prophase, *130, 131*
  sperm, 185
  transplants, 43–*44, 45*
Nutrient(s)
  absorption, 225, 227
  blood, 260
  cycling, 546
  in food chain, 530
  movement across capillary, 252
  upwelling, 565
Nutrition, 229–231

O

Ocean, fish yield, 565
Ockham, William of, 391
Ockham's razor, **391**
Odor, 375
Oil and cancer, 154
Oils, 63
Oklahoma dust bowl, 548, 550
Olfactory bulb, 375
Olfactory nerve, 375
Olfactory organs, **374**–375
Omnivores, **530**
One-gene–one-enzyme hypothesis, **112,** 121
Oocyte, **178**
  development, 178–181
  primary, 179
  secondary, 180, 185
Oogenesis, **174,** *179, 185*
  nondisjunction in, *473,* 474
Operant conditioning, **404**
Operator gene, **145**
Oparin, A. I., 12–14
Operon, **145,** 146

Opsonins, **274**
Optic chiasma, **378,** *379*
Optic disc, **378,** *379*
Optic nerve, **378**
Optic neurons, 377
Oral contraceptive pill, 188–189
Orangutang, *503*
Order, **501**
Organ, **48**–51
Organ of Corti, **382**–384
Organ systems, **48**–51
Organ transplantation, 275
Organelles, 26, **34**–48
  in eukaryotes, 48
  in unicellular organisms, 48
Organic compound, **12**
Organic ions, 350
Organic soup. *See* Primitive soup
Organism, primitive, 14–17, 20
Organophosphates, 434
Orgasm, **183**
Origin of life, 4–21, 508
Origin of man, 512–520
Origin of species, 509, 512–518
Orwell, George, 348
Osmosis, **31**
  in kidney, 307, 309
Osmotic pressure, **31**
  blood, 252, 253
  capillary, 252
Os penis, 174
Ossification, 420, **421,** 422
  joints, 426
Osteoarthritis, **426,** *427*
Osteoblast, **327**
Osteoclast, **329,** 421
Osteocyte, **418,** *419,* 420
Otolith, ear, **384**
O-type blood, 465, 468
Outer ear, 384
Ova, 177. *See also* Egg
Oval window, ear, **382,** 384
Ovarian cycle, **178,** *179, 180–181*
  and contraceptive pill, 188–189
  and temperature, *186*
Ovary, **177,** *178, 179*
  function, 336
  overactive, 336
Ovulation, 179–181, 183
  detection of, 186
  failure, 181
Oxidation, **84**

of fats, **62**
of foodstuffs, 86, 229
Oxidizing atmosphere, **12,** 18, 19
Oxygen
  atom, 583, 584
  cellular requirement, 287
  debt, **433,** 436
  deficiency, 296, *297*
  deprivation, 288
  exchange, 262, 285, 287
  life requirement, 9
  poisoning, 298
  in primitive atmosphere, 10
  receptors, 295, *296*
  requirements, 295
  transport, 261
Oxyhemoglobin, **261, 289**
Oxytocin, 66, *71,* 211–213, **324**
Ozone, 18, 566

P

PABA, 76
Pacemaker, **242**
  artificial, 246–*247*
Pacinian corpuscle, **385**
Pain
  awareness, 366
  ESB treatment, 348
P-aminobenzoic acid (PABA), 76
P-aminohippuric acid, 309
PAN, **567**
Pancreas, 225, **330–331**
Panting, 441, 444
Paralysis, 93
Paraplegics, 363
Parasites, virus, 53
Parathormone, 327, **329**–330
Parasympathetic nerves, 243, *244*
Parasympathetic nervous system, **365**
Parathion, 358
Parathyroid glands, 187, 329–330
Parathyroid hormone, 327, 329–330
Parentage, 468, 469
Parkinson's disease, 388
Parturition, **208.** *See also* Birth
Passive immunity, **271,** 273
Pasteur, Louis, 9, *10*
Pauling, Linus, 483
Pavlov, Ivan, 403
Pea genetics, 456–457, 461
Peck order, **410,** 412

Pectin, 47
Pectoral girdle, **418**
Pedigree, **455**
  albinism, 455
  color blindness, 462, *464*
  hemophilia, *466–467*
  human, 513, *514*
Pellagra, 74, 87
Pelvis, **418**, 506
Penicillin
  kidney secretion, 309
  -resistant bacteria, 485, 486
  sensitivity, 269
Penis, 170, *171, 173*, **174**
  bone, 174
  detumescence, 183
  erection, 174, 182, 183
  human, 506
  and urethra, 312
Pentoses, **58**
Pepsin, 74
Peptic ulcers, 231–233
Peptide bond, 67, 119, *120, 121*
Perennials, 546
Pericardium, **240**
Period, menstrual, 178
Periodicities, DNA, **110**, 111
Periodontal tissue, **231**
Periosteum, **420**
Peripheral vision, **378**
Peripheral nervous system, **360**, 361, 363–365, *364*
Peristalsis, *224*
  colonic, 229
  in ureter, 311
Peritonitis, **231**, 233
Permanent wave, 71
Permeability
  kidney tubules, 310
  postsynaptic membrane, 358
Permease, 144–146
Peroxyacetyl nitrate, **567**
Pesticides, 572
  as nerve poisons, 435
Petit mal, 357
Peter the Great, 357
Petroleum and cancer, 154
PGA, **102**
PGAL, **102**
pH, 585–587
  blood, 309
Phagocytic cells, 262, 265. *See also* White blood cell
Phagocytosis, 33–**34,** 39

Pharynx, 225, **284,** 286
Phase-contrast microscope, 29
Phenotype, 454–458, **455**
Phenylalanine, 113
Phenylketonuria, 113, 123
Phenylpyruvic acid, 113
Phipps, James, 272
Phosphate, in DNA, 108
Phosphoglyceraldehyde, **102**
Phosphoglyceric acid, **102**
Phospholipids, **65,** 66
Phosphorus cycle, *541*
  dietary, 76–77
Phosphorylation, **87**
Photons, **99,** 100, 101
Phototaxes, **393**
Photosynthesis, **18,** 20, 96–105, **98**
  and air pollution, 567
  carbohydrate production, 102
  cellular location, 102–103
  dark phase, **102**
  and DDT, 536
  and ecosystems, 528–529
  efficiency, 98, 103–104
  electron flow, 101, 103
  energy yield, 98
  in eukaryotes, 48
  light phase, 99–102
  location, **36,** 102–103
  noncyclic photophosphorylation, **101**
  origin of, 18
  phosphorylated intermediates, 102
  pigments, 99, 100
  in prokaryotes, 48
  reactions, *105*
  schematic, *102*
  and solar energy input, 99
  sugar formation, 102
  systems I and II, **101**
  zone of, 529
Phycobilins, 99
Phyla, **501**
Physique
  evolution of, 523
  genetic basis, 123–124
Pigment
  eye, 376
  hair, 112, 114
  muscle, 423
  photosynthetic, 99, 100
  retinal, 377, *378*
  skin, 454–455, 468–471, 523

Pill, contraceptive, 154, **188**–189
Pineal gland, **336**
Pineal tumor, 335, 336
Pinocytosis, **33**–34, 39, 252, *253*
Pioneering community, 545
Pioneer plants, 545, 547, 548
Pitch, **381,** 384
Pithecanthropus, 517
Pitocin, 209
Pituitary
  anterior, 178, **321**
  dwarf, **322**
  giant, 322–323
  gland, 170, 187, *318*, **319**–324
  hormones, 170, 212, 318–324, *320, 321*
  and hypothalamus, 336–337
  and osmoregulation, 310
  posterior, 324
  and thyroid regulation, 324
  trophic hormones, *320*
  tumor, 154, 335
Pivot joint, *426*
PKU, 113, 123
Placenta, **206**–207, *209*, 210
Plague, 122, 556
Planarian worms, 407, 408
Planets
  atmosphere ,12
  composition, 11
Plant(s), **48**
  behavior, 390
  classification, 499, 500, 501
  leguminous, 540
  pioneering invaders, 545, 547, 548
  productivity, 563
Plant cell, *25*
  mitosis, 134–135
  nutrition, 98
  telophase, 135
  vacuoles, 39, 40
  wall, *47*
Plantae, 48
Plaque
  arterial, 250
  dental, 231
Plasma, *51,* **260**
  carbon dioxide in, 290
  components, *261*
  oxygen uptake, 287, 289
  pH, 261
  proteins, 260, 261
  regulation of, 302, 303, 305, 307

Plasmablasts, **269**
Plasma cells, 261, 269, 270, *271*
Plasma membrane, *27*
    function, 30–32
    fusion at fertilization, 184
    and insulin, 330
    and mercury affects, 534
    muscle. *See* Sarcolemma
    neuron, 350, 352–353
    and prostaglandins, 336
    structure, 26, 65, 66
    and transport, 341
Plasmids, 124–*125*
    chimeras, **125**
Plastids, **37**
Plastoquinones, 101
Platelets, *51*, 261–264
Platinum, catalytic properties, 72
Pleiotropy, **484**
Pleural membrane, **284**, 290
Pleurisy, 286
Pneumonia, 108
PNS, **360**, 361, 363–365, *364*
Poikilotherms, **440**
Poison
    DDT as, 536
    enzyme, **74**, 75
    nerve, 353
    paralytic, 434–435
Poisoning
    ATP, 93
    carbon monoxide, 567
    mercury, 534–535
Poison ivy, 269
Polar body, oocyte, 179
Poliomyelitis, 272–273, 363
    and recombinant DNA, 122
Poliovirus, *53*, 272–274
Pollen, 411
Pollution, 533–537, **571**
    air, 154, 288, **566**–570, *567*
    noise, 383
    thermal, **571**
    water, 570, 572
Polygenes, 468–471, **469**
Polygenic inheritance, **123**, 124
Polymerization, **14, 58**
Polyoma virus, **159**
Polymorphism, **491, 520**
    balanced, **493**
Polypeptide chain, **67**
    coiling, 69

formation, 119, *121*
Polyribosome, **34, 119,** *121*
Polysaccharide, 60, **62**
Polysome, **119,** *121. See also* Polyribosome
Polytypic species, **520**
Pons, **361,** *363*
Population, **528**
    bomb, 555–561
    density increase, 561
    distribution, 558–561
    doubling time, 103–104
    equilibrium level, 575
    explosion, *554–577*
    genetics, 486–492
    growth, 553–577, *554*
    growth curbs, 575
    growth rate, *557*
    human, *575*
    increase rate, 103–104
    and photosynthesis, 98
    self-regulation, 573–574
    size, *557*
    structure, 556–561
    and territoriality, 409
    world, *575*
Pores, nuclear, 45
Portal veins, 228, **239**
Porter, Keith, 34
Positive feedback, **143**
Posterior chamber, **377**
Postsynaptic membrane, 358
Potassium
    active transport, 32
    in primitive atmosphere, 10
Potassium ion(s), 350, 352, 353, 358, 359
    in muscle contraction, 431
    secretion by kidney, 309
Potato virus, *53*
Potential energy, **82**
Pott, Percival, 154
Precipitation, annual, 543
Precipitins, **277**
Predator-prey relations, 544, 545
Prednisone, 275
Preening, 413
Pregnancy
    and breasts, 211
    risks, 189
    symptoms, 187
    test, 187
Pregnenolone, 333
Pregnosticon, 187

Prehormone, 328
Prepuce, **174**
Pressure
    barometric, 296, *297*
    blood, 252
    filtration, *307*
    high, 298
    low, 296
    reception, 385
Presynaptic knob, 355, 357–360
Primary consumers, **529,** *532, 533*
Primary productivity, *565*
Primate(s), *502–505*
    behavior, 399
    characteristics of, 502
    classification, *506*
    dentition, *516*
    skull shape, *516*
    territoriality, 409
Primitive organism, 14–17, 20
Primitive soup, 12, **14,** 15
Primitive streak, *198*
Primordial cloud theory, **10,** *11,* 12
Principle of recombination, **459,** *460, 461,* 487, 488
Principle of segregation, *458*
Problem solving, 404–405
Proconsul, 513
Producers, 97, **529,** 530, *532, 533,* 538
Productivity, primary, *565*
Progesterone, 180, 181, **335**–336
    and breast enlargement, 211
    and cancer, 154
    and gene activation, 341
    level, 189
    in pregnancy, 187
    and prostaglandins, 336
    synthetic, 189
Prokaryotes, **48,** 136
Prolactin, 211, 337
Promoter substance, 159
Promoters, gene, **146**
Pronator muscle, **426**
Pronuclei, 185
Prophase, *130,* **131,** 132, 134, 135
Propylthiouracil, 327
Prosimians, 513
Prostaglandins, **336**
Prostate gland, 161, **172**
Prosthetic groups, **69**
Protamines, **136**
Proteins, 66–70, **67**
    as antigens, 265

Proteins (continued)
  in bone, 420
  carrier, **307**
  chromosome, 136
  ccagulation, **68**–69
  complete, 230
  conjugated, **69**–70
  denaturation, **68**–69, *70*
  deficiency, 562
  digestion, 224, *226*
  hormone receptor, 341
  metabolism, 330
  plasma, 260, 261
  in protoplasm, 26
  secondary structure, **68,** 69
  sparing, **230**
  and species specificity, 67
  stereoisomers of, 61
  structure, 66–70
  synthesis, 115–120, 130, 407
  tertiary structure, **68,** 69
  proteinoids, **14,** *16*
Protista, **48**
Protons, **583,** 584, **585**
Protoorganism, 17, *18. See also* Primitive organism
Protoplasm, **26**–30
Protozoa, 48
  and disease, 9
Pseudopod, 28
Ptyalin, 74
Puberty
  in blind girls, 336
  and breast development, 211
  female, *178,* 335
  male, *170,* 333
Pubis, 182
Puffs, chromosome, **140**
Pulmonary artery, 264
Pulmonary circulation, 287
Pulmonary edema, 286
Pulmonary embolism, 189
Pulmonary osteoarthropathy, **288**
Pulmonary pump, **241**
Pulmonary valve, **241**
Pulmonary vein, 287
Punnett, R. C., 460
Punnett square, **460**
Punishment, 395, 404
Pupil, **376,** 377
Pus, **265**
Purines, 14, **109,** *110*
Purkinje fibers, **243**

Purpose and behavior, 391
P wave, **245**
Pyloric sphincter, 225
Pyorrhea, **231**
Pyrimidines, **109,** *110*
Pyrogens, **443**
Pyruvic acid, 89–92, 432–433

Q

QRS wave, 245
Quadrupedalism, 514–515
Quanta, **99**
Queen Victoria, 466
Queen bee, 391, 410

R

Rabies vaccine, 273
Race, **520**–523
  classification, *522*
  and sociobiology, 396
Racism, 523
Radiation, **441**
  annual, *490*
  background, **489**
  and environment, 537
  mutagenic, 122, 489, 491
  solar. *See* Solar radiation
Radioactive isotopes, **584**
Radioactive residues, 533
Radioactive waste, 537, 549
Rainfall, 543
Ramapithecus, 513–514
Rank order, 409–410
Rasputin, 466
Rats
  killer instinct, 398–399
  learning in, 408
  memory transfer, 407–408
  taxes, 393
  trial and error learning, 404
Rayon, 47
Reagins, **276**
Reasoning, **405**
Receptor, 440
  temperature, 441
Recessive gene, **458**
Recipient, **465,** 468
Recognition site, antibody, **273,** 274
Recombination, genetic, 459, 487, 488. *See also*
    Principle of recombination

Recombinant DNA, 106, 122–125
Rectum, 181, **229,** 255
Recycling, 537–544, *538*
  solid waste, 572–573
  water, 572
Red blood cell(s), *51,* **260,** 261–264
  antigens, 465, 468
  and carbon dioxide, 290
  count, 266–267
  enzymes, 262
  formation, 262
  ghosts, **26**
  lifespan, 129
  origin, 421
  osmosis in, 30–31, *32*
  shape, 262
  sickled, *484*
  structure, 263
Redi, Francesco, 7, *9*
Redirected activity, **401**
Red muscle, **423**
Reducing atmosphere, **12,** 18, 84
Reduction, **84**
Reflection, 530
Reflex arc, **365,** 366
Reflexes, **361,** *365*–366, **394,** 398
  breathing, 293
  conditioned, 394, 398, **403,** 404
  defecation, 229
  duodenocolic, **229**
  ejaculation, *175,* 183
  gastrocolic, **229**
  grasping, 402
  instinctive nature of, 402
  kneejerk, 365
  letdown, 211, *213*
  micturition, 314
  in paraplegics, 363
  in penal erection, 174
  respiratory, 292–295
  stereotype, 394, 398
  sucking, *213,* 402
  thirst, 311
  visual, 379
Refractory heartbeat, **243**
Refractory period
  male, 183
  muscle, **432**
Refractory state, nerves, **353**
Regulator gene, **144,** 145, 159
Regurgitation, newborn, 231
Reinforcement, 393, 403, 404

Relaxation, muscle, 429
Releaser, **398,** 400, 401
Releasing factors, **337**
Renal artery, **303**
Renal vein, **303**
Repair, 5–6
Repetition, 403, 404
Replication, **112,** *139*
Repressor substance, **145,** 146, 159
Reproduction, **6**
    asexual, 486
    cellular. *See* Mitosis; Meiosis
    differential, **493,** 512
    hormonal control. *See* Hormones, female; hormones, male
    in primitive organisms, 15, 17
    sexual, 166–193, 486
    and territoriality, 409
    and variation, 486, 488
Reproductive isolation, **511**
Reproductive system
    female, 174, *177, 178–182*
    male, 170, *171–174*
Reserpine, 366, *368–369*
Resistance
    DDT, 536
    environmental, **573**
    penicillin, 485, 486
Resolution, 28–**29**
Resolving power, 29
Resorption, kidney, 306, 307, 309, 310
Resource depletion, 566
Respiration, 280–299, **281.** *See also* Breathing
    aerobic, **86,** *87,* 89–92
    anaerobic, **86–**89
    artificial, 295
    cellular, 80–94, **282**
    chemical control of, 295
    in eukaryotes, 48
    evolution of, 94
    internal, **282**
    kinds of, *282*
    location of, 35
    rate regulation, 295
Respiratory center, **292,** 293, *294, 295, 296*
Respiratory chain, **91,** 92
Respiratory cycle, *91*
Respiratory reflex, 292–295
Respiratory spasms, 330
Respiratory surface, 283
Respiratory system
    functions, 302

    insect, *283*
    mouse, *283*
Respiratory tree, **283–***284*
Response, emergency, **332**
Response mechanism, 440
Responsiveness, 6
Resting period, muscle, 432
Resting potential, neuron, **352**
Restriction enzyme, 125
Retaliation, 413
Rete testis, **172**
Reticular formation, *363*
Reticulin, **46**
Reticuloendothelial system, **265,** 273
Retina, 377, *378, 379*
Reward, 395, 404
Rheumatism, **426**
Rhizobium, 540
Rhodopsin, 377
Rhythm method, **185**
Ribonucleic acid. *See* RNA
Ribonucleotide, **115,** *116*
Ribose, **115**
Ribosomal RNA, 34, 118, *119*
    and hormone synthesis, 341
Ribosome, **34,** *121*
    precursors, 43
    in prokaryotes, 48
    and protein synthesis, 119
    in translation, *120*
Ribs, **417**
    breathing movements, 290
Ribulose diphosphate, **102**
Ribulose phosphate, **102**
Rickets, 328–**329**
    in early man, 519, 523
Rickettsiae, classification, 48
Rigor mortis, 430
Ripening fruit, 47
Ritalin, 388
Ritual, 412–**413**
R-group, **66**
RNA, **115**
    and chromosomal puffs, 140
    and memory, 407
    messenger, 116, *119, 120,* 140, 341
    nitrogenous base, 115
    nucleotides, 116, 119
    polymerase, **117**
    ribosomal, 34, 118, *119,* 341
    synthesis, 115–120, 141
    transfer, **118,** *119, 120*

    translation, 119
Robertson, J. D., 26
Rods, retinal, **377,** *378*
Röntgen, Wilhelm, 490
Root, spinal nerve, 361, 363
Rough endoplasmic reticulum, 34
Round window, ear, **382**
Rous, Peyton, 157
Rous sarcoma virus, 157
rRNA. *See* Ribosomal RNA
RuDP, **102**
RuP, **102**
Runaway feedback, **143**
Ryther, J. H., 565

S

Saccharin, 154
Sahlins, Marshall, 396
St. Martin, Alexis, 222
St. Martin's fistula, **222**
Saline abortion, **216**
Saliva, 264
Salivary gland(s), 264, 318
    chromosomes, *138,* 140
Salmonella typhimurium, 157
Saltatory conduction, 355
Salt(s)
    balance, 303, 307–311
    bile, 233
    blood, 260
    bone, 419–421
    concentration gradient, 307, *309*
    intestinal absorption, 224
    membrane penetration, 30
    in seawater, 311
    in sweat, 442
SA node, 246
Sarcolemma, **428,** 430, 431
Sarcoplasm, **428**
Sarcoplasmic reticulum, **431, 432**
Satiation, **400**
Saturated fats, **64**
Scab, 262
Schick test, 269
Schizophrenia, 388
Schwann cell, *351*
Scientific method, 8
Sclera, **376**
Scrotum, 170, *171,* 173, **174**
    cancer of, 153–154
Seawater, salt content, 311

Sebaceous glands, 211
Secondary consumers, **529**, 530, *532, 533*
Secondary sex characteristics, 170
    Napoleon, 338–339
Second filial generation, **456**–457
Second law of thermodynamics, 532, 571
Secretion, kidney, 307, 309
Sedatives, 366, *368–369*
Seed production, 547
Selection. *See* Natural selection
Selection, negative, 494, **495**
Self tolerance, 268
Semen, 170, **172**, 173, 176
Semicircular canals, **384**
Semilunar valve, **241**
Seminal fluid, **172**, 336
Seminal vesicles, **172**
Seminiferous tubules, **170,** *171*
Semipermeable membrane, **31**
Sense
    balance and equilibrium, 382, 384–385
    hearing, 381–384
    olfactory, 374–375
    sight, 376–380
    taste, 375–376
    touch, 385
Sense organs, 372–387
Sensory adaptation, 402–403
Sensory nerves, 361, 363, 364
Sensory neurons, **360**
    in reflex arc, 365
Septum, deviated, 286
Serotonin, 366, 367
Sertoli cells, **170**
Serum, antilymphocyte, 275
    blood, **263**
    sickness, 273
Sewage treatment, *572*
Sex cells. *See* Gametes
Sex characteristics, secondary, 338–339
Sex chromosomes, *451,* 453, **454,** 462–463
    abnormal numbers, 471–473
    determination, 454
    differences, 317
    drive, 335, 400
    hormones, **333,** 335. *See also specific hormone in*
      question
    linkage, 462
Sexual intercourse, **182**–183
Sexuality, 166–193
Sexual reproduction, 166–193, 486, 488
Shivering, 442

Shock
    anaphylactic, 276
    and blood pressure, 250
    and edema, **253**
    and ulcer formation, 232
Shope papilloma virus, 157
Short term memory, 367, **407**
Shoulder girdle, **418**. *See also* Pectoral girdle
Shoulder joint, 426
Shrubs, 546
Siamese twins, 213
Sickle cell anemia, 482–486, *494–496*
Sickle cell disease, 483–486, 493
Sickle cell hemoglobin, 67–68, **121,** 122, 484, *485,*
    486, 493, 496
Sickle cell trait, **483,** *485,* 486, 493
Siegler, Alvin, 191
Sight, 376–380
Sigmoid growth curve, **573**
Signals, 408
Silent Spring, 537
Sinanthropus pekinensis, 518
Single bond, **587**
Sinoatrial node, **242**
Sinuses, 284, *285*
Sinusoids, **265**
Skeletal muscle, **422**. *See also* Striated muscle
Skeletal tissue, *51. See also* Bone
Skeleton, 416–422, *418*
    appendicular, **418**
    axial, **418**
    evolution of, *515,* 518
Skin, **385**
    cancer, 155–156
    cells, 151, *160*
    color, 454, 455, 468–471, 523
    division, 135
    graft, 265, 269
    and natural immunity, 264
    senses, 385
    -sensitizing antibodies, 276
    structure, *442*
    sunburnt, *156*
    sweat glands, 441
    temperature receptors, 441
    and temperature regulation, 444
    wart, 157
    wound healing, *152*
Skinner, B. F., 394–395, 398
Skinner box, 393, 395
Skull, 360, **417**
Skull

Cro-Magnon man, *519*
    evolution of, *515–516*
    human, 506
    Java man, 518
    Neanderthal man, *519*
    shape, primate, *516*
    Steinheim man, *520*
Sleep
    and LSD, 367
    rapid eye movement, 356
Sliding filament model, **429,** *430,* 431
Small intestine, 225, *227*
Smallpox
    and recombinant DNA, 122
    virus, *53,* 272. *See also* Variola
Smell, *374–375*
Smog, **288,** 568
Smoking, 288
    and cancer, 154–155
Smooth endoplasmic reticulum, 34
Smooth muscle, 51, **422,** 423, *434*
Snakebite, 273
Snake poison, 274
Sneezing, 286
Sniffing, 375
Snowshoe hare, 544
Social behavior, **408**–414
    and adaptation, 409
    bees, 410–411
Social dominance, 410
Social hierarchy, 409, **410,** 412, 413
Social insects, 396
Society, 408
Sociobiologists, 414
Sociobiology, 396, 398
Sodium
    active transport, 32
    conservation of, 311
    ions, 350, 352, 353, 357, 358, 436
    regulation, 307, *309,* 310–333
Sodium chloride, 350, 352, 442
Sodium pump, **32, 352,** *353*
Soil bacteria, 540
Solar energy
    annual, 103
    ecosystem flow, 528–531
Solar radiation, 528–*530*
    and rickets, 328–329
Solar system, origin of, 10, *11,* 12
Sol-gel transformation, **28**
Solid, diffusion in, 30
Solid waste, **572**–573

Solution, 30
Somatic cells, **170**
Somatic nervous system, *364*
Somatostatin, 123
Somatotrophic hormone. *See* Growth hormone
Somites, embryonic, *199, 200*
Soot, 153–154, 568
Sound, **381,** *383*
  reception, 381–384
  receptor, *382*
Spaceship earth, 3, 552–577
Spallanzani, Lazzaro, *9,* 176
Spark discharge apparatus, *14*
Special creation, 508
Specialization, cell, 48–51, 136
Speciation, *511,* 512
Species, **500,** 501, 520
  diversity, 547
  equilibrium, 546–548
  evolution, 447
  fixity of, **501**
  polytypic, **520**
  specificity, **67,** 115, 122
  evolution of, 516
Sperm, **170**
  abnormal, *471*
  bank, 176, 496
  chromosomes in, 454
  at fertilization, *196*
  haploid, 453
  longevity, 184
  maturation, 333
  motility, 172–173
  nucleus, 185
  penetration of egg, 46
  production, **170**–173
  tails, *41–42*
Spermatid maturation, *173*
Spermatocytes, *172, 173*
Spermatogenesis, **170,** *172, 173,* 333
  nondisjunction, *471, 474*
Spermatogonia, **170,** *172, 173*
Spermatozoa. *See* Sperm
Spermicidal solutions, **186**
Spermiogenesis, *173*
Sphincter muscle, 312, 314, **425**
Sphygmomanometer, **249**
Spina bifida, **214**
Spinal cord, *51,* 349, **361,** *362,* 363, 364, *365*
  central canal, 360
  gray matter, 361
  relay centers, 360

structure, *362*
  white matter, 361
Spinal nerves, 360, 361
Spinal reflex. *See* Reflex
Spindle, 132–134
Spine, **417.** *See also* Backbone
  split, 213
Spinnbarkheit, **186**
Spleen, 268
Spontaneous generation, **6**–10
Spores, 10
Squamous epithelium, *49*
S-shaped growth curve, **573**
Stained cells, 29, *263*
Stapes, **382**
Staphylococci, mutation, 485, 486
Starch, 58–**62**
  hydrolysis, *63*
  metabolism, 330, 331
  structure, *63*
Starr-Edwards heart valve, *243*
Starvation, 104, **230**
  deer, 545
Stationary phase, **573**
Status, 410
Steady state, **143, 439**
Steinheim man, 519, *520*
Stem cells, 268
Stereoisomers, **60**–61
Stereotyped behavior, 390–393
Stereotyped reflexes, 394, 398
Sterility
  female, 181
  male, 173, 335
Sterilization, 188, 190–191, 494
Sternum, 417
Steroids, 62, **64**–66
  and cancer, 154
  hormones, 333, *341*
  ring, *67,* **333**
Stewart, Sarah E., 159
Stickleback, reproductive behavior, 398, 409
Sticky ends, chromosome, 478
Stimulants, 366, *368–369*
Stimulus
  intensity, 354–355
  nerve cell, **352**
  subthreshold, 360
  threshold, **354**
Stirrup, ear, **382**
STM, 367, **407**
Stomach, 225

absorption in, 224
acidity, 264
and alcohol, 367
cancer, 161
fistula, **222**
ulcers, 231, 232
Strata, **508**
Stratton, Charlie, 322
Stress and ulcer formation, 232
Striated muscle, 422, **423,** *424, 428, 430*
Striped muscle, **423.** *See also* Striated muscle
Stroke, 264, 388
Stroma, chloroplast, **36,** 103
Strontium 90, 537
Structural isomers, *59,* 60
Structure, **501**
Strychnine, 366
Submucosa, **223**
Subspecies, **520.** *See also* Race
Substrate, *73*
Subthreshold stimulus, 360
Subvital gene, 486
Succession
  ecological, **545**–*547, 546*
  farmland, *546*
  lake, *547*
Suckling reflex, *213,* 402
Sucrase, 74
Sucrose, **59,** 60, 67
Suction (abortion), **216**
Sugar(s), 58–62
  blood level, 228
  in DNA. *See* Deoxyribose
  double. *See* Disaccharides
  fermentation, *88*
  formation, 14, 102
  many. *See* polysaccharides
  metabolism, 300, 331
  milk. *See* Lactose
  RNA. *See* Ribose
  single. *See* Monosaccharides
  stereoisomers of, 61
  in urine, 310
Sulfanilamide, 76
Sulfur dioxide, air content, 568
Sulfuric acid, air content, 568
Sulfur-sulfur linkage, 71
Summation, **359**–360
Sun, energy, 528–530
Sunburn, 155–*156*
Sunlight
  and ecosystems, 528

Sunlight (continued)
  and photosynthesis, 99
  and rickets, 328–329
Supinator muscle, **426**
Supporting connective tissue, *50*
Survival of the fittest, 493
Suspension, **30**
Swallowing, 384
Sweat, 441, 442
  composition, *442*
  glands, 318, 441
  and salt balance, 309
Sweden, population structure, 558, *559*
Swordfish, mercury in, 535
Sympathetic nerves, **243,** *244*
Sympathetic nervous system, 332, **364,** 365
Synapse, **355**–359
  and drugs, 366
  fatigue, 359
  and learning, 406
  transmission, 359
Synapsis, **170**
Synaptic cleft, **355,** *359*
Synaptic knob, **349,** *358, 359*
Synaptic transmission and drugs, 366, *368–369*
Syncytium, **423,** 424
Syngamy, **168,** 185. *See also* Fertilization
Synovial fluid, **426**
Synthesis, abiotic, **18,** 19
Syphilis, 363
System I, **101**
System II, **101**
Systema Naturae, 500
Systemic pump, **241**
Systole, **245,** 249
Systolic pressure, **250**

T

Tadpole, tail resorption, 38
Tanning, skin, 470
Target organ, **319**
Tarsier, *504,* 513
Taste, 375–376
Taste buds, **375**
Tatum, E. L., 112, 121
Taxes (taxis), **393,** 394
T-cells, **268,** 269, 170, 275
Tear ducts, 284
Tear glands, 284
Tears, 264, 284, **377**
Technology, 506

Tectorial membrane, 384
Teleology, **391**
Telophase, *130,* **134,** 135, 137
Temperature
  air inversion, 567, *570*
  atmospheric, 544
  control, 440–443
  core, **440**
  and ovarian cycle, *186*
  reception, 385, 441
  regulation, fetal, 205
Temporal lobe, 348
Tendon(s), **425**
427
  hand, *427*
Teratogens, **214,** 215
Territoriality, 398–401, **409**–413
  birds, 400, 401
  sticklebacks, 398, *400,* 409
Testis (testes), **170,** *171,* 174, 333
  descent, 174, 335
  hormones, 333. *See also* Androgens
  rete, **172**
  retraction, 174
  tumor, 335
Testosterone, **172, 333**
  and climacteric, 335
  structure, 67
Test-tube babies, 166
Tetanus, 273
  bacterium, 86
  muscle, **432,** *435*
Tetany, 330
Tetrahydrocannabinol, 367, *368–369*
Tetroses, **58**
Thalamus, **361,** *363*
Thalidomide, 199, 213, *214*
Theory, **8**
Thermal inversion, 567, *570*
Thermal pollution, **571**
Thermodynamic system, open, *83*
Thermodynamics, Laws of, **82,** 86, 532, 571
Thermostat, *142,* 143, 443, 444
Thiamine pyrophosphate, **88**
Thinking, 405, 506
  and drugs, 366, *369*
Thirst, 311, 400
  reflex, 311
Thompson, Francis, 547
Thoracic cavity, 284
Thoracic duct, **253**
Threshold, *360*

of learning, **383**
  motor unit, 432
  stimulus, **354**
Throat cancer, 154
Thrombin, 264
  and menstrual blood, 181
Thrombocytes, **263**
Thrombophlebitis, **189**
Thrombus, 264. *See also* Clot
Thwarting behavior, 401
Thymine, **109,** *110,* 111, *114, 115*
Thymus gland, *268,* **336**
Thyroid deficiency, 325
Thyroidectomy, 327
Thyroid gland, **324**–329
  cell receptors, 340
  in pregnancy, 187
  tumor, 327
Thyroid hormones, 324–329
  and iodine, 77
  synthesis of, *326*
Thyroid-stimulating hormone, 324, 325, 337
Thyrotrophic hormone, **321**
Thyroxine, **324**
  and radioactive waste, 537
Tissue(s), **48**–51
  classification, *49–51*
  fluid, 350
  not self, 264, 268
  periodontal, **231**
  self, 264, 268
  space, *253*
  typing, 275
Toes, embryonic, *201*
Tolerance, immunological, 275
Tone quality, **381,** 384
Tongue, *375*
Tonsillitis, 286
Tonsils, 286
Tool use, evolution of, 516, *517*
Toothache, 231
Tooth decay, 231
Touch reception, 385
Toxins, **274**
Toxoid, **274**
TPP, **88**
Tracers, atomic, **584**
Trachea, **284,** 286
Tracheal cells, 41, 287
Tracheal tubes, **283**
Training, 404
Tranquilizers, 366, *368–369*

Transcription, **117,** 115–119
Transducer, sensory, **374,** 385
Transduction and memory, **407**
Transfer RNA, **118,** *119, 120*
Transformation, experiments, *109*
Transforming principle, **108**
Transfusion, blood, 463, 465, 468
Translation, 115–*120*
Translocation, **474,** 476, 478
Transmission
  block, 435
  of Down's syndrome, 478–479
  gene, 458–463
  mechanism, 440
  nervous, 352–360
  neuromuscular, *431*
  synaptic, 355–359
Transmitter substance, 357
  antagonists, 366
  and drugs, 366, 367, *368–369*
  inhibitory, 358
  in muscle, 430
  parasympathetic neurons, 365
  release, *359*
  in summation, 359
  threshold, 358
Transport, 34, 252. *See also* Active transport
Trees, 546
Tree shrews, 513
Trench mouth, 231
Trial and error learning, **404**–405
Triceps, *425*
Trichinosis, 267
Tricuspid valve, **241**
Triglycerides, **228**
Triiodothyronine, **324**
Trioses, **58**
Triple bond, **587**
Triplet code, **114,** 119
Trivers, Robert, 396
tRNA. *See* Transfer RNA
Trophallaxis, **411**
Trophic hormones, **321,** *337,* 340
Trophoblast, *197,* 206
Tropisms, **390**
Trypsin, 74
TSH, 324, 325, 337
Tubal ligation, **188,** 191
Tuberculin test, 269
Tubulin, 40
Tumor, **152,** 153, 269
  adrenal, 335

benign, **152**
brain, 357
mammary, 158
pineal, 335, 336
pituitary, 154, 335
metastatic, 153
testis, 335
thyroid gland, 327
uterine, 154
Tuna fish, mercury in, 535
Tunnel vision, **378**
Turbinate bones, 284, *285*
Turgid cells, 40
Turner's syndrome, **473,** *474, 475*
T wave, 245
Tympanum, **381**
Tyrosine, 324

U

Ulcers, **231,** 233, 234
Ultraviolet light, carcinogenic effects, 155–156
Umbilical cord, **206,** *207*
Unbalanced ecosystems, 547–549
Underdeveloped countries, 563, 565
Unicellularity, **26**
  origin of, 20
Unicellular organisms, 48
United States
  birth rate, *560*
  fertility rate, *560*
  population structure, *559*
Unit membrane, **26**
Universal donor, **468**
Universal recipient, **468**
Unsaturated fats, **64,** 65
Uracil, 116
Uranium
  and cancer, 154
  resources, 566
Urbanization, 560–561, 564
Urea, 302, 303
Urea cycle, **302**
Ureter, **303,** 304, **311,** 312
Urethra, **172, 312,** 314
Urey, Harold C., 13
Uric acid, 312
Urinary calculi, **312**
Urinary sphincters, 312, 314
Urination, 314
Urine, 302, 303, 307, **309**
  acidity, 264

collecting duct, **304,** 305
concentration of, *308,* 310, 311
discharge, 311–314
flow regulation, 311
output, 309, *310*
storage, 312
Usher, 508
Uterine cycle, *180,* **181**
Uterine milk, 206
Uterine section, *207*
Uterine tubes, 177, **178.** *See also* Fallopian tubes
Uterus, *177,* 178, **181,** *206*
  contractions, 208
  lining. *See* Endometrium
  tumors, 154

V

Vaccination, 273
Vaccine, 272–273
Vaccinia virus, **272,** 274
Vacuoles, **39**
Vagina, 172, 174, *177,* 178, **181**
Vaginal douche, **186**
Vaginal show, 208
Vagus nerve, 293, 365
Valine, 122, 484
Valve
  aortic, **241**
  artificial, 241, *243*
  ball-and-ring, 241
  bicuspid, **241**
  bladder, 312, *314*
  lymph, 254
  pulmonary, **241**
  semilunar, **241**
  tricuspid, **241**
  venous, 251
Variation
  geographical, **520**
  sources of, 511
Varicose veins, 187
Variola, **272**
Vasectomy, **188,** 190–191
Vas deferens, 170, *171,* **172,** 174
Vasopressin, 66, **310, 324**
  disulfide bridges in, *71*
  and lactation, 211–213
  structure, 67
Veins(s), **239,** *241*
  allantoic, **206**
  compression, 250, 251

Vein(s) (continued)
  flow rate in, 250
  gastric, 228, *229*
  hepatic portal, 228
  jugular, 228
  mesenteric, 228, *229*
  portal, 228, **239**
  pulmonary, 287
  renal, **303**
  subclavian, 228
  umbilical, *209*
  valves, 251
Venom, 274
Venous pump, **251**
Ventilation, **281.** *See also* Breathing
Ventricle, **241**
  contraction, *242. See also* Systole
  electrical activity, 245
  fibrillation, 243, 249
  relaxation. *See* Diastole
Venules, **239,** 250
Vermiform appendix, **233**
Vernix caseosa, **205**
Verte' rae, 361, **417**
Vertebrates, **417**
Vesalius, 319
Vestibule
  ear, **384**
  genital, **182**
Villus (villi)
  intestinal, **227**
  chorionic, **206,** 207
Vincent's angina, 231
Virus, **19, 52**–53. *See also* Bacteriophage
  and antibodies, 274
  antigenic properties, 265
  and cancer, 157–160, 162
  cowpox, 19, 272, 274
  diseases, 53
  DNA, 136, *137*, 138
  and DNA research, 123
  genetics, 116
  infection, 47
  influenza, *53*, 274
  mumps, 335
  origin of, *19*
  polio, *53*, 272–274
  polyoma, **159**
  potato, *53*
  protein, 67
  Rous sarcoma, 157
  Shope papilloma, 157

  smallpox, *53*, 272
  vaccinia, 272, 274
  variola, **272**
  wart, 157
Visceral muscle, **423, 424.** *See also* Smooth muscle
Vision, 376–380
  peripheral, **378**
  tunnel, **378**
Visual cortex, 378
Visual purple, 377
Vitalism, 6–7
Vitalists, **7**
Vital principle, 9
Vitamin(s), 74, **75,** 76
  absorption of, 228, 229
  deficiency, 231
Vitamin A
  and eye pigment, 377
  intoxication, 39
  and lysosomes, 39
Vitamin D, 328–**329,** 421
  and skin color, 523
Vitamin K, 263
Vitreous humor, **377**
Vocal cords, **284**
Voice box. *See* Larynx
Voltage, **350**
Volume, sound, **381**
Vomiting, 385
Vulva, 177, **182**

W

Wadlow, R., 322
Wallace, Alfred, 493
Warren, Lavinia, 322
Wart, 157
Washkansky, Louis, 275
Wasp behavior, 396
Waste
  atomic, 549
  liquid, 570, 572
  nuclear, 549
  solid, 570, **572,** 573
Water
  absorption, 228, 229
  and agriculture, 563, 565
  balance, 311
  consumption, 542, 544
  cycle, 541–544, *542*
  diffusion of, 31
  intake, 311

  intestinal absorption, 224
  molecular weight, 67, 265
  molecule, 586
  pills, 311
  pollution, 570, 572
  recycling, *572*
  regulation, 309–311
  resorption in kidney, 307, 310, 311
  thermal pollution, 566
Waters, breaking of, 208
Watson, James D., 110–111
Watson, J. B., 393
Watson-Crick model, 110–111, 122, 136, 137
Wax, 63
  ear, 264
  glands, 411
Weapon use, 399
Weed killers, 103
Weight
  and age, 63
  and energy, 81
Weinberg, W., 488
Wharton's jelly, **207**
Wheeless, Clifford R., 191
White blood cell(s), *51*, 254, **260,** 261–264
  count, 266–267
  lifespan, 129
  origin, 421
  and phagocytosis, 34, 39
  type, *263*
White matter
  brain, 360, *363*
  spinal cord, 361
White muscle, **423**
Whooping cough, 273
Wilkins, M. H. F., 110
Wilson, E. O., 398
Windpipe. *See* Trachea
Witch's milk, 205
Womb. *See* Uterus
Wool, 71
Work, **82**
  caloric requirements, 230
Worker bees, 397
Wound healing, *152*
Wrist joint, 426

X

X-rays, 490
  and birth defects, 215, 217
  and cancer, 156–157

X-rays (continued)
    diffraction pattern, DNA, **110**
    and embryonic development, 215, 217
    and immunosuppression, 275

Y

Yeast, 86
    and disease, 9

and fermentation, *88*
Yellow fever vaccine, 273
Yolk sac, *200*

Z

Zeidler, Othmar, 535
Zero population growth, **556,** 558, 575

Zinc, in diet, 77
Z line, **429,** *430*
Zona pellucida, **179,** 184, *196, 197*
ZPG, **556,** 558, 575
Zygote, **168,** 169, 185, 195, *196,* **453**

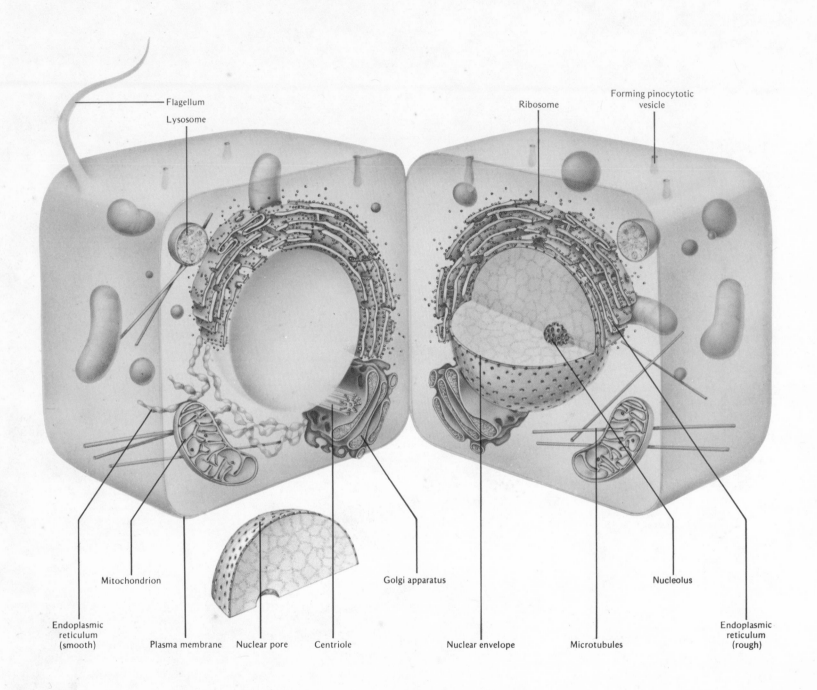

Flagellum

Lysosome

Ribosome

Forming pinocytotic vesicle

Mitochondrion

Golgi apparatus

Nucleolus

Endoplasmic reticulum (smooth)

Plasma membrane

Nuclear pore

Centriole

Nuclear envelope

Microtubules

Endoplasmic reticulum (rough)